DATE DUE

The chemistry of
the amino group

THE CHEMISTRY OF FUNCTIONAL GROUPS

A series of advanced treatises under the general editorship of
Professor Saul Patai

The chemistry of alkenes (published)
The chemistry of the carbonyl group (published)
The chemistry of the ether linkage (published)
The chemistry of the amino group (published)

The chemistry of
the amino group

Edited by

SAUL PATAI

The Hebrew University
Jerusalem, Israel

1968

INTERSCIENCE PUBLISHERS

a division of John Wiley & Sons

LONDON — NEW YORK — SYDNEY

First published by John Wiley & Sons Ltd. 1968

Library of Congress catalog card number 67–31072

SBN 470 66931 4

Made and printed in Great Britain by
William Clowes and Sons, Limited, London and Beccles

Contributing authors

Barbara E. C. Banks — Physiology Department, University College, London, England

D. V. Banthorpe — University College, London, England

Anthony R. Butler — St. Salvator's College, University of St. Andrews, Scotland

Brian C. Challis — St. Salvator's College, University of St. Andrews, Scotland

Gabriel Chuchani — Instituto Venezolano de Investigaciones Científicas, IVIC, Caracas, Venezuela

R. Daudel — Sorbonne and Centre de Mécanique Ondulatoire Appliquée, Paris, France

M. S. Gibson — Faculty of Technology, University of Manchester, England

David M. Lemal — Dartmouth College, Hanover, New Hampshire, U.S.A.

R. Bruce Martin — University of Virginia, Charlottesville, Virginia, U.S.A.

J. W. Smith — Bedford College, London, England

Paula Y. Sollenberger — University of Virginia, Charlottesville, Virginia, U.S.A.

Emil H. White — Department of Chemistry, The Johns Hopkins University, Baltimore, Maryland, U.S.A.

Y. Wolman — Department of Organic Chemistry, The Hebrew University of Jerusalem, Israel

David J. Woodcock — Department of Chemistry, The Johns Hopkins University, Baltimore, Maryland, U.S.A.

Jacob Zabicky — Institute for Fibres and Forest Products Research, Jerusalem, Israel

Foreword

The principles governing the plan and pattern of the present volume have been summarized in the Preface to the series 'The Chemistry of the Functional Groups'.

Out of the originally planned contents, three chapters failed to materialize. These should have been chapters on the 'Photochemistry of the Amino Group', on the 'Syntheses and Uses of Isotopically Labelled Amines' and on 'Enamines'.

Jerusalem, September 1967   SAUL PATAI

The Chemistry of the Functional Groups
Preface to the series

The series 'The Chemistry of the Functional Groups' is planned to cover in each volume all aspects of the chemistry of one of the important functional groups in organic chemistry. The emphasis is laid on the functional group treated and on the effects which it exerts on the chemical and physical properties, primarily in the immediate vicinity of the group in question, and secondarily on the behaviour of the whole molecule. For instance, the volume *The Chemistry of the Ether Linkage* deals with reactions in which the C—O—C group is involved, as well as with the effects of the C—O—C group on the reactions of alkyl or aryl groups connected to the ether oxygen. It is the purpose of the volume to give a complete coverage of all properties and reactions of ethers in as far as these depend on the presence of the ether group, but the primary subject matter is not the whole molecule, but the C—O—C functional group.

A further restriction in the treatment of the various functional groups in these volumes is that material included in easily and generally available secondary or tertiary sources, such as Chemical Reviews, Quarterly Reviews, Organic Reactions, various 'Advances' and 'Progress' series as well as textbooks (i.e. in books which are usually found in the chemical libraries of universities and research institutes) should not, as a rule, be repeated in detail, unless it is necessary for the balanced treatment of the subject. Therefore each of the authors is asked *not* to give an encyclopaedic coverage of his subject, but to concentrate on the most important recent developments and mainly on material that has not been adequately covered by reviews or other secondary sources by the time of writing of the chapter, and to address himself to a reader who is assumed to be at a fairly advanced post-graduate level.

With these restrictions, it is realized that no plan can be devised for a volume that would give a *complete* coverage of the subject with *no* overlap between the chapters, while at the same time preserving the readability of the text. The Editor set himself the goal of attaining *reasonable* coverage with *moderate* overlap, with a minimum of

cross-references between the chapters of each volume. In this manner, sufficient freedom is given to each author to produce readable
quasimonographic chapters.

The general plan of each volume includes the following main
sections:

(a) An introductory chapter dealing with the general and theoretical
aspects of the group.

(b) One or more chapters dealing with the formation of the functional group in question, either from groups present in the molecule,
or by introducing the new group directly or indirectly.

(c) Chapters describing the characterization and characteristics of
the functional groups, i.e. a chapter dealing with qualitative and
quantitative methods of determination including chemical and
physical methods, ultraviolet, infrared, nuclear magnetic resonance,
and mass spectra; a chapter dealing with activating and directive
effects exerted by the group and/or a chapter on the basicity, acidity
or complex-forming ability of the group (if applicable).

(d) Chapters on the reactions, transformations and rearrangements
which the functional group can undergo, either alone or in conjunction with other reagents.

(e) Special topics which do not fit any of the above sections, such as
photochemistry, radiation chemistry, biochemical formations and
reactions. Depending on the nature of each functional group treated,
these special topics may include short monographs on related functional groups on which no separate volume is planned (e.g. a chapter
on 'Thioketones' is included in the volume *The Chemistry of the Carbonyl
Group*, and a chapter on 'Ketenes' is included in the volume *The
Chemistry of Alkenes*). In other cases, certain compounds, though containing only the functional group of the title, may have special
features so as to be best treated in a separate chapter as e.g. 'Poly-
ethers' in *The Chemistry of the Ether Linkage*, or 'Tetraaminoethylenes'
in *The Chemistry of the Amino Group*.

This plan entails that the breadth, depth and thought-provoking
nature of each chapter will differ with the views and inclinations of
the author and the presentation will necessarily be somewhat uneven.
Moreover, a serious problem is caused by authors who deliver their
manuscript late or not at all. In order to overcome this problem at
least to some extent, it was decided to publish certain volumes in
several parts, without giving consideration to the originally planned
logical order of the chapters. If after the appearance of the originally

planned parts of a volume, it is found that either owing to non-delivery of chapters, or to new developments in the subject, sufficient material has accumulated for publication of an additional part, this will be done as soon as possible.

It is hoped that future volumes in the series 'The Chemistry of the Functional Groups' will include the topics listed below:

The Chemistry of the Alkenes (published)
The Chemistry of the Carbonyl Group (published)
The Chemistry of the Ether Linkage (published)
The Chemistry of the Amino Group (published)
The Chemistry of the Nitro Group (in press)
The Chemistry of Carboxylic Acids and Esters (in press)
The Chemistry of the Carbon–Nitrogen Double Bond
The Chemistry of the Cyano Group (in preparation)
The Chemistry of the Carboxamido Group (in preparation)
The Chemistry of the Carbon–Halogen Bond
The Chemistry of the Hydroxyl Group (in preparation)
The Chemistry of the Carbon–Carbon Triple Bond
The Chemistry of the Azido Group
The Chemistry of Imidoates and Amidines
The Chemistry of the Thiol Group
The Chemistry of the Hydrazo, Azo and Azoxy Groups
The Chemistry of Carbonyl Halides
The Chemistry of the SO, SO_2, $-SO_2H$ and $-SO_3H$ Groups
The Chemistry of the $-OCN$, $-NCO$ and $-SCN$ Groups
The Chemistry of the $-PO_3H_2$ and Related Groups

Advice or criticism regarding the plan and execution of this series will be welcomed by the Editor.

The publication of this series would never have started, let alone continued, without the support of many persons. First and foremost among these is Dr. Arnold Weissberger, whose reassurance and trust encouraged me to tackle this task, and who continues to help and advise me. The efficient and patient cooperation of several staff-members of the Publisher also rendered me invaluable aid (but unfortunately their code of ethics does not allow me to thank them by name). Many of my friends and colleagues in Jerusalem helped me in the solution of various major and minor matters and my thanks are due especially to Prof. Y. Liwschitz, Dr. Z. Rappoport and Dr. J. Zabicky. Carrying out such a long-range project would be quite impossible without the non-professional but none the less essential participation and partnership of my wife.

The Hebrew University,   SAUL PATAI
Jerusalem, ISRAEL

Contents

CHAPTER **1**

General and theoretical

R. Daudel

Sorbonne and Centre de Mécanique Ondulatoire Appliquée, Paris, France

I. THE NATURE OF THE CHEMICAL BOND AND THE MAIN METHODS OF CALCULATING MOLECULAR WAVE FUNCTIONS

A. Some Aspects of Molecular Structure

In order to help the readers who are not specialized in the field of quantum chemistry we will summarize in this first section some important results on the nature of the chemical bond and the main ideas which form the basis of the usual methods of calculating electronic wave functions[1].

It will be convenient to introduce the subject by the analysis of the electronic structure of an atom. Let us take, as an example, a helium atom in its first excited state (which is a triplet state). In the old

1

theory of Bohr this state would correspond to one K electron in a certain circular orbit and one L electron in another one. From the wave-mechanical point of view the electronic structure of the atom appears to be rather different.

First of all, since no experiments have as yet been devised to determine the trajectories of electrons in atoms and molecules, it is assumed that such trajectories (if they exist) cannot be known. The wave mechanics can only give us a procedure to calculate *the probability* of finding an electron at a certain instant in such and such a small volume of an atom or molecule.

However, the main forces which from the wave-mechanical viewpoint are responsible for the 'motion' of electrons and nuclei, remain essentially the same as in classical mechanics; these forces are the coulombic interactions (repulsive between electrons or between nuclei, attractive between an electron and a nucleus). The Coulomb attraction that the nucleus of an atom exerts on the electrons holds them in a very small region of space, despite the repulsive forces between the electrons. Likewise, the repulsion between the nuclei of a molecule is compensated for by the attraction of electrons for the nuclei. But furthermore, we must take into account the spin of the electrons. Let us recall that this kinetic moment is quantized, in such a way, that if its projection along a given axis is measured, only one of the two values $\pm \frac{1}{2}h/2\pi$ (where h is Planck's constant) can be found.

Obviously it is more difficult to obtain a simple geometrical picture of an atom in the framework of the wave mechanics than with the theory of Bohr.

However, such a picture is very useful, especially for chemists who like to use intuitive concepts. This is why the notion of 'loge'[2] has been introduced into the wave mechanics.

Let us go back to our helium atom in its first excited state. Let us consider a sphere of radius r (this value being completely arbitrary) with its center at the nucleus. With the help of wave mechanics it is possible to calculate the probability P of finding one electron, and one only, in this sphere. When r is very small this probability is also very small because the sphere is generally empty. When r is very large, P again will be very small because now the sphere will generally contain the two electrons (and not one only). Thus, intuitively, we must anticipate that P will possess a maximum for at least one value of r. The curve[3] of Figure 1 shows that this is true. The maximum is large as it corresponds to $P = 0.93$. The corresponding radius is $1.7\ a_0$. ($a_0 = 0.529$Å being the atomic unit of length.) We shall say

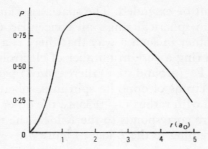

FIGURE 1. The probability P as a function of r.

that the best division of the atomic space into spherical loges is obtained when $r = 1.7\ a_0$ and that there is a probability of 93 per cent of finding one electron, and only one, in this sphere (the other one being outside). Therefore Figure 2 symbolizes the most probable organization of the electrons of helium in its first excited state. We are thus quite naturally led to associate the sphere of radius $1.7\ a_0$ with the K shell and the rest of the space with the L shell.

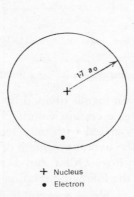

+ Nucleus
• Electron

FIGURE 2. The best decomposition in loges for the helium atom (first excited state).

It is important to point out that the K and L shells are associated with some portion of space, but not with a particular electron. It would not be convenient to speak of a K electron or an L electron from the wave-mechanical point of view, since we know that the various electrons of a system are assumed to be undistinguishable.

These results can be extended. The space associated with an atom can be cut up into spherical rings, all concentric with the nucleus, one built on the other in such a way that there is a high probability of finding in each ring a certain number of electrons[4]. For example, in the fluoride ion F^- (ground state) there is an 81 per cent probability of finding two electrons of opposite spin in a sphere with the center at the nucleus and with radius $r = 0.35\ a_0$, the other 8 electrons being outside. The sphere corresponds to the K loge, the remaining part of the space corresponding to the L loge (Figure 3).

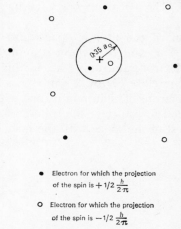

• Electron for which the projection
   of the spin is $+1/2\ \dfrac{h}{2\pi}$

○ Electron for which the projection
   of the spin is $-1/2\ \dfrac{h}{2\pi}$

FIGURE 3. The best decomposition in loges for the fluorine negative ion.

If the volume of a given loge is divided by the number of electrons which it usually contains, a certain volume v is obtained which gives an idea of the space associated with one electron in the loge. Moreover, we can evaluate the average value p of the electronic potential which is exerted in the loge. Odiot and Daudel[5] observed that for all atoms and all shells the following relation applies:

$$p^{\frac{3}{2}}v = \text{constant}$$

A kind of Boyle–Mariotte law exists between the 'electronic pressure' p and the volume v associated with one electron in an atomic loge.

Another important feature arises from the Pauli principle or in other words from the symmetry of the electronic wave functions. In wave mechanics the probability dP of finding an electron in a volume dv surrounding a point M is written as:

$$dP = |\psi_{(M)}|^2\ dv$$

where $\psi_{(M)}$ is the de Broglie wave function which satisfies the Schrö-dinger equation. For a system of electrons this writing is generalized and

$$dP_{12} = |\psi(M_1, \omega_1, M_2, \omega_2)|^2 \, dv_1 \, dv_2$$

represents the probability of finding electron 1 with a projection of the spin equal to ω_1 in a volume dv_1 surrounding the point M_1, and electron 2 with a projection of the spin ω_2 in a volume dv_2 at point M_2. As a consequence of the indistinguishability of electrons it is easy to show that ψ must be symmetric or antisymmetric with respect to a permutation of the electronic coordinates. That is to say:

$$\psi(M_1, \omega_1, M_2, \omega_2) = \pm \psi(M_2, \omega_2, M_1, \omega_1)$$

The *Pauli principle* consists of selecting the sign minus. We are led to the very important relation:

$$\psi(M_1, \omega_1, M_2, \omega_2) = - \psi(M_2, \omega_2, M_1, \omega_1) \tag{1}$$

Let us calculate the probability of finding the two electrons in the same volume dv at point M with the same projection of the spin ω. This probability will be:

$$dP = |\psi(M, \omega, M, \omega)|^2 \, dv \, dv$$

but the relation (1) becomes:

$$\psi(M, \omega, M, \omega) = - \psi(M, \omega, M, \omega)$$

which obviously shows that:

$$\psi(M, \omega, M, \omega) = 0$$

Therefore the probability of finding two electrons with the same projection of the spin in the same small volume is zero. In other words two electrons with the same projection of the spin (or, in short, 'with the same spin') do not like to occupy the same small volume of the space. Obviously this is also true for two electrons of opposite spin as there is always the Coulomb repulsion, but when the spins are the same a stricter repulsion is added to this Coulomb repulsion. This is why two electrons with opposite spins are sometimes said to be coupled. This does not mean that they like to occupy the same small volume of space. The interaction between them remains a strong repulsion but the repulsion is less severe than when they have the same spin. In conclusion it may be said that two electrons with opposite spin can be found in a smaller portion of space than two electrons possessing the same spin.

Another way of picturing this phenomenon is to compute the most probable configuration of the electrons of various atoms. The most probable configuration of an atom is the set of electronic coordinates which corresponds to the highest maximum of the modulus of the wave function. Figure 4 represents the most probable configuration of

Be (^3P)   Be (^4P)   C (^5S)

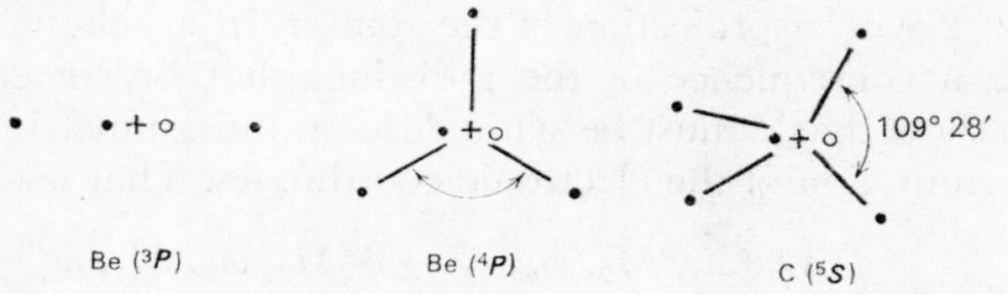

FIGURE 4. The most probable electronic configuration of small atoms.

some small atoms, computed by Linnet and Poe[6]. For beryllium in its ^3P state, the most probable configuration corresponds to two electrons of opposite spins at the nucleus and two electrons of the same spin at $2 \cdot 7\ a_0$ from the nucleus; the angle formed by these two electrons and the nucleus is therefore 180°. In the case of boron in its ^4P state, two electrons of opposite spins are found at the nucleus and three electrons with the same spin are found $2 \cdot 45\ a_0$ from the nucleus at the vertices of an equilateral triangle. In the case of carbon (^5S) four electrons possessing the same spin are found at the vertices of a regular tetrahedron. It is observed, as expected, that in the same loge, the electrons possessing the same spin tend to form the largest possible angles with the nucleus.

The notion of loge is also helpful in describing the electronic structure of molecules. In a good division of a molecule into loges, some loges usually appear which were also representative of the free atoms before bonding. Such loges are said to be *loges of the cores* while the others can be called *loges of the bonds*. As an example let us take the case of the lithium molecule Li_2. Figure 5 represents a good division

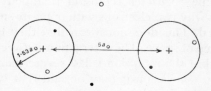

FIGURE 5. A good decomposition in loges for the lithium molecule.

of this molecule[3] into loges. In the two spherical loges which look like the K loges of the lithium atoms there is a probability of 0·96 of finding two electrons (and two only) with opposite spins. Therefore the distribution of the electrons symbolized by Figure 5 has a very high probability. The region of space outside of the two spheres where there is also a high probability of finding a pair of electrons, can be considered as the loge corresponding to a two-electron bond.

Another useful notion which gives information on the nature of the chemical bond is the *density difference function* given by the equation:

$$\delta(M) = \rho(M) - \rho^f(M)$$

where $\rho(M)$ is the actual electronic density at a given point M of a molecule, and $\rho^f(M)$ is the electronic density which would occur at this point if the density in the molecule were the simple sum of the density in the free atoms. Therefore, at any point where $\delta(M)$ is positive, an increase of electronic density results from the binding. On the other hand, in any region where $\delta(M)$ is negative, the binding has led to a decrease in the electronic density. Figure 6 shows the variation of $\delta(M)$ along the line of the nuclei of the hydrogen molecule H_2[7]. It is seen that between the nuclei $\delta(M)$ is positive. In agreement with chemical intuition the chemical binding produces an increase in the electronic density in this region.

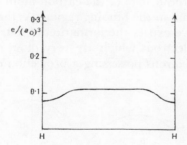

FIGURE 6. The density difference function in the hydrogen molecule.

If we consider now a molecule like O_2 we are led to a rather different conclusion. Figure 7 corresponds to that case[8]. There is no increase of electronic density between the nuclei but an annular region centered on the axis of the molecule where $\delta(M)$ is positive. The fact that there is no increase of electronic density along the nuclear axis is probably due to the presence of four electrons (two of each spin) in this vicinity. The strong repulsion between electrons possessing the same spin tends to place them outside the small space surrounding the bond axis. Therefore we can say that between the cores of small atoms near the bond axis there is only room for two electrons with opposite spins.

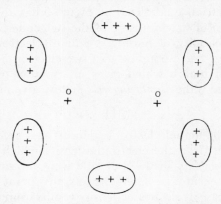

FIGURE 7. The density difference function in the oxygen molecule.

This is the reason why a molecule in which the number of electrons of the loges of the bonds is twice the number of adjacent atomic cores usually contains two-electron bonds. Let us consider methane. This molecule contains ten electrons. In a good division into loges two will be associated with the K loge of the carbon atom in such a way that eight electrons remain in the binding region. As there are four pairs of neighboring cores we expect the formation of four C-H loges associated with the two electrons which are very often near the bond axis (Figure 8). Four electrons possessing a projection of the spin $+\frac{1}{2}h/2\pi$

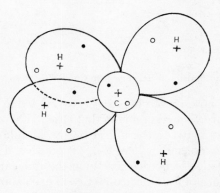

FIGURE 8. Methane.

will very often be in the binding region surrounding the carbon core. They will tend to form angles of 109° 28′ with the carbon nucleus. The same will be true for the four electrons of spin $-\frac{1}{2}\,h/2\pi$. As in the vicinity of each C—H axis there is very often one electron of positive spin (and also one of negative spin) we must expect that the angle between the C—H axes will also be 109° 28′. This explains why methane is a tetrahedral molecule.

Now let us consider the molecule NH_3 (the simplest amine!). It also contains ten electrons and two will be associated with the K loge of the nitrogen atom and eight electrons will remain in the binding region. As the total projection of the spin of such a molecule is zero this region will often contain four electrons with positive spin and four with negative spin. In the case of CH_4, the electrons possessing the same spin tended to form tetrahedral angles but as in NH_3 there are only three pairs of neighboring cores, six electrons only will be used to form three N—H bonds, while two electrons will remain as a lone pair (Figure 9). However, the angle between the N—H bonds will again be of the order of 109°, explaining why NH_3 is pyramidal. Obviously the lone pair can be considered as 'a potential bond'. If a proton is added to the system, the $NH_4{}^+$ ion is obtained (Figure 9) where the four N—H bonds produce a structure very similar to that of methane.

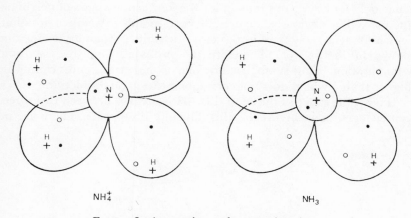

NH$_4^+$ NH$_3$

FIGURE 9. Ammoniac and ammonium ion.

When, as in CH_4, NH_3, $NH_4{}^+$, it is possible to find, between two neighboring cores, a good loge associated with a certain number n of electrons possessing a given organization of spin, it can be said that an n-electron localized bond between these cores has been defined.

But it is easy to see that in certain cases it is not possible to localize, between only two cores, a region where there is a high probability of finding n electrons with a precise organization of spins. This is the case with diborane B_2H_6. This molecule contains 16 electrons. Four of them can be associated with the two K loges of boron. Twelve electrons remain for binding purposes. Diborane consists of eight pairs of neighboring cores. Therefore it is not possible to associate a two-electron localized bond with each pair of neighboring cores. From the experimental viewpoint the four outer B—H bonds have the same behavior as normal localized two-electron bonds. Then, we are led to try a division into loges similar to the one described in Figure 10a,

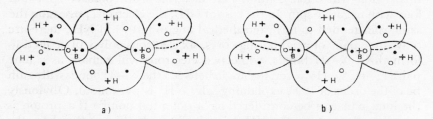

FIGURE 10. Decompositions in loges of diborane.

since only four electrons remain for the four central pairs of neighboring cores. But obviously, Figure 10b corresponds to another situation, which for symmetry reasons possesses the same probability as the one shown in Figure 10a. Therefore the probability of the electronic configuration being symbolized by one of these two figures cannot be higher than $\frac{1}{2}$ which does not correspond to a good division into loges.

If now, the space of the central B—H is divided between two three-center loges, as in Figure 11, this difficulty disappears. There are no

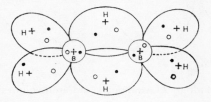

FIGURE 11. Another decomposition in loges of diborane.

a priori reasons to exclude the possibility of finding such good loges. As they are now extended over more than two cores we shall say that they correspond to delocalized bonds; more precisely to two-electron bonds delocalized over three centers. In conclusion, when it is not possible to find a good loge between two cores the loge will be extended over a greater number of cores and an n-electron bond delocalized over p centers will be considered.

B. The Main Methods of Calculating Electronic Molecular Wave Functions[9]

Assuming that the Born–Oppenheimer approximation is convenient, the main ideas forming the basis of the principal methods of calculating electronic molecular wave functions will now be summarized.

In the approximation the calculation of this function is done as if the nuclei were fixed. To obtain the space part Φ of the wave function we are led to solve a Schrödinger equation

$$\mathbf{H}\Phi = W\Phi$$

where $\mathbf{H}$ is the non-relativistic Hamiltonian and W the energy of an electronic state of the molecule. The total electronic wave function Ψ is obtained by multiplication of Φ by a convenient spin function σ and after transformation of the product $\Phi\sigma$ in order to obey the Pauli principle, we can write

$$\Psi = A\Phi\sigma \tag{2}$$

if A represents the convenient antisymmetrisor.

As it is not possible to solve the Schrödinger equation (except for very simple molecules) it is customary to use approximate solutions. An important starting point to find such solutions lies in the independent electron model. In this model the repulsion between the electrons is neglected. The corresponding part of the Hamiltonian $\mathbf{H}$ vanishes and only a certain part $\mathbf{H}^0$ remains. Thus, one must solve the equation

$$\mathbf{H}^0\Phi^0 = W^0\Phi^0 \tag{3}$$

It is easy to show that

$$\Phi^0 = \varphi_1(M_1)\varphi_2(M_2)\ldots\varphi_n(M_n) \tag{4}$$

is a solution of equation (3) if n electrons are involved in the problem, and if the various φ_i obey the equation

$$h\varphi_i = \varepsilon_i\varphi_i \tag{5}$$

where h is the Hamiltonian corresponding to the motion of one electron in the field of the nuclei of the molecule under consideration. If this solution is taken

$$W^0 = \varepsilon_1 + \varepsilon_2 + \cdots \varepsilon_n + \cdots \qquad (6)$$

The functions φ_i are called molecular orbitals. Obviously the discussion applies to the case of one atom (which may be considered to be a monoatomic molecule), and then the φ_i are atomic orbitals. The relation (6) shows that the energy associated with a given state of the independent-electron model is the sum of the energies ε_i associated with the various orbitals introduced in the wave function Φ^0.

If now we consider the total wave function obtained from equation (2) we have

$$\Psi^0 = A\Phi^0\sigma \qquad (7)$$

and it is readily seen that this function is a linear combination of Slater determinants. For this reason it is not possible to introduce in Φ^0 a given orbital more than twice (one with the α spin function corresponding to the positive projection of the spin and one with the β spin function corresponding to the negative projection), otherwise the determinants will vanish.

But the solutions of the independent model are very far from exact solutions because the repulsion between the electrons is far from being negligible. For this reason this interaction must be taken into account at least in part. To do so the general form of equation (7) can be kept, that is to say it is assumed that Φ^0 is a product of mono-electronic functions φ, but that these φ do not obey equation (5) and must be chosen by a variational procedure. The functions φ obtained in this way are again called molecular orbitals. This is the essence of the *self-consistent field method*. In fact it is very tedious to solve the exact equation corresponding to the self-consistent field method. A new approximation is usually made; it is assumed that a molecular orbital can be expanded as a linear combination of the atomic orbitals associated with the atoms constituting the molecule: this is the LCAO approximation.

Some aspects of the calculation of wave functions in this framework will be demonstrated, using the simplest amine, NH_3, as an example. The ammonia molecule has been studied by several authors[10] using the self-consistent field method with the LCAO approximation. Kaplan's results will be analyzed. Figure 12 shows how the axes are chosen. The three hydrogen nuclei $H_{(1)}$, $H_{(2)}$ and $H_{(3)}$ lie in the plane xOy, $H_{(1)}$ being on Ox. The nitrogen nucleus is on the z axis. First of all, the most important atomic orbitals must be listed. As the

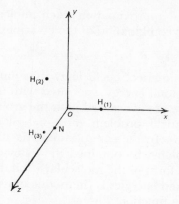

FIGURE 12. Coordinate axis for NH_3.

ground-state wave function must be computed, only those atomic orbitals which are usually used in the representation of the ground states of the free atoms will be considered. These are:

— the $1s$ orbital corresponding to the K shell for each hydrogen atom (Let a, b and c be the $1s$ orbitals of $H_{(1)}$, $H_{(2)}$ and $H_{(3)}$ respectively)

— the $1s$ orbital corresponding to the K shell of the nitrogen atom

— and the $2s$, $2p_x$, $2p_y$ and $2p_z$ orbitals associated with the L shell of the same atom.

TABLE 1. Ammonia molecular orbitals.

$\varepsilon_1 = -15 \cdot 5682$

$\varphi_1 = 1 \cdot 001(1s) - 0 \cdot 0033(2s) - 0 \cdot 0020(2p_z) - 0 \cdot 0013(h_0)$

$\varepsilon_2 = -1 \cdot 1482$

$\varphi_2 = 0 \cdot 0286(1s) + 0 \cdot 7591(2s) + 0 \cdot 1616(2p_z) - 0 \cdot 2711(h_0)$

$\varepsilon_3 = \varepsilon_4 = -0 \cdot 6625$

$\varphi_3 = 0 \cdot 6195(2p_x) + 0 \cdot 4860(h_x)$

$\varphi_4 = 0 \cdot 6195(2p_y) + 0 \cdot 4860(h_y)$

$\varepsilon_5 = -0 \cdot 4646$

$\varphi_5 = 0 \cdot 0257(1s) - 0 \cdot 4418(2s) + 0 \cdot 8956(2p_z) - 0 \cdot 2582(h_0)$

with,

$$h_0 = n_0(a + b + c)$$
$$h_x = n_x(a - \tfrac{1}{2}(b + c))$$
$$h_y = n_y(b - c)$$

n_0, n_x, n_y being normalization coefficients.

Following the LCAO approximation each molecular orbital φ_i is expanded as a linear combination of these various atomic orbitals

$$\varphi_i = k_1a + k_2b + k_3c + k_41s + k_52p_x + k_62p_y + k_72p_z + k_82s$$

where the k's are numerical coefficients that must be computed in such a way that the total energy associated with the wave function is a minimum, since we are concerned with the ground state.

Thus a mathematical problem must be solved which leads to Roothaan's equations. In the case of ammonia these equations have been solved by Kaplan who obtained the following results. Table 1 contains the explicit form of the molecular orbitals obtained with the values of the associated energies ε_i (in atomic units).

The corresponding function Φ is

$$\Phi = \varphi_1(M_1)\varphi_1(M_2)\varphi_2(M_3)\varphi_2(M_4)\varphi_3(M_5)\varphi_3(M_6)$$
$$\varphi_4(M_7)\varphi_4(M_8)\varphi_5(M_9)\varphi_5(M_{10})$$

and the total electronic wave function

$$\Psi = A\Phi\sigma$$

can be written as

$$\Psi = \det \varphi_1(M_1)\overline{\varphi_1(M_2)}\varphi_2(M_3)\overline{\varphi_2(M_4)}\varphi_3(M_5)\overline{\varphi_3(M_6)}$$
$$\varphi_4(M_7)\overline{\varphi_4(M_8)}\varphi_5(M_9)\overline{\varphi_5(M_{10})}$$

where 'det' means the Slater determinant built by permutation of the coordinates in the product of orbitals; a molecular orbital with a bar, e.g. $\bar{\varphi}$, being associated with a β spin function and an orbital without a bar being associated with an α spin function. The total energy corresponding to this wave function is $-56 \cdot 266$ a.u. There is a fair agreement with the experimental value of $-56 \cdot 596$, the error being smaller than one per cent. It is interesting to note that this energy is not the sum of the energies associated with the molecular orbitals introduced in the wave function. Such a relation only applies in the framework of the independent-electron model and disappears when the repulsion between electrons is taken into account.

Other properties can be computed from the wave function but as is well known the precision is not as good as in the case of the total energy. If the first ionization energy is calculated using Koopman's approximation, that is to say if it is taken as equal to $-\varepsilon_5$ a value of about 14 ev is obtained. Experiment gives 11 ev. Furthermore the wave function Ψ corresponds to a dipole moment of $1 \cdot 82$ D, the experimental value being $1 \cdot 46$.

II. SOME ASPECTS OF THE PROPERTIES OF AMINES IN THEIR GROUND STATE

A. A Theoretical Discussion of some Physical Properties of Aliphatic Amines in their Ground State

The general formula of aliphatic amines can be written as

where R^1, R^2 and R^3 are hydrogen atoms or alkyl groups. As has been stated the ammonia molecule NH_3 can be considered as the simplest aliphatic amine. In fact many aliphatic amines possess the most important properties of ammonia. They all contain a lone pair of electrons on the nitrogen atom and three two-electron localized bonds starting from this atom. Therefore, as in the case of NH_3, four pairs of electrons are found surrounding the nitrogen core. For this reason it must be expected that the $R\widehat{N}R$ angles will be about 109° except when steric hindrance or geometrical strain appears.

For example, in trimethylamine the $C\widehat{N}C$ angle is 108° 40′ ± 1[11]. Table 2 contains other geometrical data concerning this molecule.

TABLE 2.   Valence angles and interatomic distances in trimethylamine.

$C\widehat{N}C$ angle	108·7° ± 1[11] or 108 ± 4°[12]
C–N distance	1·472 ± 0·008 Å[11] or 1·47 ± 0·002 Å[12]
$H\widehat{C}H$ angle	108·5 ± 1·5°[11]
C–H distance	1·090 Å[11]

The electronic structure of the N—H bond in ammonia has been carefully investigated by Tavard[13] from the theoretical viewpoint. Since the electronic structure of the N—H bond in aliphatic amines is probably very similar, the main results obtained by Tavard will be reported. He calculated the electronic density in NH_3 using a very elaborate electronic wave function obtained by Moccia[14]. Figure 13 shows lines corresponding to various values of the density difference function $\delta(M)$. From this figure it appears that during the formation of the chemical bond there is some electron transfer from the hydrogen to the nitrogen (as δ is negative near the hydrogen and positive near the nitrogen in the vicinity of the bond axis). This transfer is in good agreement with chemical intuition as nitrogen is considered to be more electronegative than hydrogen. Furthermore, Figure 13 shows a region in which δ is positive and which forms an angle of about 109° with the bond axis. This result is consistent with the idea of the localization of a lone pair near this region.

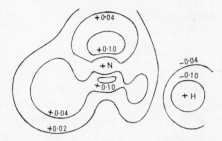

FIGURE 13. Density difference function in NH_3.

Electron diffraction experiments[15] confirmed this effect of the binding on the distribution of electron density in ammonia. Figure 14, reproduced from the paper of Iijima and coworkers[15], shows the variation of density as a function of the diffraction angle. Curve a corresponds to theoretical calculations for which it is assumed that

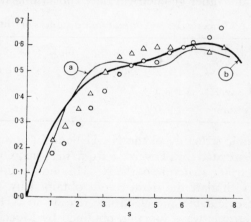

FIGURE 14. Intensity scattered as a function of the diffraction angle.

the electron density at a given point of the molecule is the sum of the electron densities of the free atoms, curve b is calculated from the actual density in the molecule when the Moccia electron wave function is used and finally the points Δ and O correspond to the experimental results. First of all, it can be seen that there is a significant difference between curve a and curve b for the small angles. This difference is a theoretical measurement of the effect of the binding on the electron density (as is the δ function). The agreement observed between curve b and the experimental results inspires confidence in this measurement.

The N—H stretching vibration corresponds to an infrared band between 3500 and 3400 cm^{-1} in the spectra of secondary aliphatic amines taken in dilute solution in carbon tetrachloride to avoid molecular association[16]. Under these conditions primary aliphatic amines show two bands near 3500 and 3400 cm^{-1} (asymmetrical and symmetrical vibrations)[16]. Among the vibration frequencies associated with ammonia two harmonic frequencies are found in approximately the same region (3506 and 3577 cm^{-1})[17]. Therefore we can anticipate that the main force constant K_{RR} will be of the same order of magnitude for NH$_3$, as well as for secondary and primary amines. This fact supports the hypothesis that the electronic structure of the N—H bonds would not be very different in all these molecules.

The force constant K_{RR} of ammonia has been evaluated as 7·17 from the infrared spectrum. Allavena calculated the same constant[18] (with the help of an electronic wave function similar to Moccia's function, but a little less precise) and obtained a value of 9. The agreement between experiment and theory is not very satisfactory and could certainly be improved if Moccia's function itself were used.

The first ionization energy of aliphatic amines, as in the case of NH$_3$, is essentially the production of an electron hole in the lone-pair region. It must therefore be expected that the first ionization energy is of the same order of magnitude in all this set of molecules. This is found to be the case since in ammonia the first ionization energy has been evaluated as 10·5[19] or 11[20] ev, the corresponding energy being estimated as 9 ev in trimethylamine[21].

The electric dipole moment of aliphatic amines also contains a contribution from the lone pair, but as the various bonds also contribute, the resulting value varies in a more significant manner in going from one molecule to another. For example, it is evaluated as 1·46 D in the case of ammonia, while values between 0·61[22] and 0·86 D[23] can be found in the literature for the dipole moment of trimethylamine.

B. Physical Properties of Aromatic Amines in their Ground State

Very little precise information is known about the geometry of aromatic amines. Some problems remain even in the case of the simplest aromatic amine, aniline. The main part of the molecule is certainly planar and the nitrogen is certainly in the plane of the aromatic ring. Figure 15 shows for example, the values of valence angles and inter-atomic distances in a parent compound[24]. It is seen that the geo-metrical organization of the aromatic ring is very similar to that of

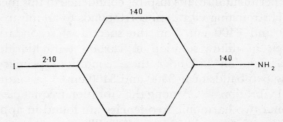

FIGURE 15. Interatomic distances in *para*-iodoaniline.

benzene, but the position of the two hydrogen atoms of the NH_2 group is not yet known exactly. Van Meerssche and Leroy[25] studied the geometrical structure of 2-bromo-4'-dimethylamino-2-cyanostil-bene by x-ray diffraction. Figure 16 shows the results concerned with the part of the molecule which contains the amino group. The authors claim that the dimethylamino group and the neighboring aromatic ring are coplanar. Furthermore, it is seen that again the geometry of the aromatic ring does not differ greatly from that of benzene. It should be noted that the length of the C—N bond connecting the amino group and the ring is only 1·34 Å. (The length of a normal C—N simple bond is 1·47 Å.)

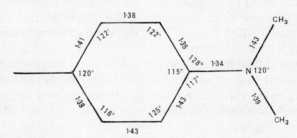

FIGURE 16. Geometry of a part of 2-bromo-4'-dimethylamino-2-cyanostilbene.

On the other hand, in their recent book Higasi and coworkers[26] remark that if the group NH_2 were in the plane of the ring, *p*-diamino-benzene should be non-polar like *p*-dichlorobenzene. This is not so, as the former has a large dipole moment of 1·5 D[27]. An explanation of this apparent discrepancy may lie in the fact that the amino group and the ring would not be coplanar in the gaseous or liquid phase, but would be so in the crystal used to observe the x-ray pattern.

No precise detailed experimental study of the electronic distribution in aromatic amines seems to be available. The only way to obtain such information is from theory. Let us consider the case of aniline. As this molecule contains a ring similar to that in benzene, data known about the latter will first be recalled. The total number of electrons is 42. If 12 electrons are accounted for as being associated with the six K loges corresponding to the carbon cores, 30 electrons remain to be associated with the loges of the bonds. The benzene molecule contains 12 pairs of neighboring cores. The number of electrons is therefore higher than twice the number of the pairs of neighboring cores. If two electrons are associated with each pair of neighboring cores, six additional electrons must still be accounted for. As the C—H bonds appear to be normal simple bonds a convenient division into loges will contain a two-electron localized bond associated with each C—H bond. The six remaining electrons have to be associated with the central hexagon. Figure 17a shows a possible division into loges but Figure 17b represents another one which for symmetry reasons has the same probability.

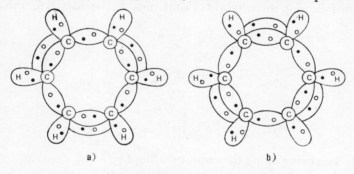

a)   b)

FIGURE 17. Decompositions in loges of benzene.

Therefore this probability will be at best equal to 0·5. This value does not correspond to a good division into loges. Figures 18a and 18b correspond to other divisions, which are poor for the same reasons. Thus the division shown in Figure 19 is usually adopted, in which there is a six-electron bond delocalized over the six carbon cores.

To calculate a corresponding electronic wave function the LCAO approximation must be introduced. In this approximation the molecular orbitals will be expanded on the atomic orbitals associated with the free atoms, that is to say a $1s$ orbital for each hydrogen atom (let us call them $h_1, h_2, \ldots, h_6$ respectively) and for each carbon atom the orbitals $1s, 2s, 2p_x, 2p_y, 2p_z$. Obviously the carbon $1s$ orbitals can be associated with the carbon K loges.

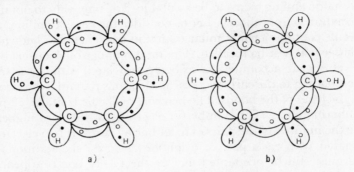

a)   b)

FIGURE 18. Other decompositions in loges of benzene.

To represent a two-electron localized bond a function must be built with important values in the corresponding loges. As the nuclei about a carbon atom ($C_{(1)}$ for example) make angles with it of approximately 120° the following three hybrid orbitals must be considered for this atom

$$tr_1 = \frac{1}{\sqrt{3}} 2s_1 + \frac{\sqrt{2}}{\sqrt{3}} 2p_{1x}$$

$$tr_1' = \frac{1}{\sqrt{3}} 2s_1 - \frac{1}{\sqrt{6}} 2p_{1x} + \frac{1}{\sqrt{2}} 2p_{1y}$$

$$tr_1'' = \frac{1}{\sqrt{3}} s_1 - \frac{1}{\sqrt{6}} 2p_{1x} - \frac{1}{\sqrt{2}} 2p_{1y}$$

the Ox axis being along the corresponding $C_{(1)}$—$H_{(1)}$ bond and the Oy axis in the plane of the ring, because the hybrid tr_1 points in the direction of the hydrogen atom, tr_1' and tr_1'' point respectively in the directions of the neighboring carbon atoms as seen in Figure 20. These hybrid orbitals are called *trigonal orbitals*. Therefore, associated with the $C_{(1)}$—$H_{(1)}$ bond is the bond orbital

$$ah_1 + btr_1$$

and with the localized $C_{(1)}$—$C_{(2)}$ bond, the bond orbital

$$ctr_1' + dtr_2''$$

and so on.

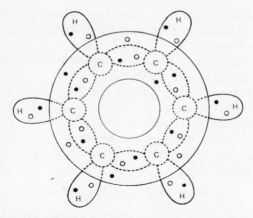

FIGURE 19. A better decomposition in loges of benzene.

Of all atomic orbitals only the $2p_z$'s remain unused. It is natural to associate them with the delocalized bond and to consider the molecular orbitals

$$\pi_i = s_{1i}2p_{1z} + s_{2i}2p_{2z} + \cdots + s_{6i}2p_{6z}$$

where the s_i's are unknown coefficients.

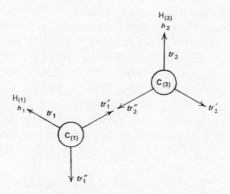

FIGURE 20. Hybridization in benzene.

As six electrons must be associated with the delocalized bonds at least three π orbitals will have to be introduced. To compute the coefficients s_i it is necessary to introduce additional approximations such as the Hückel [28] or the Pariser and Parr [29] approximation. Details of these approximations can be found elsewhere[1]. Table 3 contains the π orbitals obtained and the associated energies when the Hückel approximation is used[30].

The square of an s_i coefficient before a certain $2p_z$ is considered to be the electronic charge introduced by the π orbital in the corresponding carbon atom. It is seen, for example, that the orbital π_1 introduces a charge equal to $(0·408)^2$ on each carbon atom. As each of the three π orbitals must be used twice for the description of the ground state of benzene (as there are six electrons associated with the delocalized bond) the total electronic charge introduced in carbon 1 by the delocalized bond is

$$q_1 = 2(0·408)^2 + 2(0·577)^2 = 1$$

The same result is obtained for the other carbon atoms.

On the other hand, the quantity $s_{2i}s_{3i}$ is considered to be the contribution of the orbital π_i to the bond order[31] introduced between carbon 2 and carbon 3 by the delocalized bond. Then the bond order between carbon 1 and carbon 2 due to the presence of the delocalized bond is

$$p_{23} = 2(0·408)^2 + 2(0·500)^2 - 2(0·288)^2 = 0·666$$

The same bond order is obtained for the other pairs of adjacent carbon atoms.

Let us consider now the case of aniline. The orbitals $1s_N$, $2s_N$, $2p_{Nx}$, $2p_{Ny}$, $2p_{Nz}$ will have to be introduced to take account of the nitrogen atom in place of the orbital h_1, and the $1s$ orbitals associated with the hydrogen $H_{(7)}$ and $H_{(8)}$ belonging to the amino group must be considered. Let us call h_7 and h_8 these orbitals. Let us assume that aniline is a completely planar molecule. The $1s_N$ orbital will be associated with the nitrogen K loge. The $2s_N$, $2p_{Nx}$ and $2p_{Ny}$ orbitals will be combined in such a way as to produce three trigonal orbitals tr_N, tr'_N and tr''_N which will be used to represent the various two-electron localized bonds $C_{(1)}$—N, N—$H_{(7)}$ and N—$H_{(8)}$ (Figure 21). The

TABLE 3. Benzene molecular orbitals.

$$\varepsilon_1 = 2\beta$$
$$\pi_1 = 0·408(2p_{1z} + 2p_{2z} + 2p_{3z} + 2p_{4z} + 2p_{5z} + 2p_{6z})$$
$$\varepsilon_2 = \beta$$
$$\pi_2 = 0·500(2p_{2z} + 2p_{3z} - 2p_{5z} - 2p_{6z})$$
$$\varepsilon_3 = \beta$$
$$\pi_3 = 0·577(2p_{1z} - 2p_{4z}) + 0·288(2p_{2z} - 2p_{3z} - 2p_{5z} + 2p_{6z})$$

delocalized bond can therefore be extended over the nitrogen atom as the $2p_{Nz}$ orbital is not yet used. The π orbitals become

$$\pi_i = s_{1i}2p_{1z} + s_{2i}2p_{2z} + s_{3i}2p_{3z} + s_{4i}2p_{4z} + s_{5i}2p_{5z} + s_{6i}2p_{6z} + s_{Ni}2p_{Nz}$$

A simple calculation shows that the nitrogen atom introduces two electrons into the delocalized bond. Therefore to describe the ground state of aniline four different π orbitals must be considered, each of them being used twice.

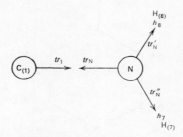

FIGURE 21. Hybridization in aniline.

Obviously, as in the case of benzene, additional approximations are necessary in calculating the coefficients s. But contrary to the case of benzene, the values of these coefficients depend on the values chosen for the parameters which characterize the atoms and the bonds. The parameters under consideration are therefore very often chosen in order to obtain a good agreement between experiment and calculation for certain properties of the molecule. In the present case, it is believed that the error introduced by assuming that aniline is completely planar is probably reduced by using this empirical choice of parameters.

A recent study of aniline using the Hückel method has been made by Fischer–Hjalmars[32]. The same author also used the more sophisticated Pariser and Parr approximation. Figure 22 represents the corresponding molecular diagrams showing the distribution of bond orders and atomic charges associated with the delocalized bonds. The presence of a bond order of the order of 0·3 along the C—N bond explains why it is shorter than a normal C—N two-electron bond. From the diagram showing the distribution of the electronic charge it is seen that in the delocalized bond there is a charge transfer of about 0·07 electrons from the nitrogen atom to the ring. The main result of this transfer is the presence of an excess of electrons at the *ortho* and *para* positions which therefore become negative.

It must also be pointed out that the charge is more negative in the *ortho* than in the *para* position when the Pariser and Parr method is used. This result seems in agreement with recent measurements of these charges based on n.m.r. experiments[33]. Finally, the contribution

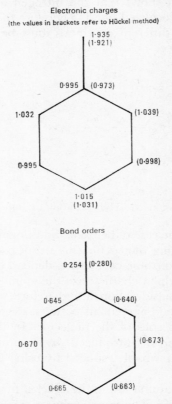

FIGURE 22. Molecular diagrams of aniline.

of the delocalized bond to the dipole moment of aniline is calculated as 1·2 D by the Hückel method and as 0·78 D by the Pariser and Parr method. The corresponding experimental estimation is 0·7 D[34].

C. Chemical Properties of Amines in their Ground State

The basic behavior of amines is certainly one of their most important chemical properties. If B denotes the neutral molecule the corresponding equilibrium constant can be written as:

$$B + H^+ \underset{K}{\rightleftharpoons} BH^+$$

and the equilibrium constant K is a measure of the basic strength of the amine.

Let us consider the equilibrium

$$A + B \underset{K}{\rightleftharpoons} C + D.$$

It is well known that K can be expressed as

$$K = \frac{f_A f_B}{f_C f_D} e^{-\Delta\varepsilon/\chi^T}$$

if the concentrations are small. In this expression the f's denote the various partition functions associated with the distribution of the various chemical species on the various translational, rotational and vibrational levels at the temperature T, and $\Delta\varepsilon$ represents the difference between the sum of the ground-state energies of molecules C and D and the sum of the ground-state energies of molecules A and B. Finally χ denotes the Boltzmann constant. Therefore

$$\Delta\varepsilon = \varepsilon_{0C} + \varepsilon_{0D} - (\varepsilon_{0A} + \varepsilon_{0B})$$

Now for a large molecule a ground-state energy ε_{0i} may be divided into various parts:

(a) the vibrational energy ε_v corresponding to the zero point energy,
(b) the energy ε_l associated with the localized bonds and the atomic cores,
(c) the energy ε_d associated with the delocalized bonds taking account of the interaction of these bonds with the cores and the localized bonds,
(d) the energy ε_{nb} associated with the interaction between non-bonded atoms, including the steric effect.

Therefore with obvious notations $\Delta\varepsilon$ can be written as

$$\Delta\varepsilon = \Delta\varepsilon_v + \Delta\varepsilon_l + \Delta\varepsilon_d + \Delta\varepsilon_{nb}$$

which leads to

$$K = \frac{f_A f_B}{f_C f_D} e^{-(\Delta\varepsilon_v + \Delta\varepsilon_l + \Delta\varepsilon_d + \Delta\varepsilon_{nb})/\chi^T}$$

But in many cases the reaction takes place in a solvent and then it is necessary to take into account the solvent effect. A simple way to do that, is to introduce a term $\Delta\varepsilon_s$ into $\Delta\varepsilon$ representing the difference between the solvation energies of the final products and the solvation energies of the initial products. But, as the solvation energy of a molecule depends on the temperature, we shall write this term as

$$\Delta \varepsilon_s(\mathrm{T})$$

Furthermore, an index s will be put on each partition function to recall that the solvent effect can affect greatly this function. (Very often, for example, the free rotation of molecules becomes impossible and is replaced by oscillation.) Finally the expression

$$K = \frac{f_A^s f_B^s}{f_C^s f_D^s} \, e^{-[\Delta \varepsilon_v + \Delta \varepsilon_l + \Delta \varepsilon_d + \Delta \varepsilon_{nb} + \Delta \varepsilon_s(\mathrm{T})]/\chi^{\mathrm{T}}} \tag{8}$$

is obtained, which shows that an equilibrium constant depends on six terms:

(a) the vibrational energy change $\Delta \varepsilon_v$
(b) the localized bond energy change $\Delta \varepsilon_l$
(c) the delocalized bond energy change $\Delta \varepsilon_d$
(d) the non-bonded atoms energy change $\Delta \varepsilon_{nb}$
(e) the solvation energy change $\Delta \varepsilon_s(\mathrm{T})$
(f) the ratio

$$f = \frac{f_A^s f_B^s}{f_C^s f_D^s}$$

of the partition functions.

To show how it is possible to use equation (8) in studying the basic strengths of amino compounds one should try to understand which are the main factors responsible for the change in base strength of the amino group from one amino acid to another. For such substances two different pK values must be considered; pK_1 corresponds to the equilibrium between

$$H_3N^+{-}CRR'{-}CO_2H \text{ and } H_3N^+{-}CRR'{-}CO_2^-;$$

and pK_2 is associated with the equilibrium between

$$H_3N^+{-}CRR'{-}CO_2^- \text{ and } H_2N{-}CRR'{-}CO_2^-.$$

Therefore the problem is the study of pK_2. Let us consider two amino acids with K_2 and K_2' as the corresponding equilibrium constants.

Taking account of equation (8)

$$\frac{K_2'}{K_2} = \frac{f'}{f} \, e^{-[\Delta\Delta \varepsilon_v + \Delta\Delta \varepsilon_l + \Delta\Delta \varepsilon_d + \Delta\Delta \varepsilon_{nb} + \Delta\Delta \varepsilon_s(\mathrm{T})]/\chi^{\mathrm{T}}} \tag{9}$$

In this equation

$$\Delta\Delta \varepsilon_i = \Delta \varepsilon_i' - \Delta \varepsilon_i$$

The theoretical prediction of the variation of the pK in the set of molecules under consideration requires the calculation of f'/f and of the five $\Delta\Delta\varepsilon$ terms which appears in the exponent.

This calculation has been made by Del Ré, Pullman and Yonezawa[35]. They assumed that the ratio f'/f does not differ significantly from one and that $\Delta\Delta\varepsilon_v$ is negligible.

Obviously $\Delta\Delta\varepsilon_l$ is a very important term because during the protonation of the amino group a new N—H bond appears and it is necessary to compute the energy of this new bond. For this calculation the Del Ré method[36] is used. In this method an orbital is associated with each bond and by the LCAO approximation this bond orbital is taken as a linear combination of the two convenient hybrids associated with the two atoms constituting the bond.

The $\Delta\Delta\varepsilon_d$ terms are calculated using the Hückel approximation and as all these approximations do not take account explicitly of the electrostatic interaction occurring between the atomic charges Q, an additional term

$$\underset{\mu<\nu}{\Delta\varepsilon}\ \frac{Q_\mu Q_\nu}{r_{\mu\nu}}$$

is calculated in which $r_{\mu\nu}$ is the interatomic distance between atom μ and atom ν. This term contains the main part of $\Delta\varepsilon_{nb}$ when no steric hindrance occurs. Furthermore estimation of the $\Delta\varepsilon_s(T)$ terms has shown that $\Delta\Delta\varepsilon_s(T)$ is negligible, and by plotting the experimental pK as a function of

$$\Delta\varepsilon_l + \Delta\varepsilon_d + \underset{\mu<\nu}{\Delta\varepsilon}\ \frac{Q_\mu Q_\nu}{r_{\mu\nu}}$$

a rather satisfactory relationship was observed between the pK's and the considered term[35].

Another theoretical discussion of the pK of amino compounds is given in the case of amino derivatives of molecules like pyridine, quinoline, isoquinoline and acridine. There is experimental evidence to show that with such molecules protonation occurs at the nitrogen belonging to the aromatic rings. Therefore the pK is not related to the protonation of the amino group. However, its study is extremely interesting because it permits a measurement of the perturbation brought to the heteroatom by the amino group. It was seen in Figure 21, that the amino group is able to increase significantly the electronic charge of certain atoms of the aromatic ring. It must be expected, therefore, that this group is able to increase the ease of protonation of the

2*

heteroatom and therefore to increase the pK_a of molecules like pyridine, quinoline and so on. This is exactly what experiment has shown.

Effectively the pK_a of pyridine at 20°c in water solution is 5·23[37] and the pK_a of p-aminopyridine is 9·17[37]. Therefore the dissociation constant of the positive ion is reduced by a factor of 10,000 as a result of the presence of an amino group, that is to say the amino derivative is much more basic than the pyridine itself.

Many authors[38] have discussed the role of $\Delta\varepsilon_d$ in the determination of the pK of such compounds. The importance of $\Delta\varepsilon_d$ is obvious, since during protonation the proton is directly bonded to the nitrogen heteroatom which belongs to the delocalized bond. The change in energy of the delocalized bond which results from this binding can be estimated by various methods. Figure 23 shows the results obtained[39] when the pK_a is plotted as a function of $\Delta\varepsilon_d$ calculated by

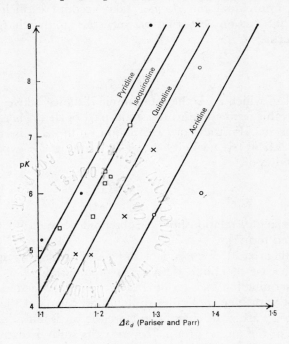

FIGURE 23. The pK as a function of $\Delta\varepsilon_d$.

the Pariser and Parr method. It is seen that the points corresponding to a set of compounds derived from a given skeleton (pyridine, iso-quinoline, quinoline or acridine) lie along a straight line.

The role of the solvent effect has been discussed by Chalvet, Daudel and Peradejordi[39, 40].

The solvation energy of a molecule can be divided into three parts: the cavitation energy ε_{sc} associated with the hole that the molecule creates in the solvent, the orientation term ε_{so} due to the fact that the molecule modifies the average orientations of neighboring molecules of the solvent and the interaction energy ε_{si} due to the intermolecular forces which appear between the solvated molecule and the solvent. With obvious notations we can write

$$\Delta \varepsilon_s = \Delta \varepsilon_{sc} + \Delta \varepsilon_{so} + \Delta \varepsilon_{si}$$

In the present case we can assume that $\Delta \varepsilon_{si}$ is the most important term because during the protonation we are going from a neutral molecule to a positive ion, therefore we must expect a rather large $\Delta \varepsilon_{si}$ term. The discussion will be focused on the estimation of $\Delta \varepsilon_{si}$. This term itself can be divided into three parts. Let us call $\Delta \varepsilon_{sia}$ the part corresponding to the anisotropic interaction like those due to the formation of hydrogen bonds between the solvated molecule and the solvent molecules. In our case this term is certainly important because a hydrogen bond can appear between the heteroatom of a molecule like pyridine and the water molecules.

But as $\Delta \varepsilon_{sia}$ represents the corresponding change of energy between the neutral and the protonated molecule, it can be assumed that this term will have about the same value for all the set of molecules under consideration, in such a way that the $\Delta \Delta \varepsilon_{sia}$ terms which appear in equation (9) will vanish in the first approximation.

The part $\Delta \varepsilon_{sid}$ due to dispersion forces will in this case be of small importance in comparison with the part $\Delta \varepsilon_{sii}$ associated with the iso-tropic interaction between electric charges and dipole moments. In conclusion, to estimate $\Delta \Delta \varepsilon_s$, account need only be taken of the $\Delta \Delta \varepsilon_{sii}$ term.

The term $\Delta \varepsilon_{sii}$ can be estimated from the formula[41]

$$\Delta \varepsilon_{sii} = -\Delta \Sigma \frac{Q_i Q_j}{2r_{ij}} \left(1 - \frac{1}{D}\right) \tag{10}$$

where the Q_i's are the apparent charges of each atom, and where r_{ij} denotes the distance between the atom i and the atom j except when i and j are identical. In the latter case r_{ii} represents a certain empirical effective radius. Finally D is the effective dielectric constant of the solvent.

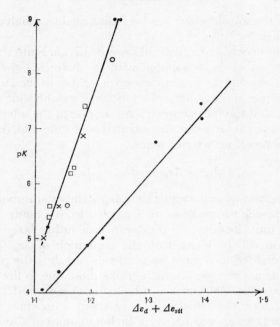

FIGURE 24. The pK as a function of $\Delta\varepsilon_d + \Delta\varepsilon_{sti}$.

Figure 24 shows what happens when the pK_a of the amino derivatives under consideration is plotted as a function of $\Delta\varepsilon_d + \Delta\varepsilon_{sti}$, this last term being derived from equation (10). Twelve points corresponding to all families of compounds are now found along the same straight line with a deviation of no more than 0·3 units of pK. The introduction of $\Delta\varepsilon_{sti}$ destroys the segregation of the various families. The precision is very satisfactory as the theory predicts the constant with an uncertainty of a factor of only 2 and there is a factor of 10,000 between the highest and the lowest values of K.

Also shown in Figure 24 are various points which lie near another straight line; they correspond to molecules containing an amino group in an *ortho* or *peri* position to the nitrogen heteroatom. In this case a special $\Delta\varepsilon_{nb}$ term appears mainly due to the interaction between the lone pair of the heteroatom and the amino group.

In conclusion, Figure 24 clearly shows the importance of $\Delta\varepsilon_d$, $\Delta\varepsilon_s(T)$ and $\Delta\varepsilon_{nb}$ in the determination of the pK of the molecules under consideration.

The basic properties of amines are also seen in other reactions. For example, amines are able to act as proton acceptors in the presence of proton donors in such a way that association compounds are formed. There is an obvious analogy between the acid–base equilibrium

$$B + H^+ \underset{K}{\rightleftharpoons} BH^+$$

and the association equilibrium

$$B + HR \underset{K'}{\rightleftharpoons} B \cdot HR$$

where HR denotes the proton donor and K' the association constant. It is, therefore, interesting to compare the pK and the association constant.

With obvious notations one can write

$$K = \frac{f_{BH^+}}{f_B f_{H^+}} e^{-[\Delta\varepsilon_v + \Delta\varepsilon_i + \Delta\varepsilon_d + \Delta\varepsilon_{nb} + \Delta\varepsilon_s(T)]/\chi^T}$$

$$K' = \frac{f_{B \cdot HR}}{f_B f_{HR}} e^{-[\Delta\varepsilon_v' + \Delta\varepsilon_i' + \Delta\varepsilon_d' + \Delta\varepsilon_{nb}' + \Delta\varepsilon_s'(T)]/\chi^T}$$

Association constants are usually measured in a non-polar solvent like carbon tetrachloride. The solvent effect represented by $\Delta\varepsilon_s'(T)$ is still further reduced.

Furthermore, the interatomic distance between the amino nitrogen and the proton is greater in the association complex B·HR than in the positive ion BH$^+$. Therefore we must expect that the effect of the electronic structure of amines will be less apparent in K' than it is in K.

Finally we can expect that K' will vary more slightly with the electronic structure than K. However there is a chance that K and K' run parallel. The following table taken from Bonnet's thesis[42] shows such an example

TABLE 4. Comparison of the pK and of values of K' at 25°c (in n-heptane).

	$10^6\,K$	$10^{10}\,K$	p-Cresol 0·65	Phenol 1·06	Naphthol 1·40
Triethylamine	435	$K' =$	55	86	117
Dibutylamine	1282		83	128	162

Such a parallel disappears if the pK depends strongly on the solvent effect and, as has been said, the association constants are measured in such conditions that this effect is usually very small.

III. SOME ASPECTS OF THE PROPERTIES OF AMINES IN THEIR ELECTRONICALLY EXCITED STATES

A. *Electronic Spectra of Amines*

The electronic spectrum of aniline has been analyzed, using wave-mechanical methods, by various authors[32, 43]. The self-consistent field method which has been presented in section I and used in section II to describe the ground state of this molecule can also be used to represent its electronically excited states. For practical reasons it is necessary to introduce the LCAO approximation and either the Hückel or the Pariser and Parr method is used. Furthermore, the 'virtual orbitals' approximation is added. Let us consider the eigenfunctions π_i of the self-consistent field operator h^{SCF} corresponding to the ground state of aniline in the LCAO approximation. They obey the equation

$$h^{SCF}\pi_i = \varepsilon_i \pi_i$$

Obviously the molecular orbitals π_i^0 which are used to represent the ground state are solutions of this equation. But there are also other functions π_i^u which obey the equation. They are called virtual orbitals for the ground state. It is possible to show that if in the total wave function of the ground state one (or more) orbital π_i^0 is replaced by one (or more) virtual orbital, a convenient approximate wave function is obtained which represents an electronically excited state of the molecule.

In Table 5 some of the values obtained by Fischer–Hjalmars using this procedure are compared with experimental results.

TABLE 5. Electronic excitation energies of aniline (in ev).

Theoretical	Experimental[44]
4·83	4·32
5·29	5·31
7·03	6·32

Taking account of the various approximations introduced into the calculation, the agreement is satisfactory. Other authors obtained analogous results. For example, Bloor and coworkers[43e] are led to the conclusion that by adjusting the core integral empirically, or by using a variable electronegativity approach, it is possible to obtain agreement between experiment for the first two electronic transitions in the vapor state, the change in ionization potential relative to benzene and the electron density *para* to the amino group.

Mataga[43d] performed an analogous analysis of the spectra of isomeric phenylenediamines, of *s*-triaminobenzene and of several amino-substituted nitrogen heterocycles such as 4-aminopyridine, 1,4-diaminotetrazine and melamine.

B. *Base Strength of Aromatic Amines in their Electronically Excited States*

Coulson and Jacobs[45] had studied, theoretically, charge migration in aniline under the effect of irradiation. They observed that the π electronic charge of nitrogen is smaller in the first excited state than in the ground state, that is to say, it should be less basic. Förster[46] has effectively observed that if a base, such as 3-aminopyrene (which from the theoretical viewpoint must have a similar behavior) is irradiated, the excited molecules have acidic properties. To show this fact Förster studied the absorption and fluorescence spectra as a function of the pH of a solution of 3-aminopyrene. The absorption spectrum gives information about the ground state of the molecule. On the other hand, the fluorescence spectrum depends on the excited states of the same molecule. It is found that up to pH $\sim$ 2, the absorption spectra are those of the $ArNH_3{}^+$ ions, whereas the fluorescence spectra correspond to $ArNH_2$. It appears that as predicted by the theory, the molecules in their excited states have less tendency to add a proton than the ground-state molecules. Furthermore, near pH 12 some new bands appear in the fluorescence spectra which can be attributed to the $ArNH^-$ ions. This attractive result shows that the excited molecules $ArNH_2{}^*$ are able to react as an acid

$$ArNH_2{}^* \rightleftharpoons ArNH^- + H^+$$

The phenomenon has been analyzed in more detail by Sandorfy[47] who calculated the distribution of the electronic charges, taking account of both the π and the σ orbitals. He found that the nitrogen which is negative in the ground state becomes positive in the first electronically excited state, which completely explains why the molecule becomes an acid.

Other amines have been studied from the same viewpoint. For example it has been possible to measure the pK of β-naphthylamine in various states. A value of 4·1 is found for the ground state, whereas a value of -2 is obtained for the first singlet state and a value of 3·3 for the first triplet state[48].

To explain the difference between the first excited singlet and triplet levels Murrell [49] pointed out that the energy of the direct donor–acceptor charge-transfer configuration lies at a higher energy than any of the actual states considered. Therefore is it normal that the actual charge transfer is greater for the excited singlet state than for the triplet, as the energy of the former is nearer the energy of the charge-transfer configuration than is the energy of the triplet state.

IV. REFERENCES

1. R. Daudel, R. Lefebvre and C. Moser, *Quantum Chemistry, Methods and Applications*, Interscience, New York, 1959.
2. R. Daudel, *Compt. Rend.*, **237**, 60 (1953).
3. R. Daudel, H. Brion and S. Odiot, *J. Chem. Phys.*, **23**, 2080 (1955).
4. S. Odiot, *Cahiers Phys.* (81), 1 (1957); (82), 23 (1957).
5. S. Odiot and R. Daudel, *Compt. Rend.*, **238**, 1384 (1954).
6. J. W. Linnett and A. J. Poe, *Trans. Faraday Soc.*, **47**, 1033 (1951).
7. M. Roux, S. Besnaïnou and R. Daudel, *J. Chim. Phys.*, **53**, 218 (1956).
8. S. Bratoz, R. Daudel, M. Roux and M. Allavena, *Rev. Mod. Phys.*, **32**, 412 (1960).
9. R. Daudel, *Advan. Quantum Chem.*, **1**, 115 (1964).
10. (a) H. Kaplan, *J. Chem. Phys.*, **26**, 1704 (1957).
    (b) J. Higuchi, *J. Chem. Phys.*, **24**, 535 (1956).
    (c) A. B. F. Duncan, *J. Chem. Phys.*, **27**, 423 (1957).
11. R. D. Lide and D. E. Mann, *J. Chem. Phys.*, **28**, 572 (1958).
12. L. O. Brockway and H. O. Jenkins, *J. Am. Chem. Soc.*, **58**, 2036 (1936).
13. C. Tavard, *Thèse Sciences*, Sorbonne, Paris, 1966.
14. R. Moccia, *J. Chem. Phys.*, **40**, 2164, 2176 and 2186 (1964).
15. T. Iijima, R. A. Bonham, C. Tavard, M. Roux and M. Cornille, *Bull. Chem. Soc. Japan*, **38**, 1757 (1965).
16. R. N. Jones and C. Sandorfy, *Chemical Application of Spectroscopy*, Interscience, New York, 1956, p. 511.
17. W. S. Benedict, E. K. Plyler and E. P. Tidwell, *J. Res. Nat. Bureau, Std.*, **61**, 123 (1958); W. S. Benedict and E. K. Plyler, *Can. J. Phys.*, **35**, 1235 (1957).
18. M. Allavena, *Cahiers Phys.*, **19**, 401 (1965).
19. J. D. Morrison and A. J. C. Nicholson, *J. Chem. Phys.*, **20**, 1021 (1952).
20. H. Sun and G. L. Weissler, *J. Chem. Phys.*, **23**, 1160 (1955).
21. I. Omura, K. Higasi and H. Baba, *Bull. Chem. Soc. Japan*, **29**, 504 (1956).
22. R. Sänger, O. Steiger and J. Gächterk, *Helv. Phys. Acta*, **5**, 200 (1932).
23. R. J. W. Lefevre and P. Russell, *Trans. Faraday Soc.*, **43**, 574 (1947).
24. H. M. Powell, G. Huse and P. W. Cooke, *J. Chem. Soc.*, 153 (1943).
25. M. Van Meerssche and G. Leroy, *Bull. Soc. Chim. Belge*, **69**, 204 (1960).
26. K. Higasi, H. Baba and A. Rembaum, *Quantum Organic Chemistry*, Wiley, New York, 1965, p. 75.
27. J. W. Smith, *Electric Dipole Moments*, Butterworths, London, 1955, p. 212; L. McClellan, *Tables of Experimental Dipole Moments*, Freeman, San Francisco, 1963.

28. E. Huckel, *Z. Physik*, **70**, 204 (1931); **76**, 628 (1932).
29. R. Pariser and R. Parr, *J. Chem. Phys.*, **21**, 466, 767 (1953).
30. C. A. Coulson and R. Daudel, *Dictionnaire des Grandeurs Théoriques Descriptives des Molécules*, V. II, C.N.R.S. Paris, 1955.
31. C. A. Coulson, *Proc. Roy. Soc. (London)*, *Ser. A*, **169**, 413 (1939).
32. I. Fischer-Hjalmars, *Arkiv Fysik*, **21**, 123 (1962).
33. R. A. Hoffman and Gronowitzs (quoted in ref. 32).
34. I. Fischer, *Acta Chem. Scand*, **4**, 1197 (1950).
35. (a) G. Del Ré, B. Pullman and T. Yonezawa, *Biochim. Biophys. Acta*, **75**, 153 (1963).
    (b) T. Yonezawa, G. Del Ré and B. Pullman, *Bull. Chem. Soc. Japan*, **37**, 985 (1964).
36. G. Del Ré, *J. Chem. Soc.*, 4031 (1958).
37. (a) A. Albert, R. Goldacre and J. Phillips, *J. Chem. Soc.*, 2240 (1948).
    (b) A. R. Osborn and K. Schofield, *J. Chem. Soc.*, 4191 (1956).
38. (a) J. Ploquin, *Compt. Rend.*, **226**, 2140 (1948).
    (b) R. Daudel, *Compt. Rend.*, **227**, 1241 (1948).
    (c) R. Daudel and O. Chalvet, *J. Chim. Phys.*, **46**, 332 (1949).
    (d) O. Chalvet, R. Daudel, M. Pages, M. Roux, N. P. Buu Hoi and R. Royer, *J. Chim. Phys.*, **51**, 548 (1954).
    (e) H. C. Longuet-Higgins, *J. Chem. Phys.*, **18**, 275 (1950).
    (f) J. J. Elliot and S. F. Mason, *J. Chem. Soc.*, 2352 (1959).
    (g) A. Pullman and T. Nakajima, *J. Chim. Phys.*, **55**, 793 (1958).
39. F. Peradejordi, *Cahiers Phys.*, **17**, 393 (1963).
40. O. Chalvet, R. Daudel and F. Peradejordi, *J. Chim. Phys.*, **59**, 709 (1962).
41. (a) M. Born, *Z. Physik.*, **1**, 45 (1920).
    (b) G. J. Hoijtink, E. De Boer, P. M. Van der Meij and W. P. Weijland, *Rec. Trav. Chim.*, **75**, 487 (1956).
42. M. Bonnet, *Thesis*, Marseille, 1960.
43. (a) J. N. Murrell, *Proc. Phys. Soc.*, **A68**, 969 (1955).
    (b) T. E. Peacock, *Mol. Phys.*, **3**, 453 (1960).
    (c) F. Peradejordi, *Calcul des Fonctions d'Onde Moléculaires C.N.R.S.*, Paris, 1958, p. 247.
    (d) N. Mataga, *Bull. Chem. Soc. Japan*, **36**, 1607 (1963).
    (e) J. E. Bloor, P. N. Daykin and P. Boltwood, *Can. J. Chem.*, **42**, 121 (1964).
44. J. R. Platt, *J. Chem. Phys.*, **19**, 101 (1951).
45. C. A. Coulson and J. Jacobs, *J. Chem. Soc.*, 1983 (1949).
46. Th. Förster, *Z. Elektrochem.*, **54**, 42, 531 (1949).
47. C. Sandorfy, *Compt. Rend.*, **232**, 841 (1951).
48. G. Jackson and G. Porter, *Proc. Roy. Soc.*, **A260**, 13 (1961).
49. J. N. Murrell, *The Theory of Electronic Spectra of Organic Molecules*, Methuen, London, 1963.

CHAPTER **2**

The introduction of the amino group

M. S. Gibson

Faculty of Technology, University of Manchester

I. INTRODUCTION

For the purposes of discussion, the term amino group is taken to include not only the primary amino group (—NH$_2$) but also substituted-amino groups bearing one or more aliphatic or aromatic groups

37

on the nitrogen atom. This chapter is concerned with methods of synthesis of aliphatic, aromatic and mixed aliphatic–aromatic amines, including primary, secondary and tertiary amines, and related quaternary compounds. Other classes of nitrogen compounds which contain an amino group, e.g. amides, are considered only in relation to preparative routes to amines.

The preparation of amines by methods involving substitution, addition and exchange reactions are discussed in sections II to VI, and reductive procedures are considered separately in section VII; molecular rearrangements are not considered as such, but are mentioned where appropriate in the text. Steric factors influence many amine syntheses, but these are considered principally in section IV, as much of the relevant information has come from studies of aminolysis of halides and related compounds.

II. INTRODUCTION OF AMINO GROUPS BY REPLACEMENT OF HYDROGEN

A. Direct Amination of Aromatic Compounds in the Presence of Lewis Acids

Examples of the amination of aromatic compounds by various electrophilic species, derived from hydroxylamine[1], and from hydrogen azide[2], have been known for some time. Aromatic compounds which have been aminated in this way range from benzene and alkylbenzenes to substances such as anthraquinone[3]. Recently, some of these reactions have been systematically examined, and the principal mechanistic features elucidated, mainly by Kovacic and coworkers.

Direct amination of toluene can be effected with hydroxylamine-O-sulphonic acid[4], alkylhydroxylamines[5] and hydroxylammonium salts[6] in the presence of aluminium chloride (at least two moles per mole of aminating agent), and with hydrogen azide in the presence of aluminium chloride or sulphuric acid[7]. These reactions lead to the corresponding toluidines. Thus toluene and hydroxylamine-O-sulphonic acid give mixed toluidines (50% yield) of the following composition:

o-toluidine (51%),
m-toluidine (13%),
p-toluidine (36%).

The preponderance of *ortho* and *para* isomers indicates attack by an active, if somewhat unselective, electrophilic species. In fact, the

Selectivity Factor, S_f, for this reaction, calculated according to equation (1)[8] is 0·74, and similar values were calculated for the other

$$S_f = \log \left(\frac{2 \times \% \ para}{\% \ meta} \right) \tag{1}$$

aminations examined. Comparison of these values of S_f for the amination reactions with those for other electrophilic substitutions led to the view that the aminating agent is likely to be the protonated or Lewis acid-complexed nitrogen compound rather than the NH_2^+ cation, and this view is reinforced when the results of competitive aminations of toluene and benzene are compared with those of other substitution reactions. Likely electrophiles in the hydroxylamine-O-sulphonic acid/aluminium chloride and hydrogen azide/sulphuric acid aminating systems are **1** and **2**:

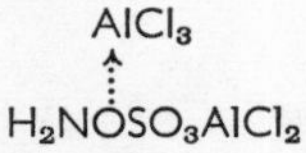

$$
\begin{array}{cc}
AlCl_3 & H^+ \\
\uparrow & \uparrow \\
H_2N\dot{O}SO_3AlCl_2 & HN{-}N{\equiv}N \\
\textbf{(1)} & \textbf{(2)}
\end{array}
$$

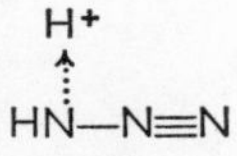

and the amination of toluene by hydrogen azide may be represented schematically by reaction (2).

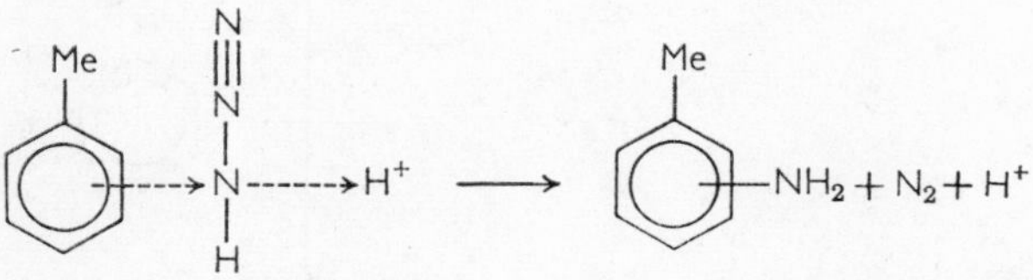

$$\tag{2}$$

Dialkylamino groups may, in principle, be introduced into aromatic rings using the dialkylchloroamine and concentrated sulphuric acid[9], except in those cases where the chloroamine undergoes preferential ring closure (Hofmann–Loffler reaction[10]) (reaction 3). However,

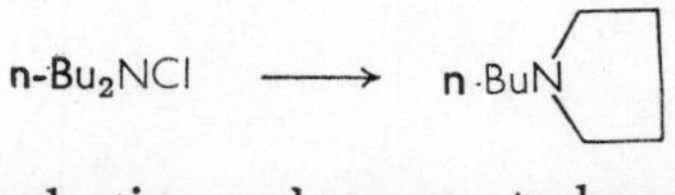

$$n\text{-}Bu_2NCl \longrightarrow n\text{-}BuN \qquad \tag{3}$$

these reactions are unselective and appear to be of preparative value only in a limited number of cases, for example benzene → N-phenyl-piperidine.

By contrast, reaction of toluene with dichloroamine or trichloroamine in the presence of aluminium chloride gives m-toluidine[11], and the same unusual meta-amination pattern was subsequently observed when the reaction was extended to include other alkylbenzenes[12]. Other products (from toluene) are o- and p-chlorotoluenes, and small amounts of m-chlorotoluene, 2-chloro-5-methylaniline and 4-chloro-3-methylaniline. The absence of substantial amounts of the last two compounds has been taken to indicate that N-chloro-m-toluidine is not an intermediate in the formation of m-toluidine, and after consideration of various possibilities, a σ-substitution mechanism has been advanced to explain the observed orientation. This involves the formation of a chloroarenonium ion, followed by addition of a nucleophilic nitrogen species, and subsequent aromatisation through loss of hydrogen chloride. The nature of the nucleophile is not certain, but it is not considered to be trichloroamine[12, 13]. The overall result may be illustrated by reactions such as (4) and (5); these also account satisfactorily for the orientation of the chlorotoluenes produced.

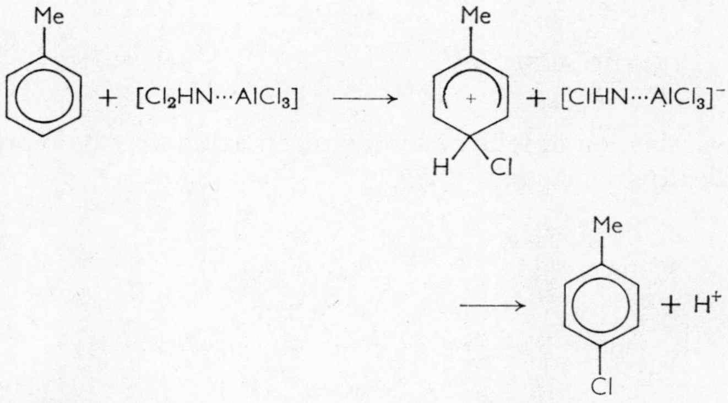

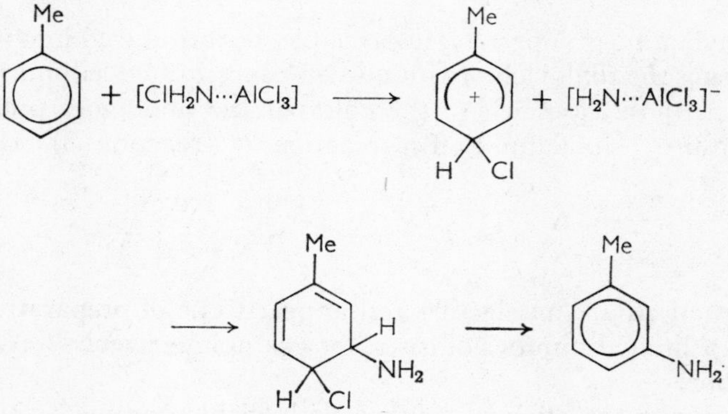

In spite of the unattractive properties of trichloroamine, this procedure may be of value for the preparation of certain m-alkylanilines;

the amination reactions based on hydroxylamine-O-sulphonic acid, hydrogen azide, etc., seem to offer little in comparison with other routes to the corresponding amines.

B. The Chichibabin and Related Reactions

Direct amination by amide ion is possible with aromatic molecules which are susceptible to nucleophilic attack. This group includes many heterocyclic compounds, particularly derivatives of pyridine, quinoline and related ring systems[14], and also a number of benzenoid compounds containing appropriate electron-attracting substituents. These reactions may be conducted at elevated temperatures in dimethylaniline or a hydrocarbon for example, or at lower temperatures in liquid ammonia. In the pyridine series, it is the electron-deficient α-position which is preferentially aminated, though γ-substitution may occur if no α-site is available (reaction 6). The second stage

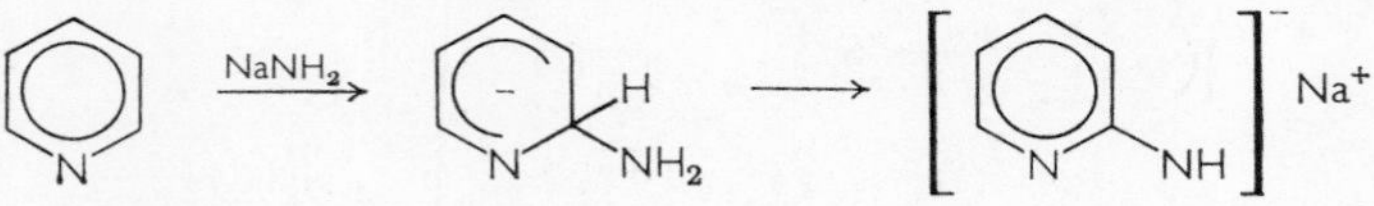

$$(6)$$

involves dehydrogenation, and ammonium and potassium nitrates have been used as additives on a number of occasions.

In some cases, conducting this type of reaction in the presence of a primary alkylamine (rather than dimethylaniline) allows effective introduction of the alkylamino group[15]. There are grounds for believing that the amino group is introduced and then displaced to a greater or lesser extent, by the alkylamino group; indeed, equilibration of 2-methylaminoquinoline and 2-aminoquinoline by means of potassamide in liquid ammonia has been demonstrated in separate experiments[16] (reaction 7).

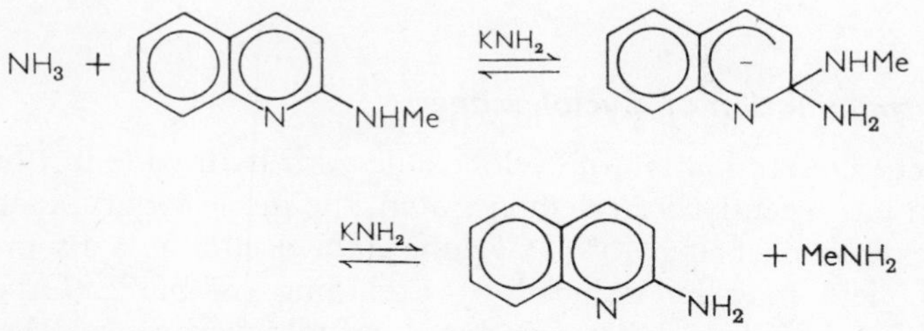

$$(7)$$

Nitrobenzene may undergo nucleophilic substitution at the *ortho* and *para* positions, but it is desirable to employ nitrobenzene in excess as some is destroyed by reduction. Thus N-o-nitrophenylpiperidine is formed from nitrobenzene and lithium piperidide (reaction 8)[17],

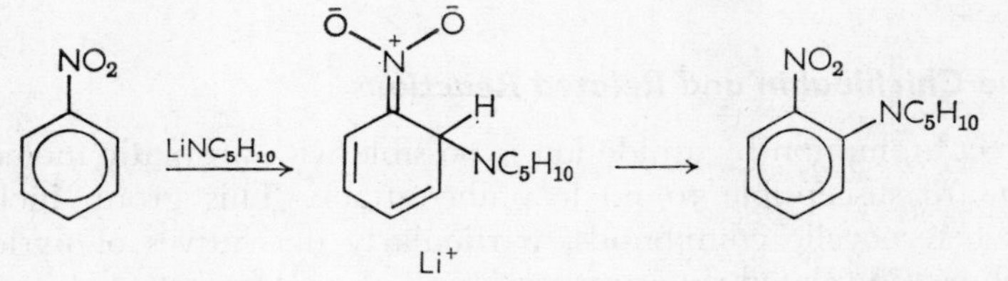

(8)

whilst nitrobenzene and sodium diphenylamide give 4-nitrotriphenyl-amine (reaction 9)[18].

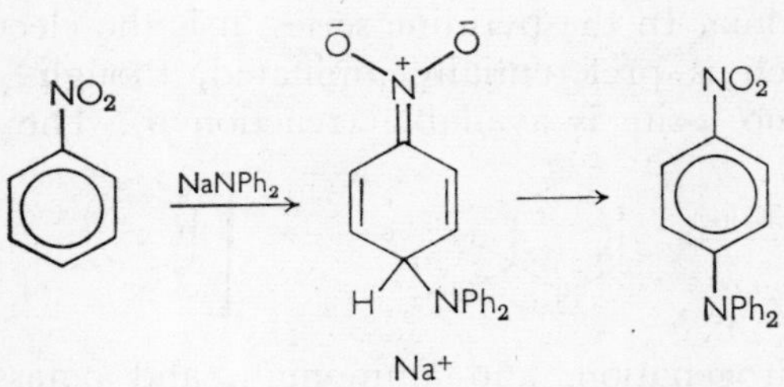

(9)

Hydroxylamine has also been employed for direct amination of a few nitro compounds (reaction 10)[19, 20].

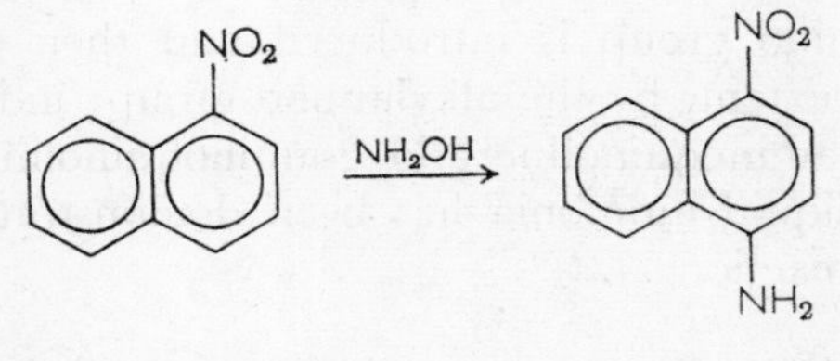

(10)

C. Photoamination of Cyclohexane

Direct photoamination of cyclohexane with hydrazine in t-butanol solution has recently been demonstrated, the most favourable yield of cyclohexylamine being 45%. Cyclohexanol occurs as a by-product, but its yield may be reduced by excluding oxygen[21]. Other by-products include bicyclohexyl and cyclohexylhydrazine, but the latter is not apparently an intermediate in the formation of cyclohexyl-amine[22].

A radical process is favoured for this reaction, with N—N bond fission as the important initiating step (reactions 11–14).

$$NH_2NH_2 \xrightarrow{h\upsilon} NH_2NH_2^* \longrightarrow 2\,NH_2^{\bullet} \tag{11}$$

$$NH_2^{\bullet} + C_6H_{12} \longrightarrow NH_3 + C_6H_{11}^{\bullet} \tag{12}$$

$$C_6H_{11}^{\bullet} + NH_2^{\bullet} \longrightarrow C_6H_{11}NH_2 \tag{13}$$

$$2\,C_6H_{11}^{\bullet} \longrightarrow C_{12}H_{22} \tag{14}$$

III. INTRODUCTION OF AMINO GROUPS BY REPLACEMENT OF HYDROXYL

For most laboratory purposes, the important reactions under this heading are the Bucherer Reaction, chiefly useful in the naphthalene series, and the Ritter Reaction, which will be considered in sections IV.B and VI.B.

The Bucherer Reaction

Naphthols and the corresponding naphthylamines are interconvertible in aqueous media containing sulphite or bisulphite ion at elevated temperatures[23]. Thus naphthols may be converted to naphthylamines by reaction with ammonia and ammonium sulphite in aqueous solution, usually in the temperature range 90–150°; the corresponding dinaphthylamine is sometimes obtained as a byproduct (reaction 15). Monoalkylamines and dialkylamines similarly

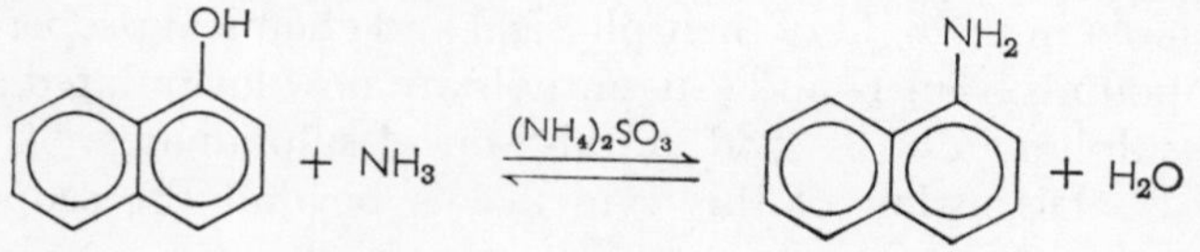

$$\tag{15}$$

give the alkylamino- and dialkylaminonaphthalenes, though higher temperatures are normally required. Naphthols also react with arylamines to give the arylaminonaphthalene, 1-naphthols reacting less readily than 2-naphthols. Similar cases have also been noted with arylamines (reaction 16).

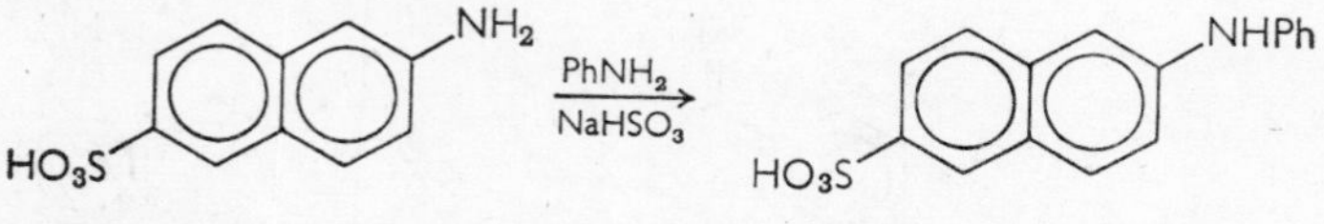

$$\tag{16}$$

Bisulphite addition compounds of naphthols have been isolated in a number of cases, and for a considerable time these were formulated as derivatives of the ketonic form of the naphthol. Until recently, the generally accepted reaction scheme was (17) with the naphthol possibly reacting in the keto form. This overall equilibrium picture

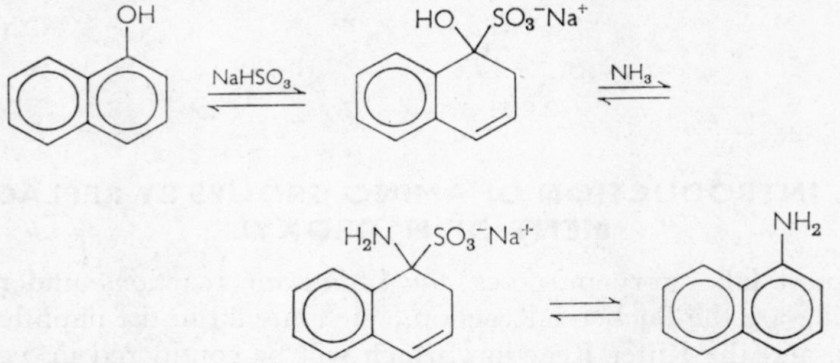

was consistent with the results of a systematic kinetic study of the reaction conducted with a number of naphthol and naphthylamine sulphonic acids and reported in 1946[24].

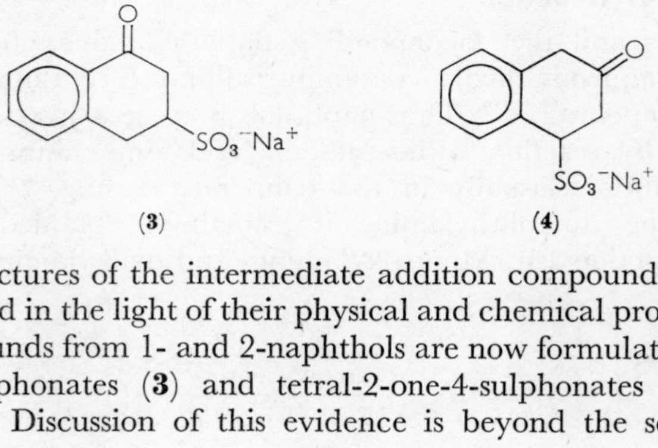

(3)   (4)

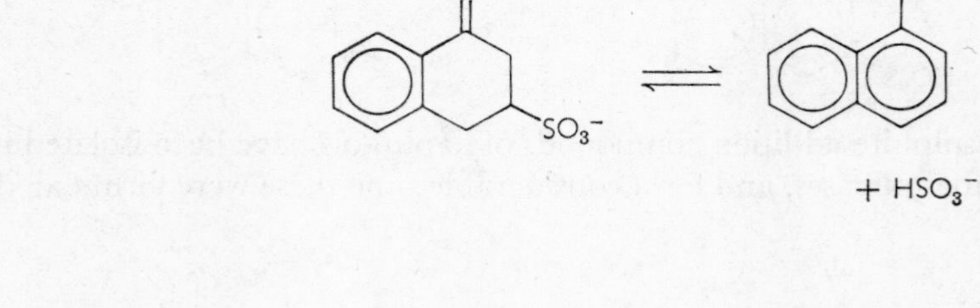

The structures of the intermediate addition compounds have been reconsidered in the light of their physical and chemical properties, and the compounds from 1- and 2-naphthols are now formulated as tetral-1-one-3-sulphonates (3) and tetral-2-one-4-sulphonates (4) respectively[25, 26]. Discussion of this evidence is beyond the scope of the present chapter but it is interesting to note that a new synthesis of 1-naphthol-3-sulphonic acids has emerged from this work. The scheme for the interconversion of naphthols and naphthylamines may be summarised by reaction (18), with the possibility of ketimine and enamine intermediates, except in the special case of reactions involving dialkylamines, when only an enamine intermediate is possible. Amine exchange reactions of 1- and 2-naphthylamines with an amine R_2NH are likely to involve further intermediates of types 5 and 6 respectively, formed by addition of the amine R_2NH to the intermediate ketimine or enamine.

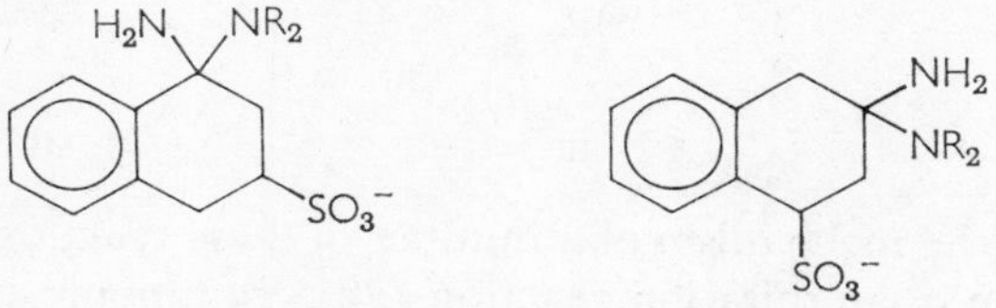

(5)   (6)

IV. INTRODUCTION OF AMINO GROUPS BY REPLACEMENT OF HALOGEN, AND RELATED REACTIONS

A. The Alkylation of Ammonia and Amines

The preparation of amines and quaternary ammonium salts through displacement of halide ion from alkyl halides by ammonia or amines is commonly referred to as alkylation of ammonia (or amines); alkyl esters of other strong acids have also been used in similar alkylation procedures. Equations (19) to (22) represent in simple terms the sequence of reactions which may occur in the interaction of ammonia or an amine with an alkyl halide.

$$NH_3 + RX \longrightarrow \overset{+}{N}H_3R \quad X^- \tag{19}$$

$$NH_2R + RX \longrightarrow \overset{+}{N}H_2R_2 \quad X^- \tag{20}$$

$$NHR_2 + RX \longrightarrow \overset{+}{N}HR_3 \quad X^- \tag{21}$$

$$NR_3 + RX \longrightarrow \overset{+}{N}R_4 \quad X^- \tag{22}$$

In principle, these reactions lead to a wide range of amines and quaternary salts, containing identical or differing R groups. Internal alkylation is also possible in appropriate cases. Thus ammonia and 1,5-dibromopentane react to give the amines (7) and (8), and the quaternary salt (**9**)[27]; ammonolysis of **10** similarly gives the symmetrical, high-melting 1-aza-adamantane (**11**) (reaction 23)[28].

$$NH_2(CH_2)_5NH_2$$

(**7**)

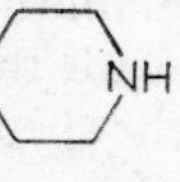

(**8**)

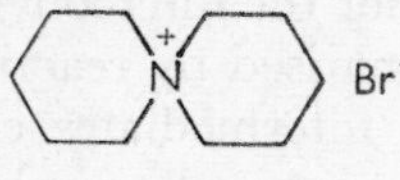

(**9**)

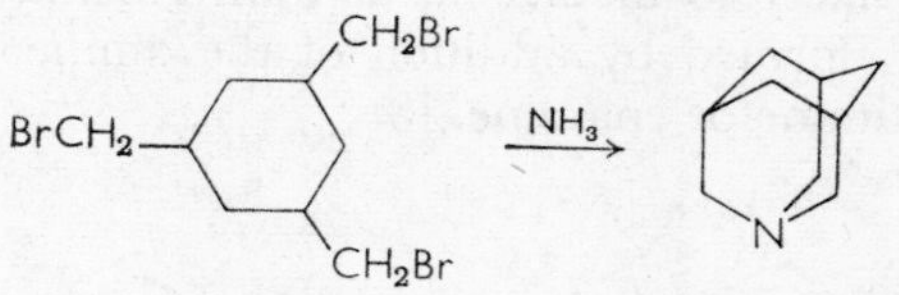

(23)

(**10**)

(**11**)

Studies of the mechanism of a number of these reactions, and particularly of the quaternisation reaction (22) (commonly known as the Menschutkin reaction), have been reviewed elsewhere[29] and only certain points are relevant here. The reactions generally exhibit the normal characteristics of S_N2 reactions with regard to kinetics, solvent polarity and structural requirements of the alkyl halide. The basic strength of the amine may provide a rough guide to its reactivity, the more strongly basic amines being frequently the most nucleophilic. Weakly basic amines such as diphenylamine can be methylated directly[30] (even to the quaternary stage under forcing conditions[31]), but in preparative work it is not uncommon to enhance nucleophilicity by conversion to an alkali-metal or Grignard derivative prior to alkylation[32]. However, steric factors are also important in relation to the nucleophilicity of amines. These may be illustrated by the relative rates of reaction of the comparably strong bases, quinuclidine and triethylamine, with methyl, ethyl and isopropyl iodides in nitrobenzene[33]. At 25°, the ratio (k_Q/k_T) of the specific rate constants for quinuclidine (**12**) (k_Q) and triethylamine (**13**) (k_T) have been determined as 57, 254 and 705 for methyl, ethyl and isopropyl iodides

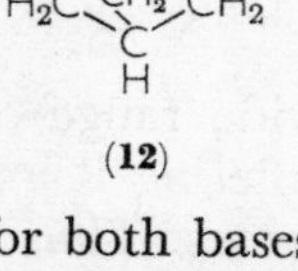

(**12**)

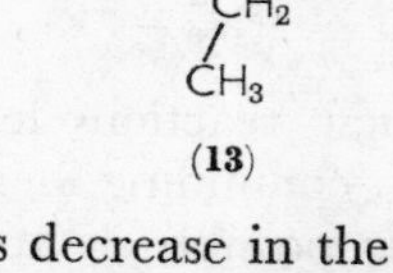

(**13**)

respectively; for both bases, the rates decrease in the expected order MeI > EtI > i-PrI. These results indicate not only that the steric requirements of quinuclidine are lower than those of triethylamine, but also that this difference in steric requirement becomes more important as the steric requirements of the alkyl halide increase. Similar effects have been observed in the reaction of pyridine and monoalkylpyridines with the same alkyl iodides[34]. Again the reaction rates drop sharply in the order MeI > EtI > i-PrI for each amine, but whereas (relative to pyridine) 3- or 4-alkyl substitution leads to slight increases in rate, 2-alkyl substitution results in decreases which become very pronounced in the case of 2-t-butylpyridine. These steric factors are of considerable importance in amine alkylations.

Limitations are also imposed by the nature of the alkyl group in the alkyl halide. Broadly speaking, the alkyl halides which react with ammonia (or amines) to give amines and quaternary ammonium salts are those which are susceptible to substitution by the S_N2 mechanism. Amines preparable by this route are consequently those in which primary or secondary alkyl groups are attached to the nitrogen atom. Attempts at alkylation of ammonia and amines using tertiary alkyl halides are normally frustrated by the occurrence of alternative elimination reactions and such problems are also apparent with other halides in which structural features favour alternative reactions (reaction 24)[35]. An important group of tertiary halides which may be

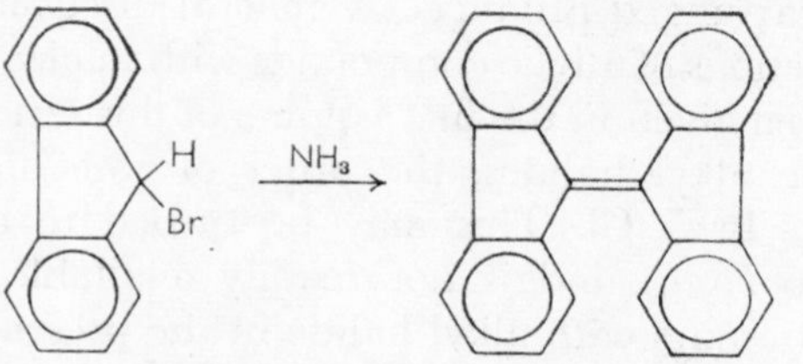

$$\text{(24)}$$

used for the alkylation of ammonia and amines are the tertiary propargylic halides (reaction 25)[36]. The ethynyl group may be reduced

$$Me_2CClC\equiv CH \xrightarrow{\ NHMe_2\ } Me_2\overset{\overset{\displaystyle NMe_2}{|}}{C}C\equiv CH \qquad \text{(25)}$$

to a vinyl or ethyl group and, since the steric requirements of the ethynyl group are small, a large variety of sterically hindered amines have become available, e.g. reactions (26) and (27)[37]. These reactions

$$Me_2CClC\equiv CH \xrightarrow[\ 2.\ H_2/Ni\ ]{1.\ t\text{-}BuNH_2} Me_2C\overset{\displaystyle Et}{\underset{\displaystyle NHBu\text{-}t}{<}} \qquad \text{(26)}$$

$$\underset{Me}{\overset{t\text{-Bu}}{>}}CClC{\equiv}CH \xrightarrow[\text{2. } H_2/Ni]{\text{1. } i\text{-PrNH}_2} \underset{Me}{\overset{t\text{-Bu}}{>}}C\underset{NHPr\text{-}i}{\overset{Et}{<}} \tag{27}$$

are relatively free from complications, although quaternisation of tertiary amines with tertiary propargylic halides may lead to the propargylic and the isomeric allenic quaternary salts (reaction 28), and may involve some destruction of the propargylic halide through dehydrohalogenation[38]. Allylic halides react with ammonia and

$$Me_2NCH_2CH{=}CH_2 + Me_2CClC{\equiv}CH \longrightarrow$$

$$\underset{CMe_2C{\equiv}CH}{\overset{CH_2CH{=}CH_2}{Me_2\overset{+}{N}<}} + \underset{CH{=}C{=}CMe_2}{\overset{CH_2CH{=}CH_2}{Me_2\overset{+}{N}<}} \tag{28}$$

with amines to give allylic amines and quaternary salts which may contain the original or the rearranged allylic group (reaction 29)[39].

$$MeCH{=}CHCH_2Cl \xrightarrow{NHEt_2} MeCH{=}CHCH_2NEt_2 \xleftarrow{NHEt_2} MeCHClCH{=}CH_2 \tag{29}$$

Primary allylic halides generally give normal products, whilst secondary allylic halides frequently give rearranged products, or mixtures of normal and rearranged products. A general discussion of the factors relevant to reactions of allylic compounds with nucleophiles, including amines, has been given in the first volume of this series[40].

Amongst the alkyl halides, the order of ease of replacement of halogen is I > Br > Cl. This may be turned to advantage, even when the iodo compound is not readily available, by conducting the reaction of amine with alkyl halide in the presence of iodide ion, so as to generate the organic iodide *in situ* by halogen exchange. The preparation of 3-(2-methylpiperid-1-yl)propanol[41] is illustrative

$$\underset{NH}{\overset{Me}{\bigcirc}} \xrightarrow[\text{NaI, EtOH}]{Cl(CH_2)_3OH,} \underset{N(CH_2)_3OH}{\overset{Me}{\bigcirc}} \tag{30}$$

(reaction 30). Whilst alkyl halides probably represent the most general type of alkylating agent for amines, various other esters have been used (particularly for the transfer of small alkyl groups) and in many cases offer practical advantages. Amongst sulphur esters, these include alkyl sulphates, and alkylbenzene- and toluene-*p*-sulphonates[42], methyl trifluoromethanesulphonate[43] and methyl sulphite[44].

Interestingly, use of methyl sulphite leads to the methanesulphonate of the methylated amine. A recently reported method, considered to be suitable for alcohols capable of yielding fairly stable carbonium ions, involves thermolysis of the corresponding dimethylsulphamate, followed by hydrolysis (reaction 31). The method may prove a useful

$$ROH \longrightarrow ROSO_2NMe \xrightarrow{\Delta} \bar{S}O_3{-}\overset{+}{N}RMe_2 \xrightarrow{H^+} NRMe_2 \qquad (31)$$

supplement to the S_N2 alkylation procedures[45]. Amines have also been alkylated by the use of alkyl phosphates[46], and in a few cases by preparation and decomposition of alkyl phosphoramidates (reaction 32)[47]. Alkyl nitrates are effective alkylating agents for aliphatic

$$(EtO)_2PONHPh \xrightarrow{\Delta} PhNEt_2 \qquad (32)$$

amines, though complications arise with aromatic amines[48]. Alkylation of amines is also possible with certain types of carboxylic esters[49], and under appropriate conditions propiolactone can be used to introduce β-carboxyethyl groups (reaction 33)[50]. Epoxides are suitable

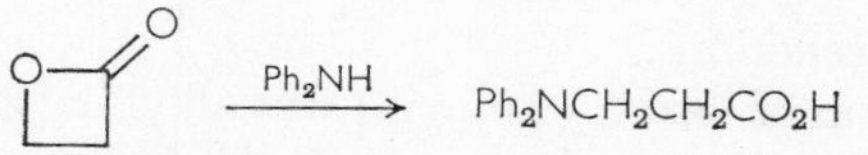

$$\xrightarrow{Ph_2NH} Ph_2NCH_2CH_2CO_2H \qquad (33)$$

for β-hydroxyalkylation of amines (reaction 34)[51]; this type of reaction is very general. being useful for the introduction of amino groups in the glycerol and sugar series[52, 53].

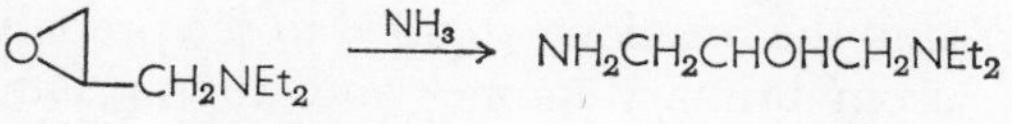

$$\xrightarrow{NH_3} NH_2CH_2CHOHCH_2NEt_2$$

It will be appreciated that many of these alkylation reactions (19) to (22) can be made reasonably selective, for example (20) to the virtual exclusion of (21) and (22), by appropriate choice of nucleophile and alkylating agent. This selectivity is not likely to occur, however, in the corresponding reactions of ammonia and unhindered primary and secondary n-alkylamines with primary alkylating agents (except in favourable cases where the desired product separates from the reaction), where mixtures of amines are to be expected. The problems attending separation of such mixtures are discussed below.

In the quaternisation reaction (22), or any sequence ending in quaternisation, separation of the ionic quaternary salt from unreacted amine(s) presents few problems as reaction is normally conducted under conditions in which step (22) is effectively irreversible. It is well to remember, however, that quaternary ammonium ions are susceptible to nucleophilic displacement reactions[31]. Ethanolamine has been shown to dealkylate tetraalkylammonium and aryltrialkyl-ammonium salts[54] at about 154° according to reaction (35). In

$$HOCH_2CH_2NH_2 + R_3\overset{+}{N}Ar \longrightarrow HOCH_2CH_2\overset{+}{N}H_2R + R_2NAr \qquad (35)$$

reactions involving a tertiary amine and a derived quaternary ammonium ion, such reaction may lead to redistribution of alkyl groups between the amine and the ion[55]. This has been demonstrated for benzyldimethylamine and the benzyltrimethylammonium ion, and for related pairs (reactions 36 and 37). The evolution of trimethyl-

$$PhCH_2NMe_2 + PhCH_2\overset{+}{N}Me_3 \longrightarrow (PhCH_2)_2\overset{+}{N}Me_2 + NMe_3 \qquad (36)$$

$$(PhCH_2)_2\overset{+}{N}Me_2 + PhCH_2NMe_2 \longrightarrow (PhCH_2)_2NMe + PhCH_2\overset{+}{N}Me_3 \qquad (37)$$

amine allows disproportionation of benzyldimethylamine to proceed in the presence of a catalytic amount of the quarternary salt. It is clearly desirable to avoid use of high temperatures for prolonged periods in these reactions.

The preparation of tertiary amines from alkyl halides and a second-ary amine, or a primary amine in which both N-hydrogen atoms are to be replaced by identical alkyl groups, can be carried out with favourable yields in many cases, as evidenced by preparations of diethyl-n-hexadecylamine (from diethylamine) and of dimethyl-n-docosylamine (from dimethylamine)[56]. Alkyl sulphonates have simi-larly been used, and the reaction extended to the preparation of cyclic tertiary amines from primary amines and appropriate terminal di-sulphonates (reaction 38)[57]. Satisfactory conditions have been reported

$$O\underset{CH_2CH_2OSO_2Ph}{\overset{CH_2CH_2OSO_2Ph}{<}} \xrightarrow{\text{n-BuNH}_2} O\bigcirc N\text{Bu-n} \qquad (38)$$

for dimethylation of primary aromatic amines using methyl sulphate[58] and for di-n-alkylation[46a] using n-alkyl phosphates; aniline and iso-propyl phosphate, however, give N-isopropylaniline, indicating the different steric requirements of the reactants.

The isolation of tertiary amines from alkylation reactions is normally straightforward; except in special cases[59], unwanted primary and secondary amines can usually be removed from the basic fraction by conversion to non-basic carboxamides (cf. ref. 46a) or sulphonamides (cf. ref. 28). If quaternisation has occurred, the desired tertiary amine may still be obtainable if the unwanted group can be removed reductively by use of lithium aluminium hydride (reaction 39)[60], or sodium and liquid ammonia[61]. With allyl or benzyl trialkylammonium

$$\langle\text{hexagon}\rangle-\overset{+}{\text{N}}\text{Me}_3 \xrightarrow{\text{LiAlH}_4} \langle\text{hexagon}\rangle-\text{NMe}_2 + \text{CH}_4 \tag{39}$$

salts, the allyl or benzyl group is removed more readily than a methyl group. It may be noted in passing that there is a tendency to solvolytic cleavage of an allyl group in certain tertiary amine hydrochlorides, and appropriate precautions should be taken during purification (reaction 40)[62]. (The other product of solvolysis was not identified, but is

$$\text{i-Bu}\overset{+}{\text{N}}\text{H}(\text{CH}_2\text{CH}{=}\text{CH}_2)_2 \xrightarrow{\text{EtOH}} \text{i-Bu}\overset{+}{\text{N}}\text{H}_2\text{CH}_2\text{CH}{=}\text{CH}_2 \tag{40}$$
$$\text{Cl}^- \qquad\qquad\qquad\qquad\qquad \text{Cl}^-$$

presumably allyl ethyl ether.)

The preparation of dialkylamines from ammonia or primary amines can be realised where steric factors militate against tertiary amine formation. Thus monoalkylation of t-butylamine with some alkyl halides and epoxides has been accomplished (reaction 41)[63].

$$t\text{-BuNH}_2 + \text{PhCH}_2\text{Cl} \longrightarrow t\text{-BuNHCH}_2\text{Ph} \tag{41}$$

More generally, ammonolysis or aminolysis of alkyl halides, and particularly primary alkyl halides, is likely to give rise to mixtures of the primary, secondary and tertiary amines (reactions 19, 20 and 21). The proportions of the three possible amines in such a mixture may be varied by alterations in the relative proportions of the reactants. Where, for example, the primary amine is the desired product, the use of liquid ammonia[27] (to ensure high concentration of ammonia), and also of sodamide (one mole per mole of alkyl halide) in liquid ammonia[64] have been recommended, though some dehydrohalogenation of the alkyl halide may occur when sodamide is used (reaction 42).

$$\text{n-C}_6\text{H}_{13}\text{Br} \longrightarrow \text{n-C}_6\text{H}_{13}\text{NH}_2 + \text{n-BuCH}{=}\text{CH}_2 \tag{42}$$
$$(74\%) \qquad\qquad (5\%)$$

In the reaction of ammonia with the alkyl halide, secondary and tertiary amine formation become more important as the concentration of ammonia is reduced. In favourable cases, e.g. the mono-, di- and trioctylamines from ammonolysis of n-octyl chloride[65], the amines may be separable by distillation. In other cases, chemical methods

3+c.a.g.

may be employed. Treatment of such a mixture of amines with benzene-sulphonyl chloride (Hinsberg's method) gives an 'alkali-soluble' amide (from the primary amine) and an 'alkali-insoluble' amide (from the secondary amine); again, phthalic or 3-nitrophthalic anhydride converts the primary and secondary amines into the phthalamic acids, but only the acid from the primary amine can be dehydrated to the corresponding phthalimide[66]. Both of these separation methods allow direct recovery of the tertiary amine from the unreacted amine fraction. Recovery of primary and secondary amines from these derivatives is considered below. Another method, suitable for separation of secondary amines, involves formation of the N-nitroso compound and subsequent regeneration[67].

B. The Use of Blocking Groups for Control of Alkylation

The difficulties associated with direct alkylation of ammonia and amines have led to the development of indirect alkylation procedures designed specifically for the preparation of primary and secondary amines. These depend on the blocking of the required number of sites in the ammonia molecule so as to limit the extent of the alkylation reaction, and fall into two categories depending on whether or not the alkylation step involves the formation of a quaternary ammonium ion.

The Délepine method[68], for primary amines, is based on the quaternisation of hexamine with an alkyl halide; the resulting hexaminium salt is then decomposed with ethanolic hydrochloric acid to give the desired amine, ammonia and the carbon fragments of the ring as diethylformal (reaction 43).

$$RX \longrightarrow \xrightarrow[\text{EtOH}]{\text{HCl}} \begin{array}{l} RNH_2 \\ 3\,NH_3 \\ 6\,CH_2(OEt)_2 \end{array} \qquad (43)$$

Primary amines are also available through the Ritter reaction in which an organic nitrile provides the blocked nitrogen atom. Though relatively poor nucleophiles, nitriles are able to capture carbonium ions; the resulting nitrilium ion reacts with water to give the amide, which can then be hydrolysed to the amine. A number of alkyl and aralkyl halides capable of yielding carbonium ions have been shown to react with nitriles in the presence of aluminium chloride[69], silver sulphate[70] or particularly antimony pentachloride (reaction 44)[71].

$$RCN + R^1Cl \xrightarrow{SbCl_5} [RC{\equiv}N{-}R^1]^+SbCl_6^- \xrightarrow{H_2O} RCONHR^1 \qquad (44)$$

For preparative purposes, Ritter's original procedure of generating the carbonium ion from the alcohol (or olefin) and sulphuric acid is preferred; under these conditions, hydrogen cyanide, produced *in situ* from sodium cyanide, can replace the nitrile[72]. Secondary and tertiary alcohols can be used[73], and the reaction, providing essentially for substitution by the S_N1 mechanism, is a valuable addition to the methods already discussed. It must be borne in mind, however, that under certain conditions carbonium ion rearrangement is possible, and has indeed been observed in some applications of this reaction[74].

For preparations of secondary amines by way of quaternisation reactions, it is necessary to employ a blocking group for two of the nitrogen valencies. Benzylidene derivatives of primary amines fulfil this function, and quaternisation of such compounds, followed by hydrolysis, gives benzaldehyde and the desired secondary amine (reaction 45)[75]. Incorporation of the primary amine into an imidate

$$PhCH{=}NR \xrightarrow{\text{MeI}} [PhCH{=}NMeR]^+I^- \longrightarrow PhCHO + RNHMe \qquad (45)$$

provides a new and interesting variant of this method, and has been employed for monomethylation of the N_α atom of tryptophan. The primary amine is condensed with γ-chlorobutyroyl chloride in pyridine, and the resulting amide cyclised to the iminolactone with silver fluoroborate. Methylation with methyl iodide, followed by hydrolysis, provides the secondary amine (reaction 46)[76]. Another route to this

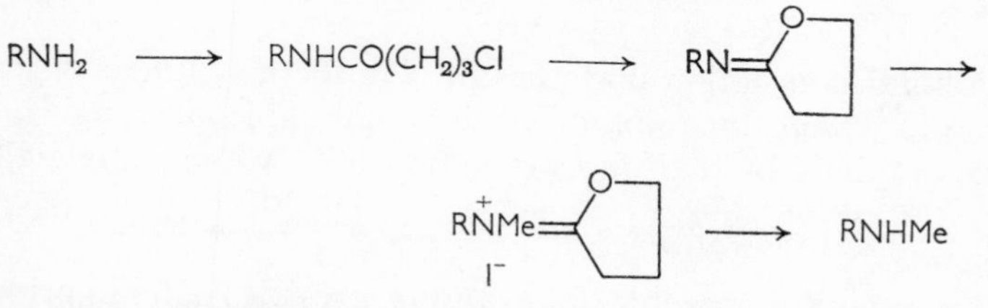

type of iminolactone (and, by implication, to secondary amines) which would appear to be particularly suitable where R is a tertiary alkyl group, is exemplified by reaction (47)[77]. It may be noted in

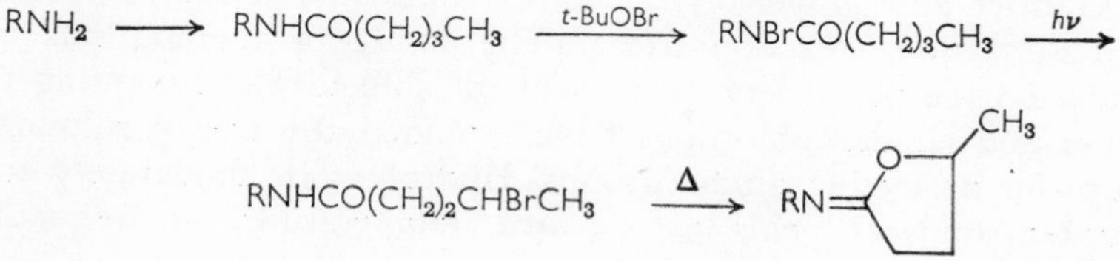

passing that, in the aromatic series, rearrangement of imidates provides a useful source of diarylamines (Chapman rearrangement)
(reaction 48)[78].

$$RC\overset{NAr}{\underset{OAr^1}{\lVert}} \longrightarrow RCONArAr^1 \longrightarrow ArNHAr^1 \qquad (48)$$

A somewhat different approach to the synthesis of primary and
secondary amines is involved in the use of blocking groups that so
reduce the nucleophilicity of ammonia or a primary amine that
alkylation does not proceed to the quaternary stage. For the conversion of ammonia to primary amines, these conditions are met in the
Gabriel synthesis in which the nucleophile is the anion of phthalimide
and the product the N-alkylphthalimide. Alkyl halides containing a
variety of other substituents, e.g. —CN, —COR, —COOR, —OR, in
the alkyl chain undergo this reaction, and the method has enjoyed
considerable popularity[79]; alkyl toluene-p-sulphonates[80] and epoxides[81] can also be used as alkylating agents. The alkylamine can be
obtained from the alkylphthalimide by hydrolysis or by the milder
hydrazinolysis procedure (reaction 49)[82]. In a recent improvement,
the use of dimethylformamide as solvent is recommended for the
alkylation reaction[83].

$$\underset{CO}{\overset{CO}{\bigotimes}}N^-K^+ \xrightarrow{RX} \underset{CO}{\overset{CO}{\bigotimes}}NR \longrightarrow RNH_2 \qquad (49)$$

A similar conception underlies the Hinsberg synthesis of secondary
amines in which the anion of a sulphonamide is alkylated[84], and the
amine obtained by hydrolysis (reaction 50). Mono- and dialkylation

$$[PhSO_2NR^1]^-K^+ \longrightarrow PhSO_2NRR^1 \longrightarrow NHRR^1 \qquad (50)$$

of toluene-p-sulphonamide, for example, provide routes to primary and
symmetrically substituted secondary amines[85], and mixed aliphatic–
aromatic amines are obtainable by use of sulphonanilides[86]. An
interesting synthesis of secondary amines, embodying features of both
the Gabriel and Hinsberg reactions, employs saccharin in a two-
stage alkylation (reaction 51)[87]. The Hinsberg method has not
attracted wide popularity because of the difficulty of liberating the
amine, and similar objections have attended the use of sulphonyl
groups for protecting amino groups. Hydrolysis by moderately concentrated mineral acids at elevated temperatures is frequently

undesirable, and other acidic procedures have been recommended for sensitive molecules[88]. A number of reductive methods (zinc and hydrochloric acid, sodium and isoamyl alcohol) have been reported for splitting sulphonamides[89], and sodium and liquid ammonia has been successfully used for this purpose in syntheses of oxytocin and related compounds[90].

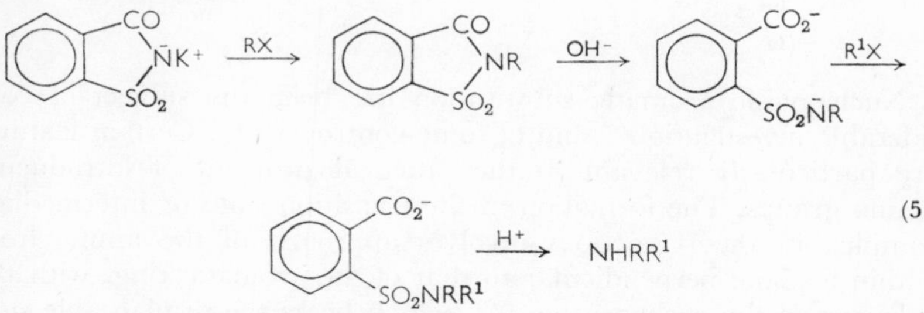

$$\text{(51)}$$

Alkylation of cyanamide (anion), followed by hydrolysis, has been used for the preparation of some symmetrically substituted secondary amines (reaction 52)[91]. It is clear from alkaloid studies that reduction

$$(\text{N—CN})^{2-} 2\,\text{Na}^+ \xrightarrow{\ \text{RX}\ } \text{R}_2\text{NCN} \longrightarrow \text{R}_2\text{NH} \qquad (52)$$

with lithium aluminium hydride may be used as an alternative to the hydrolysis stage[92].

A method which is suitable for the preparation of allylic amines involves the allylation of thiocyanate ion to give the allyl thiocyanate followed by rearrangement (by the $S_N i$ mechanism in many cases) to the isothiocyanate, and subsequent hydrolysis[40].

C. Nucleophilic Aromatic Substitution by Ammonia and Amines; the Bimolecular Mechanism [93]

Aryl halides are relatively insensitive to nucleophilic substitution, i.e. replacement of halogen as halide ion, by ammonia and amines. However, susceptibility of the halogen atom to displacement (or of other groups displaceable as anions) increases as the *ortho* and *para* positions become progressively substituted by electron-attracting groups, nitro groups being particularly effective. The reactivity of the halogen atoms in 2,4-dinitrofluorobenzene (14)[94] and in 4,4'-difluoro-3,3'-dinitrodiphenylsulphone (15)[95] are cases in point. These examples illustrate an important difference between alkyl halides and activated aryl halides towards nucleophilic substitution, namely the generally high susceptibility of fluorine to displacement in the aromatic series;

this indicates the relative importance of the bond-making step in these reactions.

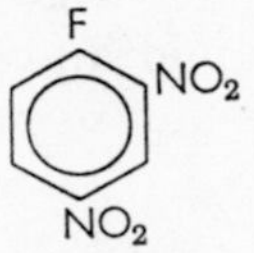

(14) 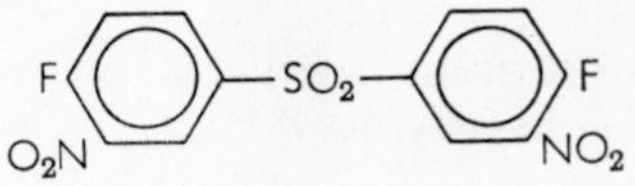   (15)

Nucleophilic aromatic substitution has been the subject of considerable investigation[93] and of some controversy[96]. Certain features are particularly relevant to the practical problem of introducing amino groups. The formation of the transition state or intermediate complex in these reactions involves approach of the amine from within a plane perpendicular to that of the aromatic ring, with the valencies of the carbon atom $C_{(1)}$ which bears the displaceable substituent becoming approximately tetrahedral. When the nucleophile is an unhindered amine, relatively little steric hindrance to this approach is offered by 2- and 6-substituents in the aromatic ring unless these are especially bulky, e.g. *t*-butyl, but as the steric requirements of the amine increase, even 2- and 6-methyl groups may exert detectable interference (cf. **16** below for the case where the activating

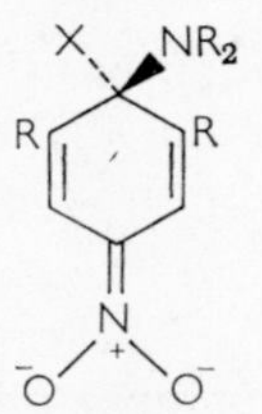

(16) 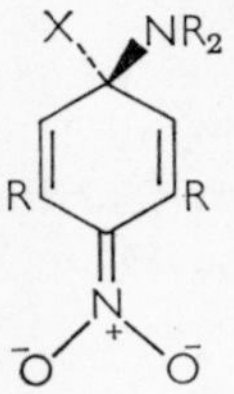   (17)

substituent is a 4-nitro group). It will be noted that for effective activation at $C_{(1)}$ by a 4-nitro group the latter should ideally be coplanar with the ring and this may be rendered difficult by the presence of alkyl groups at $C_{(3)}$ and $C_{(5)}$ (cf. **17**); 3- or 5-substitution does in fact markedly reduce ease of replacement at $C_{(1)}$[93, 97]. This second effect is often product-determining in reactions of amines with alkyldinitrobenzenes (reaction 53)[98], but is not observed with such activating substituents as the cyano group where the question of coplanarity with the ring does not arise[99]. In a number of cases, competitive displacement of substituents has been noted, as for example in reaction (54)[100].

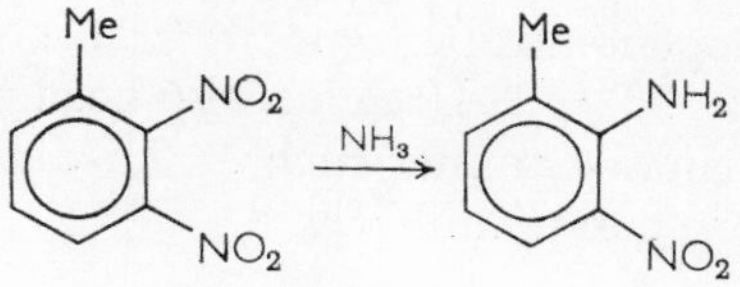

(53)

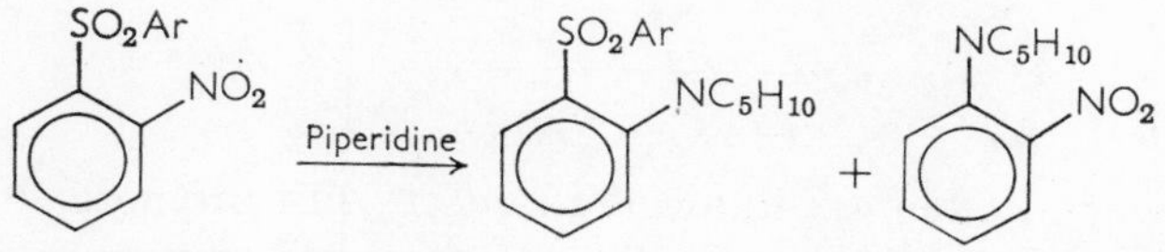

(54)

Almost comparable with the activating influence of a nitro group is that of the ring nitrogen atom in pyridine, and many examples of the introduction of amino groups by nucleophilic displacement of halogen atoms from the α- and γ-positions are known (reaction 55)[101]. Cor-

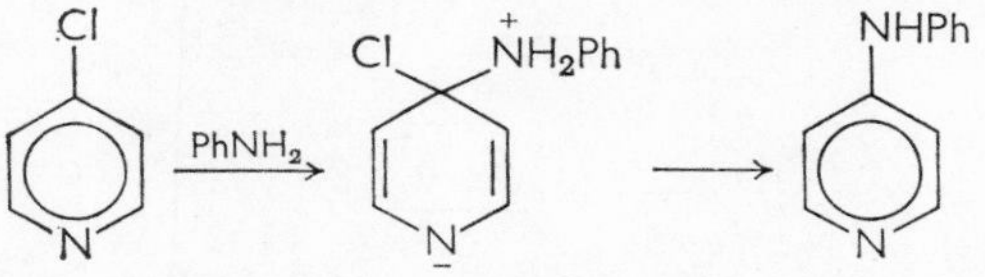

(55)

responding positions, i.e. α- or γ- with respect to the ring nitrogen atom, in related ring systems are similarly activated[102]. The synthesis of atebrine (reaction 56)[103a] is illustrative, and numerous potential antimalarial drugs have been similarly prepared from substituted 4-chloroquinolines[103b]. Site activation is likely to be increased by

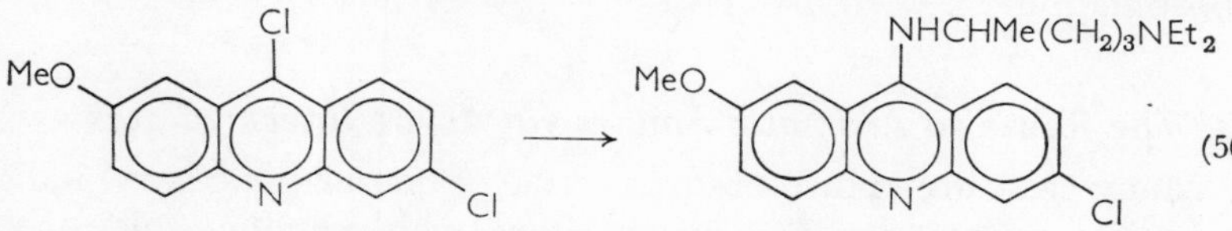

(56)

protonation or quaternisation of the ring nitrogen, and acidic catalysis has been noted in a number of cases (reaction 57)[93, 104].

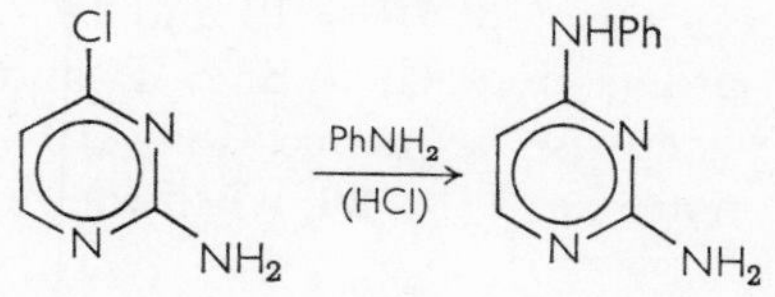

(57)

In the reactions discussed in this subsection, the displacing nucleophile has been ammonia or an amine. A number of cases have been reported, however, in which a dimethylamino group has been introduced by use of dimethylformamide; these include displacement of halogen from halogenonitrobenzenes[105a] and heterocyclic chloro compounds (reaction 58)[105b]. Urea has also been used for the preparation of aromatic primary amines such as 2,4-dinitroaniline from the corresponding chloro compounds[106].

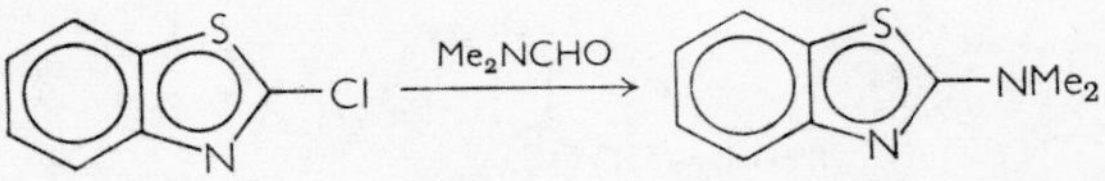

$$\qquad \qquad (58)$$

Amines can often be induced to react with aromatic halides, in which activating influences of the type discussed above are absent, by employing the device of catalysis by copper and various copper compounds (reactions 59 and 60). Diarylamines (from aromatic amines[107a]) and mixed aliphatic–aromatic amines (from aliphatic amines[107b]) are obtainable in this way. These induced reactions have

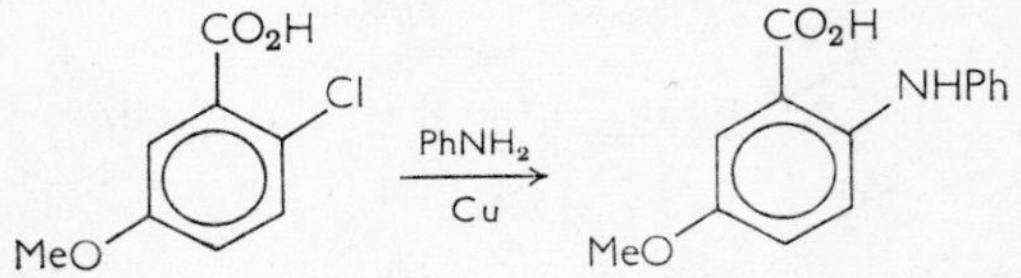

$$\qquad \qquad (59)$$

$$PhCl + MeNH_2 \xrightarrow{Cu_2Cl_2} PhNHMe \qquad (60)$$

been presented as nucleophilic substitutions of copper-complexed aryl halides in which the halogen has acquired positive character by coordination[93a].

D. The Route to Aromatic Amines via Aryne Intermediates

Many cases are known of the reaction of non-activated aryl halides, possessing at least one free *ortho* position, with metallic amides by an elimination–addition, i.e. aryne[108] or heteroaryne[109], mechanism. With the halogenobenzenes, the observed order of reactivity is Br > I > Cl ≫ F. The rates of the steps **18 → 19** and **19 → 20** are expected to decrease in the orders F > Cl > Br > I and I > Br > Cl > F respectively. The stepwise sequence **18 → 19 → 20** thus becomes effectively synchronous, i.e. **18 → 20**, when the rate of the second step

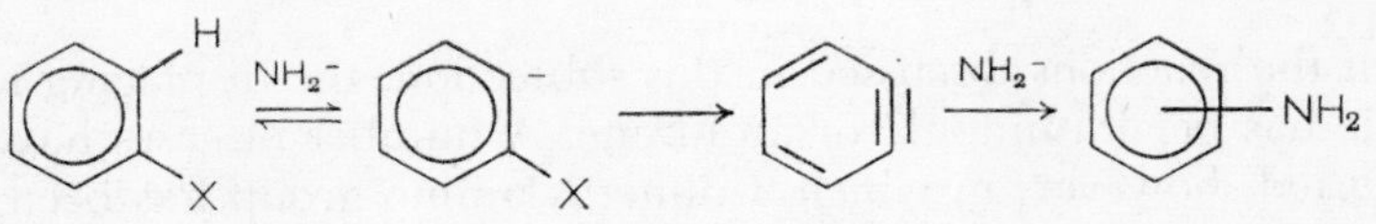

$$\qquad (18) \qquad \qquad (19) \qquad \qquad (20)$$

sufficiently exceeds the rate of the first, and this change occurs between chlorobenzene and bromobenzene[110].

With substituted halogenobenzenes, the amide ion might be expected to add preferentially to one or other end of the 'triple bond' and so lead to rearranged amines or mixtures of amines. The occurrence of such cine substitution has obvious attractions for the synthetic chemist in providing useful routes to various amines which might otherwise be difficult to obtain. An example is provided by the synthesis of 2-amino-4-methoxybiphenyl from 3-bromo-4-methoxybiphenyl (reaction 61)[111]. A similar rearrangement is exploited in a recent synthesis of the aporphine alkaloid, laureline[112].

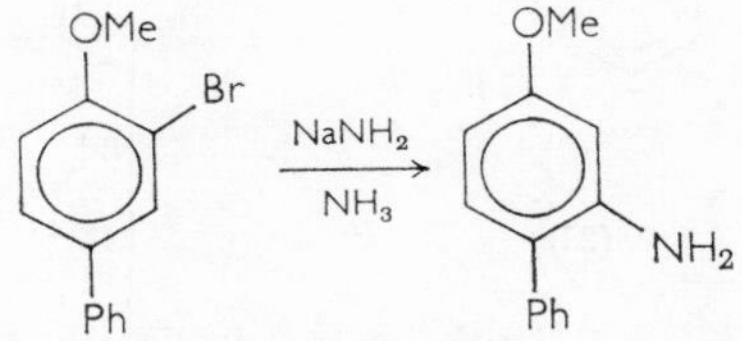

$$\text{(61)}$$

The factors underlying the orientation pattern observed with substituted halogenobenzenes have been generalised by Roberts and coworkers[113]. In the stages leading to the aryne, *ortho-* and *para-*substituted aryl halides give the 3- and 4-substituted benzynes, (**20a**) and (**20b**) respectively. Where the synchronous mechanism, **18 → 20**, applies, *meta*-substituted halides are expected to give preferentially the benzyne formed by removal of the more acidic of the hydrogen atoms *ortho* to halogen; for example, *m*-bromobenzotrifluoride gives 3-trifluoromethylbenzyne. Where the stepwise mechanism, **18 → 19 → 20**, applies, loss of halide ion from the two possible aryl anions (**19a**) and (**19b**) becomes a contributory and possibly determining factor, as in the formation of 3,4-pyridyne from 3-halogenopyridines[109].

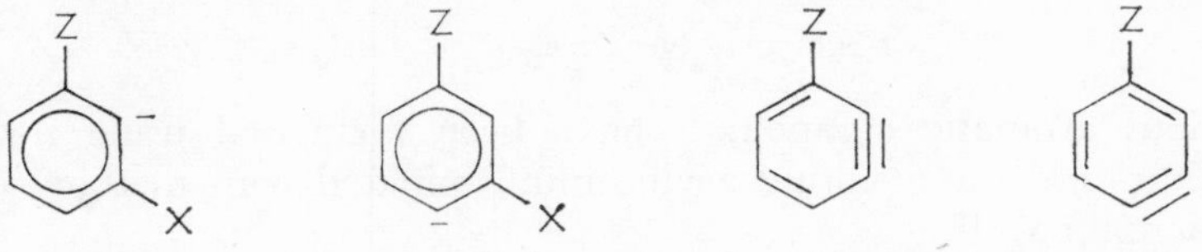

(**19a**)   (**19b**)   (**20a**)   (**20b**)

The preferred site of addition of amide ion to substituted benzynes is considered to be that giving the aryl anion which is most stabilised by the inductive effect of the substituent. With 3-substituted benzynes, amide ion would thus be expected to add preferentially to the *meta*

3*

position if Z is electron attracting, and to the *ortho* position (neglecting steric effects) if Z is electron releasing. Similarly, 4-substituted benzynes would be expected to give 4- and 3-substituted anilines respectively depending on whether Z is electron attracting or releasing. However, inductive effects are weaker with 4- than with 3-substituted benzynes, and conjugative effects may become relatively important; for example, approximately equal amounts of *m*- (**21a**) and *p*-anisidine (**21b**) are formed from *p*-bromoanisole and sodamide. These con-

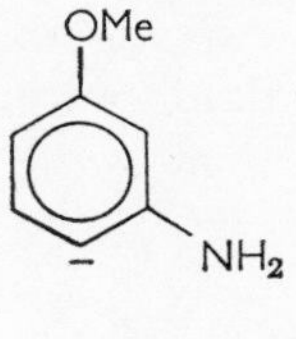

(**21a**) 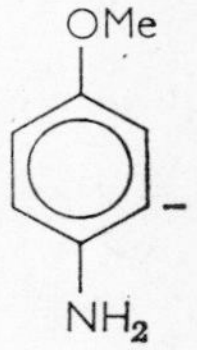   (**21b**)

siderations have been generally used for predicting and rationalising product orientation in additions of amide ion (and of alkylamide ions) to arynes[114], though the possible involvement of conjugative effects in the corresponding additions to 3-substituted benzynes has also been recently discussed[115].

V. PREPARATIONS OF AMINES INVOLVING THE USE OF GRIGNARD REAGENTS

A number of amine syntheses take advantage of the nucleophilic character of the alkyl (or aryl) group in a Grignard reagent. Thus direct amination of Grignard reagents (preferably those derived from alkyl chlorides or bromides rather than iodides) by methoxyamine leads to primary amines (reaction 62)[116, 117]. In a similar way,

$$t\text{-BuMgCl} + \text{NH}_2\text{OMe} \longrightarrow t\text{-BuNH}_2 \tag{62}$$

various aromatic compounds have been metalated using n-butyllithium and the resulting aryllithium aminated with methoxyamine (reaction 63)[118].

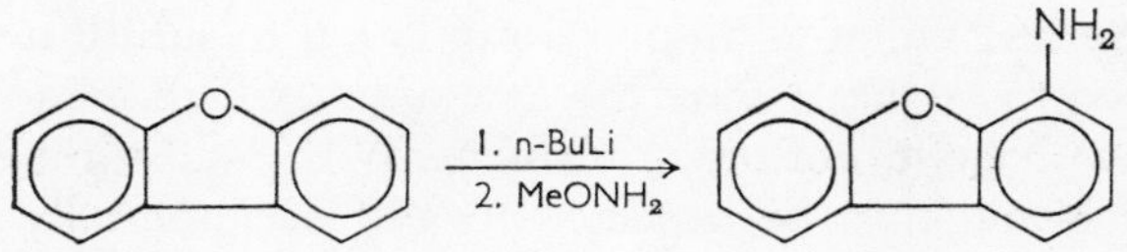

(63)

A number of α-dialkylaminoalkyl compounds, such as ethers and cyanides, react with Grignard reagents and so provide useful supplementary routes to tertiary amines. Dialkylaminomethyl ethers, prepared from the dialkylamine, formaldehyde and n-butanol, for example, react as in (64)[119]. Corresponding compounds, available

$$R_2NCH_2OBu\text{-}n + R^1MgX \longrightarrow R_2NCH_2R^1 \tag{64}$$

from cyclic secondary amines such as morpholine[120], or from aromatic aldehydes[121], react in the same way. The amino–ether grouping may also be incorporated in a cyclic structure, in which case the reaction gives rise to a hydroxy amine (reaction 65)[122].

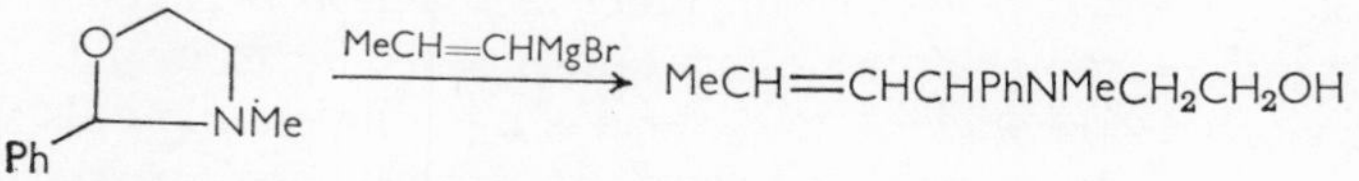

$$\xrightarrow{\text{MeCH=CHMgBr}} MeCH\!\!=\!\!CHCHPhNMeCH_2CH_2OH \tag{65}$$

α-Dialkylaminoalkyl cyanides react in a similar way to yield tertiary amines[123], although various complications have been observed[124,125]. The reaction has, however, continued to find application and has been used for the synthesis of triisopropylamine (reaction 66)[126]. Incidental examples of the reaction are also provided by

$$i\text{-}Pr_2NCHMeCN \xrightarrow{\text{MeMgCl}} i\text{-}Pr_3N \tag{66}$$

displacement of angular cyano groups in the quinolizidine and octahydropyrrocoline series (reaction 67)[127].

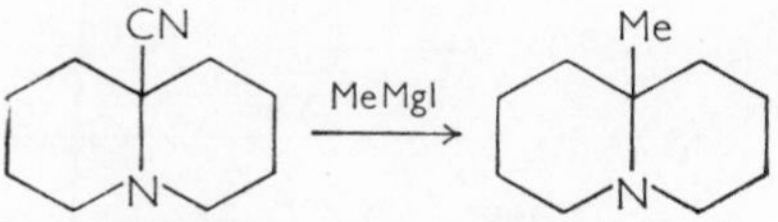

$$\xrightarrow{\text{MeMgI}} \tag{67}$$

A few tertiary amines have been prepared from N,N-dialkylformamides by reaction with an excess of Grignard reagent, though yields are frequently not high (reaction 68)[128].

$$(i\text{-}Pr)_2NCHO \xrightarrow[\text{n-BuMgBr}]{\text{Excess}} (i\text{-}Pr)_2NCH(n\text{-}Bu)_2 \tag{68}$$

VI. INTRODUCTION OF AMINO GROUPS BY ADDITION TO CARBON–CARBON DOUBLE BONDS

A. Nucleophilic Addition of Ammonia and Amines to Carbon–Carbon Double Bonds

Olefins are normally susceptible to attack by electrophiles rather than nucleophiles, and the addition of ammonia or amino compounds to carbon–carbon double bonds is only expected to occur readily in cases where the double bond bears one or more electron-attracting substituents. Typical activating substituents[129] are —CN, —COOR, —COR, —NO$_2$, —SO$_2$R. The general mechanistic features of this type of reaction have been discussed in an earlier volume of this series[130], and only problems pertinent to amination reactions will be discussed here.

These addition reactions may be typified by equation (69) where Y

$$R_2NH + CH_2{=}CHY \longrightarrow R_2NCH_2CH_2Y \qquad (69)$$

is an activating substituent. Reaction often occurs at normal or elevated temperatures without added catalysts, though catalysis is desirable and often necessary for efficient reaction between aromatic amines and, α,β-unsaturated esters and nitriles, for example. Typical catalysts are acetic acid[131], and various metallic compounds such as stannic chloride[132] or cupric acetate[133]. Catalysis by bases (sodium ethoxide, benzyltrimethylammonium hydroxide) is also employed, particularly in reactions of amines with α,β-unsaturated nitriles[131,134].

In addition reactions involving ammonia, it is sometimes difficult to obtain solely the desired product. The problem is well illustrated by McElvain and Stork's synthesis of guvacine[135], by way of the ammonia–ethyl acrylate reaction, which proceeds by reactions (70), (71) and (72). The optimum yield of the secondary amine, required for

$$NH_3 \; \underset{}{\overset{CH_2=CHCO_2Et}{\rightleftharpoons}} \; NH_2CH_2CH_2CO_2Et \qquad (70)$$

$$NH_2CH_2CH_2CO_2Et \; \underset{}{\overset{CH_2=CHCO_2Et}{\rightleftharpoons}} \; NH(CH_2CH_2CO_2Et)_2 \qquad (71)$$

$$NH(CH_2CH_2CO_2Et)_2 \; \underset{}{\overset{CH_2=CHCO_2Et}{\rightleftharpoons}} \; N(CH_2CH_2CO_2Et)_3 \qquad (72)$$

benzoylation, was about 44% (obtained using a 5:1 molar ratio of ammonia to ethyl acrylate), but the tertiary amine was formed in comparable amount (47%). The reversibility of the individual steps, however, allowed conversion of the tertiary amino compound to N,N-di(β-carbethoxyethyl)benzamide by thermal decomposition in the presence of benzoyl chloride.

Ammonia and acrylonitrile react similarly to give the corresponding primary, secondary and, to a lesser extent, tertiary amine[131]. Methylamine and ethyl acrylate give a mixture of the corresponding secondary and tertiary amines[136], or almost exclusively the tertiary amine[137], according to conditions. Methylamine and acrylonitrile, however, give mainly the corresponding secondary or tertiary amine[131], again depending on conditions. In this situation, it is customary to control the reaction as far as possible by adjusting the temperature and proportions of the reactants. With aromatic amines[138] and higher primary n-alkylamines the problems of selective reaction become less acute and it is usually possible to obtain principally the secondary or tertiary amino compound as desired[136, 139].

Secondary amines react with acrylonitrile and with ethyl acrylate to give the tertiary amino compounds. Piperidine, morpholine and to a lesser extent diethylamine react readily enough but the rates of these reactions drop with increasing size and branching of alkyl groups, and may fail with highly branched alkyl amines where branching occurs near the nitrogen atom[103b, 140]. These are reactions in which base catalysis is frequently employed[131]. Appropriate acidic conditions (with copper catalysis) have been used for the reaction of diphenylamine with acrylonitrile[141].

Attachment of an alkyl group to the double bond, either α- or β- to the activating substituent, may result in a decrease in reactivity of the unsaturated compound, but also in greater selectivity of reaction. Thus methylamine and methyl methacrylate interact to give the corresponding secondary amino compound (1:1 adduct) in very good yield (reaction 73)[142]. Esters of crotonic acid and its homologues

$$MeNH_2 + CH_2\!\!=\!\!CMeCO_2Me \longrightarrow MeNHCH_2CHMeCO_2Me \qquad (73)$$

give similar 1:1 adducts with ammonia, and primary and secondary amines (reaction 74)[143]. Interestingly, ammonia and ethyl γ-tri-

$$EtCH\!\!=\!\!CHCO_2Et + NHMe_2 \longrightarrow EtCH(NMe_2)CH_2CO_2Et \qquad (74)$$

fluorocrotonate give β-amino-γ,γ,γ-trifluorobutyramide, the inductive effect of one —CF_3 group being insufficient to reverse the direction of addition (reaction 75)[144a]; two β-CF_3 groups are however effective in this respect (reaction 76)[144b]. Esters of cinnamic acid react slug-

$$CF_3CH\!\!=\!\!CHCO_2Et + NH_3 \longrightarrow CF_3CH(NH_2)CH_2CONH_2 \qquad (75)$$

$$(CF_3)_2C\!\!=\!\!CHCO_2H + NH_3 \longrightarrow (CF_3)_2CHCH(NH_2)CO_2H \qquad (76)$$

gishly with amines to give indifferent yields of the corresponding β-amino-β-phenylpropionic esters[145].

Ammonia and amines add readily to the double bond in α,β-unsaturated ketones (reaction 77)[146] and nitro compounds (reaction 78)[147], catalysis being usually unnecessary even for aromatic amines.

$$PhCH\!\!=\!\!CHCOMe \xrightarrow{\text{Piperidine}} \underset{\underset{NC_5H_{10}}{|}}{PhCHCH_2COMe} \qquad (77)$$

$$PhCH\!\!=\!\!CHNO_2 \xrightarrow{\text{PhNH}_2} \underset{\underset{NHPh}{|}}{PhCHCH_2NO_2} \qquad (78)$$

The activating effect of a nitro group is even sufficient to promote amine addition when transmitted through a benzene ring. Thus primary and secondary amines add to o- and p-nitrostyrenes to give 1:1 adducts (reaction 79)[148], though yields decrease with increasing size of alkyl groups in the amine. Similar considerations apply to the

$$
\begin{array}{ccc}
\text{CH=CH}_2 & & \text{CH}_2\text{CH}_2\text{NEt}_2 \\
\bigcirc & \xrightarrow{\text{NHEt}_2} & \bigcirc \\
\text{NO}_2 & & \text{NO}_2
\end{array}
\tag{79}
$$

direct addition of amines to α- and γ-vinylpyridines (reaction 80)[149a,b].

$$
\begin{array}{ccc}
\text{CH=CH}_2 & & \text{CH}_2\text{CH}_2\text{NMe}_2 \\
\bigcirc & \xrightarrow{\text{NHMe}_2} & \bigcirc \\
\text{N} & & \text{N}
\end{array}
\tag{80}
$$

Styrene itself can be aminated in the presence of sodium, but the reaction here probably involves addition of the highly nucleophilic amide ion[129,150].

A few instances have been noted of allylic compounds bearing an electron-attracting substituent undergoing addition reactions with amines. These cases appear to involve shift of the double bond from the β,γ- to the α,β-position prior to reaction. For example, benzyl allyl sulphone and piperidine react to give benzyl β-piperidinopropyl sulphone (reaction 81)[151]. Allyl cyanide similarly reacts with am-

$$
\text{PhCH}_2\text{SO}_2\text{CH}_2\text{CH=CH}_2 \longrightarrow \text{PhCH}_2\text{SO}_2\text{CH}_2\text{CH(CH}_3)\text{NC}_5\text{H}_{10}
\tag{81}
$$

monia and amines at elevated temperatures to give adducts derived from crotononitrile (reaction 82)[152].

$$
\text{CH}_2\text{=CHCH}_2\text{CN} \xrightarrow{\text{NHEt}_2} \text{CH}_3\text{CH(NEt}_2)\text{CH}_2\text{CN}
\tag{82}
$$

Transamination reactions are occasionally used as an adjunct to the addition reactions described in this section. Aniline and β-diethylaminopropionitrile (as hydrochloride) undergo amine exchange at 140–150° to give β-anilinopropionitrile (reaction 83)[153]. These

$$
\text{PhNH}_2 + \text{Et}_2\overset{+}{\text{N}}\text{HCH}_2\text{CH}_2\text{CN} \longrightarrow \text{PhNHCH}_2\text{CH}_2\text{CN} + \text{Et}_2\overset{+}{\text{N}}\text{H}_2
\tag{83}
$$

reactions may proceed in some cases by an elimination–addition mechanism, i.e. via acrylonitrile, but Cymerman–Craig and coworkers have adduced evidence for a substitution mechanism[154]. In another context, the β-cyanoethyl group may be seen as a potential blocking group for —NH_2 as in recently reported syntheses of glycyl-glycine and other simple peptides[155]; removal of the blocking group may be effected using 10% ammonium hydroxide (reaction 84).

$$RN(CH_2CH_2CN)_2 \xrightarrow[\text{reflux}]{10\% \text{ NH}_4\text{OH}} RNH_2 \tag{84}$$

B. Routes to Amines Involving Electrophilic Additions to Carbon–Carbon Double Bonds

A number of routes to amino compounds take advantage of the normal susceptibility of olefins to attack by electrophiles. In general these reactions do not lead directly to amines, but rather to compounds which can be transformed to amines by subsequent reactions. A notable exception is provided by the reaction of dialkylchloro-amines with terminal olefins in acidic media (reaction 85)[156]. The

$$Et_2NCl + CH_2=CHR \xrightarrow{H^+} Et_2NCH_2CHClR \tag{85}$$

process involves addition of an aminium ion-radical, e.g. ($\dot{\,}NHEt_2$)$^+$ to the double bond followed by chain transfer. With the higher dialkylchloroamines, pyrrolidine formation (cf. section II.A) takes precedence over addition to the olefin (though not in the case of butadiene). Chlorination of the olefin is another possible complicating reaction. Nonetheless, yields of the 1:1 adduct are of the order 30–60% where the reaction is applicable.

α-Piperidyl radicals, generated from piperidine and benzoyl peroxide, also add to terminal olefins to give the alkylpiperidine[157], whilst addition of nitrogen dioxide to olefins gives products which are reducible to amino compounds[158].

Further examples of the Ritter reaction[72, 73] (cf. section IV.B) are provided by the many substituted olefins that may be protonated to give carbonium ions which can be intercepted by hydrogen cyanide or organic nitriles; cyanogen chloride can also be used as intercepting species[159], but offers no advantages. Mixtures of the two possible amides, and hence amines, are to be expected from non-terminal alkenes and from such olefinic compounds as oleic acid[160].

Olefins may also be converted to amines with yields of up to 60%, by hydroboration and subsequent reaction of the organoborane with chloroamine in alkaline solution or preferably with hydroxylamine-O-sulphonic acid in diglyme (reaction 86)[161]. The reaction is applicable

$$RCH=CH_2 \xrightarrow{B_2H_6} RCH_2CH_2-B \longrightarrow RCH_2CH_2NH_2 \tag{86}$$

to hindered olefins and is stereospecific, the boron being replaced by the NH_2 group with retention of configuration.

VII. REDUCTIVE METHODS FOR THE PREPARATION OF AMINES

A comprehensive review is available covering a large number of known reductive methods for the production of amines[162]. In principle, compounds containing an alkyl or aryl group linked to a nitrogen functional group at a higher oxidation level than —NH_2, or those containing an amino group directly linked to a carbon atom at a higher oxidation level than alkyl, or again compounds containing carbon–nitrogen multiple bonds, may all be reducible to amines.

Within the first group, reduction of nitro compounds is by far the most important, though principally in the aromatic series. Aliphatic nitro compounds may be reduced to the corresponding primary amines by catalytic hydrogenation[162-164] or by chemical means. Favoured chemical methods include reduction by iron and hydrochloric acid (with optional catalysis by ferric chloride[163]) or by lithium aluminium hydride[165]. With the latter reagent, ring enlargement has been noted with some tertiary cyclic nitroalkanes (reaction 87)[166]. Lithium aluminium hydride has also been extensively

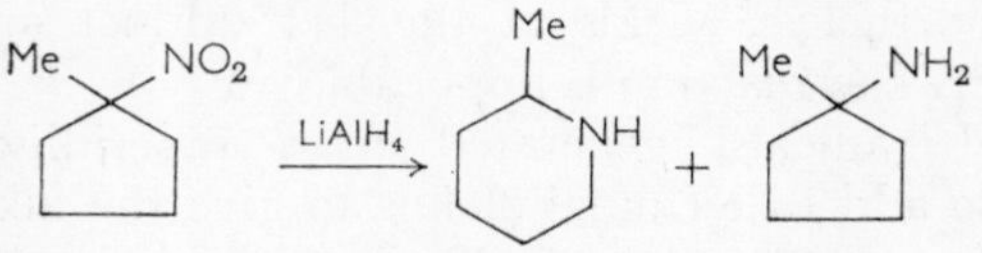

used for the reduction of β-nitrostyrenes to the important β-phenylethylamines (reaction 88)[165]. The utility of these reductions is limited

$$ArCH=CHNO_2 \xrightarrow{\text{LiAlH}_4} ArCH_2CH_2NH_2 \qquad (88)$$

by the availability of the nitro compounds, and many applications depend on the use of condensation reactions involving simple nitroalkanes to provide the required nitro compound[167].

Aromatic nitro compounds are a much more accessible class of substances, and a wide choice of methods exists for their reduction to amines. The choice of reducing agents includes (i) a variety of metals, usually employed in acidic or faintly acidic media, (ii) metallic compounds involving a low valency state of the metal, stannous chloride being frequently used and (iii) anionic sulphur compounds such as ammonium and sodium sulphides and polysulphides, and sodium hydrosulphite[162]. Catalytic hydrogenation is also widely employed[162]. Lithium aluminium hydride reduces simple aromatic nitro compounds to the azo compound[165]; sodium borohydride does not affect nitro groups except in special cases where a nitro group is displaced or eliminated as nitrite ion (reaction 89)[168]. A number of

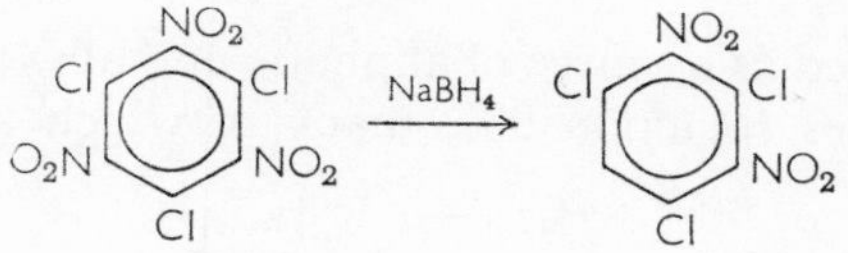

$$\text{(89)}$$

reductive procedures involving hydrogen transfer in the presence of palladium catalysts are known; hydrazine[169], and sodium borohydride in neutral or acidified solutions[170] have been used for this purpose.

A number of these methods of reduction are pleasingly selective and may be employed with molecules containing other potentially reducible groups; amongst the selective reagents are ferrous hydroxide and, particularly for the partial reduction of polynitro compounds, sodium and ammonium sulphides[162].

The substitution of a chlorine atom in the aromatic ring has been noted during some reductions in which concentrated hydrochloric acid is used; this presumably occurs by conversion of the intermediate arylhydroxylamine to the N-chloroamine and thence by rearrangement to the chloroaniline. A recent preparation of p-fluoroaniline exploits this type of substitution (reaction 90)[171].

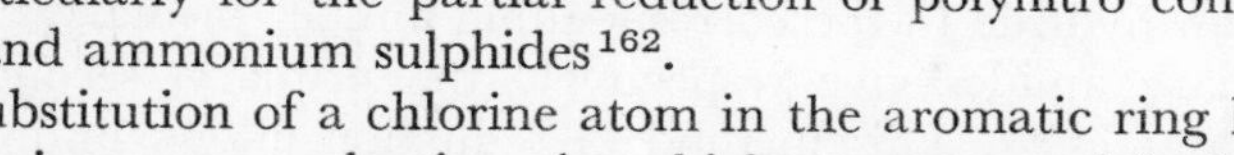

$$\text{(90)}$$

With aromatic compounds which are very reactive towards electrophiles, it may be preferable to introduce a nitroso or arylazo group rather than a nitro group; reduction to the amine can be effected in a number of ways (reaction 91)[162,172].

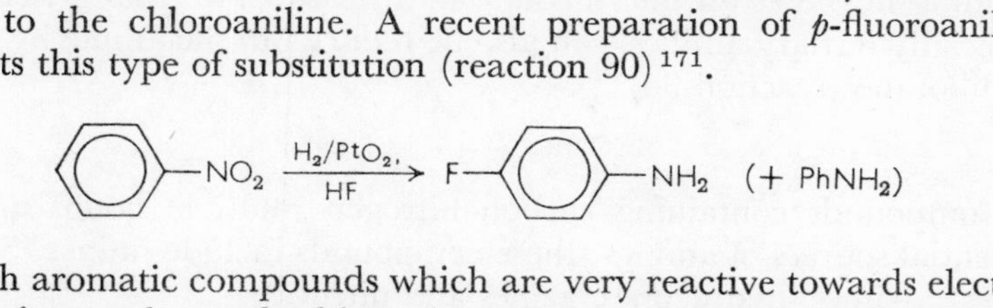

$$\text{(91)}$$

The reduction of hydrazine derivatives has not been widely exploited for the preparation of amines, but might be useful in certain cases. Benzhydrazide can be alkylated on the terminal nitrogen atom; hydrogenolysis then gives the amine (reaction 92)[173].

$$PhCONHNRR^1 \longrightarrow PhCON\overset{+}{\overset{-}{N}}RR^1Me \longrightarrow PhCONH_2 + NRR^1Me \quad (92)$$

More widely used as a source of aliphatic and alicyclic amines is the reduction of azides (reaction 93), many of which are available by

$$RN_3 + H_2 \longrightarrow RNH_2 + N_2 \quad (93)$$

stereospecific S_N2 azidolysis of various alkyl halides and toluene-p-sulphonates. The reduction may be carried out catalytically under very mild conditions[162], or by use of lithium aluminium hydride[174] or diborane[175].

The principal compounds in the second group, i.e. those containing an amino group linked to a carbon atom at a higher oxidation level than alkyl, are the carboxylic amides. These are reducible by lithium aluminium hydride to the corresponding amine; primary, secondary and tertiary amides give respectively the primary, secondary and tertiary amines[165]. This method of reduction allows indirect introduction of bridging amino groups in certain cyclic systems[176]. There is evidence that the reduction of primary amides involves the corresponding nitrile as an intermediate in at least some cases[177]. Amides, especially tertiary amides, can also be reduced to the amine by means of diborane (reaction 94)[178].

$$RCONR_2 \longrightarrow RCH_2NR_2 \quad (94)$$

Compounds containing carbon–nitrogen multiple bonds are also potential sources of amines; these compounds include imines (Schiff's bases), oximes, hydrazones, azines and nitriles.

Imines can be reduced to amines by catalytic hydrogenation, or by reduction with lithium aluminium hydride, sodium borohydride or amine–borane complexes[179]. For many purposes, it is not necessary to prepare the imine as the desired amine can, at least in principle, be obtained by catalytic reduction of a mixture of the corresponding carbonyl compound and precursoral amine (reaction 95)[180]. The

$$R_2C{=}O + R^1NH_2 \rightleftharpoons R_2C(OH)NHR^1 \rightleftharpoons R_2C{=}NR^1 \longrightarrow R_2CHNHR^1 \quad (95)$$

amine R^1NH_2 can even be replaced by compounds, e.g. nitro compounds, which are reducible to the amine under the reaction conditions[180, 181]. This reaction is termed reductive alkylation and suffers from the drawback that the amine formed may also condense with the carbonyl compound and so undergo further reductive alkylation. In practical applications of this method, undesired reactions of this type can often be minimised by suitably adjusting the concentrations of the carbonyl compound and the starting amine. A recent variant involves the reduction of a mixture of the carbonyl compound and the amine salt by sodium borohydride in aqueous solution[182].

Reductive alkylation of amines by carbonyl compounds can also be effected by procedures in which formic acid and its derivatives provide the reducing system[183]. The particular combination of formaldehyde and formic acid (Clarke–Eschweiler method) provides for the methylation of amines, usually to the tertiary stage (cf. ref. 60)[183]; the Délepine method (cf. section IV.B) can be modified to give alkyldimethylamines by use of formic acid[184]. These reactions are considered to involve decarboxylation of the intermediate aminocarbinyl formate (reaction 96)[185].

$$R_2C{=}O \rightleftharpoons R_2C(NHR^1)OH \longrightarrow$$
$$R_2C(NHR^1)OCH{=}O \longrightarrow R_2CHNHR^1 + CO_2 \quad (96)$$

Oximes, hydrazones and azines are obtainable from carbonyl compounds and can be reduced to primary amines (sometimes accompanied by secondary amines) in various ways (reaction 97)[162,186].

$$RCH{=}NOH \longrightarrow RCH_2NH_2 \quad (97)$$

Rearrangements have been noted in the reduction of some oximes with lithium aluminium hydride[187].

The catalytic reduction of nitriles is also frequently employed as a route to primary amines (reaction 98)[162,188]. This involves an imine

$$RC{\equiv}N \longrightarrow RCH_2NH_2 \quad (98)$$

as an intermediate stage and may, as in the reductive alkylation procedures, lead to the primary amine together with appreciable amounts of the secondary amine. For the preparation of primary amines, it has become common practice to conduct hydrogenation in the presence of an excess of ammonia, or alternatively in the presence of acetic anhydride, when the primary amide is produced. Lithium aluminium hydride is also widely used for the reduction of nitriles[165].

VIII. REFERENCES

1. R. N. Keller and P. A. S. Smith, *J. Am. Chem. Soc.*, **66**, 1122 (1944); *J. Am. Chem. Soc.*, **68**, 899 (1946).
2. G. M. Hoop and J. M. Tedder, *J. Chem. Soc.*, 4685 (1961).

3. A. C. Robson and S. Coffey, *J. Chem. Soc.*, 2372 (1954).

4. P. Kovacic and R. P. Bennett, *J. Am. Chem. Soc.*, **83**, 221 (1961).

5. P. Kovacic and J. L. Foote, *J. Am. Chem. Soc.*, **83**, 743 (1961).

6. P. Kovacic, R. P. Bennett and J. L. Foote, *J. Am. Chem., Soc.*, **84**, 759 (1962).

7. P. Kovacic, R. L. Russell and R. P. Bennett, *J. Am. Chem. Soc.*, **86**, 1588 (1964).

8. H. C. Brown and C. R. Smoot, *J. Am. Chem. Soc.*, **78**, 6255 (1956).

9. (a) H. Bock and K. L. Kompa, *Angew. Chem.*, **77**, 807 (1965).
   (b) H. Bock and K. L. Kompa, *Chem. Ber.*, **99**, 1347 (1966), and the two following papers.

10. M. E. Wolff, *Chem. Rev.*, **63**, 55 (1963).

11. (a) P. Kovacic, R. M. Lange, J. L. Foote, C. T. Goralski, J. J. Hiller and J. A. Levisky, *J. Am. Chem. Soc.*, **86**, 1650 (1964).
    (b) P. Kovacic, C. T. Goralski, J. J. Hiller, J. A. Levisky and R. M. Lange, *J. Am. Chem. Soc.*, **87**, 1262 (1965).

12. P. Kovacic, J. A. Levisky and C. T. Goralski, *J. Am. Chem. Soc.*, **88**, 100 (1966).

13. P. Kovacic and J. A. Levisky, *J. Am. Chem. Soc.*, **88**, 1000 (1966).

14. M. T. Leffler, *Organic Reactions*, Vol. I, John Wiley and Sons, New York, 1942, Chap. 4.

15. F. W. Bergstrom, H. G. Sturz and H. W. Tracy, *J. Org. Chem.*, **11**, 239 (1946).

16. N. G. Luthy, F. W. Bergstrom and H. S. Mosher, *J. Org. Chem.*, **14**, 322 (1949).

17. R. Huisgen and H. Rist, *Ann. Chem.*, **594**, 159 (1955).

18. F. W. Bergstrom, I. M. Granara and V. Erickson, *J. Org. Chem.*, **7**, 98 (1942).

19. J. Meisenheimer and E. Patzig, *Chem. Ber.*, **39**, 2533 (1906).

20. C. C. Price and S. Voong, *Organic Syntheses*, Coll. Vol. III, John Wiley and Sons, New York, 1955, p. 664; and refs. cited.

21. Y. Ogata, Y. Izawa and H. Tomioka, *Tetrahedron*, **22**, 483 (1966).

22. Y. Ogata, Y. Izawa, H. Tomioka and T. Nishazawa, *Tetrahedron*, **22**, 1557 (1966).

23. N. L. Drake, *Organic Reactions*, Vol. I, John Wiley and Sons, New York, 1942, Chap. 5.

24. (a) W. A. Cowdrey and C. N. Hinshelwood, *J. Chem. Soc.*, 1036 (1946).
    (b) W. A. Cowdrey, *J. Chem. Soc.*, 1041 (1946), and the two following papers.

25. A. Rieche and H. Seeboth, *Ann. Chem.*, **638**, 43 (1960), and the six following papers.

26. H. Seeboth and A. Rieche, *Ann. Chem.*, **671**, 70 (1964), and the following paper.

27. J. v. Braun, *Chem. Ber.*, **70**, 979 (1937).

28. R. Lukeš, V. Galík and J. Bauer, *Chem. Listy*, **48**, 858 (1954); *Collection Czech. Chem. Commun.*, **19**, 712 (1954).

29. (a) C. K. Ingold, *Structure and Mechanism in Organic Chemistry*, Bell, London, 1953, Chap. 7.
    (b) A. Streitwieser, *Chem. Rev.*, **56**, 571 (1956).

2. The Introduction of the Amino Group

30. S. Patai and S. Weiss, *J. Chem. Soc.*, 1035 (1959).
31. E. D. Hughes and D. J. Whittingham, *J. Chem. Soc.*, 806 (1960).
32. F. E. King, T. J. King and I. H. M. Muir, *J. Chem. Soc.*, 5 (1946).
33. H. C. Brown and N. R. Eldred, *J. Am. Chem. Soc.*, **71**, 445 (1949).
34. H. C. Brown and A. Cahn, *J. Am. Chem. Soc.*, **77**, 1715 (1955).
35. L. A. Pinck and G. E. Hilbert, *J. Am. Chem. Soc.*, **68**, 377 (1946).
36. G. F. Hennion and K. W. Nelson, *J. Am. Chem. Soc.*, **79**, 2142 (1957).
37. (a) G. F. Hennion and R. S. Hanzel, *J. Am. Chem. Soc.*, **82**, 4908 (1960).
    (b) N. R. Easton, R. D. Dillard, W. J. Doran, M. Livezey and D. E. Morrison, *J. Org. Chem.*, **26**, 3772 (1961).
    (c) C. Ainsworth and N. R. Easton, *J. Org. Chem.*, **26**, 3776 (1961).
38. G. F. Hennion and C. V. DiGiovanna, *J. Org. Chem.*, **30**, 3696 (1965); *J. Org. Chem.*, **31**, 1977 (1966).
39. W. G. Young, I. D. Webb and H. L. Goering, *J. Am. Chem. Soc.*, **73**, 1076 (1951).
40. R. H. DeWolfe and W. G. Young in *The Chemistry of Alkenes* (Ed. S. Patai), Interscience, New York, 1964, Chap. 10.
41. R. O. Clinton, U. J. Salvador and S. C. Laskowski, *J. Am. Chem. Soc.*, **71**, 3366 (1949).
42. C. M. Suter, *The Organic Chemistry of Sulphur*, John Wiley and Sons, New York, 1944, Chaps. 1 and 5.
43. (a) T. Gramstad and R. N. Haszeldine, *J. Chem. Soc.*, 4069 (1957).
    (b) J. Burdon and V. C. R. McLoughlin, *Tetrahedron*, **21**, 1 (1965).
44. W. Voss and E. Blanke, *Ann. Chem.*, **485**, 258 (1931); cf. W. E. Bissinger, F. E. Kung and C. W. Hamilton, *J. Am. Chem. Soc.*, **70**, 3940 (1948).
45. E. H. White and C. A. Elliger, *J. Am. Chem. Soc.*, **87**, 5261 (1965).
46. (a) J. H. Billman, A. Radike and B. W. Mundy, *J. Am. Chem. Soc.*, **64**, 2977 (1942).
    (b) D. G. Thomas, J. H. Billman and C. E. Davis, *J. Am. Chem. Soc.*, **68**, 895 (1946).
    (c) T. L. Fletcher, M. E. Taylor and A. W. Dahl, *J. Org. Chem.*, **20**, 1021 (1955).
    (d) J. R. Cox and O. B. Ramsay, *Chem. Rev.*, **64**, 317 (1964).
    (e) R. F. Hudson, *Structure and Mechanism in Organo-Phosphorus Chemistry*, Academic Press, London and New York, 1965.
47. (a) B. P. Lugovkin and B. A. Arbuzov, *Zh. Obshch. Chem.*, **22**, 2041 (1952).
    (b) J. I. G. Cadogan, *J. Chem. Soc.*, 1079 (1957) and refs. cited; cf. H. E. Baumgarten and R. A. Setterquist, *J. Am. Chem. Soc.*, **81**, 2132 (1959).
48. E. S. Lane, *J. Chem. Soc.*, 2006 (1956); cf. R. Boshan, R. T. Merrow and R. T. Van Dolah, *Chem. Rev.*, **55**, 485 (1955).
49. (a) H. E. Zaugg, P. F. Helgren and A. D. Schaefer, *J. Org. Chem.*, **28**, 2617 (1963).
    (b) T. Kametani, K. Kigasawa, T. Hayasaka, M. Hiiragi, H. Ishimaru and S. Asagi, *J. Heterocyclic Chem.*, **3**, 129 (1966).
50. T. L. Gresham, J. E. Jansen, F. W. Shaver, R. A. Bankert and F. T. Fiedorek, *J. Am. Chem. Soc.*, **73**, 3168 (1951).
51. (a) A. F. McKay and H. H. Brownell, *J. Org. Chem.*, **15**, 648 (1950).
    (b) C. L. Browne and R. E. Lutz, *J. Org. Chem.*, **17**, 1187 (1952).

(c) A. R. Graham, A. F. Millidge and D. P. Young, *J. Chem. Soc.*, 2180 (1954).

52. H. Gilman, C. S. Sherman, C. C. Price, R. C. Elderfield, J. J. Maynard, R. H. Reitsema, L. Tolman, S. P. Massie, F. J. Marshall and L. Goldman, *J. Am. Chem. Soc.*, **68**, 1291 (1946).

53. A. B. Foster and D. Horton, *Advan. Carbohydrate Chem.*, **14**, 213 (1959).

54. S. Hünig and W. Baron, *Chem. Ber.*, **90**, 395 (1957); cf. D. A. Archer and H. Booth, *J. Chem. Soc.*, 322 (1963).

55. (a) H. R. Snyder, R. E. Carnahan and E. R. Lovejoy, *J. Am. Chem. Soc.*, **76**, 1301 (1954).
    (b) H. R. Snyder, E. L. Eliel and R. E. Carnahan, *J. Am. Chem. Soc.*, **72**, 2958 (1950).

56. J. v. Braun and R. Clar, *Chem. Ber.*, **73**, 1417 (1940).

57. D. D. Reynolds and W. O. Kenyon, *J. Am. Chem. Soc.*, **72**, 1597 (1950).

58. S. Hünig, *Chem. Ber.*, **85**, 1056 (1952).

59. C. F. Culberson and P. Wilder, *J. Am. Chem. Soc.*, **82**, 4939 (1960).

60. A. C. Cope, E. Ciganek, L. J. Fleckenstein and M. A. P. Meisinger, *J. Am. Chem. Soc.*, **82**, 4651 (1960).

61. (a) E. Grovenstein, S. Chandra, C. E. Collum and W. E. Davis, *J. Am. Chem. Soc.*, **88**, 1275 (1966).
    (b) E. Grovenstein and R. W. Stevenson, *J. Am. Chem. Soc.*, **81**, 4850 (1959).

62. G. F. Hennion and R. H. Ode, *J. Org. Chem.*, **31**, 1975 (1966).

63. (a) J. N. Tilley and A. A. R. Sayigh, *J. Org. Chem.*, **28**, 2076 (1963).
    (b) N. Bortnick, L. S. Luskin, M. D. Hurwitz, W. E. Craig, L. J. Exner and J. Mirza, *J. Am. Chem. Soc.*, **78**, 4039 (1956).

64. (a) R. N. Shreve and L. W. Rothenberger, *Ind. Eng. Chem.*, **29**, 1361 (1937), and refs. cited.
    (b) G. F. Hennion and E. G. Teach, *J. Am. Chem. Soc.*, **75**, 1653 (1953).
    (c) M. S. Kharasch, W. Nudenberg and E. K. Fields, *J. Am. Chem. Soc.*, **66**, 1276 (1944), and refs. cited.

65. O. Westphal and D. Jerchel, *Chem. Ber.*, **73**, 1002 (1940).

66. J. W. Alexander and S. M. McElvain, *J. Am. Chem. Soc.*, **60**, 2285 (1938).

67. J. S. Buck and C. W. Ferry, *Organic Syntheses*, Coll. Vol. II, John Wiley and Sons, New York, 1943, p. 290.

68. S. J. Angyal, *Organic Reactions*, Vol. VIII, John Wiley and Sons, New York, 1954, Chap. 4.

69. G. W. Cannon, K. K. Grebber and Y. Hsu, *J. Org, Chem.*, **18**, 516 (1953).

70. J. Cast and T. S. Stevens, *J. Chem. Soc.*, 4180 (1953).

71. H. Meerwein, P. Laasch, R. Mersch and J. Spille, *Chem. Ber.*, **89**, 209 (1956).

72. (a) J. J. Ritter and P. P. Minieri, *J. Am. Chem. Soc.*, **70**, 4045 (1948).
    (b) J. J. Ritter and J. Kalish, *J. Am. Chem. Soc.*, **70**, 4048 (1948).

73. (a) F. R. Benson and J. J. Ritter, *J. Am. Chem. Soc.*, **71**, 4128 (1949).
    (b) H. Plant and J. J. Ritter, *J. Am. Chem. Soc.*, **73**, 4076 (1951).

74. R. Jaquier and H. Christol, *Compt. Rend.*, 556 (1954).

75. (a) H. Decker and P. Becker, *Ann. Chem.*, **395**, 362 (1913).
    (b) A. L. Morrison and H. Rinderknecht, *J. Chem. Soc.*, 1478 (1950).
    (c) J. N. Baxter and J. Cymerman-Craig, *J. Chem. Soc.*, 1940 (1953).

76. H. Peter, M. Brugger, J. Schreiber and A. Eschenmoser, *Helv. Chim. Acta*, **46**, 577 (1963).
77. R. S. Neale, N. L. Marcus and R. C. Schepers, *J. Am. Chem. Soc.*, **88**, 3051 (1966).
78. J. W. Schulenberg and S. Archer, *Organic Reactions*, Vol. XIV, John Wiley and Sons, New York, 1965, Chap. 1.
79. S. Gabriel, *Chem. Ber.*, **20**, 2224 (1887), and later papers by Gabriel and coworkers in *Chem. Ber.*
80. E. J. Sakellarios, *Helv. Chim. Acta*, **29**, 1675 (1946).
81. S. Gabriel and H. Ohle, *Chem. Ber.*, **50**, 819 (1917).
82. H. R. Ing and R. H. F. Manske, *J. Chem. Soc.*, 2348 (1926).
83. J. C. Sheehan and W. A. Bolhofer, *J. Am. Chem. Soc.*, **72**, 2786 (1950).
84. O. Hinsberg, *Chem. Ber.*, **23**, 2962 (1890); *Ann. Chem.* **265**, 178 (1891).
85. D. Klamann, G. Hofbauer and F. Drahowzal, *Monatsh. Chem.*, **83**, 870 (1952).
86. (a) F. E. King, R. M. Acheson and A. B. Yorke-Long, *J. Chem. Soc.*, 1926 (1948).
    (b) F. E. King, R. J. S. Beer and S. G. Waley, *J. Chem. Soc.*, 92 (1946).
87. K. Abe, *J. Pharm. Soc. Japan*, **75**, 153 (1955), and the three following papers.
88. D. I. Weisblat, B. J. Magerlein and D. R. Myers, *J. Am. Chem. Soc.*, **75**, 3630 (1953).
89. D. Klamann and G. Hofbauer, *Chem. Ber.*, **86**, 1246 (1953).
90. C. Ressler and V. du Vigneaud, *J. Am. Chem. Soc.*, **76**, 3107 (1954); cf. V. du Vigneaud, C. Ressler, J. M. Swan, C. W. Roberts and P. G. Katsoyannis, *J. Am. Chem. Soc.*, **76**, 3115 (1954); V. du Vigneaud and O. K. Behrens, *J. Biol. Chem.*, **117**, 27 (1937).
91. W. Traube and A. Englehardt, *Chem. Ber.*, **44**, 3149 (1911); cf. E. B. Vliet, *Organic Syntheses*, Coll. Vol. I., John Wiley and Sons, New York, 1941, pp. 201 and 203.
92. (a) V. Prelog, B. C. McKusick, J. R. Merchant, S. Julia and M. Wilhelm, *Helv. Chim. Acta*, **39**, 498 (1956).
    (b) A. C. Currie, G. T. Newbold and F. S. Spring, *J. Chem. Soc.*, 4693 (1961).
93. (a) J. F. Bunnett and R. E. Zahler, *Chem. Rev.*, **49**, 273 (1951).
    (b) J. F. Bunnett, *Quart. Rev.* (*London*), **12**, 1 (1958).
    (c) S. D. Ross in *Progress in Physical Organic Chemistry*, Vol. I (Eds. S. G. Cohen, A. Streitwieser and R. W. Taft), Interscience, New York, 1963, Chap. 2.
94. F. Sanger, *Biochem. J.*, **39**, 507 (1945).
95. H. Zahn, *Angew. Chem.*, **67**, 561 (1955).
96. J. F. Bunnett and R. H. Garst, *J. Am. Chem. Soc.*, **87**, 3875 (1965), and refs. cited.
97. B. Capon and N. B. Chapman, *J. Chem. Soc.*, 600 (1957).
98. (a) J. Kenner and M. Parkin, *J. Chem. Soc.*, **117**, 852 (1920).
    (b) K. Ibbotson and J. Kenner, *J. Chem. Soc.*, **123**, 1260 (1923).
99. W. C. Spitzer and G. W. Wheland, *J. Am. Chem. Soc.*, **62**, 2995 (1940).
100. J. D. London and T. D. Robson, *J. Chem. Soc.* 242 (1937).

101. H. E. Mertel in *Pyridine and its Derivatives*, a monograph in *The Chemistry of Heterocyclic Compounds* series, Vol. II (Ed. E. Klingsberg), Interscience, New York, 1961, Chap. 6.

102. N. B. Chapman and D. Q. Russell-Hill, *J. Chem. Soc.*, 1563 (1956), and refs. cited.

103. (a) F. Mietzsch and H. Mauss to I. G. Farbenindustrie, *Ger. Pat.* 553,072 (1930).

(b) D. S. Tarbell, N. Shakespeare, C. J. Claus and J. F. Bunnett, *J. Am. Chem. Soc.*, **68**, 1217 (1946); cf. D. E. Pearson, W. H. Jones and A. C. Cope, *J. Am. Chem. Soc.*, **68**, 1225 (1946); R. C. Fuson, W. E. Parham and L. J. Reed, *J. Am. Chem. Soc.*, **68**, 1239 (1946).

104. C. K. Banks, *J. Am. Chem. Soc.*, **66**, 1127 (1944).

105. (a) D. S. Deorha and H. L. Sharma, *J. Indian Chem. Soc.*, **40**, 819 (1963); M. Wakae and K. Hamano, *Bull. Chem. Soc. Japan*, **36**, 230 (1963).

(b) J. J. D'Amico, S. T. Webster, R. H. Campbell and C. E. Twine, *J. Org. Chem.*, **30**, 3618 (1965); L. Joseph and A. H. Albert, *J. Heterocyclic Chem.*, **3**, 107 (1966).

106. (a) W. H. Bentley and W. Blythe and Co., *Brit. Pat.* 263,552 (1925).

(b) K. H. T. Pfister to Rohm and Haas Co. *U.S. Pat.* 1,752,998 (1925).

107. (a) F. Ullmann and H. Kipper, *Chem. Ber.*, **38**, 2120 (1905), and refs. cited; P. E. Weston and H. Adkins, *J. Am. Chem. Soc.*, **50**, 859 (1928).

(b) E. C. Hughes, F. Keatch and V. Elersich, *Ind. Eng. Chem.*, **42**, 787 (1950).

108. (a) R. Huisgen and J. Sauer, *Angew. Chem.*, **72**, 91 (1960).

(b) J. F. Bunnett, *J. Chem. Educ.*, **38**, 278 (1961).

(c) H. Heaney, *Chem. Rev.*, **62**, 81 (1962), and refs. cited.

109. T. Kaufmann, *Angew. Chem.*, **77**, 557 (1965), and refs. cited.

110. J. D. Roberts, D. A. Semenow, H. E. Simmons and L. A. Carlsmith, *J. Am. Chem. Soc.*, **78**, 601 (1956).

111. C. K. Bradsher, F. C. Brown and P. H. Leake, *J. Org. Chem.*, **22**, 500 (1957).

112. M. S. Gibson and J. M. Walthew, *Chem. Ind.* (*London*), 185 (1965).

113. J. D. Roberts, C. W. Vaughan, L. A. Carlsmith and D. A. Semenow, *J. Am. Chem. Soc.*, **78**, 611 (1956).

114. J. F. Bunnett and B. F. Hrutfiord, *J. Org. Chem.*, **27**, 4152 (1962).

115. G. B. R. de Graaff, H. J. Den Hertog and W. C. Melger, *Tetrahedron Letters*, 963 (1965).

116. N. I. Sheverdina and K. A. Kocheshkov, *Zh. Obshch. Chim.*, **8**, 1825 (1938).

117. R. Brown and W. E. Jones, *J. Chem. Soc.*, 781 (1946).

118. H. Gilman and J. W. Morton, *Organic Reactions*, Vol. VIII, John Wiley and Sons, New York, 1954, Chap. 6.

119. G. M. Robinson and R. Robinson, *J. Chem. Soc.*, **123**, 532 (1923).

120. J. P. Mason and M. Zief, *J. Am. Chem. Soc.*, **62**, 1450 (1940).

121. A. T. Stewart and C. R. Hauser, *J. Am. Chem. Soc.*, **77**, 1098 (1955).

122. J. Ficini and H. Normant, *Bull. Soc. Chim. France*, 1454 (1957).

123. P. Bruylants, *Bull. Soc. Chim. Belge.*, **33**, 467 (1924); *Bull. Acad. Roy. Belge*, **11**, 261 (1925).

124. T. Thomson and T. S. Stevens, *J. Chem. Soc.*, 2607 (1932).

125. M. Bockmühl and G. Ehrhart, *Ann. Chem.*, **561**, 52 (1949).
126. F. Kuffner and W. Koechlin, *Monatsh. Chem.*, **93**, 476 (1962).
127. N. J. Leonard and A. S. Hay, *J. Am. Chem. Soc.*, **78**, 1984 (1956).
128. F. Kuffner and E. Polke, *Monatsh. Chem.*, **82**, 330 (1951), and refs. cited.
129. H. Bestian, *Ann. Chem.*, **566**, 211 (1950).
130. S. Patai and Z. Rappoport in *The Chemistry of Alkenes*, (Ed. S. Patai), Interscience, New York, 1964, Chap. 8.
131. H. A. Bruson, *Organic Reactions*, Vol. V, John Wiley and Sons, New York, 1949, Chap. 2, and refs. cited.
132. P. L. Southwick and R. T. Crouch, *J. Am. Chem. Soc.*, **75**, 3413 (1953), and refs. cited.
133. S. A. Heininger, *J. Org. Chem.*, **22**, 1213 (1957), and refs. cited.
134. (a) A. Pohland and H. R. Sullivan, *J. Am. Chem. Soc.*, **77**, 2817 (1955).
    (b) J. M. Stewart and C. H. Chang, *J. Org. Chem.*, **21**, 635 (1956).
135. S. M. McElvain and G. Stork, *J. Am. Chem. Soc.*, **68**, 1049 (1946).
136. D. W. Adamson, *J. Chem. Soc.* (supplement), 144 (1949); cf. R. W. Holley and A. D. Holley, *J. Am. Chem. Soc.*, **71**, 2127 (1949).
137. (a) R. Mozingo and J. H. McCracken, *Organic Syntheses*, Coll. Vol. III, John Wiley and Sons, New York, 1955, p. 258.
    (b) A. Dobrowsky, *Monatsh. Chem.*, **83**, 443 (1952).
138. J. T. Braunholtz and F. G. Mann, *J. Chem. Soc.*, 1817 (1953).
139. (a) R. C. Fuson, W. E. Parham and L. J. Read, *J. Am. Chem. Soc.*, **68**, 1239 (1946).
    (b) S. M. McElvain and K. Rorig, *J. Am. Chem. Soc.*, **70**, 1826 (1948).
    (c) N. J. Leonard, F. E. Fischer, E. Barthel, J. Figueras and W. C. Wildman, *J. Am. Chem. Soc.*, **73**, 2371 (1951).
    (d) M. Lipp, F. Dallacker and H. Rey, *Chem. Ber.*, **91**, 2239 (1958).
140. O. Hromatka, *Chem. Ber.*, **75**, 131 (1942).
141. R. C. Cookson and F. G. Mann, *J. Chem. Soc.*, 67 (1949).
142. D. R. Howton, *J. Org. Chem.* **10**, 277 (1945).
143. D. W. Adamson, *J. Chem. Soc.*, 885 (1950).
144. (a) M. M. Walborsky and M. Schwarz, *J. Am. Chem. Soc.*, **75**, 3241 (1953).
    (b) I. L. Kunyants, *Intern. Symposium Fluorine Chem.*, Birmingham, 1959.
145. C. B. Pollard and G. C. Mattson, *J. Am. Chem. Soc.*, **78**, 4089 (1956).
146. (a) N. H. Cromwell, *Chem. Rev.*, **38**, 83 (1946).
    (b) K. Bowden, I. M. Heilbron, E. R. H. Jones and B. C. L. Weedon, *J. Chem. Soc.*, 39 (1946).
    (c) P. Reynaud and J. Matti, *Compt. Rend.*, **236**, 2156 (1953).
    (d) R. Baltzly, E. Lorz, P. B. Russell and F. M. Smith, *J. Am. Chem. Soc.*, **77**, 624 (1955).
147. (a) R. L. Heath and J. D. Rose, *J. Chem. Soc.*, 1486 (1947).
    (b) A. T. Blomquist and T. H. Shelley, *J. Am. Chem. Soc.*, **70**, 147 (1948).
    (c) A. Dornow and F. Boburg, *Ann. Chem.*, **578**, 94 (1952).
    (d) F. Brower and H. Burkett, *J. Am. Chem. Soc.*, **75**, 1082 (1953).
148. W. J. Dale and G. Buell, *J. Org. Chem.*, **21**, 45 (1956).
149. (a) H. E. Reich and R. Levine, *J. Am. Chem. Soc.*, **77**, 4913, 5434 (1955).
    (b) A. J. Matuszko and A. Taurins, *Can. J. Chem.*, **32**, 538 (1954); G. Magnus and R. Levine, *J. Am. Chem. Soc.*, **78**, 4127 (1956); A. R. Phillips, *J. Am. Chem. Soc.*, **78**, 4441 (1956).

150. R. Wegler and G. Pieper, *Chem. Ber.*, **83**, 1 (1950).

151. (a) H. J. Backer and G. J. de Jong, *Rec. Trav. Chim.*, **67**, 884 (1948).

(b) H. J. Backer and R. van der Ley, *Rec. Trav. Chim.*, **70**, 564 (1951).

152. T. Kurihara and K. Ro, *J. Pharm. Soc. Japan*, **75**, 1267 (1955).

153. A. F. Bekhli, *Zh. Obshch. Chim.*, **21**, 86 (1951); *Zh. Obshch. Chim.*, **28**, 204 (1958).

154. (a) J. Cymerman-Craig, M. Moyle, A. J. C. Nicholson and R. L. Werner, *J. Chem. Soc.*, 3628 (1955).

(b) R. J. Bates, J. Cymerman-Craig, M. Moyle and R. J. Young, *J. Chem. Soc.*, 388 (1956).

(c) J. Cymerman-Craig, M. Moyle and L. F. Johnson, *J. Org. Chem.*, **29**, 410 (1964).

155. P. Buckus and A. I. Buckiene, *Zh. Obshch. Chim.*, **33**, 3775 (1963).

156. (a) R. S. Neale, *J. Am. Chem. Soc.*, **86**, 5340 (1964).

(b) R. S. Neale and R. L. Hinman, *J. Am. Chem. Soc.*, **85**, 2666 (1963).

157. W. H. Urry, O. O. Juveland and F. W. Stacey, *J. Am. Chem. Soc.*, **74**, 6155 (1952).

158. J. C. D. Brand and I. D. R. Stevens, *J. Chem. Soc.*, 629 (1958).

159. E. M. Smolin, *J. Org. Chem.*, **20**, 295 (1955).

160. E. T. Roe and D. Swern, *J. Am. Chem. Soc.*, **77**, 5408 (1955).

161. (a) H. C. Brown, W. R. Heydkamp, E. Breuer and W. S. Murphy, *J. Am. Chem. Soc.*, **86**, 3565 (1964).

(b) M. W. Rathke, N. Inoue, K. R. Varma and H. C. Brown, *J. Am. Chem. Soc.*, **88**, 2870 (1966).

162. R. Schröter in Houben-Weyl's *Methoden der Organischen Chemie*, Vol. XI/1, 4th ed. (Ed. E. Müller), George Thieme Verlag, Stuttgart, 1957, Chap. 4.

163. K. Johnson and E. F. Degering, *J. Am. Chem. Soc.*, **61**, 3194 (1939).

164. D. C. Iffland and F. A. Cassis, *J. Am. Chem. Soc.*, **74**, 6284 (1952).

165. N. G. Gaylord, *Reduction with Complex Metal Hydrides*, Interscience, New York, 1956.

166. H. J. Barber and E. Lunt, *J. Chem. Soc.*, 1187 (1960).

167. Cf. C. F. H. Allen, A. Bell and J. W. Gates, *J. Org. Chem.*, **8**, 373 (1943); G. E. Ullyot, J. J. Stehle, C. I. Zirkle, R. L. Shriner and F. J. Wolfe, *J. Org. Chem.*, **10**, 429 (1945).

168. (a) T. Severin, R. Schmitz and H. L. Temme, *Chem. Ber.*, **96**, 2499 (1963).

(b) L. A. Kaplan, *J. Am. Chem. Soc.*, **86**, 740, (1964).

169. A. Furst, R. C. Berlo and S. Hooton, *Chem. Rev.*, **65**, 51 (1965).

170. (a) T. Neilson, H. C. S. Wood and A. G. Wylie, *J. Chem. Soc.*, 371 (1962).

(b) H. C. Brown and K. Sivasankaran, *J. Am. Chem. Soc.*, **84**, 2828 (1962).

171. D. A. Fidler, J. S. Logan and M. M. Boudakian, *J. Org. Chem.*, **26**, 4014 (1961).

172. Cf. L. F. Fieser, *Organic Syntheses*, Coll. Vol. II, John Wiley and Sons, New York, 1943, p. 39.

173. (a) R. L. Hinman and M. C. Flores, *J. Org. Chem.*, **24**, 660 (1959).

(b) S. Wawzonek, J. Chua, E. L. Yeakey and W. McKillip, *J. Org. Chem.*, **28**, 2376 (1963).

174. A. K. Bose, J. F. Kistner and L. Farber, *J. Org. Chem.*, **27**, 2925 (1962).

175. A. Hassner and L. A. Levy, *J. Am. Chem. Soc.*, **87**, 4203 (1965); cf. H. C. Brown, *Hydroboration*, Benjamin, New York, 1962, Chap. 17 for a general account of reductions by diborane.

176. Cf. R. W. Guthrie, A. Phillipp, Z. Valenta and K. Wiesner, *Tetrahedron Letters*, 2945 (1965); J. W. ApSimon and O. E. Edwards, *Can. J. Chem.*, **40**, 896 (1962).

177. M. S. Newman and T. Fukunaga, *J. Am. Chem. Soc.*, **82**, 693 (1960).

178. H. C. Brown and P. Heim, *J. Am. Chem. Soc.*, **86**, 3566 (1964).

179. J. H. Billman and J. W. McDowell, *J. Org. Chem.*, **27**, 2640 (1962), and refs. cited.

180. W. S. Emerson, *Organic Reactions*, Vol. IV, John Wiley and Sons, New York, 1948, Chap. 3.

181. Cf. R. A. Heacock and O. Hutzinger, *Can. J. Chem.*, **40**, 128 (1962).

182. K. A. Schellenberg, *J. Org. Chem.*, **28**, 3259 (1963).

183. M. L. Moore, *Organic Reactions*, Vol. V, John Wiley and Sons, New York, 1949, Chap. 7.

184. M. Sommelet and J. Guioth, *Compt. Rend.*, **175**, 1149 (1923), and refs. cited.

185. A. Lukasiewicz, *Tetrahedron*, **19**, 1789 (1963), and refs. cited.

186. J. C. Jochims, *Monatsh. Chem.*, **94**, 677 (1963).

187. (a) M. N. Rerick, C. H. Trottier, R. A. Daignault and J. D. Defoe, *Tetrahedron Letters*, 629 (1963).

   (b) A. E. Petrarca and E. M. Emery, *Tetrahedron Letters*, 635 (1963).

188. (a) M. Freifelder, *J. Am. Chem. Soc.*, **82**, 2386 (1960).

   (b) F. E. Gould, G. S. Johnson and A. F. Ferris, *J. Org. Chem.*, **25**, 1658 (1960).

CHAPTER 3

Detection, determination, and characterisation of amines

Jacob Zabicky

Institute for Fibres and Forest Products Research, Jerusalem, Israel

I. INTRODUCTION

The wide distribution of amines in nature, their importance in industry as raw materials, intermediates, and finished products, and their use in the laboratory causes continuous interest in the analytical problems connected with them. These will be focussed at two levels in the present chapter; at the purely functional level attention will be paid to the nitrogen atom and its immediate vicinity, and at the molecular level the interactions between the amino group and the rest of the molecule will be considered. The alternative approach, centered on the discussion of the available methods, techniques, and instrumentation, is essential when one has to carry out an actual analysis, and constitutes the main part of the literature on the subject, in books and treatises on general analysis, organic analysis, and organic chemistry, as well as in original papers and reviews. Some outstanding titles are listed in references 1 to 16.

(1) 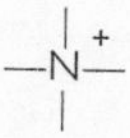   (2)

This chapter will deal with compounds in which structures **1** or **2** appear at least once, where the nitrogen atoms are singly bonded to carbon or hydrogen atoms, and furthermore, when these carbon atoms are linked to hydrogen or carbon atoms. Thus structures such as pyrrole (**3**), morpholine (**4**), and their salts will be considered, whereas pyridine (**5**) and 1-aminoethanol will be excluded, in spite of the fact that many of the methods applied in analysing the former compounds are useful for the latter too.

(3) 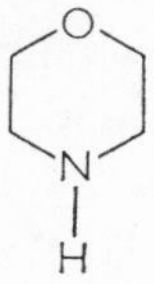   (4)   (5)

II. FEATURES OF THE GROUP

The amino group is endowed with a set of physical and chemical properties which depend mainly on the possession of certain structural elements. These do by no means belong to amines exclusively, but they may be used for the partial or total solution of many analytical problems.

A. The Nitrogen Atom

I. Pyrolyses and other chemical methods

On strong heating, many nitrogen-containing organic compounds yield hydrogen cyanide and a few yield cyanogen. Ammonia is evolved from compounds such as urea, guanidine, and some of their derivatives. Trimethylamine is released by compounds such as choline and betaine[9]. If the pyrolysis is carried out in the presence of calcium oxide the evolution of ammonia can be detected easily[9, 17].

The reductive pyrolysis of organic compounds in the presence of metals leads to the formation of the metal cyanide, which can be detected as Prussian blue $(Fe_4[Fe(CN)_6]_3)$ or by the copper(II) acetate–benzidine test[18]. Several metals and salts have been recommended for the fusion of the organic compound, e.g. potassium[19, 20], sodium[21, 22], magnesium mixed with potassium carbonate[23], zinc mixed with potassium carbonate[24], and a mixture of dextrose with sodium carbonate[25]. When the compound contains sulphur the metal thiocyanate is also produced, which can be detected by ferric chloride[26, 27].

Another type of reductive process is the combustion in hydrogen atmosphere in the presence of a nickel catalyst, where at temperatures over 1000° nitrogen is converted into ammonia and subsequently determined[28–30]. A similar method adapted for ultramicroanalysis has been proposed[31].

Nitrogen and hydrogen can be determined simultaneously by burning in dry carbon dioxide in the presence of a copper–copper oxide–silver catalyst, when water and nitrogen are liberated and subsequently determined[32].

A very sensitive test for nitrogen consists of heating the sample in admixture with manganese dioxide or other oxidants such as lead dioxide, red lead oxide (Pb_3O_4), cobalt(III) oxide, nickel(III) oxide, or manganese(III) oxide. Nitrous acid is evolved and can be detected by the Griess reaction[9, 33].

Many determinations of nitrogen based on oxidative pyrolyses of organic samples are variations, adaptations, and refinements of the original Dumas method[34], consisting of burning the organic compound in oxygen. Adaptations suited to ultramicroanalysis (nitrogen content of the order of 1 µg) have been reported[35, 36]. The Dumas method has been amply reviewed[15, 30, 37, 38].

The Kjeldahl method[39, 40] converts the bonded nitrogen into ammonium bisulphate which is then determined as ammonia. Many variations have been introduced in order to refine the method or adapt it to diverse types of organic materials. It was originally devised to determine the nitrogen of amino or amido groups and it cannot be used for other functions without due precautions or previous treatment of the sample, and in certain cases it fails to render quantitative results, for example when applied to compounds with partial structure C—N—N—C, probably because elementary nitrogen is lost. Ultramicro scale determinations have been carried out by this method[41, 42]. Ample reviews and discussions can be found elsewhere[15, 30, 43, 44].

Various classes of nitrogen-containing organic compounds were found to give colour reactions with p-dimethylaminobenzaldehyde in the presence of sulphuric acid in toluene solution[45].

2. Nitrogen by mass spectrometry

High-resolution mass spectrometers are capable of recording m/e values down to several decimal places, and the intensity of the peaks with an accuracy of fractions of 1%. This allows one to carry out elegant, though by no means straightforward, determinations of the elementary analysis and molecular weight of organic compounds.

The m/e value of the molecular ion—the ion resulting from losing one electron of the original molecule—is equal to the molecular mass M of the compound. It can be computed from equation (1), where n_i is the number of atoms of the nuclide with mass number i present in the molecule, and d_i is the difference between the atomic mass of the nuclide and its mass number. The fractional part of M can be calculated from equation (2), where I is an arbitrarily large positive integer. By finding the values of n_i which best fit equations (1) and (2), the elementary analysis, and therefore the nitrogen content of the compound, have in fact been determined. This can be accomplished by trial and error or by computer. In Table 1 the isotopes most frequently encountered in organic compounds are listed.

$$M = \sum_i n_i(i + d_i) \tag{1}$$

$$\text{Fractional part of } M = \text{Fractional part of } \left[I + \sum_i n_i d_i\right] \tag{2}$$

An alternative approach correlates the natural abundance of the isotopes (see Table 1) with the relative intensities of the peaks at M, $M + 1$, $M + 2$, etc. In order to illustrate the method consider a molecule with n carbon atoms with molecular ion of mass M. The intensity of the peak at $M + 1$ will be close to $(n \times 1 \cdot 1)\%$ of the

TABLE 1. Naturally occurring isotopes frequently encountered in organic compounds[a].

Element	Mass number[b]	Atomic mass[c]	Difference[d]	Natural abundance (%)
Hydrogen	1	1·007825	0·007825	99·985
	2	2·01410	0·01410	0·015
Carbon	12	12·00000	0·00000	98·89
	13	13·00335	0·00335	1·11
Nitrogen	14	14·00307	0·00307	99·63
	15	15·00011	0·00011	0·37
Oxygen	16	15·99491	−0·00509	99·759
	17	16·99914	−0·00086	0·037
	18	17·99916	−0·00084	0·204
Fluorine	19	18·99840	−0·00160	100
Phosphorus	31	30·97376	−0·02624	100
Sulphur	32	31·97207	−0·02793	95·0
	33	32·97146	−0·02854	0·76
	34	33·96786	−0·03214	4·22
	36	35·96709	−0·03291	0·014
Chlorine	35	34·96885	−0·03115	75·53
	37	36·96590	−0·03410	24·47
Bromine	79	78·9183	−0·0817	50·54
	81	80·9163	−0·0837	49·46
Iodine	127	126·9044	−0·0956	100

[a] From reference 46.
[b] i in equation (1).
[c] Based on the arbitrarily assigned atomic mass of 12·00000 to the nuclide $^{12}_{6}C$, in the physical scale.
[d] d_i in equation (1).

intensity of the peak at M, because for every carbon atom in the molecule there is a probability of 1·1% of it being ^{13}C. The contribution of ^{2}H to the $M + 1$ peak is much smaller because of the low natural abundance of this isotope, however ^{15}N and other nuclides may have a noticeable effect, especially when more than one atom of these elements is present in the molecule. For compounds of high molecular weight, containing many carbon atoms, statistical considerations have to be made in order to calculate the M to $M + 1$ intensity ratio. The $M + 2$ peak can be used in the same way relative to the $M + 1$ peak. By comparing the calculated and experimental values of these ratios the elementary composition of the molecular ion can be found.

The amino group is a source of instability in the molecular ion which may cause its rearrangement before reaching the collector of the mass spectrometer. If such rearrangements involve splitting of all the molecular ions into two particles neither the molecular weight nor the elementary analysis will be accessible by mass spectrometry. This splitting will occur less frequently in cyclic compounds, where the bond rearrangements will not cause fragmentation of the molecule.

The molecular ions of amines frequently abstract a hydrogen atom from a neutral species thus acquiring stability. This will make the molecular peak appear at $M + 1$. Such processes can be detected by studying the relative intensities of many peaks in the spectrum, as the relative intensities of peaks resulting from hydrogen abstraction will be pressure-dependent, as their corresponding ions result from bimolecular processes [47-51].

The application of mass spectrometry in the determination of molecular weights has been reviewed [47, 48, 52-54] and tables for the elementary analysis of compounds containing C, H, N, and O, according to both mass spectrometry methods previously described, have been compiled [55].

3. Nuclear activation

Activation analysis is a method that in some cases allows one to determine small amounts of nitrogen, of the order of 1 μg or less. It consists of irradiating the sample in a suitable generator of particles or rays, most frequently in a nuclear reactor, and measuring the decay of the new radioactive nuclides produced. The radioactivity A induced in such a manner can be calculated from equation (3), where N is the number of target atoms, σ the cross section of the reaction, Φ the irradiation flux, t the irradiation time, and λ the decay constant of the new nuclide [56].

$$A = N\sigma\Phi(1 - e^{-\lambda t}) \tag{3}$$

In a nuclear activation reaction the following factors have to be considered besides the variables of equation (3): threshold energy, mode of decay of the product, interference by other elements present in the sample yielding the same products as the element investigated, or products with overlapping decay spectra, etc. For example in the determination of compound **6** the $^{14}N(n, 2n)^{13}N$ reaction was used [57].

$$O{=}P(CH_2CH_2NH_2)_3$$
$$(6)$$

This is a 'fast neutrons' reaction, with threshold energy of 10·5 мev, and is applicable only if sufficient high-energy neutrons are available. The other two neutron reactions which could be possibly used, namely $^{14}N(n, p)\,^{14}C$ and $^{15}N(n, \gamma)\,^{16}N$, are not very useful in this case because of their low yields: with the former ^{14}C has a very low decay constant λ, resulting in a low saturation factor, which is the expression enclosed in brackets in equation (3), whereas with the latter reaction the natural abundance (see Table 1) which is directly related to N, and the thermal neutron cross section σ of ^{15}N, are small[58]*.

In Table 2 nuclear reactions are listed in which nitrogen is converted into a radioactive nuclide.

TABLE 2. Nuclear activation reactions of nitrogen.

Reaction[a,b,c]	Half-life of product[d]	Possible interferences[a,b,c]
$^{14}N(n, p)^{14}C$	5770 years	$^{17}O(n, \alpha)^{14}C$ and $^{13}C(n, \gamma)^{14}C$
$^{14}N(n, 2n)^{13}N$	10·0 min	$^{13}C(p, n)^{13}N$ and $^{16}O(p, \alpha)^{13}N$
$^{14}N(\gamma, n)^{13}N$	10·0 min	$^{13}C(p, n)^{13}N$ and $^{16}O(p, \alpha)^{13}N$
$^{14}N(d, n)^{15}O$	2·03 min	$^{16}O(d, dn)^{15}O$, $^{14}N(p, \gamma)^{15}O$, and $^{15}N(p, n)^{15}O$
$^{14}N(p, \gamma)^{15}O$	2·03 min	$^{16}O(p, pn)^{15}O$
$^{15}N(p, n)^{15}O$	2·03 min	$^{16}O(p, pn)^{15}O$
$^{14}N(p, \alpha)^{11}C$	20·5 min	
$^{15}N(n, \gamma)^{16}N$	7·35 sec	

[a] From references 58 and 59.
[b] In a reaction $X(a, b)Y$, nuclide X irradiated with a releases b yielding nuclide Y.
[c] n–neutron, p–proton, γ–γ-ray, d–deuterium nucleus, α–α-particle.
[d] From reference 46.

B. The Free Electron Pair

The fact that in amines the nitrogen atom has two free electrons readily available for reaction is the basis of a large number of detection and determination methods. Furthermore, in many cases in which the products are apparently the result of a reaction with N—H groups, the presence of the free electron pair is essential for the reaction to take place. The latter cases however will be treated in section II.C together with other properties of the N—H moiety.

Table 3 summarises the various ways in which the free electron pair (p electrons) of amines can react with other compounds or func-

* The nuclides ^2H, ^{12}C, ^{13}C, ^{14}N, ^{15}N, ^{16}O, ^{17}O, and ^{18}O can be determined after proton or deuteron activation, in gaseous samples, at pressures of a few mm Hg, by neutron time of flight spectroscopy[610]. In the case of nitogen the reactions are $^{14}N(d, n)\,^{15}O$ and $^{15}N(p, n)^{15}O$.

tions. Types 1 and 2 of this table can be considered to differ from one another in the degree of coordination of the p electrons with a proton, types 2 to 7 are distinguished by the nature of the electron acceptor, and type 8 is a contribution (hypothetical or actual) to the structure of amines where the amino group is attached to an unsaturated system. It should be noticed that the molecule tends to be planar at the nitrogen site in type 8, in contrast with the nearly tetrahedral geometry in the remaining seven types.

TABLE 3. Coordination of the p electrons of amines.

Type of coordination	Section in which considered
1. Hydrogen bonding	II.E, IV.C.4
2. Ammonium salts of protonic acids	II.B.1
3. Ammonium compounds with Lewis acids	II.B.2, IV.C.4
4. Complexes with metallic ions	II.B.3, IV.C.2
5. Quaternary ammonium compounds	II.B.4, III
6. Amine oxides	
7. Complexes with π acids	II.B.1.b
8. Electron transfers in unsaturated molecules	IV.B

A further way in which the free electrons may react is by losing only one of them, thus yielding a radical–cation, for example by electron impact in a mass spectrometer, where reaction (4) probably takes place (see also sections II.A.2 and IV.A.2).

$$\diagdown N: + e \longrightarrow \diagdown N^{\cdot +} + 2e \qquad (4)$$

I. Reaction with protonic acids

a. Basicity tests. Amines are weak bases capable of reacting with acids. Several methods of detecting this reaction have been proposed. The simplest is the solubility test of a sample in dilute mineral acid, used in qualitative organic analysis[8, 16]. In another test the sample is dissolved in hydrochloric acid, evaporated to dryness, the residue redissolved, and the solution tested for chloride with silver nitrate[9]. Some basicity tests make use of equilibria involving the formation of coloured precipitates, such as the complex of dimethylglyoxime (7) with nickel(II) ions[60], or 8-hydroxyquinoline (8) with zinc ions[61]. If in equations (5) and (6) the concentration of protons is adjusted to a value at which the solubility product of the metal complex is nearly reached, then the addition of a base will lower the proton concentration and enhance the complex concentration thus causing its precipitation.

$$Ni^{2+} + 2XH \rightleftharpoons NiX_2 + 2H^+ \quad (XH = 7) \tag{5}$$

$$Zn^{2+} + 2XH \rightleftharpoons ZnX_2 + 2H^+ \quad (XH = 8) \tag{6}$$

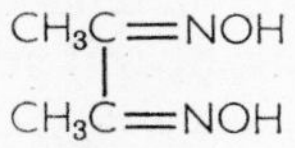

CH₃C=NOH
|
CH₃C=NOH

(7)

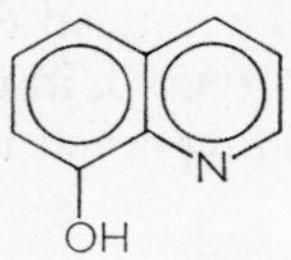

(8)

Many aliphatic amines have pK_a values in the range of 9 to 11, and aromatic ones in the range of 4 to 5. Accordingly, an aqueous solution buffered at pH 5·5 will dissolve alkylamines but not arylamines[62].

b. Precipitation of salts. Amines are lipophilic substances, especially when no acidic groups are present in the same molecule, and therefore they will dissolve in organic solvents such as chloroform, ether, benzene, etc., whereas their water solubility is small, except for a few with low molecular weight[63]. On the other hand, ammonium salts are lipophobic owing to their ionic character. These properties can be applied to the separation of amines from other organic materials dissolved in solvents of low polarity. Precipitation by acids from these solutions, according to equation (7), has been applied qualitatively and quantitatively. The extraction of ion pairs by organic solvents is sometimes possible owing to the hydrophobic character of the radicals attached to the ammonium moiety[64].

$$R^1R^2R^3N + HX \rightleftharpoons R^1R^2R^3NH^+X^- \tag{7}$$
$$(R^1, R^2, R^3 = H, alkyl, aryl)$$

A great variety of salts derived from amines have been used for analytical purposes, a partial list of which appears in Table 4. It should be noticed that the nitro compounds listed in this table are both protonic and π acids, and therefore their derivatives may be ammonium salts, or under certain circumstances π complexes. With the commercial availability of reagent **18**, the preparation of tetraphenylborates has become an important method for the determination of organic bases[95].

c. Aqueous titrations. Amines in aqueous solution interact with water yielding two formal species: a hydrate in which water is hydrogen bonded to the nitrogen atom, and an ammonium ion. The former can

TABLE 4. Salts derived from amines and used for analysis[a].

Halides[65]
Perchlorates[65]
Antimony(III) iodide complexes (**9, 10**, M = Sb)[66, 67]
Bismuth(III) iodide complexes (**9, 10**, M = Bi)[68, 69]
Chloroaurates[65]
Metaphosphates[70]
Phosphomolybdates[71]
Phosphotungstates[71, 72]
Silicotungstates[71, 73–76]
Fluorosilicates[77]
Reinecke salts (**11**)[78, 79]
Chloroplatinates[65, 79]
Oxalates[65]
Imidazole-4,5-dicarboxylates (**12**)[80]
3,5-Dinitrobenzoates[81]
2,4-Dinitrobenzoates[82]
3,5-Dinitro-*o*-toluates[83]
3,5-Dinitro-*p*-toluates[84]
Picrates (**13**)[65, 79, 85, 86]
Picrolonates (**14**)[79, 87–89]
Flavianates (**15**)[90]
Styphnates (**16**)[79, 91]
2-Nitro-1,3-diketohydrindene salts (**17**)[92]
Tetraphenylborates (**18**)[79, 93–97]

[a] Formulae **11** to **18** belong to the reagents.

Amine . MI_3 . HI   2 Amine . MI_3 . HI $NH_4[Cr(NH_3)_2(SCN)_4]$
(**9**)   (**10**)   (**11**)

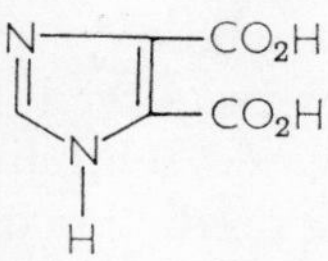

(**12**) 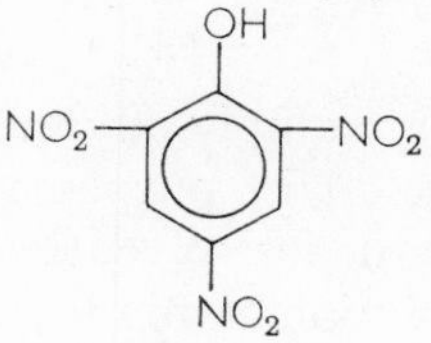   (**13**) 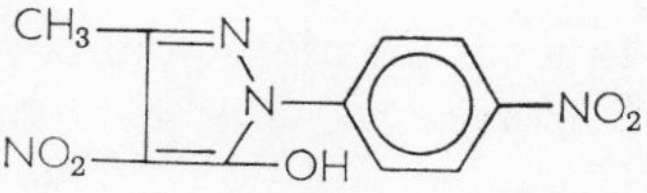   (**14**)

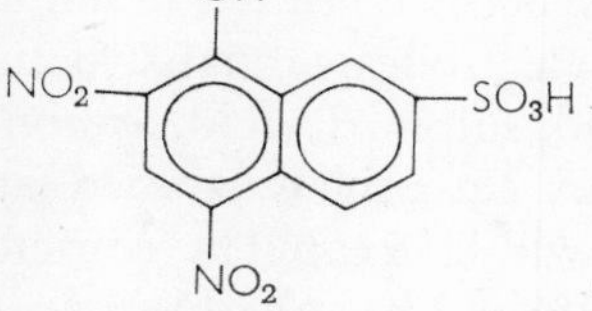

(**15**) 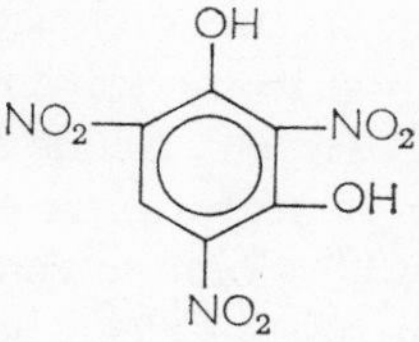   (**16**)

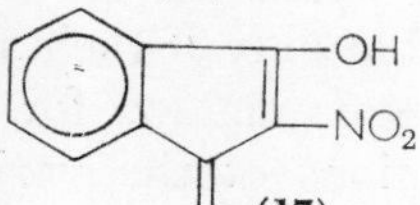

$Na^+ B(C_6H_5)^-_4$

(**17**)   (**18**)

be considered as the most abundant species when amines are dissolved in water[98-100], whereas the latter is the result of the acid–base dissociation of the former according to equation (8).

$$\diagdown N \cdots H - OH \rightleftharpoons \diagdown \overset{+}{N} - H + OH^- \qquad (8)$$

The equilibrium favours the undissociated form on the left hand side of equation (8), but it can be displaced to the right by addition of a strong acid which will react with the hydroxide ions. Titrations with standard mineral acids can be carried out easily in aqueous solution when the amine has $pK_a \geq 9$, while for lower values it is difficult to distinguish the end-point. By adding neutral salts to the solution, the potentiometric break is enhanced, allowing better detection of the end-point[101, 102].

The low solubility of most amines in water is usually corrected by adding organic solvents such as ethanol, methanol, or dioxane. Another potential source of error arises from the possibility of association in solution, for example if the ammonium ion and a neutral amine molecule become associated the potentiometric break will be produced before the equivalent amount of acid has been added, as was found to be the case with aminophenols[103].

Amino acids in the zwitterion form[104, 105] can be titrated with mineral acids only if the acid moiety is a weak acid such as a carboxyl group, as in equation (9), which is a special case of titration of ammonium salts of weak acids.

$$NH_3^+ \sim CO_2^- + H^+ \rightleftharpoons NH_3^+ \sim CO_2H \qquad (9)$$

Aqueous titrations of amines are amply discussed elsewhere[14, 15, 37].

d. Nonaqueous titrations. Three main purposes may be served by carrying out titrations in nonaqueous solvents: increased solubility, change of the pH scale, and resolution of mixtures. The prediction of a potentiometric titration curve in an arbitrary solvent is a difficult task, in which many factors intervene, such as dielectric constant, definition of acid and base in relation to the solvent, electrodes, actual structure of conjugate acids and bases, etc. Acetic acid, sulphuric acid, acetonitrile, and alcohol–water mixtures have been extensively studied and were reviewed elsewhere[106]. Some solvents will be treated here briefly:

Acetic acid. Bases with $pK_a \geq 2$ can be titrated in anhydrous acetic acid[107], where they show a good potentiometric break. On the other hand, in this solvent bases become distinguishable only if their pK_a values differ widely, and it is therefore not recommended for the determination of mixtures in which concentrations of the various components are required[108]. Standardised perchloric acid in anhydrous acetic acid is the titrant most frequently used. Very weak bases such as diphenylamine were titrated with this solvent–titrant system by differential thermometry[109]. It should be noted that approximately 0·6% of water is contained in a 0·1N perchloric acid solution prepared from commercially available reagents, and if strictly anhydrous conditions are required, it can be removed by treatment with the required amount of acetic anhydride[110].

Acetic anhydride. This is also an excellent solvent for titrating organic bases, giving good end-points when they have $pK_a \geq 0.5$[107], although acetylation may occur to some extent in certain cases. Standard perchloric acid[107, 111, 112] or fluorosulphuric acid[113] in acetic anhydride have been used as titrants. Mixtures of acetic acid with acetic anhydride were also proposed as solvents for the titration of amines[107, 114, 115].

Acetonitrile is a solvent in which excellent end-points and resolution of mixtures can be attained[14, 116, 117]. It has, however, the disadvantage of enabling the titration of many amides which may be mixed with amines in the sample. Perchloric acid is usually the titrant. Hydrogen bromide was used for thermometric titrations of amines in this solvent[118].

Nitromethane has properties similar to those of acetonitrile[119–121].

Dioxan gives good end-points. It can be used for titrating bases such as pyridine, but titrations of aniline were not satisfactory in this solvent. The titrant is perchloric acid[122].

Glycols give good resolution in the titration of amines with close pK_a values[123, 124].

Many other solvents have been used for the determination of pure and mixed amines. The subject is amply reviewed and discussed elsewhere[14, 15, 37, 125–128].

2. Reaction with Lewis acids

A limited application of Lewis acids has been made in analyses of amines; boron trifluoride[129], titanium tetrachloride, and tin tetrachloride[130] were used to titrate heterocyclic bases in aprotic solvents. Other applications connected with steric effects in amines will be discussed in section IV.C.4.

3. Complexes with metal ions

Amines can coordinate their free electron pair into empty orbitals of some metal ions, much in the same way as ammonia does in inorganic

4*

complexes. This has found some application in the detection and determination of amines although the scope of each reaction is limited by factors such as the type of amine analysed and interferences from other classes of compounds.

Aqueous solutions of copper(II) salts give a green to blue colour or precipitate in the presence of water-soluble amines[16]. In organic solvents, and preferably in the absence of water, aliphatic amines form a complex with excess of copper(II) chloride. Although the structure of this complex is unknown, it has composition $2CuCl_2$.amine, becoming therefore useful for quantitative analysis[131]. Amino acids yield complexes with the same salts, which are stabilised by a chelate structure **19**. After isolation of the complex, it can be analysed for its copper content gravimetrically[132, 133] or colourimetrically[133-135].

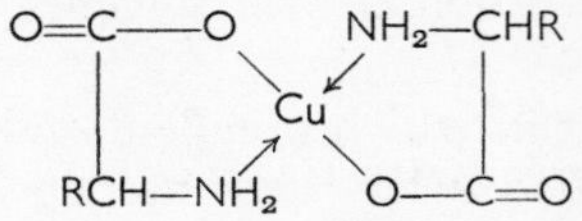

(19)

Ferric complex ions have been used for very sensitive tests of detection and identification of certain classes of amines. Green to blue complexes are formed by the reaction of primary aromatic amines with sodium pentacyanoaquoferriate, as in equation (10)[9, 136].

$$Na_2[Fe(CN)_5H_2O] + C_6H_5NH_2 \longrightarrow Na_2[Fe(CN)_5C_6H_5NH_2] + H_2O \qquad (10)$$

When a secondary amine is treated with sodium nitroprusside $(Na_2[Fe(CN)_5NO])$ in the presence of acetaldehyde, acetone, pyruvic acid, etc., blue to violet compounds of unknown structure are obtained[9, 137, 138]. This is the so-called Rimini test[139]. A colourimetric method for determining secondary amines is based on this reaction[140].

In the absence of oxidants, cobalt(II) nitrate in a basic medium yields deeply coloured complexes with certain classes of amines, such as histamine (**20**), which can thus be determined colourimetrically. The formation of such complexes does not seem to depend on the presence of the imidazole (**21**) partial structure, although **21** itself yields a violet precipitate[4, 141].

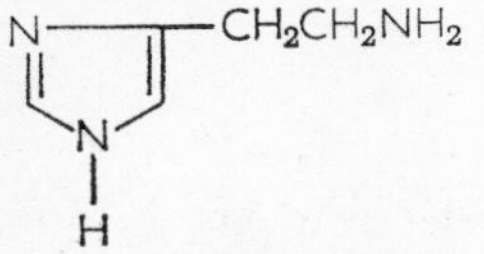

(20)

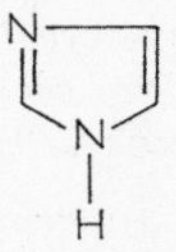

(21)

Copper and cobalt complexes of amino acids have found applications in the solution of stereochemical problems, as mentioned in section IV.C.2.

4. Quaternisation

Amines react with alkyl halides or certain esters attaining a higher degree of alkylation. Primary and secondary amines give, usually, mixtures of amines and ammonium salts while tertiary amines undergo a one-step reaction (equation 11), the product of which is relatively easy to separate and purify. Some reagents which raise the degree of substitution by one only, are treated in section II.C.3, and similarly sultones react according to equation (12), the best results being obtained with primary amines[142].

$$R^1R^2R^3N + R^4X \longrightarrow R^1R^2R^3R^4N^+X^- \tag{11}$$

$$RNH_2 + \begin{matrix} CH_2CH_2CH_2 \\ | \qquad\qquad | \\ O\text{------}SO_2 \end{matrix} \longrightarrow R\overset{+}{N}H_2CH_2CH_2CH_2SO_3^- \tag{12}$$

Quaternisation reactions can be carried out with alkyl halides such as methyl iodide[143,144], and esters of strong acids such as methyl 2,4-dinitrobenzenesulphonate[145], methyl p-toluenesulphonate[146], methyl picrate (O-methyl derivative of **13**)[147], and methyl sulphate[148]. The reaction can take place with or without solvent and at temperatures ranging from room temperature up to that of reflux[79,143-145]. If the quaternary salt is obtained quantitatively it can be filtered off and weighed, or otherwise determined according to section III.

Some amines in chloroform solution yield the corresponding N-(dichloromethyl)ammonium chloride[149]. This reaction may lead to abnormal products, as is the case with **22**, from which **23** is derived[150].

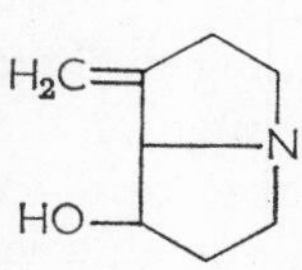

(**22**) 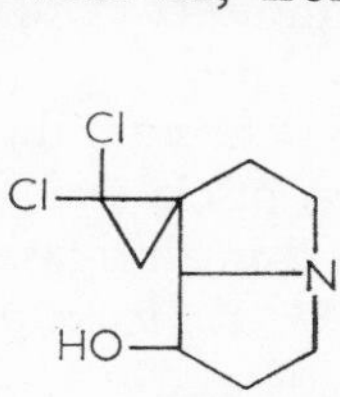   (**23**)

For heterocyclic bases a method of determination was proposed based on quaternisation with ethyl iodide or methyl sulphate, followed by treatment with basic potassium permanganate. Volatile amines are thus formed which can be collected in standardised mineral acid and back titrated[148].

Certain complex structures yield with alkylating reagents atypical products, for example acronycidine (**24**) on treatment with ethyl iodide yields the *N*-ethyl analogue of isoacronycidine (**25**)[151]. More complex results were reported for the action of the same reagent on the similar compound dictamnine (**34**)[151] (section II.C.2.b).

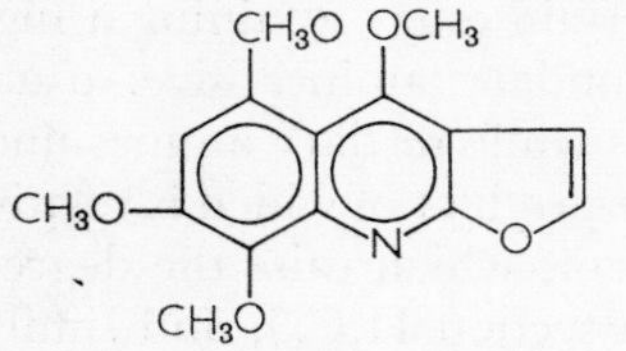

(**24**) 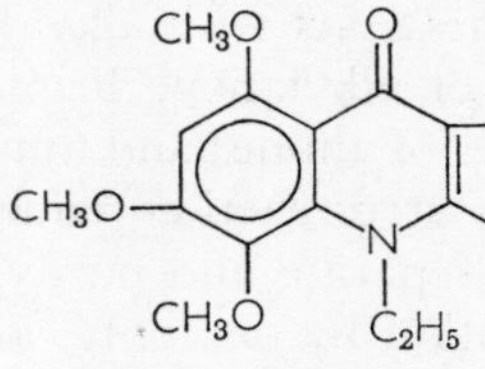   (**25**)

A method based on quaternisation was proposed for the detection of β-hydroxyethylamines[9, 152], which can be possibly extended to other β-substituted ethylamines. It consists of heating the amine with sodium chloroacetate. A betaine is first formed (equation 13), which decomposes on further heating, yielding acetaldehyde (equation 14). The latter can be detected with a morpholine–sodium nitroprusside solution (compare with the Rimini test in section II.B.3).

$$R_2NCH_2CH_2OH + ClCH_2CO_2Na \longrightarrow \begin{array}{c} \overset{+}{R_2N}CH_2CH_2OH \\ | \\ CH_2CO_2^- \end{array} + NaCl \quad (13)$$

$$\begin{array}{c} \overset{+}{R_2N}CH_2CH_2OH \\ | \\ CH_2CO_2^- \end{array} \longrightarrow R_2\overset{+}{N}HCH_2CO_2^- + CH_3CHO \quad (14)$$

5. Some oxidation methods

Oxidations of amines usually yield degradation products, some of which provide indirect evidence for the presence of different types of amines.

A solution of potassium dichromate in sulphuric acid has been used for the detection of classes of amines according to the colour changes observed[153], and in thin-layer chromatography for the identification of arylamines[154]. Ceric sulphate in sulphuric acid can be used for spotting alkaloids and probably other bases in thin-layer chromatography[155]. A standard solution of ceric sulphate was used for determining *p*-methylaminophenol[156].

Lead dioxide in dilute acetic acid or alcohol reacts with aromatic amines giving characteristic colours[157]. The same oxide dissolved in phosphoric acid can be used for colourimetric determination of aromatic amines[158].

The following oxidants were reported to give colour reactions with aromatic amines, which may be used for identification purposes under controlled conditions: chloric acid[159], ammonium persulphate–silver nitrate[160], benzoyl peroxide[161], and sodium hypochlorite[162]. Amines with partial structures **26** and **27**, where R^1, R^2 = H, alkyl, give the haloform reaction with hypochlorites in basic media[163].

$$—CH_2CH_2NR^1R^2$$

(**26**)

$$\begin{matrix} —CH_2 \\ \diagdown \\ CHNR^1R^2 \\ \diagup \\ CH_3 \end{matrix}$$

(**27**)

N-Iodosuccinimide with primary and tertiary amines yields stable brown solutions whereas with secondary amines the colour fades in a few minutes; *N*-bromosuccinimide yields orange precipitates with tertiary amines[16, 164], and gives colour reactions which may be of aid in the classification of other types of amines[165].

6. Spectral properties of the electron pair

a. Electronic spectrum. The *p* electrons of amines have an absorption band at the fringes of the far ultraviolet region, due to a transition from a *p* orbital to an antibonding *σ* orbital of nitrogen. Thus trimethyl-amine[166] and piperidine (**28**)[167] have a band at 200 mμ, ε_{max} 4000. It has been suggested[166] that this band can be used to distinguish primary, secondary, and tertiary amines, but this region is frequently obscured by other chromophores and is therefore rather difficult to interpret in more complicated cases[168]. A band at ∼214 mμ, shows that the amino group is not attached to an unsaturated carbon atom, as otherwise the band is displaced towards longer wavelengths[169]. Further spectral properties associated with transitions of the *p* electrons of amines are discussed in section IV, in connection with molecular structure.

b. Vibrational spectrum. Bands in the 2800 cm^{-1} region are associated with the *p* electrons of certain amines, because they disappear on coordination of these electrons[170]. Some of the structures in which these bands arise are the following: *N*-methyl and *N,N*-dimethyl (not in *N*-ethyl) with medium to strong bands[171–174], in cyclic compounds when two or more adjacent C—H groups are *trans* to the electron pair as in 10-methyl-*trans*-quinozilidine (**29**), with a group of small bands[175], and in other compounds such as di- and triethylamine, piperidine (**28**), *N*-ethylpiperidine, morpholine (**4**), etc., with a group of small bands[176].

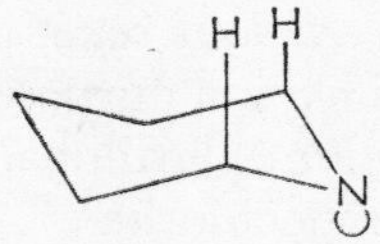

(28) 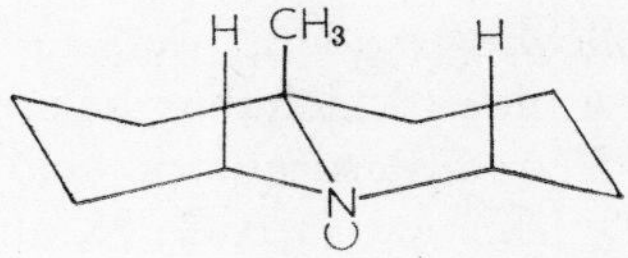   (29)

The N-methyl bands may be of aid in the further characterisation of the molecule[172] as they appear at 2810–2820 cm^{-1} in aromatic amines and at 2780–2805 cm^{-1} in aliphatic or alicyclic amines. A dimethylamino group attached to an aromatic system has a band near 2800 cm^{-1}, and when attached to an aliphatic or alicyclic system it has two bands, a strong one at 2765–2775 cm^{-1} and the second one at 2810–2825 cm^{-1}.

C. The N—H Bond

Among the most important reactions of ammonia and of amines are ammonolyses and condensations. Both start with an attack of the p electrons on an electrophilic centre, followed by the loss of one or two protons, and other transformations depending on the electrophile. These reactions are the basis of many analytical methods for amines. Other methods are based on the instability of the bond, on its ability to form hydrogen bonds, and on its distinctive vibrational and nuclear magnetic resonance spectra.

I. Active hydrogen

The N—H partial structure confers upon primary and secondary amines a set of properties which can be connected mainly with two phenomena: active hydrogen and hydrogen bonding. Both are inter-related, although the methods of determining the former are usually 'chemical', while the latter is investigated mainly through physical properties as shown in section II.E.

Contrasting with the high thermochemical stability of the N—H bond (bond energies at 0°κ, in kcal/mole: C—H 98·2, C—N 78, N—H 92·2[177]) it is most unstable kinetically, and is capable of undergoing extremely fast proton interchange reactions in protonic solvents such as water or alcohols[178], and even in the gas phase this interchange is probably very fast[179].

Various methods have been used for detecting and determining active hydrogen in general, and they can be applied with diverse degrees of success to amines.

a. *The Zerewitinoff method.* This consists of allowing the active hydrogen compound to react with a Grignard reagent (equations 15 and 16) and measuring the evolved hydrocarbon[180-182].

$$RNH_2 + 2\,CH_3MgI \longrightarrow RN(MgI)_2 + 2\,CH_4 \tag{15}$$

$$R_2NH + CH_3MgI \longrightarrow R_2NMgI + CH_4 \tag{16}$$

An interesting example of a positive Zerewitinoff reaction is provided by sempervirine (**30**), in spite of its lack of N—H bonds. It was proposed that the active hydrogen is present in the methylene group marked with (*) in **31**, as **31** is a resonance form of **30**[183].

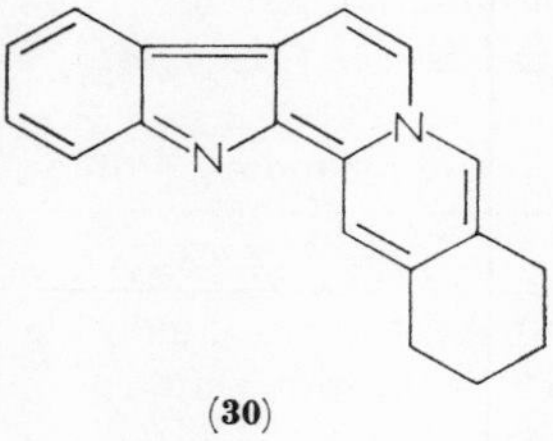

(**30**) 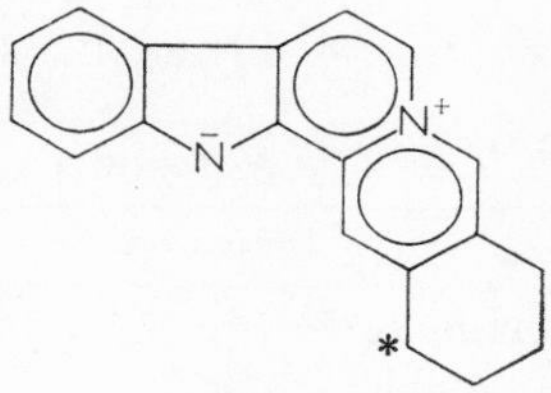   (**31**)

b. *Lithium aluminium hydride.* In the presence of active hydrogen compounds (X—H) $LiAlH_4$ decomposes according to equation (17), and the evolved hydrogen is measured[184, 185].

$$X—H + \tfrac{1}{4}\,LiAlH_4 \longrightarrow X—(LiAl)_{1/4} + H_2 \tag{17}$$

c. *Isotopic exchange methods.* Amines exchange their active hydrogen among themselves and with other active hydrogen compounds, in particular with water. This provides a method for detecting and determining active hydrogen by equilibrating with pure deuterium oxide, and then estimating the amount of O—H bond formed by means of density measurements[186], i.r. spectrophotometry[187], n.m.r. spectroscopy[188], or alternatively, by using tritium-labelled water and measuring the radioactivity of the amine after equilibration[189].

d. *N—H and C—H spin–spin coupling.* That the amino group is able to exchange protons in aqueous solutions was also shown by nuclear magnetic resonance. In solutions of high pH methylamine shows a singlet corresponding to the three methyl protons, although a triplet could be expected from coupling of these protons with the two on the nitrogen atom. The singlet was interpreted as a result of the rapid exchange of the protons on the amino group, which does not allow their alignment with the external field, and therefore the spin states of the amino protons are averaged to zero. On the other hand, in solutions of low pH, where the methylammonium ion is prevalent, proton exchange with the solvent occurs at a slow rate and a quartet is observed for the methyl group, resulting from spin–spin coupling with the three ammonium protons[190].

The analysis of active hydrogen is extensively discussed elsewhere[15, 180, 185, 191, 192].

2. Acylation

Table 5 summarises some of the derivatives obtained by acylation (equation 18), which have found application in analysis. Acyl derivatives are widely used for identification purposes[8, 16].

$$R^1R^2NH + Acyl\text{-}X \longrightarrow R^1R^2N\text{-}Acyl + XH \tag{18}$$
$$(R^1, R^2 = H, \text{ alkyl, aryl; } X = \text{halogen, OH, alkoxy, O–acyl, etc.})$$

TABLE 5. *N*-Acyl derivatives of primary and secondary amines.

Derivatives	Type of reagent
Acetamides[184, 193–204, a]	Anhydride or acyl chloride
Formamides[162, 205–207]	Acid or mixed acetic–formic anhydride
Trifluoroacetamides[162, 208–210, b]	Anhydride
p-Nitrophenylacetamides[211, 212]	Acyl chloride
Benzamides[213–219, a]	Anhydride or acyl chloride
3,5-Dinitrobenzamides[220–222]	Acyl chloride
p-Phenylazobenzamides[223]	Acyl chloride
Phthalimides[224, a, c]	Anhydride
Pyromellitic diimides[225, 226, a, c]	Dianhydride
N- and *N,N*-substituted 3-nitrophthalamic acids[227, a]	Anhydride
3-Nitrophthalimides[227, a, c]	Anhydride
N,N-Diphenylureas[228]	Acyl halides
Benzenesulphonamides[229–231, a]	Acyl chloride
p-Toluenesulphonamides[72, 232–234]	Acyl chloride
Benzylsulphonamides[233]	Acyl chloride
Methanesulphonamides[235]	Acyl chloride
p-Bromobenzenesulphonamides[236]	Acyl chloride
m-Nitrobenzenesulphonamides[237]	Acyl chloride
Sulphonebisacetamides[238]	Ethyl ester
p-Phenylazobenzenesulphonamides[239]	Acyl chloride
2,4-Dinitrobenzenesulphenamides[240]	Sulphenyl chloride
o-Nitrobenzenesulphenamides[241, 242]	Sulphenyl chloride

[a] See text.
[b] Derived from esters of amino acids.
[c] Primary amines only.

a. Acetylation. A great variety of methods for the preparation of acetamides from acetic anhydride have been suggested, both in the absence of catalyst[16] and in the presence of catalysts such as acids[193–195], pyridine[16, 196, 197], and sodium hydroxide[8].

Although this reaction is of great value in the characterisation of primary and secondary amines, the reagents can cause rearrangements obscuring structural elucidation. Thus for example, harmaline (**32**) undergoes acetylation at $N_{(2)}$ yielding compound **33**[198, 199].

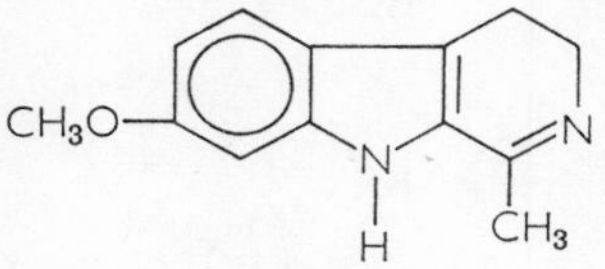

(**32**) 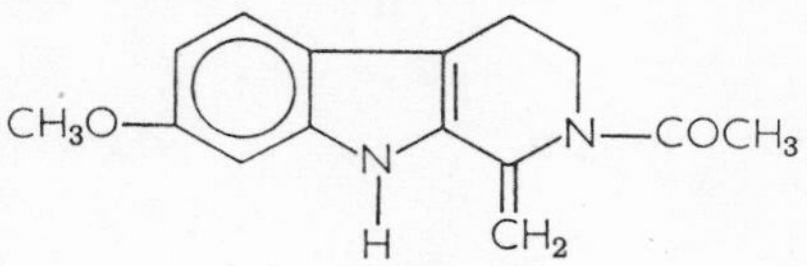   (**33**)

Acetylation with acetic anhydride in anhydrous pyridine can be used for the determination of primary and secondary amines according to equations (19) and (20).

$$RNH_2 + (CH_3CO)_2O \longrightarrow RNHCOCH_3 + CH_3CO_2H \qquad (19)$$

$$R_2NH + (CH_3CO)_2O \longrightarrow R_2NCOCH_3 + CH_3CO_2H \qquad (20)$$

The amount of amine reacted is determined by titration of the liberated acid[196] or by the amount of water necessary for hydrolysing the excess acetic anhydride[197].

Acetylation proceeds at different rates for alcohols, phenols, aliphatic amines, and aromatic amines. It is possible therefore to determine the latter in the presence of all the others[200], or sometimes even in alcoholic solution[201]. Similar determinations based on the use of acetyl chloride were proposed[202, 203], but the anhydride methods seem to be more advantageous[15].

Further discussions of this reaction and lists of acetamides can be found elsewhere[8, 10, 14–16, 37, 203, 204].

b. Benzoylation. This reaction has found its main application in the characterisation of compounds as 'acylatable'[184], and in the identification of primary and secondary amines.

Benzoylation of certain compounds may yield atypical products, for example dictamnine (**34**) gives no reaction with acetic anhydride but yields *N*-benzoylnordictamnine (**35**) on treatment with benzoyl chloride[213], and myosmine (**36**) yields compound **37** on treatment with benzoic anhydride[214].

Further discussions of benzoylation and lists of benzamides can be found elsewhere[8, 10, 14–16, 204, 215–219].

c. Phthaloylations. Phthalic anhydride[224] and pyromellitic anhydride (**38**)[225, 226] are reagents for the determination of primary and secondary amines.

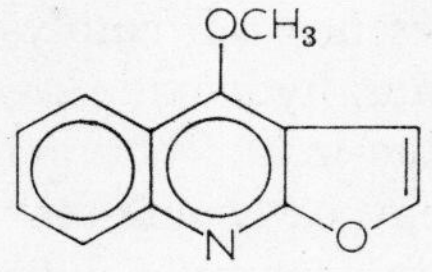

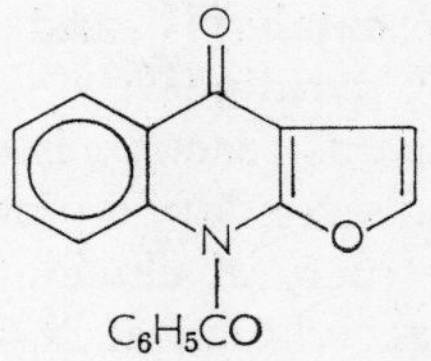

(34)

(35)

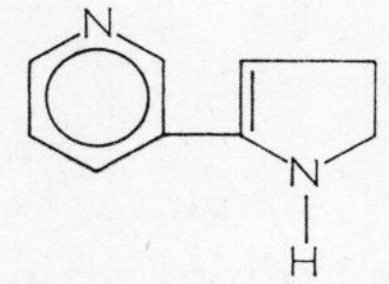

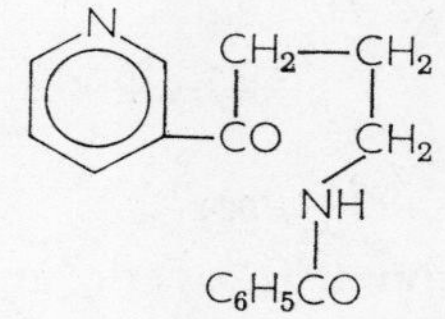

(36)

(37)

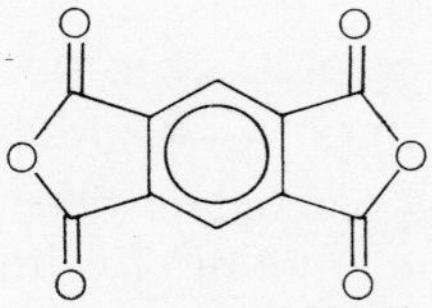

(38)

3-Nitrophthalic anhydride reacts with primary and secondary amines according to equations (21) and (22) [227], on mild heating. On further heating at about 145° dehydration (equation 23) takes place in the case of primary amines, while the phthalamic acids (**40**) derived from secondary amines are stable. Compounds **40** and **41** can be easily distinguished by their chemical properties.

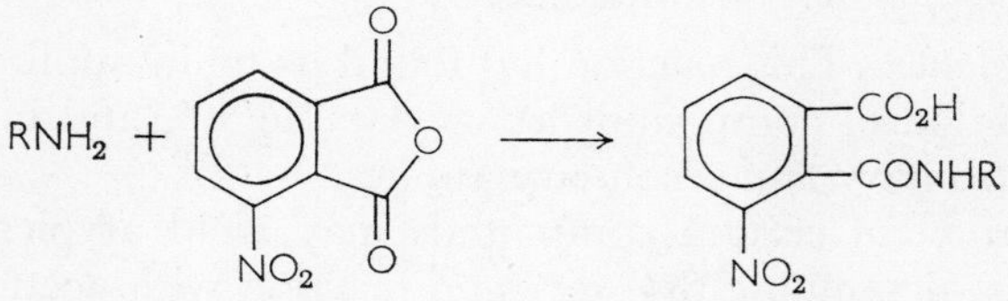

$$RNH_2 + \quad \longrightarrow \qquad\qquad\qquad (21)$$

(39)

$$R_2NH + \quad \longrightarrow \qquad\qquad\qquad (22)$$

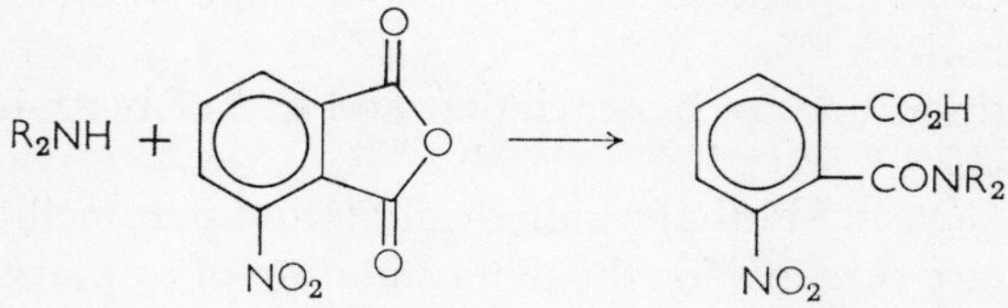

(40)

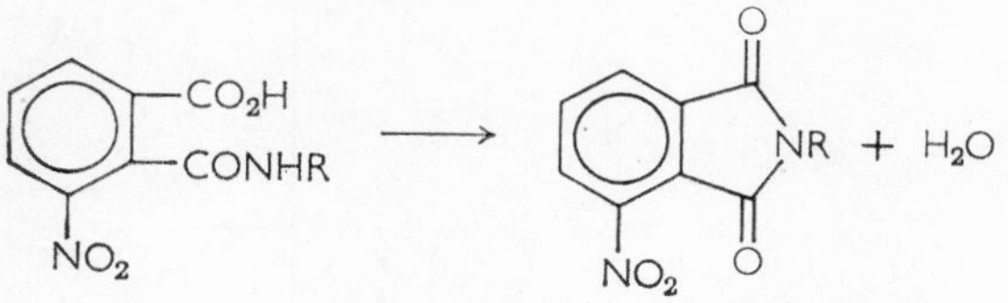

$$\text{(23)}$$

(41)

d. Sulphonylations. Sulphonyl chlorides (see Table 5) yield, with primary and secondary amines, the corresponding sulphonamides, which can be used for identification purposes. An application of these reactions is the Hinsberg method[229] which is amply discussed elsewhere[8, 16, 204, 230].

e. Nitrosation of secondary amines. Nitrous acid yields nitrosamines with aromatic and aliphatic secondary amines (equation 24). The mechanism probably involves dinitrogen trioxide as intermediate[243, 244].

$$R_2NH + HO-N=O \longrightarrow R_2N-N=O + H_2O \qquad (24)$$

This reaction has been applied in the determination of secondary amines, and their mixtures with other classes of amines. Although the latter react with nitrous acid in a different way (sections II.C.10 and II.F.3) the one-to-one stoichiometry is preserved. The determinations can be carried out by direct titration of the organic sample with standardised sodium nitrite[245], back titration of the excess reagent[246, 247], or by measuring the nitrosamine spectrophotometrically[248, 249].

An indirect method was proposed consisting of hydrogenation of the nitrosamine to a hydrazine and oxidation of the latter (reactions 25 and 26), the nitrogen thus liberated being measured[250].

$$R_2N-NO \xrightarrow{\text{Zn,HCl}} R_2N-NH_2 \qquad (25)$$

$$R_2N-NH_2 \xrightarrow{\text{K}_3[\text{Fe(CN)}_6]} N_2 \qquad (26)$$

Again, abnormal behaviour in nitrosation is observed with certain complex molecules. Thus, pelletierine (42) fails to undergo *N*-nitrosation, although it yields the corresponding acetamide and benzamide[251], and on the other hand, the tertiary amine codeine (43) yields *N*-nitrosonorcodeine by displacement of the *N*-methyl group[252].

Analysis of secondary amines via nitrosation is further discussed elsewhere[14, 15, 37].

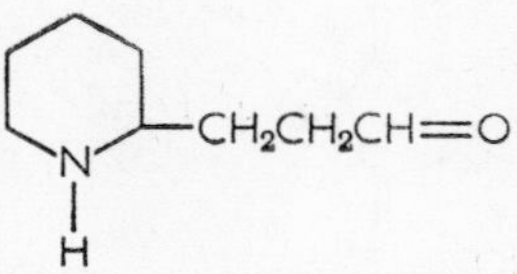

(42) 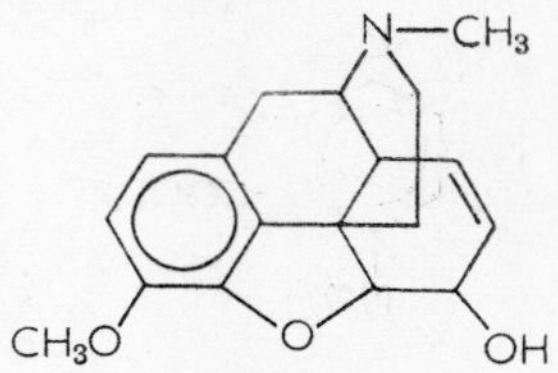　　　　　　　　　　　　　　　　(43)

3. Nitroarylations and nitrobenzylations

Certain aryl and benzyl halides react with primary and secondary amines yielding the corresponding secondary and tertiary amines, in addition to hydrogen halide, much in the same fashion as described in section II.B.4. However, by virtue of the electronegative nature of the reagents, the reaction does not proceed further up to quaternisation as with the previous case, neither does it yield quaternary ammonium compounds with tertiary amines.

Many authors include these reactions among acylations, based on the similarity of the processes and the nomenclature of some reagents, for example from picric acid (13) one prepares picryl chloride which yields a picramide according to equation (18).

The reactions of amines with p-nitrobenzyl chloride[253], and of amino acids with 5-fluoro-2,4-dinitroaniline[254], were recommended for identification purposes. So was picryl chloride[255], which can be also used in quantitative analysis, by determining the hydrogen chloride released in the reaction[255, 256].

2,4-Dinitrophenyl chloride or fluoride with amines give a strong colour which under certain conditions can be tentatively correlated with the type of amine, i.e. yellow for primary amines and orange to brown for other types[9]. The derivative is useful for identification purposes[257].

2,4-Dinitrophenyl fluoride has found wide application in biochemical analysis, in problems related to amino acids, peptides, and proteins, such as the chromatographic separation and identification of amino acids[258, 259], and the analysis of end amino acids of peptidic chains, (equations 27 and 28)[260–265]. The same reagent was used for the determination of primary and secondary amines in general[266, 267] and of α-amino acids in particular[268, 269], by measuring the derivative spectrophotometrically after extraction with a suitable solvent or after chromatographic separation[270]. The spectra of the derivatives present bands at 350 and 390 mμ, providing a way of distinguishing between primary and secondary amines, as the ratio of the extinction coefficient of the former band to that of the latter varies from 0·4 to 0·8 for primary amines, and from 2·1 to 2·4 for secondary amines[271]. It was pointed out[15] that the ratio is 2·1 to 2·4 for α-amino acids, that is, similar to the case of secondary amines.

$$2,4\text{-}(NO_2)_2C_6H_3F + NH_2CHRCO\text{-Peptide residue} \xrightarrow[\text{2. Acid}]{\text{1. Base}}$$

$$2,4\text{-}(NO_2)_2C_6H_3NHCHRCO\text{-Peptide residue} \quad (27)$$

$$2,4\text{-}(NO_2)_2C_6H_3NHCHRCO\text{-Peptide residue} \xrightarrow{H_2O(Acid)} 2,4\text{-}(NO_2)_2C_6H_3NHCHRCO_2H$$
$$+ \text{Peptide degradation products} \quad (28)$$

The applications of 2,4-dinitrophenyl fluoride in organic analysis are amply reviewed elsewhere[260, 272].

4. The Folin reaction

In this method primary aliphatic and aromatic amines yield coloured products by nucleophilic displacement of the sulphonic group of 1,2-naphthoquinone-4-sulphonic acid (44) (equation 29)[273]. This reaction has also been applied in colourimetric determinations[274, 275].

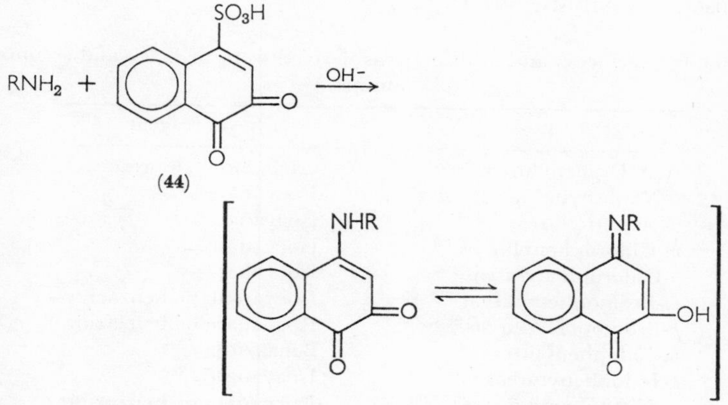

$$(29)$$

5. Ureas, thioureas, hydantoins, and thiohydantoins

Primary and secondary amines react with aryl isocyanates, as in equations (30) and (31), or with azides of arylcarboxylic acids, after their rearrangement to isocyanates, according to equation (32).

$$ArN{=}C{=}O + NH_2R \longrightarrow ArNHCONHR \quad (30)$$

$$ArN{=}C{=}O + NHR_2 \longrightarrow ArNHCONR_2 \quad (31)$$

$$ArCON_3 \longrightarrow ArN{=}C{=}O + N_2 \quad (32)$$

When the arylurea derivative of an α-amino acid is treated with acid a hydantoin is formed, according to equation (33), which can be used for detection and identification[276].

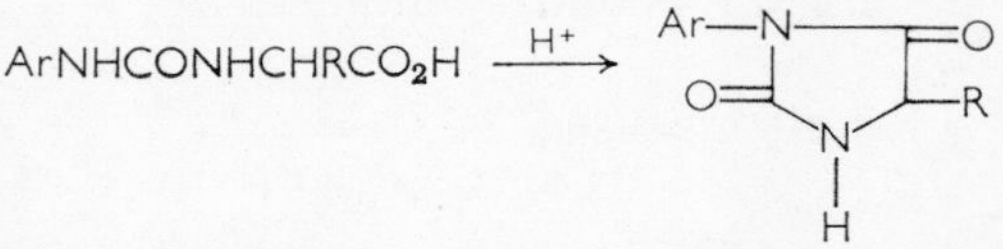

$$ArNHCONHCHRCO_2H \xrightarrow{\ H^+\ } \quad\text{(structure)}\qquad (33)$$

It is sometimes advantageous to use aryl isothiocyanates instead of aryl isocyanates yielding the corresponding thioureas, as these reagents are more resistant to decomposition by water[8]. Both aryl isocyanates and aryl isothiocyanates were found to yield atypical products with certain enamines, as shown for example in (34), where a vinylogue of an arylurea is obtained from an aryl isocyanate[277].

$$\text{(structure)} + OCNC_6H_5 \longrightarrow \text{(structure)}-CONHC_6H_5 \qquad (34)$$

Table 6 summarises the ureas and thioureas derived from amines and used in analysis.

TABLE 6. Arylureas and arylthioureas derived from primary and secondary amines.

Derivatives	Type of reagent
N,N-Diphenylureas[228]	Carbamoyl chloride
α-Naphthylureas[278]	Isocyanate
β-Naphthylureas[279]	Isocyanate
m-Chlorophenylureas[280]	Isocyanate
p-Chlorophenylureas[281]	Isocyanate
m-Bromophenylureas[280, 282]	Isocyanate or benzazide
p-Bromophenylureas[283, 284]	Isocyanate or benzazide
m-Iodophenylureas[285]	Benzazide
p-Iodophenylureas[286]	Isocyanate
m-Nitrophenylureas[287, 288]	Isocyanate or benzazide
p-Nitrophenylureas[289]	Isocyanate or benzazide
3,5-Dinitrophenylureas[290, 291]	Isocyanate or benzazide
2,6-Dinitro-p-tolylureas[292]	Benzazide
Phenylthioureas[293]	Isothiocyanate
α-Naphthylthioureas[294]	Isothiocyanate
β-Naphthylthioureas[293]	Isothiocyanate
p-Biphenylylthioureas[293]	Isothiocyanate
o-Tolylthioureas[293]	Isothiocyanate
p-Chlorophenylthioureas[293]	Isothiocyanate
m-Nitrophenylthioureas[295]	Isothiocyanate

An important application of phenyl isothiocyanates is the analysis of end amino acids in peptides and proteins. This is represented by equations (35) to (39), and involves derivatives of the end amino acids such as thiazolines (**45**), phenylthioureas (**46**), and phenylthiohydantoins (**47**) [296-301].

$$C_6H_5NCS + NH_2CHRCO\text{—Peptide residue} \xrightarrow{\text{PH 8-9}}$$

$$C_6H_5NHCSNHCHRCO\text{—Peptide residue} \quad (35)$$

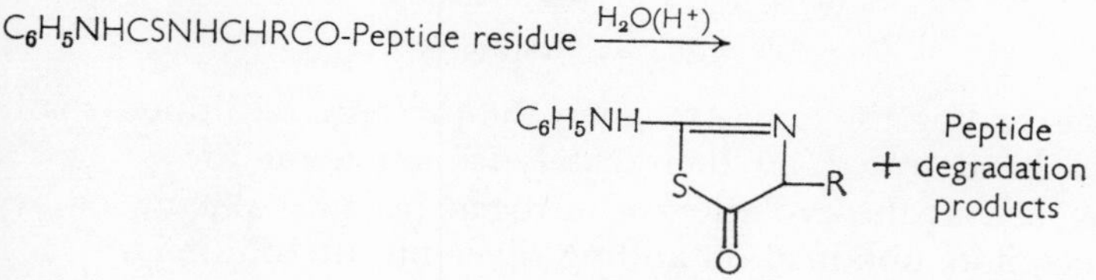

$$C_6H_5NHCSNHCHRCO\text{-Peptide residue} \xrightarrow{H_2O(H^+)}$$

Peptide + degradation products   (36)

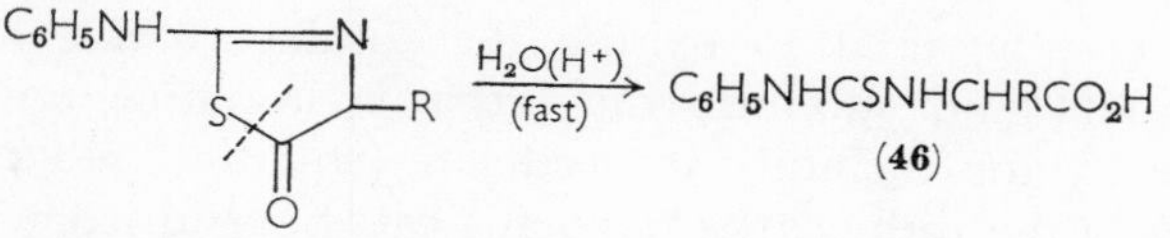

(**45**)

$$\xrightarrow[\text{(fast)}]{H_2O(H^+)} C_6H_5NHCSNHCHRCO_2H \quad (37)$$

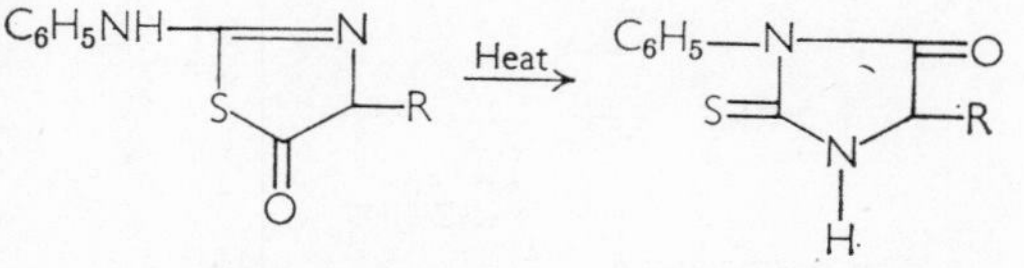

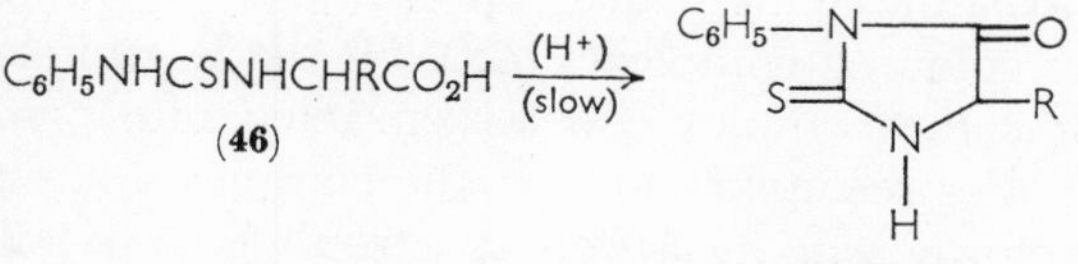

(**46**)

(**45**)

$$\xrightarrow{\text{Heat}}$$

(**45**)   (**47**)   (38)

$$C_6H_5NHCSNHCHRCO_2H \xrightarrow[\text{(slow)}]{(H^+)}$$

(**46**)

(**47**)   (39)

The applications of phenyl isothiocyanate to biochemical analysis are amply discussed elsewhere [260, 272].

The reaction of amines with phenyl isothiocyanate is of the second order and proceeds at different rates for different amines, thus providing a means of determining the components of a mixture, by following kinetically the decrease of the amine concentration in the reaction mixture[302, 303]. Among aliphatic amines secondary ones react faster than primary ones, but alkylarylamines react slower than primary arylamines. In general aromatic amines react faster than aliphatic amines[302].

6. Reactions with carbon disulphide

Primary and secondary amines yield dithiocarbamic acids with carbon disulphide (equations 40 and 41).

$$RNH_2 + CS_2 \longrightarrow RNHCS_2H \tag{40}$$

$$R_2NH + CS_2 \longrightarrow R_2NCS_2H \tag{41}$$

After discarding the excess reagent, the presence of dithiocarbamic acids can be detected by their catalytic action on the process in equation (42), as the evolution of nitrogen becomes patent, or by a black precipitate obtained on adding silver nitrate solution[9].

$$2\,NaN_3 + I_2 \longrightarrow 2\,NaI + 3\,N_2 \tag{42}$$

Iodine can be used to oxidise the dithiocarbamic acids to an isothiocyanate and sulphur; the latter is insoluble and therefore detectable by the turbidity it produces (equations 43 and 44)[304]. Alternatively the dithiocarbamic acids can be oxidised with iron(III) chloride or mercury(II) chloride, as in equation (45), and the hydrogen sulphide easily detected[305].

$$2\,RNHCS_2H \xrightarrow{\ I_2\ } RNH\overset{\overset{\displaystyle S}{\|}}{C}SS\overset{\overset{\displaystyle S}{\|}}{C}NHR \tag{43}$$

$$RNH\overset{\overset{\displaystyle S}{\|}}{C}SS\overset{\overset{\displaystyle S}{\|}}{C}NHR \xrightarrow[\ 2.\ I_2\ ]{1.\ NaOC_2H_5} 2\,RNCS + 2\,S \tag{44}$$

$$RNHCS_2H \longrightarrow RNCS + H_2S \tag{45}$$

Dithiocarbamic acids with copper(II) salts in the presence of ammonia give coloured complexes (48a). If the R groups are sufficiently hydrophobic the complex can be extracted into benzene for better detection[9]. For secondary amines the complex 48a can be extracted into chloroform and measured spectrophotometrically[306, 307]. With nickel(II) salts coloured precipitates (48b) can be obtained[9, 308].

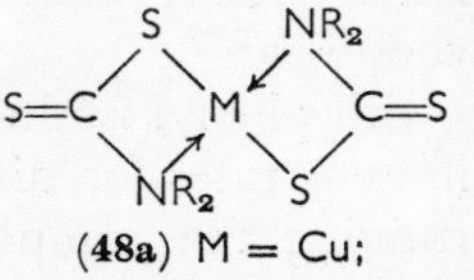

$$S=C\overset{\displaystyle S}{\underset{\displaystyle NR_2}{\diagup\diagdown}}M\overset{\displaystyle NR_2}{\underset{\displaystyle S}{\diagup\diagdown}}C=S$$

(48a) M = Cu;
(48b) M = Ni

Reactions (40) and (41) can be followed by direct titration of the dithiocarbamic acids with alkali[309, 310], or by precipitating complex **48b** and determining the nickel in the precipitate[311]. Secondary amines were determined coulometrically as the dithiocarbamate anions react with mercury(II) cations generated at the electrode, as in equation (46)[312].

$$2\ R_2NCS_2{}^- + Hg^{2+} \longrightarrow (R_2NCS_2)_2Hg \tag{46}$$

The analysis of amines via dithiocarbamic acids is amply discussed elsewhere[14–16, 37].

7. Reaction with sulphur trioxide

Aromatic amines can be determined in dioxane solution by adding excess of sulphur trioxide (equation 47), then destroying the excess reagent with water and titrating the sulphuric acid thus produced[313].

$$ArNH_2 + SO_3 \longrightarrow ArNHSO_3H \tag{47}$$

8. Reactions of primary amines with aldehydes and ketones

The general reaction described by equation (48), when carried out with certain reagents and under controlled conditions, provides good methods of detection and determination of primary amines, although some amines of other types may interfere.

$$R^1NH_2 + O{=}C\!\!\begin{array}{c} R^2 \\ \diagdown \\ R^3 \end{array} \longrightarrow R^1N{=}C\!\!\begin{array}{c} R^2 \\ \diagup \\ R^3 \end{array} + H_2O \tag{48}$$

$$(\mathbf{49})$$

$$(R^1,\ R^2,\ R^3 = H,\ alkyl,\ aryl)$$

a. Benzaldehyde and similar reagents. Primary aliphatic and aromatic amines can be determined with benzaldehyde in nonaqueous solution. Excess of reagent is used and destroyed afterwards with hydrogen cyanide[314] (equation 49), and the water produced according to (48) is measured by the Karl Fischer method[315].

$$C_6H_5CHO + HCN \longrightarrow C_6H_5CH(CN)OH \tag{49}$$

2-Ethylhexaldehyde[119] and vanillin (**50**) react, in a way similar to benzaldehyde, with primary aromatic amines; the Schiff base (**49**) derived from **50** can be determined spectrophotometrically[316]. With salicylaldehyde (**51**) the use of hydrogen cyanide is avoided, and instead the excess reagent is titrated with standardised sodium

methoxide[317], or otherwise the Schiff base is measured spectrophotometrically[318].

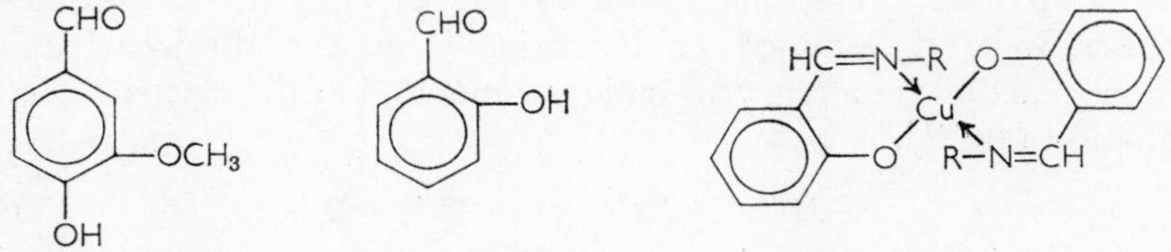

(50)   (51)   (52)

A variation of the salicylaldehyde method, applicable only to
primary aliphatic amines with an unbranched α-position, consists of
carrying out the reaction in the presence of copper(II) chloride and a
base such as triethanolamine, whence complex **52** is formed, which
can be extracted and measured spectrophotometrically[310].

2,4-Pentanedione undergoes tautomerisation according to equation
(50), and therefore resembles salicylaldehyde in its reactions with
primary amines[319, 320].

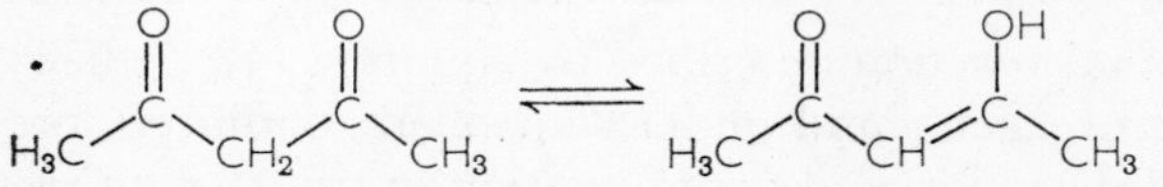

(50)

p-Dimethylaminobenzaldehyde is the reagent used in very sensitive
tests for the detection and possibly identification of primary amines
as coloured Schiff bases[9, 321]. Other amines, notably diphenylamine
and pyrrole derivatives in which the NH group and at least one hydrogen atom of the pyrrole ring are preserved, also give colour reactions
with this reagent[322, 323]. The reaction with the pyrrole derivatives is
due to condensation of the aldehyde with a methylene group, produced in the heterocyclic compound by tautomerisation according to
equation (51). Condensations with pyrroles and other atypical amines
can be avoided when detecting primary amines, by controlling the
pH of the solution[9].

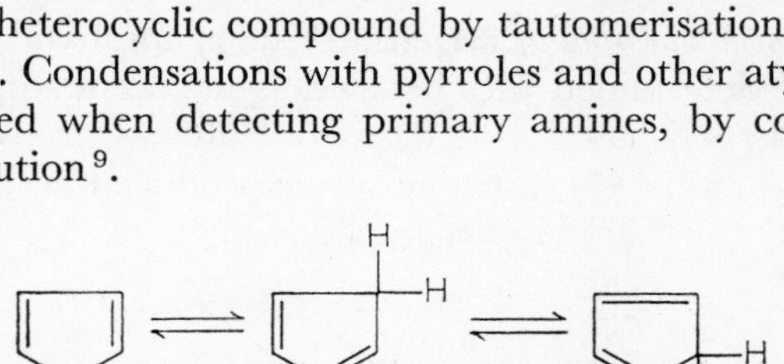

(51)

b. *Glutaconic aldehyde.* This compound yields with primary amines
polymethyne dyes[324-326] such as **56**. The reagent is however un-

stable and has to be prepared not more than a few days before use, either by irradiating aqueous pyridine with ultraviolet light[327] according to equation (52) or by treatment of N-(4-pyridyl)pyridinium chloride hydrochloride with alkali[328] according to (53). Products **53** and **54** of these reactions are enolate salts derived from glutaconic aldehyde (**55**), and are able to condense with primary amines in the presence of acid (equation 54)[329].

$$C_6H_5N + 2\,H_2O \xrightarrow{hv} O{=}CHCH{=}CHCH{=}CH{-}O^-NH_4^+ \qquad (52)$$
$$(53)$$

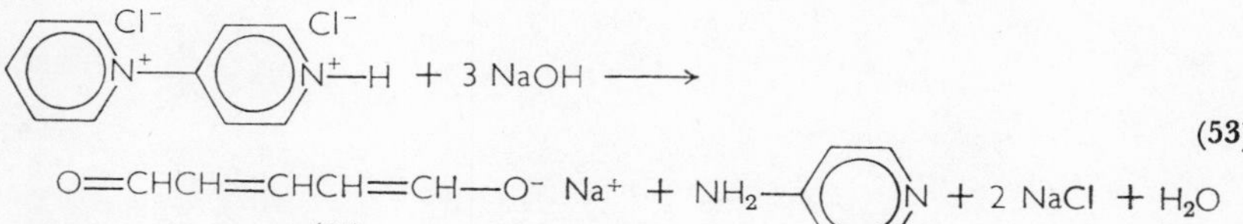

$$(53)$$

$$O{=}CHCH{=}CHCH{=}CH{-}O^-\ Na^+ + NH_2{-}\ \bigcirc\!\!N + 2\,NaCl + H_2O$$
$$(54)$$

$$O{=}CHCH{=}CHCH_2CH{=}O + ArNH_2 + H^+ \longrightarrow$$
$$(55)$$

$$Ar\overset{+}{N}H{=}CHCH{=}CHCH{=}CH{-}NHAr \quad (54)$$
$$(56)$$

Glutaconic aldehyde can react with compounds other than primary amines[325], for example it undergoes condensations at C-positions with pyrrole derivatives that tautomerise according to equation (51).

The condensations of amines with aldehydes have been reviewed and discussed elsewhere[9, 14–16, 37].

9. Reaction of α-amino acids with ninhydrin

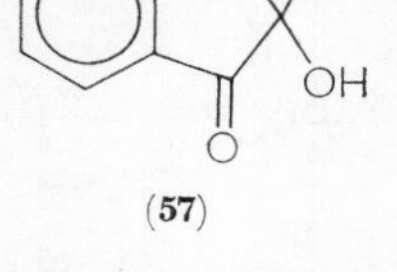

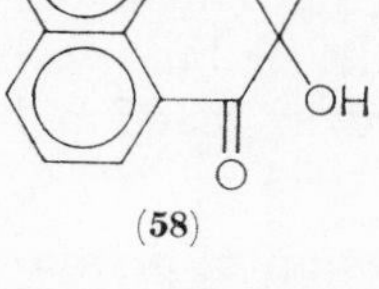

Ninhydrin (hydrate form, **57**)[330, 331] and similar reagents such as perinaphthalenetrione (hydrate form, **58**)[332, 333] are capable of oxidising α-amino acids in neutral solution. This provides a very sensitive detection method, as dyes such as **59** are probably produced in a process involving equations (55) to (57) in the case of ninhydrin, and similar ones for the other reagents.

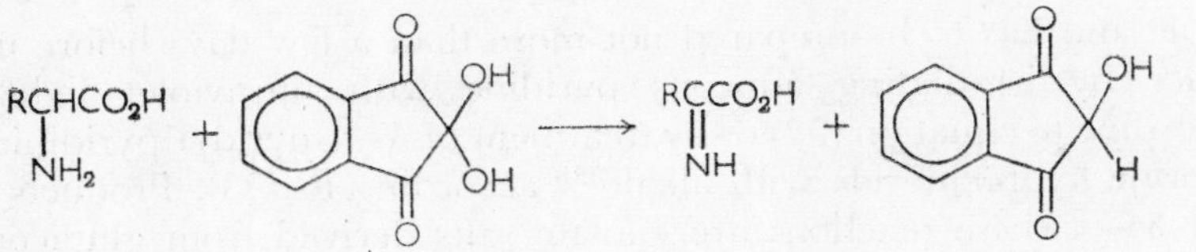

$$\underset{\substack{|\\NH_2}}{RCHCO_2H} + \text{(indanetrione hydrate)} \longrightarrow \underset{\substack{||\\NH}}{RCCO_2H} + \text{(2-hydroxyindanone)} \qquad (55)$$

$$\underset{\substack{||\\NH}}{RCCO_2H} + H_2O \longrightarrow RCH{=}O + NH_3 + CO_2 \qquad (56)$$

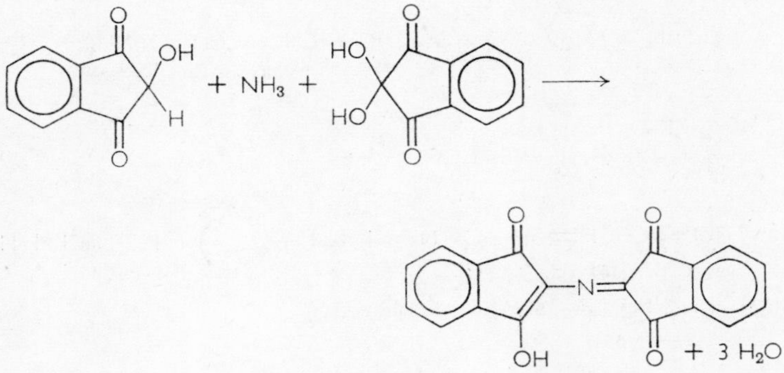

$$\text{(2-hydroxyindanone)} + NH_3 + \text{(indanetrione hydrate)} \longrightarrow$$

$$\text{(bis-indanedione imine)} \quad OH \qquad O + 3\ H_2O \qquad (57)$$

$$(59)$$

The colour reaction is transient but it can be fixed with special reagents and applied for detection and identification of individual amino acids, by making use of polychromic methods, especially after a chromatographic separation[334–341].

The determination of amino acids based on the ninhydrin reaction can be carried out in several ways: by spectrophotometry of the coloured solutions[342–345], or spectral reflectance after chromatographic separation of the amino acids and colour development[339, 346], by spectrophotometry of ninhydrin–zinc chloride complexes[347], by determining the aldehyde obtained in equation (56), either chemically[342] or by gas chromatography[162], or by determining ammonia, carbon dioxide, or both, produced in equation (56)[348–354].

The ninhydrin reaction has been amply discussed and reviewed elsewhere[9, 14–16, 133, 192, 355–359].

10. Diazotisation of primary amines

With nitrous acid, primary amines are diazotised (equation 58), secondary amines undergo N-nitrosation (section II.C.2.e), and tertiary aliphatic or alicyclic amines fail to react while tertiary aromatic amines may undergo C-nitrosation.

$$R{-}NH_2 + HNO_2 + H^+ \longrightarrow R{-}\overset{+}{N}{\equiv}N + H_2O \qquad (58)$$

$$(R = \text{alkyl, aryl})$$

a. The van Slyke method. Primary aliphatic amines, including amino alcohols and amino acids, react with nitrous acid (equation 58), followed immediately by decomposition (equations 59 or 60). The amines are determined by measuring the amount of nitrogen evolved[360, 361]. Ammonia reacts as a primary amine, and has to be determined independently when present. Amines of high molecular weight or complex structure fail to give quantitative results[37]. The method has been amply reviewed[14, 15, 37, 356].

$$R-\overset{+}{N}\equiv N + H_2O \longrightarrow ROH + H^+ + N_2 \tag{59}$$

$$R-\overset{+}{N}\equiv N \longrightarrow Olefin + H^+ + N_2 \tag{60}$$

b. Aryldiazonium salts. Primary aromatic amines can be detected, after diazotisation and coupling, as deeply coloured azo dyes are produced in the presence of phenols or aromatic amines in solutions of controlled pH[8, 14]. Quantitative analyses via diazotisation can be performed in various ways; for example by titrating with standardised sodium nitrite[3, 362], or by spectrophotometry of the diazonium salt itself[363] or its coupled product[364].

Several methods have been proposed based on the decomposition of diazonium salts. The volume of nitrogen can be measured after treating aryldiazonium compounds with copper(I) chloride[14], potassium iodide[365], or titanium(III) chloride[366, 367]. Otherwise, the excess catalyst can be determined, as in the case of titanium(III) chloride[14], or chromium(II) chloride[368]. The diazonium compound strongly heated with hydriodic acid yields iodine which can be titrated[369].

The differential kinetic method (section II.C.5) is also applicable to mixtures of primary aromatic amines, which after diazotisation can be determined by measuring the first-order rates of evolution of nitrogen, catalysed by copper(I) chloride[370].

These methods are further discussed elsewhere[8, 14, 15, 65].

II. Spectral properties of the N—H bond

a. Vibrational spectrum. Two complementary spectroscopic methods are available, namely infrared and Raman spectroscopy. Unfortunately the latter has not yet acquired the popularity and widespread application of the former. Raman spectra can be used as fingerprints of compounds for identification, and for the detection of functional groups, among which amino groups appear with characteristic bands arising from N—H vibrations. The N—H group of primary and secondary amines has characteristic infrared absorption bands which are summarised in Table 7.

TABLE 7. Absorption bands of N—H in amines[a].

Region[b] (cm^{-1})	Origin of band
1600	Deformation modes
3400	Stretching (fundamental)
5000	Combination of stretching and deformation modes
6700	Stretching (1st overtone)
10000	Stretching (2nd overtone)

[a] Data from references 170, 371, and 372.
[b] The influence of hydrogen bonding is not included.

Considerable discussion notwithstanding, the absorption band in the 1600 cm^{-1} region has been assigned to N—H deformation modes[371]. For primary amines a medium to strong band appears almost always at 1590–1650 cm^{-1}; in aromatic amines this can be confused with a band from an aromatic ring vibration near 1600 cm^{-1}. In the case of secondary amines this band is much weaker.

The most useful region for analytical purposes is that of the fundamental stretching modes. In the absence of association, primary amines present two bands, which for aromatic amines appear at about 3400 and 3500 cm^{-1}, while in the case of aliphatic amines they appear at frequencies, lower by about 100 cm^{-1}. The lower frequency band is assigned to the symmetric and the higher to the asymmetric stretching mode, both being of medium strength[373-378]. This pair of bands can be identified more easily if the correlation in equation (61) is used[377]. Secondary amines show, in the absence of association phenomena, a single band in the 3300–3500 cm^{-1} region[379-384]. The intensities in the fundamental stretching region are much higher for aromatic and some heterocyclic primary and secondary amines than for aliphatic amines, and the former can be determined in this region[372, 385, 386].

$$\nu_{\mathrm{sym}} = 0.876\nu_{\mathrm{asym}} + 345.5 \tag{61}$$

Primary amines have absorption bands in the 5000 cm^{-1} region, probably arising from the additive combination of bending and stretching modes[387]. For aromatic amines it appears at 5050–5100 cm^{-1} and has been proposed for the characterisation and determination of these compounds[388, 389]. The absorptions of aliphatic amines are of about half the intensity of those of aromatic amines, appearing at somewhat lower frequencies, and it has been suggested that they too can be used for quantitative analysis[372].

Primary aromatic amines have two bands assigned to the first overtone of the stretching mode, the symmetric one appearing at about 6700 cm^{-1} and the asymmetric one at about 6900 cm^{-1}, the former being some six times more intense than the latter[389]. Secondary aromatic amines have only one band in this region, near the site of the symmetric band of primary aromatic amines but of about half the intensity[372, 388, 390]. Primary and secondary aliphatic amines absorb at about 6500 cm^{-1}, which is out of the range of the aromatic ones[390], and allows the determination of mixtures of both classes[388]. It has been suggested that this and the combination region are better than the fundamental region for the determination of aliphatic amines, owing to the wider choice of solvents that can be used for this purpose[372].

Finally, in the second overtone region only very weak bands can be observed for aromatic amines, the most outstanding being the symmetric band at about 9800 cm^{-1}, of limited analytical value[389].

It has been pointed out that a possible way of distinguishing between primary, secondary, and tertiary amines is by treatment with acid and looking for the bands arising from the ammonium salt, as these are distinctive for each ammonium type[170, 176]. The problem of distinguishing between OH and NH groups in the fundamental stretching region was also considered[391]. A comparison of absorption bands and intensities of functions containing N—H bonds can be found elsewhere[392].

b. Nuclear magnetic resonance spectrum. The resonance bands due to NH groups are summarised in Table 8. The main factors affecting the location and shape of the N—H bands are hydrogen bonding[393, 394], the quadrupole moment of ^{14}N, and the fast proton exchange occurring in certain solvents[394]. Two main effects stem from the quadrupole moment of ^{14}N, namely a splitting of the band into a triplet, by coupling with the three states of ^{14}N, and the bands becoming broad, even sometimes disappearing[395, 396]. The fast proton exchange occurring in certain solvents tends to sharpen the N—H band (see however reference 397, and also section II.C.1).

TABLE 8. Chemical shifts of NH groups[a].

Type of compound	Chemical shift[b]
Aliphatic and alicyclic amines	2·9–5·0
Aromatic amines	3·6–4·7
Amides	5·0–8·0

[a] Data from reference 393.
[b] In parts per million (p.p.m.) of the δ scale.

The values in Table 8 are only general guides for the assignment of resonance bands, and wide discrepancies have been observed, for example histidine (**60**) dissolved in trifluoroacetic acid presents a band at $\delta -1$, assigned to the nitrogen-bonded imidazole proton[398].

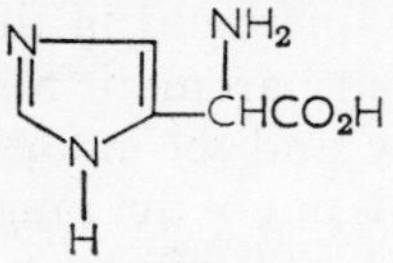

(**60**)

D. The N—C Bond

I. Chemical methods

The N—C bond of amines is very resistant to chemical attack and is usually cleaved only under drastic conditions.

a. *The Herzig–Meyer method*[399–402]. When an amine and hydriodic acid are heated strongly, dealkylation occurs (equations 62 and 63).

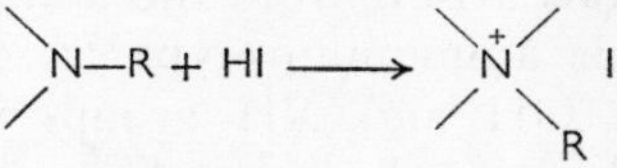

$$\diagdown\!\!\!\overset{\diagup}{N}\!\!-\!R + HI \longrightarrow \overset{H}{\underset{R}{\diagdown\!\!\!\overset{\diagup}{N}{}^{+}}}\ I^{-} \tag{62}$$

$$\overset{H}{\underset{R}{\diagup\!\!\!\overset{\diagdown}{N}{}^{+}}}\ I^{-} \longrightarrow \diagdown\!\!\!\overset{\diagup}{N}\!\!-\!H + RI \tag{63}$$

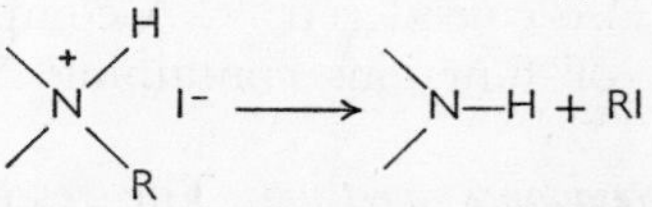

The alkyl iodide can be determined by gravimetry, iodometry, or gas chromatography, and identified by the latter method or by preparing derivatives recommended for alkyl halides[2, 10, 16].

Alkoxy groups undergo dealkylation more readily than amines, and in the presence of the former, after addition of the hydriodic acid the sample should be subjected first to distillation at 150°, in order to determine the alkoxy groups, and then to pyrolysis at 360°, where the Herzig–Meyer method is operative[403].

The method is further discussed elsewhere[15, 356, 404].

b. *Fusion with benzoyl peroxide*[9]. On melting an amine with benzoyl peroxide, the *N*-alkyl substituents become oxidised according to equation (64).

$$\diagdown\!\!\!\overset{\diagup}{N}\!\!-\!CH_2R + (C_6H_5CO)_2O_2 \longrightarrow \diagdown\!\!\!\overset{\diagup}{N}\!\!-\!COC_6H_5 + C_6H_5CO_2H + RCHO \tag{64}$$

In the case of *N*-methyl compounds (R = H in equation 64) formaldehyde is liberated, which gives a violet colour with a chromotropic acid (**61**)–sulphuric acid solution[405]. When R = CH_3 or a

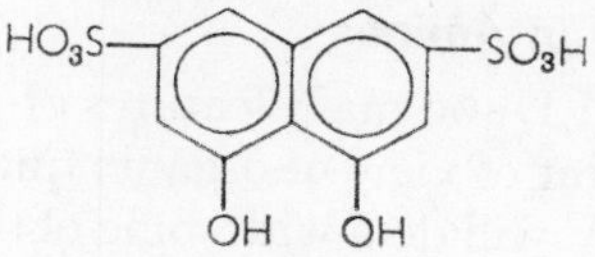

(61)

somewhat higher homologue, the aldehyde produced can be detected by the colour it produces with a solution of sodium nitroprusside and morpholine, which is similar to the Rimini test described in section II.B.3.

Compounds containing alkoxy or propenyl groups interfere with this method, as they too yield aldehydes on fusion with benzoyl peroxide.

c. β-Substituted ethylamines[9]. The process in equations (65) and (66) takes place when compounds of general structure $X—CH_2CH_2—Y$ are fused with moist zinc chloride[406]. The acetaldehyde of equation (66) can be detected by the sodium nitroprusside–morpholine reagent.

$$X—CH_2CH_2—Y + 2\,H_2O \longrightarrow HO—CH_2CH_2—OH + XH + YH \qquad (65)$$

$$\left(X, Y = OH, OR, N\diagup_{\diagdown}, SH, SR, \text{halogen, etc.}\right)$$

$$HOCH_2CH_2OH \longrightarrow CH_3CHO + H_2O \qquad (66)$$

2. Vibrational spectrum

Table 9 summarises the regions in which bands assigned to C—N vibrations appear. In the case of aliphatic amines[407] the band is difficult to locate both because of its low intensity and the fact that it appears in a region where other bands also usually do[371]. The bands of aromatic amines[407, 408] can be used for confirmation of proposed structures, although the possibility of confusion with bands belonging to other groups should be born in mind.

TABLE 9. Absorption bands of C—N stretching vibrations[a].

Type of compound	Region (cm^{-1})	Intensity
Aliphatic amines	1020–1220	Weak to medium
Aromatic amines		
Primary	1250–1340	Strong
Secondary	1280–1350	Strong
Tertiary	1310–1360	Strong

[a] Data from references 170 and 371.

5+c.a.g.

E. Hydrogen Bonding in Amines

In sections II.A to II.D the main features of the amino group were examined from the point of view of organic qualitative and quantitative analysis. Section IV will deal with some of the structural problems posed by amines. It is convenient to bridge these two aspects of analysis with a brief examination of hydrogen bonding in the amino group, which has a deep influence on a wide range of physical and chemical properties used in the elucidation of molecular structure. The importance of hydrogen bonding can be more fully appreciated by considering Table 10, in which nearly all these properties are listed.

TABLE 10. Influence of hydrogen bonding on the properties of a system[a].

Property	Intramolecular hydrogen bond	Intermolecular hydrogen bond
Structure of compound	Usually found in *ortho, cis, peri*, diequatorial, and equatorial-axial positions	Usually found in *meta, para, trans*, and diaxial positions
Geometry of bond	Usually bent	Usually straight
Molecular weight	Normal	Increased
Density		Increased
Molecular volume	Decreased	Decreased
Molecular refraction	Increased	
Viscosity	Decreased	Increased
Self diffusion		Decreased
Parachor	Decreased	Decreased
Surface tension	Decreased	Increased
2nd Virial coefficient (Berthelot equation)	Normal	Absolute value increased
Vapour pressure	Normal	Decreased
Boiling point	Normal	Increased
Melting point	Normal	Increased
Solvent power	Normal	Increased if solute becomes hydrogen bonded to solvent
Thermal conductivity	Normal	Increased
Acoustic conductivity	Normal	Increased
Electric conductivity	Normal	Increased if hydrogen-bonded network is formed
Dielectric constant	Normal for solids and variable according to molecular shape for liquids	Increased for solids and variable according to molecular shape for liquids
Dipole moment	Lower than calculated for structure. Concentration independent	Higher than calculated for structure. Increases with concentration

TABLE 10—(*Cont.*).

Property	Intramolecular hydrogen bond	Intramolecular hydrogen bond
Reaction rate	Often influenced	Often influenced
Reaction mechanism	Often influenced	Often influenced
Optical rotation	Can be large if spiro structures are formed	Usually none
Electronic absorption bands	Shifted	Shifted
Phototropy	Often presented	Seldom presented
Vibrational spectrum		
Stretching frequencies (ν_s)	Shifted down	Shifted down
Bending frequencies (ν_b)	Shifted up	Shifted up
Concentration dependence of ν_s and ν_b	None	Present
Size of $\Delta\nu_s$	Determined by functional groups and size of ring formed	Determined by functional groups
Effect of phase change on $\Delta\nu_s$	Small	Large
Intensity of stretching band	Somewhat increased	Very much increased
Half-width of stretching band	Somewhat increased	Very much increased
Nuclear magnetic resonance		
Chemical shifts	Shifts to lower field	Shifts to lower field
Concentration dependence of chemical shifts	Small	Large

a From G. C. Pimentel and A. L. McClellan, *The Hydrogen Bond*, W. H. Freeman and Company, San Francisco, 1960.

Amino groups are capable of becoming associated among themselves or with other functions by two types of hydrogen bonding, formally represented as **62** and **63**. In the former the amine acts as a weak protonic acid, and must therefore be primary or secondary (or an ammonium compound other than quaternary). On the other hand, in **63** it acts as a weak base, sharing its p electrons, and can be any type of amine (but not an ammonium compound).

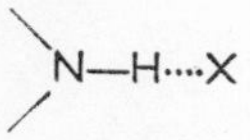

N—H····X

(**62**)

—N····H—X

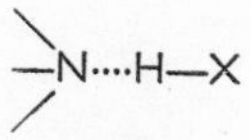

(**63**)

Hydrogen bonds of types **62** and **63** are relatively weak and there-fore amenable to investigation by i.r. spectroscopy[179, 384, 409–412]. Special attention should be paid to the stretching frequencies of the N—H and X—H bonds, the shape and intensity of the absorption bands, and the effect of temperature, concentration, and deuterium exchange.

F. Miscellaneous Chemical Properties

I. Preliminary tests for detection and classification

A great number of methods have been proposed hitherto in order to ascertain whether a nitrogen-containing compound is an amine or not. It is possible to use some of these tests to classify amines into aromatic, aliphatic, primary, secondary, or tertiary. As the chemical process involved in most of them is vaguely known, results should be taken only as a helpful guide. Such tests include many of the oxidations described in section II.B.5 and the ones listed below.

a. Potassium thiocyanate test. Salts of amines (not quaternary ammon-ium compounds) yield hydrogen sulphide on heating at 200–250° with potassium thiocyanate[413]. Compounds which liberate water on heating also give a positive test. The process is probably similar to the am-monium thiocyanate rearrangement (equation 67)[414, 415].

$$NH_4SCN \longrightarrow \overset{NH_2}{\underset{NH_2}{\diagup}} C{=}S \longrightarrow NH_2CN + H_2S \qquad (67)$$

b. Chloranil test. Chloranil (**64**) yields coloured products in the presence of amines. Amino acids do not react, but some amides such as anilides do. Aliphatic amines react more readily than aromatic amines, and the colour may be used for classification as follows[416, 417]: red—primary, violet—secondary, and emerald—tertiary. This re-action can be used for chromatographic development[418].

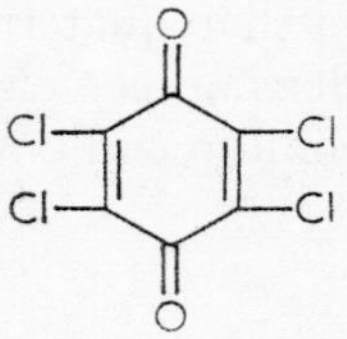

(**64**)

c. Reaction with triphenylmethyl dyes. All amines yield on melting with a mixture of 'fluorescein chloride' (65) and anhydrous zinc chloride, red water-soluble dyes of structure 66. Their fluorescence under u.v. illumination may be used to classify amines as follows: yellow–green fluorescence shows the presence of an NH—alkyl partial structure in 66, orange fluorescence shows the N(alkyl)$_2$ partial structure, and if an *N*-aryl group is present in 66 no fluorescence is observed[419, 420]. Pyrroles, indoles, and carbazoles give blue fluorescence[9].

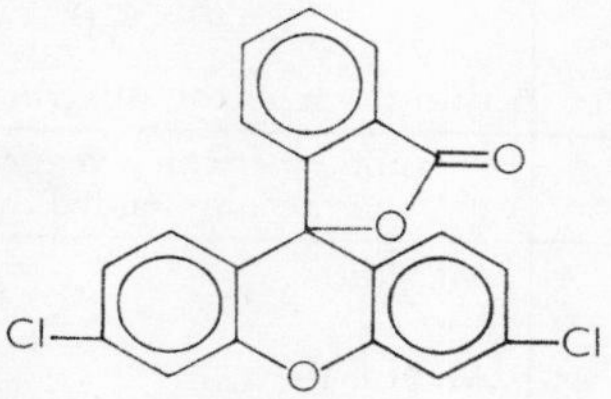

(65)

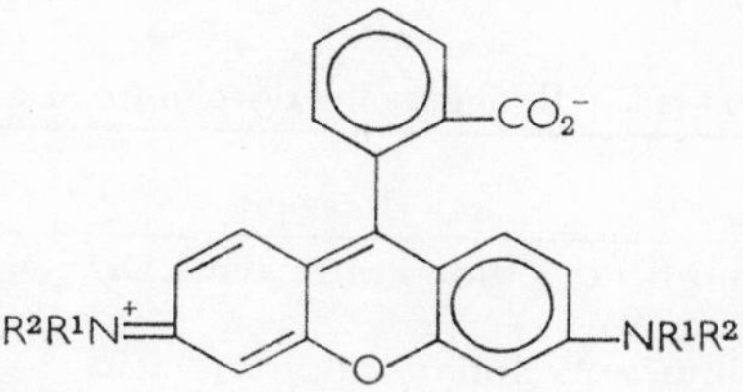

(66)

Bromophthalein magenta E (67) yields coloured derivatives with aliphatic amines in anhydrous solutions. The colour can be correlated with structure as follows: purple (λ_{max} 530–540 mμ)—primary, blue (λ_{max} 570–580 mμ)—secondary, and red (λ_{max} 520–530 mμ)—tertiary[421].

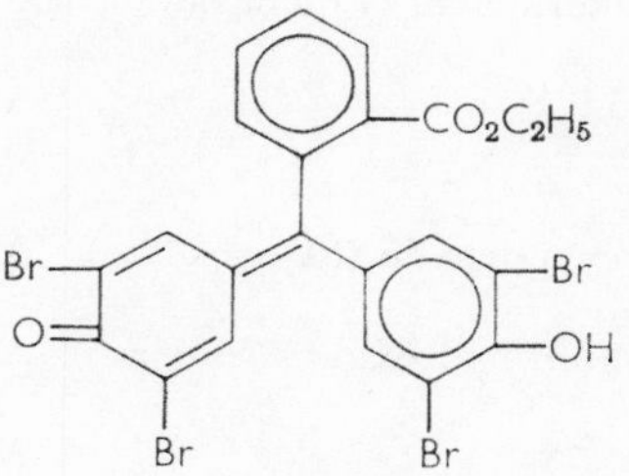

(67)

d. Lignin test. The lignin present in newsprint reacts swiftly at room temperature with primary and secondary arylamines, in the presence of strong acid, yielding yellow to orange spots. Aliphatic and alicyclic primary and secondary amines give the reaction on heating. Tertiary amines, amino acids, and amides do not react[16].

e. o-Diacetylbenzene test. Primary amines, aliphatic and aromatic, undergo colour reactions with this compound. Secondary and tertiary amines, and some primary amines such as glucosamine give negative results[422].

f. Reaction with polycarboxylic acids. Tertiary amines yield red to blue coloured solutions when heated briefly with malonic (**68**), citric (**69**), or aconitic acid (**70**), in acetic anhydride solution[423]. Aconitic anhydride (**71**) can also be used for this test[417].

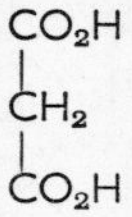

(**68**) 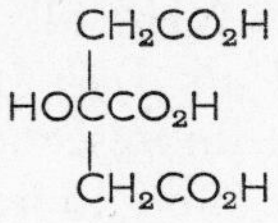   (**69**) 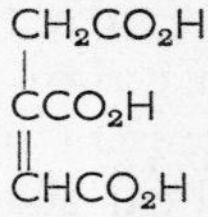   (**70**) 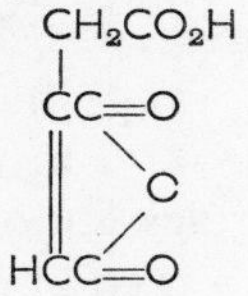   (**71**)

TABLE 11. Reagents for development of amines in thin-layer chromatography[a].

Reagent	Amine for which reagent is recommended
Bismuth(III) iodide/acetic acid (Dragendorff reagent)	Alkaloids
Cerium(IV) sulphate/sulphuric acid	Alkaloids
Chloranil (**64**)	Aliphatic amines
Chloroplatinic acid/potassium iodide	Alkaloids, *N*-heterocyclic compounds
Chromic acid/sulphuric acid	Aromatic amines
Cinnamic acid (**72**)/hydrochloric acid	Indole derivatives
p-Dimethylaminobenzaldehyde/hydrochloric acid	Indole derivatives, ergot alkaloids
p-Dimethylaminobenzaldehyde-2,4-pentanedione	Aminosaccharides
Formaldehyde/hydrochloric acid (Prochazka reagent)	Indole derivatives
Glucose/phosphoric acid	Aromatic amines
Iodine solution	General reagent for organic compounds
Malonic acid (**68**)/salicylaldehyde (**51**)	Amines, *N*-heterocyclic compounds
Mercury(II) iodide/sodium hydroxide (Nessler reagent)	Alkaloids
4-Methylumbeliferone (**73**)	*N*-heterocyclic compounds
Ninhydrin (hydrate form, **57**)	Amino acids, amines
Ninhydrin/copper(II) nitrate (polychromatic reagent)	Amino acids
Nitric acid	Alkaloids, amines
Potassium iodate	Sympaticomimetic amines[b]
Sodium 1,2-naphthoquinone-4-sulphonate (Folin reagent, sodium salt of **44**)	Amino acids
Sodium nitroprusside/acetaldehyde (Rimini test)	Secondary aliphatic and alicyclic amines
Sulphuric acid (charring reagent)	Alkaloids, amines
Vanillin (**50**)	Amino acids, amines

[a] Data from references 16, 424, and 425.
[b] E.g. (phenylethyl)amine and derivatives.

2. Development of chromatograms

In Table 11 a list is given of the most usual developing reagents recommended for amines in thin-layer chromatography, many of which can be used for the same purposes also in column and paper chromatography.

$$C_6H_5CH{=}CHCO_2H$$

(72)

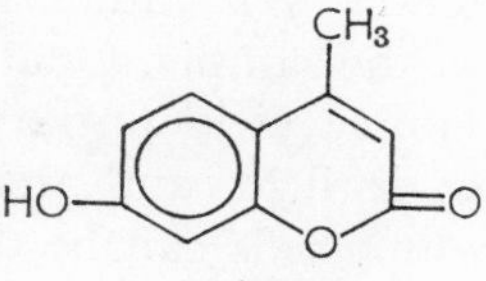

(73)

3. Ring substitution in aromatic amines

Aromatic amines are usually very reactive towards electrophilic reagents yielding derivatives substituted at carbon atoms. Reactions such as nitration[8] and bromination[8, 14–16, 37] can be applied to the amine as such, or to the amine previously protected at the nitrogen site. Although the derivatives from these reactions are useful for identification and possibly quantitative purposes, the site and stoichiometry of the substitution are specific for every system, and no generalisation can be made. Aryldiazonium salts undergo coupling reactions with aromatic amines to yield azo dyes, which can be used for detection and determination. In Table 12 some of the aryldiazonium salts used in analysis are listed.

TABLE 12. Diazonium salts for coupling with aromatic amines.

p-Acetylbenzenediazonium chloride[4]
p-Aminobenzenediazonium chloride[4]
p-Arsonobenzenediazonium chloride[426]
Benzenediazonium salts[427]
2-Carboxy-4-nitrobenzenediazonium resinesulphonate[426]
2,5-Dichlorobenzenediazonium resinesulphonate[426]
m-Nitrobenzenediazonium chloride[14]
p-Nitrobenzenediazonium resinesulphonate[426]
p-Nitrobenzenediazonium salts[4, 427]
p-Nitrobenzenediazonium tetrafluoroborate[418, 428]
p-Phenylazobenzenediazonium tetrafluoroborate[429]
p-Sulphobenzenediazonium salts[4, 427, 430]
p-Toluenediazonium chloride[14]

III. QUATERNARY AMMONIUM COMPOUNDS

The analytical methods based on chemical reactions of compounds with structure **2**, in which the nitrogen atom is bonded to four carbon atoms, can be grouped into two classes, namely methods preserving the cationic part of the molecule, and methods in which this part is destroyed.

The latter class is rather limited as it involves, usually, drastic treatment of the sample. Quaternary ammonium compounds treated with morpholine yield tertiary amines (equation 68) which can be fractionally distilled and determined[431]. Simple quaternary ammonium compounds can be determined by repeatedly heating with sodium hydroxide and collecting the liberated tertiary amines, for example tetramethylammonium salts yield trimethylamine by this method[432].

$$R_4N^+ + O\!\!-\!\!NH \longrightarrow R_3N + O\!\!-\!\!N^+\!\!\big\langle\!\!\begin{array}{c} R \\ H \end{array} \tag{68}$$

In the most important methods the ammonium ion is not destroyed. Quantitative analyses can be carried out by determining either the cation or the anion, whereas in detection or identification problems, of course, the cation has to be traced.

Quaternary ammonium hydroxides are strong bases, comparable with alkali hydroxides, and therefore can be easily titrated with standardised mineral acid. This affords a method of determining quaternary salts, by using ion exchangers and titrating the resulting hydroxide[433, 434]. Ammonium salts other than quaternary will yield weak bases distinct from the strong quaternary ones. The converse ion exchange can be also carried out, thus alkaloid salts were converted into magnesium salts, and the metal cation determined[435].

Salts of weak acids such as carboxylates[436], picrates[437], or carbonates[438] can be titrated with mineral acids. Ammonium halides can be converted into acetates by treatment with mercury(II) acetate in glacial acetic acid, and determined by subsequent titration with perchloric acid[439, 440]. Salts of strong acids, such as sulphates and halides, can be determined by measuring the anion[441], or alternatively the quaternary cation can react to form complexes which precipitate, or remain in aqueous solution, or have to be extracted into organic solvents. These complexes can be measured directly, or indirectly by measuring the excess reagent. In Table 13 appears a list of reagents used for the determination of quaternary ammonium compounds by complex formation.

TABLE 13. Reagents for the determination of quaternary ammonium compounds.

Reinecke salt (**11**) [3, 442]
Phosphotungstic acid [443]
Potassium dichromate [444]
Picric acid (**13**) [445, 446]
Dipicrylamine [447–449]
Potassium iodide [450]
Potassium cadmium iodide ($CdI_2 . 2 KI$) [451]
Sodium tetraphenylborate (**18**) [95, 452, 453]
Alipal CO-436 [14, a]
Igepon T-77 [14, b]
Sodium lauryl sulphate [454, 455]
Sodium dodecyl sulphate [456]
Sodium sulphosuccinate [457]
Eosin A (**74**) [458, c]
Bromophenol blue (**75**) [453, 459–463, c]
Bromothymol blue (**76**) [449, 461, 462, c]
Bromocresol purple (**77**) [464, c]
Chloranil (**64**) [465, c]
Aconitic anhydride (**71**) [465, c]

---

[a] Trade name for ammonium nonyl phenoxy polyethoxy sulphate of mol. wt. 504.
[b] Trade name for oleyl methyl tauride.
[c] Can be used in colour reactions for detection.

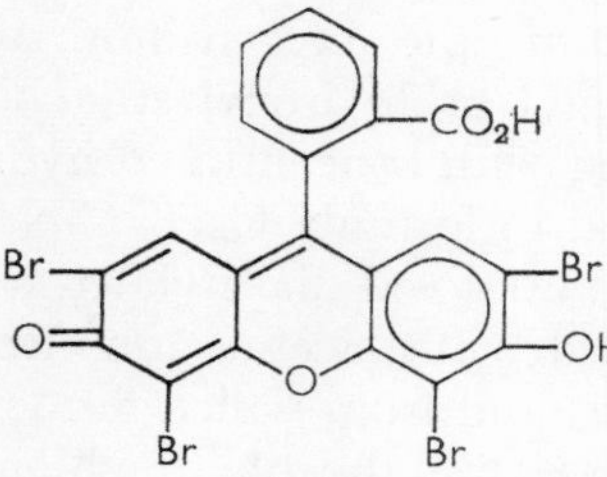

(**74**)

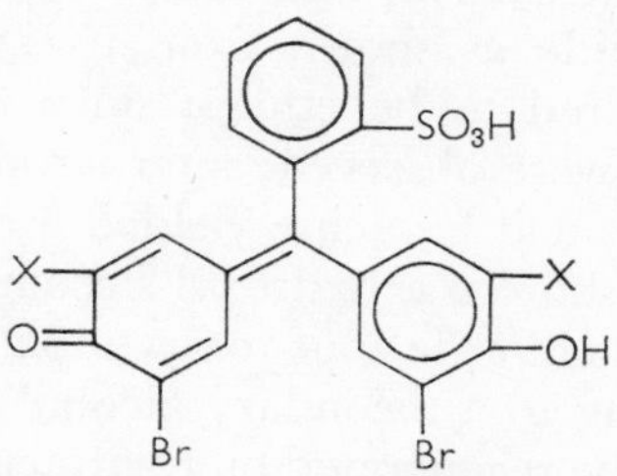

(**75**) X = Br   (**76**) X = i-Pr   (**77**) X = CH₃

IV. ELUCIDATION OF STRUCTURES

The chemist confronted with the problem of assigning a structural formula to a compound has a great variety of methods to aid him in his pursuit. His choice will usually depend on the amount of background information available, the relation between the amount of work to be invested to the amount of information expected, and the adaptability of the method to the particular problem. Thus for example, if he has to find the structure of a crystalline organic solid, and assuming that all the requirements are met, a single method that

5*

could solve the problem completely is x-ray diffraction. This however will probably be his last resource, as the method requires specialised training and can be extremely time consuming and expensive. Usually information is obtained from elementary analysis and physical properties such as melting point, boiling point, pK_a, etc., and from i.r., u.v., n.m.r., and mass spectra. The compound is subjected to chemical reactions and the above proceedings repeated with the derivatives. At progressive stages it is possible to propose more and more detailed structures, until finally the elucidation is complete or no further advancement can be expected. It may then be necessary to resort to crystallographic methods for confirmation or completion of the solution of the problem. To round up the case a total or partial synthesis of the compound may be attempted, assuming that the structure of the starting material, intermediates, and products is unambiguously known.

For example, the elucidation of the structure of festucine (**78**) went through the following steps[466]: elementary analysis and mass spectroscopy showed formula $C_8H_{14}N_2O$ and molecular weight 154; n.m.r. indicated the presence of one N-methyl group, the two nitrogen atoms were shown to be basic by preparing the dihydrochloride, pK_a values 2·5–3·0 and 8·25; the alkaloid underwent monoacetylation and mononitrosation, and the acetyl derivative was an N,N-disubstituted acetamide as shown from its i.r. spectrum; the oxygen atom was considered to be etheric since no carbonyl or hydroxyl stretching bands were observed; after strong heating with concentrated hydrochloric acid festucine yielded a compound of formula $C_8H_{15}ClN_2O$, which showed a hydroxyl stretching band (and was diacetylated with acetyl chloride but only monoacetylated with acetic anhydride, pointing to a secondary alcohol structure) and from which festucine could be regenerated on treatment with base; and the dihydrochloride of festucine did not undergo hydrogenation, pointing to a saturated structure. From its physical and chemical properties, analogous to those of other known bases, it was concluded that festucine was a cyclic ether with pyrrolizidine skeleton, and the definitive structure **78** was assigned after carrying out an x-ray diffraction study[467].

(**78**)

In the following sections an outline will be presented of the solution of some important structural problems in which the properties of the amino group are taken into account. It should be pointed out that the methods mentioned below, especially those based on spectroscopy and crystallography, are capable of yielding much more detailed information than that stemming from the interactions of the amino group with the rest of the molecule alone, but the discussion of such applications lies beyond the scope of the present chapter.

A. Saturated Vicinity of the Nitrogen Atom

It was shown in section II.D.1 that chemical methods for the analysis of groups attached to the nitrogen atom require drastic treatment. On the other hand i.r., n.m.r., and mass spectroscopy afford excellent methods for such studies. In sections II.B.6.b, II.C.11.a, and II.D.2 certain structural features of saturated amines yielding absorption bands in the infrared region were mentioned already, and no further treatment of i.r. spectroscopy will be made here.

I. Nuclear magnetic resonance spectrum

N-Methyl groups show a band at about δ 2·15 p.p.m.; this is shifted to higher fields when —CH_2—, —CHR—, or —CRR^1— groups are interposed between the nitrogen atom and the methyl group, making the chemical shift almost indistinguishable from a hydrocarbon methyl group. In some cases it may be helpful to locate the band of the interposed methylene or methyne group at lower field (see below), and observe the splitting into a quartet for the free amine and a more complicated pattern when the amine is converted into an ammonium salt (see section II.C.1). The methyl bands of aryl-N—CH_3 and acyl-N—CH_3 systems[468] are shifted towards lower fields, which is of great aid in the assignment of structures.

Protons of N-methylene and N-methyne groups give rise to bands at slightly lower fields than those of N-methyl groups, and they too become shifted to lower fields by N-acetylation.

Examples illustrating the chemical shifts of various saturated groups attached to the amino group are given in Table 14. The bands arising from N—H groups were treated in section II.C.11.b. Detailed discussions on this method can be found elsewhere[393, 469–472].

TABLE 14. Nuclear magnetic resonance bands of alkyl groups in amines[a].
(Groups containing the relevant protons are indicated by *).

Structure	Chemical shift[b]
$CH_3*NR^1R^2$ (R^1, R^2 = H, alkyl)	~2·15
$(CH_3)_2*NCH_2C_6H_5$	2·17
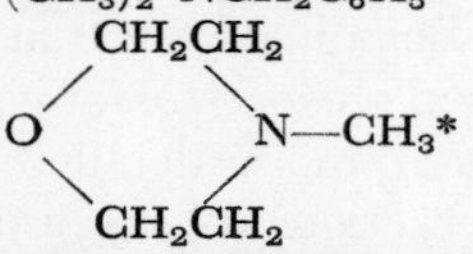 $\begin{array}{c} CH_2CH_2 \\ O \qquad N-CH_3* \\ CH_2CH_2 \end{array}$	2·20
$CH_3*CH_2NR^1R^2$ (R^1, R^2 = H, alkyl)	~0·95
$CH_3*CH_2CH_2CH_2NR^1R^2$	~0·93
$(CH_3)_2*CHNH_2$	1·01
$(CH_3)_3*CNH_2$	1·15
$(CH_3)_2*NC_6H_5$	2·90
$(CH_3)_2*NC_6H_4CHO$-p	3·05
Alkyl—CH_2*NH_2	2·42–2·63
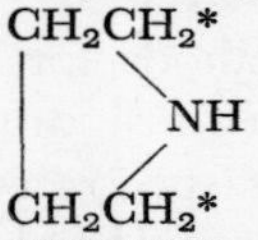 $\begin{array}{c} CH_2CH_2* \\ \quad NH \\ CH_2CH_2* \end{array}$	2·74
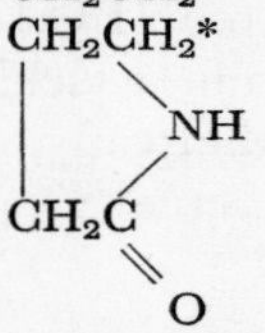 $\begin{array}{c} CH_2CH_2* \\ \quad NH \\ CH_2C \\ \parallel \\ O \end{array}$	3·40
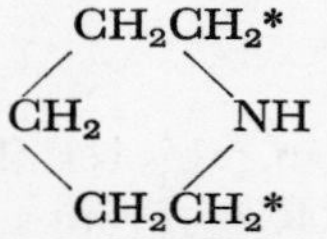 $\begin{array}{c} CH_2CH_2* \\ CH_2 \qquad NH \\ CH_2CH_2* \end{array}$	2·69
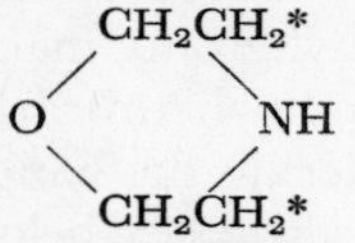 $\begin{array}{c} CH_2CH_2* \\ O \qquad NH \\ CH_2CH_2* \end{array}$	2·83
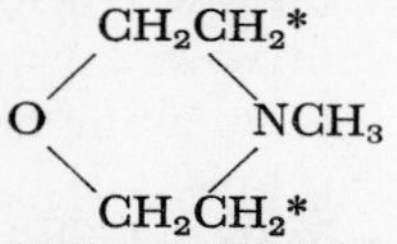 $\begin{array}{c} CH_2CH_2* \\ O \qquad NCH_3 \\ CH_2CH_2* \end{array}$	2·28
$(CH_3)_2CH*NH_2$	2·87
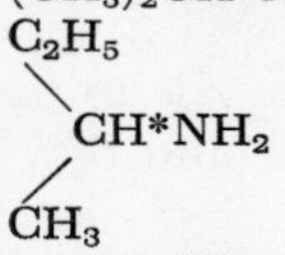 $\begin{array}{c} C_2H_5 \\ \quad CH*NH_2 \\ CH_3 \end{array}$	2·78
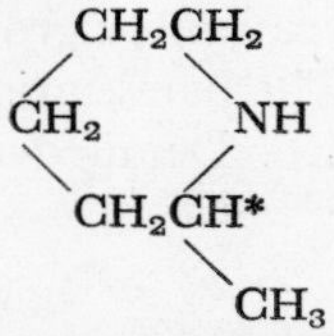 $\begin{array}{c} CH_2CH_2 \\ CH_2 \qquad NH \\ CH_2CH* \\ \qquad CH_3 \end{array}$	2·57 ± 0·10

[a] Data from references 468–470.
[b] In parts per million (p.p.m.) of the δ scale; δ 0·00 for tetramethylsilane.

2. Mass spectrum

Mass spectroscopy provides a powerful means for the elucidation of the structure of alkylamines. The radical-ions derived from these compounds tend to become stabilised by undergoing scission of α,β-carbon–carbon bonds, as in equation (69).

$$[R^3{-}CH_2{-}NR^1R^2]^{\cdot+} \longrightarrow R^{3\cdot} + [CH_2{=}\overset{+}{N}R^1R^2 \longleftrightarrow \overset{+}{C}H_2{-}NR^1R^2] \quad (69)$$

$$(79)$$

$$(R^1, R^2, R^3 = H, \text{alkyl})$$

Cation **79** is capable of undergoing further cleavage if R^1 or R^2 have a hydrogen atom in the α-, β-, γ-, or δ-positions, according to equation (70), where M represents a neutral molecule derived from R^2 by scission and loss of one hydrogen atom[473–476]. The same process of course can take place on cation **80**.

$$CH_2{=}\overset{+}{N}\overset{\displaystyle R^1}{\underset{\displaystyle R^2}{\Big<}} \longrightarrow M + CH_2{=}\overset{+}{N}\overset{\displaystyle R^1}{\underset{\displaystyle H}{\Big<}} \quad (70)$$

$$\textbf{(79)} \qquad\qquad\qquad \textbf{(80)}$$

Cleavage according to equation (69) has been applied in the study of amino acid sequences in oligopeptides, by reducing the peptide to an amino alcohol (**81**) (equation 71) and determining its mass spectrum.

$$\overset{R^1}{\underset{\underset{O}{\|}}{NH_2{-}CH{-}C}}{-}NH{-}\overset{R^2}{\underset{\underset{O}{\|}}{CH{-}C}}{-}NH{-}\overset{R^3}{CH}{-}CO_2H \xrightarrow{\text{LiAlH}_4}$$

$$NH_2{-}\overset{R^1}{CH}{-}CH_2NH{-}\overset{R^2}{CH}{-}CH_2NH{-}\overset{R^3}{CH}{-}CH_2OH \quad (71)$$

(with scission labels d, e, f above and a, b, c below)

$$\textbf{(81)}$$

In **81** scissions at bonds a, b, or c, yield two different ions, as two heteroatoms can be responsible for the scission. If the reduction is also carried out with lithium aluminium deuteride (LiAlD$_4$), the assignment of the peaks in the mass spectrum will be more facile, as a displacement of $+2$ mass units will be observed for every reduced carbonyl group included in the ion, for example in **81** the molecular ion and the ions resulting from scissions at d, e, or f will be displaced by $+6$ mass units, the ion on the left of a scission at b will be displaced by $+2$ units, and that on the right side by $+4$ units[50, 52].

B. Unsaturated Vicinity of the Nitrogen Atom

I. Ultraviolet spectrum

The amino group, when attached to an unsaturated system, shifts the absorption bands in the ultraviolet and visible regions towards longer wavelengths (see Table 15) and enhances their intensity.

TABLE 15. Bathochromic shifts due to an $-NR_2$ (R = H, alkyl) group attached to an unsaturated system[a].

System	Shift (mμ)
$C{=}C{-}NR_2$	40
$C{=}C{-}C{=}C{-}NR_2$	65
$C_6H_5{-}C{=}C{-}NR_2$	58
$O{=}C{-}C{=}C{-}NR_2$	95
$HO_2C{-}C{=}C{-}NR_2$	80
$C_6H_5{-}NR_2$	51 and 43

[a] Data from references 477–479.

The main absorption bands of unsaturated amines are associated with electronic shifts such as those depicted in equations (72) and (73).

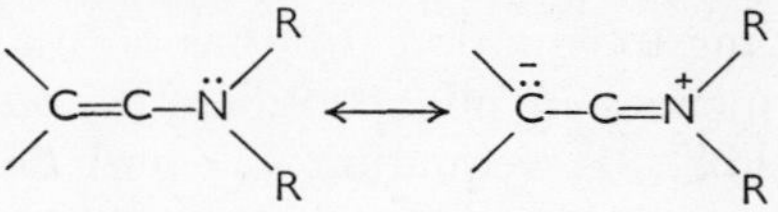

$$\text{(72)}$$

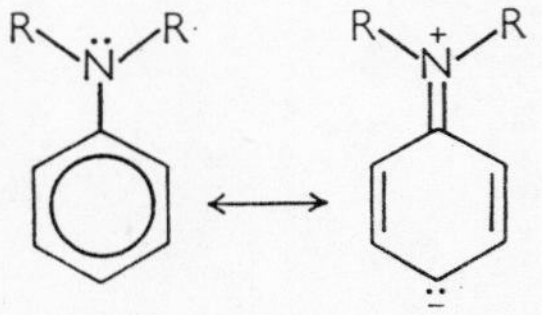

$$\text{(73)}$$

When the p electrons of the amino group are coordinated, as occurs in ammonium compounds, the spectra resemble those of the hydrocarbon skeleton, for example the spectrum of the anilinium ion is similar to that of benzene, and differs much from that of aniline[480, 481], and that of the N,N-dimethylindolium ion (82) is similar to that of styrene and indene (83) but not to that of indole (84) as shown in Table 16[482].

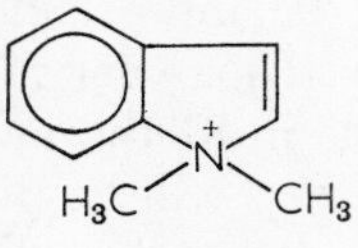

(82) 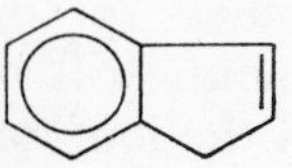   (83) 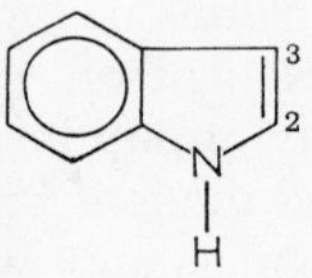   (84)

TABLE 16. Ultraviolet spectra of some amines.

Compound	λ_{max} (mμ)	ε_{max}
Indole (84)[487]	218	27500
	271	6170
	278	6000
	287	4570
2,3-Dimethylindole (cf. 84)[487]	228	30200
	283	7080
	290	6300
Ibogamine (85)[485]	230	33100
	285	7260
Alloyohimbine (86)[486]	225	32900
	290	5540
Ajmalicine (87)[486]	226	44400
	291	6760
N-Methyl-Δ^{8-9}-octahydroisoquinolin-7-one (90)[496]	222	5100
3-(Dimethylaminomethyl)cyclohex-2-en-1-one (91)[496]	225	5950
4-(2-Dimethylaminoethyl)cyclohex-2-en-1-one (92)[496]	224	4150
4-(Dimethylaminomethyl)cyclohex-2-en-1-one (93)[496]	225	6140
Δ^{1-9}-Octahydronaphthalene-2-one (94)[496]	241	10000

The absorption spectrum of a compound can be associated with a partial structure of the molecule (sometimes the whole molecule), and it can be expected that all molecules embodying that part as their only chromophore will have the same spectrum. This principle has been applied in the elucidation of structures of natural products [483, 484]. Thus for example, it was found that the indole alkaloids have two absorption bands, at about 225 mμ, $\varepsilon_{max} \sim 25000$, and at 270–290 m$\mu$, $\varepsilon_{max} \sim 6000$, as in ibogamine (85)[485], alloyohimbine (86)[486], and ajmalicine (87)[486], shown in Table 16. The enhanced absorption of 87 at 226 mμ is due to superimposition of the absorption of the acrylic system also present in the molecule. Some care has to be taken when deciding which is the chromophore responsible for the absorption spectrum of a compound. Thus 2,3-dimethylindole (cf. 84)[487] is a better choice than indole (84)[487] itself, as a model for the spectra of alkaloids 85, 86, and 87, as shown in Table 16.

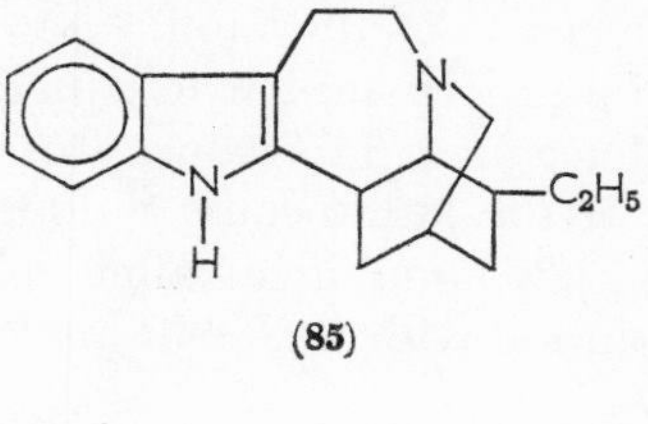

(85)

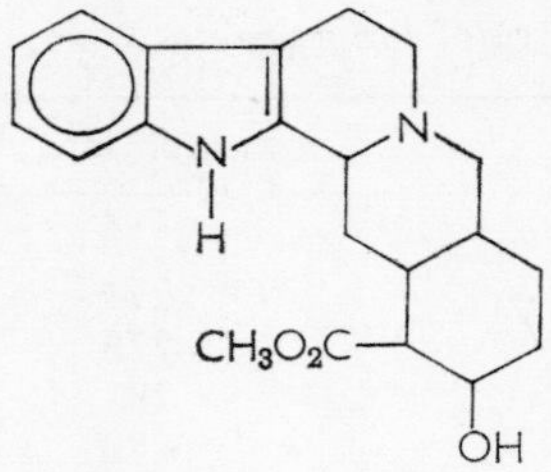

(86) 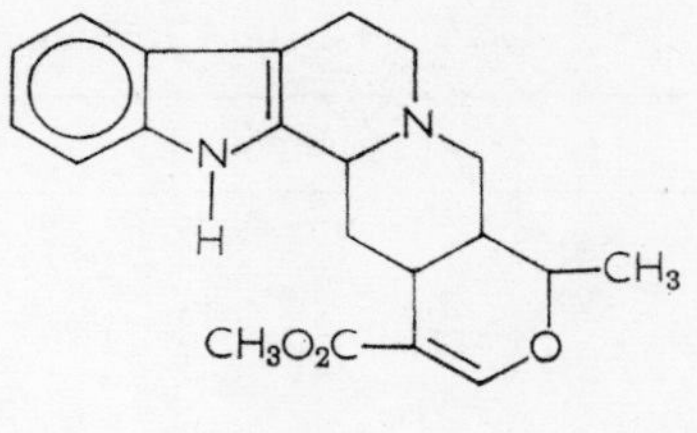   (87)

Amino groups attached to vinyl systems can undergo tautomerisation as in equation (74), or take part in processes as in equation (75).

$$\text{(74)}$$

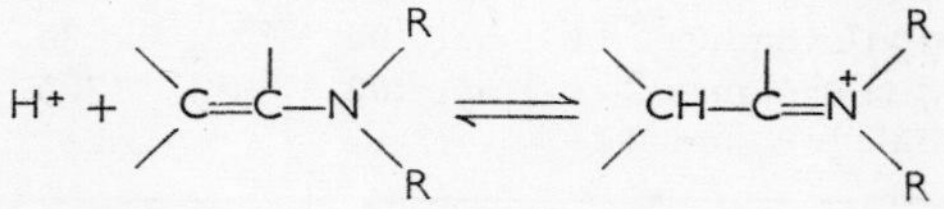

$$\text{(75)}$$

Equations (51) and (76) are examples of tautomerisation equilibria. In the former the *N*-protonated form is predominant, while in the latter it is the *C*-protonated form[488].

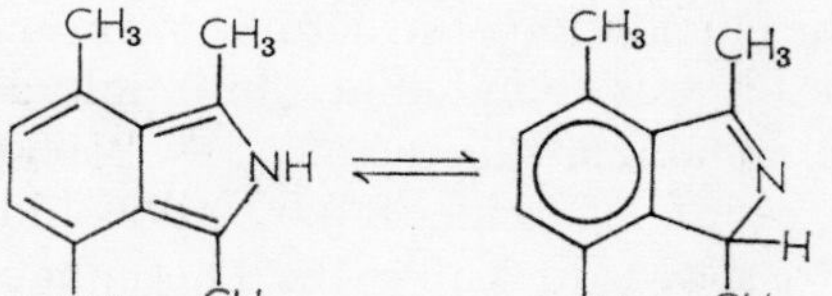

$$\text{(76)}$$

Reaction (74) is the main reason for the instability of vinylamines, as it facilitates solvolysis in the azomethine form. It has been studied by following the changes in the spectrum of the system[489, 490].

Usually an unsaturated amine will change its absorption spectrum on passing from a basic or neutral solution to an acid solution. This is due either to the conversion of the amine into an ammonium salt or to the occurrence of a process similar to equation (75). The former cause allows one to determine pK_a values from spectrophotometric data[479]. An example of the latter cause is found among retinenes: compound **88** has λ_{max} 365 mμ in alkaline solution, and acquires structure **89** in acid solution with λ_{max} 440 mμ[491, 492].

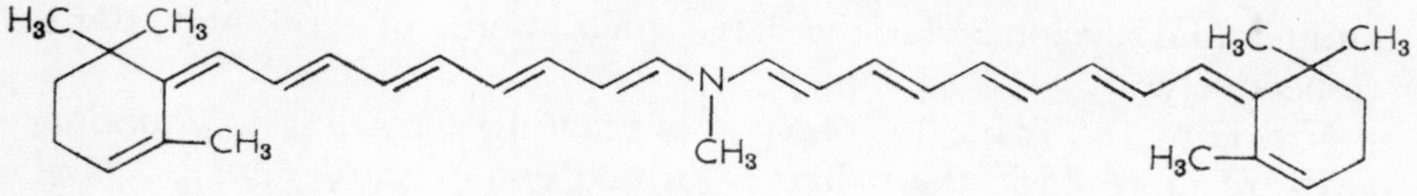

(88)

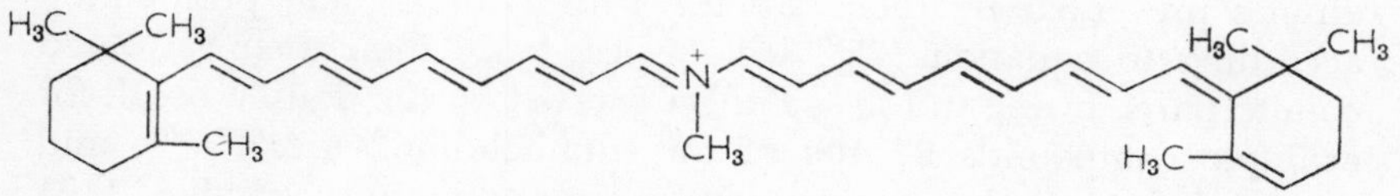

(89)

It should be pointed out that the process suggested by equation (75) has been recently subjected to criticism and revision[490, 493–495].

The p electrons of amines can also influence the spectrum of a distant chromophore. In Table 16 the absorption bands of some compounds (90–93) are shown in which 'long range conjugation' (equation 77) possibly takes place, resulting in a displacement towards lower wavelength and smaller ε_{max} values as compared with the model compound 94[496].

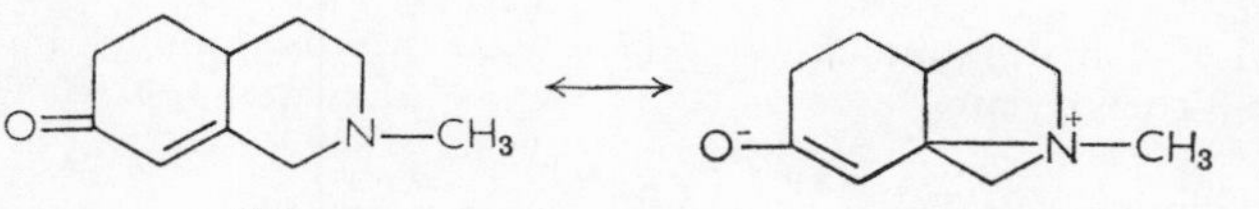

(90)   (77)

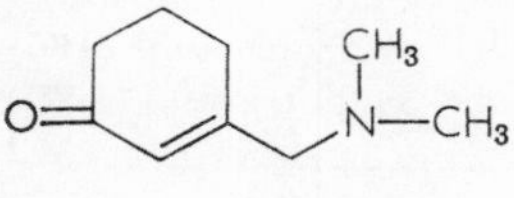

(91) 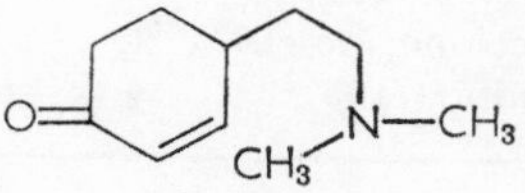   (92)

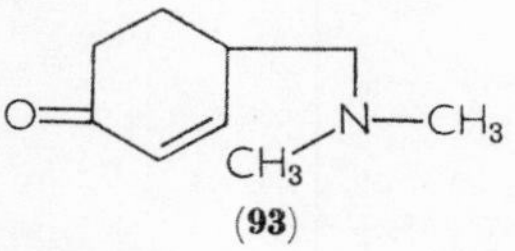

(93) 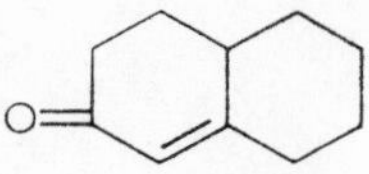   (94)

Fuller discussions on the applications of u.v. spectroscopy in structural elucidation can be found elsewhere[168, 479, 483, 484].

2. Basicity of unsaturated amines

Aromatic amines are less basic than aliphatic ones owing to the delocalisation of the p electrons in the resonance hybrid (e.g. equation 73) and to the higher $-I$ effect of unsaturated structures, for example aniline and cyclohexylamine have pK_a values of 4·58 and 10·64 respectively[497].

Vinylamines (Table 17) capable of reacting according to equation (75) are more basic than their saturated counterparts[498], e.g. compound **95** as compared with **96**[499]. On the other hand, unsaturated amines prevented by their structures from undergoing protonation according to equation (75), will be less basic than their saturated counterparts, owing to the $-I$ effect exerted by the double bond, for example compounds **97–100** of the quinuclidine series[500, 501], and neostrychnine (**101**) as compared with dihydrostrychnine (**101** without the isolated double bond)[502].

TABLE 17.   Basicity of unsaturated amines and their saturated counterparts.

Unsaturated amines	←pK_a	pK_a→	Saturated amines
N,2-Dimethyl-Δ^2-pyrroline (**95**)[499]	11·94	10·26	N,2-Dimethylpyrrolidine (**96**)[499]
Δ^2-Dehydroquinuclidine (**97**)[500, 501]	9·82	10·95	Quinuclidine (**98**)[500, 501]
Methyl Δ^2-dehydroquinuclidine-3-carboxylate (**99**)[500, 501]	7·17	9·40	Methyl quinuclidine-3-carboxylate (**100**)[500, 501]
Neostrychnine (**101**)[502]	3·8 ⎫		
Strychnine (**108**)[502]	7·4 ⎬	7·45	Dihydrostrychnine (cf. **101**)[502]
Retronecine (**102**)[503]	8·94	10·22	Platinecine (**103**)[503]
Deoxyretronecine (**104**)[503]	9·55	10·91	Retronecanol (**105**)[503]
Heliotridine (**106**)[503]	10·60	11·48	Heliotridane (**107**)[503]

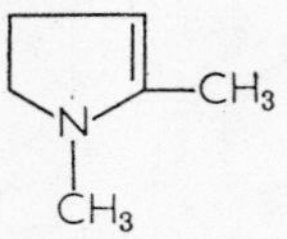

(95) 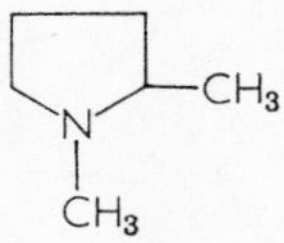   (96) 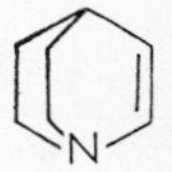   (97) 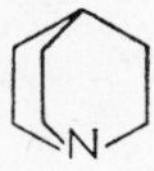   (98)

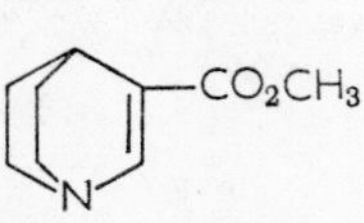

(99) 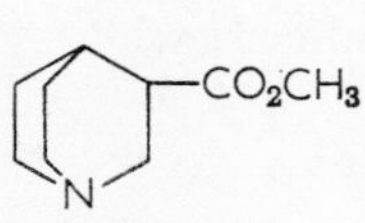   (100)

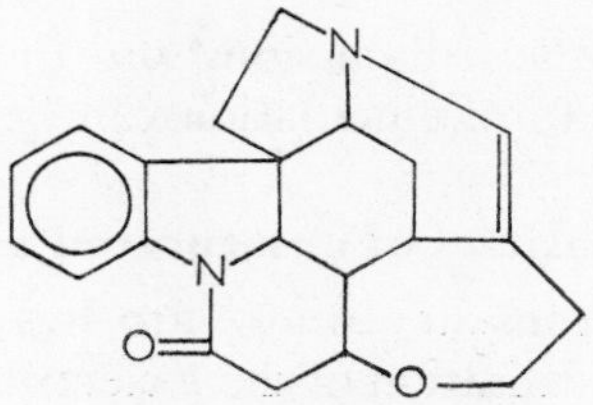

(101)

The comparison between pK_a values of unsaturated and saturated amines may be of aid in placing the unsaturation. Thus the senecio alkaloids **102**, **104**, and **106** were correctly assumed to be allylamines because their pK_a values are lower than those of the saturated alkaloids **103**, **105**, and **107** (Table 17). The vinylamine structure was discarded in these, as basicities higher than those of the corresponding saturated compounds would be expected[503]. The opposite should be true in the case of neostrychnine (**101**), a vinylamine unable to react according to equation (75), and therefore less basic than strychnine (**108**), an allyl-amine in which the $-I$ effect of the unsaturation on the amino group is weakened by distance (Table 17)[502].

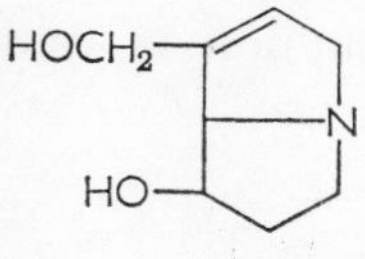

(102) 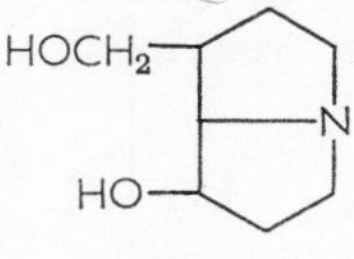   (103) 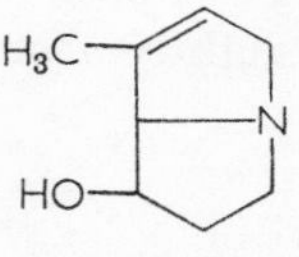   (104) 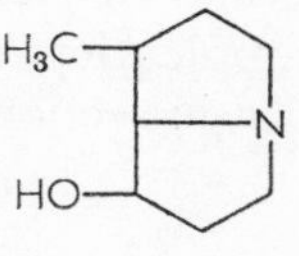   (105)

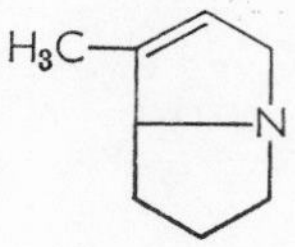

(106) 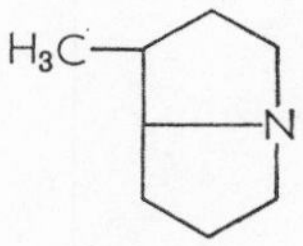   (107)

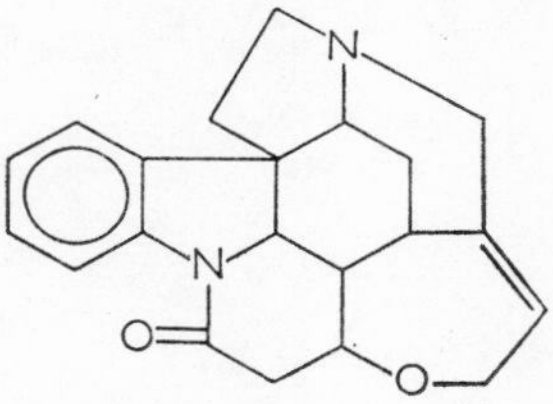

(108)

C. Stereochemical Aspects of Amines

I. Asymmetric nitrogen atom

A nitrogen atom can become a centre of asymmetry in compounds such as ammonium salts. The resolution into two optical isomers of benzylmethylphenylpropylammonium salts[504], and other similar compounds[505,506], was accomplished at the turn of the century, by precipitation of diasteroisomeric salts. A study on aziridinium salts was carried out recently[507].

The nitrogen atom of amines is an asymmetric centre, at least in theory, if three different groups are attached to it, owing to the non-planarity of the group. The fact that such amines cannot be resolved, is attributed to the mobility of the p electrons at the heteroatom[508,509]. More recently it was shown that in certain cases, such as the aziridines

109 and **110**, the inversion at the nitrogen atom is sufficiently slow near room temperature to show two distinct *C*-methyl bands in the n.m.r. spectra, which should coincide were that inversion much faster[510, 511], as is the case with larger heterocycles[510]. This however is still far from actual resolution into optical isomers. In fact, the asymmetric nitrogen has to be held in a rigid structure in order to avoid inversion. In such cases the asymmetric nitrogen is usually accompanied by asymmetric centres at carbon atoms, for example compounds **101–108** and **120–127**. To the author's knowledge, amines with asymmetric centres exclusively at nitrogen atoms have not yet been resolved, although racemic mixtures of compounds such as **111** are known[512]. Tröger's base (**112**) was actually resolved[513], but this example departs slightly from the restricted definition of amines given in section I.

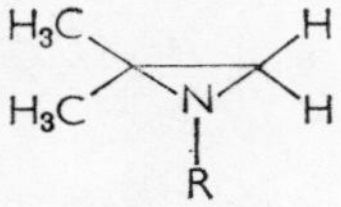

(109) 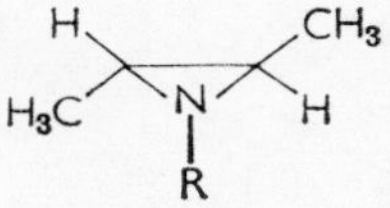   (110) 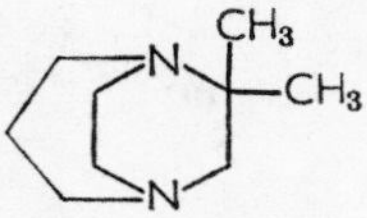   (111)

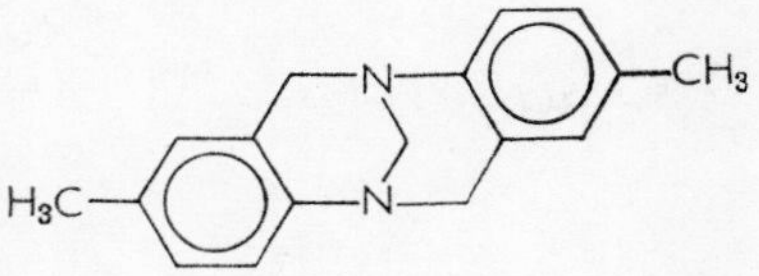

(112)

2. α-Amino acids

Among amines where the amino group is attached to an asymmetric centre α-amino acids and peptides are by far the most widely investigated class from the point of view of stereochemistry. The main results will be presented here, and more detailed discussions can be found elsewhere[514–517]. It was found that most α-amino acids derived from peptidic and proteinic materials belong to the L series, that is they have absolute configuration **113**, related to L(+)-lactic acid (**114**), when **113** and **114** are drawn according to the Fischer–Klyne convention[518]. Evidence for these assignments came from optical rotation studies in series of derivatives of hydroxy acids and amino acids[519, 520], α-azido- and α-halogenopropionic acid derivatives[521], which can be

correlated with the derivatives of lactic acid and alanine[522,523], and
D-2-glucosaminic acid as compared with D-amino acids[524-529].

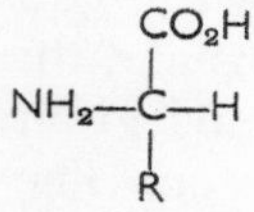

$$
\begin{array}{c}
CO_2H \\
| \\
NH_2—C—H \\
| \\
R \\
\mathbf{(113)}
\end{array}
\qquad\qquad
\begin{array}{c}
CO_2H \\
| \\
HO—C—H \\
| \\
R \\
\mathbf{(114)}
\end{array}
$$

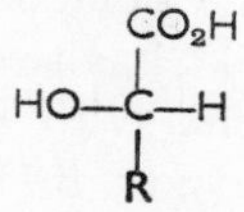

It was shown that the optical rotation changes observed on treating
neutral natural amino acids with acid or by converting them into
hydantoins (**115**), were of the same sign and about the same magnitude
for almost all the cases studied. This was interpreted as being due to all
these compounds having a unique configuration[524-531]. Assignments
of configuration based exclusively on optical rotatory evidence have
led in some cases to the wrong conclusions[532]. It is fortunate therefore
that the configurations proposed above were confirmed by several
independent methods, including x-ray diffraction[515].

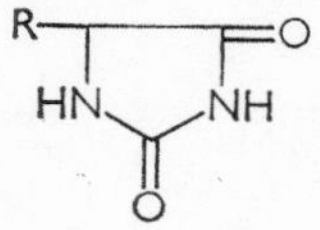

$$
\begin{array}{c}
R—\!\!\!\boxed{}\!\!\!=O \\
HN\diagdown\ \diagup NH \\
\| \\
O \\
\mathbf{(115)}
\end{array}
$$

More recently, widespread use has been made of optical rotatory
dispersion as a means for studying configurations. α-Amino acids were
investigated as such[533-535] or as derivatives, notably alkyl dithio-
carbamates (**116**)[536,537], copper(II) complexes[538,539], and cobalt(III)
complexes[540]. From these studies useful though not always straightforward correlations have been obtained, between the shape and
location of the dispersion curves and the configuration of the compounds.

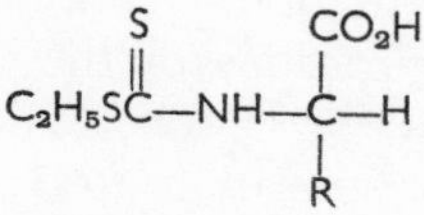

$$
\begin{array}{c}
S \qquad\quad CO_2H \\
\| \qquad\qquad | \\
C_2H_5SC—NH—C—H \\
| \\
R \\
\mathbf{(116)}
\end{array}
$$

Rotatory-dispersion curves of polypeptides ideally depend almost
exclusively on the conformation of the peptide-links backbone, and
not on the amino acid composition[541], although in special cases the
contribution of individual amino acids may be important[542-544].

3. Examples of configuration assignment

Certain reactions take place under strict steric requirements as is the case of the pinacol-type rearrangement shown in equation (78). Compound **117** has two asymmetric carbon atoms and therefore four diasteroisomers. With racemate α (**118**) the phenyl group migrates, and with racemate β (**119**), the p-chlorophenyl group. The conformations shown in **118** and **119** are the less strained, where the smallest group H is staggered between the two bulky aromatic substituents. The relative positions of the groups depicted in **118** and **119** are preserved in the transition state, thus determining the nature of the rearranged product, and from the latter the geometry of the amino alcohol can be deduced[545–547]. The same could be found from kinetic data, as the migration aptitude of a phenyl group is larger than that of a p-chlorophenyl group[548]. Similar studies were carried out also with diasteroisomeric 2-amino-1,2-diphenylethanol[549].

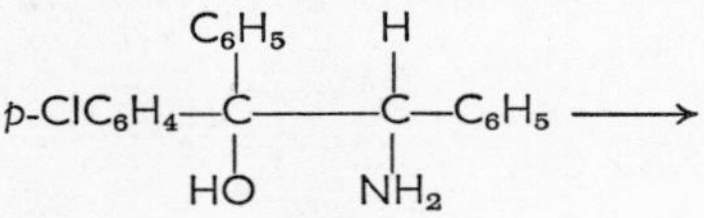

$$p\text{-ClC}_6\text{H}_4\text{—C—C—C}_6\text{H}_5 \quad\text{or}\quad \text{C}_6\text{H}_5\text{—C—C—C}_6\text{H}_5 \qquad (78)$$

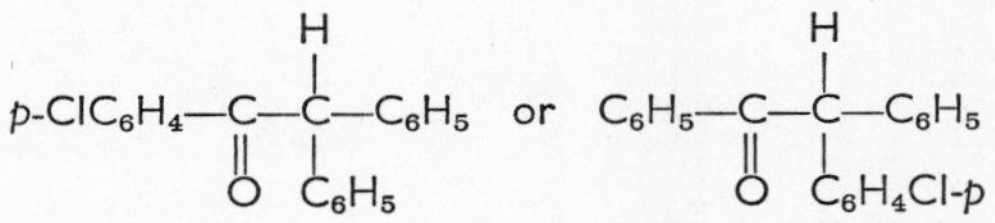

(**117**)

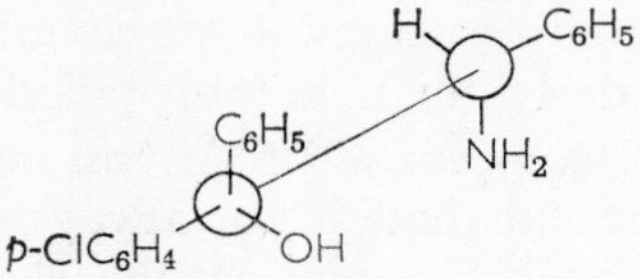

(**118**) 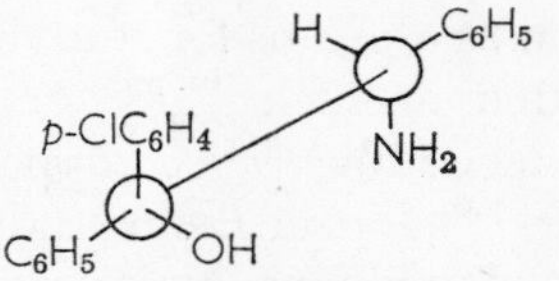   (**119**)

Evidence gathered from chemical reactions, optical rotation, and basicity, as well as conformational considerations were applied in order to assign the configurations of the two families of cinchona alkaloids listed in Table 18[550–552]. The main steps of the elucidation were the following: the configuration at $C_{(3)}$ and $C_{(4)}$ (cf. **120**) is the same for the eight alkaloids; the decreasing order of optical rotations in the two series was taken as a guide-line for proposing parallel configurations at $C_{(8)}$ and $C_{(9)}$ (cf. **120**); the highest positive optical rotation and the negative one in each series were considered to have opposite configurations at both $C_{(8)}$ and $C_{(9)}$; quinidine and cinchonine were known to yield a cyclic ether by reaction of the hydroxy

TABLE 18. Physical properties of cinchona alkaloids[a].

Compound	$[\alpha]_D$	pK_a
Quinidine (**120**)	$+254°$	7·95
*Epi*quinidine (**122**)	$+102°$	8·32
*Epi*quinine (**124**)	$+43°$	8·44
Quinine (**126**)	$-158°$	7·73
Cinchonine (**121**)	$+224°$	
*Epi*cinchonine (**123**)	$+120°$	
*Epi*cinchonidine (**125**)	$+63°$	
Cinchonidine (**127**)	$-111°$	

[a] Data on optical rotation from reference 551, and on basicity from references 550 and 552.

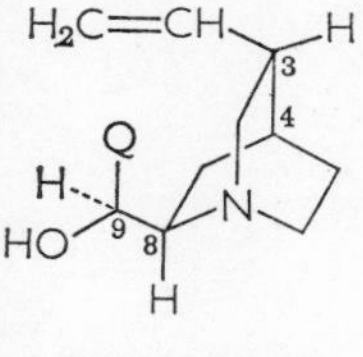

(**120**) Q = **128**
(**121**) Q = **129**

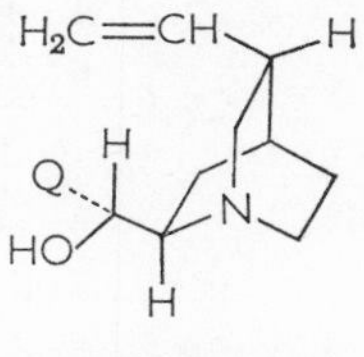

(**122**) Q = **128**
(**123**) Q = **129**

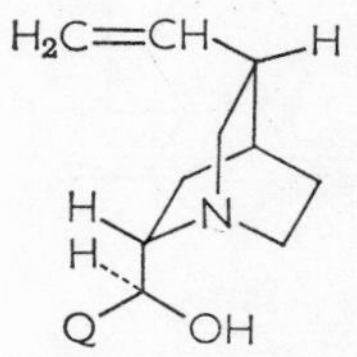

(**124**) Q = **128**
(**125**) Q = **129**

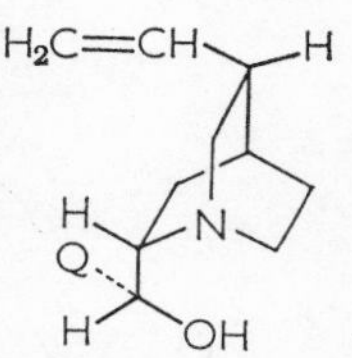

(**126**) Q = **128**
(**127**) Q = **129**

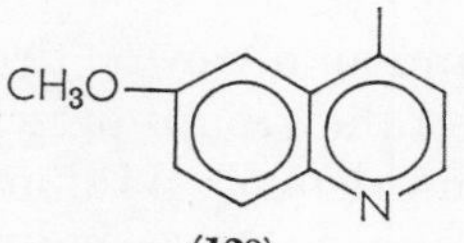

(**128**) 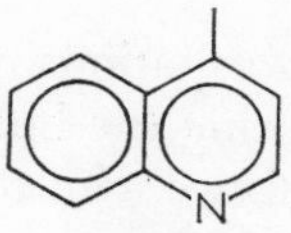   (**129**) 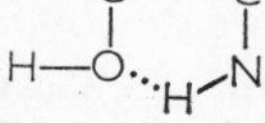   (**130**)

and vinyl groups; an alkaloid could be converted into the *epi* form and vice versa by oxidation of the alcohol followed by reduction; the fact that in a pair, alkaloid-*epi*alkaloid, the latter is the more basic was attributed to the stabilisation of the ammonium cation by hydrogen bonding, as the conformation attained on forming the hydrogen-bonded structure **130** is sterically less hindered in the *epi* forms.

The configurations of four 10-hydroxydihydrodesoxycodeines (**131**) appear in Table 19. The assignments were made by taking into account the stabilisation of the ammonium ion by hydrogen bonding[553], by acquiring a structure such as **130**. Lysergic acid (pK_a 7·8) and isolysergic acid (pK_a 8·4) are two isomers of structure **132**. They were respectively assigned configurations where the carboxyl group could not and could establish hydrogen bonding with the ammonium cation on forming the salt[554–556].

TABLE 19. Configuration of
10-hydroxydihydrodesoxycodeines (131)[a].

R	pK_a	Assignment[b]
H	8·72	*trans*
H	9·17	*cis*
CH$_3$	7·71	*trans*
CH$_3$	9·41	*cis*

[a] Data from reference 553.
[b] C$_{(10)}$—OH relative to C$_{(9)}$—NR.

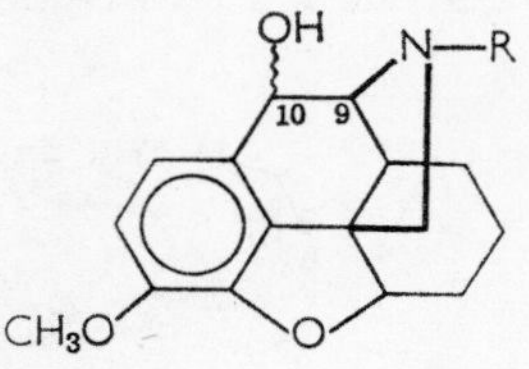

(131) 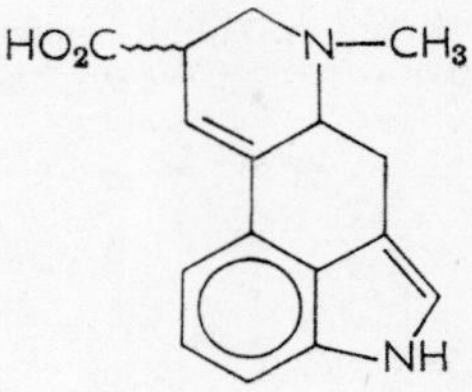   (132)

4. Conformations, steric hindrance, and other stereochemical problems

The preferential conformations of a compound can be deduced in certain cases by considering the bulkiness of the groups present in the molecule, as was done above with the amino alcohols (118) and (119), and the alkaloids (120–127). In other instances more indirect considerations have to be made. For the needs of the present section, the changes in configuration resulting from inversion of p electrons at nitrogen atoms will be regarded as changes in conformation.

The equilibrium of conformers of *trans-N*,2-dimethyl-3-isopropyl-aziridine, shown in equation (79), was found by n.m.r. spectroscopy to favour the left hand side by a ratio of about 4 to 1, at temperatures from $-55\cdot5$ to $9°$. The method could not be applied at somewhat higher temperatures[557].

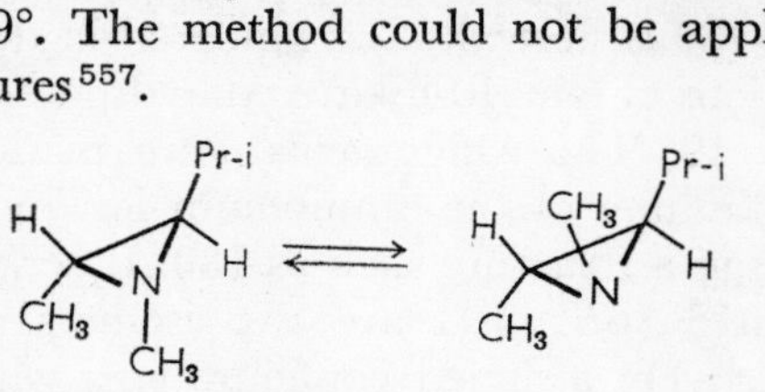

(79)

N.m.r. was also used to show that at low temperatures there is considerable freezing of ring conformation in *N*-methylmorpholine

(**133**) and N,N'-dimethylpiperazine (**134**) as compared with morpholine and cyclohexane[558].

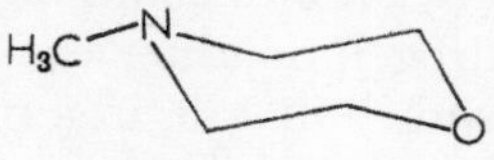

(**133**) 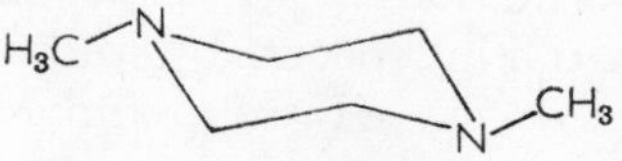   (**134**)

From optical rotatory dispersion studies, 3-benzoyl-3-chloro-N-methylpiperidine was shown to have conformation **135**, with the carbonyl group pointing away from the amino function, in neutral solution, and conformation **136**, with the carbonyl and ammonium moieties pointing towards each other, in acid solution[559].

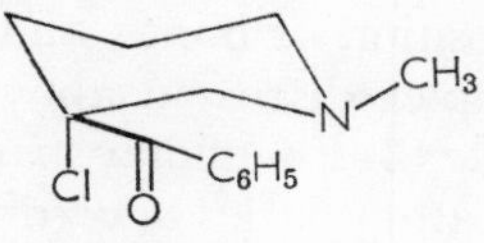

(**135**) 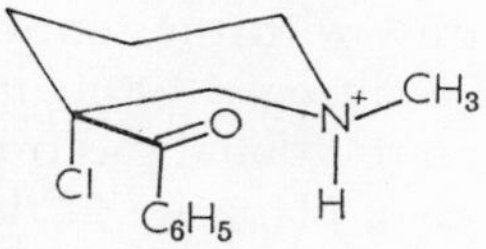   (**136**)

The preferential conformation of a substituent at the nitrogen atom of piperidine seems to depend on its size and the state of solvation of the p electrons[560]. Thus it was found that the steric requirements of the unsolvated pair are less than those of a proton or a methyl group[561, 562], for example in equation (80) equilibrium favours form **137** over **138** at room temperature in benzene solution ($\sim 88\%$ for R = H, and $\sim 94\%$ for R = CH$_3$), as was shown by dipole-moment measurements[561]. On the other hand, the solvated electron pair has larger steric requirements than an N-hydrogen, as shown for example from the kinetics of equilibration of epimers of methyl N-methyl-decahydroquinoline-4-carboxylate (**139**)[563]. Results from the n.m.r. spectroscopy of piperidine and N-methylpiperidine in methanol (a

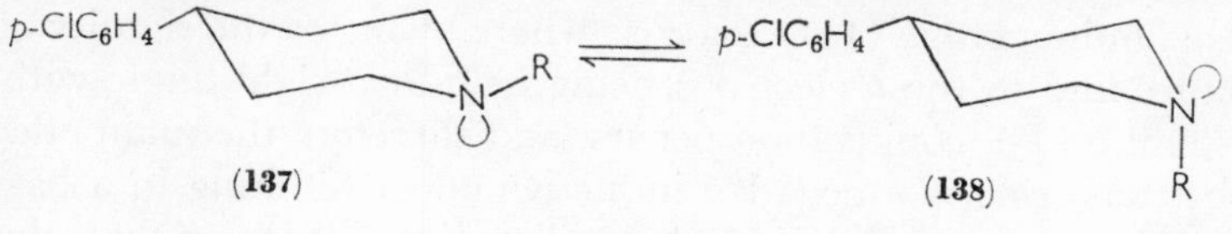

(**137**)   (**138**)   (80)

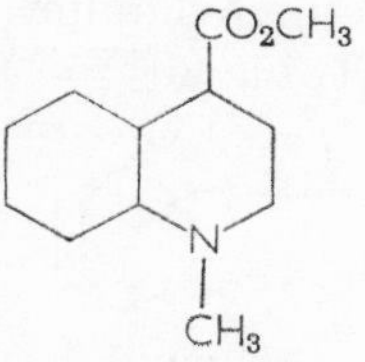

(**139**)

strongly solvating solvent), also support these hypotheses on the steric requirements of the p electrons of amines[564].

Conformations of o,o'-disubstituted diphenylamines could be determined from i.r. spectra by studying the hydrogen bonding established between the N—H group and X or Y, as shown in **140**, in a manner similar to the one applied for o,o'-disubstituted diazoaminobenzenes[565].

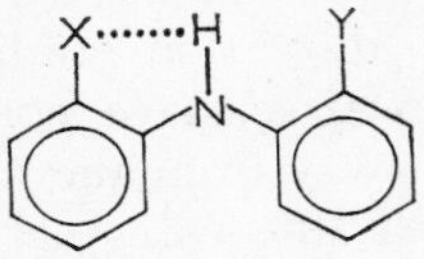

(140)

In contrast with the tetrahedral disposition of bonds and electron pair at the nitrogen atom in the examples discussed above, it was shown from rotational spectra that pyrrole (**3**) is a planar structure[566]. The degree of planarity at the nitrogen atom varies according to the substituents attached to it. From its rotational spectrum aniline was shown to be non-planar at the nitrogen atom[567], nevertheless arylamines are more planar at that site than alkylamines, owing to the contributions of planar forms to the resonance hybrid (e.g. equation 73). The degree of planarity at the nitrogen atom can be changed by steric hindrance, as can be deduced from the u.v. spectra of substituted N,N-dimethylanilines[568, 569], their basicity[497], and their dipole moments[570]. Thus methyl groups in the *ortho* positions will hinder the dimethylamino group from attaining coplanarity with the ring, resulting in u.v. spectra that are intermediate between that of dimethylaniline and that of benzene. The p electrons by participating less in p-π conjugation, will tend to adopt the tetrahedral geometry of ammonia, resulting in their increased availability (basicity) and a lowered dipole moment.

Benzoquinuclidine (**141**) is a case where owing to the rigidity of the structure and to the p electrons being held at a right angle with the π system, no p-π conjugation occurs, and therefore the main effect of the benzene ring is a negative inductive effect resulting in a basicity higher than that of N,N-dimethylaniline but still lower than that of quinuclidine, as shown in Table 20[571,572].

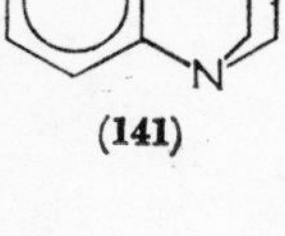

(141)

TABLE 20. Basicity of some amines.

Compound	pK_a
N,N-Dimethylaniline [497]	5·06
Benzoquinuclidine (141) [571, 572]	7·79
Quinuclidine (98) [571, 572]	10·65
Ammonia [576]	9·25
Methylamine [575]	10·63
Dimethylamine [575]	10·78
Trimethylamine [575]	9·80

Steric hindrance to coplanarity, may also shift u.v. absorption bands to longer wavelengths (Brunings–Corwin effect) [573], for example in dyes such as 142 (λ_{max} 446 mμ) as compared with 143 (λ_{max} 479 mμ). It was suggested that the resonance energy was more affected by steric hindrance in the ground state than in the electronically excited state, in compounds of this type, yielding E_1–E_0 values for the transition which are smaller, and therefore corresponding to a longer λ_{max}, in the case of 143 [574].

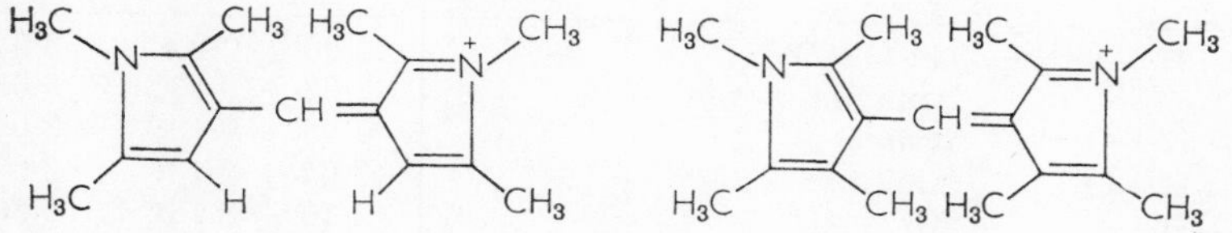

(142)   (143)

The basicity of an amine can be considered as the result of electronic and steric effects acting on the functional nitrogen atom. The increasing pK_a values observed on progressively methylating ammonia stop at dimethylamine and drop for trimethylamine [575, 576] (Table 20). The increase in pK_a parallels the increasing inductive effect by methyl groups, and the decrease observed for trimethylamine was attributed to steric effects [577-579]. More recent evidence from hydrogen bonding of ammonia and the three methylated amines with methanol in the gas phase, shows strict inductive order for the series [179].

Steric hindrance by substituents at the nitrogen atom of amines has been determined by preparing certain Lewis salts and measuring their dissociation constants and heats of formation. Quinuclidine (98) was found to yield the most stable adduct with trimethylboron [580], while the adduct in the case of triethylamine is unstable [581]. These results show that the approach to the nitrogen atom is hindered to a certain extent by the ethyl groups, which are free to rotate and flip around the nitrogen atom, while in quinuclidine these movements are absent, leaving free the approach to the p electrons[580].

An important consequence of the formation of a hydrogen bond A—H$\cdots$B is the shortening of the interatomic distance A–B, as compared with the sum of the van der Waals radii of A and B. This has to be considered when studying the stereochemistry of amines and ammonium compounds, especially in the solid phase, as very significant distortions and steric effects may result. Table 21 summarises hydrogen bonds involving nitrogen atoms, and their lengths are compared with the van der Waals distances. The contraction taking place on hydrogen-bond formation is the more noteworthy as the van der Waals radius of hydrogen is not taken into account for this comparison.

TABLE 21. Hydrogen-bond lengths and van der Waals distances in crystalline compounds[a].

A—H$\cdots$B	A—H type	A–B distance (Å)	van der Waals A–B distance (Å)[b]
O—H$\cdots$N	All O—H	$2 \cdot 80 \pm 0 \cdot 09^c$	3·05
N$^+$—H$\cdots$O	Ammonium	$2 \cdot 88 \pm 0 \cdot 13^d$	3·05
N—H$\cdots$O	Amide	$2 \cdot 93 \pm 0 \cdot 10^e$	3·05
N—H$\cdots$O	Amine	$3 \cdot 04 \pm 0 \cdot 13^f$	3·05
N—H$\cdots$N	All N—H	$3 \cdot 10 \pm 0 \cdot 13^g$	3·30
N$^+$—H$\cdots$F$^-$	NH$_4$F	2·69	3·00
N—H$\cdots$F	NH$_3$.BF$_3$	$3 \cdot 01 \pm 0 \cdot 3$	3·00
N$^+$—H$\cdots$Cl$^-$	NH$_2$(CH$_2$)$_6$NH$_2$.2 HCl	3·01, 3·07	3·45
N—H$\cdots$Cl$^-$	Adenine (**144**).HCl.$\frac{1}{2}$ H$_2$O	3·11, 3·21	3·45
N—H$\cdots$Cl	4,5-Diamino-2-chloropyrimidine (**145**)	3·51, 3·52	3·45
N$^+$—H$\cdots$Br$^-$	NH$_2$(CH$_2$)$_{10}$CO$_2$H.HBr.$\frac{1}{2}$ H$_2$O	3·30, 3·44	3·60
N—H$\cdots$I$^-$	Cleavamine methiodide (**146**)[582]	3·4	3·80

[a] From G. C. Pimentel and A. L. McClellan, *The Hydrogen Bond*, W. H. Freeman and Company, San Francisco, 1960.

[b] Estimated from 1·65 Å proposed by Pimentel and McClellan for the van der Waals radius of nitrogen[100], and those given by Pauling for the other atoms[99].

[c] Average of 21 data.

[d] Average of 41 data.

[e] Average of 35 data.

[f] Average of 30 data.

[g] Average of 37 data.

In spite of the A–B contraction, the A—H bond undergoes elongation in these cases. The location of the hydrogen atom participating in a hydrogen bond has been accomplished by several methods, for

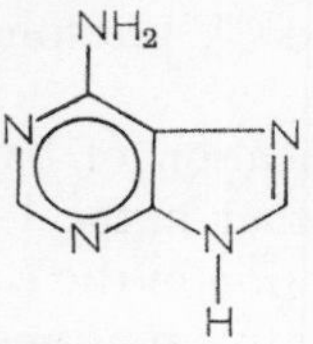

(144)

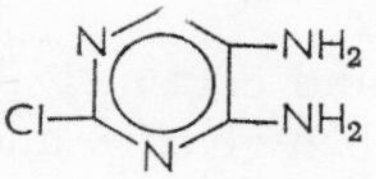

(145)

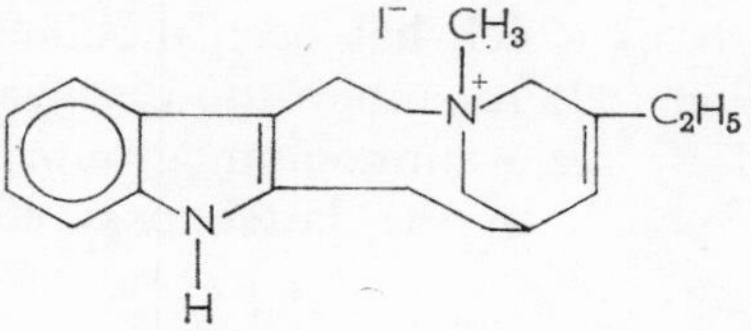

(146)

example x-ray diffraction which although rather ineffective for this purpose owing to the low scattering amplitudes of hydrogen atoms, yielded in some instances reliable results, e.g. for compound 145[583]. Electron diffraction was also used for the same purpose[584,585]. The neutron scattering amplitude of hydrogen is comparable with that of carbon or oxygen, allowing therefore the localisation of hydrogen atoms by neutron diffraction. Furthermore, by this method hydrogen and deuterium atoms can be distinguished, thus adding to its usefulness[100]. Finally, N—H distances have been calculated for amines, ammonium compounds, and other functions, based on the band widths measured by n.m.r. spectroscopy[586-593].

D. Multifunctional Amines

The presence of additional functional groups in an amine can of course be detected by spectroscopic and chemical methods specially devised for these groups. Functional groups affect strongly the properties of the amino group if they are near the nitrogen atom, but their influence decreases steeply with distance. For example the pK_a values of the senecio alkaloids 102–107 in Table 18 illustrate the effect of the electronegativity of hydroxyl groups, as these lower the pK_a of the compound. Extensive studies have been carried out to establish the influence of substituents on the basicity of amines of structure $X—(CH_2)_n—NH_2$. It was shown that the pK_a value is strongly affected for functional groups X, compared with X = H, when $n = 0, 1$, only slightly for n = 2, and remains practically unaffected for n ≥ 3. Some of the groups studied were hydrogen[594-596], carboxylic acid and ester[597], sulphonic acid[598], phosphonic acid[599], phenyl[600], etc.

An early and extremely important application of basicity studies was made when glycine, and by analogy other natural amino acids, was established as a zwitterion. The dissociation of the conjugate acid of this compound could take place according to equations (81) or (82). The former is similar to the dissociation of the conjugate acid of methylamine (pK_a 10·6) which has been modified by an α-carboxy group, while equation (82) resembles the dissociation of acetic acid (pK_a 4·8) modified by an α-ammonium group. The pK_a value of glycine being 2·3, points to the latter possibility as the correct one[104, 105].

$$NH_3{}^+CH_2CO_2H \rightleftharpoons NH_2CH_2CO_2H + H^+ \qquad (81)$$

$$NH_3{}^+CH_2CO_2H \rightleftharpoons NH_3{}^+CH_2CO_2{}^- + H^+ \qquad (82)$$

The effect on the basicity of aniline (pK_a 4·58[497]) has been studied for a wide list of substituents in the *ortho*, *meta*, and *para* positions. It was found that the pK_a value of a substituted aniline can be estimated from that of aniline by the use of additive terms depending only on the nature and position of the substituent[552]. Thus the calculated and experimental values of polysubstituted anilines showed good agreement for a wide range of substituents and substitution patterns[601–604], thus providing a method for structural assignment based on a fundamental property of the amino group. For example, on bromination of 6-acetylamino-1,2,3,4-tetrahydronaphthalene followed by hydrolysis of the acetyl group, two monobromo compounds with pK_a 3·05 were obtained. They were assigned structures **147** and **148**, while structure **149** was discarded, as the calculated values, based on aniline and assuming that the tetramethylene chain acts as two methyl groups, are 3·18 for **147** and **148**, and 4·09 for **149**[601, 602]. Fuller discussions of substituent effects in amines can be found elsewhere[552, 605].

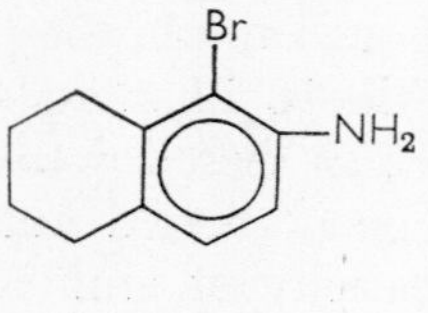

(147) 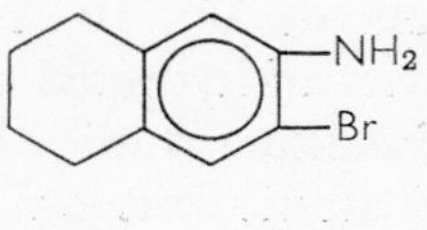   (148) 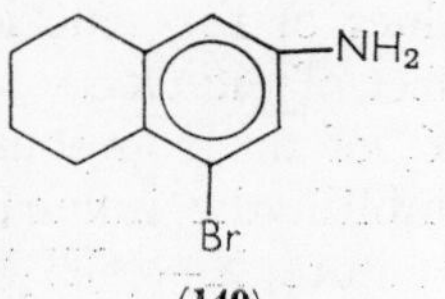   (149)

Correlations between chemical shifts of *N*-protons in substituted anilines[606–608], and methyl protons of substituted *N*,*N*-dimethylanilines[609] have been found, which may be of aid in the assignment of substitution sites of similar compounds.

V. DEDICATION AND ACKNOWLEDGMENT

The author wishes to dedicate this chapter to Dr. José Herrán in recognition of many years of guidance, and to acknowledge the invaluable assistance of his wife in the preparation of the manuscript.

VI. REFERENCES

1. J. Mitchell, Jr., I. M. Kolthoff, E. S. Proskauer, and A. Weissberger, (Eds.), *Organic Analysis*, Vols. I–IV, Interscience Publishers, New York, 1953–1960.
2. E. Müller (Ed.), *Methoden der Organischen Chemie (Houben-Weil)*, Vol. II, Georg Thieme Verlag, Stuttgart, 1953.
3. F. Wild, *Estimation of Organic Compounds*, Cambridge University Press, Cambridge, 1953.
4. F. D. Snell and C. T. Snell, *Colorimetric Methods of Analysis*, Vol. IV, 3rd ed., D. Van Nostrand, Princeton, 1954.
5. D. Glick (Ed.), *Methods of Biochemical Analysis*, Vols. I–XIII, Interscience Publishers, New York, London, and Sydney, 1954–1965.
6. E. A. Braude and F. C. Nachod (Eds.), *Determination of Organic Structures by Physical Methods*, Academic Press, New York, 1955.
7. F. C. Nachod and W. D. Phillips (Eds.), *Determination of Organic Structures by Physical Methods*, Vol. II, Academic Press, New York and London, 1962.
8. R. L. Shriner, R. C. Fuson, and D. Y. Curtin, *The Systematic Identification of Organic Compounds*, 4th ed., John Wiley and Sons, New York, 1956.
9. F. Feigl, *Spot Tests in Organic Analysis*, 6th ed., Elsevier Publishing Company, Amsterdam, London, New York, and Princeton, 1960.
10. M. Frankel and S. Patai, *Tables for Identification of Organic Compounds*, Chemical Rubber Publishing Company, Cleveland, 1960.
11. C. N. Reiley (Ed.), *Advances in Analytical Chemistry and Instrumentation*, Vols. I–IV, Interscience Publishers, New York and London, 1960–1965.
12. T. Higuchi and E. Brochman-Hansen (Eds.), *Pharmaceutical Analysis*, Interscience Publishers, New York and London, 1961.
13. K. W. Bentley (Ed.), *Elucidation of Structures by Physical and Chemical Methods*, Vol. XI of *Technique of Organic Chemistry*, (Ed., A. Weissberger), Interscience Publishers, New York and London, 1963.
14. S. Siggia, *Quantitative Organic Analysis Via Functional Groups*, 3rd ed., John Wiley and Sons, New York, 1963.
15. N. D. Cheronis and T. S. Ma, *Organic Functional Group Analysis by Micro and Semimicro Methods*, Interscience Publishers, New York, London, and Sydney, 1964.
16. N. D. Cheronis, J. B. Entrikin, and E. M. Hodnett, *Semimicro Qualitative Organic Analysis*, 3rd ed., Interscience Publishers, New York, London, and Sydney, 1965.

17. L. E. Brown and C. L. Hoffpanir, *Anal. Chem.*, **23**, 1035 (1951).
18. G. Kainz and F. Schoeller, *Mikrochim. Acta*, 333 (1954).
19. J. L. Lassaigne, *Compt. Rend.*, **16**, 387 (1843).
20. J. L. Lassaigne, *Ann. Chim. (Paris)*, **48**, 367 (1843).
21. E. A. Kehrer, *Chem. Ber.*, **35**, 2523 (1902).
22. S. H. Tucker, *J. Chem. Educ.*, **22**, 212 (1945).
23. R. H. Baker and C. Barkenbus, *Ind. Eng. Chem. Anal. Ed.*, **9**, 135 (1937).
24. H. Middleton, *Analyst*, **60**, 154 (1935).
25. C. L. Wilson, *Analyst*, **63**, 332 (1938).
26. O. Jacobsen, *Chem. Ber.*, **12**, 2316 (1879).
27. E. Täuber, *Chem. Ber.*, **32**, 3150 (1899).
28. H. ter Meulen, *Bull. Soc. Chim. Belges*, **49**, 103 (1940).
29. A. Lacourt and Ch. F. Chang, *Bull. Soc. Chim. Belges*, **49**, 167 (1940).
30. S. Jacobs, in reference 5, Vol. XIII, pp. 241 ff.
31. G. Tolg, *Z. Anal. Chem.*, **205**, 40 (1964).
32. B. Wurzschmitt, *Mikrochim. Acta*, **36/37**, 614 (1951).
33. F. Feigl and R. Amaral, *Anal. Chem.*, **30**, 1148 (1958).
34. J. B. A. Dumas, *Ann. Chim. (Paris)*, **2**, 198 (1830).
35. K. Hozumi and W. J. Kirsten, *Anal. Chem.*, **34**, 434 (1962).
36. W. J. Kirsten, *Z. Anal. Chem.*, **181**, 1 (1961).
37. E. F. Hillenbrandt, Jr. and C. A. Pentz, in reference 1, Vol. III, pp. 129 ff.
38. W. Merz and W. Pfab, *Microchem. J.*, **10**, 346 (1966).
39. J. Z. Kjeldahl, *Z. Anal. Chem.*, **22**, 366 (1883)
40. J. Z. Kjeldahl, *Compt. Rend. Trav. Lab. Carlsberg*, **2**, 12 (1883).
41. W. L. Doyle and J. H. Omoto, *Anal. Chem.*, **22**, 603 (1950).
42. J. A. Kuck, A. Kingsley, D. Kinsey, F. Sheehah, and G. F. Swigert, *Anal. Chem.*, **22**, 604 (1950).
43. R. B. Bradstreet, *Chem. Rev.*, **27**, 331 (1940).
44. R. B. Bradstreet, *Anal. Chem.*, **26**, 185 (1954).
45. C. Menzie, *Anal. Chem.*, **28**, 1321 (1956).
46. R. C. Weast (Ed.), *Handbook of Chemistry and Physics*, 46th ed., The Chemical Rubber Company, Cleveland, 1965.
47. J. H. Beynon, *Mikrochim. Acta.*, 437 (1956).
48. F. W. McLafferty. *Anal. Chem.*, **28**, 306 (1956).
49. F. W. McLafferty, *Anal. Chem.*, **29**, 1782 (1957).
50. K. Biemann, F. Gapp, and J. Seibl, *J. Am. Chem. Soc.*, **81**, 2274 (1959).
51. K. Biemann, J. Seibl, and F. Gapp, *Biochem. Biophys. Res. Commun.*, **1**, 307 (1959).
52. K. Biemann, in reference 13, pp. 259 ff.
53. F. W. McLafferty, in reference 7, pp. 93 ff.
54. P. de Mayo and R. I. Reed, *Chem. Ind. (London)*, 1481 (1956).
55. J. H. Beynon and A. E. Williams, *Mass and Abundance Tables for Use in Mass Spectrometry*, Elsevier Publishing Company, Amsterdam, London, and New York, 1963.
56. H. H. Ross in *Guide to Activation Analysis*, (Ed. W. S. Lyon, Jr.), D. Van Nostrand, Princeton, 1964, pp. 1 ff.
57. J. T. Gilmore and D. E. Hull, *Anal. Chem.*, **34**, 187 (1962).
58. E. Ricci, in reference 56, pp. 116 ff.

59. R. C. Koch, *Activation Analysis Handbook*, Academic Press, New York and London, 1960.
60. F. Feigl and C. P. J. Silva, *Ind. Eng. Chem. Anal. Ed.*, **14**, 316 (1942).
61. L. Velluz and M. Pesez, *Ann. Pharm. Franç.*, **4**, 10 (1946).
62. A. E. Petrarca, *J. Org. Chem.*, **24**, 1171 (1959).
63. A. Seidell and W. F. Linke, *Solubilities of Inorganic and Organic Compounds*, Vols. I, II, and suppl., 3rd ed., D. Van Nostrand, New York, 1940, 1941, and 1952.
64. G. Schill, R. Modin, and B.-A. Persson, *Acta Pharm. Suecica*, **2**, 119 (1965).
65. E. von Hulle, in reference 2, pp. 692 ff.
66. S. Besson and J. J. Brignon, *Bull. Soc. Sci. Nancy*, **12**, 61 (1953).
67. S. Besson and J. J. Brignon, *Ann. Pharm. Franç.*, **11**, 535 (1953).
68. W. Poethke and H. Trabert, *Pharm. Zentralhalle*, **94**, 214 (1955).
69. M. Bobtelsky and M. M. Cohen, *Anal. Chim. Acta*, **22**, 270 (1960).
70. W. Schlömann, *Chem. Ber.*, **26**, 1020 (1893).
71. M. Bobtelsky and I. Barzily, *Anal. Chim. Acta*, **28**, 82 (1963).
72. E. Fischer and M. Bergmann, *Ann. Chem.*, **398**, 96 (1913).
73. J. Abildgaard and H. Baggesgaard-Rasmussen, *Arch. Pharm.*, **268**, 356 (1930).
74. E. Graf and E. Fiedler, *Naturwiss.*, **39**, 556 (1952).
75. T. Ogawa, *J. Chem. Soc. Japan*, **76**, 739 (1955).
76. T. Ogawa, *Denki Kagaku*, **25**, 377 (1957); *Chem. Abstr.*, **52**, 4668 (1958).
77. C. A. Jacobson, *J. Am. Chem. Soc.*, **53**, 1011 (1931).
78. P. Duquénois and M. Faller, *Bull. Soc. Chim. France*, **6**, 988 (1939).
79. N. W. Rich and L. G. Chatten, *J. Pharm. Sci.*, **54**, 995 (1965).
80. H. Pauli and E. Ludwig, *Z. Physiol. Chem.*, **121**, 165 (1922).
81. C. A. Buehler, E. J. Currier, and R. Lawrence, *Ind. Eng. Chem. Anal. Ed.*, **5**, 277, (1933).
82. C. A. Buehler and J. D. Calfee, *Ind. Eng. Chem. Anal. Ed.*, **6**, 351, (1934).
83. P. P. T. Sah and C. H. Tien, *J. Chinese Chem. Soc.*, **4**, 490, (1936).
84. P. P. T. Sah and K. H. Yuin, *J. Chinese Chem. Soc.* **5**, 129, (1937).
85. K. G. Cunningham, W. Dawson, and F. S. Spring, *J. Chem. Soc.*, 2305 (1951).
86. A. H. Cross, D. McClaren, and S. G. E. Stevens, *J. Pharm. Pharmacol.*, **11** (suppl.), 103T (1959).
87. H. Mathes and O. Rammstedt, *Z. Anal. Chem.*, **46**, 565 (1907).
88. D. D. van Slyke, *J. Biol. Chem.*, **12**, 127 (1912).
89. E. Abderhalden and A. Weil, *Z. Physiol. Chem.*, **78**, 150 (1912).
90. J. Wachsmuth, *J. Pharm. Belg.*, **8**, 283 (1953).
91. R. Opfer-Schaum, *Mikrochemie*, **31**, 324 (1944).
92. G. Wanag and A. Lode, *Chem. Ber.*, **70**, 547 (1937).
93. F. E. Crane, Jr., *Anal. Chem.*, **28**, 1794 (1956).
94. F. E. Crane, Jr., *Anal. Chem.*, **30**, 1426 (1958).
95. H. Flaschka and A. J. Barnard, Jr., in reference 11, Vol. I, pp. 1 ff.
96. K. Beyermann, *Z. Anal. Chem.*, **212**, 199 (1965).
97. K. Beyermann, *Z. Anal. Chem.*, **215**, 316 (1966).
98. T. S. Moore and T. F. Ninmill, *J. Chem. Soc.*, **101**, 1635 (1912).
99. L. Pauling, *The Nature of the Chemical Bond*, 3rd ed., Cornell University Press, Ithaca, 1962.

100. G. C. Pimentel and A. L. McClellan, *The Hydrogen Bond*, W. H. Freeman, San Francisco and London, 1960.
101. F. E. Critchfield and J. B. Johnson, *Anal. Chem.*, **30**, 1247 (1958).
102. R. P. Courgnaud and B. Trémillon, *Bull. Soc. Chim. France*, 752 (1965).
103. G. Chuchani, J. A. Hernández, and J. Zabicky, *Nature*, **207**, 1385 (1965).
104. E. Q. Adams, *J. Am. Chem. Soc.*, **38**, 1503 (1916).
105. N. Bjerrum, *Z. Physik. Chem.*, **104**, 147 (1923).
106. R. G. Bates, *Determination of pH. Theory and Practice*, 2nd ed., John Wiley and Sons, New York, London, and Sydney, 1964, pp. 201 ff.
107. M. Gutterson and T. S. Ma, *Mikrochim. Acta*, 1 (1960).
108. J. S. Fritz, *Anal. Chem.*, **22**, 1028 (1950).
109. H. J. Keily and D. N. Hume, *Anal. Chem.*, **36**, 543 (1964).
110. O. W. Kolling, *J. Am. Chem. Soc.*, **79**, 2717 (1957).
111. R. Belcher, J. Berger, and T. S. West, *J. Chem. Soc.*, 2882 (1959).
112. D. B. Cowell and B. D. Selby, *Analyst*, **88**, 974 (1963).
113. R. C. Paul and S. S. Pahil, *Anal. Chim. Acta*, **30**, 466 (1964).
114. V. Vajgand and T. Pastor, *J. Electroanal. Chem.*, **8**, 40 (1964).
115. V. Vajgand and T. Pastor, *J. Electroanal. Chem.*, **8**, 49 (1964).
116. J. S. Fritz, *Anal. Chem.*, **25**, 407 (1953).
117. W. Huber, *Z. Anal. Chem.*, **216**, 260 (1966).
118. E. J. Forman and D. N. Hume, *Talanta*, **11**, 129 (1964).
119. F. E. Critchfield and J. B. Johnson, *Anal. Chem.*, **28**, 432 (1956).
120. C. A. Streuli, *Anal. Chem.*, **31**, 1652 (1959).
121. O. W. Kolling and J. L. Lambert, *Inorg. Chem.*, **4**, 170 (1965).
122. J. S. Fritz, *Anal. Chem.*, **22**, 578 (1950).
123. S. Palit, *Ind. Eng. Chem. Anal. Ed.*, **18**, 246 (1946).
124. J. Ruch and F. E. Critchfield, *Anal. Chem.*, **33**, 1569 (1961).
125. M. E. Auerbach, *Drug Standards* **19**, 127 (1951).
126. J. S. Fritz, *Acid–Base Titrations in Nonaqueous Solvents*, G. Frederick Smith Chemical Company, Columbus, 1952.
127. J. A. Riddick, *Anal. Chem.*, **24**, 41 (1952).
128. A. H. Beckett and E. H. Tinley, *Titration in Nonaqueous Solvents*, 3rd ed., The British Drug Houses, Poole, 1960.
129. M. C. Henry, J. F. Hazel, and W. M. McNabb, *Anal. Chim. Acta.*, **15**, 187 (1956).
130. R. C. Paul, J. Singh, and S. S. Sandhu, *J. Indian Chem. Soc.*, **36**, 305 (1959)
131. H. M. Hershenson and D. N. Hume, *Anal. Chem.*, **29**, 16 (1957).
132. P. A. Kober and K. Sugiura, *J. Biol. Chem.*, **13**, 1 (1912).
133. A. J. Woiwod, *Biochem. J.*, **45**, 412 (1949).
134. J. R. Spies and D. C. Chambers, *J. Biol. Chem.*, **191**, 787 (1951).
135. M. Trop, A. Pinsky, and R. Beitner, *Israel J. Chem.*, **2**, 319 (1964).
136. V. Anger, *Mikrochim. Acta*, **2**, 3 (1937).
137. N. A. Sheppard, *J. Am. Chem. Soc.*, **38**, 2507 (1916).
138. F. Feigl and V. Anger, *Mikrochim. Acta*, **1**, 138 (1937).
139. E. Rimini, *Ann. Farmacoter. Chim.*, 193 (1898); *Chem. Zentr.*, **69**II, 132 (1898).
140. C. F. Cullis and D. J. Waddington, *Anal. Chim. Acta*, **15**, 158 (1956).
141. W. Zimmermann, *Z. Physiol. Chem.*, **186**, 260 (1930).

142. C. F. H. Allen, C. F. Murphy, and W. E. Yoerger, *Anal. Chem.*, **37**, 156 (1965).
143. T. Sharp, *J. Chem. Soc.*, 1353 (1938).
144. K. B. Prasad and G. A. Swan, *J. Chem. Soc.*, 2024 (1958).
145. A. I. Kiprianov and A. I. Tolmachev, *Zh. Obsch. Khim.*, **27**, 142 (1957).
146. C. S. Marvel, E. W. Scott, and K. L. Amstutz, *J. Am. Chem. Soc.*, **51**, 3638 (1929).
147. M. Kohn and F. Grauer, *Monatsh. Chem.*, **34**, 1751 (1913).
148. O. Sackur, *Bull. Soc. Chim. France*, 270 (1949).
149. M. E. von Klemperer and F. L. Warren, *Chem. Ind. (London)*, 1553 (1955).
150. C. C. J. Culvenor, L. W. Smith, and W. G. Woods, *Tetrahedron Letters*, 2025 (1965).
151. T. J. Batterham and J. A. Lamberton, *Australian J. Chem.*, **18**, 859 (1965).
152. M. I. Rosen, *Anal. Chem.*, **27**, 114 (1955).
153. H. Agulhon and P. Thomas, *Bull. Soc. Chim. France*, **11**, 69 (1912).
154. D. Waldi, in *Dünnschicht Chromatographie*, (Ed. E. Stahl), Springer Verlag, Berlin, Göttingen, and   Heidelberg, 1962, pp. 446 ff.
155. C. Kump and H. Schmid, *Helv. Chim. Acta*, **45**, 1090 (1962).
156. T. P. Sastri and G. G. Rao, *Z. Anal. Chem.*, **163**, 263 (1958).
157. C. Lauth, *Compt. Rend.*, **111**, 975 (1890).
158. J. Jan, J. Kolseck, and M. Perpar, *Z. Anal. Chem.*, **153**, 4 (1956).
159. R. L. Datta and J. K. Choudhury, *J. Am. Chem. Soc.*, **38**, 1079 (1916).
160. C. Cândea and E. Macovski, *Bull. Soc. Chim. France*, **4**, 1398 (1937).
161. I. de Paolini, *Atti Accad. Sci. Torino* (Classe Sci. Fis. Mat. Nat.), **65**, 201 (1930); *Chem. Abstr.*, **25**, 1182 (1931).
162. B. Potteau, *Bull. Soc. Chim. France*, 3747 (1965).
163. C. J. de Wolff, *Pharm. Weekblad*, **89**, 40 (1954).
164. P. F. Kruse, K. L. Grist, and T. A. McCoy, *Anal. Chem.*, **26**, 1319 (1954).
165. M. Z. Barakat, N. Wahba, and M. M. El-Sadr, *Analyst*, **79**, 715 (1954).
166. E. Tannenbaum, E. M. Coffin, and A. J. Harrison, *J. Chem. Phys.*, **21**, 311 (1953).
167. L. W. Pickett, M. E. Corning, G. M. Wieder, D. A. Semenov, and J. M. Buckley, *J. Am. Chem. Soc.*, **75**, 1618 (1953).
168. J. C. D. Brand and A. I. Scott, in reference 13, pp. 61 ff.
169. N. J. Leonard and D. M. Locke, *J. Am. Chem. Soc.*, **77**, 437 (1955).
170. K. Nakanishi, *Infrared Absorption Spectroscopy-Practical*, Holden Day, San Francisco, and Nankodo, Tokyo, 1962.
171. S. Oseko, *Yakugaku Zasshi*, **77**, 120 (1957).
172. R. D. Hill and G. D. Meakins, *J. Chem. Soc.*, 760 (1958).
173. J. T. Braunholtz, E. A. V. Ebsworth, F. G. Mann, and N. Sheppard, *J. Chem. Soc.*, 2780 (1958).
174. C. C. Watson, *Spectrochim. Acta*, **16**, 1322 (1960).
175. F. Bohlmann, *Chem. Ber.*, **91**, 2157 (1958).
176. K. Nakanishi, T. Goto, and M. Ohashi, *Bull. Chem. Soc. Japan*, **30**, 403 (1957).
177. J. G. Aston, in reference 6, pp. 557 ff.
178. J. Hine, *Physical Organic Chemistry*, 2nd ed., McGraw-Hill Book Company, New York, San Francisco, Toronto, and London, 1962.
179. D. J. Millen and J. Zabicky, *J. Chem. Soc.*, 3080 (1965).

180. G. F. Wright, in reference 1, Vol. I, pp. 155 ff.
181. T. Zerewitinoff, *Chem. Ber.*, **47**, 2417 (1914).
182. H. Roth, *Mikrochemie*, **11**, 140 (1932).
183. R. B. Woodward and B. Witkop, *J. Am. Chem. Soc.*, **71**, 379 (1949).
184. E. D. Olleman, *Anal. Chem.*, **24**, 1425 (1952).
185. T. Higuchi, in reference 1, Vol. II, pp. 123 ff.
186. W. H. Hamill, *J. Am. Chem. Soc.*, **59**, 1152 (1937).
187. W. R. Harp and R. C. Effert, *Anal. Chem.*, **32**, 794 (1960).
188. P. J. Paulsen and W. D. Cooke, *Anal. Chem.*, **36**, 1721 (1964).
189. J. F. Eastham and V. F. Raaen, *Anal. Chem.*, **31**, 555 (1959).
190. E. Grunwald, A. Loeverstein, and S. Meiboom, *J. Chem. Phys.*, **27**, 630 (1957).
191. F. G. Arndt, in reference 1, Vol. I, pp. 197 ff.
192. K. H. Overton, in reference 13, pp. 20 ff.
193. A. E. Smith and K. J. P. Orton, *J. Chem. Soc.*, **93**, 1242 (1908).
194. S. Siggia and I. R. Kevenski, *Anal. Chem.*, **23**, 117 (1951).
195. J. S. Fritz and G. H. Schenle, *Anal. Chem.*, **31**, 1808 (1959).
196. C. L. Ogg, W. L. Porter, and C. O. Willito, *Ind. Eng. Chem. Anal. Ed.*, **17**, 394 (1945).
197. J. Mitchell, Jr., W. Hawkins, and D. M. Smith, *J. Am. Chem. Soc.*, **66**, 782 (1944).
198. W. H. Perkin and R. Robinson, *J. Chem. Soc.*, **115**, 933 (1919).
199. H. Nishikawa, W. H. Perkin, and R. Robinson, *J. Chem. Soc.*, **125**, 657 (1924).
200. C. A. Reynolds, F. H. Walker, and E. Cochran, *Anal. Chem.*, **32**, 983 (1960).
201. L. Lumière and H. Barbier, *Bull. Soc. Chim. France*, **35**, 625 (1906).
202. V. R. Olson and H. B. Feldman, *J. Am. Chem. Soc.*, **59**, 2003 (1937).
203. N. O. V. Sonntag, *Chem. Rev.*, **52**, 237 (1953).
204. G. A. Swan, in reference 13, pp. 457 ff.
205. Ya. L. Gol'dfarb and L. M. Smogonskiĭ, *Zh. Obshch. Khim.*, **12**, 255 (1942).
206. G. P. Clemo and G. A. Swan, *J. Chem. Soc.*, 603 (1945).
207. G. P. Clemo and G. A. Swan, *J. Chem. Soc.*, 198 (1948).
208. A. Darbre and K. Blau, *J. Chromatog.*, **17**, 31 (1965).
209. W. M. Lamkin and C. W. Gehrke, *Anal. Chem.*, **37**, 383 (1965).
210. F. Marcucci, E. Mussini, F. Poy, and P. Gagliardi, *J. Chromatog.*, **18**, 487 (1965).
211. H. P. Ward and E. F. Jenkins, *J. Org. Chem.*, **10**, 371 (1945).
212. E. A. Smirnov, *Zh. Obshch. Khim.*, **20**, 696 (1950).
213. Y. Asahina, T. Ohta, and M. Inubuse, *Chem. Ber.*, **63**, 2045 (1930).
214. E. Späth, A. Wenusch, and E. Zajic, *Chem. Ber.*, **69**, 393 (1936).
215. C. Schotten, *Chem. Ber.*, **17**, 2544 (1884).
216. L. von Udránszki and E. Baumann, *Chem. Ber.*, **21**, 2744 (1888).
217. O. Hinsberg and L. von Udránszki, *Ann. Chem.*, **254**, 252 (1889).
218. C. Schotten, *Chem. Ber.*, **23**, 3430 (1890).
219. G. R. Clemo and S. P. Popli, *J. Chem. Soc.*, 1406 (1951).
220. B. C. Saunders, *Biochem. J.*, **28**, 580 (1934).
221. B. W. Town, *Biochem. J.*, **35**, 578 (1941).

222. K. Teichert, E. Mutschler, and H. Rochelmeyer, *Deut. Apotheker-Z.*, **100**, 283 (1960).
223. E. O. Woolfolk and E. H. Roberts, *J. Org. Chem.*, **21**, 436 (1956).
224. P. J. Elving and B. Warshowski, *Anal. Chem.*, **19**, 1006 (1947).
225. S. Siggia, J. G. Hanna, and R. Culmo, *Anal. Chem.*, **33**, 900 (1961).
226. S. Siggia and J. G. Hanna, *Anal. Chem.*, **37**, 600 (1965).
227. J. W. Alexander and S. M. McElvain, *J. Am. Chem. Soc.*, **60**, 2285 (1938).
228. D. E. Rivett and J. F. K. Wilshire, *Australian J. Chem.*, **18**, 1667 (1965).
229. O. Hinsberg, *Chem. Ber.*, **23**, 2962 (1890).
230. P. E. Fanta and C. S. Wang, *J. Chem. Educ.*, **41**, 280 (1964).
231. E. Maas, *Chem. Ber.*, **41**, 1635 (1908).
232. O. Hinsberg and J. Kessler, *Chem. Ber.*, **38**, 906 (1905).
233. C. S. Marvel and H. B. Gillespie, *J. Am. Chem. Soc.*, **48**, 2943 (1926).
234. E. W. McChesney and W. K. Swann, *J. Am. Chem. Soc.*, **59**, 1116 (1937).
235. C. S. Marvel, M. D. Helfrick, and J. P. Belsley, *J. Am. Chem. Soc.*, **51**, 1272 (1929).
236. C. S. Marvel and F. E. Smith, *J. Am. Chem. Soc.*, **45**, 2696 (1923).
237. C. S. Marvel, F. L. Kingsbury, and F. E. Smith, *J. Am. Chem. Soc.*, **47**, 166 (1925).
238. J. P. Alden and B. Houston, *J. Am. Chem. Soc.*, **56**, 413 (1934).
239. E. O. Woolfolk, W. E. Reynolds, and J. L. Mason, *J. Org. Chem.*, **24**, 1445 (1959).
240. J. H. Billman, J. Garrison, R. Anderson, and B. Wolnak, *J. Am. Chem. Soc.*, **63**, 1920 (1941).
241. J. H. Billman and E. O'Mahoney, *J. Am. Chem. Soc.*, **61**, 2340 (1939).
242. J. E. Dunbar and J. H. Rogers, *Tetrahedron Letters*, 4291 (1965).
243. C. K. Ingold, *Structure and Mechanism in Organic Chemistry*, Cornell University Press, Ithaca, 1953, p. 398.
244. J. Packer and J. Vaughan, *A Modern Approach to Organic Chemistry*, Oxford University Press, Oxford, 1958, pp. 147 ff, 628.
245. T. Sabalitschka and H. Schrader, *Z. Anal. Chem.*, **34**, 45 (1921).
246. H. R. Lee and D. C. Jones, *Ind. Eng. Chem.*, **16**, 930 (1924).
247. H. R. Lee and D. C. Jones, *Ind. Eng. Chem.*, **16**, 948 (1924).
248. S. J. Clark and D. J. Morgan, *Mikrochim. Acta*, 966 (1956).
249. D. J. Morgan, *Mikrochim. Acta*, 104 (1958).
250. M. Yokoo, *Chem. Pharm. Bull. (Tokyo)*, **6**, 64 (1958).
251. K. Hess, *Chem. Ber.*, **50**, 368 (1917).
252. E. Speyer and L. Walther, *Chem. Ber.*, **63**, 852 (1930).
253. E. Lyons, *J. Am. Pharm. Assoc.*, **21**, 224 (1932).
254. E. D. Bergman and M. Bentov, *J. Org. Chem.*, **26**, 1480 (1961).
255. B. Linke, H. Preissecker, and J. Stadler, *Chem. Ber.*, **65**, 1282 (1932).
256. G. Spencer and J. E. Brinley, *J. Soc. Chem. Ind. (London)*, **64**, 53 (1945).
257. E. J. van der Kam, *Rec. Trav. Chim.*, **45**, 722 (1926).
258. G. Biserte, J. W. H. Holleman, J. Holleman-Dehove, and P. Sautière, *J. Chromatog.*, **2**, 225 (1959).
259. G. Biserte, J. W. H. Holleman, J. Holleman-Dehove, and P. Sautière, *J. Chromatog.*, **3**, 85 (1960).
260. H. Fraenkel-Conrat, J. I. Harris, and A. L. Levy, in reference 5, Vol. II, pp. 359 ff.

261. F. Sanger, *Biochem. J.*, **39**, 507 (1945).
262. F. Sanger, *Biochem. J.*, **40**, 261 (1946).
263. F. Sanger, *Biochem. J.*, **45**, 562 (1949).
264. R. Porter and F. Sanger, *Biochem. J.*, **42**, 287 (1948).
265. F. C. Green and L. McKay, *Anal. Chem.*, **24**, 726 (1952).
266. F. C. McIntire, L. M. Clements, and M. Sproul, *Anal. Chem.*, **25**, 1757 (1953).
267. B. J. Camp and J. A. More, *J. Am. Pharm. Assoc. Sci. Ed.*, **49**, 158 (1960).
268. B. B. Brodie and S. Udenfriend, *J. Biol. Chem.*, **158**, 705 (1945).
269. G. Koch, and W. Weidel, *Z. Physiol. Chem.*, **303**, 213 (1956).
270. J. M. Dellacha and A. V. Fontanive, *Experientia*, **21**, 351 (1965).
271. D. T. Dubin, *J. Biol. Chem.*, **235**, 783 (1960).
272. M. Brenner, A. Niederweiser, and G. Pataki, in reference 154, pp. 403 ff.
273. O. Folin and H. Wu, *J. Biol. Chem.*, **51**, 377 (1922).
274. E. G. Frame, J. A. Russell, and A. E. Wilhelmi, *J. Biol. Chem.*, **149**, 255 (1943).
275. D. H. Rosenblatt, P. Hlinka, and J. Epstein, in reference 12, p. 424.
276. A. J. Patten, *Z. Physiol. Chem.*, **39**, 350 (1903).
277. W. Ried and W. Kaeppeler, *Ann. Chem.*, **673**, 132 (1964).
278. H. E. French and A. F. Wirtel, *J. Am. Chem. Soc.*, **48**, 1736 (1926).
279. P. P. T. Sah, *J. Chinese Chem. Soc.*, **5**, 100 (1937).
280. P. P. T. Sah and C. S. Wu, *J. Chinese Chem. Soc.*, **4**, 513 (1936).
281. C. H. Kao, N. Y. Fang, and P. P. T. Sah, *J. Chinese Chem. Soc.*, **3**, 137 (1935).
282. P. P. T. Sah and L. H. Chang, *Rec. Trav. Chim.*, **58**, 8 (1939).
283. P. P. T. Sah, S. H. Chang, and H. H. Lei, *J. Chinese Chem. Soc.*, **2**, 225 (1934).
284. P. P. T. Sah, C. H. Kao, and S. M. Wang, *J. Chinese Chem. Soc.*, **4**, 193 (1936).
285. P. P. T. Sah and S. C. Chen, *J. Chinese Chem. Soc.*, **14**, 74 (1946).
286. P. P. T. Sah and Y. K. Wang, *Rec. Trav. Chim.*, **59**, 364 (1940).
287. K. C. Meng and P. P. T. Sah, *J. Chinese Chem. Soc.*, **4**, 75 (1936).
288. K. J. Karrman, *Svensk Kem. Tidskr.*, **60**, 61 (1948).
289. P. P. T. Sah, *Rec. Trav. Chim.*, **59**, 231 (1940).
290. T. Curtius and A. Riedel, *J. Prakt. Chem.*, **76**, 238 (1907).
291. P. P. T. Sah and T. S. Ma, *J. Chinese Chem. Soc.*, **2**, 159 (1934).
292. P. P. T. Sah, *Rec. Trav. Chim.*, **58**, 1008 (1939).
293. E. N. Brown and N. Campbell, *J. Chem. Soc.*, 1699 (1937).
294. C. M. Suter and E. W. Moffett, *J. Am. Chem. Soc.*, **55**, 2496 (1933).
295. P. P. T. Sah and H. H. Lei, *J. Chinese Chem. Soc.*, **2**, 153 (1934).
296. P. Edman, *Acta Chem. Scand.*, **4**, 277 (1950).
297. P. Edman, *Acta Chem. Scand.*, **4**, 283 (1950).
298. P. Edman, *Nature*, **177**, 667 (1956).
299. P. Edman, *Acta Chem. Scand.*, **10**, 761 (1956).
300. J. Sjöquist, *Arkiv Kemi*, **11**, 129 (1957).
301. J. Sjöquist, *Biochem. Biophys. Acta*, **41**, 20 (1960).
302. J. G. Hanna and S. Siggia, *Anal. Chem.*, **34**, 547 (1962).
303. S. Siggia and J. G. Hanna, *Anal. Chem.*, **33**, 896 (1961).
304. J. von Braun, *Chem. Ber.*, **55**, 3536 (1922).

305. A. Skita, *Chem. Ber.*, **56**, 1014 (1923).
306. M. Weiser and M. K. Zacherl, *Mikrochim. Acta*, 577 (1957).
307. G. R. Umbreit, *Anal. Chem.*, **33**, 1572 (1961).
308. F. R. Duke, *Ind. Eng. Chem. Anal. Ed.*, **17**, 196 (1945).
309. F. E. Critchfield and J. B. Johnson, *Anal. Chem.*, **28**, 430 (1956).
310. F. E. Critchfield and J. B. Johnson, *Anal. Chem.*, **28**, 436 (1956).
311. L. Nebbia and F. Guerrieri, *Chim. Ind. (Milan)*, **35**, 896 (1953); *Chem. Abstr.*, **48**, 9869 (1954).
312. E. P. Przybylowicz and L. B. Rogers, *Anal. Chim. Acta*, **18**, 596 (1958).
313. A. P. Terentev, N. B. Kupletskaya, and E. V. Andreeva, *Zh. Obshch. Khim.*, **26**, 881 (1956).
314. W. M. Bryant, J. Mitchell, Jr., and D. M. Smith, *J. Am. Chem. Soc.*, **62**, 3504 (1940).
315. W. Hawkins, D. M. Smith, and J. Mitchell, Jr., *J. Am. Chem. Soc.*, **66**, 1662 (1944).
316. E. N. Deeb, *Drug Standards*, **26**, 175 (1958).
317. J. B. Johnson, *Anal. Chem.*, **28**, 1977 (1956).
318. A. J. Milun, *Anal. Chem.*, **29**, 1502 (1957).
319. F. E. Critchfield and J. B. Johnson, *Anal. Chem.*, **29**, 477 (1957).
320. F. E. Critchfield and J. B. Johnson, *Anal. Chem.*, **29**, 1174 (1957).
321. S. N. Chakravarti and M. B. Roy, *Analyst*, **62**, 603 (1937).
322. H. Fischer and F. Meyer-Betz, *Z. Physiol. Chem.*, **75**, 232 (1911).
323. E. Salkowski, *Biochem. Z.*, **103**, 185 (1920).
324. F. Feigl, V. Anger, and R. Zappert, *Mikrochemie*, **16**, 74 (1934).
325. V. Anger and S. Ofri, *Mikrochim. Acta*, 626 (1964).
326. V. Anger and S. Ofri, *Mikrochim. Acta*, 770 (1964).
327. H. Freytag and W. Neudert, *J. Prakt. Chem.*, **135**, 15 (1932).
328. W. Koenig and H. Greiner, *Chem. Ber.*, **64**, 1049 (1931).
329. F. Feigl and V. Anger, *J. Prakt. Chem.*, **139**, 180 (1934).
330. E. Abderhalden and H. Schmidt, *Z. Physiol. Chem.*, **72**, 37 (1911).
331. W. Grassmann and K. von Arnim, *Ann. Chem.*, **509**, 288 (1934).
332. R. Moubasher, *J. Biol. Chem.*, **175**, 18 (1948).
333. R. Moubasher, *J. Chem. Soc.*, 1137 (1949).
334. T. L. Hardy, D. O. Holland, and J. H. C. Nayler, *Anal. Chem.*, **27**, 971 (1955).
335. E. D. Moffat and R. I. Lytle, *Anal. Chem.*, **31**, 926 (1959).
336. E. Mutschler and H. Rochelmeyer, *Arch. Pharm.*, **292**, 449 (1959).
337. E. Nürnberger, *Arch. Pharm.*, **292**, 610 (1959).
338. M. Brenner and A. Niederweiser, *Experientia*, **16**, 378 (1960).
339. A. J. Woiwod, *J. Chromatog.*, **3**, 278 (1960).
340. R. W. Frei and M. M. Frodyma, *Anal. Biochem.*, **9**, 310 (1964).
341. Y. P. Lee and T. Takahashi, *Anal. Biochem.*, **14**, 71 (1966).
342. R. Moubasher and W. A. Awad, *J. Biol. Chem.*, **179**, 915 (1949).
343. S. Moore and W. H. Stein, *J. Biol. Chem.*, **211**, 893 (1954).
344. S. Ishii, *J. Biochem. (Tokyo)*, **43**, 531 (1956).
345. S. Moore, D. H. Spackman, and W. H. Stein, *Anal. Chem.*, **30**, 1185 (1958).
346. M. M. Frodyma and R. W. Frei, *J. Chromatog.*, **17**, 131 (1965).
347. S. Moore and W. H. Stein, *J. Biol. Chem.*, **176**, 367 (1948).

348. D. D. van Slyke, R. T. Dillon, and D. A. MacFadyer, *J. Biol. Chem.*, **141**, 627 (1941).

349. D. D. van Slyke, D. A. MacFadyer, and P. Hamilton, *J. Biol. Chem.*, **141**, 671 (1941).

350. R. Moubasher and A. Sina, *J. Biol. Chem.*, **180**, 681 (1949).

351. R. Moubasher, A. Sina, W. A. Awad, and A. E. M. Othman, *J. Biol. Chem.*, **184**, 693 (1950).

352. P. Linko, *Suomen Kemistilehti*, **28**, 96 (1956).

353. R. Riemschneider, R. Koka, and H. Kiesler, *Monatsh. Chem.*, **94**, 1131 (1963).

354. I. R. Kennedy, *Anal. Biochem.*, **11**, 105 (1965).

355. R. Moubasher and A. Schoenberg, *Chem. Rev.*, **50**, 272 (1952).

356. H. Roth, in reference 2, pp. 663 ff.

357. D. J. McCaldin, *Chem. Rev.*, **60**, 39 (1960).

358. R. West, *J. Chem. Educ.*, **42**, 386 (1965).

359. M. Friedman and C. W. Sigel, *Biochemistry*, **5**, 478 (1966).

360. D. D. van Slyke, *J. Biol. Chem.*, **9**, 185 (1911).

361. D. D. van Slyke, *J. Biol. Chem.*, **12**, 275 (1912).

362. R. S. Soxena and C. S. Bhatnegar, *Naturwiss.*, **44**, 583 (1957).

363. A. Watanabe and M. Kamada, *Yakugaku Zasshi*, **72**, 972 (1952).

364. F. J. Bandelin and C. R. Kemp, *Ind. Eng. Chem. Anal. Ed.*, **18**, 470 (1946).

365. F. Gasser, *Chemiker Z.*, **51**, 206 (1950).

366. H. Rathsburg, *Chem. Ber.*, **54**, 3183 (1921).

367. W. E. Shaefer and W. W. Becker, *Anal. Chem.*, **19**, 307 (1947).

368. R. S. Bottei and N. H. Furman, *Anal. Chem.*, **27**, 1182 (1955).

369. R. Aldrovandi and F. DeLorenzi, *Ann. Chim. (Rome)*, **42**, 298 (1952).

370. S. Siggia, J. G. Hanna, and N. M. Serencha, *Anal. Chem.*, **35**, 575 (1963).

371. L. J. Bellamy, *The Infra-red Spectra of Complex Molecules*, 2nd ed., Methuen, London, 1958.

372. R. F. Goddu, in reference 11, Vol. I, pp. 347 ff.

373. R. E. Richards, *Trans. Faraday Soc.*, **44**, 40 (1948).

374. M. St. C. Flett, *Trans. Faraday Soc.*, **44**, 767 (1948).

375. M. St. C. Flett, *J. Chem. Soc.*, 1441 (1948).

376. D. E. Hathway and M. St. C. Flett, *Trans. Faraday Soc.*, **45**, 818 (1949).

377. L. J. Bellamy and R. F. Williams, *Spectrochim. Acta*, **9**, 341 (1957).

378. W. J. Orville-Thomas, A. E. Parsons, and C. P. Ogden, *J. Chem. Soc.*, 1047 (1958).

379. R. E. Richards and W. R. Burton, *Trans. Faraday Soc.*, **45**, 874 (1949).

380. B. Witkop, *J. Am. Chem. Soc.*, **72**, 614 (1950).

381. L. Marion, D. A. Ramsay, and R. N. Jones, *J. Am. Chem. Soc.*, **73**, 305 (1951).

382. B. Witkop and J. B. Patrick, *J. Am. Chem. Soc.*, **73**, 1558 (1951).

383. B. Witkop and J. B. Patrick, *J. Am. Chem. Soc.*, **73**, 2188 (1951).

384. N. Fuson, M. L. Josien, R. L. Powell, and E. Utterback, *J. Chem. Phys.*, **20**, 145 (1952).

385. R. A. Russell and H. W. Thompson, *J. Chem. Soc.*, 483 (1955).

386. P. J. Kruger and H. W. Thompson, *Proc. Roy. Soc. (London)*, **A243**, 143 (1957).

387. W. Kaye, *Spectrochim. Acta*, **6**, 257 (1954).

388. K. B. Whetsel, W. E. Roberson, and M. W. Krell, *Anal. Chem.*, **30**, 1594 (1958).
389. K. B. Whetsel, W. E. Roberson, and M. W. Krell, *Anal. Chem.*, **30**, 1598 (1958).
390. K. B. Whetsel, W. E. Roberson, and M. W. Krell, *Anal. Chem.*, **29**, 1006 (1957).
391. D. Hadži and M. Škrbljak, *J. Chem. Soc.*, 843 (1957).
392. R. F. Goddu and D. A. Delker, *Anal. Chem.*, **32**, 140 (1960).
393. N. F. Chamberlain, *Anal. Chem.*, **31**, 56 (1959).
394. C. S. Springer, Jr., and D. W. Meek, *J. Phys. Chem.*, **70**, 481 (1966).
395. J. N. Sholery, *Discussions Faraday Soc.*, **19**, 215 (1955).
396. L. H. Piette, J. D. Ray and R. A. Ogg, Jr., *J. Chem. Phys.*, **26**, 1341 (1957).
397. R. A. Ogg, Jr., *J. Chem. Phys.*, **22**, 560 (1954).
398. F. A. Bovey and G. V. D. Tiers, *J. Am. Chem. Soc.*, **81**, 2870 (1959).
399. J. Herzig and H. Meyer, *Chem. Ber.*, **27**, 319 (1894).
400. J. Herzig and H. Meyer, *Monatsh. Chem.*, **15**, 613 (1894).
401. J. Herzig and H. Meyer, *Monatsh. Chem.*, **16**, 599 (1895).
402. J. Herzig and H. Meyer, *Monatsh. Chem.*, **18**, 379 (1897).
403. R. Belcher, M. K. Bhatty, and T. S. West, *J. Chem. Soc.*, 2393 (1958).
404. A. Elek, in reference 1, Vol. I, pp. 667 ff.
405. E. Eegrive, *Z. Anal. Chem.*, **110**, 22 (1937).
406. A. Würtz, *Ann. Chem.*, **108**, 68 (1858).
407. J. Colthup, *J. Opt. Soc. Am.*, **40**, 397 (1950).
408. R. B. Barnes, R. C. Gore, R. W. Stafford, and v. Z. Williams, *Anal. Chem.*, **20**, 402 (1948).
409. S. N. Vinogradov and R. H. Linnell, *J. Chem. Phys.*, **23**, 93 (1955).
410. R. A. Russell and H. W. Thompson, *Proc. Roy. Soc.* (*London*), **A234**, 318 (1956).
411. G. M. Barrow, *J. Am. Chem. Soc.*, **80**, 86 (1958).
412. D. J. Millen and J. Zabicky, *Nature*, **196**, 889 (1962).
413. F. Feigl and H. E. Feigl, *Mikrochim. Acta*, 85 (1954).
414. H. Krall, *J. Chem. Soc.*, **103**, 1124 (1913).
415. W. Gluud, K. Keller, and W. Klempt, *Angew. Chem.*, **39**, 1071 (1926).
416. J. Sivadjian, *Bull. Soc. Chim. France*, **2**, 623 (1935).
417. S. Sass, J. J. Kaufman, A. A. Cardenas, and J. J. Martin, *Anal. Chem.*, **30**, 529 (1958).
418. A. L. LeRosen, R. T. Moravek, and J. K. Carlton, *Anal. Chem.*, **24**, 1335 (1952).
419. F. Feigl, V. Anger, and R. Zappert, *Mikrochemie*, **16**, 67 (1934).
420. F. Feigl, V. Anger, and R. Zappert, *Mikrochemie*, **16**, 70 (1934).
421. J. L. Valentine, J. B. Entrikin, and M. W. Hanson, *J. Chem. Educ.*, **41**, 569 (1964).
422. F. Weigand, H. Weber, E. Maekawa, and G. Eberhardt, *Chem. Ber.*, **89**, 1994 (1956).
423. S. Ohkuma, *Yakugaku Zasshi*, **75**, 1124 (1955).
424. A. S. Curry, in reference 5, Vol. VII, pp. 39 ff.
425. D. Waldi, in reference 154, pp. 496 ff.
426. J. L. Lambert and V. E. Cates, *Anal. Chem.*, **29**, 508 (1957).
427. W. Büchler, *Helv. Chim. Acta*, **47**, 639 (1964).

428. C. L. Hilton, *Rubber Age*, **84**, 263 (1958).
429. E. Sawicki, T. W. Stanley, and T. R. Hauser, *Chemist-Analyst*, **48**, 30 (1959).
430. P. Ehrlich, *Z. Klin. Med.*, **5**, 285 (1882).
431. A. Barber, C. C. T. Chinnick, and P. A. Lincoln, *J. Appl. Chem. (London)*, **5**, 594 (1955).
432. J. J. Bikerman, *Z. Anal. Chem.*, **90**, 335 (1932).
433. F. O. Gunderson, R. Heiz, and R. Klevstrand, *J. Pharm. Pharmacol.*, **5**, 608 (1953).
434. J. Knabe, *Deut. Apotheker-Z.*, **96**, 874 (1956).
435. E. Sjostrom and W. Rittner, *Z. Anal. Chem.*, **153**, 321 (1956).
436. P. C. Markunas and J. A. Riddick, *Anal. Chem.*, **24**, 312 (1952).
437. J. R. Clark and S. M. Wang, *Anal. Chem.*, **26**, 1230 (1954).
438. N. Y. C. Chang, *Anal. Chem.*, **30**, 1095 (1958).
439. C. W. Pifer and E. G. Wollish, *Anal. Chem.*, **24**, 300 (1952).
440. K. K. Kunov and M. N. Das, *Anal. Chem.*, **31**, 1358 (1959).
441. G. Kainz and M. Polun, *Mikrochemie*, **35**, 189 (1950).
442. J. B. Wilson, *J. Assoc. Offic. Agr. Chemists*, **35**, 455 (1952).
443. P. A. Lincoln and C. C. T. Chinnick, *Analyst*, **81**, 100 (1956).
444. I. Renard, *J. Pharm. Belg.*, **7**, 403 (1952).
445. S. Weiner, *Chemist-Analyst*, **42**, 9 (1953).
446. T.-F. Chin and J. L. Lach, *J. Pharm. Sci.*, **54**, 1550 (1965).
447. G. Schill and B. Danielson, *Anal. Chim. Acta*, **21**, 248 (1959).
448. G. Schill and B. Danielson, *Anal. Chim. Acta*, **21**, 341 (1959).
449. G. Schill, *Acta Pharm. Suecica*, **2**, 177 (1965).
450. R. Reiss, *Arzneimittel Forsch.*, **6**, 77 (1956).
451. B. Budensky and E. Vanickova, *Chem. Listy*, **50**, 1241 (1956).
452. J. Gautier, J. Renault, and F. Pallerin, *Ann. Pharm. Franç.*, **13**, 725 (1955).
453. D. M. Patel and R. A. Anderson, *Drug Standards*, **26**, 189 (1958).
454. T. E. Furlong and P. R. Elliker, *J. Dairy Sci.*, **36**, 225 (1953).
455. M. Dolezil and J. Bulander, *Chem. Listy*, **51**, 225 (1957).
456. E. D. Carkuff and W. F. Boyd, *J. Am. Pharm. Assoc. Sci. Ed.*, **43**, 240 (1954).
457. G. R. F. Rose and C. H. Bayley, *Nat. Res. Council Can.*, *NRC Bull.* (2875), (1952); *Chem. Abstr.*, **47**, 6662 (1953).
458. W. K. Moseley, *Milk Plant Monthly*, **38**, 76 (1949).
459. M. E. Auerbach, *Ind. Eng. Chem. Anal. Ed.*, **15**, 492 (1943).
460. M. E. Auerbach, *Ind. Eng. Chem. Anal. Ed.*, **16**, 739 (1944).
461. J. B. Wilson, *J. Assoc. Offic. Agr. Chemists*, **34**, 343 (1951).
462. L. D. Metcalfe, *Anal. Chem.*, **32**, 70 (1960).
463. J. van Steveninck and M. Maas, *Rec. Trav. Chim.*, **84**, 1166 (1965).
464. J. Fogh, P. O. H. Rasmussen, and K. Skadhauge, *Anal. Chem.*, **26**, 392 (1954).
465. S. Sass, J. J. Kaufman, A. A. Cardenas, and J. J. Martin, *Anal. Chem.*, **30**, 530 (1958).
466. S. G. Yates and H. L. Tookey, *Australian J. Chem.*, **18**, 53 (1965).
467. J. A. McMillan and R. E. Dickerson, unpublished results, cited in reference 466.
468. J. B. Stothers, in reference 13, pp. 175 ff.

469. L. M. Jackman, *Applications of Nuclear Magnetic Resonance in Organic Chemistry*, Pergamon Press, Oxford, 1962.

470. N. S. Bhacca, D. P. Hollis, L. F. Johnson, E. A. Pier, and J. N. Shoolery, *NMR Spectra Catalog*, Vols. I and II, Varian Associates, Palo Alto, 1961 and 1962.

471. W. Gordy in *Chemical Applications of Spectroscopy*, (Ed. W. West), Vol. IX of *Technique of Organic Chemistry*, (Ed. A. Weissberger), Interscience Publishers, New York and London, 1956, pp. 146 ff.

472. J. A. Pople, W. G. Schneider, and H. J. Bernstein, *High Resolution Nuclear Magnetic Resonance*, McGraw-Hill Book Company, New York, 1959.

473. G. Spiteller and M. Spiteller-Friedmann, *Angew. Chem. Intern. Ed. Engl.*, **4**, 383 (1965).

474. C. Djerassi and C. Fenselau, *J. Am. Chem. Soc.*, **87**, 5747 (1965).

475. C. Djerassi and C. Fenselau, *J. Am. Chem. Soc.*, **87**, 5752 (1965).

476. F. W. McLafferty, *Chem. Commun.*, 78 (1966).

477. K. Bowden, E. A. Braude, and E. R. H. Jones, *J. Chem. Soc.*, 948 (1946).

478. K. Bowden and E. A. Braude, *J. Chem. Soc.*, 1068 (1952).

479. E. A. Braude, in reference 6, pp. 131 ff.

480. L. A. Flexser, L. P. Hammett, and A. Dingwall, *J. Am. Chem. Soc.*, **57**, 2103 (1935).

481. L. A. Flexser and L. P. Hammett, *J. Am. Chem. Soc.*, **60**, 885 (1938).

482. R. L. Hinman and J. Lang, *J. Org. Chem.*, **29**, 1449 (1964).

483. A. I. Scott, *Interpretation of the Ultraviolet Spectra of Natural Products*, Pergamon Press, Oxford, London, Edinburgh, New York, Paris, and Frankfurt, 1964.

484. A. W. Sangster and K. L. Stuart, *Chem. Rev.*, **65**, 69 (1965).

485. C. A. Burckhardt, R. Goutarel, M. M. Janot, and E. Schlittler, *Helv. Chim. Acta*, **35**, 642 (1952).

486. N. Neuss, *Physical Data of Indole and Dihydroindole Alkaloids*, Eli Lilly and Co., Indianapolis, 1961.

487. G. Pappalardo and T. Vitali, *Gazz. Chim. Ital.*, **88**, 564 (1958); through *Organic Electronic Spectral Data, Volume IV, 1958–1959*, (Eds. J. P. Phillips and F. C. Nachod), Interscience Publishers, New York and London, 1963.

488. C. O. Bender and R. Bonnett, *Chem. Commun.*, 198 (1966).

489. C. A. Grob, *Helv. Chim. Acta*, **33**, 1789 (1950).

490. D. A. Nelson and J. J. Worman, *Chem. Commun.*, 487 (1966).

491. R. Hubbard and G. Wald, *Science*, **115**, 60 (1952).

492. F. D. Collins, *Nature*, **171**, 469 (1953).

493. E. J. Stamhuis and W. Maas, *J. Org. Chem.*, **30**, 2156 (1965).

494. E. J. Stamhuis, W. Maas, and H. Wynberg, *J. Org. Chem.*, **30**, 2160 (1965).

495. J. Elguero, R. Jacquier, and G. Tarrago, *Tetrahedron Letters*, 4719 (1965).

496. C. B. Clarke and A. R. Pinder, *J. Chem. Soc.*, 1967 (1958).

497. N. F. Hall and M. R. Sprinkle, *J. Am. Chem. Soc.*, **54**, 3469 (1932).

498. N. J. Leonard and V. W. Gash, *J. Am. Chem. Soc.*, **76**, 2781 (1954).

499. R. Adams and J. E. Mahan, *J. Am. Chem. Soc.*, **64**, 2588 (1942).

500. C. A. Grob, A. Kaiser, and E. Renk, *Chem. Ind. (London)*, 598 (1957).

501. C. A. Grob, A. Kaiser, and E. Renk, *Helv. Chim. Acta*, **40**, 2170 (1957).

502. V. Prelog and O. Häfliger, *Helv. Chim. Acta*, **32**, 1851 (1949).

503. R. Adams, M. Carmack, and J. E. Mahan, *J. Am. Chem. Soc.*, **64**, 2593 (1942).

504. W. J. Pope, and S. T. Peachley, *J. Chem. Soc.*, **75**, 1127 (1899).

505. M. B. Thomas and H. O. Jones, *J. Chem. Soc.*, **89**, 280 (1906).

506. J. Meisenheimer and W. Theilacker in *Stereochemie*, (Ed. K. Freudenberg), Franz Deuticke, Leipzig and Vienna, 1933, pp. 963 ff.

507. A. T. Bottini, B. F. Dowden, and R. L. VanEtten, *J. Am. Chem. Soc.*, **87**, 3250 (1965).

508. R. L. Shriner, R. Adams, and C. S. Marvel in *Organic Chemistry, an Advanced Treatise*, Vol. II, 2nd ed. (Chief Ed., H. Gilman), John Wiley and Sons, New York, 1949, pp. 214 ff.

509. G. W. Wheland, *Advanced Organic Chemistry*, 2nd ed., John Wiley and Sons, New York, 1954.

510. A. T. Bottini and J. D. Roberts, *J. Am. Chem. Soc.*, **80**, 5203 (1958).

511. A. Loewenstein, J. F. Neumer, and J. D. Roberts, *J. Am. Chem. Soc.*, **82**, 3599 (1960).

512. S. M. McElvain and L. W. Bannister, *J. Am. Chem. Soc.*, **76**, 1126 (1954).

513. V. Prelog and P. Wieland, *Helv. Chim. Acta*, **27**, 1127 (1944).

514. A. Neuberger in *Advances in Protein Chemistry*, Vol. IV, (Eds. M. L. Anson and J. T. Edsall), Academic Press, New York, 1948, pp. 297 ff.

515. W. Klyne, in reference 6, pp. 73 ff.

516. G. G. Lyle and R. E. Lyle, in reference 7, pp. 1 ff.

517. C. Djerassi, *Optical Rotatory Dispersion*, McGraw-Hill Book Company, New York, Toronto, and London, 1960.

518. W. Klyne, *Chem. Ind. (London)*, 1022 (1951).

519. K. Freudenberg and F. Rhino, *Chem. Ber.*, **57**, 1547 (1924).

520. K. Freudenberg and M. Meister, *Ann. Chem.*, **518**, 86 (1935).

521. K. Freudenberg, W. Kuhn, and I. Bumann, *Chem. Ber.*, **63**, 2380 (1930).

522. W. A. Cowdrey, E. D. Hughes, and C. K. Ingold, *J. Chem. Soc.*, 1243 (1937).

523. W. A. Cowdrey, E. D. Hughes, C. K. Ingold, S. Masterman, and A. D. Scott, *J. Chem. Soc.*, 1252 (1937).

524. P. A. Levene, *J. Biol. Chem.*, **63**, 95 (1925).

525. P. A. Levene, T. Mori, and L. A. Mikeska, *J. Biol. Chem.*, **75**, 337 (1927).

526. P. A. Levene, L. W. Bass, A. Rothen, and R. E. Steiger, *J. Biol. Chem.*, **81**, 687 (1929).

527. O. Lutz and B. Jirgensons, *Chem. Ber.*, **63**, 448 (1930).

528. O. Lutz and B. Jirgensons, *Chem. Ber.*, **64**, 1221 (1931).

529. O. Lutz and B. Jirgensons, *Chem. Ber.*, **65**, 784 (1932).

530. G. W. Clough, *J. Chem. Soc.*, **107**, 1509 (1915).

531. G. W. Clough, *J. Chem. Soc.*, **113**, 526 (1918).

532. P. Brewster, F. Hiron, E. D. Hughes, C. K. Ingold, and P. A. D. S. Rao, *Nature*, **166**, 178 (1950).

533. J. A. Schellman, in reference 517, pp. 210 ff.

534. J. C. Craig and S. K. Roy, *Tetrahedron*, **21**, 391 (1965).

535. J. P. Jennings, W. Klyne, and P. M. Scopes, *J. Chem. Soc.*, 294 (1965).

536. B. Sjöberg, A. Fredga, and C. Djerassi, *J. Am. Chem. Soc.*, **81**, 5002 (1959).

537. S. Yamada, K. Ishikawa, and K. Achiwa, *Chem. Pharm. Bull. (Tokyo)*, **13**, 892 (1965).

538. T. Yasui, *Bull. Chem. Soc. Japan*, **38**, 1746 (1965).
539. T. R. Emerson, D. F. Ewing, W. Klyne, D. G. Neilson, D. A. V. Peters, L. H. Roach, and R. J. Swan, *J. Chem. Soc.*, 4007 (1965).
540. J. Fujita, T. Yasui, and Y. Shimura, *Bull. Chem. Soc. Japan*, **38**, 654 (1965).
541. G. C. Schellman and J. A. Schellman, *Compt. Rend. Trav. Lab. Carlsberg, Ser. Chim.*, **30**, (27), (1958).
542. E. Katchalski, J. Kurtz, G. D. Fasman, and A. Berger, *Bull. Res. Council Israel*, **5A**, 264 (1956).
543. W. F. Harrington and M. Sela, *Biochem. Biophys. Acta*, **27**, 24 (1958).
544. G. F. Bryce and F. R. N. Gurd, *J. Biol. Chem.*, **241**, 122 (1966).
545. P. I. Pollak and D. Y. Curtin, *J. Am. Chem. Soc.*, **72**, 961 (1950).
546. D. Y. Curtin and P. I. Pollak, *J. Am. Chem. Soc.*, **73**, 992 (1951).
547. D. Y. Curtin, E. E. Harris, and P. I. Pollak, *J. Am. Chem. Soc.*, **73**, 3453 (1951).
548. W. E. Bachmann and F. H. Moser, *J. Am. Chem. Soc.*, **54**, 1124 (1932).
549. J. W. Huffman and R. P. Elliott, *J. Org. Chem.*, **30**, 365 (1965).
550. V. Prelog and O. Häfliger, *Helv. Chim. Acta*, **33**, 2021 (1950).
551. R. B. Turner and R. B. Woodward in *The Alkaloids*, Vol. III, (Eds. R. H. F. Manske and H. L. Holmes), Academic Press, New York, 1953, pp. 1 ff.
552. H. C. Brown, D. H. McDaniel, and O. Häfliger, in reference 6, pp. 567 ff.
553. H. Rapoport and S. Masamune, *J. Am. Chem. Soc.*, **77**, 4330 (1955).
554. A. Stoll, A. Hoffmann, and F. Troxler, *Helv. Chim. Acta*, **32**, 506 (1949).
555. A. Stoll, *Chem. Rev.*, **47**, 197 (1950).
556. A. L. Glenn, *Quart. Rev.* (*London*), **8**, 192 (1954).
557. A. T. Bottini, R. L. VanEtten, and A. J. Davidson, *J. Am. Chem. Soc.*, **87**, 755 (1965).
558. R. K. Harris and R. A. Spragg, *Chem. Commun.*, 314 (1966).
559. E. E. Smissman and G. Hite, *J. Am. Chem. Soc.*, **82**, 3375 (1960).
560. E. L. Eliel, *Angew. Chem. Intern. Ed. Engl.*, **4**, 761 (1965).
561. R. J. Bishop, L. E. Sutton, D. Dineen, R. A. Y. Jones, and A. R. Katritzky, *Proc. Chem. Soc.* (*London*), 257 (1964).
562. N. L. Allinger, J. G. D. Carpenter, and F. M. Karkowski, *Tetrahedron Letters*, 3345 (1964).
563. K. Brown, A. R. Katritzky, and A. J. Waring, *Proc. Chem. Soc.* (*London*), 257 (1964).
564. J. B. Lambert and R. G. Keske, *J. Am. Chem. Soc.*, **88**, 620 (1966).
565. V. C. Farmer, R. L. Hardie, and R. H. Thomson, *Hydrogen Bonding Paper Symposium, Ljubljana, 1957*, publ. 1959, pp. 475 ff.; *Chem. Abstr.*, **54**, 20376 (1960).
566. W. S. Wilcox and J. H. Goldstein, *J. Chem. Phys.*, **20**, 1656 (1952).
567. D. G. Lister and J. K. Tyler, *Chem. Commun.*, 152 (1966).
568. W. R. Remington, *J. Am. Chem. Soc.*, **67**, 1838 (1945).
569. H. B. Klevens and J. R. Platt, *J. Am. Chem. Soc.*, **71**, 1714 (1949).
570. L. E. Sutton, in reference 6, pp. 373 ff.
571. B. M. Wepster, *Rec. Trav. Chim.*, **71**, 1159 (1952).
572. B. M. Wepster, *Rec. Trav. Chim.*, **71**, 1171 (1952).
573. K. J. Brunings and A. H. Corwin, *J. Am. Chem. Soc.*, **64**, 593 (1942).

574. L. G. S. Brooker, F. L. White, R. H. Sprague, S. G. Deut, Jr., and G. Van Zandt, *Chem. Rev.*, **41**, 325 (1947).
575. D. H. Everett and W. F. K. Wynne-Jones, *Proc. Roy. Soc. (London)*, **A177**, 499 (1941).
576. R. G. Bates and G. D. Pinching, *J. Am. Chem. Soc.*, **72**, 1393 (1950).
577. H. C. Brown, H. Bartholomay, Jr., and M. D. Taylor, *J. Am. Chem. Soc.*, **66**, 435 (1944).
578. H. K. Hall, Jr., *J. Am. Chem. Soc.*, **79**, 5441 (1957).
579. G. Briegleb, *Z. Elektrochem.*, **53**, 350 (1949).
580. H. C. Brown and S. Sujishi, *J. Am. Chem. Soc.*, **70**, 2871 (1948).
581. H. C. Brown and M. D. Taylor, *J. Am. Chem. Soc.*, **69**, 1332 (1947).
582. N. Camerman and J. Trotter, *Acta Cryst.*, **17**, 384 (1964).
583. N. E. White and C. J. B. Clews, *Acta Cryst.*, **9**, 586 (1956).
584. B. K. Vaĭnsteĭn, *Tr. Inst. Kristallogr., Akad. Nauk SSSR*, **10**, 49 (1954); *Chem. Abstr.*, **50**, 1404 (1956).
585. B. K. Vaĭnsteĭn, *Dokl. Akad. Nauk SSSR*, **99**, 81 (1954); *Chem. Abstr.*, **49**, 6685 (1955).
586. H. S. Gutowsky, G. B. Kistiakovsky, G. E. Pake, and M. E. Purcell, *J. Chem. Phys.*, **17**, 972 (1949).
587. H. S. Gutowsky, G. E. Pake, and R. Bersohn, *J. Chem. Phys.*, **22**, 643 (1954).
588. R. Bersohn and H. S. Gutowsky, *J. Chem. Phys.*, **22**, 651 (1954).
589. C. M. Deeley and R. E. Richards, *Trans. Faraday Soc.*, **50**, 560 (1954).
590. D. Pendred and R. E. Richards, *Trans. Faraday Soc.*, **51**, 468 (1955).
591. R. A. Kromhout and W. G. Moulton, *J. Chem. Phys.*, **23**, 1673 (1955).
592. W. G. Moulton and R. A. Kromhout, *J. Chem. Phys.*, **25**, 34 (1956).
593. R. E. Richards, *Quart. Rev. (London)*, **10**, 480 (1956).
594. H. S. Harned and B. B. Owen, *J. Am. Chem. Soc.*, **52**, 5079 (1930).
595. P. Damsgaard-Sørensen and A. Unmack, *Z. Physik. Chem.*, **A172**, 389 (1935).
596. C. W. Hoerr, M. R. McCorkle, and A. W. Ralston, *J. Am. Chem. Soc.*, **65**, 328 (1943).
597. A. Neuberger, *Proc. Roy. Soc. (London)*, **A158**, 68 (1937).
598. P. Rumpf, *Bull. Soc. Chim. France*, **5**, 871 (1938).
599. P. Rumpf and V. Chavanne, *Compt. Rend.*, **224**, 919 (1947).
600. J. F. J. Dippy, *Chem. Rev.*, **25**, 151 (1939).
601. F. Kieffer and P. Rumpf, *Compt. Rend.*, **230**, 1874 (1950).
602. F. Kieffer and P. Rumpf, *Compt. Rend.*, **230**, 2302 (1950).
603. M. Gillois and P. Rumpf, *Bull. Soc. Chim. France*, 112 (1954).
604. P. Legatte and G. E. Dunn, *Can. J. Chem.*, **43**, 1158 (1965).
605. J. F. King, in reference 13, pp. 317 ff.
606. H. Suhr, *Z. Elektrochem.*, **66**, 466 (1962).
607. L. K. Dyall, *Australian J. Chem.*, **17**, 419 (1964).
608. T. Yonemoto, W. F. Reynolds, H. M. Hutton, and T. Schaefer, *Can. J. Chem.*, **43**, 2668 (1965).
609. I. D. Rae and L. K. Dyall, *Australian J. Chem.*, **19**, 835 (1966).
610. M. Peisach, *Chem. Commun.*, 632 (1966).

CHAPTER **4**

Basicity and complex formation

J. W. SMITH

Bedford College, London

I. INTRODUCTION

Owing to the presence of a lone pair of electrons on the nitrogen atom, all amines can potentially act as bases, accepting protons from Lowry–Brönsted acids. In the case of normal acids in aprotic solvents this results in the formation of substituted ammonium ions, probably with hydrogen bonds between the proton and the anion, but with some weak proton donors such as alcohols it may lead only to the production of hydrogen bonded complexes (equations 1 and 2).

$$AH + NH_2R \rightleftharpoons A^- \cdots H\overset{+}{N}H_2R \tag{1}$$

$$AH + NH_2R \rightleftharpoons AH \cdots NH_2R \tag{2}$$

The reaction in a so-called 'ionising' solvent, however, is very different. It is generally written formally as in equation (3)

$$SH^+ + RNH_2 \rightleftharpoons S + RNH_3^+ \tag{3}$$

where S represents a solvent molecule, but this is an over-simplified representation of a chemical reaction in which solvent participation is both fundamental and complex. Thus change of solvent is tantamount to a change of reaction under study.

The lone pair of electrons can be broadly described as occupying an sp^3 hybrid orbital of the nitrogen atom, but the precise character of this orbital depends upon the groups linked to the nitrogen atom. Thus the energy of the N—H bond produced, and the thermodynamic stability of the positive ion, depend on the nature of these groups. As a result of this and other factors the strengths of amines as bases vary considerably.

In aqueous solution the strength of a base B can be defined in terms of the equilibrium constant K_b of the reaction (4).

$$B + H_2O \rightleftharpoons BH^+ + OH^- \tag{4}$$

$$K_b = \frac{c_{BH^+} \cdot c_{OH^-}}{c_B} \cdot \frac{f_{BH^+} \cdot f_{OH^-}}{f_B} \tag{5}$$

At low concentrations the activity coefficient term f_B can be taken as unity, and $f_{BH^+} = f_{OH^-} = f_{\pm}$, so

$$K_b = \frac{c_{BH^+} \cdot c_{OH^-}}{c_B} \cdot f_{\pm}^2 \tag{6}$$

For convenience in expression, and by analogy with the pH and pK_a scales, K_b may be written in terms of its negative logarithm, i.e. $pK_b = -\log_{10} K_b$.

The alternative and more usual scheme, however, is to describe the strength of a base in terms of the dissociation constant K_a of its conjugate positive ion, i.e. the constant of reaction (7).

$$BH^+ + H_2O \rightleftharpoons B + H_3O^+ \tag{7}$$

$$K_a = \frac{c_B \cdot c_{H_3O^+}}{c_{BH^+}} \cdot \frac{f_B \cdot f_{H_3O^+}}{f_{BH^+}} \qquad (8)$$

Having activity coefficient terms for univalent ions in both the numerator and denominator, at low concentrations this reduces to equation (9).

$$K_a = c_B \cdot c_{H_3O^+} / c_{BH^+} \qquad (9)$$

The product $K_a K_b$ is given by equation (10).

$$K_a K_b = \frac{c_B \cdot c_{H_3O^+}}{c_{BH^+}} \cdot \frac{c_{BH^+} \cdot c_{OH^-}}{c_B} \cdot f_\pm^2 = c_{H_3O} \cdot c_{OH^-} \cdot f_\pm^2 = K_w \qquad (10)$$

Since at 25°c the value of K_w, the ionic activity product of water, is about 10^{-14}g. $\text{ion}^2/\text{litre}^2$, it follows that $pK_a + pK_b = -\log K_w = 14$. The lower the value of pK_b and hence the higher the value of pK_a the stronger is the base.

The values of pK_a for bases can be determined by essentially the same methods as are used for acids. These have been listed, and the pK_a values for a large number of amines tabulated, by Perrin[1].

II. FACTORS DETERMINING THE STRENGTHS OF AMINES AS BASES

A. Thermodynamic Considerations

The standard free-energy change $\Delta G°$ attending the dissociation of an aminium ion is given by

$$-\Delta G° = RT \ln K_a \qquad (11)$$

so it follows that

$$pK_a = \Delta G°/2.303 \, RT \qquad (12)$$

where the standard states are those of unit activity for the amine, its cation, and the solvated proton and the pure state for the solvent. The pK_a value, therefore, is determined by the temperature and by the free-energy difference between the (imaginary) states of the system when the solution contains (a) the amine at unit activity, H_3O^+ ions at unit activity, and solvent, and (b) the aminium ions at unit activity and solvent. Again since

$$\Delta G° = \Delta H° - T\Delta S° \qquad (13)$$

this free-energy difference depends upon the enthalpies and entropies of the two states of the system. Any factors which can modify these properties in either state will therefore influence the base strength of an amine.

The dissociation of a molecular acid HA is accompanied by net generation of charges, and hence there are very large net hydrational changes. The increased order in the system which this brings about leads to large changes in the entropy of the system. Also, because the extent of this ordering is temperature dependent, it leads to large changes in the heat capacity. Typically for the dissociation of a molecular acid in water (equation 14)

$$HA + H_2O = H_3O^+ + A^- \tag{14}$$

$\Delta S°$ is about -20 cal/deg/mole and $\Delta C_p°$ is about -40 cal/deg/mole. The dissociation of an amine cation, on the other hand, is isoelectric, so $\Delta S°$ and $\Delta C_p°$ may be zero or have low values, positive or negative.

The analysis of the factors determining the value of pK_a is complicated by the fact that $\Delta G°$, $\Delta H°$, and $\Delta S°$ all vary with the temperature:

$$d\Delta G°/dT = -\Delta S°, \quad d\Delta H°/dT = \Delta C_p°, \quad d\Delta S°/dT = \Delta C_p°/T.$$

Even $\Delta C_p°$ is temperature variant. Since $\Delta G°$ is essentially the quantity measured, it follows that the precision with which $\Delta H°$, $\Delta S°$, and $\Delta C_p°$ can be determined decreases with the increase in the number of differentials upon which they depend. Thus even if $\Delta G°$ can be determined to ± 0.5 cal/mole, the error in $\Delta H°$ may be ± 20 cal/mole and in $\Delta S°$ ± 0.1 cal/deg/mole, whilst the error in $\Delta C_p°$ will be even greater. Present accuracy does not permit the temperature dependence of $\Delta C_p°$ to be determined. The apparent alternative route of measuring $\Delta H°$ directly by heat of neutralization measurements is not useful, as it is as yet impossible to achieve the accuracy required with solutions sufficiently dilute for the purpose.

Meaningful values of the thermodynamic quantities at a particular temperature T can only be determined, therefore, by precision measurements over a range of graduated temperatures on either side of T. Then if $\Delta C_p°$ can be regarded as effectively constant over the temperature range used,

$$\Delta H_T° = \Delta H_0° + T\Delta C_p° \tag{15}$$

$$\Delta S_T° = \Delta S_0° + \Delta C_p° \ln T \tag{16}$$

Introducing these values into equation (13) gives

$$\Delta G_T° = \Delta H_0° + T\Delta C_p° - T(\Delta S_0° + \Delta C_p° \ln T) \tag{17}$$

and

$$pK_a = \frac{\Delta G_T^\circ}{2 \cdot 303\ RT} = \frac{\Delta H_0^\circ}{2 \cdot 303\ RT} + \frac{\Delta C_p^\circ}{R} \log T - \frac{\Delta S_0^\circ - C_p^\circ}{2 \cdot 303\ RT} \qquad (18)$$

or

$$pK_a = A/T + (\Delta C_p^\circ/R) \log T + B \qquad (19)$$

where $A = -\Delta H_0^\circ/2 \cdot 303\ R$ and $B = -(\Delta S_0^\circ - \Delta C_p^\circ)/2 \cdot 303\ R$. The results of measurements at different ionic strengths are extrapolated to zero ionic strength so as to obtain the value of pK_a at each temperature. The resulting data are then analysed using equation (19) and the best values of A, $\Delta C_p^\circ/R$, and B deduced. Thence the values of ΔH_T°, ΔS_T°, and ΔC_p° are derived[2].

Unfortunately only a few amines have been studied with such precision. A number of authors, however, have deduced approximate values of ΔS° and ΔH° from measurements at two temperatures by using the relationship $(d\Delta G^\circ/dT) = -\Delta S^\circ$, the value of ΔH° being obtained by application of equation (13). The values so obtained are very useful for comparative purposes, but their limitations must be realised when interpreting small differences.

The comparison of data for different amines is made more complicated by the fact that they often refer to measurements with different solvents or solvent mixtures and at different temperatures. The effects of these factors on pK_a will therefore be considered first.

B. The Nature of the Solvent

In general the net effects of the presence of any solute species in a solution should be assessed in terms both of solute–solvent interactions and of the influence of the solute on solvent–solvent interactions, which in water are particularly strong and specialised[3]. For ionic species close-range nearest-neighbour interactions with solvent molecules are strong. This interaction depends on the charge type and effective size of the ion, the factors which determine the intensity of its field, and also on its chemical nature. It may involve orbital overlap, an interaction akin to hydrogen bonding, or intense charge–dipole interaction. For aminium ions or amine molecules in aqueous solution hydrogen-bonding interactions undoubtedly occur with nearest-neighbour water molecules. Whatever type of interaction is involved in this primary hydration 'shell' it is sterically limited in extent but gives rise to a substantial loss of enthalpy (since attractive forces are satisfied), entropy (through loss of freedom of the solvent molecules), and usually, of heat capacity (through loss of freedom and through the fact that freedom tends to be regained with rise of temperature)[3].

The orderliness of arrangement can be regarded as transmitted outwards from the primary zone, so that there is 'correlation' between the first and second shells. The field of the ion is still strong, and highly polarised water molecules in the first shell readily form hydrogen bonds with others further from the ion. Proceeding outwards, however, the influence of the ion must decline and restriction of freedom of the solvent molecules decreases until orientational rather than translational freedom is lost. As a result, the secondary hydration zone is less well defined than the primary zone, but it contributes to the loss of entropy and of enthalpy. Its effect on heat capacity is probably composite. Since weak restriction of freedom will undergo release with increase of temperature there may be a positive contribution.

The order generated in this way locally around an ion does not fit particularly well into the peculiar orderliness of the structure of the water. It is primarily centro-symmetrical, whereas the water structure is not, and the result of conflict between the incompatible ordering influences seems to be a zone of disorder which makes a substantial positive contribution to the entropy. This is the reason why Frank and Evans[4] found that ions in aqueous solution have 'too much' entropy.

Non-polar solutes have the effect of promoting the ordered arrangement of water molecules about themselves, lowering both the enthalpy and entropy. Thus methane is only very slightly soluble in water because of the very large entropy loss which dissolution involves. This effect is common not only to all non-polar solutes but also to non-polar parts of bifunctional solute molecules with hydrophilic groups. It is called 'hydrophobic hydration'[4] because it involves no specific solute–solvent interactions, but is the effect of an inert body on solvent–solvent interactions in its vicinity. It is the cause of the apparently ice-like structure which seems to exist at the interface between water and gas or water and a hydrocarbon.

This hydrophobic hydration gives a very large contribution to the heat capacity, apparently because it 'melts' with rise of temperature. Since it conflicts with the centro-symmetrical association it is probably weakened or destroyed by the field of a charge on an adjacent part of the molecule. As a result it is more important for the neutral amine molecule than for its conjugate cation, and so it should lead to a negative contribution to ΔS° and a positive contribution to ΔC_p°, as well as making $\Delta H°$ less positive. It may be noted that it is a general phenomenon that the factors which lead to a decrease in $\Delta H°$ also decrease $\Delta S°$, and hence there is a partial compensation in the values of $\Delta G°$ and hence of pK_a.

These effects have been discussed in detail by Ives and Marsden[3], upon whose views the majority of this section has been based. They will be referred to again later in the discussion of various groups of amines.

In general, the apparent basic character of an amine is influenced by the base strength of the solvent, the greater its proton affinity the lower the pK_a value. In addition, however, it depends upon the relative extents to which the solvation of the amine molecule and the H_3O^+ ion on the one hand, and the amine cation on the other, affect the entropy and enthalpy of the system. There is no simple formula, therefore, relating the relative strengths in one solvent to those of the same amines in another solvent, and it is not surprising that the basic strengths of amines in anisole and chlorobenzene solutions, compared by measuring their equilibria with the indicators 2,6-dinitrophenol and bromophenol blue, do not parallel completely their strengths in water[5]. On the other hand, they show a much closer relationship with the catalytic constants for the decomposition of nitramide in anisole[6]. Similar behaviour has been observed with various bases in isoamyl alcohol[7].

The basic strengths of many amines have been measured in mixed solvents, particularly in mixtures of water with an alcohol. In such media a certain degree of solvent sorting occurs, with the result that the solvation layers around the various species tend to differ in composition from the bulk of the solution, thereby confusing the position still further. In spite of these complications, however, the pK_a values for different amines in these hydroxylic media differ numerically from, but are closely parallel with, their values in water. The parallelism arises doubtless because of the variation of $\Delta H°$, which depends predominantly on the relative proton affinities of the base and solvent. The actual numerical values, however, have not quite the same significance as for aqueous solutions, since the relationship $pK_a + pK_b = 14$ no longer holds.

C. Temperature

Since $\Delta G° = 2.303\ RT\ pK_a$ it follows that $(d\Delta G°/dT) = -\Delta S° = 2.303R\ pK_a + 2.303\ RT(d(pK_a)/dT)$ and hence

$$-\frac{(pdK_a)}{dT} = \frac{\Delta S°}{2.303RT} + \frac{pK_a}{T} \tag{20}$$

Thus (dpK_a/dT) is negative so long as $-\Delta S°$ is less than 4·576 pK_a. As $\Delta S°$ never has such high negative values the temperature coefficient of pK_a is always negative.

Perrin[8] has pointed out that the temperature variation of the pK_a values of univalent organic cations can be represented satisfactorily by the function

$$-\frac{d(pK_a)}{dT} = \frac{pK_a - 0·9}{T} \pm 0·004 \tag{21}$$

and for bivalent cations by

$$-\frac{d(pK_a)}{dT} = \frac{pK_a}{T} \tag{22}$$

These corrections tacitly assume that for univalent cations $\Delta S°$ is about $-4·1$ cal/deg/mole, whilst for divalent cations it is zero. Since for many amines $\Delta S°$ has relatively low values no great error is introduced by employing these relations, provided that the temperature range over which correction is made is relatively small.

D. Structural Features

The nature of the remainder of the molecule to which the amino group is linked affects the relative energies of the amine and its cation. Factors which decrease the free energy of the cation more than that of the amine tend to stabilize the former and hence to increase pK_a, whilst if the base is stabilized to a greater extent than the ion the pK_a value is reduced. The formation of the amine cation requires the donation of the lone pair on the nitrogen atom to a proton, so any factor which tends to increase the electron density or availability at the nitrogen atom will increase its proton affinity, that is will increase the strength of the base, whilst any factor which tends to decrease this electron density has the reverse effect. The modifications to the base strength produced by various groupings in the molecule are generally classified under three headings.

I. Inductive (polar) effects

When a hydrogen atom linked to a carbon atom in an amine molecule is replaced by a halogen atom or a nitro group, or by any other group attracting electrons, a dipole is set up with its negative end directed away from the carbon atom. This exerts a $-I$ inductive effect, that is electron density is drawn away from the carbon atom making it more electronegative. It therefore tends to draw electron density away from its neighbouring atoms, and eventually this leads to a decrease in the electron density on the nitrogen atom. The new dipole also exerts a direct field effect, tending to polarise the orbitals of all the other electrons in the molecule. The result of these effects is a reduction of the base strength of the amine.

A methyl group, having a weak dipole in the reverse sense, with its positive end away from the carbon atom, should have a slight tendency to exert a $+I$ effect, with resulting increase of base strength.

These inductive effects fall off very rapidly with distance, particularly in saturated compounds, so it is with substituents relatively near to the amine group that they are most marked. When the substituent is linked to an aromatic ring or other conjugated system with a highly polarizable π-electron system, however, these inductive effects extend over much further distances.

2. Mesomeric (resonance) effects

When the amino group is linked directly to a conjugated system the lone-pair electrons tend to become delocalized and to participate in the conjugated system. This reduces their availability for accepting a proton and hence lowers the base strength. If another group is now introduced into the system in such a position that its mesomeric effect can interact with that of the amino group there is a further effect. A nitro group or other group with a $-M$ effect will withdraw electron density from the conjugated system by the mesomeric mechanism and so effectively increase the $+M$ mesomeric effect of the amino group and decrease the base strength still further. On the contrary, a halogen atom, a hydroxyl group, or any other group with a $+M$ effect donates electron density to the system, opposing the mesomeric effect of the amino group and so increasing the base strength.

Since in general the inductive and mesomeric effects of different substituents present in the molecule are roughly additive, attempts have been made to correlate the pK_a values of aromatic amines with the Hammett σ functions of the substituents[9] and those of aliphatic amines with the corresponding Taft σ^* functions[10]. These relationships and their applications have been fully reviewed by Clark and Perrin[11].

3. Steric effects

When bulky groups are present in close proximity to an amino group, steric effects are to be expected. If they are present, for instance in an aromatic amine, in positions where they tend to prevent the amino group from assuming the conformation most favourable for mesomerism, they should increase the basic strength of the amine relative to that of the corresponding compound in which obstructing groups are absent. Such behaviour is observed with *ortho*-substituted *N,N*-dialkylanilines. In some other cases, however, the presence of such bulky groups tends to decrease the basicity of the amine. Although this general behaviour has been known for some time, the mechanism through which the effect is exerted has been the subject of considerable controversy. Brown and Cahn[12] suggested that the $-NH_3^+$ group in the cation is appreciably larger than the $-NH_2$ group of the free amine and is comparable in size with a methyl group. On this basis the presence of other large groups in the vicinity may cause steric interference which increases the potential energy of the cation relative to that of the free base. This would be unfavourable to cation formation and so would lower the strength of the base.

This theory, however, has not received universal acceptance. For one thing it is by no means certain that the anilinium ion has greater spatial requirements than the free amine, in fact the reverse seems more likely to be true. Although the ammonium ion is tetrahedrally symmetrical, the HÑH bond angle in ammonia is smaller than the tetrahedral angle. This suggests that the lone-pair electrons exert a greater repulsive force on the hydrogen atoms than does one bound hydrogen atom on another. The relatively small HÔH bond angle $(104\frac{1}{2}°)$ in the water molecule and the structures of some of the inter-halogen compounds also suggest that a lone pair has considerable spatial requirement. It seems much more likely that in the majority of cases the reason for decreased base strength is to be sought in solvation phenomena, as discussed in section II.B, rather than in actual physical interference of the groups.

As in some instances more than one mechanism may be acting, these effects will be discussed in conjunction with the classes of compounds in which they arise, particularly the alkylamines, *ortho*-substituted anilines and *N,N*-dialkylanilines, and polynuclear aromatic amines.

III. ALIPHATIC AMINES

A. The Methylamines

The basic strength of methylamine is appreciably greater than that of ammonia. This is sometimes attributed to the $+I$ inductive effect of the methyl group, leading to an increased electron availability at the nitrogen atom[13]. It would perhaps be better to regard this as a mutual polarisation of the methyl and amino groups. At the same time it is of importance to note, as will be discussed later, that the difference in $\Delta G°$ and hence in pK_a arises more from the difference between the entropy changes associated with the dissociation of the ions than from the difference between the enthalpy changes.

The substitution of a second hydrogen atom of the ammonia molecule by a methyl group to give dimethylamine leads to only a small further increase in base strength, whilst the substitution of the third hydrogen atom leads to a considerable decrease, trimethylamine being only slightly stronger as a base than ammonia.

This peculiar order of base strengths was attributed by Brown and his coworkers to what they call 'B-strain'[14, 15]. This they define as the strain introduced into a molecule as a result of changes in the normal bond angles of an atom brought about by the steric requirements of bulky groups attached to that atom. Thus they suggested that in trimethylamine the three methyl groups are crowded around the small nitrogen atom, and that their steric requirements are met by a spreading of the CÑC bond angles to a value greater than the tetrahedral angle. On this view the addition of a proton to the lone pair, which would tend to reduce the bond angle to the tetrahedral value, is resisted by the molecule. Very similar ideas have been expressed by Fyfe[16]. These views, however, seem to be inconsistent with the observations of Lide and Mann[17], who have deduced from the microwave absorption spectrum that the CÑC bond angle in trimethylamine is $108 \cdot 7 \pm 1°$.

The apparent anomaly is clarified to a considerable extent, however, by reference to the precision measurements of Everett and Wynne-Jones on ammonia[18, 19], methylamine, dimethylamine, and trimethylamine[2], which covered a wide range of temperatures and of ionic strengths. These permitted the evaluation of $\Delta C_p°$, $\Delta S°$, $\Delta G°$, and $\Delta H°$ for the dissociation of each of the cations. The final results for 25° and zero ionic strength are included in Table 1.

For the dissociation of the ammonium ion $\Delta S°$ is small and $\Delta C_p°$ is zero. Unlike methane, which is repelled from water due to the large negative ΔS value involved, the ammonium ion must be very strongly attracted by water. As a result there is only a very small decrease in entropy on dissociation corresponding with the slight changes in the orderliness of arrangement attending the replacement of the NH_4^+ ion by an H_3O^+ ion and a water molecule by an ammonia molecule. The equilibrium is determined principally, therefore, by $\Delta H°$, which in turn depends mainly on the relative base strengths of ammonia and water.

TABLE 1.   Thermodynamic functions for the dissociation of amine
cations in aqueous solution at 25°c.

Amine	ΔC_p° (cal/deg/ mole)	ΔS° (cal/deg/ mole)	ΔG (kcal/ mole)	ΔH° (kcal/ mole)	pK_a
Ammonia [18, 19]	0·0	−0·60	12·608	12·425	9·24
Methylamine [21]	+7·5	−4·7	14·484	13·088	10·63
Dimethylamine [21]	+19·9	−9·5	14·687	11·859	10·78
Trimethylamine [21]	+41·0	−15·2	13·358	8·815	9·80
Ethylamine [22, 16]	—	−3·1	14·504	13·58	10·61
	—	−11·4	14·55	11·15	10·67
Diethylamine [22, 16]	—	−7·2	14·916	12·77	10·94
	—	−15·7	14·97	10·3	10·98
Triethylamine [16]	—	−16·3	14·55	9·7	10·67
n-Propylamine [22]	—	−1·7	14·36	13·85	10·53
n-Butylamine [22]	—	−1·3	14·45	14·07	10·60
Ammonia in 60% MeOH [23]	0·0	+2·65	11·77	12·56	8·64
Methylamine in 60% MeOH [23]	−12	−1·73	13·306	12·79	9·76
[a]Dimethylamine in 60% MeOH [23]	0	−4·6	13·41	12·03	
[a]Trimethylamine in 60% MeOH [23]	Positive	−6·6	11·93	9·97	

[a] Data for ionic strength 0·10.

As the hydrogen atoms of the ammonia molecule are successively replaced by methyl groups the value of ΔS° changes systematically to more negative values. Trotman–Dickenson [20] explained this as arising from the progressive decrease in the number of hydrogen atoms on the amine cation available for hydrogen bonding to solvent molecules, thus decreasing the constraint produced in the solvent through this cause. At the same time ΔC_p° acquires a progressively larger positive value, this being associated with a large decrease in ΔH°. For the change from ammonia to methylamine the entropy effect predominates and leads to a large increase in ΔG° and hence in pK_a, but for the subsequent changes from methylamine to dimethylamine and from dimethylamine to trimethylamine almost similar increases occur in $-\Delta S^\circ$, but with each step ΔC_p° shows increasing increments, with associated decreases in ΔH°. The result is that a maximum value of ΔG° is reached at dimethylamine.

In addition to the effect produced by the reduction of the number of hydrogen bonds to the cation, and the further ordered structure produced near the aminium group, the successive additions of methyl groups would be expected to yield increasing areas of surface over which hydrophobic hydration can occur in the neutral amine molecule, and to reduce slightly the hydrogen bonding of its amino group. It is difficult to assess the area over which hydrophobic hydration can occur but it is not unreasonable to suppose that this may be very much greater over the almost complete hemisphere formed by three methyl groups than over the smaller surfaces presented by two or one methyl groups. Being close to the centre of positive charge this hydration atmosphere will be almost wholly dispersed in the cation, so the heat capacity of the system in the dissociated state will be higher than that for the cation–solvent system, since it contains a term representing the heat required to disperse this solvent layer. This has a corresponding effect on the enthalpy of the amine–solvent system, since a considerable amount of heat would be required to bring the solvent atmosphere around the methyl groups isothermally to the more random state which exists around these groups in the cation. These effects and theories regarding hydrophobic hydration are discussed fully by Ives and Marsden[3].

It is interesting to note that the effects become modified in aqueous methanol solution, where for ammonia $\Delta S°$ becomes positive, possibly through the ammonium ion exerting a greater sorting effect in its hydrogen bonding than does the hydroxonium ion. Although $\Delta C_p°$ has the rather unexpected value of -12 cal/deg/mole for methylamine, it is very low for the other amines, indicating that no phenomenon analogous to hydrophobic hydration occurs to an appreciable extent in this solvent mixture. The changes in $\Delta S°$, therefore, may reflect the decrease in the number of hydrogen atoms in the cation available for hydrogen-bond formation. The $\Delta G°$ values for dimethylamine and triethylamine must be lower at zero ionic strength than at $I = 0.10$, so the net result is that in this medium methylamine seems to be the strongest of the bases.

B. Higher Alkylamines

Lengthening of the alkyl chain makes only relatively slight differences to the pK_a values of primary amines (Table 2). After a very small increase in ethylamine they remain almost constant up to n-decylamine, after which there seems to be a slight decrease. Chain branching also produces very little effect. Amongst secondary amines (Table 3) there is evidence of the values rising to a maximum at di-n-butylamine, after which they assume an almost constant value.

TABLE 2.   pK_a Values of primary alkylamines in aqueous solution at 25°c.

Amine	pK_a
Ammonia	9·24[24, 19], 9·25[25]
Methylamine	10·62[26], 10·63[21], 10·64[25]
Ethylamine	10·64[22], 10·67[16]
n-Propylamine	10·53[22], 10·69[27]
n-Butylamine	10·60[22], 10·61[28, 29], 10·66[30]
n-Pentylamine	10·63[29], 10·61[31]
n-Hexylamine	10·64[29]
n-Heptylamine	10·66[29]
n-Octylamine	10·65[29], 10·57[31]
n-Nonylamine	10·64[29]
n-Decylamine	10·64[29]
n-Undecylamine	10·63[29]
n-Dodecylamine	10·63[29]
n-Tridecylamine	10·63[29]
n-Tetradecylamine	10·62[29]
n-Pentadecylamine	10·61[29]
n-Hexadecylamine	10·61[29]
n-Heptadecylamine–n-Docosylamine	10·60[29]
i-Propylamine	10·63[28]
i-Butylamine	10·43[29]
t-Butylamine	10·68[31]
i-Pentylamine	10·60[29], 10·64[28]
t-Pentylamine	10·72[31]
Methyl(diethylmethyl)amine	10·63[31]
Triethylmethylamine	10·59[31]

TABLE 3.   pK_a Values of secondary alkylamines in aqueous solution at 25°c.

Amine	pK_a
Dimethylamine	10·73[25], 10·77[32], 10·78[21], 10·81[33]
Diethylamine	10·94[22], 10·98[28, 16], 11·04[33]
Di-n-propylamine	11·00[31]
Di-n-butylamine	11·25[31], 11·31[28]
Di-n-pentylamine	11·18[28]
Di-n-hexylamine	11·01[29]
Di-n-octylamine	11·01[29]
Di-n-dodecylamine–Di-n-octadecylamine	11·00[29]
Di-i-propylamine	11·05[28]
Di-i-butylamine	10·50[34], 10·82[28]
Di-s-butylamine	11·01[34]
Di-i-pentylamine	11·00[28]
N-Methyl-2-aminoheptane	10·82[31]

The effect of chain length is more pronounced in the tertiary alkyl-amines (Table 4), where triethylamine is much more basic than trimethylamine, whilst N,N-dimethylethylamine and N,N-diethyl-methylamine lie intermediately. There are some anomalies amongst the data for the higher members of this series, and further measurements on them would be of interest.

TABLE 4. pK_a Values of tertiary alkylamines in aqueous solution at 25°c.

Amine	pK_a
Trimethylamine	9·74[25], 9·75[36], 9·76[35], 9·80[21, 32], 9·81[33]
Triethylamine	10·65[35], 10·67[16], 10·71[32]
Tri-n-propylamine	10·65[28]
Tri-n-butylamine	9·93[28], 10·87[37]
N,N-Dimethylethylamine	9·99[35]
N,N-Diethylmethylamine	10·29[35]
N,N-Dimethyl-n-propylamine	9·99[35]
N,N-Dimethyl-n-butylamine	10·02[35]
N,N-Dimethyl-i-propylamine	10·30[35]
N,N-Dimethyl-i-butylamine	9·91[35]
N,N-Dimethyl-s-butylamine	10·40[35]
N,N-Dimethyl-t-butylamine	10·52[35]

Studies of the entropy changes attending the dissociation of the cations of these amines have led to somewhat conflicting results (Table 1). From measurements at only two temperatures Evans and Hamann[22] found the value of $-\Delta S°$ decreased systematically from methylamine to n-butylamine, this being accompanied by a progressive increase in $\Delta H°$. They also found $-\Delta S°$ for the dissociation of the diethylammonium ion to be less than for the dimethylammonium ion. Fyfe[16], on the other hand, inferred from measurements at four temperatures that the values of $-\Delta S°$ for the ethylammonium, diethylammonium, and triethylammonium ions were all much higher than for the corresponding methylammonium ions.

The difference between these two series of amines is likely to arise principally from the hydrophobic hydration effects. If Evans and Hamann's results are correct they suggest that in ethylamine the extent of hydrophobic hydration may not be much greater than in methylamine, or may persist in part in the cations, since the β-carbon atom is an appreciable distance from the charge centre near the nitrogen atom. Fyfe's measurements, on the contrary, suggest that this hydrophobic hydration atmosphere is dispersed in the cation. Further measurements on these compounds are obviously required.

C. Substituted Alkylamines

The introduction of a trimethylsilyl group into an alkyl group increases the base strength, the pK_a values of both trimethylsilyl-methylamine and di(trimethylsilylmethyl)amine being appreciably higher than the values for methylamine and dimethylamine, respectively. The effect diminishes, however, the further the trimethylsilyl group is removed from the amino group[27] (Table 5).

Although chlorinated alkylamines generally hydrolyse or cyclise rapidly in solution, the pK_a values of some of these compounds have been measured. As is to be expected from their strong $-I$ effects,

TABLE 5. pK_a Values of substituted alkylamines in aqueous solution.

Amine	pK_a/Temperature (°c)
Trimethylsilylmethylamine	10·96/25[27]
β-Trimethylsilylethylamine	10·99/25[27]
γ-Trimethylsilylpropylamine	10·75/25[27]
Di(trimethylsilylmethyl)amine	11·40/25[27]
β,β-Difluoroethylamine	7·52/25[38]
β,β,β-Trifluoroethylamine	8·30/25[39]
γ,γ,γ-Trifluoro-n-propylamine	8·70/25[39]
β,β,β-Trichloroethylamine	5·47/20[40]
γ,γ,γ-Trichloro-n-propylamine	9·65/20[40]
δ,δ,δ-Trichloro-n-butylamine	9·93/20[40]
γ-Bromo-n-propylamine	8·93/21[41]
N,N-Di(β-chloroethyl)ethylamine	5·45/25[42]
N,N-Di(β-chloroethyl)-n-propylamine	6·68/25[42]
N,N-Di(β-chloroethyl)-i-propylamine	6·98/25[42]
N,N-Di(β-chloroethyl)-n-butylamine	6·61/25[42]
β-Hydroxyethylamine (ethanolamine)	9·50/25[43], 9·51/25[44, 45]
γ-Hydroxy-n-propylamine	9·85/25[46], 9·96/25[47]
δ-Hydroxy-n-butylamine	10·35/20[48]
Ethylenediamine (pK_{a2})	9·87/25[49], 9·93/25[50], 9·95/25[51], 9·98/25[52]
Propylenediamine (pK_{a2})	10·30/25[49], 10·47/25[51], 10·65/25[52]
Butylenediamine (pK_{a2})	10·75/25[51], 10·84/25[52]
Pentylenediamine (pK_{a2})	10·25/25[49]
Hexylenediamine (pK_{a2})	10·93/25[50]
Ethylenediamine (pK_{a1})	6·80/25[49], 6·85/25[50], 6·90/25[51]
Propylenediamine (pK_{a1})	8·29/25[49], 8·48/25[51]
Butylenediamine (pK_{a1})	8·78/25[49], 9·20/25[51]
Pentylenediamine (pK_{a1})	9·13/25[49]
Hexylenediamine (pK_{a1})	9·83/25[50]

halogens decrease the base strength of an amine, the effect diminishing the further the halogen is removed from the amino group. Although the available data are somewhat restricted, they indicate that fluorine has an appreciably greater effect than chlorine.

The presence of a hydroxyl group also decreases the base strength, and again there is a marked effect of the position of the substituent.

The introduction of a second amino group similarly decreases the pK_a value of ethylamine, but with the introduction of further methylene groups between the amino groups the base strength increases progressively and becomes higher than that of the corresponding alkylamine. This is attributable to the statistical factor, since in these compounds there are two possible sites for acceptance of the proton, as well as to the fact that with increasing separation the mutually polarizing effects of the two groups are diminished.

The protonated amino group naturally reduces the pK_a value more than the neutral amino group. Data are shown as first dissociation constant (pK_{a1}) values for the doubly charged diamine cation in Table 5. With increasing number of methylene groups separating the two amino groups the effect of the positive charge gradually diminishes, but again, for statistical reasons the pK_{a1} values are always smaller than for the corresponding alkylamines.

These effects are illustrated by the thermodynamic data shown in Table 6. In interpreting them it must be borne in mind that the

TABLE 6. Thermodynamic data for the dissociation of alkylenediamine and hydroxy-amine cations in aqueous solution at 25°c.

	ΔC_p° (cal/deg/ mole)	ΔS° (cal/deg/ mole)	ΔG° (kcal/mole)	ΔH° (kcal/mole)
Ethylenediamine 1[50, 51]	17·5	+5·1	9·343	10·87
	—	+3	9·4	10·3
2[50, 51]	9·6	−5·8	13·545	11·82
	—	−7	13·5	11·5
Propylenediamine 1[51]	—	+8	11·6	13·9
2[51]	—	−3	14·3	13·3
Butylenediamine 1[51]	—	−2	12·6	12·0
2[51]	—	−9	14·5	12·0
Hexylenediamine 1[50]	8·2	+1·3	13·411	13·82
2[50]	8·4	−3·3	14·912	13·91
Ethanolamine [43, 44]	−1·17	−2·94	12·955	12·08
	—	−5	13·0	11·5
β-Methoxyethylamine [44]	—	−4	12·9	11·7

entropy changes attending the first and second stages of the dissociation of the divalent cation include terms $+R \ln 2$ and $-R \ln 2$, respectively, for the statistical effect. Similarly the $\Delta G°$ terms include corresponding contributions of $-RT \ln 2$ and $+RT \ln 2$, respectively.

Everett and Pinsent[50] explained the results for ethylenediamine on the grounds that $\Delta G°$ for the first stage is low because of the high initial potential energy of the divalent cation through charge repulsion. This repulsion will stiffen the molecule, probably with a greater effect on the entropy than on the heat capacity. The cooperative effects of the two charges will cause much stronger solvent orientation than two separate charges. As the reaction to give the univalent cation and a proton is associated with freeing these solvent molecules, it will be accompanied by increases in both the entropy and heat capacity. The second stage of the dissociation follows much of the normal pattern for an alkylamine. In hexamethylenetetramine the effects of the repulsive forces between the charges seem to have virtually disappeared, the values of $\Delta S°$ and $\Delta G°$ for the two stages differing only by the statistical factors. Everett and Pinsent inferred, therefore, that the zone of influence of a single positive charge extends over about 5 Å.

The negative value of $\Delta C_p°$ for ethanolamine is of some interest, as it suggests that the presence of the hydroxyl group leads to hydrogen-bonding hydration in both the cation and free amine, thus effectively nullifying the effect of hydrophobic hydration.

The replacement of one of the hydrogen atoms in the alkyl chain of an alkylamine by a phenyl group causes a marked decrease in pK_a, the effect being greatest for benzylamine and diminishing with increased separation of the phenyl group from the amino group (Table 7). This can be attributed to the withdrawal of electron density from the nitrogen atom through the $-I$ inductive effect of the phenyl group. This arises through the fact that the sp^2 hybridisation of the orbitals of the ring carbon atoms causes its electronegativity to be greater than that of the methylene carbon atom with sp^2-hybridised orbitals[53]. Hyperconjugation of the methylene group to the aromatic ring may also play a role. The electron-donating character of the methyl group has been suggested as explaining the fact that the α-methyl derivatives of these compounds are stronger bases than the unmethylated compounds[54].

The electron-withdrawing effect of sp^2-hybridised carbon atoms also accounts for the fact that the presence of a double bond in the hydro-

TABLE 7. pK_a Values of phenyl-substituted and unsaturated amines in aqueous solution at 25°c.

Amine	pK_a
PhCH$_2$NH$_2$	9·35[55], 9·37[56]
Ph(CH$_2$)$_2$NH$_2$	9·83[55]
Ph(CH$_2$)$_3$NH$_2$	10·20[55]
Ph(CH$_2$)$_4$NH$_2$	10·39[55]
Ph(CH$_2$)$_5$NH$_2$	10·49[55]
PhCH$_2$NHMe	9·58[55]
PhCH$_2$NHEt	9·68[55]
PhCH$_2$NHPr	9·62[55]
Ph(CH$_2$)$_2$NHMe	10·14[55]
PhCMe$_2$NH$_2$	10·27[31]
PhCH$_2$CMe$_2$NH$_2$	10·03[31]
PhCHMeCH$_2$NH$_2$	9·80[31]
Allylamine	9·49[31]
Diallylamine	9·29[31]
Triallylamine	8·31[31]
Diallylmethylamine	10·11[34]

carbon chain decreases the basic strength of an amine. This is illustrated by the fact that the pK_a values of allylamine, diallylamine, and triallylamine decrease in that order, and are all lower than the values for the corresponding n-propylamines.

D. Cycloalkylamines

The base strengths of the amino-substituted cycloparaffins are of some interest as they vary with the size of the ring (Table 8). The pK_a value of cyclohexylamine in water is about equal to that for an ordinary alkylamine. In 50% ethanol, however, the values for cyclopropylamine and cyclobutylamine are much smaller than those of cyclopentylamine and cyclohexylamine. This apparently higher electron-attracting effect is in accordance with the view[59] that the external bonds of small ring compounds have more s character, and the ring carbon atoms greater electronegativity than in macrocyclic compounds[57]. The N,N-dimethyl derivatives of these cycloalkylamines also show a similar pattern.

With increasing ring size the pK_a in 80% methyl cellosolve appears to pass through a slight maximum for an eight-membered ring. It has been suggested that with a ring of this size steric factors are, on the average, most favourable for proton addition[58].

TABLE 8. pK_a Values of cycloalkylamines at 25°c.

Amine	pK_a in water	pK_a in 50% ethanol	pK_a in 80% methyl cellosolve
Cyclopropylamine		8·66[57]	
Cyclobutylamine		9·34[57]	
Cyclopentylamine		9·95[57]	
Cyclohexylamine	10·64[28] 10·68[51]	9·83[57]	9·82[58]
Cycloheptylamine			9·99[58]
Cyclooctylamine			10·01[58]
Cyclononylamine			9·95[58]
Cyclodecylamine			9·85[58]
Cycloundecylamine			9·71[58]
Cyclododecylamine			9·62[58]
Cyclotridecylamine			9·63[58]
Cyclotetradecylamine			9·54[58]
Cyclopentadecylamine			9·54[58]
Cycloheptadecylamine			9·57[58]
Cyclooctadecylamine			9·54[58]
Cyclopropyldimethylamine		7·70[57]	
Cyclobutyldimethylamine		8·77[57]	
Cyclopentyldimethylamine		8·93[57]	
Cyclohexyldimethylamine		9·17[57]	

IV. AROMATIC AMINES

A. Aniline

Aromatic amines are much weaker bases than are their aliphatic analogues, aniline having a pK_a value in water of 4·60 as compared with about 10·63 for methylamine or 10·66 for cyclohexylamine. Thus whilst the latter amines are stronger bases than ammonia, aniline is very much weaker. This difference can be ascribed to a small extent to the electronegativity of the phenyl group but is mainly due to the very large $+M$ mesomeric effect of the amino group when linked to a conjugated system.

The maximum mesomeric effect of the amino group could be exerted only if the hydrogen atoms of the amino group became coplanar with the ring. For this to occur it would be necessary for the hybridisation of the orbitals of the nitrogen atom to change from the approximately sp^3 tetrahedral character found in ammonia to the trigonal sp^2 type. Analogy with the energy associated with the inversion process in the ammonia molecule suggests that about 6 kcal/mole would be required for this change in hybridisation before the full gain in resonance energy could be achieved[60]. There is a considerable weight of evidence from various physical measurements that the atoms in aniline are not coplanar, and it appears that a substantial resonance energy can be achieved without coplanarity and hence with less energy expenditure. The result seems to be a compromise, whereby the HŃH bond angle is increased as compared with that in ammonia but without the group achieving complete coplanarity with the ring.

This explains why aniline is so much weaker as a base than is ammonia. When the anilinium $PhNH_3^+$ ion is formed, the electron pair, which is partially used to conjugate with the π-electron system of the ring and so lead to an increase in the resonance energy of the system, can no longer fulfil this function. The result is that, compared with ammonia or methylamine, the energy of the conjugate cation is very much higher than that of the base.

B. Meta- *and* Para-*substituted Anilines*

The introduction of substituents in the positions *meta* and *para* to the amino group of aniline leads to marked changes in the pK_a value which reveal clearly the two effects (inductive and mesomeric) which are operative in the molecules concerned. The effects of *ortho* substituents are more complicated and therefore they will be discussed separately. The pK_a values for a range of monosubstituted anilines are shown in Table 9.

The introduction of a *meta*-halogen atom reduces the pK_a of aniline considerably, the effects of fluorine, chlorine, bromine, and iodine being roughly equal. On the other hand a *para*-halogen atom has much less effect, a fluorine substituent even increasing it slightly. This can be accounted for by the fact that the $-I$ inductive effect of the *meta* halogen atom has a greater influence on the electron density on the nitrogen atom than has the inductive effect of a *para* halogen atom. At the same time, however, the $+M$ mesomeric effect of a *para* halogen atom interacts through the aromatic system in such a manner as to oppose the $+M$ mesomeric effect of the amino group. No similar mesomeric interaction can occur with a substituent in the *meta* position, which therefore has a much greater effect than a *para* halogen in decreasing the basic strength. It is interesting to note that if, as is suggested by the figures for the *m*-halogenoanilines, the four halogen atoms exert nearly equal $-I$ effects from the *para* position, their $+M$ effects must follow the order $F > Cl > Br > I$.

TABLE 9. pK_a Values of monosubstituted anilines in aqueous solution at 25°C.

The following are average values, with preference given to consistent data, based on the results given in the references shown in the last column.

Substituent	ortho	pK_a meta	para	References
(H)	(4·60)	(4·60)	(4·60)	28, 61, 62, 63, 64, 65, 66, 67, 68, 69, 70
CH_3	4·45	4·72	5·10	28, 61, 66, 70
F	3·20	3·55	4·60	61, 66, 71
Cl	2·64	3·50	3·98	28, 61, 67, 70, 71
Br	2·53	3·54	3·86	61, 70
I	2·60	3·61	3·78	61, 70
OH	4·74	4·30	5·65	70
OCH_3	4·52	4·21	5·34	28, 61, 66, 70
OC_2H_5	4·47	4·18	5·25	28, 66
NH_2	4·74	4·98	6·16	70, 72, 73
NH_3^+	0·6	2·24	2·67	70, 73
NMe_3^+	—	1·98	2·15	73
NO_2	−0·26	2·47	1·11	61, 66, 68, 70, 71, 74
CN	0·95	2·76	1·74	66, 68, 70
COMe	2·22	3·59	2·19	70
CO_2Me	—	3·55	—	70
SO_3^-	2·46	3·74	3·22	70, 75
SO_2NH_2	1·0	2·84	—	70
SO_2CH_3	—	2·70	1·47	76
CF_3	—	3·49	2·54	77
$SiMe_3$	—	4·64	4·36	78
C_6H_5	3·78	4·22	4·27	28, 79

Analogous behaviour, leading to rather more exaggerated results, is shown in the presence of hydroxyl, methoxyl, or ethoxyl substituents. These groups have only relatively weak $-I$ effects, but very strong $+M$ effects. The result is that when they occupy positions *meta* to an amino group they decrease the pK_a value slightly, but when in a position *para* to it they increase the pK_a value considerably.

In the case of a nitro substituent there is a $-M$ effect as well as a $-I$ effect, so in the *para* position it conjugates with the $+M$ mesomeric effect of the amino group, causing a still greater decrease in electron density at the nitrogen atom. Thus whilst *m*-nitroaniline is a much weaker base than aniline, *p*-nitroaniline is a very much weaker base still. Analogous effects are shown to a less marked degree by the cyano, acetyl, methylsulphonyl, and trifluoromethyl groups. This behaviour of the methylsulphonyl group has been interpreted[76, 80] as suggesting that the sulphur atom uses its vacant $3d$ orbitals in acquiring a fairly strong $-M$ mesomeric effect. That the trifluoromethyl group acts in this same way is particularly interesting. Its fairly strong $-M$ effect, revealed by the fact that it exerts a greater effect when in the *para* position, cannot be explained by the utilisation of d orbitals, and can be interpreted on valence-bond theory only by assuming that there are important contributions from structures of the type[77]:

$$\overset{+}{H_2N}=\underset{}{\langle\quad\rangle}=CF_2 \; F^-$$

The anilinesulphonic acids exist as zwitterions $^-O_3SC_6H_4NH_3{}^+$ and so can be regarded as substituted anilinium ions. Their pK_a values indicate that the base strength is reduced by the presence of the $SO_3{}^-$ group[81–83]. This has been interpreted as indicating that the $SO_3{}^-$ group has still a strong $-I$ effect, even though it carries a negative charge[84]. Its effect, however, is much greater when in the *para* position than in the *meta* position, suggesting that it also exerts a $-M$ effect, leading to conjugative interaction with the $+M$ effect of the amino group less strong in character than that of the nitro group.

Methyl and other alkyl groups have electron-donating effects and hence they tend to increase the pK_a value. The fact that this electron donation is primarily of a mesomeric character is indicated by the fact that *p*-toluidine is a stronger base than is *m*-toluidine.

Similar effects of the same kind are produced by the introduction of a second amino group. The pK_a values here are increased slightly through the statistical factor, since there are two sites for the acceptance of a proton, but nevertheless *p*-phenylenediamine is a much stronger base than aniline. This can be attributed to the opposition of the mesomeric effects of the two amino groups which increases the electron density at each nitrogen atom. The introduction of an $NH_3{}^+$ group, on the other hand, diminishes the base strength strongly. Here the statistical factor decreases the pK_a slightly, but the main contributing factor is the interaction between the groups. The fact that the difference between the first and second dissociation constants is so great both for the *m*-phenylenediaminium and the *p*-phenylenediaminium ions suggests that the inductive effect is the main factor.

C. Ortho-substituted Anilines

The effects of *ortho* substituents upon the basic strength of aniline are anomalous as compared with those of the corresponding *meta* and *para* substituents. Thus although an alkyl group in the *meta* or *para* positions raises the pK_a, an alkyl group in the *ortho* position always decreases it.

Wepster[62] has estimated the magnitude of the anomaly thus created by calculating the base strength to be expected on the assumption that the electrical effect of each alkyl group in the *meta* position raises the pK_a value by 0·15 units and that of each alkyl group in a *para* or *ortho* position raises it by 0·4 units. This equation of the effects of *ortho* and *para* substituents was justified by observations on analogous derivatives of pyridine. The difference between this calculated pK_a value and the observed value is denoted by δpK_a in Table 10. The pattern which emerges is fairly clear. Mono *ortho* substitution by a methyl, ethyl, or isopropyl group leads to a δpK_a value of about 0·6, but substitution by a *t*-butyl group leads to a value of 1·3. Two *ortho* methyl groups have about the same effect as one *t*-butyl group or double that of one methyl group, whilst two *ortho* *t*-butyl groups have rather more than twice the effect of one. The effects of the *ortho* groups, therefore, appear to be roughly additive.

The effect of an *ortho* methyl group in causing *o*-toluidine to have such a low pK_a value was attributed by Bennett and Mosses[85] to the strong inductive effect produced in adjacent groupings by the dipole of the methyl group. Later views have generally attributed it to steric effects but the exact nature of the steric effects concerned has been the subject of considerable controversy.

It is certain that the effect does not arise simply through a modification of the resonance in the amine itself caused by the presence of an alkyl group. Such steric inhibition of resonance would in fact tend to increase the base strength, as it would increase the energy of the base relative to that of the ion. The decrease in pK_a would suggest, therefore, a steric promotion of resonance rather than its inhibition, but there is no evidence of this. Other physical properties such as the ultraviolet absorption spectra and molecular refractions of the compounds concerned indicate little or no change in the mesomeric effect of the amino group produced by the presence of *ortho* substituents.

It was suggested by Brown and Cahn[11] that the NH_3^+ group in the anilinium ion is larger than the amine group and comparable in size with a methyl group. On this view, if there is another group of

TABLE 10. pK_a Values of alkyl-substituted anilines in water or 50 volume percent ethanol at 25°c.

Substituent	Water		50 vol. % EtOH	
	pK_a	δ pK_a	pK_a	δ pK_a
None	4·60[62]	(0)	4·26[86]	(0)
2-Methyl	4·44[62]	0·6	4·09[62]	0·6
3-Methyl	4·68[87]	0·1		
4-Methyl	5·11[62]	−0·1	4·74[62]	−0·1
2-Ethyl	4·37[62]	0·6	4·04[62]	0·6
3-Ethyl	4·70[66]	0·0		
2-n-Propyl	4·36[87]	0·6		
2-i-Propyl	4·42[62]	0·6	4·06[62]	0·6
3-i-Propyl	4·67[66]	0·1		
2-n-Butyl	4·26[87]	0·7		
2-t-Butyl	3·78[62]	1·2	3·38[62]	1·3
3-t-Butyl	4·66[66]	0·1		
4-t-Butyl	4·95[62]	0·0	4·62[62]	0·0
2,3-Dimethyl	4·72[62]	0·4	4·42[86]	0·4
2,4-Dimethyl	4·84[87]	0·6	4·61[88]	0·4
2,5-Dimethyl	4·57[62]	0·6	4·23[62]	0·6
2,6-Dimethyl	3·89[62]	1·5	3·49[62]	1·6
3,4-Dimethyl	5·22[79]	−0·1		
3,5-Dimethyl	4·91[62]	0·0	4·61[62]	−0·1
2,4-Di-t-butyl			3·80[74]	1·3
2,5-Di-t-butyl			3·58[74]	1·2
2,6-Di-t-butyl			1·80[74]	3·3
3-5-Di-t-butyl	4·97[74]	−0·1	4·73[74]	−0·2
2,4,6-Trimethyl	4·37[62]	1·4	4·00[62]	1·3
2,4,6-Tri-i-propyl			4·04[62]	1·4
2,4,6-Tri-t-butyl			< 2[62]	
2,6-Dimethyl-4-t-butyl			3·88[74]	1·6
2,4-Dimethyl-6-t-butyl			3·40[74]	2·1
2-Methyl-4,6-di-t-butyl			3·35[74]	2·1
2,3,5,6-Tetramethyl	4·30[62]	1·4	4·09[62]	1·3

similar or larger size *ortho* to it the groups will suffer steric interference. The resulting strain in the ion would increase its potential energy, thereby opposing its formation and increasing the base strength. This type of strain caused by steric interference of atoms or groups which are attached to different atoms has been denoted 'F Strain' by Brown and his coworkers.

On this theory the effect of a substituent alkyl group should increase with increasing bulk, and this was apparently confirmed by the fact that the pK_a value is reduced from 4·60 in aniline to 4·44 in *o*-toluidine and 3·78 in *o-t*-butylaniline. As has been mentioned in section II.D.3 the premises of this theory are by no means generally acceptable. In addition Wepster[62] has pointed out that since the most favourable conformation of the amine has the protons as nearly as possible in the plane of the ring, it is impossible to accept that the addition of a proton would produce an appreciable increase of steric strain even if its spatial requirements are greater than that of a lone pair of electrons. He allows that another configuration might be more favourable for the ion, but this would imply a decrease of steric strain corresponding with an increase in base strength.

Wepster also rejects the suggestion made by Beale[63] that the relatively low basicities of 2-methyl- and 2,6-dimethylaniline arise from an electrostatic interaction between the partially unscreened nuclei of the methyl hydrogen atoms and the lone pair. This, he suggests, amounts to the assumption of a hydrogen bond, and since a $N \cdots H—N$ bond is already very weak a $N \cdots H—C$ bond must be still weaker. Wepster also points out that the spatial configuration of 2-methylaniline is altogether unsuited for hydrogen-bond formation and that the explanation cannot be extended to 2-*t*-butylaniline and to 2,6-di- and 2,4,6-tri-*t*-butylaniline.

The suggestion made by Wepster is that the relative solvation energies of the anilinium ions may be the factor determining their relative strengths, as the solvation energies of the uncharged amines may be neglected. With increasing size of the alkyl groups in the *ortho* position the solvation shell round the NH_3^+ group becomes smaller and smaller, and hence renders the ion more and more unstable. This must result in an increasing rise of the value of δpK_a and so can be described as steric hindrance to solvation.

Brown and his coworkers had rejected steric hindrance to hydration as a main cause of these *ortho* effects, because they considered that a more gradual increase in the anomalies should be found[89]. Wepster points out that substitution of a hydrogen atom by a methyl group removes at least one water molecule from the solvation shell, so that for instance in the ion of 2-*t*-butylaniline the solvation energy will be decreased considerably as compared with the 2-isopropylammonium ion.

Doubtless the factors which determine this *ortho* effect in the alkyl-anilines are also responsible, in part at least, for the similar effects in other *ortho*-substituted anilines, although in some cases the dipolar character of the substituent and strong inductive effects may tend to confuse the position. Unfortunately there have been only two series of precision measurements from which thermodynamic data can be deduced with certainty. For the three sulphonic acids the ΔC_p° values are 18·6, 7·7, and 7·1 cal/deg/mole for the *ortho*, *meta*, and *para* compounds respectively, whilst the corresponding values of ΔS° are 3·32, 0·57, and 0·37 cal/deg/mole[81]. For aniline itself ΔS° has the low positive value of $+0.28$ cal/deg/mole which is increased to 8·1 cal/deg/mole for *o*-chloroaniline, whilst ΔH° is decreased from 7·105 to 6·007 kcal/mole[21]. The entropy change is the reverse of that to be expected if the hydrophobic hydration of the amine were a controlling factor, but the values for the *ortho* compounds accord with Wepster's views on solvent exclusion. In each case, however, other factors may enter, so further studies are required to ascertain whether similar effects arise with the *o*-alkylanilines.

D. Secondary and Tertiary Alkyl- and Arylanilines

Replacement of the amino hydrogen atoms of aniline by methyl groups to yield methylaniline and dimethylaniline leads to successive increases in base strength (Table 11). This contrasts with the order of strength of the methylamines, where there is actually a decrease in basic strength on passing from dimethylamine to trimethylamine. The increases in pK_a observed as the hydrogen atoms of aniline are replaced by ethyl groups are even more pronounced. These changes are far too great to be accounted for by the inductive effect, and there is no doubt that steric effects of some kind are playing an important role, but unfortunately there are again no data from which the entropy, heat capacity, and enthalpy changes accompanying the dissociation of the cation can be derived, so the precise steric mechanism operating is uncertain.

Brown has suggested[95] that the conformation of N,N-diethylaniline with the two ethyl groups so arranged as to minimise interaction with the phenyl group would cause steric interference between the ethyl groups themselves. Rotation of one ethyl group towards the ring to remove this interference would cause some interference with the *ortho* hydrogen atoms, an overcrowding effect which could be relieved only by some twisting of the diethylamino group from its position for maximum resonance, with resulting increase in energy of the amine relative to its conjugate cation and hence an increase in its base strength.

Further substitution of the hydrogen atoms of the methyl groups to give N,N-diisopropylaniline leads to a further increase in base strength.

7*

TABLE 11.   pK_a Values of N-alkyl- and N,N-dialkylanilines at 25°c.

Derivative	Water	pK_a Value in 50 vol % EtOH	75 vol % EtOH
N-Methyl-	4·85[28,62,90]	4·29[74]	—
N-Ethyl-	5·11[28]	4·71[74]	4·25[91]
N-n-Propyl-	5·02[28]		
N-n-Butyl-	5·12[31]		4·02[91]
N-n-Hexyl-	5·33[31]		4·02[91]
N-n-Octyl-			4·02[91]
N-i-Propyl	5·77[31], 5·50[92]		
N-i-Butyl	4·43[31]		
N-t-Butyl-	7·00[8]		
N-t-Pentyl-	6·75[93], 6·35[94]		
N-s-Butyl	5·22[94]		
N-Neopentyl-	4·17[31]		
N-t-Hexyl-	6·34[31]		
N,N-Dimethyl-	5·06[28], 5·07[68], 5·12[62], 5·15[90]	4·39[86]	
N,N-Diethyl-	6·56[28]		5·41[91]
N,N-Di-n-propyl-	5·59[28]		
N,N-Di-n-butyl-	6·21[31]	4·78[68]	4·50[91]
N,N-Di-n-hexyl-			4·39[91]
N,N-Di-n-octyl-			4·37[91]
N,N-Diisopropyl-	7·37[31]		
N,N-di-t-butyl-		4·00[93]	
N-Ethyl-N-methyl-	5·98[28]		
N-Methyl-N-n-propyl-	5·64[28]		
N-i-Butyl-N-methyl-	5·20[54]		
N-s-Butyl-N-methyl-	6·03[54]		
N-t-Butyl-N-methyl-	7·39[31]		
N-Ethyl-N-n-propyl-	6·34[54]		
N-t-Butyl-N-ethyl-	8·02[31]		
N-Cyclopentyl-	5·30[54]		
N-Cyclohexyl-	5·60[54]		
N-Cyclopentyl-N-methyl-	6·71[54]		
N-Cyclohexyl-N-methyl-	6·35[54]		

This feature makes the relatively low value found[93] by Vexlearschi and Rumpf for N,N-di-t-butylaniline in 50% ethanol rather puzzling and indicates once again the necessity for further work in this field, especially as N-t-butyl-N-ethylaniline is reported as having a pK_a value of 8·02, the highest value observed for an aromatic amine.

In relation to these values the relative base strengths of the mono-alkylanilines present some interesting features. Again, replacement of the methyl hydrogen atoms by methyl groups to give N-ethylaniline, N-i-propylaniline, and N-t-butylaniline leads to successive increases in base strength. On the other hand, replacement of the β-hydrogen atoms in N-ethylaniline to give N-n-propylaniline, N-i-butylaniline, and N-neopentylaniline is accompanied by decreases in pK_a, whilst a further decrease still is observed in N-t-hexylamine (α-ethyl-α-methyl-N-propylamine) [93]. It appears, therefore, that closely packed groups near to the nitrogen atom increase the base strength, whilst when further away from the nitrogen atom they decrease it. The latter effect was attributed by Vexlearschi and Rumpf to electrostatic interaction diminishing the stability of the ion relative to the amine [93], but it is difficult to see why such an effect should be so strong as to cause the large effects observed.

A further anomalous feature is that, apart from a slight deviation for N-n-propylaniline, the pK_a values of the N-alkylanilines increase progressively with chain length when measured in aqueous solution. In 75 vol % ethanol, however, N-n-butyl-, N-n-hexyl-, and N-n-octyl-aniline are reported as having equal pK_a values which are appreciably less than the value for N-ethylaniline in the same solvent.

It is quite possible that in these series steric inhibition of resonance does play some part in determining the pK_a values. This would account for the large effects of compact groups near to the nitrogen atom, but extensive series of precision measurements will be required before the problems can be solved completely.

Replacement of one of the amino hydrogen atoms in aniline by a phenyl group to give diphenylamine leads to a decrease of the pK_a value to 0·79 [98], whilst the pK_a value of triphenylamine is too small for measurement. These facts can be associated with the further stabilisation of the amine with respect to the ion through the further incorporation of the lone pair of electrons into π-electron systems of the rings.

E. Derivatives of Methylaniline and Dimethylaniline

In so far as the *meta* and *para* derivitives of N-methylaniline and N,N-dimethylaniline are concerned the pK_a values follow very closely the same pattern as for the derivatives of aniline (Table 12). It is in the *ortho*-substituted compounds that the most interesting features arise. N,N-Dimethyl-o-toluidine has an appreciably higher basic strength than N,N-dimethylaniline. This is to be expected since steric obstruction by the *ortho* methyl group prevents the dimethylamino

TABLE 12.   pK_a Values for derivatives of N,N-dimethylaniline at 25°c
in water (w) or in 50% ethanol (ae).

Substituent	ortho	pK_a meta	para	References
None	w 5·12			28, 62, 68, 96
	ae 4·39			86
Methyl-	w 6·11	w 5·24	w 5·50	28, 62, 68
	ae 5·15	ae 4·60	ae 4·88	78, 86
Fluoro-			ae 3·96	97
Chloro-		w 3·79	w 4·34	68, 96
		ae 3·05	ae 3·29	78, 97
Bromo-	w 4·31		w 4·23	31, 68
		ae 3·04	ae 3·47	78
Methoxy-	ae 5·42		ae 5·09	97
Cyano-	w 0·95	w 2·75	w 1·78	68
Trifluoromethyl-		w 3·27	w 2·67	77
Trimethylsilyl-		ae 4·35	ae 3·94	78
Nitroso-			ae 3·48	97
Nitro-		w 2·63	w 0·61	68

group from assuming near coplanarity with the ring and hence decreases the mesomeric effect. On the other hand N-methyl-o-toluidine, like o-toluidine, is a weaker base than its parent amine, but 2,6-dimethyl-N-methylaniline is a stronger base than N-methylaniline. This can be accounted for because the NHMe group of N-methyl-o-toluidine should be able to attain near coplanarity with the ring whilst the methyl groups are remote from one another, but when there are two *ortho* methyl groups this is impossible.

On these lines a second methyl group in the *ortho* position might be expected to lead to an increased obstruction to the coplanarity of the dimethylamino group with the ring, but, in fact, 2,6-dimethyl-N,N-dimethylaniline is a weaker base than N,N-dimethyl-o-toluidine. This observation led Thomson[88] to the conclusion that resonance must not be such an important factor in determining base strength as had previously been supposed.

In order to obtain a comparative picture of the relative effects of various substituents on the pK_a value Wepster[62] has used the same method as for the derivatives of aniline, but with an additional term to allow for the steric inhibition of resonance of the base. Thus he adds to the pK_a value of dimethylaniline contributions of $+0·4$ units for each *para* or *ortho* alkyl group and $0·15$ units for each *meta* alkyl group. The extra term is calculated from the extinction coefficient ε of the

'C' band in the ultraviolet spectrum at about 250 mμ. This is taken as being 15,500 when mesomerism is fully exerted and zero when it is excluded. This would correspond with a change in pK_a of about 3 units, so the effect of the inhibition of resonance on the pK_a value is evaluated as $3(15,500 - \varepsilon)/15,500$. The differences between the pK_a values calculated in this way and the observed values give the δpK_a values recorded in Table 13. Qualitatively the order of the results follows closely those for the primary amines, but differ from them in magnitude by a factor of 2 to 3.

The interpretation of the pK_a values of this series of compound has led to even more controversy than for the derivatives of aniline. Brown and Cahn[12] pointed out that in N,N-dimethyl-o-toluidine resonance is already partly suppressed in the base, so addition of a

TABLE 13.   pK_a Values for alkyl-substituted N,N-dimethylanilines at 25°c.

Substituent	Water	pK_a in 50 vol % EtOH	δ pK_a
N,N-Dimethylaniline	5·12[62]	4·39[86]	(0)
2-Methyl-	6·11[62]	5·15[86]	1·4
3-Methyl-	5·24[28]	4·60[78]	0·0
4-Methyl-	5·50[28]	4·88[78]	−0·1
2,3-Dimethyl-		5·25[86]	1·6
2,4-Dimethyl-		5·28[88]	1·6
2,5-Dimethyl-		5·19[88]	1·6
2,6-Dimethyl-	5·3[78]	4·81[62]	3·0
3,5-Dimethyl-		4·48[88]	0·2
2-Ethyl-		5·20[62]	1·6
4-Ethyl-		4·63[99]	0·2
2-i-Propyl-		5·05[86]	1·9
4-n-Propyl-		4·37[99]	
4-i-Propyl-		4·71[99]	0·1
4-n-Butyl-		4·56[99]	0·2
4-i-Butyl		4·13[99]	
2-t-Butyl-		4·28[62]	3·4
4-t-Butyl-		4·59[99]	
2,5-Di-t-butyl-		4·32[100]	
2,4,6-Trimethyl-		5·19[74]	3·2
2,3,5,6-Tetramethyl-		4·75[62]	3·3
6-t-Butyl-2,4-dimethyl-		2·93[74]	5·5
4,6-Di-t-butyl-2-methyl-		2·77[74]	5·7
2-t-Butyl-4-nitro-		1·63[100]	
2,5-Di-t-butyl-4-nitro-		1·91[100]	
3,5-Di-t-butyl-4-nitro-		2·93[74]	

proton involves a smaller loss of resonance energy than it does in N,N-dimethylaniline. At the same time it was assumed that the total energy change is increased by an additional steric strain accompanying the formation of the ion, with the net result that the basic strength is increased somewhat, but not by so much as it would have been in the absence of this strain. They suggest that in 2,6-dimethyl-N,N-dimethylaniline the resonance is reduced still further in the base, but that the steric strain in the formation of the ion is increased still more and becomes the predominating factor, leading to a reduction in pK_a. On the basis of these considerations Brown and Cahn predicted that as the bulk of the *ortho* substituent is increased the base strength should first increase with the inhibition of resonance and then decrease through strain in the ion.

Once again, however, the assumption of increased strain in the ion as compared with the amine itself does not seem to be justified. Wepster[62] has suggested that in 2-t-butyl-N,N-dimethylaniline steric repulsions cause the lone pair to be strongly oriented towards the t-butyl group, and that little effective reorientation is to be expected in the ion. He has also pointed out that the same argument cannot be used for 2-methyl-N,N-dimethylaniline. The steric strain in this molecule should be about the same as in the ion of 2-methylaniline, since the dimethylamino group is of similar size to the isopropyl group, the steric requirements of which are not much greater than those of the methyl group. Hence Wepster considers that Brown's premises cannot furnish a consistent explanation for the differences in δpK_a between primary and tertiary amines.

Wepster explains this difference between primary and tertiary amines on the grounds of the strong orientation of the solvent shells, especially in the presence of only one hydrogen atom in the ions. The greater the angle of twist of the dimethylamino group around the $C_{(Ar)}$—N bond the more strongly is the N—H bond oriented towards the group in the *ortho* position and the more the solvation energy will decrease. In nonaqueous media amines do not appear to exhibit the anomaly or do so to a less marked degree. For instance in hexane 2,6-dimethyl-N,N-dimethylaniline behaves as a much stronger base than 2-methyl-N,N-dimethylaniline, suggesting a pK_a value in water several units higher than the observed value[5].

As would be expected, the pK_a values of alkyl-substituted derivatives of N-methylaniline vary from that of the parent amine in a similar manner to those of the corresponding aniline derivatives when there is not more than one *ortho* substituent, but follow more closely the

TABLE 14. pK_a Values of alkyl derivatives of N-methylaniline and N-ethylaniline in 50 vol % ethanol at 25°C.

Compound	pK_a	δpK_a
N-Methylaniline	4·29[62]	(0)
2-t-Butyl-	3·35[74]	1·3
2,4-Di-t-butyl-	3·75[74]	1·3
2,4,6-Trimethyl-	5·77[74]	1·3
2-Methyl-4,6-di-t-butyl-	4·49[74]	2·5
2,4,6-Tri-t-butyl-	3·57[74]	3·8
N-Ethylaniline	4·71[74]	(0)
2,4,6-Trimethyl-	5·67[74]	1·8
2,4,6-Tri-t-butyl-	3·20[74]	4·5

values for the N,N-dimethylaniline derivatives when both *ortho* positions are occupied. These effects are illustrated in Table 14.

F. Amino Derivatives of Polynuclear Hydrocarbons

The relative pK_a values of amines derived from polynuclear hydrocarbons present a fairly consistent pattern, although all its features cannot be fully explained. The pK_a values of *m*- and *p*-aminobiphenyl are somewhat lower than that of aniline, while that of *o*-aminobiphenyl is much lower still (Table 15). Some lowering in all three cases is to be expected through the inductive effect of the second ring, but the extra effect for the *ortho* compound must arise from some steric effect, possibly steric exclusion of hydration of the cation. It is noteworthy that ionisation of the *o*-aminobiphenyl cation is accompanied by a much greater increase in entropy than for its *meta* and *para* isomers. Closely analogous behaviour is shown by the aminofluorenes.

The amines of condensed ring systems are all less basic than aniline, and in the absence of possible steric effects there is a tendency for their pK_a values to decrease with increasing number of condensed rings. The measurements of Elliott and Mason[101] indicate that in the unhindered amines the decrease in pK_a arises because, although both the enthalpy and entropy changes accompanying dissociation of the cations decrease, the enthalpy effect outweighs the entropy effect. These results are in accord with the view that increase in the size of the conjugate system is accompanied by increased delocalization of the lone-pair electrons.

1-Naphthylamine, however, is appreciably less basic than 2-naphthylamine, and in general, compounds in which the amino group

TABLE 15.　pK_a Values for polynuclear amines in water (w), in 50 vol %
ethanol (AE), or 20 wt % dioxan (AD), and enthalpy and entropy
changes on dissociation.

Amine	Solvent temperature	pK_a	ΔH (kcal/ mole)	ΔS (cal/deg/ mole)
Aniline	w (25°)	4·60[21]	7·1	+0·28
	AE (20°)	4·19[101]	5·9	+1·0
2-Aminobiphenyl	AE (20°)	3·03[101]	5·7	+5·2
3-Aminobiphenyl	AE (20°)	3·82[101]	5·2	+0·2
4-Aminobiphenyl	AE (20°)	3·81[101]	5·6	+1·5
1-Naphthylamine	w (25°)	3·92[102]		
	AE (20°)	3·40[101]	5·9	+4·6
2-Naphthylamine	w (25°)	4·16[102]		
	AE (20°)	3·77[101]	4·8	−0·9
1-Aminoanthracene	AE (20°)	3·22[101]	5·2	+2·9
1-Aminophenanthrene	AE (20°)	3·23[101]	5·6	+4·2
2-Aminophenanthrene	AE (20°)	3·60[101]	4·4	−1·3
3-Aminophenanthrene	AE (20°)	3·59[101]	4·4	−1·3
9-Aminophenanthrene	AE (20°)	3·19[101]	5·6	+4·3
1-Aminofluorene	w (25°)	3·87[103]	3·87	
2-Aminofluorene	w (25°)	4·64[103]	4·64	
	AE (20°)	4·21[101]	5·7	+0·3
3-Aminofluorene	w (25°)	4·82[103]	7·1	
4-Aminofluorene	w (25°)	3·39[103]		
3-Aminopyrene	AE (20°)	2·91[101]	3·9	−0·1
4-Aminoacenaphthene	w (25°)	4·50[105]		
5-Aminoacenaphthene	w (25°)	4·58[105]		
1-Aminobenzo(C)- phenanthrene	w (26°)	2·78[104]		
2-Aminobenzo(C)- phenanthrene	w (26°)	3·66[104]		
3-Aminobenzo(C)- phenanthrene	w (26°)	3·56[104]		
4-Aminobenzo(C)- phenanthrene	w (26°)	3·32[104]		
5-Aminobenzo(C)- phenanthrene	w (26°)	3·10[104]		
6-Aminobenzo(C)- phenanthrene	w (26°)	3·22[104]		
1-Dimethylaminonaphthalene	AD (25°)	4·43[105]		
2-Dimethylaminonaphthalene	AD (25°)	4·36[105]		
4-Dimethylamino- acenaphthene	AD (25°)	4·92[105]		
5-Dimethylamino- acenaphthene	AD (25°)	5·07[105]		

occupies a *peri* position, as in 1-aminoanthracene and 1- and 2-amino-phenanthrene, have lower pK_a values than their isomers. Mason[106] has obtained infrared evidence for steric hindrance to conjugation in these amines, so the increased values of $\Delta S°$ for these compounds and also for *o*-aminobiphenyl may be due to the rotation entropy of the amino group not being lost in the free amine, as it is for an unhindered amine. The dissociation enthalpies of the cations with *peri* hydrogen atoms are only slightly higher than for the unhindered cations, so the entropy effect outweighs that of the enthalpy, leading to a net decrease in $\Delta G°$ and hence in pK_a. It seems, therefore, as though the steric effect may hinder free rotation of the amino group without greatly reducing its conjugation with the aromatic system, an effect which would be expected to increase $\Delta H°$. Alternatively the particular positioning of the *peri* hydrogen atoms may cause a large effect through solvent exclusion from the conjugate cation, but this would also be expected to give a negative contribution to $\Delta S°$. Similar behaviour is to be expected, and is observed, whenever the amino group is close to a hydrogen atom linked to another ring.

It is of interest that, through steric inhibition of resonance of the free amine, N,N-dimethyl-1-naphthylamine is a slightly stronger base than N,N-dimethyl-2-naphthylamine, but the obstruction in ace-naphthene is insufficient to make N,N-dimethyl-4-aminoacenaphthene a stronger base than N,N-dimethyl-5-aminoacenaphthene.

The pK_a values of the nuclear-substituted naphthylamines present some features of interest (Table 16). Substituents in the same ring as the amino group produce effects similar to those in the aniline series, but the presence of the nitro group has a greater effect on 1- than on 2-naphthylamine, thus accentuating the difference in base strength. The pronounced difference between the pK_a values for 1- and 3-nitro-2-naphthylamines is attributable to the double-bond character of the 1-2 bond being greater than that of the 2–3 bond[107]. A sub-stituent in the opposite ring to the amino group has a much smaller effect, attributable mainly to induction. Here the relative effects are greater with 1- than with 2-naphthylamine.

V. COMPLEXES WITH LEWIS ACIDS AND WITH METAL IONS

A. Addition Products with Lewis Acids

Besides accepting protons from Lowry–Brönsted acids, amines can also add on to Lewis acids, that is to say to compounds which can

TABLE 16. pK_a Values of substituted naphthylamines in aqueous solution at 25°C[75, 103, 107].

1-Naphthylamine Substituent	2	3	4	pK_a 5	6	7	8
CH$_3$		3·96					
Cl		2·66		3·34	3·48	3·48	
Br		2·67	3·21				
OH		3·30		3·96	3·97	4·20	
OMe		3·26			3·90	4·07	
CO$_2$Me		3·12					
NO$_2$	−1·74	2·07	0·54	2·73	2·89	2·83	2·79
CN		2·26					
SO$_3$	1·71	3·20	2·81	3·69	3·80	3·66	5·03

2-Naphthylamine Substituent	1	3	4	5	6	7	8
Cl			3·38			3·71	
Br			3·40				
I			3·41				
OH				4·07		4·25	
OMe			4·05		4·64	4·19	
CO$_2$Me			3·38				
NO$_2$	−0·85	1·48	2·43	3·01	2·62	3·10	2·73
CN			2·66				
SO$_3$	2·35		3·70	3·96	3·74	3·95	3·89

accept an electron pair. One of the best known examples of such addition products is that formed between ammonia and boron trifluoride.

$$H_3N: + BF_3 \longrightarrow H_3N{\rightarrow}BF_3$$

Exactly analogous compounds are formed, however, by amines, products of the types $R_3N{\rightarrow}BF_3$, $R_3N{\rightarrow}BH_3$, and $R_3N{\rightarrow}BMe_3$ being characteristic. It has been found that some of the factors which may modify the basic character of amines can be more readily studied through observations of the equilibrium constants of these reactions with Lewis acids than through measurements on aqueous solutions. In this way a great deal of information can be obtained regarding the inductive, mesomeric, and steric effects upon the basic character of amines. Measurements of the degree of dissociation of the complexes are made in the vapour phase at a series of temperatures, so as to obtain values of both the equilibrium constant and its temperature coefficient. As a result $\Delta G°$, $\Delta H°$, and $\Delta S°$ can all be calculated.

The validity of comparisons made in this way was first demonstrated by Brown and Bartholomay[13], who investigated the base strengths of ammonia and the methylamines in the vapour phase by using trimethylboron as a reference acid. The total pressures set up by equimolecular mixtures in the vapour state at various temperatures were measured, and the dissociation constants of the addition products calculated from the results. These were found to place the bases in the same relative order as had been found for aqueous solution (Table 17).

TABLE 17.   Thermodynamic data for the dissociation of the complexes of amines with trimethylboron.

Amine complexed with trimethylboron	Reference	K_{100} atm	ΔG°_{100} (kcal/mole)	ΔH° (kcal/mole)	ΔS° (cal/deg/ mole)
Ammonia	13	4·62	−1·13	13·75	39·9
Methylamine	13	0·0360	2·46	17·64	40·6
Dimethylamine	13	0·0214	2·88	19·26	43·6
Trimethylamine	13	0·472	0·56	17·62	45·7
Ethylamine	109	0·0705	1·97	18·00	43·0
Diethylamine	110	1·22	−0·15	16·31	44·1
Triethylamine	110	Too highly dissociated for measurement			
n-Propylamine	109	0·0598	2·09	18·14	43·0
n-Butylamine	109	0·0470	2·27	18·41	43·2
n-Pentylamine	109	0·0415	2·36	18·71	43·9
n-Hexylamine	109	0·0390	2·40	18·53	43·2
i-Propylamine	111	0·368	0·74	17·42	44·7
s-Butylamine	111	0·373	0·73	17·26	44·3
t-Butylamine	111	9·46	−1·67	12·99	39·3

They suggested that the relative degree of dissociation exhibited by these addition compounds might result from two opposing effects of the methyl group, the $+I$ inductive effect which tends to increase the stability of the addition product, and the steric effect which tends to decrease its stability. Thus they considered that the steric strain resulting from the presence of three methyl groups is sufficient to counteract the increased inductive effect and hence to decrease markedly the strength of trimethylamine as a base. This was essentially the same explanation as Brown and his coworkers had suggested to account for the order of base strengths observed in aqueous solution.

The dissociation constants at 100°c (K_{100}) of the trimethylboron complexes of aliphatic primary amines, deduced from such measurements[13,108,109], suggested that there is a large increase in base strength between ammonia and methylamine, but further increase in the length of the alkyl chain produces only small extra effects. This is similar to the behaviour with protonic acids. The rather higher $\Delta S°$ values for ethylamine and higher amines were attributed to the effect of the large BMe_3 group in preventing free rotation of the ethyl group around the C—N bond and thus lowering the entropy in the complex. Dissociation is therefore accompanied by a greater gain in entropy than is the case with methylamine. Although it has been adversely criticised when applied to the dissociation of amine cations, this steric effect may well explain the relative dissociation constants of these complexes.

Branching of the alkyl group causes a considerable decrease in the stability of the addition compound[111]. This arises from a decrease in $\Delta H°$ and a small increase in $\Delta S°$. The latter, suggests that there is increased steric interference in the complex. It is very striking that although the change from ethylamine to i-propylamine or s-butyl-amine causes more than a five-fold increase in the dissociation constant and a change in $\Delta G°$ of over 1·2 kcal, an even greater change is produced when all three α-hydrogen atoms are replaced by methyl groups to give t-butylamine. For this amine $\Delta H°$ is much lower and so, strangely, is $\Delta S°$, so it seems evident that some other effect is coming into play here.

An interesting feature of these complexes is that the order of the apparent strengths of the primary, secondary, and tertiary amines depends on the sizes of the groups linked to the nitrogen and to the boron atom. Thus with BMe_3 and BF_3 the methylamines follow the order $NH_3 < NMe_3 < NH_2Me < NHMe_2$, as in aqueous solution, but the ethylamines follow the order $NR_3 < NH_3 < NHR_2 < NH_2R$, and this latter order is also followed by the methylamines in reaction with triethylboron. The order $NR_3 < NHR_2 < NH_3 < NH_2R$ is followed by the i-propylamines with BF_3, BMe_3, and BEt_3, by the ethylamines with BEt_3, $B(Pr-i)_3$, and perhaps $B(Bu-t)_3$, and methyl-amines with $B(Bu-t)_3$. When both components contain bulky groups the order is $NR_3 < NHR_2 < NH_2R < NH_3$. It is clear, therefore that in the formation of these complexes steric strain is of very great importance[112].

B. Complexes of Amines with Metal Ions

The ability to coordinate to metal ions is another well-known characteristic of amine groups. This tendency can to a limited extent be correlated with the dipole moment of the ligand molecule, but steric factors also play an extremely important part in determining the complexing ability. Presumably on this account primary amines usually have less tendency towards complex formation than ammonia itself, while complex formation occurs less readily still with secondary and tertiary amines.

The stabilities of the complexes formed by amines become more marked when the molecule contains two amino groups and hence chelate compounds can be formed. This is exemplified by the well-known series of coordination compounds formed by metals with ethylenediamine and its derivatives. The instability constants of these compounds have been extensively studied and they are fully discussed in books on Inorganic Chemistry.

As in the case of the ammonia complexes the number of amine molecules which can be coordinated depends on the electronic configuration of the metal concerned. With the Ag^+ ion two groups are coordinated to give a linear structure about the silver atom which apparently uses sp-hybrid orbitals. On the other hand Pd^{2+} and Pt^{2+} ions show a coordination number of four and their bonding orbitals are of the dsp^2-hybrid type, so they yield square-planar complexes with amines. The Ni^{2+}, Cr^{3+}, and Co^{3+} ions, along with many others, have a coordination number of six and so form octahedral complexes with amines.

The resemblance between these compounds and those formed by ammonia is exemplified by the fact that they form non-ionising compounds of the type $[Pd(CN)_2(NR_3)]$ [113] and $[PtCl_2(NH_2R)_2]$. The instability constants of the second group of compounds are about the same whether they contain ammonia, methylamine, or ethylamine, and in each case the *trans* isomer is more stable than the *cis* [114]. The methylamine complex of this type will undergo a number of reactions without loss of amine. Thus on boiling with potassium nitrite it gives $[Pt(NO_2)_2(NH_2Me)_2]$, while on treatment with silver nitrate followed by potassium bromide, potassium iodide, or potassium thiocyanate it yields $[PtBr_2(NH_2Me)_2]$, $[PtI_2(NH_2Me)_2]$, and $[Pt(SCN)_2(NH_2Me)_2]$, respectively [115].

The silver complexes of amines of the general type $[Ag(RNH_2)]^+$ provide some features of interest in relation to the base strengths of the amines in protonic solvents. Some comparative data are shown in Table 18. Although there are slight discrepancies between the data from the different sources [116-118] there is an evident parallelism [119].

Similar results are obtained with secondary and tertiary amines, and Trotman-Dickenson [20] has pointed out that when the logarithms

TABLE 18. Comparison of the logarithms of
stability constants ($\log \beta_2$) for formation of silver
complexes with pK_a values in aqueous solution.

Amine	$\log \beta_2$	pK_a
n-Butylamine	7·48	10·65
Ethylamine	7·30	10·65
i-Butylamine	7·24	10·43
Benzylamine	7·14	9·62
Ammonia	7·12	9·24
Methylamine	6·68	10·63
Ethanolamine	6·68	9·60
2-Methoxyethylamine	6·34	9·45
p-Toluidine	3·48	5·10
o-Toluidine	3·61	4·45
Aniline	3·47	4·60
2-Naphthylamine	3·23	4·16
m-Nitroaniline	1·7	2·47
p-Nitroaniline	1·6	1·11

of the dissociation constants of these ions are plotted against the pK_a
values for the amines the points for primary, secondary, and tertiary
amines lie on three separate straight lines. This is as would be expected,
since the complexes of the tertiary amines cannot be stabilised at all
by hydrogen bonding to water molecules. On the other hand, those of
secondary amines can form one hydrogen bond from each amine
group whilst those of primary amines can form two such bonds. The
slopes of the lines obtained are about 0·25, indicating a much smaller
range of base strength when measured against the silver ion as reference
acid than when measured against water.

Finally, it may be mentioned that silver provides an exception to
the rule regarding the higher stability of ethylenediamine complexes.
This arises because the ethylenediamine molecule cannot span the
silver ion so as to give the necessary linear arrangement of the bonds.

VI. REFERENCES

1. D. D. Perrin, *Dissociation Constants of Organic Bases in Aqueous Solution.*
   London, Butterworths, 1965.
2. D. H. Everett and W. F. K. Wynne-Jones, *Trans. Faraday Soc.*, **35**, 1380
   (1939).
3. D. J. G. Ives and P. D. Marsden, *J. Chem. Soc.*, 649 (1965); D. J. G. Ives,
   Personal Communication.
4. H. S. Frank and M. W. Evans, *J. Chem. Phys.*, **13**, 507 (1945).
5. R. P. Bell and J. W. Bayles, *J. Chem. Soc.*, 1518 (1952).

6. R. P. Bell and A. F. Trotman-Dickenson, *J. Chem. Soc.*, 1288 (1949).
7. J. N. Brönsted, A. Delbanco, and A. Tovborg-Jenson, *Z. Physik. Chem.*, **169A**, 361 (1934).
8. D. D. Perrin, *Australian J. Chem.*, **17**, 484 (1964).
9. L. P. Hammett, *Chem. Rev.*, **17**, 125 (1935).
10. R. W. Taft, *J. Am. Chem. Soc.*, **74**, 3120 (1952); **75**, 4231 (1953).
11. J. Clark and D. D. Perrin, *Quart. Rev. (London)*, **18**, 295 (1964).
12. H. C. Brown and A. Cahn, *J. Am. Chem. Soc.*, **72**, 2939 (1950).
13. H. C. Brown and H. Bartholomay, Jr., *J. Chem. Phys.*, **11**, 43 (1943).
14. H. C. Brown, H. Bartholomay, Jr., and M. D. Taylor, *J. Am. Chem. Soc.*, **66**, 435 (1944).
15. H. C. Brown, *J. Am. Chem. Soc.*, **67**, 374, 503 (1945).
16. W. S. Fyfe, *J. Chem. Soc.*, 1347 (1955).
17. D. R. Lide, Jr., and D. E. Mann, *J. Chem. Phys.*, **28**, 572 (1958).
18. D. H. Everett and W. F. K. Wynne-Jones, *Proc. Roy. Soc.*, **169A**, 190 (1938).
19. D. H. Everett and D. A. Landsman, *Trans. Faraday Soc.*, **50**, 1221 (1954).
20. A. F. Trotman-Dickenson, *J. Chem. Soc.*, 1293 (1949).
21. D. H. Everett and W. F. K. Wynne-Jones, *Proc. Roy. Soc.*, **177A**, 499 (1941).
22. A. R. Evans and S. D. Hamann, *Trans. Faraday Soc.*, **47**, 34 (1951).
23. D. H. Everett and W. F. K. Wynne-Jones, *Trans. Faraday Soc.*, **48**, 531 (1952).
24. P. Damsgaard-Sörensen and A. Unmack, *Z. Physik. Chem.*, **172A**, 389 (1935).
25. H. S. Harned and B. B. Owen, *J. Am. Chem. Soc.*, **52**, 5079 (1930).
26. J. Buchanan and S. D. Hamann, *Trans. Faraday Soc.*, **49**, 1425 (1953).
27. L. H. Sommer and J. Rockett, *J. Am. Chem. Soc.*, **73**, 5130 (1951).
28. N. F. Hall and M. R. Sprinkle, *J. Am. Chem. Soc.*, **54**, 3469 (1932).
29. C. W. Hoerr, M. R. McCorkle, and A. W. Ralston, *J. Am. Chem. Soc.*, **65**, 328 (1943).
30. G. Vexlearschi, *Compt. Rend.*, **228**, 1655 (1949).
31. G. Girault-Vexlearschi, *Bull. Soc. Chim. France*, 589 (1956).
32. H. S. Hamann and W. Strauss, *Trans. Faraday Soc.*, **51**, 1684 (1955).
33. W. C. Somerville, *J. Phys. Chem.*, **35**, 2412 (1931).
34. H. K. Hall, Jr., *J. Am. Chem. Soc.*, **79**, 5441 (1957).
35. J. Hansen, *Svensk. Kem. Tidskr.*, **67**, 256 (1955).
36. R. G. Bates and G. D. Pinching, *J. Res. Nat. Bur. Std.*, **43**, 519 (1949).
37. R. G. Pearson and F. V. Williams, *J. Am. Chem. Soc.*, **76**, 258 (1954).
38. F. Swarts, *Bull. Acad. Roy. Belg.*, **186**, 762 (1904).
39. A. L. Henne and J. J. Stewart, *J. Am. Chem. Soc.*, **77**, 1901 (1955).
40. A. N. Nesmeyanov, L. I. Zakharkin, and R. K. Freidlina, *Izv. Akad. Nauk SSSR, Otd. Khim. Nauk*, 841 (1954); *Chem. Abstr.*, **53**, 1111 (1959).
41. G. Girault and P. Rumpf, *Compt. Rend.*, **243**, 663 (1956).
42. B. Cohen, E. R. van Artsdalen, and J. Harris, *J. Am. Chem. Soc.*, **70**, 281 (1948).
43. R. G. Bates and G. D. Pinching, *J. Res. Nat. Bur. Std.*, **46**, 349 (1951).
44. J. R. Lotz, B. P. Block, and W. C. Fernelius, *J. Phys. Chem.*, **63**, 541 (1959).

45. S. P. Datta and A. K. Grzybowski, *J. Chem. Soc.*, 1091 (1959).
46. J. R. Dudley, *J. Am. Chem. Soc.*, **73**, 3007 (1951).
47. K. Schwabe, W. Graichen, and D. Spiethoff, *Z. Physik. Chem.*, **20**, 68 (1959).
48. G. Girault and P. Rumpf, *Compt. Rend.*, **246**, 1705 (1958).
49. A. Gero, *J. Am. Chem. Soc.*, **76**, 5158 (1954).
50. D. H. Everett and B. R. W. Pinsent, *Proc. Roy. Soc.*, **215A**, 416 (1952).
51. C. R. Bertsch, W. C. Fernelius, and B. P. Block, *J. Phys. Chem.*, **62**, 444 (1958).
52. G. Schwarzenbach, *Helv. Chim. Acta*, **16**, 522 (1933).
53. R. S. Mulliken, *J. Am. Chem. Soc.*, **72**, 4500 (1950).
54. G. Vexlearschi and P. Rumpf, *Compt. Rend.*, **229**, 1152 (1949).
55. R. A. Robinson and A. K. Kiang, *Trans. Faraday Soc.*, **52**, 327 (1956).
56. W. H. Carothers, C. F. Bickford, and G. J. Hurwitz, *J. Am. Chem. Soc.*, **49**, 2908 (1928).
57. J. D. Roberts and V. C. Chambers, *J. Am. Chem. Soc.*, **73**, 5030 (1951).
58. V. Prelog, M. F. El-Neweiky and O. Häfliger, *Helv. Chim. Acta.*, **33**, 365 (1950).
59. A. D. Walsh, *Trans. Faraday Soc.*, **45**, 179 (1949).
60. C. A. Coulson, *Valence*, Clarendon Press, Oxford, 1952, p. 246.
61. A. I. Biggs and R. A. Robinson, *J. Chem. Soc.*, 388 (1961).
62. B. M. Wepster, *Rec. Trav. Chim.*, **76**, 357 (1957).
63. R. N. Beale, *J. Chem. Soc.*, 4494 (1954).
64. K. J. Pederson, *Kgl. Danske Videnskab. Selskab. Mat. Fys. Medd.*, **14**, 9 (1932).
65. A. L. Bacarella, E. Grunwald, H. P. Marshall, and E. L. Purlee, *J. Org. Chem.*, **20**, 747 (1955).
66. A. Bryson, *J. Am. Chem. Soc.*, **82**, 4858 (1960).
67. J. C. James and J. G. Knox, *Trans. Faraday Soc.*, **46**, 254 (1950).
68. M. M. Fickling, A. Fischer, B. R. Mann, J. Packer, and J. Vaughan, *J. Am. Chem. Soc.*, **81**, 4226 (1959).
69. B. Gutbezahl and E. Grunwald, *J. Am. Chem. Soc.*, **75**, 559 (1953).
70. J. M. Vandenbelt, C. Henrich, and S. G. Vanden Berg, *Anal. Chem.*, **26**, 726 (1954).
71. M. Kilpatrick and C. A. Arenberg, *J. Am. Chem. Soc.*, **75**, 3812 (1953).
72. G. M. Kharkharova, *Sb. Statei Obshch. Khim.*, **2**, 1663 (1953); *Chem. Abstr.*, **49**, 5340 (1955).
73. A. V. Willi, *Z. Physik. Chem. (Frankfurt)*, **26**, 42 (1960); **27**, 233 (1961).
74. J. Burgers, M. A. Hoefnagel, P. E. Verkade, H. Visser, and B. M. Wepster, *Rec. Trav. Chim.*, **77**, 491 (1958).
75. A. Bryson, *Trans. Faraday Soc.*, **47**, 528 (1951).
76. H. Kloosterziel and H. J. Backer, *Rec. Trav. Chim.*, **71**, 295 (1952).
77. J. D. Roberts, R. L. Webb, and E. A. McElhill, *J. Am. Chem. Soc.*, **72**, 408 (1950).
78. R. A. Benkeser and H. R. Krysiak, *J. Am. Chem. Soc.*, **75**, 2421 (1953).
79. F. Kieffer and P. Rumpf, *Compt. Rend.*, **230**, 1874 (1950).
80. F. G. Bordwell and G. D. Cooper, *J. Am. Chem. Soc.*, **74**, 1058 (1952).

81. R. N. Diebel and D. F. Swinehart, *J. Phys. Chem.*, **61**, 233 (1957).
82. R. D. McCoy and D. F. Swinehart, *J. Am. Chem. Soc.*, **76**, 4708 (1954).
83. R. O. McLaren and D. F. Swinehart, *J. Am. Chem. Soc.*, **73**, 1822 (1951).
84. H. Zollinger, W. Bückler, and C. Wittwer, *Helv. Chim. Acta*, **36**, 1711 (1953).
85. G. M. Bennett and A. N. Mosses, *J. Chem. Soc.*, 2364 (1930).
86. R. van Helden, P. E. Verkade, and B. M. Wepster, *Rec. Trav. Chim.*, **74**, 39 (1954).
87. H. Kloosterziel, *Thesis Groningen* 1952; *Chem. Weekblad*, **51**, 143 (1955).
88. G. Thomson, *J. Chem. Soc.*, 1113 (1946).
89. H. C. Brown, G. K. Barbaras, H. L. Berneis, W. H. Bonner, R. B. Johannesen, M. Grayston, and K. L. Nelson, *J. Am. Chem. Soc.*, **75**, 1 (1953).
90. A. L. Bacarella, E. Grunwald, H. P. Marshall, and E. L. Purlee, *J. Org. Chem.*, **20**, 747 (1955).
91. J. A. C. Th. Brouwers, S. C. Bijlama, P. E. Verkade, and B. M. Wepster, *Rec. Trav. Chim.*, **77**, 1080 (1958).
92. G. Baddeley, J. Chadwick, and H. T. Taylor, *J. Chem. Soc.*, 2405 (1954).
93. G. Vexlearschi and P. Rumpf, *Compt. Rend.*, **233**, 1630 (1951).
94. V. Wolf and D. Ramin, *Ann. Chem.*, **626**, 47 (1959).
95. H. C. Brown, D. H. McDaniel and O. Häfliger in *Determinations of Organic Structures by Physical Methods* (Eds. E. A. Braude and F. C. Nachod), Academic Press, New York, 1955, p. 607.
96. H. P. Marshall and E. Grunwald, *J. Am. Chem. Soc.*, **76**, 2000 (1954).
97. W. C. Davies and H. W. Addis, *J. Chem. Soc.*, 1622 (1937).
98. M. A. Paul, *J. Am. Chem. Soc.*, **76**, 3236 (1954).
99. W. C. Davies, *J. Chem. Soc.*, 1865 (1938).
100. B. M. Wepster, *Rec. Trav. Chim.*, **75**, 1471 (1956).
101. J. J. Elliott and S. F. Mason, *J. Chem. Soc.*, 2352 (1959).
102. A. Bryson, *J. Am. Chem. Soc.*, **82**, 4862 (1960).
103. P. H. Grantham, E. K. Weisburger, and J. H. Weisburger, *J. Org. Chem.*, **26**, 1008 (1961).
104. M. S. Newman and J. Blum, *J. Am. Chem. Soc.*, **86**, 1835 (1964).
105. A. Fischer, G. J. Sutherland, R. D. Topsom, and J. Vaughan, *J. Chem. Soc.*, 5948 (1965).
106. S. F. Mason, *J. Chem. Soc.*, 3619 (1958); 1281 (1959); J. J. Elliott and S. F. Mason, *J. Chem. Soc.*, 1275 (1959).
107. A. Bryson, *Trans. Faraday Soc.*, **45**, 257 (1949).
108. H. C. Brown, *J. Am. Chem. Soc.*, **67**, 1452 (1945).
109. H. C. Brown, M. D. Taylor, and S. Sjishi, *J. Am. Chem. Soc.*, **73**, 2464 (1951).
110. H. C. Brown and M. D. Taylor, *J. Am. Chem. Soc.*, **69**, 1332 (1947).
111. H. C. Brown and G. K. Barbaras, *J. Am. Chem. Soc.*, **75**, 6 (1953).
112. Lawton, *Thesis Purdue* 1952; see H. C. Brown, *J. Chem. Soc.*, 1248 (1956).
113. F. Feigl and G. B. Heisig, *J. Am. Chem. Soc.*, **73**, 5631 (1951).
114. A. A. Grinberg and E. S. Postnikova, *Dokl. Akad. Nauk SSSR.*, **153**, 340 (1963).

115. I. I. Chernyaev, N. N. Zheligovskaya, and T. N. Leonava, *Zh. Neorg. Khim.*, **9**, 347 (1964).
116. E. Larsson, *Z. Physik. Chem.*, **169**, 207 (1934).
117. H. T. S. Britton and W. G. Williams, *J. Chem. Soc.*, 796 (1935).
118. R. J. Bruehlman and F. H. Verhoek, *J. Am. Chem. Soc.*, **70**, 1401 (1948).
119. B. E. Douglas and D. H. McDaniel, *Concepts and Models of Inorganic Chemistry*, Blaisdell Pub. Co., New York, Toronto, London. 1965, p. 399.

CHAPTER **5**

Directing and activating effects

GABRIEL CHUCHANI

Instituto Venezolano de Investigaciones Científicas, IVIC, Caracas, Venezuela

I. INTRODUCTION

Quantitative studies on the polar effects of amine substituents in chemical reactions have not been extensively undertaken due to several factors which are inherent in the processes involved. In the first place, the basic nature of these substituents can facilitate interaction with any electrophile present in the media, thus producing the corresponding ammonium derivatives, which have an entirely different character from that of the parent base. Furthermore, the very strong effect of the amine substituents on reactivity can sometimes impede the application of standard kinetic techniques. Moreover, amines behave as exceptions and anomalies to many important generalizations. The polar effects of the groups under consideration in this chapter have not been evaluated as a series, as have, for example, the alkyls or halogens, but as a comparison with other commonly known groups. Positively charged nitrogen substituents are necessarily discussed, since many reaction conditions inevitably produce these entities and also in order to contrast their electronic displacements with the corresponding free amines. A discussion of the effects of variation of N-alkyl substituent size is included in order to bring out the fact that electrical or steric factors can affect reactivity. Basicity, an important property of these substituents, has been to a large extent minimized here as a factor which can give weight to the explanation of known phenomena.

Although a number of interesting reactions in which the amine substituents exert important effects were omitted, it is hoped that the processes cited in this chapter will give ample insight into the general behavior of these substituents.

II. THE POLAR EFFECT OF SUBSTITUENTS

The modern structural theory of organic chemistry has incorporated the concept of relative electronegativity of atoms [1] and used it to interpret the electrical dissymmetry of valence bonds in molecules as an explanation of certain physical and chemical properties. Those properties are usually affected by changes in the skeletal geometry of the systems and the presence of substituents often influences their reactivity relative to the parent compound. The amino groups generally attract electrons along strong σ bonds or weaker fluid π bonds by the inductive effect $(-I)$, while their lone pair of electrons or p electrons can also interact through delocalization with the molecular orbital of a system containing π bonds by the resonance $(+R)$ or the conjugative $(+T)$ effect.

Many researchers, in particular the 'English school', separate the electronic effect present in the ground state of molecules from those occurring only in the transition state or activated complex. The former is termed 'mesomeric effect' determined by permanent polarization, while the latter 'electromeric effect' is determined by the polarizability [2]. However, these effects will be here included together and defined broadly as 'resonance' or 'conjugative effects', with $+R$ indicating those groups which supply electron density to unsaturated or conjugated systems, and $-R$ those groups which attract electron density from such systems. In like manner, the concept 'inductive effect' will be designated so as to comprise both the 'inductive effect' determined by permanent polarization and the 'inductometric effect' determined by the polarizability, with $-I$ referring to those groups which withdraw electron density mainly in saturated systems and $+I$ to those groups which release electron density to similar systems.

Some important generalizations [2] are given of the polar effects of substituted and unsubstituted nitrogen as substituents in organic molecules, if possible with at least one example of physical evidence (see sections II.A and II.B). This could be useful to visualize, both amply and comparatively, the magnitude of their modes of transmission, and to deduce, at the same time, that any alteration of these general inferences brings new phenomenological concepts which are interesting and often even desirable.

A. The Inductive Effect

The capacity of the amino groups to attract electrons by the $-I$ effect is due to the greater electronegativity of nitrogen relative to carbon. The unequal sharing of electrons between these two atoms can be transmitted by the successive polarization of σ bonds as in homologous carbon chains (**1**) and the small dipoles produced from such electronic displacement decrease progressively away from the polar group

$$R_2\overset{\delta^-}{N} \longleftarrow \overset{\delta^+}{C} \longleftarrow \overset{\delta\delta^+}{C} \longleftarrow \overset{\delta\delta\delta^+}{C}$$

(**1**)

The electrostatic interaction arising from the influence of an electronegative group is not limited to transmission along σ bonds of saturated molecules, but can also be transmitted by a direct inductive effect, known as field effect, across space or through solvents between the group to the reaction center. This type of transmission was first considered separable from the inductive effect[3], but subsequent attempts to divide the two effects experimentally have either proved difficult or shown that one effect predominates. Discussions of these topics are available[4-11], and it has been argued[12] that the inductive effect is in reality a field effect and that propagation by successive polarization of σ bonds, i.e. σ inductive effect may not be important at positions separated from a substituent by more than one or two bonds. However, it was suggested[13] that an absolute separation of these two effects appears to be unlikely and that the field effect is an important feature of the polar effect.

The comparative magnitude of the inductive effect $(-I)$ of the amino groups in relation to other elements in the periodic table is shown by the following sequence:

$$F > OH(OR) > NH_2(NR_2)$$

This order for the isoelectronic series is demonstrated satisfactorily by dipole-moment measurements. The simplest organic molecules containing these substituents, i.e. MeF, MeOH and $MeNH_2$, have moments of 1·81, 1·68 and 1·28 D respectively[14]. The strengths of saturated acids may be increased by the $-I$ effect, responsible for lowering the free energy of ionization of the anion relative to the corresponding acid. The strength of these acids has been studied as supporting evidence for the above relationship[15], but the electrostatic interaction with the basic amine nitrogen could to some degree lead to false conclusions (see section IV.A). Experimental results[16] relating proton 'chemical shifts' and the ionic character of the chemical bond to the proton in methyl and ethyl derivatives of a number of substituent groups illustrate the above order of electron-withdrawal power. Furthermore, these data were used to assign effective values of the electronegativities of the substituent groups.

The presence of π electrons in multiple nitrogen to carbon bonds facilitates electron attraction and results in a greater $-I$ effect than for the corresponding σ-bonded saturated group. At the same time, compared with an unsubstituted amino group, the electron-withdrawing effectiveness is partially offset by the opposite inductive effect of alkyl groups when the latter replace hydrogens. Hence, the following order would be expected:

$$C\equiv N > C=NR > C-NH_2 > C-NHR > C-NR_2$$

Unsubstituted amines have a higher dipole moment than the most highly alkyl-substituted ones[17], but lower than the corresponding nitriles[14]. Actually, the μ values in the vapor phase, i.e. $MeC\equiv N$, $\sim 4\cdot 0$ D, ($MeCH=NH$ not reported due to instability), $MeCH_2NH_2$, $1\cdot 22$ D, $MeCH_2NHEt$, $0\cdot 92$ D and $MeCH_2NEt_2$, $0\cdot 66$ D, provide a reasonable support for the above generalization. These large differences of course, are valid only for simple molecules of small angular structure when determined under similar experimental conditions.

The effect of a positively charged nitrogen is best understood when compared with neutral nitrogen. The bonding electrons of onium substituents are supposed to be very tightly held. Consequently their electron-withdrawing effect should be greater than that of amino groups. The actual difference should depend on the presence of multiple bonds and on the opposing effect of alkyl groups which replace hydrogen, thus:

$$=\overset{+}{N}R_2 > \overset{+}{N}H_3 > \overset{+}{N}H_2R > \overset{+}{N}HR_2 > \overset{+}{N}R_3 > NR_2$$

The literature does not provide conclusive evidence for this sequence. However, some studies[18-20] of these groups give a different pattern of substituent effects. This has been attributed to hydration through hydrogen bonding, and will be discussed in subsequent sections.

Because alkyl groups oppose the effective transmission of the inductive effect of the amine substituent, any increment of the carbon chain must tend to diminish their $-I$ effect. This phenomenon should be appreciable with branched-chain carbon. A similar reasoning can be applicable to dialkyl-substituted amines. Nevertheless, only scattered information is found in the literature to substantiate any generalization of the electronic interactions with the reaction sites. Moreover, since the phenyl group attracts electrons, the N-phenyl group exhibits an electron demand, and the order may be represented as follows:

$$NHPh > NH_2 > NHR$$

The $-I$ effect of amine substituents can be outweighed by the $+R$ electron release, provided that their unshared electrons are capable of conjugation with aromatic systems. However, the $+R$ effect may be entirely repressed by electrostatic induction of charge displacements $(-I)$, when the positive ionic center is bonded directly to the benzene nucleus. The $\overset{+}{N}R_3$ group deactivates the conjugated ring with respect to electrophilic attack and may, therefore, be represented by 2, which resembles the progressive decrease of transmission with

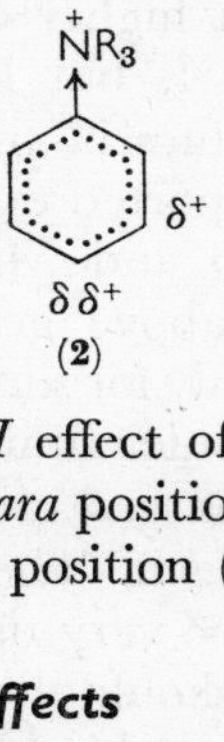

(2)

increasing distance of the $-I$ effect of the amine substituent in 1. Structure 2 suggests that the *para* position possesses a relatively higher electron density than the *meta* position (see section IX).

B. Resonance or Conjugative Effects

The resonance or conjugative effect $(+R)$ results from overlap of the unshared electron pairs of an amino group with the p orbital of an adjacent unsaturated carbon supplying its p electrons along π bonds in such systems. Theoretical considerations of the dipole moments of p electrons in aniline[21] reveal that the inductive effect of the amino group is as important as the resonance interaction of this substituent with the aromatic π-electron system, and this is in agreement with experimental findings. An interesting argument has been raised with regard to the most effective transmission of electron displacement when the inductive effect of the benzene ring is compared with the resonance effect in aromatic amines[22]. For example, aniline is 10^6 times weaker as a base than cyclohexylamine. This large difference cannot solely be ascribed to the inductive effect of the phenyl group, but rather to delocalization of the unshared electron pair of the amine nitrogen over the aromatic ring. There is no experimental evidence to show how much each of these effects contributes to the basic properties of anilines.

Although the chemical properties of the conjugated amines (**3**) are

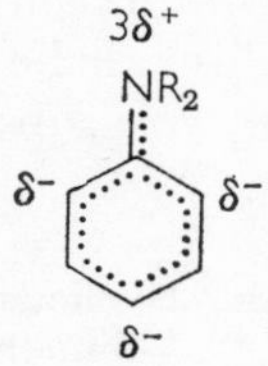

(**3**)

in many ways similar to aliphatic amines, the direction of the $+R$ effect markedly alters their relative reactivity for example in the ease of substitution by electrophilic reagents. This mode of transmission may be enhanced when the amine substituent is in direct conjugation with electron-attracting groups such as —NO_2, —COOMe, etc. Such conjugation delocalizes the p electron of nitrogen and confers a formal positive charge on the amino group. Base strengths, ultraviolet excitations and dipole moments[23, 24] give some idea of this effect when the groups are oriented *ortho* or *para* to each other. The resonance interaction may be written as in **4**. Since the *meta* conjuga-

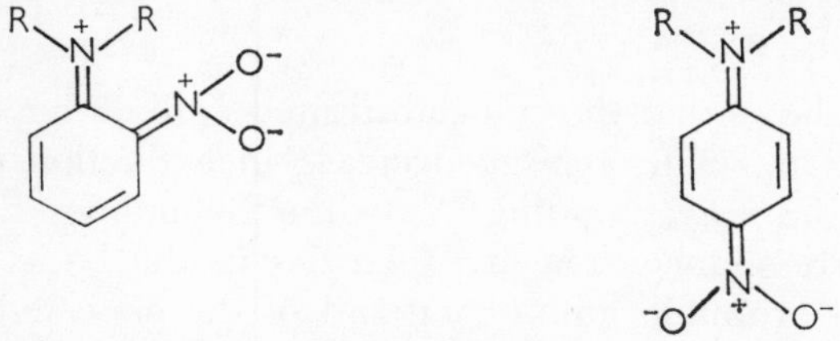

(**4**)

tive interaction would require an unstable long bond no significant contribution to resonance hybrids is expected from polar forms **5**.

There is no definite evidence concerning the geometry of the amine substituent in aromatic amines[25-27], although the resonance effect will be more pronounced when the nitrogen and its two hydrogens or alkyls are coplanar with the conjugated system. Any departure from planarity caused by bulky adjacent groups gradually inhibits the $+R$ effect. The phenomenon of steric inhibition of resonance has been

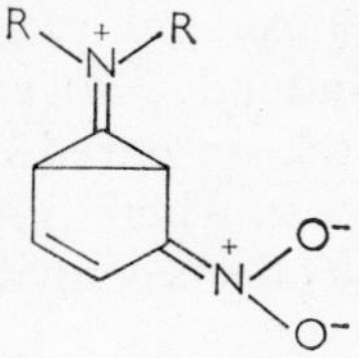

(5)

substantiated[27-29] by studies on basicity[30-33], dipole moment[34-39] and ultraviolet spectra[40-43]. This can be illustrated by the extinction coefficients and basicities of *ortho*-substituted anilines and N,N-dimethylanilines (see Table 1).

TABLE 1. Extinction coefficients and basicities of anilines and N,N-dimethylanilines[29].

	$\varepsilon_{max}{}^a$	$pK_a{}^b$
Aniline	9130	4.26
2-Me—	8800	4·09
2-Et—	8030	4·04
2-i-Pr—	7780	4·06
2-t-Bu—	7850	3·38
N,N-Dimethylaniline	15500	4·39
2-Me—	6360	5·15
2-Et—	4950	5·20
2-i-Pr—	4300	5·05
2-t-Bu—	630	4·28

^{a} At about 250 mμ.
^{b} 50% by vol ethanol.

This table shows that the ring-substituted anilines have almost the same absorption, while a large decrease in extinction coefficient is observed in the corresponding N,N-dimethylanilines due to steric inhibition of resonance. The irregularities in the basic strengths of both series of aromatic amines, instead of the expected increase in pK_a values, was attributed to steric hindrance to solvation and cannot be used as a valuable source of information on the steric inhibition of resonance. However, in comparing the base strength of the ring-alkylated dimethylanilines with the corresponding anilines, the former are stronger bases than the latter, since the o-alkyl groups cause a greater departure from coplanarity of the NMe_2 than for the NH_2 group, and result in a smaller $+R$ effect on the benzene ring. The influence of hydration of hindered aromatic bases reported recently[44] permits the observation of this fact from the estimates of their inductive effect and steric inhibition of resonance. Another good illustration of this effect of inhibition of resonance is the work of Smith[39] on dipole moments of various aromatic amines.

The relative magnitude of electron release $(+R)$ of the amine substituents as compared with other common polar groups is in the following order:

$$NH_2(NR_2) > OH(OR) > F$$

The resonance effect generally decreases with increasing electronegativity[2], that is, the more electronegative (electron attracting) a given group, the more reluctant it is to release electrons. Several pieces of evidence[45-47] provide support for this sequence and others will be discussed in subsequent sections of this chapter.

The $+I$ effect of alkyl groups replacing hydrogen atoms of an amine will increase the electron density on the nitrogen and consequently should impart a greater tendency for covalency with the adjacent atom. The π overlap with the amine nitrogen may be repressed when a phenyl group replaces at least one of the hydrogens or may be suppressed as in a positively charged ammonium group. Thus the electron-release effects vary as follows:

$$NR_2 > NHR > NH_2 > NHPh > \overset{+}{N}R_3(\overset{+}{N}H_3)$$

$$NHPr > NHEt > NHMe > NHPh$$

$$NPr_2 > NEt_2 > NMe_2 > NPh_2$$

The bathochromic effect $NEt_2 > NHEt > NH_2$ in adjacent ethylenic systems[48] and dipole-moment values[17] ($PhNEt_2$, $1 \cdot 84$ D $>$ $PhNHEt$, $1 \cdot 70$ D $> PhNH_2$, $1 \cdot 41$ D $> Ph_2NH$, $1 \cdot 02$ D) are in accord with the order in the first series, resulting from the interaction of the p electrons of the substituents with the π bonds of the system. The variation in dipole moments of N-substituted anilines[17] (the second series) are relatively small and this is mainly due to the steric inhibition discussed previously. Nevertheless, an increase in the dipole-moment values favoring the $+R$ effect is indicated for the last sequence[49], thus $PhNBu_2$, $1 \cdot 9$ D $> PhNEt_2$, $1 \cdot 84$ D $> PhNMe_2$, $1 \cdot 64$ D $> Ph_3N$, $0 \cdot 71$ D.

The resonance effect of the aforementioned substituents becomes evident when directly conjugated to a reaction center, as is readily seen when the groups are located *para* to each other in the benzene ring. The $-I$ effect, though relatively small in this case, is more efficient when the substituent is *meta* with respect to the reaction center. Steric and other factors complicate the interpretations of the significance of both effects in *ortho* positions.

III. CORRELATION BETWEEN STRUCTURE AND REACTIVITY

The success of the Hammett empirical relation has been the subject of continuing efforts to correlate the effects of structure on chemical reactivity and has led to the development of similar linear free-energy or structure–reactivity relationships in other systems. Due to the many intervening factors involved, the applications and utility of these correlations are, as yet, only approximate in the case of the amino group. Despite this disadvantage, it is highly useful for describing reaction mechanisms and for calculating a large number of rate and equilibrium constants still to be measured or experimentally difficult to determine. The present chapter is not intended to present a thorough literature coverage nor to account for a critical examination of the scope, limitations or the essential principles of structure–reactivity relationships, which are excellently reported in recent reviews [19, 50–55]. Instead, attention is directed to correlating and to interpreting the polar effect of various amino and ammonium groups, defined by sigma values, in each type of reaction rate or equilibrium.

A. The Hammett Substituent Constants

The semiquantitative linear relationship between the effects of *meta* and *para* substituents (k), compared to the unsubstituted compound (k_0), on the rates or equilibrium constants of aromatic side-chain reactions is represented by the Hammett equation [56] (1):

$$\log k - \log k_0 = \rho\sigma \tag{1}$$

where σ, the *substituent constant*, is independent of reaction type or conditions. A positive σ value indicates a net electron withdrawal relative to hydrogen and a negative σ indicates a weaker electron attractor than hydrogen. The *reaction constant*, ρ, is a function of the changes in electron density at the reaction center depending on the reaction conditions. Positive ρ values indicate that electron-attracting substituents promote the reaction under consideration, and vice versa for negative ρ values. A table of σ values for amino and ammonium groups have been compiled (see Table 2). The validity of the Hammett σ constants has been restricted to substituents in the *meta* and *para* positions, since frequent deviations in reactivities arise to varying extents from superimposed steric and field factors on the polar effects of *ortho* substituents. The *ortho*-substituent constant for NH_2 was found, in a special case, to have a σ value of -0.26 and to release

TABLE 2. Representative substituent constants.

Group	σ_{meta}	σ_{para}	σ_{meta}^{+}	σ_{para}^{+}	σ_{meta}^{o}	σ_{para}^{o}	σ_{meta}^{n}	σ_{para}^{n}	$\sigma_I^{(al)}$	$\sigma_{I\,meta}^{(ar)}$	$\sigma_{I\,para}^{(ar)}$	$\sigma_{R\,meta}$	$\sigma_{R\,para}$	$\sigma_{R\,meta}^{o}$
NH_2	−0·161[a]	−0·660[a]	−0·16[h]	−1·3[h]	−0·14[k]	−0·38[k]	−0·04[l]	−0·17[k]	0·13[l]	0·04[k]	(0·04)[q]	−0·25[n]	−0·76[n]	−0·48[k]
$NHMe$	—	−0·84[a]	—	−1·59[i]	—	—	—	—	—	—	—	—	—	—
$NHEt$	−0·302[b]	−0·592[b]	—	—	—	—	—	—	—	—	—	—	—	—
$NHBu$	−0·240[b]	−0·607[c]	—	—	—	—	—	—	—	—	—	—	—	—
$NHPh$	−0·344[a]	−0·512[c]	—	−1·4[h]	—	—	—	—	—	—	—	—	—	—
NMe_2	−0·211[b]	−0·83[a] −0·600[b]	—	−1·7[h]	−0·15[k]	−0·44[k]	−0·05[l]	−0·17[k]	0·10[k]	0·04[k]	(0·04)[q]	—	−0·83[o]	−0·54[k]
NEt_2	—	−0·87[d] −0·425[e]	—	—	—	—	—	—	—	—	—	—	—	—
$\overset{+}{N}H_3$	0·634[a]	0·485[f]	—	—	m	—	—	—	0·60[k]	—	—	—	—	—
$\overset{+}{N}H_2Me$	0·985[b]	—	—	—	m	—	—	—	0·60[p]	—	—	—	—	—
$\overset{+}{N}H_2Et$	0·985[b]	—	—	—	m	—	—	—	0·60[p]	—	—	—	—	—
$\overset{+}{N}HMe_2$	—	0·802[g]	—	—	m	—	—	—	0·70[p]	—	—	—	—	—
$\overset{+}{N}Me_3$	0·88[a] 0·904[b]	0·82[a] 0·859[b]	0·36[h]	0·41[h] 0·859[j]	m	—	0·86[l]	0·80[k]	0·92[k]	0·90[k]	(0·90)[q]	0·14[n]	—	—

[a] Reference 60.
[b] Reference 50.
[c] M. Charton, J. Org. Chem., **28**, 3121 (1963).
[d] P. J. Bray, J. Phys. Chem., **22**, 1787 (1954).
[e] K. B. Whetsel, Spectrochim. Acta, **17**, 614 (1961).
[f] Reference 85a.
[g] Reference 59.
[h] Reference 67; N. C. Deno and W. L. Evans, J. Am. Chem. Soc., **79**, 5804 (1957).
[i] H. H. Jaffé, J. Org. Chem., **23**, 1790 (1958).
[j] H. H. Jaffé and R. W. Gardner, J. Am. Chem. Soc., **80**, 319 (1958).
[k] Reference 55.
[l] Reference 53.
[m] Treatment not applied to charged substituents.
[n] Reference 19.
[o] P. J. Krueger and H. W. Thompson, Proc. Roy. Chem. Soc., **250**, 22 (1959).
[p] Reference 75.
[q] It is assumed $\sigma_I^{(ar)}$ is identical for meta and para substituents.

more electrons towards the *ortho* than the *meta* position[57]. Another limitation is the enhanced σ value of electron-attracting substituents in direct conjugation with strong electron-releasing groups[56]. Jaffé[50] evaluated and denoted these enhanced σ values as σ^*, while others[55], in accord with current conventions, denoted them by a nucleophilic substituent constant σ^- (see section III.B).

A closer inspection of Table 2 and a large body of other experimental data leads to the conclusion that uniform σ_{para} and σ_{meta} values for each amine substituent do not appear to exist. Some of the difficulties may result from the susceptibility of the electron pair of the amine nitrogen to interaction with the reaction media. Frequent deviations from the linear free-energy equation are observed with these groups and appropriate interpretation with factors already known is difficult. Some idea of the complexity of this situation can be gained from the fact that the 17 σ values reported for p-$N(CH_3)_2$ range from -0.206 to -1.049[58, 59].

Although σ constants for amino substituents are only approximate, they fit the previously indicated sequence of electronic displacement when representative σ values are compared[50, 59, 60].

$$NH_2, \quad (NMe_2) > OH, \quad (OMe) > F$$

$$\sigma_{para}: -0.66, \quad (-0.83) \quad > -0.37 \ (-0.268) > 0.062$$

$$\sigma_{meta}: -0.16, \quad (-0.211) > 0.121(0.115) \quad > 0.337$$

The σ_{para} values reflect the corresponding order of resonance effect of these groups located in a *para* position, while the σ_{meta} values indicate merely a decrease in the $+R$ transmission or an increase in the $-I$ effect when at a *meta* position. A similar order of the ability of these groups to release electrons was obtained from measurements of the acidity constants of *meta-* and *para*-substituted benzoic acids and phenols[61].

The differences in the σ values available for various amino substituents are sometimes not large enough for the results to be interpreted. For example the σ_{para} values of NMe$_2$ (-0.83), NHMe (-0.84) and NH$_2$ (-0.66), based on the ionization constants of substituted benzoic acids give no significant indication of their $+R$ effects. This could be rationalized by the greater inhibition to coplanarity of the NMe$_2$ groups as compared with the NHMe group. In any further evaluation of the substituent constants shown in Table 2, it must be ascertained, whether these data were recorded under identical or sufficiently similar reaction conditions. As an illustration of this restriction, other σ_{para} values reported for the groups just considered as well as for other NHR groups give an anomalous order for their efficiency of the $+R$ transmission in aromatic systems. Likewise, the differences $(\sigma_{meta} - \sigma_{para})$ of these substituents do not reveal the relative sequence of electron-releasing resonance.

The σ_{meta} values for NHMe (-0.302), NMe_2 (-0.211) and NH_2 (-0.161) apparently follow the $+R$ sequence of these substituents, in spite of the existing $-I$ effect, when located at the *meta* position of the aromatic ring. Again, the NMe_2 group shows a similar discrepancy in the sequence of the two effects $(-I, +R)$, probably due to the fact that each of these values was evaluated from different reaction conditions[50].

Examination of the σ_{meta} values for NHR groups determined under identical conditions of equilibria from the data of Jaffé[50] and of Wepster and collaborators[59], leads to the series:

$$NHBu > NHMe > NHEt > NH_2$$

$$\sigma_{meta}: \begin{cases} -0.344 \quad -0.302 \quad -0.240 \quad -0.161 \quad \text{(ref. 50)} \\ -0.223 \quad -0.196 \quad -0.142 \quad -0.070 \quad \text{(ref. 59)} \end{cases}$$

In both cases, the NHMe group impedes the substantiation of the relative importance of resonance effect of these substituents as established in section II.B, even though the inductive mechanism becomes significant. An identical σ_{meta} sequence of these groups was obtained from the ionization constants of the corresponding *meta* derivatives of arylphosphonic acids[62]. The Hammett equation has also been successfully used to calculate a σ value for the resonance inhibited p-NMe_2 group[11]. In most cases σ_{meta} is consistently less negative than σ_{para}.

As pointed out before, positively charged nitrogen groups, have no resonance interaction with the aromatic ring and their effective $-I$ transmission is greater in the *meta* than in the *para* position[6, 63]. This can be seen from Table 2, where the σ_{meta} value for $\overset{+}{N}R_3$ is more positive than the corresponding σ_{para} value. In addition, these substituents are also susceptible to reaction conditions. For example, the σ_{meta} values for $\overset{+}{N}Me_3$ show large variations[63] as do the $\sigma_{meta} - \sigma_{para}$ differences. This suggests that solvation may be an important factor.

The rather scattered results reported under similar conditions[59] for ammonium groups support the currently accepted $-I$ order, that is m-$\overset{+}{N}H_3$ $(0.800) > m$-$\overset{+}{N}Me_3$ (0.780); and p-$\overset{+}{N}HMe_2$ $(0.802) >$

p-$\overset{+}{\text{N}}\text{Me}_3$ (0·712). The fact that the $-I$ value for p-$\overset{+}{\text{N}}\text{Me}_3$ (0·695) is greater than p-$\overset{+}{\text{N}}\text{H}_3$ (0·666) can be ascribed to hydrogen bonding through hydration of the latter group in aqueous media.

The σ_{para} values have been favorably correlated with *ortho*-substituted benzene derivatives in which substituents and reaction centers are not adjacent[64]. The ionization constants of 2-amino- and 2-dimethylaminobenzohydroxamic acid and their rates of reaction with sarin (isopropyl diethylphosphonofluoridate)[64, 65] indicated the absence of steric effects due to the amino groups. Thus their electrical effects as well as their order of magnitude were of the same type as in the *para* position.

B. Electrophilic and Nucleophilic Substituent Constants

The extension of the Hammett equation to electron-deficient side-chain reactions was carried out by Brown and Okamoto[66–68] from the solvolysis of phenyldimethylcarbinyl chloride. These σ^+ values are found to be applicable only to substituents capable of important resonance interaction or charge delocalization $(+R)$ with an electron-deficient reaction site, and excellent correlations for electrophilic aromatic substitutions and other reaction processes have been reported[52, 55, 69]. However, the σ^+ treatment has scarcely been applied to highly polar groups (such as the amino group), obviously due to their solvation and interaction with the media. The solvolysis of unsubstituted and N-substituted p-amino derivatives of benzyl chloride and phenyldimethylcarbinyl chloride has not been used as yet to estimate σ^+ parameters probably because of the instability or high reactivity of these compounds. In electrophilic aromatic substitutions, similar factors frequently affect and impede the evaluation of the reactivity of these groups. The many inconsistencies in the σ^+ values have resulted in serious criticisms[70] of the original assumption which considered the transition state in a substitution reaction to be independent of the reagent.

The difference in activating power in electrophilic aromatic substitution between amino substituents and other representative groups is well supported by the σ^+_{para} values, thus: NH_2 $(-1·3)$ > OH $(-0·92)$ > F $(-0·07)$. Although the correlation of the σ^+ parameters for several electrophilic reactivities is by no means perfect, the representative σ^+_{para} values given in Table 2 indicate the following order: NMe_2 $(-1·7)$ > NHMe $(-1·59)$ > NHPh $(-1·4)$ > NH_2 $(-1·3)$. This is in partial agreement with the sequence of conjugative abilities

$(+R)$ of these groups with aromatic systems. The value of σ^+ for p-NHPh appears rather high. The σ^+_{meta} and σ_{meta} values of the NH_2 group (see Table 2) are identical, in support of the lesser importance of resonance effects for most substituents located at the *meta* position of the aromatic nucleus[51].

The surprisingly small value of σ^+ for m-$\overset{+}{N}Me_3$ has caused erroneous prediction of the effect of this group in various electrophilic aromatic substitutions[70]. An explanation[71] of this abnormal value was suggested by the observed difference of the entropy of activation in solvolytic reactions of m-$Me_3\overset{+}{N}C_6H_4CMe_2Cl$ relative to the parent compound $C_6H_5CMe_2Cl$. This indicates that further studies are needed on this group.

The enhanced Hammett σ values denoted as σ^- are applicable to reactions involving direct conjugation between an electron-attracting substituent with strong electron-releasing groups. In these reactions, the σ^-_{meta} values, like the σ^+_{meta} values, are in most cases closely related to the σ_{meta} constants. The electron-releasing nature of amino substituents usually limits the determination of nucleophilic substituent parameters in aromatic systems, except for positive ionic centers.

Thus p-$\overset{+}{N}Me_3$ exhibits the following σ^- values[53]: ionization of phenols = 0·73, ionization of aniline = 0·70 and ring substitution = 1·11.

C. Additional Substituent Constants

Groups which are susceptible to variations in solvation or resonance during the ionization of benzoic acids were found to give poor $\rho\sigma$ correlations, hence their corresponding effects are not considered 'normal'. On the other hand, when conjugation of any substituent is prevented or minimized with the reaction site, it is regarded as exerting a normal effect. Taft[72, 73] derived his σ° values, as 'normal substituent constants', while Bekkum, Verkade and Wepster[59] calculated σ^n values. However, the rates of saponification of several substituted ethyl phenylacetates imply that the σ^n values still show the polar effects of the substituent on the resonance interaction of the reaction site with the aromatic ring. Further revaluation led to new substituent constants known as the σ_G values[74]. The values of σ°, σ^n and σ_G as a rule differ only slightly from the σ_{meta} values of Hammett, but this does not apply to the amino groups, for which the $+R$ effects vary from one scale to another, and the resonance effect apparently occurs even in the *meta* position, possibly on account of an inductive relay

8*

involving the ring and the amine substituents. Some regularities in the polar effects are indicated, since both p-NMe$_2$ > p-NH$_2$, and m-NMe$_2$ > m-NH$_2$ invariably follow the order of the $+R$ effect of these groups in conjugated systems. Moreover, resonance becomes less important in the *meta* position, since it is less negative, than at the *para* position.

An important linear free-energy correlation for aliphatic reactivities was first gained from a study of 4-substituted bicyclo [2·2·2]octane-1-carboxylic acids and esters[7], in which the substituent exerts its influence largely through the inductive direct field effect and can also provide a measure of the contribution of induction to the Hammett substituent constant. The resulting scale (the σ' values) was the basis of further research in aliphatic systems. Taft noticed[11] a constant ratio between the σ' value of a group and his polar substituent constant σ^* and additional σ' values were thus calculated for a larger number of substituents. This fact led to the introduction of a new and important scale named the inductive substituent constant $\sigma_I^{(al)}$, as determined from the aliphatic series. The inductive substituent constants of the ammonium groups of Table 2, have an opposite order to the generalization made in section II.A, thus:

$$\overset{+}{N}Me_3 > \overset{+}{N}HMe_2 > \overset{+}{N}H_2Me = \overset{+}{N}H_3$$

$$\sigma_I^{(al)}: \quad 0{\cdot}92 \quad\quad 0{\cdot}70 \quad\quad 0{\cdot}60 \quad\quad 0{\cdot}60$$

Once more, hydration through hydrogen bonding in aqueous media may be the cause of the above deviation, though it should be noticed that these $\sigma_I^{(al)}$ values were not determined in identical experimental conditions[55, 75]. The values of σ_I have also been evaluated from aromatic compounds and were found to be similar to $\sigma_I^{(al)}$[73]. However, this treatment is limited since it suggests that the inductive effect of a substituent is identical for both the *meta* and the *para* positions. The resonance substituent constant σ_R gives negative values for the NH$_2$ group in the *meta* position (Table 2), which implies that conjugation is still significant. The sequences deduced from $\sigma_{R_{para}}$ parameters indicate that p-NMe$_2$ > p-NH$_2$ and NH$_2$ > OH > F (see reference 19) for π overlap. A contradictory order for the $+R$ effect is indicated by the σ_R values derived from chemical shifts of 2,6-dimethyl-1-substituted benzenes[76], since NH$_2$ ($-0{\cdot}76$) > NHMe ($-0{\cdot}39$) > NMe$_2$ ($-0{\cdot}21$), suggesting that the two methyl groups *ortho* to the methylamino or dimethylamino group greatly inhibit its resonance interaction with the ring by preventing it from attaining coplanarity. At present the experimental facts do not permit a common scale of resonance effects independent of reaction types and conditions[55]. However, a general scale of $\sigma_{R_{meta}}$ is operative for various *meta* substituents, while the NH_2 and NMe_2 groups can not be included unless consideration is restricted to reaction series of slight electron demand. In view of this fact, a precise scale for σ_R, proposed as σ_R^o is used to define systems without resonance interaction between the substituent and the reaction site, and represents the estimates for the substituents capable of conjugation with the aromatic ring[77]. The $\sigma_{R_{meta}}$ values are approximately half the σ_R^o values for *para* substituents as seen in Table 2 for the NH_2 group.

The fact that sigma parameters are inapplicable to reactions involving highly polarizable substituents[78], such as amino groups, and other related phenomena, inevitably necessitates improvements in correlating substituent effects, and the proposed structure–reactivity relations are still only partially successful. An excellent and critical examination of the various linear free-energy equations is found in a recent and valuable publication of Ritchie and Sager[55].

IV. EQUILIBRIA

The $-I$ or $+R$ electronic transmission of positive or negative amino-nitrogen substituents can markedly affect the equilibria, particularly for organic acids or bases. However, other commonly known factors or effects can also modify these processes and many related examples will be discussed.

A. The Ionization Constants of Aliphatic Amino Carboxylic Acids

The carboxyl group can be described by contributing nonequivalent hybrid structures (6). Energetically equivalent hybrid structures de-

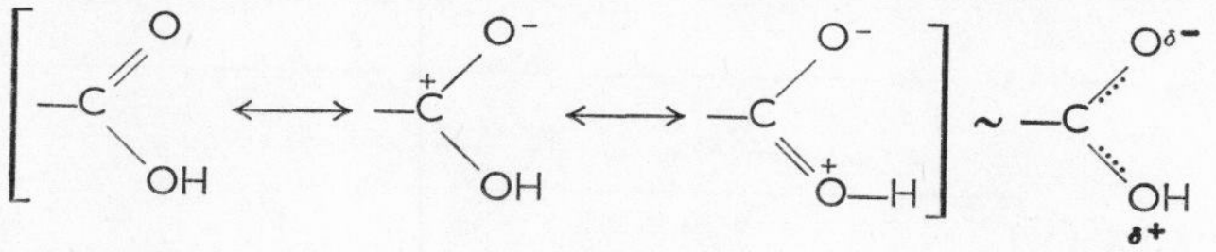

(6)

scribe the carboxylate anion (7). Consequently, the resonance energy

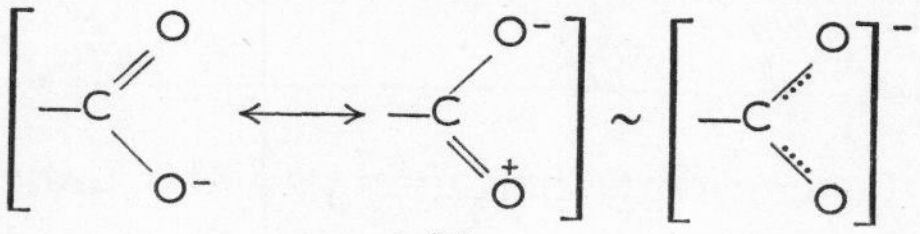

(7)

of **7** should be greater than that of **6**. The $-I$ effect of a substituent is considered to stabilize more the resonance structure **7** and thus assist the dissociation of the carboxylic acid. All common groups with positive inductive effect $(+I)$ are acid weakening.

Actually, an amine substituent and the carboxyl group within simple molecules may simultaneously ionize to form an inner salt known as 'zwitterion', where the pK_a value of the basic group is greater than, or close to, the pK_a value of the acidic group. An example is the glycine inner salt (**9**), which is a weak acid owing to the formal negative charge on the carboxyl group, while its conjugate acid (**8**) is a stronger acid because of the $-I$ effect of positive nitrogen.

$$\overset{+}{N}H_3CH_2COOH \xrightarrow[K_a = 4 \times 10^{-3}]{-H^+} \overset{+}{N}H_3CH_2COO^- \xrightarrow[K_a = 1.6 \times 10^{-9}]{-H^+} NH_2CH_2COO^-$$

$\qquad$ (**8**) $\qquad\qquad\qquad\qquad\qquad$ (**9**)

The effect of the neutral amine substituent on the ionization of substituted carboxylic acids is complicated by the presence of the zwitterion equilibrium. Hence the effect of these groups will be chiefly considered as that of the positive charged nitrogen center. Table 3 indicates that the operative inductive effects of these onium substituents greatly increase the acid strength when compared to XCH_2COOH (X = H, acetic acid $pK_a = 4.76$). It is also observed that the effect of the $\overset{+}{N}H_3$ substituent on the ionization constants decreases rapidly with increasing distance from the reaction center, and, at the same time, approaches the pK_a value of acetic acid. The effective electron-withdrawing power of these onium groups shown in the order of carboxylic acid acidity is (Table 3):

$$H_3\overset{+}{N}CH_2COOH < Me\overset{+}{N}H_2CH_2COOH < Me_2\overset{+}{N}HCH_2COOH < Me_3\overset{+}{N}CH_2COOH$$

TABLE 3. Values of pK_a for aliphatic acids containing positive nitrogen substituents[18].

X	$\overset{+}{N}H_3$	$\overset{+}{N}H_2Me$	$\overset{+}{N}HMe_2$	$\overset{+}{N}Me_3$
XCH_2COOH	2.31	2.16	1.95	1.83
$X(CH_2)_2COOH$	3.60			
$X(CH_2)_3COOH$	4.23			
$X(CH_2)_4COOH$	4.27			
$X(CH_2)_5COOH$	4.43			

constant and the constants pertaining to the individual ionization processes.

This is contrary to the sequence established in section II.A, and is attributable to hydrogen bonding[79], thus:

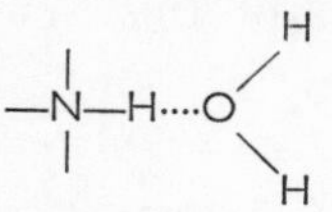

$$-\overset{|}{\underset{|}{N}}-H\cdots O\overset{\displaystyle H}{\underset{\displaystyle H}{\diagup}}$$

This is deduced from Hall's assumption of the steric effect on the solvation of tertiary ammonium ions[80]. The ionization constants of *N*-(substituted-phenyl)glycines reported recently[81] provide interesting facts on the nature of the substituent, which in conjunction with spectral data, have been used to calculate the zwitterion equilibrium constant and the constants pertaining to the individual ionization process.

B. The Ionization Constants of Aromatic Amino Carboxylic Acids

The location of a substituent in the ring of benzoic acid has been shown to affect the properties of the COOH group. The effects in the *ortho* position are rather complex and generally these are the strongest acids, even when a substituent has both $+I$ and $+R$ effects. This acid-strengthening effect can be the result of inductive transmission owing to the short distance, and because the resonance interaction $(+R)$ is reduced since the carboxyl group is forced out of coplanarity with the ring. In the *meta* position, the inductive mechanism promotes the ionization of substituted benzoic acids relatively more than in the *para* position, as the $-I$ effect decreases with increasing distance from the acidic group. *Para* substituents with $+I$ and $+R$ effects as a rule decrease the acid strength, whereas those with $-R$ effects enhance it.

The ionization of *o*-aminobenzoic acid represents a special case, since the NH_2 group can participate in the reaction. The equilibrium constant and the extended Hammett relationship of this system[82] reveal that the interaction between COOH (or COO^-) and NH_2 (or $\overset{+}{N}H_3$) is sufficiently strong to make it difficult to separate the zwitterion from the non-zwitterion species, or rather, to discriminate between ionizations from the COOH and $\overset{+}{N}H_3$ groups. The proposed ionic equilibria are shown in equation (2).

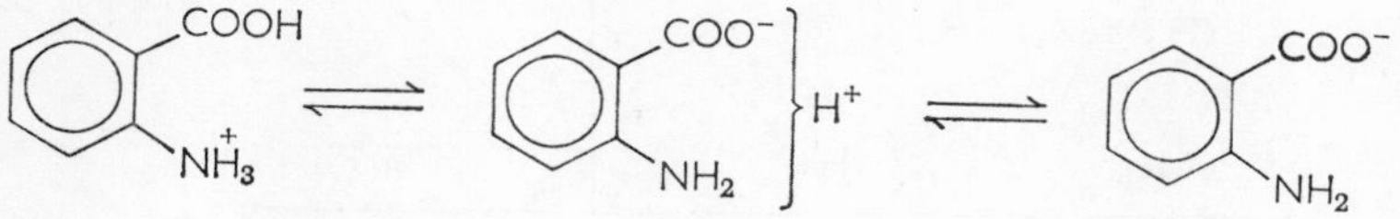

$$\text{(2)}$$

Similarly *m*-aminobenzoic acid exists largely in the zwitterionic form in aqueous solution[83–85] and in several low-dielectric solvents[86], Accordingly, the ionization of this acid could be represented by equation (3).

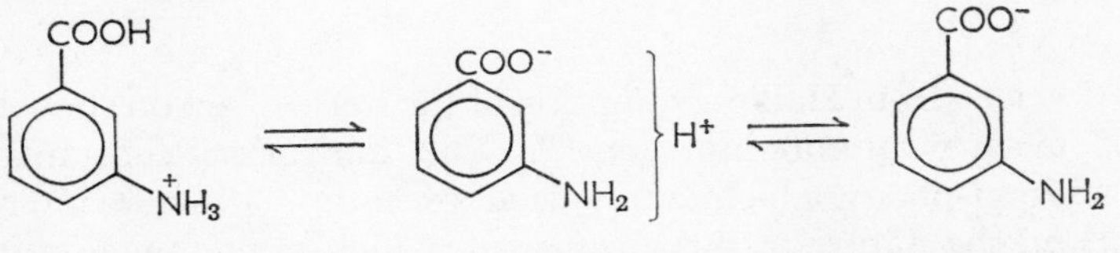

$$\tag{3}$$

Normally, the *para*-amino substituents additionally weaken the acidity of the carboxyl group[86–89] due to resonance (equation 4)

$$\tag{4}$$

which produces a higher negative charge on the carboxyl oxygen atoms, thus increasing their attraction for hydrogen ions, and, as a consequence, interfering with the resonance stabilization of the carboxyl anion.

This acidity order, *ortho* > *meta* > *para*, is shown to be irregular (Table 4) for neutral amine substituents. The evaluation of the inductive effects, including the zwitterion form and significant resonance contributions of the groups for each position of the ring, is difficult to rationalize. As an illustration, it is surprising to note that *ortho*-substituted compounds are less acidic, even if one considers the steric inhibition of resonance and their existence as dipolar ions.

TABLE 4.   Values of pK_a for amino-substituted benzoic acids in water[18].

	pK_a of XC_6H_4COOH		
X	*ortho*	*meta*	*para*
(H)	(4·20)	(4·20)	(4·20)
NH_2	4·98	4·79	4·92
NHMe	5·33	5·10	5·04
NMe_2	8·42	5·10	5·03
$\overset{+}{N}H_3$	2·04	—	—
$\overset{+}{N}H_2Me$	1·89	—	—
$\overset{+}{N}HMe_2$	1·4	—	—
$\overset{+}{N}Me_3$	1·37	3·45	3·43

With regard to acid strengthening, the inductive sequence $NH_2 >$ $NHMe > NMe_2$ can, to some extent, be applied to the *meta* positions. However, the sequence $NMe_2 > NHMe > NH_2$ of $+R$ effects is more evident for *ortho* than for *para* substituents. These groups in the *ortho* position do not appear to be bulky enough for steric inhibition of resonance, although reversal of the order expected from the $-I$ mechanism may result from the actual tendency of the *ortho* isomers to exist as inner salts. The high pK_a value of o-dimethylaminobenzoic acid (8·42) is attributed to hydrogen bonding, an important feature in the zwitterion[18]. Further, steric inhibition to resonance in the NMe_2 group and the relief of steric strain should assist the formation of a stable hydrogen-bonded structure as follows:

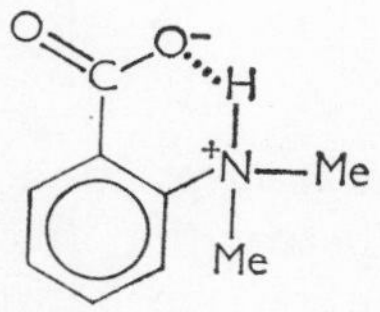

The inductive effect of positively charged nitrogen substituents in benzoic acid must increase the acid strength more at the *meta* than at the *para* position[18, 63]. However, this is not the case for the $\overset{+}{N}Me_3$ group (see Table 4), although the *ortho* isomer is the strongest acid, as expected. The acidity *meta* > *para* for the trimethylammonium group, is obtained for the ionization constants of these acids in 50% ethanol[63] and in dioxane–water solutions[90]. As with the acetic acid derivatives (section IV.A) the same type of effect of onium groups (in addition to the carboxyl group being prevented from adopting coplanarity with the ring), can be noted with the o-benzoic acids, that is $\overset{+}{N}Me_3 > \overset{+}{N}HMe_2 > \overset{+}{N}H_2Me > \overset{+}{N}H_3$.

The variation of the electronic effects of amino substituents on the ionization constants of benzoic acids has been shown to be the result of steric inhibition of resonance, originating in the presence of adjacent groups[91-95]. Another interesting phenomenon is the fact that the effect of the NH_2 group in position 4 of biphenylcarboxylic acid is transmitted in a predictable manner, but it is quantitatively less than in benzoic acids[96].

The effect of two or more substituents upon the acidity of benzoic acid has been shown in many cases to be additive whenever supplementary steric and resonance interactions are absent[18, 93, 94, 97, 98].

Recent work[99] indicates serious departure from the additivity relationship and the application of this principle has not yet been properly examined for amino groups in the presence of other substituents.

C. The Ionization Constants of Aminophenols

The acidity of phenols, compared with aliphatic alcohols, can be accounted for by the combined $-I$ effect of the benzene ring and the delocalization of the electrons of oxygen in the phenoxide ion, as shown by the resonance structures in equation (5)[100]. In general, the

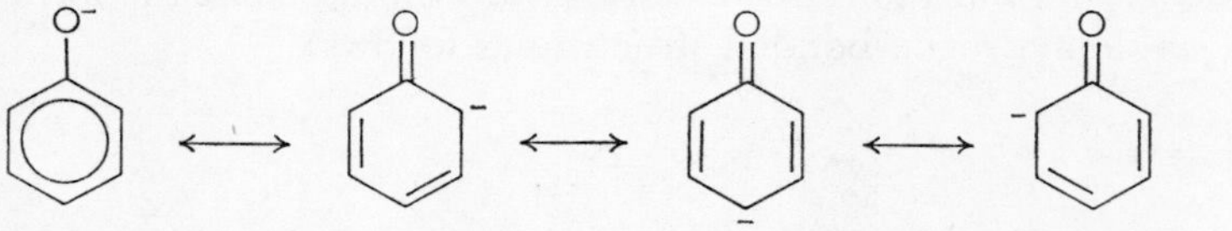

$$\tag{5}$$

effect of substituents in the *meta* position of phenol is similar to that found in benzoic acids, whereas the effect in the other positions depends on the resonance interaction between the substituent and the reaction center. Thus groups with $+I$ and $+R$ effects decrease acidity, while those with $-I$ and $-R$ effects show lower pK_a values.

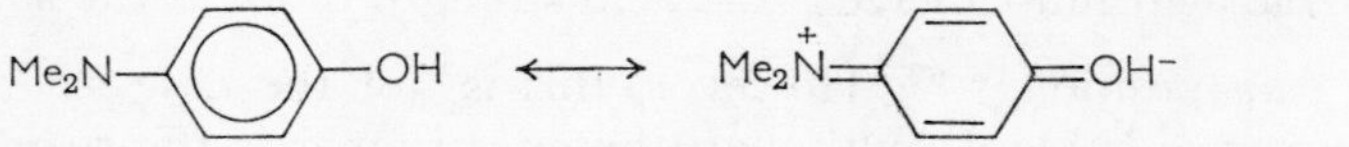

$$\tag{6}$$

This is explained by the fact that the phenol anion has a greater capacity for direct conjugation than the benzoate anion[100].

The information given in Table 5 permits few conclusions to be drawn. The inductive mechanism for $\overset{+}{N}Me_3$ is in agreement with the general principles discussed for benzoic acids. However, it is interesting to note that aminophenols can undergo association as half-salt cations in aqueous solution[101]. Many studies have considered the effect of

TABLE 5. The ionization constants of amino-substituted phenols in water.

Group	pK_a			
	ortho	*meta*	*para*	Reference
NH_2	9·71	9·87	10·30	18
NMe_2	—	9·85	10·22	61
$\overset{+}{N}Me_3$	7·42	7·98	8·20	90

amino groups in derivatives of aminophenols, aminonaphthols and related condensed aromatic compounds, but they cannot be discussed due to limitations of space [340-342].

D. The Ionization Constants of Nitrogen Bases

The effects of substituents on the conjugate acids of aliphatic amines are much the same as with aliphatic carboxylic acids. In Table 6, a

TABLE 6.   Values of pK_a for conjugate acids of aliphatic amines [18].

	$X = NH_2$	$X = \overset{+}{N}H_3$
$XCH_2\overset{+}{N}H_3$	—	—
$X(CH_2)_2\overset{+}{N}H_3$	9·98	6·97
$X(CH_2)_3\overset{+}{N}H_3$	10·65	8·59
$X(CH_2)_4\overset{+}{N}H_3$	10·84	9·31
$X(CH_2)_5\overset{+}{N}H_3$	11·05	9·74

positively charged group $(X = \overset{+}{N}H_3)$ increases the acidity of conjugate acids studied, while neutral group (NH_2) affects them less. The $-I$ mechanism for both cases increases the pK_a values with increasing distance from the reaction site, and the inductive effect of the $\overset{+}{N}H_3$ is greater than the corresponding NH_2 group.

The base-weakening resonance in aromatic amines (3) was discussed in section II.B. Substituents capable of $-I$ effect increase the proton removal of conjugate acids of aniline more strongly from the nearer position. Groups with $+I$ and $+R$ effects in the *ortho* and *para* positions result in an increase in pK_a values, whereas those groups with $-R$ effects decrease these values. The low pK_a value of $o\text{-}NH_2$ relative to $m\text{-}NH_2$ (see Table 7), presumably reflects steric factors impeding

TABLE 7.   Values of pK_a for anilinium derivatives [18].

Group	*ortho*	*meta*	*para*
NH_2	4·47	4·88	6·08
NMe_2	—	—	7·96[a]
$\overset{+}{N}H_3$	1·3	2·65	3·29
$\overset{+}{N}Me_3$	—	2·26	2·51

[a] Reference *e* of Table 2.

some resonance interaction between this group and the acidic onium center. However, the value for p-NMe_2 is larger than for p-NH_2 due to a $+R$ effect. The accepted sequence *ortho* > *meta* > *para* for electron-withdrawing substituents holds well for ammonium groups, except $\overset{+}{N}Me_3 > \overset{+}{N}H_3$ in $-I$ transmission. The latter observation can be simply accounted for in terms of hydrogen bonding with the solvent.

The polar effects of substituents in rigid hetero-aromatic bases such as pyridine can be estimated with greater ease than for the corresponding benzene derivatives. This is largely due to the fact that the basic center is not affected by the steric inhibition of resonance phenomena[18]. As in aniline, electron-releasing groups $(+I, +R)$ increase the base strength of pyridine, whereas electron-withdrawing groups $(-I, -R)$ have the opposite effect. The data for the conjugate acids of 2- and 4-aminopyridine (Table 8), indicate that proton removal from

TABLE 8.   Values of pK_a for aminopyridine in water[a].

	2-NH_2	3-NH_2	4-NH_2
Pyridine	6·86	5·98	9·17

[a] A. Albert, R. Goldacre and J. Phillips, *J. Chem. Soc.*, 2240 (1948).

these is more difficult due to direct conjugation of the two basic sites (equation 7). The resonance interaction is responsible for base strengthening and is, evidently, more effective in the 4-position.

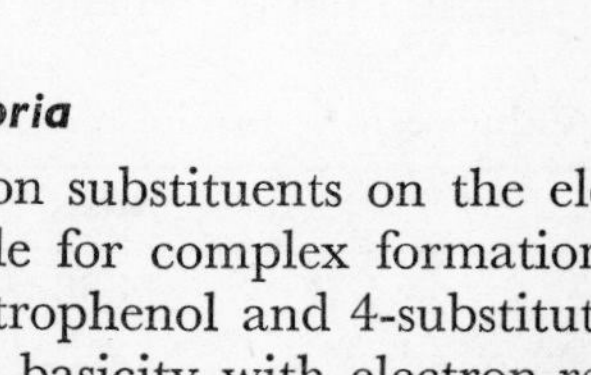

$$(7)$$

E. Miscellaneous Equilibria

The effect of common substituents on the electron density at the nitrogen atom available for complex formation (through hydrogen bonding) between p-nitrophenol and 4-substituted benzalanilines[102], indicates an increasing basicity with electron-releasing power of the substituents. The changes in equilibrium constant with 4-substituents show good correlations with the σ^+ values, and the 4′-substituted series with the Hammett σ values. The enhancement of basicity of benzalanilines by the 4-NMe$_2$ group can be attributed to the following resonance contribution involving direct conjugation with the substituent (equation 8).

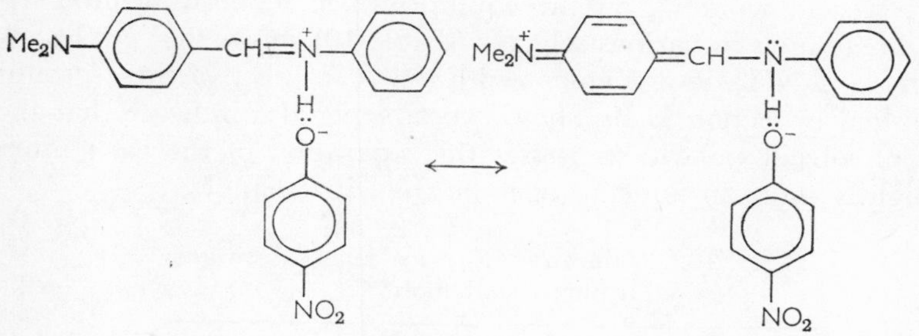

$$(8)$$

The influence of amino groups on the equilibrium constants of triphenylmethanol (ROH) in strong acid media[103, 104] have shown

$$ROH + H^+ \longrightarrow R^+ + HOH$$

large deviations from the correlation of pK_{R^+} with the Hammett σ values. The reason is the great tendency of the unshared electron pair to stabilize an electron-deficient reaction center through resonance interaction, as represented by the following contributing resonance structure **10**[105].

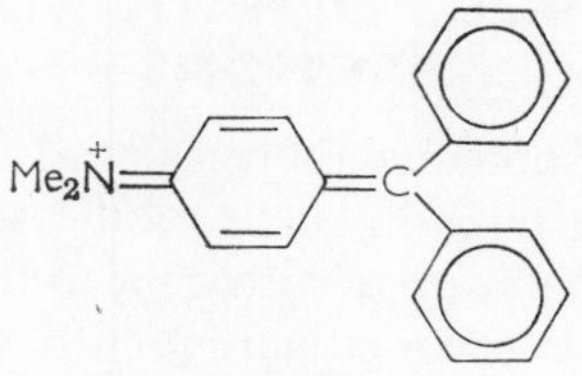

(10)

TABLE 9.   Values of pK_{R^+} for substituted triphenylmethanols[103].

Substituent	pK_{R^+}
4-NH$_2$	+4·6
4,4'-(NH$_2$)$_2$	+5·38
4,4',4''-(NH$_2$)$_3$	+7·57
4-NMe$_2$	+4·75
4,4'-(NMe$_2$)$_2$	+6·90
4,4',4''-(NMe$_2$)$_3$	+9·36
4,4'-($\overset{+}{N}$Me$_3$)$_2$	−7·52
4,4',4''-($\overset{+}{N}$Me$_3$)$_3$	−10·15

The data listed in Table 9 indicate that resonance stabilization is greater as *para*-amino groups are successively introduced into each ring, and that p-NMe$_2$ is greater than p-NH$_2$ in π delocalization $(+R)$. Recent work[106], on the equilibrium of monosubstituted triphenylmethanols in various solvents (Table 10) shows that p-NEt$_2$ > p-NMe$_2$ > p-NH$_2$, in agreement with the $+R$ effect of these groups established in section II.B. These experiments also indicate that the effect of solvent polarity represses the separation of the carbonium ion slightly more in aqueous acetone than in ethanol.

TABLE 10. Values of pK_{R^+} for monosubstituted triphenylmethanols[106].

Substituent	$pK_{R^+}{}^a$	$pK_{R^+}{}^b$	$pK_{R^+}{}^c$
4-NH$_2$	2·61	1·92	2·34^d
4-NMe$_2$	3·54	2·29	2·67
4-NEt$_2$	4·78	3·32	3·74

^{a} In water at 25°.
^{b} In 50% acetone by volume at 25°.
^{c} In 50% ethanol by volume at 25°.
^{d} The only data at 20°.

V. NUCLEOPHILIC ALIPHATIC DISPLACEMENT REACTIONS

Only a few types of reactions have been found useful for studying the polar effect of amino groups in nucleophilic substitutions. This fact may be due to the basic nature of nitrogen, to the instability of highly reactive substrates, and to the media frequently used in these reactions. The main mechanisms[107-109] to be considered are shown in reactions (9 and 10). With saturated carbon, the unimolecular mechanism S_N1, is a two-stage process (reaction 9), and the bimolecular mechanism S_N2 (reaction 10), is a one-stage process with inversion of configuration.

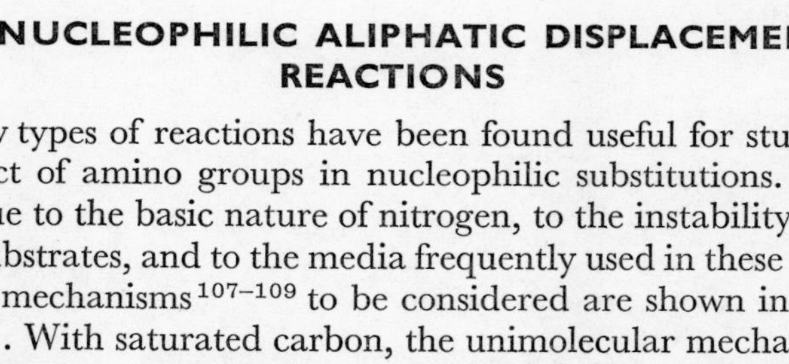

$$\overset{|}{\underset{|}{C}}-X \longrightarrow \left[\overset{|}{\underset{|}{C}}{}^{\delta+}\cdots X^{\delta-} \right] \longrightarrow \overset{|}{C}{}^+ + :X^- \quad \text{(slow)}$$

transition state

(9)

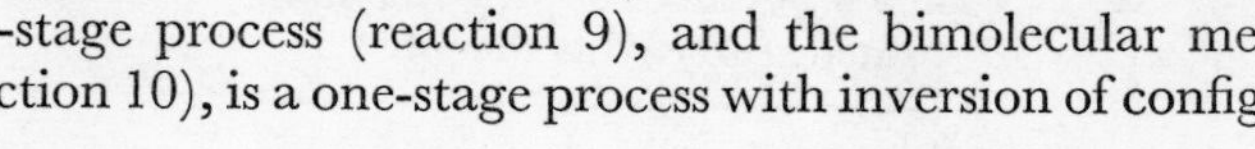

$$\overset{|}{C}{}^+ + :Y^- \longrightarrow \overset{|}{\underset{|}{C}}-Y \quad \text{(fast)}$$

$$Y: + \overset{|}{\underset{|}{C}}-X \longrightarrow Y\cdots \overset{|}{\underset{|}{C}}\cdots X \longrightarrow Y-\overset{|}{C} + :X \quad (10)$$

The general course of aliphatic displacement at unsaturated carbon involves multistage mechanisms, which consist of a nucleophilic addition to a π system followed by an elimination step (reaction 11).

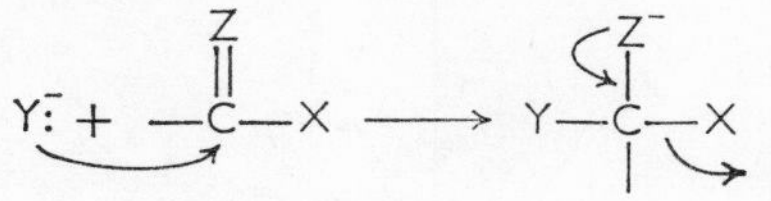

$$Y\colon\!\!\bar{} + \underset{|}{\overset{\|}{\underset{|}{-}\overset{Z}{C}-}}X \longrightarrow Y-\overset{Z^-}{\underset{|}{\underset{|}{C}}}-X \longrightarrow Y-\overset{Z}{\underset{|}{\overset{\|}{C}}}- + \colon\!X^- \tag{11}$$

A. Neighboring-group Participation

I. Haloamines

The presence of an amino group in n-alkyl halides is a cause of rapid solvolysis accompanied by ring closure. This phenomenon is shown by substituents capable of stabilizing the transition state by becoming bonded or partially bonded to the reaction center. The rate increases are explained by neighboring-group participation[110], the substituent providing anchimeric assistance[111] to the reaction center[112].

The rearrangement of primary to secondary halide in β-haloamines has been postulated to involve a cyclic 'ethylene immonium' ion intermediate[113, 114]. A similar mechanism is also suggested for the ring expansion of some heterocyclic amines (equation 12)[115]. The

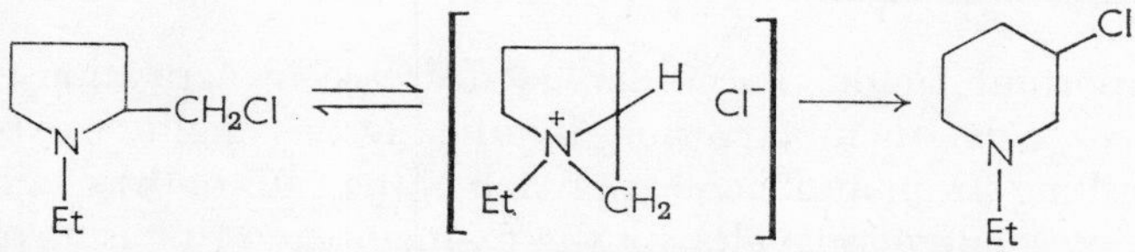

$$\tag{12}$$

fast solvolysis of β-amino primary chlorides is explained by the formation of a cyclic intermediate or by participation of the neighboring amino group in the rate-determining step of the reaction[109]. The cyclization of 1-amino-2-chloropropane proceeds more than one hundred times faster than the estimated solvolysis of 2-chloropropane in water at $25°c$[109]. The concept of 'ethylene immonium' ions is supported by kinetic studies on β-haloamines and by work on the dimerization of bis-β-haloalkylamines[116–122].

The isomerization of primary halides generally involves an S_N2 mechanism, whereas that of secondary halides follows the S_N1 mechanism. This type of intramolecular attack of the ω-nitrogen atom on the halogenated carbon to form 3-, 4- and 5-membered cyclic ammonium ions, is shown in reaction (13).

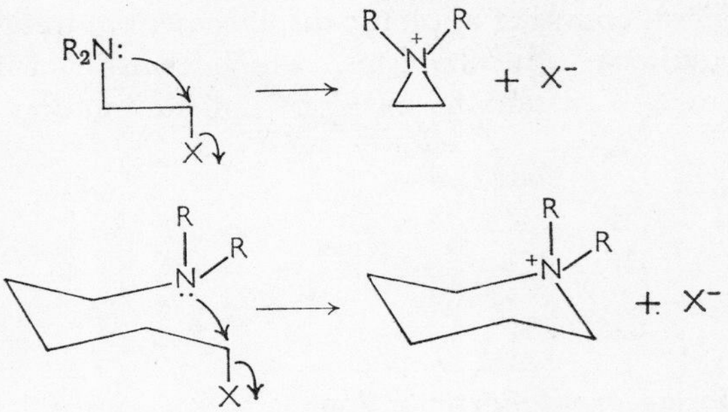

$$(13)$$

Data for the ring closure of ω-bromoalkylamines[123] are summarized in Table 11. It is evident that the distance separating the two groups

TABLE 11.   Rates of cyclization of ω-bromoalkylamines in water at $25°C$[125].

Compound	k (per min)	E (kcal)	Log PZ (per sec)
$H_2N(CH_2)_2Br$	0·036	24·9	14·8
$H_2N(CH_2)_3Br$	0·0005	23·2	11·4
$H_2N(CH_2)_4Br$	30	16·8	12
$H_2N(CH_2)_5Br$	0·5	18·3	11
$H_2N(CH_2)_6Br$	0·0015	24·9	13·3

is an important factor. The observed decrease in activation energy from three- to five-membered rings is indicative, as might be expected, of diminishing strain and facile approach of the ends of the chain. The higher activation-energy value for the six-membered ring is surprising, while that for the seven-membered ring is justified from the 'surface-tension effect'[124, 125]. This postulates that any dissolved molecule tends to adopt a spatial disposition with the lowest surface of contact in relation to the media. Therefore, the free rotation of the seven-membered amine in solution may be partially restricted so that ring closure requires a higher energy of activation. The frequency factors (PZ) account for the conformation and the relative facility by which the ends of the chain may meet to yield the corresponding cyclic ammonium ions[125].

No regular trend in reactivity is found as the nucleophilicity of the NH_2 group decreases with decreasing distance from the bromine. Nucleophilicity is roughly correlated to basicity, and increasingly basic amino groups in ω-haloamines have been shown to increase solvolysis rates[119], e.g. *N,N*-dimethyl-2-halogenoethylamines cyclize more rapidly than the corresponding unsubstituted compounds:

$$CH_3CHBrCH_2NMe_2 > CH_3CHBrCH_2NH_2$$

$$PhCHBrCH_2NMe_2 > PhCHBrCH_2NH_2$$

However, when the nucleophilicity of the nitrogen, at a fixed position in the haloamine, is reduced, the intramolecular S_N2 process (13) is supplemented by another closely related one. For example β-anilino-ethyl bromide (**11**) loses a proton from the NH group, and the resulting anion undergoes cyclization (reaction 14)[126]. For amines of still lower

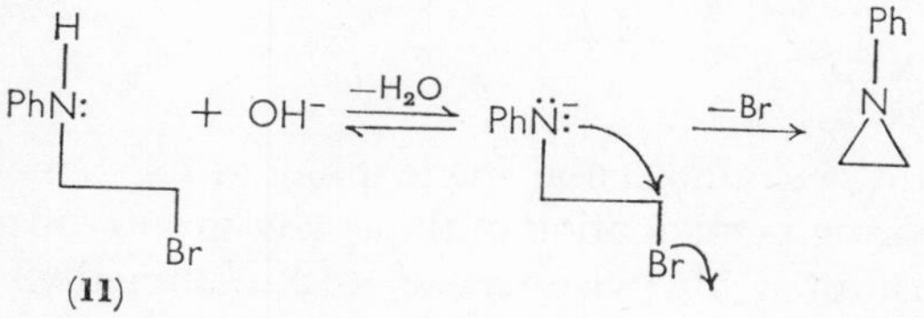

(14)

(**11**)

nucleophilicity, for example when the phenyl group of **11** is replaced by ROOC[127] or PhSO$_2$[128], the neighboring-group participation is no longer operative, and the cyclization mechanism occurs entirely according to (14). The S_N1 hydrolysis of *N,N*-di(2-chloroethyl)aniline (**12**) gives neither dimerization nor a cyclic ammonium compound[129],

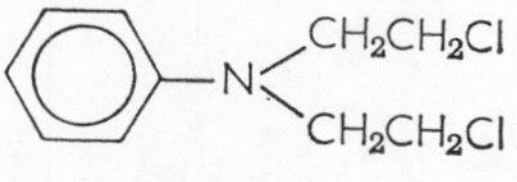

(**12**)

probably owing to the delocalization of the *p* electrons of nitrogen in the aromatic ring.

The rates of formation of cyclic ammonium ions from compounds of the type $RR^1NCH_2CH_2X$[130, 131] have been found to depend on the nature of R and R^1, and increased in the order i-Pr > Et > Me. This sequence is attributed to some extent to the electron density at the nitrogen but is mainly of steric origin. Yet when alkyl groups are branched at different positions of the closing chain, no steric effects are observed[125, 132] as in the case of 4-bromobutylamine derivatives[133] (see Table 12).

It is frequently found that at least four processes can occur in γ-haloamines, that is fragmentation, nucleophilic displacement, 1,2-elimination and neighboring-group participation[134, 135]. Their relative importance is determined mainly by structure, stereochemistry

TABLE 12.   Relative rates of cyclization of
chain-branched 4-bromobutylamines[133].

Compound	$k_{rel.}$
$H_2NCH_2CH_2CH_2CH_2Br$	1
$H_2NCH_2CMe_2CH_2CH_2Br$	158
$H_2NCH_2CEt_2CH_2CH_2Br$	594
$H_2NCH_2C(i\text{-}Pr)_2CH_2CH_2Br$	9190
$H_2NCH_2CPh_2CH_2CH_2Br$	5250

and media. However, further discussion is outside the scope of this chapter.

2. Amino esters

An interesting intramolecular nucleophilic displacement is the first-order neighboring participation of the NMe_2 group during the hydrolysis of 4-substituted phenyl esters of γ-$(N,N$-dimethylamino)butyric and valeric acids (reaction 15)[136]. The rate constant for the formation

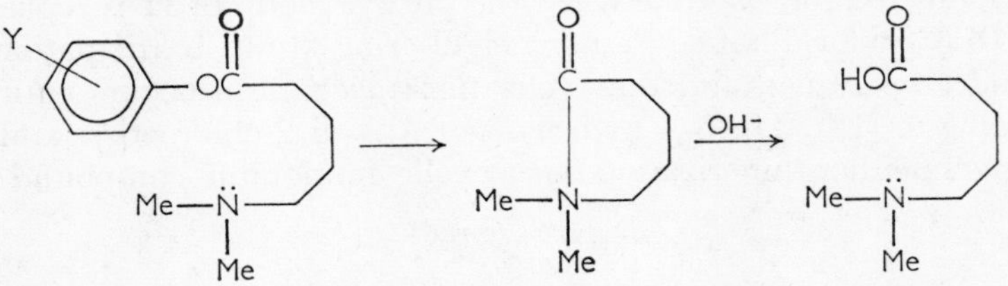

$$(15)$$

of the five-membered ring is twice that for the formation of the six-membered ring. A plausible explanation of this may be the suggested coiled conformation[136] of the ester, resulting in greater steric hindrance, of the phenyl group with the NMe_2 group, in the six- than in the five-membered ring. Yet in the base-catalyzed lactamization of the methyl esters of ω-amino-α-(toluene-p-sulphonamido)butyric and valeric acids at 25°c[137] the formation of the six-membered cyclic compound is as expected, faster (reaction 16).

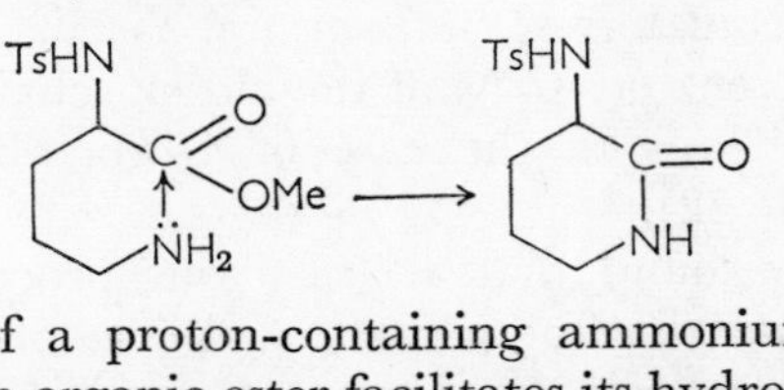

$$(16)$$

The presence of a proton-containing ammonium group on the alcoholic part of an organic ester facilitates its hydrolysis considerably more than the corresponding trialkylammonium group[138–141]. Thus the second-order rate constant for **13** is 240 times greater than that of **14**.

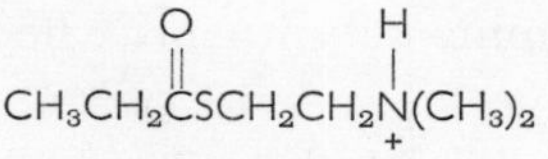

$$CH_3CH_2\overset{\overset{O}{\|}}{C}SCH_2CH_2\overset{+}{N}(CH_3)_2$$
$$\overset{H}{}$$

$$CH_3CH_2\overset{\overset{O}{\|}}{C}SCH_2CH_2\overset{+}{N}(CH_3)_3$$

(13) 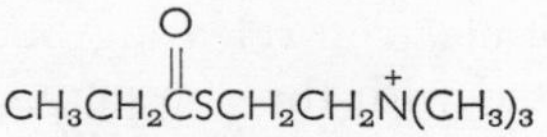   (14)

Some evidence appears to support intramolecular general acid–specific base catalysis through electrostatic bonding as in **15**[139, 141–143], recent work[144, 145], especially that on the relative order of hydrolysis, selenoester > thiolester > ester, suggests intramolecular nucleophilic catalysis through anchimeric assistance of the dimethylamino group as in **16**[145].

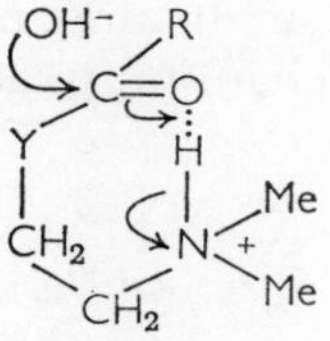

(15) 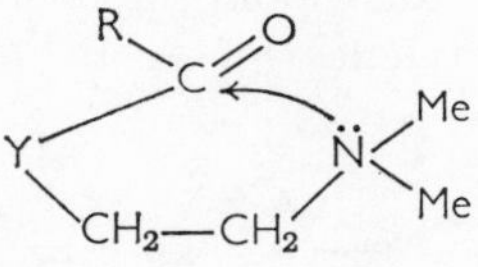   (16)

$(Y = O, S, Se)$

The reaction of hindered secondary amines with 2-chloroethyl carbonates and acetates to form tetrasubstituted ethylenediamines[146] is another process which appears to take place by bimolecular alkyl cleavage with neighboring-group participation of the amino group (reaction 17). Other examples of neighboring amino-group participa-

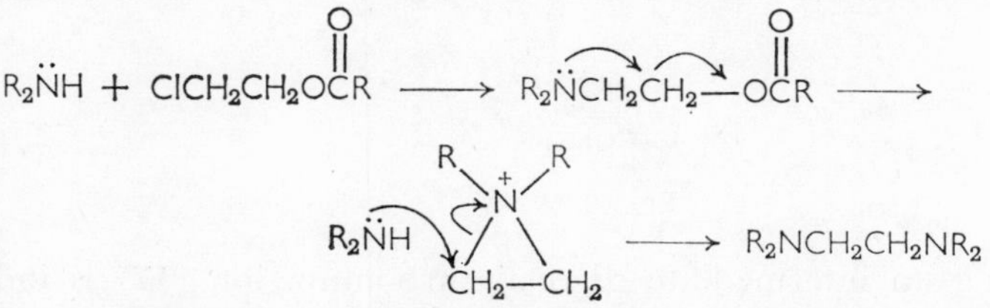

$$R_2\overset{..}{N}H + ClCH_2CH_2O\overset{\overset{O}{\|}}{C}R \longrightarrow R_2\overset{..}{N}CH_2CH_2-O\overset{\overset{O}{\|}}{C}R \longrightarrow$$

(17)

$$R_2\overset{..}{N}H \longrightarrow R_2NCH_2CH_2NR_2$$

tion are the Prevost reaction[147], the thermal decompositions of α-aminothiol esters[148], nitrogen mustards[149] and some amides and peptides[150, 151].

VI. NUCLEOPHILIC AROMATIC DISPLACEMENT REACTIONS

Nucleophilic substitution in aromatic systems has been discussed in many comprehensive reviews[152–157]. These reactions are usually

dependent on activation of the ring by electron-withdrawing substituents and consequently little information is found on the polar
effects of electron-releasing neutral amino substituents when directly
conjugated with the π system.

The most widely used illustration for the S_N1 mechanism is the
uncatalyzed decomposition of diazonium salts. Such substrates carrying amino substituents have scarcely been investigated, except for the
magnetochemical study on the photodecomposition of p-(N,N-
dimethylamino)benzenediazonium chloride[158]. The result of this and
other work[159] indicated a free-radical mechanism.

An interesting example of aromatic unimolecular solvolysis is the
interaction of *para*-substituted α-bromostyrenes with 80% ethanol to
yield the corresponding acetophenones and small amounts of the
phenylacetylenes[160]. A two-step S_N1–$E1$ mechanism (reaction 18)

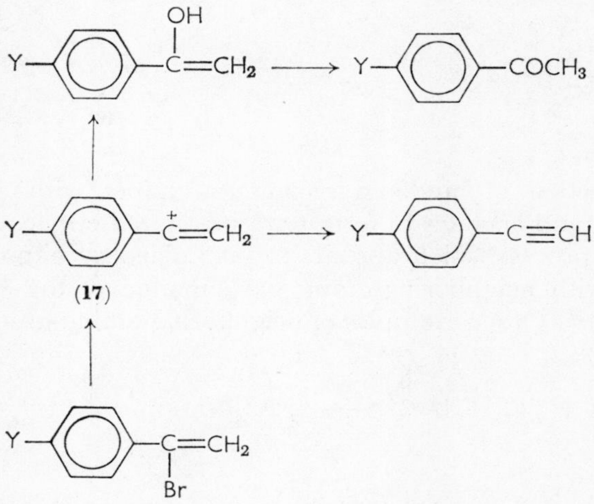

(18)

involving an intermediate digonal carbonium ion (**17**) is indicated
where the p-NH$_2$ group accelerates the reaction more than all the
other substituents studied, and is 10^8 times faster than with the unsubstituted compound. The effect of the amino group is rather surprising, since the resonance interaction through delocalization of the
type indicated by **18**, should retard solvolysis.

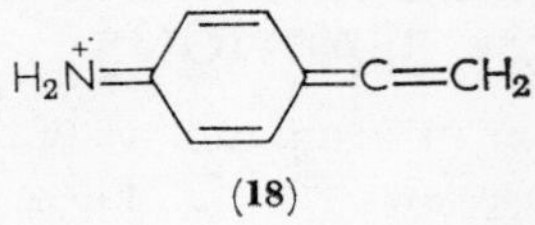

(**18**)

The S_N2 displacement (reaction 19) involves, most frequently, an intermediate complex (**19**), with the hybridization of carbon at the

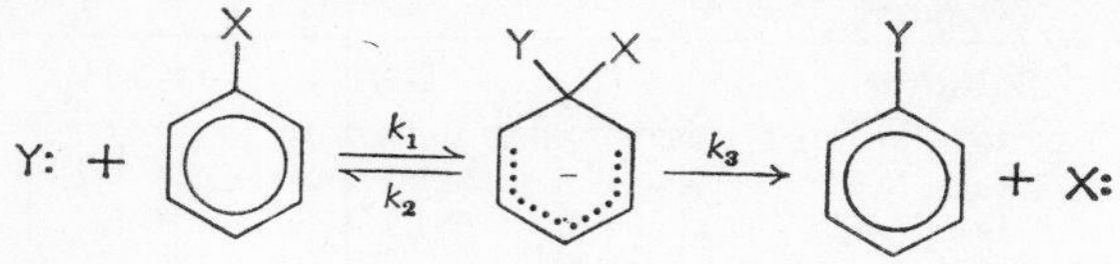

$$\text{Y: } + \quad \underset{k_2}{\overset{k_1}{\rightleftharpoons}} \quad \xrightarrow{k_3} \quad + \text{ X:} \tag{19}$$

(**19**)

reaction center changing from sp^2 to sp^3. The tetrahedral configuration in **19** again is stabilized by electron-attracting groups. Either step might be rate determining, depending on the relative magnitude of k_2 and k_3.

Both 4-amino and 4-dimethylamino substitution in 2-nitrobromo-benzenes greatly retards the reaction with piperidine[161], as expected. The 4-NMe$_2$ compound reacts 10 times faster than the 4-NH$_2$ compound even though the order of the $+R$ effect is NMe$_2$ > NH$_2$ in electron release to aromatic rings. Moreover, in this case the unsubstituted compound reacts 10^5 times faster than the *para*-amino compound. In contrast, the unactivated compounds, i.e. chlorobenzene and *p*-aminochlorobenzene, exhibit almost identical reaction rates in piperidine[162], as also do bromobenzene and *p*-aminobromobenzene towards sodium methoxide at 150°C[156]. However, a *para*-amino group lowers the reaction rates of *o*-fluoronitrobenzenes with sodium ethoxide[163] about one hundredfold.

The rate of halogen displacement has been correlated with the polar effects of substituents *para* to the reaction center[164–171], the activating effect of the *o*-nitro group being considered constant, although in some cases resonance interaction is possible with the substituent, as shown in structures of the type **20**. The relative rates of reactions for a

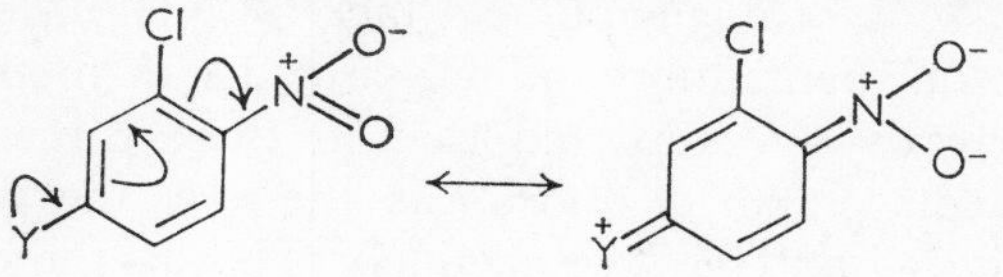

(**20**)

series of 4- and 5-substituted 2-nitrochlorobenzenes with piperidine in benzene[172] have shown a greater enhancement of velocities when the substituent and the NO$_2$ group are *para* to each other. The data in Table 13 reflect a retarding effect of the 4-amino substituent in direct conjugation with the reaction center, whereas from position 5 the $+R$ effect is transmitted to the site of displacement by a secondary

TABLE 13.   Kinetic data for the reaction of
substituted 2-nitrochlorobenzenes with piperidine
in benzene[172].

Substituent	H	5-NH$_2$	4-NH$_2$
k_S/k_H	1	0·86	0·001
E (kcal/mole)	13·9	13·7	19·6

inductive effect and is almost compensated for by the negative inductive effect of the substituent. Similar considerations in the reaction of 4-substituted 2,6-dinitrochlorobenzenes and 5-substituted 2,4-dinitrochlorobenzenes with sodium methoxide in methanol[173] have been correlated with the activating influence of the substituents in these systems. The effects of the nitro group in positions 2 and 4 are not equivalent, and they differ in magnitude and also with the nature of the nucleophilic reagent.

The steric effect of the amino group adjacent to the site of substitution is rather small[174], and, therefore, the order of the reaction rates of nucleophilic displacement is o-NH$_2$, > p-NH$_2$, since the resonance interaction from the *para* position is more marked than from the *ortho* position.

The magnitude of the activating effect of positively charged nitrogen in nucleophilic replacement[167,175,176] was found to be weaker than that of the NO$_2$ group. This fact is unexpected, because ammonium groups deactivate electrophilic aromatic substitutions more strongly than nitro groups. Apparently $\overset{+}{N}R_3$ > NO$_2$ in the polarization of the ring for *deactivation* processes, whereas $\overset{+}{N}R_3$ < NO$_2$ in polarizability of the ring for *activation* processes[175].

A detailed study of nucleophilic attack by NH$_3$, MeNH$_2$, Me$_2$NH and EtONa on pentafluoroaniline, and its N-methyl and N,N-dimethyl derivatives[177] reported that in these three substrates the ratio of the replacement of fluorine in the *meta* and *para* positions is approximately 7, 1 and 0·07, respectively. This fits the steric inhibition of resonance in the order NMe$_2$ > NHMe > NH$_2$, and is also supported by ultraviolet spectra, that is C$_6$F$_5$NMe$_2$ (ε 10,000) > C$_6$F$_5$NHMe (ε 16,800) > C$_6$F$_5$NH$_2$ (ε 18,000). The NMe$_2$ group is forced greatly out of the plane of the ring, reducing the resonance interaction and leading to a relatively rapid replacement of the *para* halogen. The NH$_2$ group is less hindered, and owing to conjugation with the ring the deactivation of the *para* position is more than that of the *meta* position. All three substrates gave very low yield of *ortho* replacement. For the reactions of substituted pentafluorobenzenes with EtONa in EtOH [178] the NMe$_2$ group deactivates *para* replacement more than any other group studied.

VII. NUCLEOPHILIC ADDITION AND SUBSTITUTION AT CARBOXYL GROUPS

Nucleophilic substitution at carbonyl carbon, takes place in bimolecular carboxylic ester hydrolysis with acyl-oxygen fission (**21**) by the base-catalyzed ($B_{AC}2$) and acid-catalyzed ($A_{AC}2$) mechanisms; in some instances, the fission of alkyl to oxygen bond (**22**) can also occur.

Generally, electron-withdrawing groups located either in R^1 or in R^2 facilitate the $B_{AC}2$ process, whereas the polar effect of substituents in acid-catalyzed ester hydrolysis is known to be rather small. However, most groups with $+I$ effect will retard $A_{AC}2$ reactions. Detailed information on ester hydrolysis may be found in standard references [182–186].

A. Aliphatic Esters

Only a few rates of hydrolysis of aliphatic esters containing amino groups have been studied. Structural factors and reaction conditions are important in determining whether hydrolysis occurs through neighboring amino-group participation (see section V.A.2) or by nucleophilic attack at the carbonyl carbon.

The kinetic results for hydrolysis of Et$_3\overset{+}{N}$CH$_2$COOEt, as compared with that of ethyl acetate, indicate that the introduction of $\overset{+}{N}$Et$_3$ increases the rate of alkaline hydrolysis 200-fold, and decreases that of acid hydrolysis 2000-fold [187]. These data suggested a simple electrostatic picture in which the presence of a net charge in the ester will be the major factor in determining the reaction velocity, but were not sufficient for deciding the finer points of the mechanism. Similarly, the acid and alkaline hydrolysis of other quaternary amino esters has shown that the main effect of the positive charge can reasonably be assumed to be electrostatic [6, 188, 189].

The alkaline hydrolysis of some esters containing nitrogen substituents[190] (Table 14) indicates little difference whether the substituent

TABLE 14.   Alkaline hydrolysis of aliphatic esters[190].

Y in YCH_2COOEt	X in CH_3COOCH_2X	$k_2{}^a$
H	—	0·111
—	H	0·185
NH_2	—	0·83
$\overset{+}{N}Me_3$	—	66
—	$\overset{+}{N}Me_3$	60

[a] Second-order rate coefficient M^{-1} sec^{-1} at 25°c and zero ionic strength.

is in the alcohol or the acyl side of the ester. In the basic hydrolysis of glycine ethyl ester hydrochloride[190] the reaction seems to occur only with the uncharged molecule, as k values did not vary with pH range (9·70–11·5). Ammonium groups enhance saponification rates considerably more than the neutral NH_2 group. This can be accounted for either by the electrostatic effect or by the order $\overset{+}{N}Me_3 > NH_2$ in inductive transmission.

B. Aromatic Esters

The polar effects of most *meta* and *para* substituents on the second-order rates of hydrolysis of aromatic esters[191-195] have demonstrated the marked influence on the energy of activation with little affect to the entropy term. In contrast, *ortho* substituents alter both the activation energy and the entropy factor. As a general rule, the alkaline hydrolysis of aromatic carboxylic esters is accelerated by substituents with $-I$, $-R$ effects and is retarded by substituents with $+I$, $+R$ effects.

The relative rates of S_N2 hydrolysis in Table 15 show strong re-

TABLE 15.   Relative rates of the second-order alkaline hydrolysis of aromatic amino esters[183].

	Y		
	m-NH_2	p-NH_2	H
$YC_6H_4COOEt^a$	—	0·023	1
$YC_6H_4COOEt^b$	0·574	0·0293	1
$AcOC_6H_4Y^c$	—	0·510	1

[a] In 85% aqueous EtOH at 25°c.
[b] In 60% aqueous Me_2CO at 25°c.
[c] In 60% aqueous Me_2CO at 0°c.

tardation by p-NH$_2$ in the aroyl side, and a much weaker effect by m-NH$_2$ in the aroyl side or by a p-NH$_2$ group in the phenyl ester. Evidently, resonance interaction between the amino and carbethoxy groups in p-H$_2$NC$_6$H$_4$COOEt is responsible for the low rates. The o-NH$_2$ group also influences the alkaline hydrolysis of ethyl o-amino-benzoate[196], but the decrease in the rates arises partially from steric effects.

Good agreement is found in the rate ratios of p-NH$_2$-substituted and unsubstituted ethyl benzoates expressed as k_S/k_H[195] (Table 16). Three very similar velocity ratios were determined in aqueous ethyl alcohol,

TABLE 16. Relative rates (k_S/k_H) of hydrolysis of ethyl p-aminobenzoates.

	Kindler[191]	Ingold and Nathan[192]	Tommila[194]	Jones and Robinson[195]
p-NH$_2$	0·0233	0·0231	0·030	0·0244

and a slightly higher value in aqueous acetone. In this connection, it should be noted that these reactions and the substituent effects were shown to be sensitive to changes in the solvents[196–199].

The introduction of alkyl substituents adjacent to the 4-amino group in aromatic esters prevents coplanarity, and the decreases in the rate of hydrolysis[88] caused by the latter. For example, compound **24** is hydrolyzed approximately 34 times more slowly than its parent

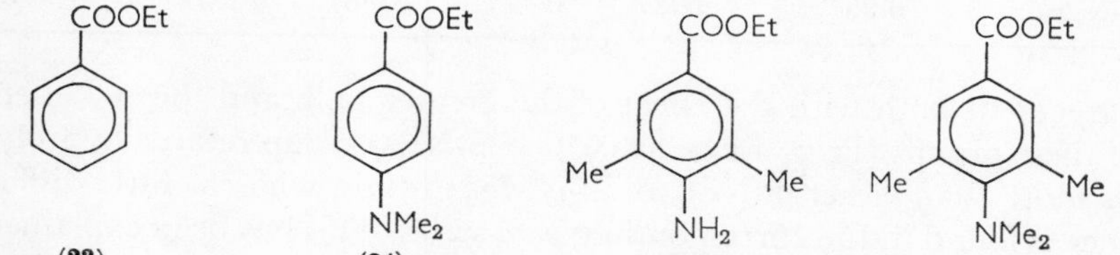

(23)	(24)	(25)	(26)
0·052	0·00152	0·019	0·206

k in l/mole/min

compound **23**, as a result of resonance interaction with the NMe$_2$ group ($+R$ effect), while the alkaline hydrolysis of **26** is much faster than that of **24** and of **25**. Comparing the substrates **25** and **26**, obviously the small size of the NH$_2$ group still allows conjugation with the carbethoxy group, thus giving a greater stabilization and a retarding hydrolysis of the ester, while this does not occur with the strongly hindered bulky NMe_2 group. The electrical influence of the dimethylamino group on the rate of hydrolysis of the ester groups, located in either *meta* or *para* position, has been found to be greatly affected by steric inhibition of resonance by one *ortho*-methyl group[200]. The effect of the NMe_2 group is greater at the *para* position. The influence at the *meta* position is attributed to a second-order field effect, arising from the negative charges placed on the *ortho*- and *para*-carbon atoms by resonance, or to an inductive effect. Steric inhibition of resonance, in the hydrolysis of ethyl 1-naphthoates with 4-NMe_2 groups has been decided[201].

Increasing the size of the *N*-alkyl group decreases the rate of hydrolysis[199], in the order $NMe_2 > NEt_2 > NPr_2$ (Table 17). This is in

TABLE 17.　Hydrolysis rates and σ values of alkyl N,N-dialkylaminobenzoates[199].

Y	YC_6H_4COOMe (80% MeOH) $k_{25} \times 10^4$	σ_{25}	YC_6H_4COOEt (80% EtOH) $k_{25} \times 10^4$	σ_{25}	YC_6H_4COOPr (80% PrOH) $k_{25} \times 10^4$	σ_{25}
H	4·21	0	8·76	0	8·94	0
m-NMe_2	2·20	−0·140	4·31	−0·136	4·36	−0·132
p-NMe_2	0·116	−0·774	0·226	−0·703	—	—
m-NEt_2	1·45	−0·231	2·79	−0·220	2·78	−0·215
p-NEt_2	0·064	−0·902	0·131	−0·808	—	—
m-NPr_2	1·25	−0·261	2·20	−0·262	2·19	−0·259
p-NPr_2	0·0572	−0·927	0·127	−0·815	0·127	−0·783

agreement both with the effect of the greater bulk and the $+I$ effect of these groups (Pr > Et > Me). The m-NPr_2 group retards hydrolysis more than either m-NEt_2 or m-NMe_2 groups, whereas little difference is found in the corresponding *para* series. This has been explained by the increased importance of the bulk effect in the *meta* position. However, this sequence may also be supported by the $-I$ effect of these groups *meta* to the reaction center. Furthermore, any dialkylamino group enhances basic hydrolysis more at the *meta* than at the *para* position, as expected. Table 17 shows a decrease in the value of the Hammett σ constants for all *para*-NR_2 groups, when they were determined by hydrolysis of the methyl, ethyl or propyl esters. Since the σ_{meta} values remain unchanged, it was suggested that the diminution in resonance stabilization is due to increased steric hindrance in structure **27** as the size of R^1 is increased. The increasing negative σ_{para} values with increasing size of R in the NR_2 group may be evidence for steric hindrance of solvation, rather than for a decrease of direct conjugation because of the greater bulk of R.

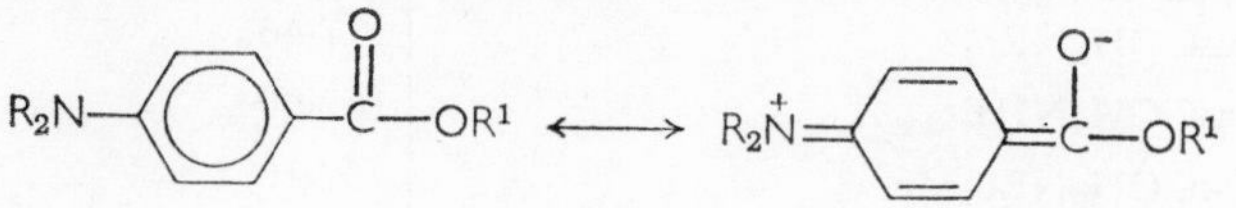

(**27**)

The effect of *para* substitution in methyl benzoate is not completely independent of the effect of simultaneous substitution in the alkyl portion[202], as is shown by the hydrolysis rate constants of p-$NH_2C_6H_4COOR$ listed in Table 18[199].

TABLE 18.   Hydrolysis rate constants of
p-$H_2NC_6H_4COOR$[202].

R	$k(M^{-1} min^{-1})$ (60% vol. aqueous dioxane at 35°c)
Me	0·0612
Et	0·0194
n-Bu	0·0104
i-Bu	0·00934

Recent work[203] on the alkaline hydrolysis of protolytic esters has considered the possibility of hydrogen-bond formation between some ammonium groups and the carbonyl carbon of the ester, e.g. **28** (see section V.A.2). However, when such esters contain a rigid acetylenic group in the alcohol side, hydrogen bonding is prevented and hydro-

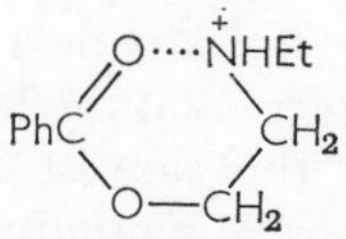

(**28**)

lysis rates decrease considerably, as illustrated in Table 19. In every case $\overset{+}{N}R_3 > NH_2$ for the $B_{AC}2$ type of hydrolysis, as expected.

The basic hydrolysis constant of ring-substituted phenylacetic esters has shown that even though the methylene group of the acid interrupts the resonance interaction between the substituent and the reaction center[204,205], the electron-withdrawing $-I$ effect of the

9+c.a.g.

TABLE 19. Second-order hydrolysis rate constants of
esters $PhCOOCH_2R$ [203].

R	k (1/mole min) (constant pH in 5% EtOH—H_2O at 25°c)
$C{\equiv}CCH_2NEt_2$	7·46
$C{\equiv}CCH_2\overset{+}{N}HEt_2$	167·00
$C{\equiv}CCH_2\overset{+}{N}Et_2Me$	32·10
CH_2NEt_2	2·48
$CH_2\overset{+}{N}HEt_2$	2500·00
$CH_2\overset{+}{N}Et_3$	69·80

$\overset{+}{N}Me_3$ group enhances hydrolysis rates more from the *meta* than from the *para* positions, and that the effect of p-NMe_2 is greater than that of p-NH_2, as expected.

C. Esterification and Reactions of Carboxyl Derivatives

I. Aromatic esterification

The influence of amino substituents in the few reported esterification reactions of aromatic acids, indicates that the polar effect of these groups is similar to that in hydrolysis. Thus the reaction of substituted benzoic acids with diphenyldiazomethane[206] is retarded by the p-NH_2 group more than by any other studied group at the *para* position, as anticipated from the resonance interaction between the NH_2 and $COOH$ groups. The reactivity of the m-$\overset{+}{N}Me_3$ is greater than p-$\overset{+}{N}Me_3$ for this type of esterification[63], as expected. Rate data for the reaction of *trans*-cinnamic acids[207] and phenylacetic acids[208] with diphenyldiazomethane for a wide range of substituents at the *para* position of the rings suggest that the large $+R$ effect of the amino nitrogen is responsible for the slowest esterification velocity. The limited success of various theories in obtaining parameters of structure–reactivity relationships for cinnamic and related systems, is indicative that more data are needed for their development and refinement[207].

2. Phthalides

Several studies of the polar effect of various groups on the velocity of ring opening of phthalide and its derivatives[209–212] showed considerable lowering of the hydrolysis rates by NH_2 groups when compared with other common substituents. The rate coefficients listed in Table 20[211] can be interpreted in terms of direct conjugation

TABLE 20.   Rate coefficients of basic hydrolysis of substituted phthalides[211].

k (1/mole min) at 25°c					
H	4-NH_2	5-NH_2	6-NH_2	7-NH_2	3-Me,4-NH_2
11·7	8·75	0·47	6·7	0·60	3·3

between the amino group ($+R$ effect) in positions 5 and 7 with the carbonyl group of the lactone (29). The NH_2 group in position 4 or

(29)

6 appears to have mainly an inductive effect ($-I$). The presence of an additional 3-methyl group in 4-aminophthalide again represses the rate and an increase in the size of the 3-alkyl group inhibits further the opening of the ring[209]. These results suggest a mechanism analogous to that for the alkaline hydrolysis of benzoic esters, although the *ortho effect* is smaller in the case of the phthalides[211].

3. Aromatic amides

In the hydrolysis of substituted benzamides with aqueous barium hydroxide[183], electron-attracting groups ($-I$, $-R$) accelerate the reaction, whereas electron-releasing groups ($+I$, $+R$) retard it. Thus the relative k values for the $B_{AC}2$ hydrolysis in Table 21 imply that

TABLE 21.   Relative rates of hydrolysis of substituted benzamides[183].

Condition and Mechanisms	ortho	NH_2 meta	para	H
$B_{AC}2$ (aq. BaOH)	—	0·93	0·20	1
$A_{AC}2$ (n/2 HCl)	0·085	0·81	0·85	1

a m-NH$_2$ group by a $-I$ effect accelerates the reaction compared with a p-NH$_2$ group, since in the retarding $+R$ effect p-NH$_2$ is more effective. The acid hydrolysis $A_{AC}2$ of aromatic amides containing the amino group (Table 21) is only weakly affected by this (or any other substituent) at the *meta* or *para* positions[183], while o-NH$_2$ markedly retards the reaction, due most probably (apart from steric factors) to hydrogen bonding.

In phthalimides, the 4-NH$_2$ group accelerates the basic hydrolysis[213] more than the 3-NH$_2$ group. This result is ascribed to hydrogen bonding between the 3-NH$_2$ with its adjacent carbonyl group, which in turn affects the other carbonyl group too. Substituted N-alkyl phthalimides show one sequence MeN > EtN > HN in reaction velocity, suggesting the presence of both steric and electronic effects.

VIII. ELECTROPHILIC ALIPHATIC REACTIONS

The π electrons of unsaturated carbons are susceptible to electrophilic attack, generally, by a two-step ionic mechanism as in (20). Alkenes react faster than alkynes since the p electrons in the latter are more

$$Y^+ + \ \ \ \ \ \ \searrow C = C \nearrow \ \ \ \longrightarrow \ \ Y-\overset{|}{\underset{|}{C}}-\overset{|}{\underset{|}{C}}{}^+ \ \xrightarrow{\ Z:^-\ } \ Y-\overset{|}{\underset{|}{C}}-\overset{|}{\underset{|}{C}}-Z \tag{20}$$

$$Y^+ + \ \ \searrow C = C \nearrow \ \ \longrightarrow \ \ \overset{\overset{Y^+}{\diagup\diagdown}}{C \quad C} \ \xrightarrow{\ Z:^-\ } \ -\overset{\overset{Y}{|}}{\underset{|}{C}}-\overset{|}{\underset{\underset{Z}{|}}{C}}-$$

tightly bound. Saturated carbon is also attacked by electrophilic reagents, either by the two-step mechanism S_E1 (reaction 21) or by

$$-\overset{|}{\underset{|}{C}}-Z \ \xrightarrow{\text{Slow}} \ -\overset{|}{\underset{|}{C}}{:}^- \ + \ Z^+$$

$$-\overset{|}{\underset{|}{C}}{:}^- \ + \ Y^+ \ \xrightarrow{\text{Fast}} \ -\overset{|}{\underset{|}{C}}-Y \tag{21}$$

the single stage mechanism S_E2 (reaction 22). Details can be found

$$-\overset{|}{\underset{|}{C}}-Z \ + \ Y^+ \ \longrightarrow \ \left[-\overset{|}{\underset{|}{C}}\overset{Y^{\delta+}}{\underset{Z^{\delta+}}{\cdots}}\overset{\delta-}{} \right] \ \longrightarrow \ -\overset{|}{\underset{|}{C}}-Y \ + \ Z^+ \tag{22}$$

transition state

in references 182, 214, 215.

A. Addition of Acids

The reaction of halogen acids with allylamines (**30**)[216] has been shown to produce equal quantities of **31** and **32** (reaction 23), and

$$R_2NCH_2CH{=}CH_2 + HX \longrightarrow R_2NCH_2CHXCH_3 + R_2NCH_2CH_2CH_2X \quad (23)$$
$$\text{(30)} \qquad\qquad\qquad\qquad \text{(31)} \qquad\qquad\qquad \text{(32)}$$

two independent addition mechanisms were proposed. The isomer **32** is believed to be formed by addition of the acid to the ionized salt of **30**, and **31** by addition to the free base. The formation of **32** may be justified by the very strong electron-withdrawing character of the ammonium group which can be transmitted across the interfering methylene group and render the terminal carbon more positive. Since the NR_2 group also exerts electron attraction, two possible intermediates, **33** and **34**, may be considered to explain the apparent anomaly of the formation of **31**.

$$\begin{array}{c} R \\ | \\ R{-}N{:}^{\delta+} \\ \diagup \quad \ddots \quad \frown_{\delta-} \\ CH_2{-}CH{=}CH_2 \\ \text{(33)} \end{array}$$

$$R_2\ddot{N}CH_2\ddot{C}\bar{H}{-}\overset{+}{C}H_2 \longleftrightarrow R_2\ddot{N}CH_2\overset{+}{C}H{-}\ddot{C}H_2{-} $$
$$\text{(34)}$$

Recent work on addition of acids to unsaturated amines[217] (equations 24 and 25) seems to oppose the previous hypothesis, since high

$$Me_2NCH_2CH{=}CHCH_3 + HCl \longrightarrow Me_2NCH_2CH_2CHClCH_3 \ (98\%) \quad (24)$$
$$\text{(35)}$$

$$Me_2NCH_2CH_2C{\equiv}CH + HCl \longrightarrow Me_2NCH_2CH_2CCl{=}CH_2 \ (80{\cdot}5\%) \quad (25)$$

yields of 'normal' products were obtained. The formation of product **35** is additionally favored since the intermediate carbonium ion is stabilized by the $+I$ effect of the terminal CH_3 group.

The electrophilic attack on unsaturated ammonium compounds (equation 26) reveals that the strong inductive effect of the positively

$$\overset{+}{Me_3N}CH{=}CH_2 + HI \longrightarrow Me_3\overset{+}{N}CH_2CH_2I \quad (26)$$

charged group assists the formation of highest electron density on the nearest unsaturated carbon[216, 218]. This phenomenon also occurs with alkynes even when they contain only uncharged amino substituents (equation 27)[217].

$$Me_2NCH_2C{\equiv}CCH_3 + HCl \longrightarrow Me_2NCH_2CH{=}CClCH_3 \quad (27)$$

The hydration of acetylenic amines in dilute acids consistently gives those amino ketones, e.g. **36**, in which the oxygen is attached to the carbon farthest from the amino substituent (equation 28)[219].

$$RC{\equiv}C(CH_2)_nNMe_2 \xrightarrow[\text{HgSO}_4]{\text{H}_2\text{SO}_4} RCOCH_2(CH_2)_nNMe_2 \quad\quad (28)$$
$$\text{(36)}$$

$$n = 1, 2, 3$$

This is consistent with the $-I$ effect of the nitrogen atom which decreases with increasing distance from the unsaturated center. The size of the R group does not appear to affect the reaction.

It is interesting to mention that possible conjugation between the acetylenic π electrons with the p electrons of the nitrogen substituents results in delocalization. Such a change in the transmission of the effect of the amino groups is illustrated in the bromination of the vinyl acetylenic amine (**37**), which occurs predominantly at the triple

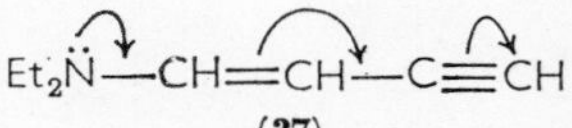

$$Et_2\overset{\curvearrowright}{N}-CH{=}CH-C{\equiv}CH$$
$$\text{(37)}$$

bond[220], even though the reactivity of the olefinic bond is greater. The electronic polarization and some properties of **37** agree with the observed magnitude of its dipole moment[221].

B. Decarboxylation

In carboxylic acids, electron-withdrawing groups in the α-position facilitate decarboxylation as anion[182,222,223]. Many amino acids undergo this type of reaction in the presence of an acylating agent. In the case of an α-NH$_2$ group[224-226], decarboxylation occurs through

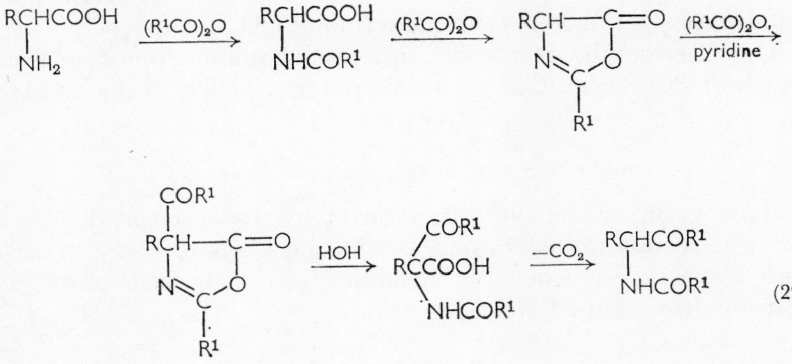

$$(29)$$

an oxazolone intermediate and involves subsequent additional reactions with the acylating agent (reaction 29)[225-227]. In certain instances the acylated oxazolone intermediate has been isolated[228].

The $-I$ effect of an α-NHR group in decarboxylative acylation of amino acids[229] also allows the reaction to proceed by an analogous route[230]. However, with α-NR$_2$ groups the corresponding tertiary amino ketones are not obtained, but undergo further reaction to give N,N-disubstituted amides (reaction 30)[227, 231]. Tertiary amino acids,

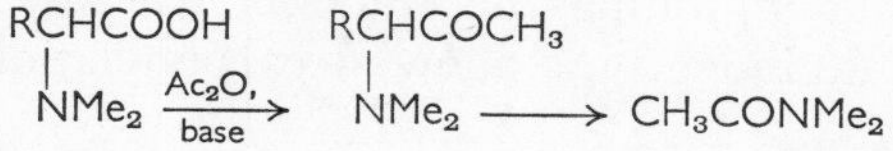

$$\underset{\substack{| \\ NMe_2}}{RCHCOOH} \xrightarrow[\text{base}]{Ac_2O,} \underset{\substack{| \\ NMe_2}}{RCHCOCH_3} \longrightarrow CH_3CONMe_2 \qquad (30)$$

as well as acids having an aryl, heterocyclic or aryloxy group in the α-position, seem to react through a different mechanism (reaction 31)

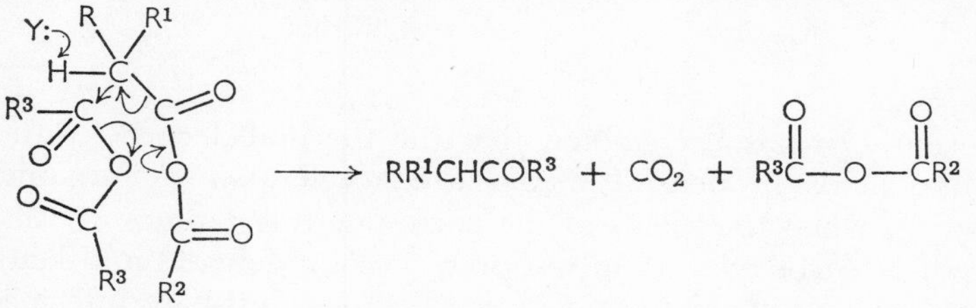

$$\longrightarrow RR^1CHCOR^3 + CO_2 + R^3\overset{O}{\underset{||}{C}}-O-\overset{O}{\underset{||}{C}}R^2 \qquad (31)$$

The abstraction of the α-hydrogen atom of the acid seems to permit formation of a quasi six-membered ring and subsequent disproportionation into the actual product[227, 232]. Although the correct mechanistic picture for these reactions has been the subject of many discussions[232, 233], it is reasonable to assume that the acylating agent must participate prior to, or simultaneously with, the decarboxylation in a concerted action[227].

Under different experimental conditions, α-amino acid derivatives can undergo decomposition according to equation (32) where X is

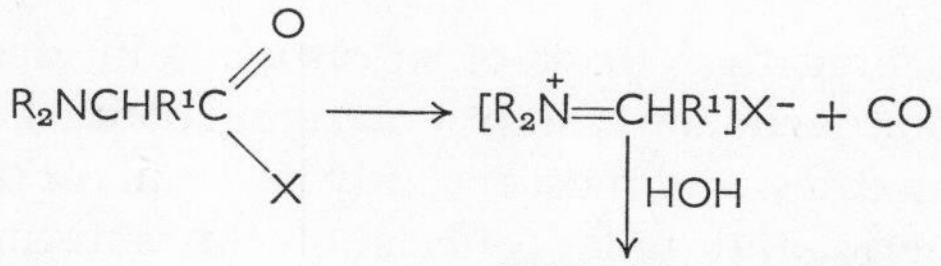

$$R_2NCHR^1\overset{O}{\underset{\diagdown X}{\overset{\diagup}{C}}} \longrightarrow [R_2\overset{+}{N}{=}CHR^1]X^- + CO \qquad (32)$$
$$\downarrow HOH$$
$$R_2NH \cdot XH + R^1CHO$$

an electronegative group. The presence of the NR$_2$ group in the α-position has a strong effect on decarbonylation reactions, while NHR and NH$_2$ groups do not affect this process[234]. Generally, these reactions are catalyzed by acids and inhibited by bases. The catalytic effect of an acid on decarbonylation can be achieved through an attack at the electronegative group as in (33). This suggests that the

$$\diagdown\ddot{N}{-}\underset{|}{C}{-}C\diagup^{O} \qquad\qquad (33)$$
$$\diagup \qquad \diagdown X \xleftarrow{\quad\cdots\cdots\cdots\quad} Y^+(H^+)$$

primary α-amino acid derivative, as in the case of the corresponding acid chloride, must first undergo protonation at the nitrogen atom (reaction 34), thus inhibiting the reaction, whereas acid derivatives with a tertiary amino group allow decarbonylation to take place (reaction 35).

$$H_2NCHRCOCl + H^+ \longrightarrow H_3\overset{+}{N}CHRCOCl \qquad (34)$$

$$R_2NCHRC\diagup^{O}_{\diagdown Cl} + H^+ \longrightarrow R_2NCHRC\diagup^{O}_{\diagdown \overset{+}{Cl}{-}H} \qquad (35)$$

Recently, it has been noted that the thermal decarboxylations of α-amino acids in neutral media are accelerated by organic peroxides[235], when the yields of the corresponding amines are also improved. Decarboxylation in the presence of aldehydes and ketones is also relatively fast, giving, in most cases, a Schiff-base intermediate, prior to formation of the expected amine.

The ease of thermal decarboxylation has been related to the strength of the acid, that is, the greater the acidity, the faster it loses CO_2[236]. Therefore, the $+R$ effect of the NH_2 group at an *ortho* or *para* position in aromatic carboxylic acids is expected to inhibit this type of decomposition. The actual results show the opposite to be true and this contradiction is explained on the basis of intramolecular catalytic action of the amino groups in the aminobenzoic acids.

IX. ELECTROPHILIC AROMATIC SUBSTITUTION

In studies concerning the effects of substituents in electrophilic aromatic substitution, particularly in orientational control of the entering group, in reaction rates, and most recently in details of the substitution process itself, steric and polar effects of the substituent and the reagent appear to be factors simultaneously affecting the reaction.

Owing to the highly polar and basic nature of aromatic amines they frequently appear as exceptions to rules and generalizations. Superimposed effects of the amino group arising from experimental conditions, and other related phenomena, have restricted the evaluation of the polar effect of the amino group in comparison with other substituents. Details and excellent discussions on electrophilic aromatic substitution in general, appeared in several recent publications [15b, 52, 70, 237-240].

A. General Considerations

A great body of evidence supports the formulation of structure **38**, as a key intermediate in aromatic substitution, although it is by no

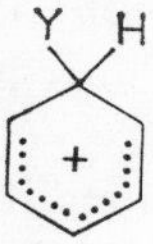

(38) 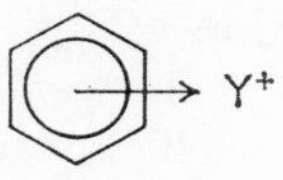   **(39)**

means the only one. The formation of this σ complex, in which both the electrophile and the hydrogen atom are attached to a new sp^3-hybridized tetrahedral carbon is the rate-determining step, followed by rapid decomposition to the product. The formation of a π complex (**39**), may take place usually prior to, although in some cases after, the formation of the σ complex, although some scepticism exists as to whether the π complex is essential in the reaction path. **38** may be formulated in terms of the hybrid contributing structures **40**. Electron

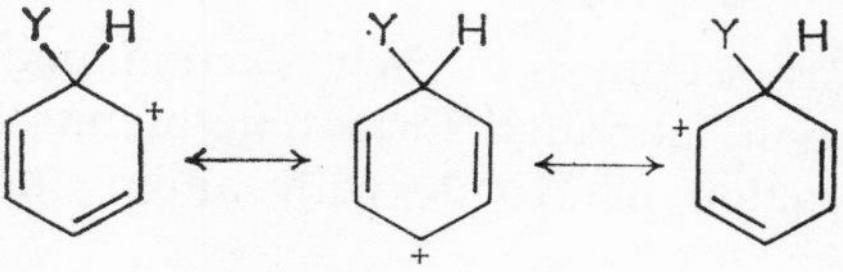

(40)

release by the amino substituents will assist the stabilization of the positive charge in the *ortho* and *para* positions, and will increase their reactivity. In contrast, with positively charged nitrogen substituents, such as the trimethylammonium group, owing to the strong $-I$ effect the *meta* position is more favored for substitution than the *ortho* and *para* positions.

The orientational effect is often expressed in terms of yields of isomers, and the rate data as overall rate of reaction relative to that of benzene. Another convenient quantitative expression of rates is the partial rate factor, f, which measures the reactivity of the position concerned, relative to one of the six equivalent positions in benzene. The relative reactivity of the *meta* and *para* positions of monosubstituted

9*

benzenes can also alter according to the nature of the attacking re-agent, a fact described as *selectivity* of the electrophile.

B. Monosubstituted Aromatic Amines

I. Reaction with nitrogen electrophiles

a. Nitration. The great variety of phenomena occurring simultane-ously under different conditions in the nitration of anilines[241] has been discouraging for any quantitative work on mechanism and polar effects. However, numerous qualitative observations permit some interpretation of the directive effect of these groups in the benzene ring.

Aniline itself requires special nitration conditions if side-reactions are to be avoided[242, 243]. The fact that a non-acidic nitrating agent, such as N_2O_5, yields *N*-nitration, suggested that ring nitration of the aromatic amine by acidic nitrating agents is in general proceeded by *N*-nitration and subsequent rearrangement. This assumption was later disproved by study of the acid-catalyzed rearrangement of phenylnitroamine and the nitration of aniline with HNO_3, under comparable conditions, with the following results[243]:

Reaction	o-, %	m-, %	p-, %
Rearrangement of phenylnitroamine	93	0	7
Nitration of aniline	6	34	59

The proportions of isomers in each reaction, point to two different types of orientation, although formation of nitroamines also takes place in the reaction of HNO_3 with aniline derivatives in acid media[244, 245].

The introduction of *N*-alkyl groups into aniline so as to prevent nitroamine formation results in rather complex mixtures of products with HNO_3, at different concentrations and media[15b, 241]. As an example, in the nitration of *N,N*-dimethylaniline, decrease in acid concentration decreases *meta*, and increases *para* substitution. Some 2,4-dinitro derivative is formed by nitration of the *para*-nitro com-pound, and the use of glacial acetic acid solvent directs about one third of the initial substitution to the *ortho* position. Yet, the reaction of PhNMe$_2$ with liquid N_2O_4 yields mostly *para* substitution with traces of the *meta* isomer[246].

In diphenylamine the benzene ring is still very strongly activated. Thus nitration in acetic anhydride (Table 22), gives a mixture of 2- and 4-nitrodiphenylamine as the main nitration product[247].

Table 22.   Nitration of diphenylamine in acetic acid at 25°C[247].

Isomer proportions (%)		Overall reactivity	Partial rate factors for positions	
ortho	para	$(C_6H_6 = 6)$	ortho	para
71	29	4,430,000	831,000	575,000

Nitrations of aromatic compounds containing positively charged nitrogen groups often yield approximately equal amounts of the *meta*- and *para*-substituted products. This might be the result of two concurrent reactions in which the conjugate acid gives the *meta* compound, while the free amine gives the *para* compound. An interesting and valuable quantitative study of the reaction mechanism and substituent effects of several ammonium groups in aromatic nitration has recently been described[248,249]. This report has shown (see Table 23) that

Table 23.   Isomer proportions and rate of nitration of aniline in concentrated sulfuric acid at 25°C[248].

H_2SO_4	(%)	85	89.4	92.4	94.8	96.4	98.0	100
ortho	(%)	6	3	—	—	—	—	—
meta	(%)	34	45	53	57	58	62	64
para	(%)	59	52	47	43	42	38	36
k_o		—	2.61	2.01	1.47	1.08	0.668	0.655
k_m		—	0.587	0.533	0.419	0.313	0.207	0.210
k_p		—	1.36	0.945	0.632	0.454	0.254	0.236

All rate coefficients in 1/mole sec.

while the product composition depends on acidity of the medium, the production of *meta* and *para* derivatives does not fit the assumption of substitution in the conjugate acid and free amine. At high acidities the *meta/para* ratio is not very sensitive to the acidity, and the effect of the $\overset{+}{N}H_3$ group in 98% sulfuric acid leads to almost equal deactivation at both the *meta* and *para* positions, so that a contribution to the reaction through the free amine seems to be practically absent. The orientational results, the consideration of encounter rates, the effect of changing the reaction medium from sulfuric acid to deuterosulfuric acid and a study of the *p*-chloroanilinium ion[248], are evidence which firmly support the above argument.

The rates of nitration of *N*-mono-, di- and trimethylanilinium ions also suggest a similar reaction mechanism[249]. The results collected in

TABLE 24. Product composition, rate coefficient and partial rate factors for nitration of N-methylated anilinium ions[249].

C_6H_5Y Y	$p(\%)$	$10^2 k_m$ (l/mole /sec)	$10^2 k_p$ (l/mole/sec)	$10^8 f_m$	$10^8 f_p$
$\overset{+}{N}H_3$	38	21	26	162	195
$\overset{+}{N}H_2Me$	30	7·4	6·4	57	49
$\overset{+}{N}HMe_2$	22	1·6	0·93	12·3	7·1
$\overset{+}{N}Me_3$	11	0·55	0·14	4·2	1·0

Table 24 for these amines reveal an approximate linearity in which the gradual replacement of the hydrogen atoms of the $\overset{+}{N}H_3$ substituent by methyl groups decreases the rate of substitution at a given position by nearly equal factors. The anilinium ion reacts fastest in *meta* nitration, and, more remarkable, also in *para* nitration; the smaller deactivating effect of the $\overset{+}{N}H_3$ group compared with methylated ammonium groups is attributed to hydrogen bonding of the unmethylated ion. An additional observation is the small difference in polar effect of these substituents, as derived from partial rate factors, when contrasted with their common deactivation effect on the benzene ring.

The *meta*-orienting effect of positively charged nitrogen decreases when a carbon chain of increasing length is interposed between the onium substituent and the aromatic ring[250, 251]. This classical fact is well illustrated in the sequence below:

Compound	% *meta* derivative
$Ph\overset{+}{N}Me_3$	100
$PhCH_2\overset{+}{N}Me_3$	88
$PhCH_2CH_2\overset{+}{N}Me_3$	19
$PhCH_2CH_2CH_2\overset{+}{N}Me_3$	5

Apart from the effect of the distance of the positive nitrogen[250, 251], hyperconjugation as in **41**[252] may be an additional factor accounting for the decreasing amount of the *meta* isomer.

More details on the nitration of anilines can be obtained from a recently published book[253].

$$\langle\text{ring}\rangle\overset{H^+}{\underset{H}{\overset{|}{=}}}CCH_2\overset{+}{N}Me_3$$

(41)

b. *Nitrosation.* The weak electrophilic properties of NO$^+$ limit its nuclear substitution to highly activated rings, such as those of phenols and anilines. The process often leads to diazotization of weakly basic primary aromatic amines and to N-nitrosation of N-alkylanilines (see references 15b and 70). However, excess of nitrous acid or the presence of strong sulfuric acid enables direct electrophilic attack on the ring of some secondary aromatic amines[254]. The effect of the di-alkylamino group results initially in *para* substitution, but subsequent reactions and formation of by-products can occur[255, 256]. Bulky N-alkyl groups in anilines have been shown to impede ring nitrosation due to steric inhibition of resonance[257, 258].

c. *Diazonium Coupling.* The very weakly electrophilic diazonium ion reacts only with aromatic amines and phenols. No information is available on the polar effect of amino groups. Details of the process can be found in the literature[258-265].

2. Electrophilic sulphonation

The scarcity of quantitative results available on the relative rates of sulfonation of aromatic compounds is due to the uncertainty about the exact nature of the electrophile, the reversibility of the process and the isomerizations occurring during the reaction. In addition to these factors, the sulphonated products are, as a rule, difficult to isolate, thus impeding an adequate analysis of isomer proportions and partial rate factors. A good review of the subject appeared recently[70].

Sulfonation of aromatic compounds containing amino groups yield different isomer ratios, depending upon the experimental conditions employed. At low temperature, aniline gives predominantly *ortho* and *para* derivatives[266], whereas dimethylaniline gives approximately equal amounts of *meta* and *para* isomers[267]. However, at high temperatures ('baking process') most aromatic amines give largely the *para* product[268-273]. The unexpected results with aniline are attributed to the formation of a phenylsulfamic acid intermediate[266], which is responsible for the activation of the *ortho* and *para* positions. The dimethylamino group apparently forms no complex with the SO$_3$ of sulfuric acid, and the equal yield of *meta* and *para* sulfonates arises from the equilibrium mixture of the free NMe_2 group with its corresponding $\overset{+}{N}HMe_2$ salt (reaction 36) [266, 267].

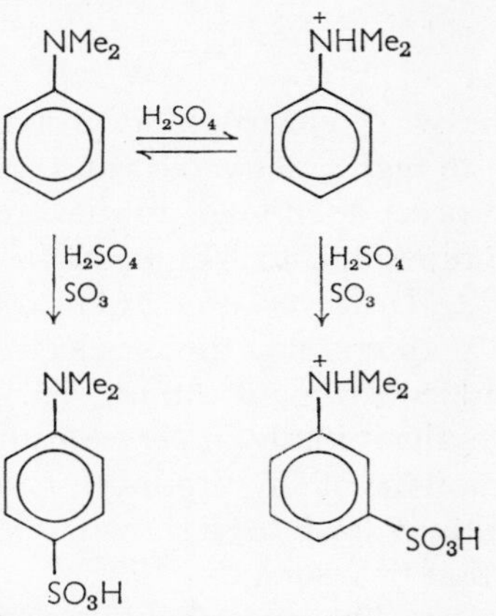

(36)

The phenyltrimethylammonium ion on sulfonation under conditions believed to give kinetic control, produces 8% *ortho*, 78% *meta* and 14% *para* substitution [274]. The rather surprising relatively high yield of *ortho* and *para* isomers apparently refutes the above mechanistic view according to which *para* sulfonation in strong acid media requires dissociation to free amines and reaction in the latter form [266, 267]. This work seems to confirm the conclusion [248, 249], that substitution occurs overwhelmingly through the ammonium ion.

3. Electrophilic halogenations

In aromatic halogenation processes with molecular halogen, either alone or in the presence of a carrier or catalyst, or with positively charged halogenating species, there are great differences in relative reactivity of these species. Discussion of this subject is covered extensively by some of the standard works [15b, 52, 70, 237].

a. Bromination. The powerful electron release of the NMe_2 group to the *ortho* and *para* positions of the benzene ring has been assessed by bromination of dimethylaniline in acetic acid [275], which was found to be 10^9 times faster than that of anisole. Previous work indicated a reverse order in activity, i.e. $OMe > NMe_2$, for the relative rates of aqueous bromination of various anilines and phenols [276]. In this earlier investigation, the acidity of the reaction mixture was not controlled, so that the amino substituents, were probably present to a greater extent as the onium salts. This assumption seems to be confirmed by the kinetic data obtained on bromination of N,N-dialkylanilines in strong acid media, which establish that the reaction velocity decreases sharply with increasing acidity[277].

The isotope effect in the bromination of N,N-dimethylaniline in aqueous acid[278] implies dissimilar mechanisms for the *ortho* and the *para* substitution. The rate-determining step could well involve, among other factors, fission or lengthening of the C—H bond in *ortho* bromination, but this is unlikely in *para* bromination. The dimethylamino group and bromine are believed to form an intermediate complex in these reactions, which appears to decrease the velocity of substitution with additional electrophile. Recent quantitative measurements, in electrophilic bromination[279], have reached the conclusion that the NMe_2 group activates the *para* position of the ring ten times more than the *ortho* position.

The change of the orientational control by amino substituents in acid media or on complex formation is obviously supported by the bromination of N,N-dimethylaniline in sulfuric acid containing silver sulfate, which produces 60% of 3-bromo- and 20% 4-bromo-N,N-dimethylaniline[280]. This result, again seemed to suggest that the appreciable *para* product in acidic halogenation of aromatic amines may be due to an ion-pair mechanism or to partial dissociation to the free base[281]. Yet the formation of *para* products in the presence of excess of strong Lewis or Brönsted acid with aniline, could hardly indicate the presence of free amine, and there is evidence that electrophilic bromine attack, at the *meta* as well as at the *para* positions, occurs on deactivated anilinium ions[248, 249].

A very interesting observation is that incorporation of anilines into coordination compounds does not affect the reactivity of the former towards bromination nor their orientation in the substitution[282]. Palladium(II) complexes of the type $[Pd(aniline)_2Br_2]$ show that the NH_2 group favors *ortho* and *para* bromination even in strongly acid solution, where any complication due to partial dissociation should be absent. In order to account for these results, it seems that the coordination process may involve the p electrons of the nitrogen atom in some manner fundamentally different from that in aromatic ammonium compounds, i.e. that coordination does not impart a positive charge to the nitrogen of anilines.

The successive introduction of N-phenyl groups to aniline (diphenylamine and triphenylamine), was shown to inhibit *ortho* substitution, and the bromine tends to occupy one or more of the available *para* positions[283, 284]. Some electrophilic halogenations of monosubstituted benzenes containing positively charged nitrogen substituents have been studied[280]. The slow rate of bromination of trimethylanilinium perchlorate[285] shows that the $\overset{+}{N}Me_3$ and the NO_2 groups have about the same effect in deactivating the aromatic ring. This fact contradicts the assumption quoted in section VI[175], namely that $^+NR_3 > NO_2$ in deactivation.

b. Chlorination and iodination. In chlorinating *N,N*-dimethyl-aniline[286], the *ortho/para* ratio and the yields depend strongly on the chlorinating agent. Similar processes on other aromatic amines[287, 288] indicate that an aniline, being an ambident[289] nucleophile, is able to form an *N*-chloro intermediate, with subsequent rearrangement to ring-chlorinated products. Chlorination in the presence of strong mineral[280] or Lewis acids[281] provides similar information as in bromination (see section IX.B.3.a), where the presence of the ammonium or complex amino groups deactivates the ring to give rise mainly to *meta* and some *para* substitution.

The rate of iodination of aniline[290], in the presence of potassium iodide and acids, decreases with increasing iodide concentration and rises with increasing pH. These results fit a mechanism in which positive iodine (I^+) is the attacking electrophile of free aniline. The same kinetic data would also agree with a reaction between hypo-iodous acid and anilinium ion, but the *para* isomer is the main product, with practically no formation of *ortho* or *meta* isomers. Therefore, the alternative, involving attack on an ammonium ion, is improbable, while it can be assumed that the $+R$ effect of the NH_2 group permits *para* substitution by I^+. Iodination of *para*-substituted anilines by iodine in aqueous methanol[291], and by iodine monochloride in aqueous hydrochloric acid[292], similarly show the activating effect of the NH_2 group, but the attack is at the *ortho* position, since the *para* position is blocked.

No kinetic isotope effect was observed in the iodination of some aromatic amines[293] although the same study confirms the activating sequence of substituents in the order $NMe_2 > NHMe > NH_2$.

Many studies indicate qualitatively that electrophilic iodination of aromatic amines in the presence of persulfate[294], mercuric oxide[295], acetic acid[296], iodine bromide[297], complexing reagents[298], and without solvent and catalyst[299], yield *ortho* and/or *para* products as expected from the $+R$ effect of the amino group.

4. Additional electrophilic reactions

a. Carbon electrophiles. The relative reactivities of Friedel-Crafts alkylations vary with the structure of the aromatic system undergoing the reaction. Substituents with $+I$ and $+R$ effects facilitate, whereas those with $-I$ and $-R$ effects inhibit alkylation. The strong effects of amino substituents are greatly diminished in Friedel-Crafts alkylations because the nitrogen atom is complexed with the Lewis acid catalyst. The deactivating effect of the amine complex group attached to the benzene nucleus is responsible for the scarcity of quantitative measurements regarding orientation and reactivity. Similarly, Friedel-Crafts acylations of anilines fail to give ring substitution due to these inactive complexes. However, these reactions sometimes result in acylation of the substituent, especially on primary aromatic amines.

Alkylations of aromatic amines can be satisfactorily achieved with stable carbonium ions, even in the presence of high concentrations of mineral or Lewis acid catalysts. The diphenylmethyl and triphenylmethyl cations have been shown to attack predominantly at the *para* position of anilines [300-306], and the prominent features of these reactions are a prior formation of the resonance-stabilized carbonium ion in a fast equilibrium, followed by a slow and selective rate-determining bimolecular reaction between these cations and the activated *para* position of the ring (reaction 37) [307-311]. The results in

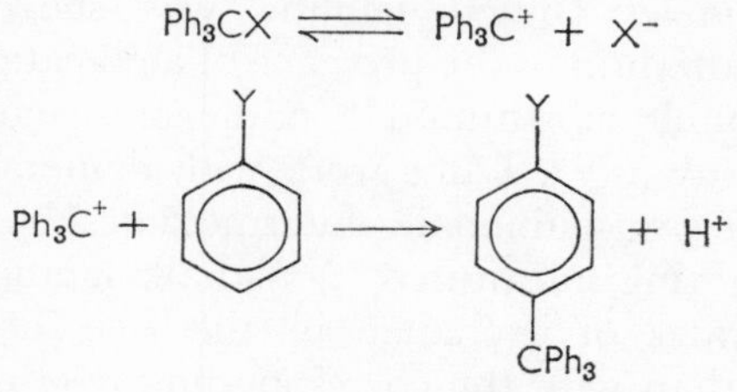

$$Ph_3CX \rightleftharpoons Ph_3C^+ + X^-$$

$$(Y = \text{activating group})$$

systems involving competition of aromatic amines for a triphenylmethyl cation [312] indicate the sequence $NMe_2 > NHMe > NH_2$ in directive power, again substantiating the $+R$ effect of these substituents.

The formation of *N,N*-dimethyl-4-tricyanovinylaniline from tetracyanoethylene and *N,N*-dimethylaniline in chloroform [179, 180] is regarded as a vinylic substitution on the ethylene, with the aromatic amine acting as a nucleophile, or in other terminology as an aromatic electrophilic substitution. The initial formation of a π complex is followed by transition to a carbanionic intermediate[180] which decomposes slowly to the final product by base-catalyzed proton removal and subsequent elimination of the cyano group (reaction 38). The

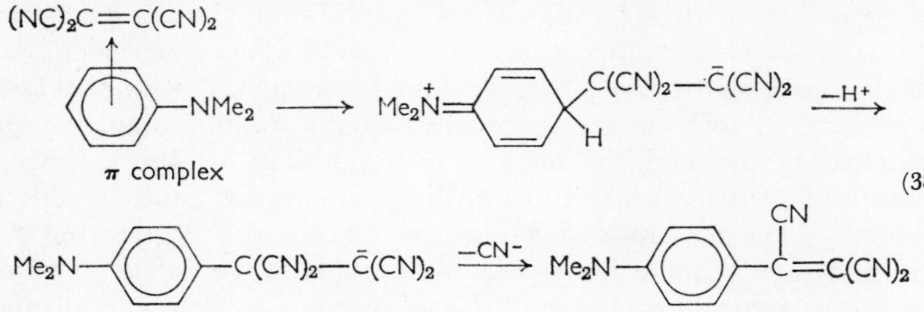

$$\tag{38}$$

interaction of tricyanovinyl chloride with a series of dialkylanilines[181] is, in contrast, not base catalyzed and prior elimination of chloride ion is followed by proton removal. The very low values of activation energies for the studied dialkylamino groups suggest a multistage process. The order of reactivity of the amino substituents was found to be $NBu_2 \sim NEt_2 \sim NPr_2 > NMe_2$ reflecting partially the $+R$ effect.

The limitations and scope of reactions involving other electrophilic carbon reagents with anilines have been extensively described elsewhere[313].

b. *Exchange reactions.* Dimethylaniline was shown to exchange hydrogen with deuterium in the presence of acid catalysts, the process involving electrophilic substitution[314] and exclusive orientation to the *ortho* and *para* positions[315]. Later work with deuterated alcohol suggested *ortho-* and *para*-quinonoid intermediates[316,317]. A series of experiments with ring-substituted *N,N*-dialkylanilines and D_2SO_4 revealed lower rates of exchange in the case of *ortho*-halo-*N,N*-dimethylanilines than with the corresponding *meta* and *para* isomers. The inhibition increases with the atomic size of the halogens, as the order $o\text{-Br} > o\text{-Cl} > o\text{-F}$ prevents coplanarity. This argument has been confirmed by the high reactivity in the exchange reaction of *N*-methylindoline, where the nitrogen atom is in the plane of the benzene ring[318].

The rates of tritiation of *N,N*-dialkylanilines have been determined in solutions containing a mild excess of acid[319]. Under such conditions, the anilinium ion may preferentially be attacked at the *meta* position, but quantitative data appear to conform to an equilibrium of the free amine with its conjugate acid (reaction 39). Further, most

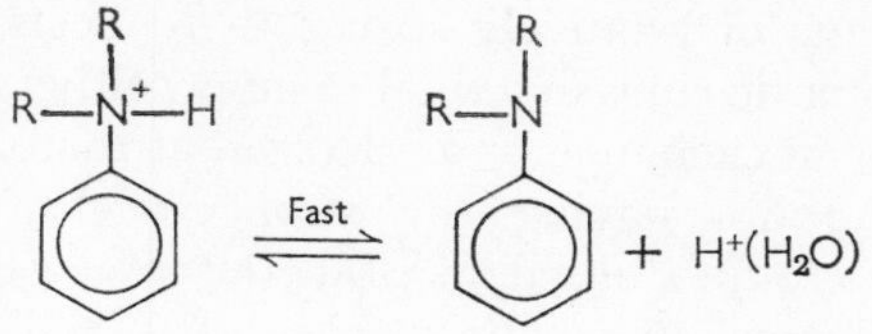

$$(39)$$

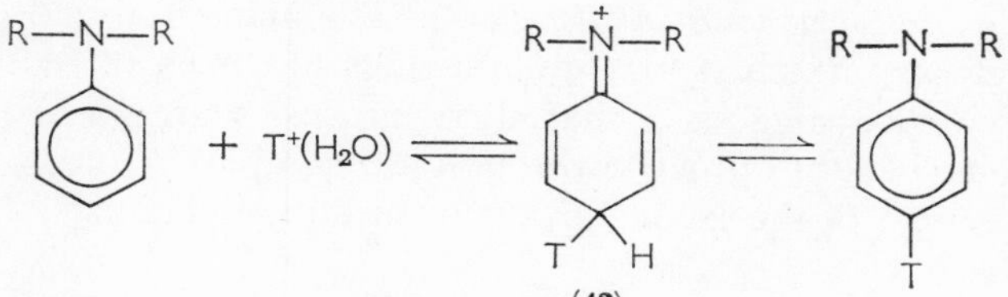

$$(40)$$

(42)

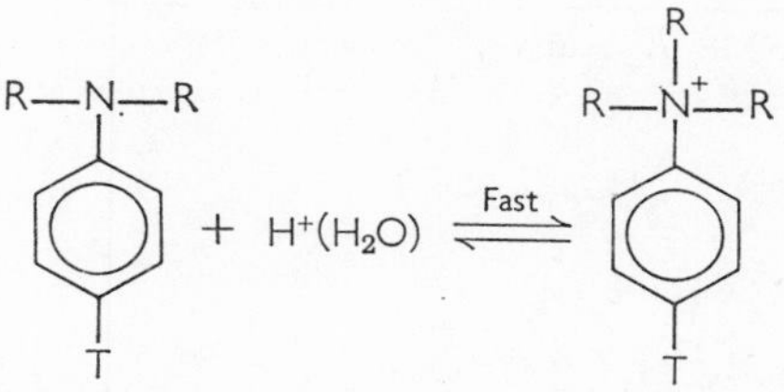

ortho-substituted compounds failed to undergo exchange, as demonstrated before[318], due to the steric inhibition resonance.

The rates for the N,N-dialkyl series are probably controlled by the entropy of activation, since steric factors, both in the electrophile and in the intermediate (42), have strong influence on the reaction (reaction 40). Accordingly, the $+I$ effect of the dialkyl group going from NMe_2 to NBu_2 groups is expected to stabilize 42, yet steric strain of the more bulky substituent may inhibit reaching the co-planarity necessary for this stabilization. Solvation, another important factor in these reactions, can be centered on the amine nitrogen of 42, but steric crowding may increasingly interfere in this process with increasing size of the substituents. An exception has been observed in the case of N,N-diethylaniline.

The partial rate factors for nuclear deuterium exchange of N,N-dimethylaniline indicate that in basic solutions the reactivity decreases with increasing distance from the dimethylamino group ($f_{ortho} > f_{meta} > f_{para}$), while in acid solutions the meta atoms are replaced at a slower rate, or not at all[320, 321]. These facts are explained in terms of the inductive effect which determines the more enhanced rate of exchange of the ortho atom. Moreover, it has also been shown that substituents of relatively strong $+R$ effects may influence the activity of aromatic rings in a dual manner, either as electron releasers in acid hydrogen exchange or as electron attractors in base exchange. The NMe_2 group, having the strongest electron-releasing effect among all the groups studied, exhibits the weakest electron-withdrawing effect in deuterium exchange with a base.

c. Removal of a substituent by proton. The acid-catalyzed cleavage of some groups attached to aromatic rings has been of recent interest. The evidence suggests[70] that these processes are electrophilic substitutions, likely to proceed through a σ complex.

The effect of the NMe_2 group in protodesilylation[322], (Table 25)

TABLE 25.   Partial rate factors for protodemetalylations.

Reaction of NMe_2 derivatives	$f_m^{NMe_2}$	$f_p^{NMe_2}$	Reference
Desilylation $(Me_2NC_6H_4SiMe_3, H^+)$	—	$3 \cdot 0 \times 10^7$	322
Degermylation $(Me_2NC_6H_4GeEt_3, H^+)$	—	3×10^6	324
Destannylation $(Me_2NC_6H_4Sn(C_6H_{11})_3, H^+)$	—	2×10^4	326
Reaction of $\overset{+}{N}Me_3$ derivatives	$f_m^{\overset{+}{N}Me_3}$	$f_p^{\overset{+}{N}Me_3}$	Reference
Desilylation $(Me_3\overset{+}{N}C_6H_4SiMe_3, H^+)$	—	$3 \cdot 84 \times 10^{-4}$	323
Degermylation $(Me_3\overset{+}{N}C_6H_4GeEt_3, H^+)$	$12 \cdot 6 \times 10^{-4}$	$10 \cdot 6 \times 10^{-4}$	325
Destannylation $(Me_3\overset{+}{N}C_6H_4Sn(C_6H_{11})_3, H^+)$	—	$68 \cdot 0 \times 10^{-4}$	326

reveals the great activating power of this group, except in strongly acidic media where the deactivating effect of the $\overset{+}{N}Me_3$ group dominates[323]. Similar results have been observed in protodegermylation[324], and the trimethylammonium group deactivates the *para* position more than the *meta*, as expected, but surprisingly the difference is only slight[325], supporting the previous conclusion[248, 249] that the *para* position is not highly deactivated. The rate of cleavage of p-$Me_2NC_6H_4Sn(C_6H_{11})_3$ is found to double by doubling the acid concentration, thus suggesting that the amino nitrogen group is present as a free base[326].

The effect of amine substituents in protodehalogenations has been examined qualitatively[327]. Iodine is replaced by a proton more easily in p-iodo-N,N-dimethylaniline than in p-iodoaniline. On the other hand, NH_2 is more efficient than NMe_2 in halogen displacement of the corresponding 2,4,6-triiodoanilines, as a result of the greater steric inhibition of resonance of the bulky dimethylamino group by the *ortho* iodines. More recently, the kinetics of deiodination of p-iodoaniline (reaction 41) in aqueous acid media[328] was shown to be

$$+ H^+ \qquad + 2I^- \qquad + I_3^- \qquad (41)$$

first order in the aromatic amine, but independent of the concentration of hydrogen and of iodine ions. Furthermore, the rate in H_2O is 6 times faster than in D_2O. These results imply the formation of the same transition state in iodination at low acidity[290] and in deiodination. The latter process also suggests that the proton transfer from solvent to aniline is the rate-determining S_E2 reaction, where the concentration of free amine depends on the acidity function H_0, and not on the concentration of hydrogen ions.

C. Polysubstituted Aromatic Amines

In the presence of two or more substituents in the benzene ring, existing data about the magnitude and polar nature of these substituents usually permit a satisfactory prediction of their relative influence on orientation in additional substitution. However, only meager quantitative work has been reported for amino groups, due probably, to experimental difficulties involved when common inorganic electrophiles are used.

Some data relating to orientation of disubstituted benzenes are shown in Table 26. Thus the halogenation[329, 330] and tritylation[330–334] of aniline derivatives results in substitution chiefly as expected from the $+R$ effect. References on nitration and sulphonation have not been included in Table 26, since these processes often lead to oxidation products, making the orientational comparison in these compounds doubtful.

Although qualitative observations of *meta*-disubstituted compounds suggest that an *ortho*-, *para*-directing substituent usually takes control

TABLE 26.   Directive effects of disubstituted benzenes.

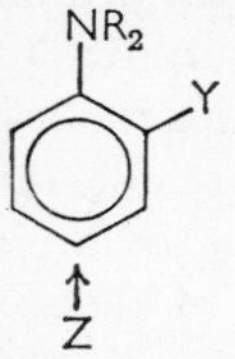

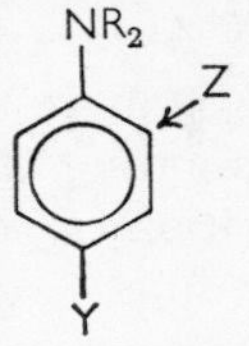

Z = Halogen (attack on *ortho*- or *para*-disubstituted benzenes)
NR$_2$ = NH$_2$, NMe$_2$ Ref. 329, 330
Y = OR, OH, NHAc, R, Ph, halogens and *m*-orienting groups.

Z = Trityl (attack on *ortho*-disubstituted benzenes only)
NR$_2$ = NH$_2$ Ref. 331–334
Y = OH, NHAc, OR, R, Cl, NO$_2$

over a deactivating group[329], some doubt as to the relative dominance may arise when both substituents are strongly activating.

Electrophilic attack on benzene containing three or more substituents can give variations from the expected results owing to steric factors or to conjugation between an activator and a deactivator. The additivity principle, which has been excellently applied for most polysubstituted benzenes, is restricted by these factors[335].

D. Comparative Reactivities

Quantitative data comparing the effects of amino substituents with other common groups in electrophilic substitution, are scattered throughout the literature. An indirect comparison of the scale of reactivities listed in Table 27 clearly illustrates the very powerful

TABLE 27.   Estimated partial rate factors for *para* halogenation of monosubstituted benzenes[329].

C$_6$H$_5$Y		Y			
	NMe$_2$	NHAc	OH	OMe	F
f_p	3×10^{19}	$1 \cdot 2 \times 10^9$	5×10^{11}	7×10^9	$6 \cdot 3$

activating effect of the dimethylamino group and the expected order of $+R$ effect, NR$_2$ > OR > F, discussed in section II.B.

Complete series of monosubstituted aromatic amines have not been studied quantitatively in electrophilic substitution, under identical conditions, except in the case of anilinium ions[249] (see section IX.B.1.a). Some of the free bases have been compared qualitatively using (see section IX.B.4.a) the non-kinetic 'competition method' with the triphenylmethyl cation[312]. In the absence of steric factors the expected sequence was obtained:

$$NMe_2 > NHMe > NH_2 > OH$$

Recently, studies of cross-migration of the triphenylmethyl group from $PhNHCPh_3$ and $PhOCPh_3$ to several monosubstituted benzenes gave the same activation sequence[336]:

$$NMe_2 > NH_2 > OH > OMe$$

Similar results were obtained in intramolecular competition of disubstituted benzenes with common inorganic electrophilic reagents[337] and the trityl cation[330-333]. The amino group again showed its predominance in the directive effect more than any other common substituent:

$$NH_2 > OH > NHAc > OMe > Me$$

These results corroborate the generalization $N > O > F$ in orientational control, indicating a decrease in the effect due to an increase in the number of unshared electrons as the group number increases[237,338,339].

X. ACKNOWLEDGMENTS

The author wishes to thank Drs. M. Tamers and O. H. Wheeler for the English revision of this work.

XI. REFERENCES

1. L. Pauling, *Nature of the Chemical Bond*, 3rd. ed., Cornell University Press, Ithaca, N.Y., 1960, pp. 88–102.
2. C. K. Ingold, *Structure and Mechanism in Organic Chemistry*, G. Bell and Sons Ltd., London, 1956, pp. 60–92.
3. C. K. Ingold, *Chem. Rev.*, **15**, 225 (1934).
4. J. G. Kirkwood and F. H. Westheimer, *J. Chem. Phys.*, **6**, 506 (1938).
5. F. H. Westheimer, *J. Am. Chem. Soc.*, **61**, 1977 (1939).
6. F. H. Westheimer and M. W. Shookhoff, *J. Am. Chem. Soc.*, **62**, 269 (1940).
7. J. D. Roberts and W. T. Moreland, Jr., *J. Am. Chem. Soc.*, **75**, 2167 (1953).
8. C. Tanford, *J. Am. Chem. Soc.*, **79**, 5348 (1957).
9. S. Siegel and J. M. Komarmy, *J. Am. Chem. Soc.*, **82**, 2547 (1960).

10. H. H. Jaffé, *J. Chem. Phys.*, **21**, 415 (1953).

11. R. W. Taft in *Steric Effects in Organic Chemistry* (Ed. M. S. Newman), John Wiley and Sons, New York, 1956, pp. 556–675.

12. M. J. S. Dewar and P. J. Grisdale, *J. Am. Chem. Soc.*, **84**, 3539, 3548 (1962).

13. K. Bowden, *Can. J. Chem.*, **41**, 2781 (1963).

14. R. J. W. Le Févre, *Dipole Moments*, 3rd ed., Methuen, London, 1953, pp. 132–135.

15. This order is in part derived from the acid dissociation constants of α-anilinoacetic acid and α-phenoxyacetic acid.
    (a) W. A. Waters, *Physical Aspects of Organic Chemistry*, 4th ed., Routledge and Paul, London, 1950.
    (b) see also P. B. D. de la Mare and J. H. Ridd, *Aromatic Substitution, Nitration and Halogenation*, Butterworths, 1959, pp. 5–25.

16. B. P. Dailey and J. N. Shoolery, *J. Am. Chem. Soc.*, **77**, 3977 (1955).

17. E. G. Cowley, *J. Chem. Soc.*, 3557 (1952).

18. H. C. Brown, D. H. McDaniel and O. Häfliger in *Determination of Organic Structures by Physical Methods* (Eds. E. A. Braude and F. C. Nachod), Academic Press, New York, 1955, pp. 567–662.

19. R. W. Taft, Jr., N. C. Deno and P. S. Skell, *Annual Review of Physical Chemistry*, Vol. 9 (Ed. H. Eyring), Annual Reviews, Palo Alto, 1958, pp. 287–314.

20. See reference 248 and 249.

21. I. Fischer, *Arkiv Fysik*, **5**, 377 (1952); *Chem. Abstr.*, **46**, 8432 (1952).

22. (a) B. M. Wepster, *Rec. Trav. Chim.*, **71**, 1171 (1952).
    (b) E. S. Gould, *Mechanism and Structure in Organic Chemistry*, H. Holt and Co., New York, 1959, pp. 199–243.

23. L. N. Ferguson, *The Modern Structural Theory of Organic Chemistry*, Prentice-Hall, New Jersey, 1964, pp. 306–452.

24. G. W. Wheland, *Resonance in Organic Chemistry*, John Wiley, New York, 1955, pp. 200–243.

25. (a) B. M. Wepster, *Rec. Trav. Chim.*, **72**, 661 (1953).
    (b) E. G. McRae and L. Goodman, *J. Chem. Phys.*, **29**, 334 (1958).
    (c) P. J. Krueger, *Nature*, **194**, 1077 (1962).

26. G. S. Hammond and M. F. Hawthorne in *Steric Effects in Organic Chemistry* (Ed. M. S. Newman), John Wiley, New York, 1956, pp. 164–200.

27. B. M. Wepster in *Progress in Stereochemistry*, Vol. 2 (Eds. W. Klyne and P. B. D. de la Mare), Butterworths, London, 1958, pp. 99–153.

28. B. M. Wepster, *Rec. Trav. Chim.*, **76**, 335, 357 (1957).

29. J. Burgers, M. A. Hoefnagel, P. E. Verkade, H. Wisser and B. M. Wepster, *Rec. Trav. Chim.*, **77**, 491 (1958).

30. G. Thomson, *J. Chem. Soc.*, 1113 (1946).

31. H. C. Brown and A. Cahn, *J. Am. Chem. Soc.*, **72**, 2939 (1950).

32. R. van Helden, P. E. Verkade and B. M. Wepster, *Rec. Trav. Chim.*, **73**, 39 (1954).

33. D. H. McDaniel and H. C. Brown, *J. Am. Chem. Soc.*, **77**, 3756 (1955).

34. R. H. Birtles and G. C. Hampson, *J. Chem. Soc.*, 10 (1937).

35. I. Fischer, *Acta Chem. Scand.*, **4**, 1197 (1950).

36. I. Fischer, *Nature*, **165**, 239 (1950).

37. J. W. Smith and S. M. Walshaw, *J. Chem. Soc.*, 4527 (1957).
38. A. E. Lutskii and B. P. Kondratenko, *Zh. Fiz. Khim.*, **33**, 2017 (1959); *Chem. Abstr.*, **54**, 14839 (1960).
39. J. W. Smith, *J. Chem. Soc.*, 81 (1961).
40. W. R. Remington, *J. Am. Chem. Soc.*, **67**, 1838 (1945).
41. P. Grammaticakis, *Bull. Soc. Chim. France*, 134 (1949).
42. M. Godfrey and J. N. Murrell, *Proc. Roy. Soc.*, **278**, 71 (1964).
43. P. Rumpf and G. Girault, *Compt. Rend.*, **238**, 1892 (1954).
44. F. E. Condon, *J. Am. Chem. Soc.*, **87**, 4494 (1965).
45. (a) L. Doub and J. M. Vandenbelt, *J. Am. Chem. Soc.*, **69**, 2714 (1947).
(b) C. Sandorfy, *Bull. Soc. Chim. France*, 615 (1949).
46. P. L. Corio and B. P. Dailey, *J. Am. Chem. Soc.*, **78**, 3043 (1956).
47. W. F. Forbes, W. A. Mueller, A. S. Ralph and J. F. Templeton, *Can. J. Chem.*, **35**, 1049 (1957).
48. K. Bowden, E. A. Braude and E. R. H. Jones, *J. Chem. Soc.*, 948 (1946).
49. H. Schindlbauer and G. Hajek, *Chem. Ber.*, **96**, 2601 (1963).
50. H. H. Jaffé, *Chem. Rev.*, **53**, 191 (1953).
51. J. E. Leffler and E. Grunwald, *Rates and Equilibria of Organic Reactions*, John Wiley, New York, 1963, pp. 171–262.
52. L. M. Stock and H. C. Brown in *Advances in Physical Organic Chemistry*, Vol. I (Ed. V. Gold), Academic Press, London, 1963, pp. 35–154.
53. P. R. Wells, *Chem. Rev.*, **63**, 171 (1963).
54. S. Ehrenson in *Progress in Physical Organic Chemistry*, Vol. II (Eds. S. G. Cohen, A. Streitwieser, Jr. and R. W. Taft), Interscience, New York, 1964, pp. 195–251.
55. C. D. Ritchie and W. F. Sager in *Progress in Physical Organic Chemistry*, Vol. II (Eds. S. G. Cohen, A. Streitwieser, Jr. and R. W. Taft), Interscience, New York, 1964, pp. 323–400.
56. L. P. Hammett, *Physical Organic Chemistry*, McGraw-Hill, New York, 1940, pp. 184–228.
57. H. Negita, H. Yamamura and H. Shiba, *Bull. Chem. Soc. Japan*, **28**, 271 (1955); *Chem. Abstr.*, **52**, 2533i (1958).
58. H. H. Jaffé, *J. Org. Chem.*, **23**, 1790 (1958).
59. H. van Bekkum, P. E. Verkade and B. M. Wepster, *Rec. Trav. Chim.*, **78**, 815 (1959).
60. D. H. McDaniel and H. C. Brown, *J. Org. Chem.*, **23**, 420 (1958).
61. F. G. Bordwell and P. J. Boutan, *J. Am. Chem. Soc.*, **78**, 854 (1956).
62. M. I. Kabachnik, *Dokl. Akad. Nauk SSSR*, **110**, 393 (1956); *Chem. Abstr.*, **51**, 5513c (1957).
63. J. D. Roberts, R. A. Clement and J. J. Drysdale, *J. Am. Chem. Soc.*, **73**, 2181 (1951).
64. M. Charton, *Can. J. Chem.*, **38**, 2493 (1960).
65. M. A. Stolberg and W. A. Mosher, *J. Am. Chem. Soc.*, **79**, 2618 (1957).
66. H. C. Brown and Y. Okamoto, *J. Am. Chem. Soc.*, **79**, 1913 (1957).
67. H. C. Brown and Y. Okamoto, *J. Am. Chem. Soc.*, **80**, 4979 (1958).
68. Y. Okamoto and H. C. Brown, *J. Org. Chem.*, **22**, 485 (1957).
69. H. C. Brown and L. M. Stock, *J. Am. Chem. Soc.*, **84**, 3298 (1962).
70. Leading reference: R. O. C. Norman and R. Taylor, *Electrophilic Substitution in Benzenoid Compounds*, Elsevier, Amsterdam, 1965, pp. 288–312.

71. Y. Okamoto and H. C. Brown, *J. Am. Chem. Soc.*, **80**, 4976 (1958).
72. R. W. Taft, *J. Phys. Chem.*, **64**, 1805 (1960).
73. R. W. Taft and I. C. Lewis, *J. Am. Chem. Soc.*, **81**, 5343 (1959).
74. R. O. C. Norman, G. K. Radda, D. A. Brimacombe, P. D. Ralph and E. M. Smith, *J. Chem. Soc.*, 3247 (1961).
75. M. Charton, *J. Org. Chem.*, **29**, 1222 (1964).
76. R. R. Fraser, *Can. J. Chem.*, **38**, 2226 (1960).
77. R. W. Taft, Jr., S. Ehrenson, I. C. Lewis and R. E. Glick, *J. Am. Chem. Soc.*, **81**, 5352 (1959).
78. G. Kohnstam and L. Williams in *The Chemistry of the Ether Linkage* (Ed. S. Patai), Interscience Publishers, London, 1966, Chapter 6.
79. Leading references: 18 and 19.
80. H. K. Hall, Jr., *J. Am. Chem. Soc.*, **79**, 5441, 5444 (1957).
81. A. Bryson, N. R. Davies and E. P. Serjeant, *J. Am. Chem. Soc.*, **85**, 1933 (1963).
82. P. Leggate and G. E. Dunn, *Can. J. Chem.*, **43**, 1158 (1965).
83. E. J. Cohn, T. L. McMeekin, J. T. Edsall and M. H. Blanchard, *J. Am. Chem. Soc.*, **56**, 784 (1934).
84. H. C. Saraswat and U. D. Tripathi, *Bull. Chem. Soc. Japan*, **38**, 1555 (1965).
85. (a) A. V. Willi and W. Meier, *Helv. Chim. Acta*, **39**, 318 (1956).
    (b) A. Bryson and R. W. Matthews, *Australian J. Chem.*, **14**, 237 (1961).
86. R. R. Miron and D. M. Hercules, *Anal. Chem.*, **33**, 1770 (1961).
87. F. H. Westheimer, *J. Am. Chem. Soc.*, **62**, 1892 (1940).
88. F. H. Westheimer and R. P. Metcalf, *J. Am. Chem. Soc.*, **63**, 1339 (1941).
89. H. Gilman and G. E. Dunn, *J. Am. Chem. Soc.*, **73**, 3404 (1951).
90. A. V. Willi, *Z. Phys. Chem.*, **26**, 42 (1960).
91. J. P. Schaeffer and T. J. Miraglia, *J. Am. Chem. Soc.*, **86**, 64 (1964).
92. D. Peltier, *Bull. Soc. Chim. France*, 994 (1958).
93. D. Peltier and M. Conti, *Compt. Rend.*, **244**, 2811 (1957).
94. D. Peltier and A. Pichevin, *Compt. Rend.*, **245**, 436 (1957).
95. D. Peltier and M. Kerdavid, *Compt. Rend.*, **243**, 2086 (1956).
96. E. Berliner and E. A. Blommers, *J. Am. Chem. Soc.*, **73**, 2479 (1951).
97. J. Shorter and F. J. Stubbs, *J. Chem. Soc.*, 1180 (1949).
98. W. Simon, A. Mörikofer and E. Heilbronner, *Helv. Chim. Acta*, **40**, 1918 (1957).
99. J. F. J. Dippy and S. R. C. Hughes, *Tetrahedron*, **19**, 1527 (1963).
100. A. Albert and E. P. Serjeant, *Ionization Constants of Acids and Bases*, Methuen, London, 1962, pp. 121–149.
101. G. Chuchani, J. A. Hernández and J. Zabicky, *Nature*, **207**, 1385 (1965).
102. J. Weinstein and E. McIninch, *J. Am. Chem. Soc.*, **82**, 6064 (1960).
103. N. C. Deno, J. J. Jaruzelski and A. Schriesheim, *J. Org. Chem.*, **19**, 155 (1954).
104. N. C. Deno and A. Schriesheim, *J. Am. Chem. Soc.*, **77**, 3051 (1955).
105. H. Walba and G. E. K. Branch, *J. Am. Chem. Soc.*, **73**, 3341 (1951).
106. R. Reynaud, *Compt. Rend.*, **260**, 6933 (1965).
107. C. K. Ingold, *Structure and Mechanism in Organic Chemistry*, Cornell University Press, 1953, 308–418.
108. Reference 22 b, pp. 250–313.

109. J. Hine, *Physical Organic Chemistry*, 2nd ed., McGraw Hill, New York, 1962, pp. 123–185.
110. S. Winstein and R. E. Buckles, *J. Am. Chem. Soc.*, **64**, 2780 (1942).
111. S. Winstein, C. R. Lindegren, H. Marshall and L. L. Ingraham, *J. Am. Chem. Soc.*, **75**, 147 (1953).
112. B. Capon, *Quart. Rev. (London)*, **18**, 45 (1964).
113. J. F. Kerwin, G. E. Ullyot, R. C. Fuson and C. L. Zirkle, *J. Am. Chem. Soc.*, **69**, 2961 (1947).
114. E. M. Schultz and J. M. Sprague, *J. Am. Chem. Soc.*, **70**, 48 (1948).
115. R. C. Fuson and C. L. Zirkle, *J. Am. Chem. Soc.*, **70**, 2760 (1948).
116. (a) P. D. Bartlett, S. D. Ross and C. G. Swain, *J. Am. Chem. Soc.*, **69**, 2971 (1947).
    (b) P. D. Bartlett, J. W. Davis, S. D. Ross and C. G. Swain, *J. Am. Chem. Soc.*, **69**, 2977 (1947).
117. S. D. Ross, *J. Am. Chem. Soc.*, **69**, 2982 (1947).
118. (a) W. E. Hanby and H. N. Rydon, *J. Chem. Soc.*, 513 (1947).
    (b) W. E. Hanby, G. S. Hartley, E. O. Powell and H. N. Rydon, *J. Chem. Soc.*, 519 (1947).
119. N. B. Chapman and D. J. Triggle, *J. Chem. Soc.*, 1385 (1963).
120. B. Cohen, E. R. Van Artsdalen and J. Harris, *J. Am. Chem. Soc.*, **70**, 281 (1948).
121. P. D. Bartlett, S. D. Ross and C. G. Swain, *J. Am. Chem. Soc.*, **71**, 1415 (1949).
122. P. L. Levins and Z. B. Papanastassiou, *J. Am. Chem. Soc.*, **87**, 826 (1965).
123. G. Salomon, *Helv. Chim. Acta*, **16**, 1361 (1933).
124. G. Salomon, *Trans. Faraday Soc.*, **32**, 153 (1936).
125. E. L. Eliel in *Steric Effects in Organic Chemistry* (Ed. M. S. Newman), John Wiley, New York, 1956, pp. 61–163.
126. H. W. Heine and B. L. Kapur, *J. Am. Chem. Soc.*, **77**, 4892 (1955).
127. A. Hassner and C. Heathcock, *J. Org. Chem.*, **29**, 3640 (1964).
128. W. J. Gensler and B. A. Brooks, *J. Org. Chem.*, **31**, 568 (1966).
129. W. C. J. Ross, *J. Chem. Soc.*, 183 (1949).
130. B. Hansen, *Acta Chem. Scand.*, **16**, 1945 (1962).
131. B. Hansen, *Acta Chem. Scand.*, **17**, 1483 (1963).
132. C. A. Grob and F. A. Jenny, *Tetrahedron Letters*, 25 (1960).
133. R. F. Brown and N. M. van Gulick, *J. Org. Chem.*, **21**, 1046 (1956).
134. C. A. Grob and F. Ostermayer, *Helv. Chim. Acta*, **45**, 1119 (1962).
135. C. A. Grob, *Gazz. Chim. Ital.*, **92**, 902 (1962).
136. T. C. Bruice and S. J. Benkovic, *J. Am. Chem. Soc.*, **85**, 1 (1963).
137. E. F. Curragh and D. T. Elmore, *J. Chem. Soc.*, 2948 (1962).
138. T. Wieland and H. Hornig, *Ann. Chem.*, **600**, 12 (1956).
139. B. Hansen, *Acta Chem. Scand.*, **12**, 324 (1958).
140. A. Ågren, U. Hedsten and B. Jonsson, *Acta Chem. Scand.*, **15**, 1532 (1961).
141. E. Schätzle, M. Rottenberg and M. Thürkauf, *Helv. Chim. Acta*, **42**, 1708 (1959).
142. W. Davis and W. C. J. Ross, *J. Chem. Soc.*, 3056 (1950).
143. J. Butterworth, D. D. Eley and G. S. Stone, *Biochem. J.*, **53**, 30 (1953).
144. W. P. Jencks, *Ann. Rev. Biochem.*, **32**, 641 (1963).
145. S. H. Chu and H. G. Mautner, *J. Org. Chem.*, **31**, 308 (1966).

146. L. Wientraub and R. Terrell, *J. Org. Chem.*, **30**, 2470 (1965).
147. A. Ferreti and G. Tesi, *Tetrahedron Letters*, 2975 (1964).
148. S. Searles, Jr., and S. Nukina, *J. Am. Chem. Soc.*, **87**, 5656 (1965).
149. Leading reference: A. Streitwieser, Jr., *Chem. Rev.*, **56**, 676 (1956).
150. R. W. Holley and A. D. Holley, *J. Am. Chem. Soc.*, **74**, 3069, 5445 (1952).
151. M. Jutisz and W. Ritschard, *Biochem. Biophys. Acta*, **17**, 548 (1955).
152. S. D. Ross in *Progress in Physical Organic Chemistry*, Vol. I (Eds. S. G. Cohen, A. Streitwieser, Jr. and R. W. Taft), Interscience, New York, 1963, pp. 31–74.
153. J. Sauer and R. Huisgen, *Angew. Chem.*, **72**, 294 (1960).
154. J. F. Bunnett, *Theoretical Organic Chemistry*, Butterworths, London, 1959, pp. 144–157.
155. J. F. Bunnett, *Quart. Rev. (London)*, **12**, 1 (1958).
156. J. F. Bunnett and R. E. Zahler, *Chem. Rev.*, **49**, 273 (1951).
157. (a) Reference 107, pp. 797–815.
     (b) Reference 109, pp. 384–401.
158. E. A. Boudreaux and E. Boulet, *J. Am. Chem. Soc.*, **80**, 1588 (1958).
159. J. H. Gorvin, *J. Chem. Soc.*, 1693 (1951).
160. C. A. Grob and G. Cseh, *Helv. Chim. Acta*, **47**, 194 (1964).
161. E. Berliner and L. C. Monack, *J. Am. Chem. Soc.*, **74**, 1574 (1952).
162. G. M. Badger, J. W. Cook and W. P. Vidal, *J. Chem. Soc.*, 1109 (1947).
163. C. W. L. Bevan, *J. Chem. Soc.*, 655 (1953).
164. J. Miller, *J. Chem. Soc.*, 3550 (1952).
165. D. T. Downing, R. L. Heppolette and J. Miller, *Chem. Ind. (London)*, 1260 (1953).
166. J. Miller, *J. Am. Chem. Soc.*, **76**, 448 (1954).
167. J. F. Bunnett, H. Moe and D. Knutson, *J. Am. Chem. Soc.*, **76**, 3936 (1954).
168. J. Miller and V. A. Williams, *J. Am. Chem. Soc.*, **76**, 5482 (1954).
169. J. F. Bunnett and R. F. Snipes, *J. Am. Chem. Soc.*, **77**, 5422 (1955).
170. E. L. Eliel and K. W. Nelson, *J. Org. Chem.*, **20**, 1657 (1955).
171. R. L. Heppolette, J. Miller and V. A. Williams, *J. Chem. Soc.*, 2929 (1955).
172. W. Greizerstein, R. A. Bonelli and J. A. Brieux, *J. Am. Chem. Soc.*, **84**, 1026 (1962).
173. M. Liveris, P. G. Lutz and J. Miller, *J. Am. Chem. Soc.*, **78**, 3375 (1956).
174. J. Miller and V. A. Williams, *J. Chem. Soc.*, 1475 (1953).
175. J. F. Bunnett, F. Draper, Jr., P. R. Ryason, P. Noble, Jr., R. G. Tonkyn and R. E. Zahler, *J. Am. Chem. Soc.*, **75**, 642 (1953).
176. (a) B. A. Bolto and J. Miller, *Chem. Ind. (London)*, 640 (1953).
     (b) B. A. Bolto, M. Liveris and J. Miller, *J. Chem. Soc.*, 750 (1956).
     (c) R. L. Heppolette and J. Miller, *J. Chem. Soc.*, 2329 (1956).
177. J. G. Allen, J. Burdon and J. C. Tatlow, *J. Chem. Soc.*, 6329 (1965).
178. J. Burdon, W. B. Hollyhead, C. R. Patrick and K. V. Wilson, *J. Chem. Soc.*, 6375 (1965).
179. B. C. McKusick, R. E. Heckert, T. L. Cairns, D. D. Coffman and H. F. Mower, *J. Am. Chem. Soc.*, **80**, 2806 (1958).
180. (a) Z. Rappoport, *J. Chem. Soc.*, 4498 (1963).
     (b) Z. Rappoport and A. Horowitz, *J. Chem. Soc.*, 1348 (1964).
181. Z. Rappoport, P. Greenzaid and A. Horowitz, *J. Chem. Soc.*, 1334 (1964).

182. Reference 22 b, pp. 314–364.
183. Reference 107, pp. 751–796.
184. Reference 109, pp. 275–301.
185. M. S. Newman in *Steric Effects in Organic Chemistry* (Ed. M. S. Newman), John Wiley, New York, 1956, pp. 201–248.
186. A. G. Davies and J. Kenyon, *Quart. Rev. (London)*, **9**, 203 (1955).
187. R. P. Bell and F. J. Lindars, *J. Chem. Soc.*, 4601 (1954).
188. G. Aksnes and J. E. Prue, *J. Chem. Soc.*, 103 (1959).
189. R. P. Bell and M. Robson, *Trans. Faraday Soc.*, **60**, 893 (1964).
190. R. P. Bell and B. A. W. Coller, *Trans. Faraday Soc.*, **60**, 1087 (1964).
191. K. Kindler, *Ann. Chem.*, **450**, 1 (1926); **452**, 90 (1927); **464**, 278 (1928); *Chem. Ber.* **69**, 2792 (1936).
192. C. K. Ingold and W. S. Nathan, *J. Chem. Soc.*, 222 (1936).
193. D. P. Evans, J. J. Gordon and H. B. Watson, *J. Chem. Soc.*, 1430 (1937).
194. E. Tommila and C. N. Hinshelwood, *J. Chem. Soc.*, 1801 (1938).
195. B. Jones and J. Robinson, *J. Chem. Soc.*, 3845 (1955).
196. E. Tommila, J. Paasivirta and K. Setälä, *Suomen Kemistilehti*, **33B**, 187 (1960); *Chem. Abstr.*, **55**, 7996a (1961).
197. E. Tommila, A. Nurro, R. Murén, S. Merenheimo and E. Vuorinen, *Suomen Kemistilehti*, **32B**, 115 (1959); *Chem. Abstr.*, **53**, 21080i (1959).
198. E. Tommila and I. Palenius, *Acta Chem. Scand.*, **17**, 1980 (1963).
199. C. C. Price and W. J. Belanger, *J. Am. Chem. Soc.*, **76**, 2682 (1954).
200. C. C. Price and D. C. Lincoln, *J. Am. Chem. Soc.*, **73**, 5838 (1951).
201. A. Fischer, H. M. Fountain and J. Vaughan, *J. Chem. Soc.*, 1310 (1959).
202. C. K. Hancock and C. P. Falls, *J. Am. Chem. Soc.*, **83**, 4214 (1961).
203. A. Agren, *Acta Pharm. Suecica*, **2**, 387 (1965); *Chem. Abstr.*, **64**, 4887f (1966).
204. R. O. C. Norman and P. D. Ralph, *J. Chem. Soc.*, 5431 (1963).
205. J. G. Watkinson, W. Whatson and B. L. Yates, *J. Chem. Soc.*, 5437 (1963).
206. J. D. Roberts, E. A. McElhill and R. Armstrong, *J. Am. Chem. Soc.*, **71**, 2923 (1949).
207. J. D. S. Ritter and S. I. Miller, *J. Am. Chem. Soc.*, **86**, 1507 (1964).
208. R. M. O'Ferrall and S. I. Miller, *J. Am. Chem. Soc.*, **85**, 2440 (1963).
209. A. Tasman, *Rec. Trav. Chim.*, **46**, 653, 922 (1927).
210. J. Véne and J. Tirouflet, *Bull. Soc. Chim. France*, 211, 217 (1954).
211. J. Tirouflet, *Bull. Soc. Sci. Bretagne*, Spec. No. **26**, 89, 101 (1951); *Chem. Abstr.*, **47**, 8694i, 8695b (1953).
212. J. Tirouflet, *Compt. Rend.*, **236**, 1796 (1953).
213. (a) R. Dabard and J. Tirouflet, *Bull. Soc. Chim. France*, 565 (1957).
    (b) J. Tirouflet, R. Dabard and E. Laviron, *Bull. Soc. Chim. France*, 570 (1957).
214. References 107, pp. 197–220.
215. References 109, pp. 214–232.
216. M. S. Kharasch and C. F. Fuchs, *J. Org. Chem.*, **10**, 159 (1945).
217. A. T. Babayan, A. A. Grigoryan and G. T. Martirosyan, *J. Gen. Chem. USSR (Eng. Transl.)*, **29**, 390 (1959).
218. E. Schmidt, *Ann. Chem.*, **267**, 300 (1892).
219. (a) M. Koulkes, *Bull. Soc. Chim. France*, 39 (1954).
    (b) M. Koulkes, *Compt. Rend.*, **241**, 1789 (1955).

220. A. A. Petrov and I. A. Maretina, *Zh. Obshch. Khim.*, **29**, 2458 (1959).
221. A. A. Petrov, K. S. Mingaleva, I. A. Maretina and V. D. Nemirovskii, *J. Gen. Chem. USSR (Eng. Transl.)*, **30**, 2230 (1960).
222. Reference 109, pp. 302–316.
223. B. R. Brown, *Quart. Rev. (London)*, **5**, 131 (1951).
224. P. A. Levene and R. E. Steiger, *J. Biol. Chem.*, **74**, 689 (1927); **79**, 95 (1928).
225. H. D. Dakin and R. West, *J. Biol. Chem.*, **78**, 91, 745 (1928).
226. G. H. Cleland and C. Niemann, *J. Am. Chem. Soc.*, **71**, 841 (1949).
227. J. A. King and F. H. McMillan, *J. Am. Chem. Soc.*, **77**, 2814 (1955).
228. J. Attenburrow, D. F. Elliott and G. F. Penny, *J. Chem. Soc.*, 310 (1948).
229. (a) R. H. Wiley, *Science*, **111**, 259 (1950).
     (b) R. H. Wiley and O. H. Borum, *J. Am. Chem. Soc.*, **72**, 1626 (1950).
230. J. W. Cornforth and D. F. Elliott, *Science*, **112**, 534 (1950).
231. J. A. King and F. H. McMillan, *J. Am. Chem. Soc.*, **73**, 4451 (1951).
232. J. A. King and F. H. McMillan, *J. Am. Chem. Soc.*, **73**, 4911 (1951).
233. R. H. Wiley, *Science*, **114**, 448 (1951).
234. Leading reference: V. I. Maksimov, *Tetrahedron*, **21**, 687 (1965).
235. G. Chatelus, *Bull. Soc. Chim. France*, 2523 (1964).
236. A. S. Sultanov, *J. Gen. Chem. USSR (Eng. Transl.)*, **16**, 1835 (1946).
237. Reference 107, pp. 223–305.
238. K. L. Nelson, *J. Org. Chem.*, **21**, 145 (1956).
239. E. Berliner in *Progress of Physical Organic Chemistry*, Vol. II (Eds. S. G. Cohen, A. Streitwieser, Jr. and R. W. Taft), Interscience, New York, 1964, pp. 253–321.
240. H. Zollinger in *Advances in Physical Organic Chemistry*, Vol. II (Ed. V. Gold), Academic Press, London, 1964, pp. 163–200.
241. J. Glazer, E. D. Hughes, C. K. Ingold, A. T. James, G. T. Jones and E. Roberts, *J. Chem. Soc.*, 2657 (1950).
242. J. Bishop Tingle and F. C. Blanck, *J. Am. Chem. Soc.*, **30**, 1395 (1908).
243. Leading reference: E. D. Hughes and G. T. Jones, *J. Chem. Soc.*, 2678 (1950).
244. E. Macciota, *Gazz. Chim. Ital.*, **71**, 81 (1941).
245. H. H. Hodgson, *J. Soc. Dyers Colourist*, **60**, 151 (1944); *Chem. Abstr.*, **38**, 5491 (1944).
246. P. P. Shorygin, A. V. Topchiev and V. A. Anan'ina, *J. Gen. Chem. USSR (Eng. Transl.)*, **8**, 986 (1938).
247. M. J. S. Dewar and D. S. Urch, *J. Chem. Soc.*, 3079 (1958).
248. M. Brickman and J. H. Ridd, *J. Chem. Soc.*, 6845 (1965).
249. M. Brickman, J. H. P. Utley and J. H. Ridd, *J. Chem. Soc.*, 6851 (1965).
250. (a) F. R. Goss, C. K. Ingold and I. S. Wilson, *J. Chem. Soc.*, 2440 (1926).
     (b) F. R. Goss, W. Hanhart and C. K. Ingold, *J. Chem. Soc.*, 250 (1927).
251. C. K. Ingold and I. S. Wilson, *J. Chem. Soc.*, 810 (1927).
252. L. N. Ferguson, *Chem. Rev.*, **50**, 47 (1952).
253. A. V. Topchiev, *Nitration of Hydrocarbons*, Pergamon, 1959, pp. 74–143.
254. L. Bangley, *Helv. Chim. Acta*, **21**, 1579 (1938).
255. (a) H. H. Hodgson and A. Kershaw, *J. Chem. Soc.*, 277 (1930).
     (b) H. H. Hodgson and D. E. Nicholson, *J. Chem. Soc.*, 470 (1941).

256. J. C. Earl and A. W. Mackney, *J. Proc. Roy. Soc. N.S. Wales*, **67**, 231 (1933); *Chem. Abstr.*, **28**, 1340 (1934).

257. W. J. Hickinbottom, *J. Chem. Soc.*, 946 (1933).

258. T. C. Van Hoek, P. E. Verkade and B. M. Wepster, *Rec. Trav. Chim.*, **77**, 559 (1958).

259. P. Friedlander, *Monatsh. Chem.*, **19**, 627 (1898).

260. W. J. Hickinbottom and E. W. Lambert, *J. Chem. Soc.*, 1383 (1939).

261. J. H. Gorvin, *J. Chem. Soc.*, 1693, 1697 (1951).

262. S. Senent Pérez and J. M. Recio Pascual, *Anales Real Soc. Españ. Fís. Quím. (Madrid)*, **49B**, 175, 183 (1953).

263. J. M. Recio Pascual, *Anales Real Soc. Españ. Fís. Quím. (Madrid)*, **50B**, 837 (1954).

264. J. M. Tedder, *J. Chem. Soc.*, 4003 (1957).

265. W. Bradley and J. D.Thompson, *Chimia*, **15**, 147 (1961).

266. E. R. Alexander, *J. Am. Chem. Soc.*, **68**, 969 (1946).

267. E. A. Shilov and A. N. Kurakin, *Zh. Obshch. Khim.*, **18**, 2092 (1948); *Chem. Abstr.*, **43**, 3803e (1949).

268. I. S. Uppal and K. Venkataraman, *J. Soc. Chem. Ind.*, **57**, 410 (1938); *Chem. Abstr.*, **33**, 2116 (1939).

269. G. V. Shirolkar, I. S. Uppal and K. Venkataraman, *J. Indian Chem. Soc.*, **17**, 443 (1940).

270. J. J. Jacobs, Jr., D. F. Othmer and A. Hokanson, *Ind. Eng. Chem.*, **35**, 321 (1943).

271. E. R. Alexander, *J. Am. Chem. Soc.*, **69**, 1599 (1947).

272. E. R. Alexander, *J. Am. Chem. Soc.*, **70**, 1274 (1948).

273. Z. Skrowaczewska, *Trav. Soc. Sci. Lettres Wroclaw Ser. B*, (61), 5 (1953); *Chem. Abstr.* **48**, 7568i (1954).

274. J. C. D. Brand and A. Rutherford, *J. Chem. Soc.*, 3927 (1952).

275. P. W. Robertson, P. B. D. de la Mare and B. E. Swedlund, *J. Chem. Soc.*, 782 (1953).

276. A. W. Francis, *J. Am. Chem. Soc.*, **47**, 2340, 2588 (1925).

277. R. P. Bell and E. N. Ramsden, *J. Chem. Soc.*, 161 (1958).

278. P. G. Farrel and S. F. Mason, *Nature*, **183**, 250 (1959); **197**, 590 (1963).

279. (a) J. E. Dubois, P. Alcais and G. Barbier, *Compt. Rend.*, **254**, 3000 (1962).
    (b) J. E. Dubois and R. Uzan, *Tetrahedron Letters*, 2397 (1964); 309 (1965).

280. J. H. Gorvin, *J. Chem. Soc.*, 1237 (1953).

281. B. R. Suthers, P. H. Riggins and D. E. Pearson, *J. Org. Chem.*, **27**, 447 (1962).

282. R. L. Jetton and M. M. Jones, *Inorg. Chem.*, **1**, 309 (1962).

283. L. Galatis and J. Megaloikonomos, *Prakt. Akad. Athenon*, **9**, 20 (1934); *Chem. Abstr.*, **29**, 6886 (1935).

284. T. N. Baker, III, W. P. Doherty, Jr., W. S. Kelley, W. Newmeyer, J. E. Rogers, Jr., R. E. Spalding and R. I. Walter, *J. Org. Chem.*, **30**, 3714 (1965).

285. P. B. D. de la Mare and I. C. Hilton, *J. Chem. Soc.*, 997 (1962).

286. T. H. Chao and L. P. Cipriani, *J. Org. Chem.*, **26**, 1079 (1961).

287. R. S. Neale, R. G. Schepers and M. R. Walsh, *J. Org. Chem.*, **29**, 3390 (1964).

288. P. Haberfield and D. Paul, *J. Am. Chem. Soc.*, **87**, 5502 (1965).

289. N. Kornblum, R. A. Smiley, R. K. Blackwood and D. C. Iffland, *J. Am. Chem. Soc.*, **77**, 6269 (1955).
290. E. Berliner, *J. Am. Chem. Soc.*, **72**, 4003 (1950).
291. E. Berliner and F. Berliner, *J. Am. Chem. Soc.*, **76**, 6179 (1954).
292. E. Berliner, *J. Am. Chem. Soc.*, **78**, 3632 (1956).
293. E. Shilov and F. Weinstein, *Nature*, **182**, 1300 (1958).
294. K. Elbs and H. Volks, *J. Prakt. Chem.*, **99**, 269 (1919).
295. L. Jurd, *Australian J. Sci. Research*, **2A**, 111 (1949).
296. W. Militzer, E. Smith and E. Evans, *J. Am. Chem. Soc.*, **63**, 436 (1941).
297. A. G. Sharpe, *J. Chem. Soc.*, 3713 (1953).
298. B. V. Tronov and S. F. Kolesnikova, *Soobshch. Nauchn-Isled. Rabot. Chlenov Primorsk. Otd. Vses. Khim. Obshchestva*, 46 (1953); *Chem. Abstr.*, **49**, 8173g (1955).
299. H. H. Hodgson and E. Marsden, *J. Chem. Soc.*, 1365 (1937).
300. R. Cantarel, *Compt. Rend.*, **225**, 638 (1947).
301. F. Seel and L. Suchanek, *Chem. Ber.*, **83**, 438 (1950).
302. F. Ullman and A. Münzhuber, *Chem. Ber.*, **36**, 404 (1903).
303. W. J. Hickinbottom, *J. Chem. Soc.*, 1700 (1934).
304. D. B. Clapp, *J. Am. Chem. Soc.*, **61**, 523 (1939).
305. V. A. Izmail'skii and D. K. Surkov, *J. Gen. Chem. USSR (Eng. Transl.)*, **13**, 848 (1943).
306. C. A. MacKenzie and G. Chuchani, *J. Org. Chem.*, **20**, 336 (1955).
307. D. Bethell and V. Gold, *J. Chem. Soc.*, 1905, 1930 (1958).
308. V. Gold and T. Riley, *J. Chem. Soc.*, 2973 (1960).
309. H. Hart and F. A. Cassis, *J. Am. Chem. Soc.*, **76**, 1634 (1954).
310. G. Chuchani, H. Díaz and J. Zabicky, *J. Org. Chem.*, **31**, 1573 (1966).
311. N. Barroeta, G. Chuchani and J. Zabicky, *J. Org. Chem.*, **31**, 2330 (1966).
312. V. Kese and G. Chuchani, *J. Org. Chem.*, **27**, 2032 (1962).
313. G. A. Olah, Ed., *Friedel-Crafts and Related Reactions*, Vol. I–IV, Interscience, London, 1963.
314. C. K. Ingold, C. G. Raisin and C. L. Wilson, *J. Chem. Soc.*, 1637 (1936).
315. A. P. Best and C. L. Wilson, *J. Chem. Soc.*, 28 (1938).
316. M. S. Kharasch, W. G. Brown and J. McNab, *J. Org. Chem.*, **2**, 36 (1938).
317. W. G. Brown, M. S. Kharasch and W. R. Sprowls, *J. Org. Chem.*, **4**, 442 (1939).
318. W. G. Brown, A. H. Widiger and N. J. Letang, *J. Am. Chem. Soc.*, **61**, 2597 (1939).
319. B. B. P. Tice, I. Lee and F. H. Kendall, *J. Am. Chem. Soc.*, **85**, 329 (1963).
320. A. I. Shatenshtein and Y. I. Ranneva, *J. Gen. Chem. USSR (Eng. Transl.)*, **31**, 1317 (1961).
321. A. I. Shatenshtein, *Tetrahedron*, **18**, 95 (1962).
322. C. Eaborn, *J. Chem. Soc.*, 4858 (1956).
323. F. B. Deans and C. Eaborn, *J. Chem. Soc.*, 2299 (1959).
324. C. Eaborn and K. C. Pande, *J. Chem. Soc.*, 297 (1961).
325. C. Eaborn and K. C. Pande, *J. Chem. Soc.*, 5082 (1961).
326. C. Eaborn and J. A. Waters, *J. Chem. Soc.*, 542 (1961).
327. R. B. Sandin and J. R. L. Williams, *J. Am. Chem. Soc.*, **69**, 2747 (1947).
328. H. S. Choguill and J. H. Ridd, *J. Chem. Soc.*, 822 (1961).
329. Reference 15b, pp. 131–148.

330. G. Chuchani, *Acta Cient. Venezolana*, Supl. 1, 200 (1963).
331. G. Chuchani, *J. Chem. Soc.*, 1753 (1959).
332. G. Chuchani, *J. Chem. Soc.*, 325 (1960).
333. G. Chuchani and J. Zabicky, *J. Chem. Soc.*, (*C*), 297 (1966).
334. R. A. Benkeser and R. B. Gosnell, *J. Am. Chem. Soc.*, **78**, 4914 (1956).
335. Reference 15b, pp. 88–92.
336. G. Chuchani and V. Rodríguez-Uzcanga, *Tetrahedron*, **22**, 2665 (1966).
337. A review of the reaction of disubstituted aromatic compounds with inorganic electrophiles, on the one hand, and their tritylation on the other, is given in reference 330.
338. C. K. Ingold and E. H. Ingold, *J. Chem. Soc.*, 1310 (1926).
339. E. L. Holmes and C. K. Ingold, *J. Chem. Soc.*, 1328 (1926).
340. M. Rapoport, C. K. Hancock and E. A. Meyers, *J. Am. Chem. Soc.*, **83**, 3489 (1961).
341. H. G. Hanson, *Acta Chem. Scand.*, **16**, 1956 (1962).
342. A. Bryson and R. W. Matthews, *Australian J. Chem.*, **16**, 401 (1963).

CHAPTER **6**

Substitution at an amino nitrogen

BRIAN C. CHALLIS and ANTHONY R. BUTLER

St. Salvator's College, University of St. Andrews, Scotland

277

I. INTRODUCTION

Amines are powerful nucleophiles owing to the presence of a lone pair of electrons on the nitrogen.

All substitutions at this nitrogen atom result from nucleophilic attack by the amine on electrophilic centres in positively charged or neutral species. The only exception is condensation of free radicals derived from aromatic amines by hydrogen abstraction, reactions which are not considered in detail in this review. Amines are such powerful nucleophiles that they will react readily with even feebly electrophilic centres such as that in carbon dioxide.

The general mechanism of substitution is a two-step process shown in equation (1) comprising formation (step a) and decomposition (step b) of a quaternary ammonium ion (**1**). This is a formal represen-

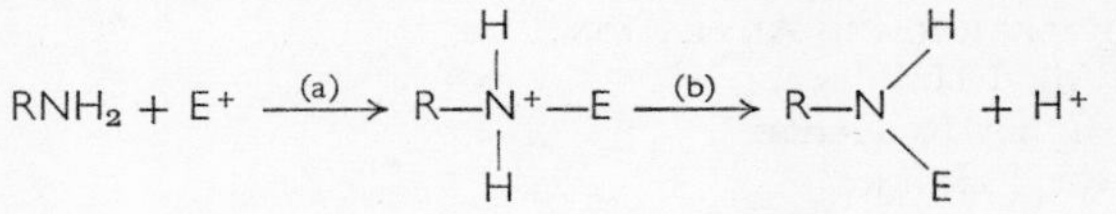

$$\text{RNH}_2 + \text{E}^+ \xrightarrow{\text{(a)}} \text{R}-\overset{\displaystyle H}{\underset{\displaystyle H}{\text{N}^+}}-\text{E} \xrightarrow{\text{(b)}} \text{R}-\overset{\displaystyle H}{\underset{\displaystyle E}{\text{N}}} + \text{H}^+ \qquad (1)$$

(**1**)

tation of an S_N2 reaction (such as the aminolysis of alkyl halides) and in these cases the ammonium ion is a high-energy configuration on the reaction coordinate, i.e. a transition state. However, in other cases (such as the aminolysis of esters and of aryl halides), this ammonium ion can acquire considerable stability and exists as a tetrahedral intermediate. It is then possible to make either step (a) or step (b) rate controlling and the factors effecting this change are discussed.

The displaced group in substitution at the amino nitrogen of primary and secondary amines is the *N*-proton. Apart from reactions leading to quaternisation of the nitrogen, as in *N*-oxide formation, tertiary amines do not normally react with many reagents that readily attack primary amines, as substitution in this instance involves fission of a carbon–nitrogen bond. The energy required for this is too large for the reaction to occur readily. Also, quaternary ammonium salts, including the conjugate acid of the amine, do not, in general, undergo substitution reactions.

Among primary and secondary amines, it might be supposed that nucleophilicity should parallel basicity, but this is rarely observed exactly. Several factors contribute to this difference, one of the most important being steric. Hall, for example, has found that Taft sigma* parameters correlate with the basicity of alkylamines[1], but not with their reactivity as nucleophilic reagents[2]. Instead, the data from several substitution reactions can be fitted[2] quite successfully to the Swain–Scott equation[3]. Both substitution and protonation involve quaternisation of the nitrogen atom at some stage and bring about steric interactions between existing substituents on nitrogen (B strain in Brown's terminology[4]) and with the incoming reagent (F strain[4]). Both should be more important for substitution with bulky reagents than with protonation, and there is good evidence that increased branching in alkylamines lowers their reactivity (mainly by increased F strain) much more than their basicity[2].

Quaternisation of the nitrogen atom during substitution or protonation should increase the acidity of the amino hydrogens, and stabilisation by polar solvents should then lower the transition-state energy. The unexpected inversion of basicity in going from secondary to tertiary amines has been ascribed to this effect[5]. There is little evidence for rate accelerations by similar solvation in substitution reactions, although exhaustive investigations have not been carried out. Probably the factor is a small one as kinetic data for substitution reactions in many different solvents can be correlated with a single Swain–Scott equation[2].

The unshared electron pair also activates the nucleus in aromatic amines and the alpha hydrogens in alkylamines. Thus attack by electrophilic reagents at these alternative sites can compete with substitution on nitrogen. These side-reactions are discussed in relation to halogenation and oxidation in which they may become dominant under certain conditions.

The multiplicity of substitution reactions in amine chemistry renders a complete coverage virtually impossible. We have therefore concentrated on just five important reactions from which it is possible to establish and summarise the important factors influencing the mechanism of substitution.

II. ACYLATION

Primary and secondary amines react with many carbonyl compounds to form amides. In these reactions the normal order of carbonyl reactivity is observed. Thus acid halides are the most reactive, anhydrides and esters react less readily, acids and amides react with difficulty, while other carbonyl compounds such as aldehydes and ketones give different products. Tertiary amines are generally unreactive, although salt-like addition products have been isolated from reaction with acid chlorides. The reaction of primary and secondary amines with esters has been studied most extensively and this reaction is considered first because many mechanistic conclusions can be applied to the other reagents.

A. Esters

Although not particularly good acylating reagents, esters will react with amines to give amides (equation 2). From the synthetic stand-

$$RCO_2R + RNH_2 \longrightarrow RCONHR + ROH \qquad (2)$$

point the reaction with anhydrous liquid ammonia is frequently used for preparing amides. The addition of ammonium salts catalyses the reaction, and as the efficacy depends upon the anion, this is probably a case of base catalysis[6]. Anhydrous amines react in a similar way and this reaction is also catalysed by amine salts[7]. Methanol is a good solvent and, with unreactive systems, alkoxide ions are effective base catalysts.

The mechanism of the reaction has been subjected to extensive investigation. The importance of amine basicity was soon recognised from the fact that straight-chain primary amines react faster than ammonia. The reaction is also subject to steric hindrance, for secondary amines, although stronger bases than ammonia, react more slowly. Early work by Betts and Hammett[8] and Watanabe and DeFonso[9] indicated that the reaction is not a simple bimolecular one. General base catalysis was first clearly demonstrated by Bunnett and Davis[10] in the reaction between n-butylamine and ethyl formate in alcoholic solution, and by Jencks and Carriuolo[11] in the aminolysis of phenyl acetate in aqueous solution, and rate equations for these reactions have the following form (equation 3) where k_n and k_{gb} are the rate constants for nucleophilic and general base catalysed aminolysis respectively. This suggests that the reaction proceeds simultaneously by

$$\text{Rate} = k_n[RNH_2][\text{ester}] + k_{gb}[RNH_2]^2[\text{ester}] +$$
$$+ k_{OH^-}[RNH_2][OH^-][\text{ester}] \quad (3)$$

two pathways: direct nucleophilic attack and base-catalysed substitution by the amine on the ester. The basic catalyst can either be a second molecule of amine or a lyate anion.

It seems probable that the mechanism of ester aminolysis is similar to that of the reaction between amines and aryl halides, a reaction which also exhibits general base catalysis (see section III.C). The reaction would then involve the formation of a tetrahedral intermediate (**2**) by addition of amine across the carbonyl double bond in an equilibrium step, which can then decompose via two independent pathways, one catalysed by base (k_4) and the other not (k_3). Whether

$$RNH_2 + RC-OR \underset{k_2}{\overset{k_1}{\rightleftharpoons}} RC-OR \overset{k_3}{\underset{k_4}{\rightarrow}} Products$$

SCHEME 1. Ester aminolysis.

formation (k_1) or decomposition ($k_3 + k_4$) of the tetrahedral intermediate is rate controlling depends on the structure of the ester and the amine.

It is quite clear that the incidence of general base catalysis is greatly affected by the nature of the leaving group. Results obtained for the ammonolysis of substituted phenyl acetates by Bruice and Mayahi[12] are shown in Table 1. It is only with poor leaving groups, like p-

TABLE 1. Ammonolysis of substituted phenyl acetates.

Substituent	k_n (l/mole min)	k_{gb} (l/mole min)
p-NO$_2$	29·2	
m-NO$_2$	10·5	
p-Cl	1·03	0·800
H	0·245	0·722
p-CH$_3$	0·130	0·510

chlorophenyl, that base catalysis can be detected. Probably a catalysed reaction occurs to a certain extent in all cases, but with nitroamines unassisted attack (k_3) is so much larger that the catalysed reaction cannot be detected over the range of amine concentrations employed in this study. Activation energies for catalysed aminolyses are generally smaller than those for unaided attack[13], so that general base catalysis is more readily observed at lower temperatures[14]. For esters containing a good leaving group, both catalysed (k_4) and uncatalysed (k_3) decomposition of the intermediate should be fast, and the formation of the tetrahedral intermediate becomes rate determining. On the other hand, with poor leaving groups the catalysed decomposition becomes kinetically important.

There is no direct evidence for tetrahedral intermediates with acyl esters. However, their formation seems very probable by analogy with ester hydrolysis[15], and Bruice and coworkers[16] have produced substantial kinetic evidence for such an intermediate in reactions between amines and thiol esters. Also, a two-step mechanism obviates the statistical improbability of a termolecular collision in the catalysed reaction.

The exact nature of the base catalysis is not clear, but protonation of the leaving group by the conjugate acid of the catalysing base seems most reasonable (Scheme 2). This interpretation is consistent

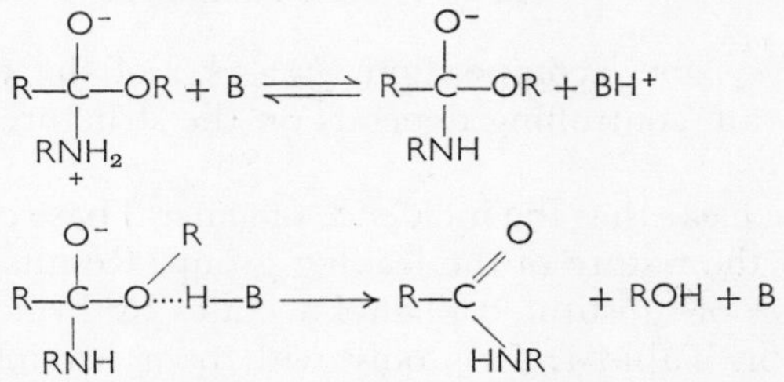

$$R-\underset{\underset{RNH_2}{\overset{+}{|}}}{\overset{\overset{O^-}{|}}{C}}-OR + B \; \rightleftharpoons \; R-\underset{\underset{RNH}{|}}{\overset{\overset{O^-}{|}}{C}}-OR + BH^+$$

$$R-\underset{\underset{RNH}{|}}{\overset{\overset{O^-}{|}}{C}}-O\cdots H-B \; \longrightarrow \; R-\overset{O}{\underset{HNR}{C}} \; + ROH + B$$

SCHEME 2. General base catalysis in ester aminolysis.

with the low Hammett rho factor reported by Bruice and Mayahi[12] (see Table 1). Since leaving tendency is a measure of anion stability, this factor will become far less important if simultaneous protonation occurs, and will, therefore, lead to the observed insensitivity to electronic effects. The low reactivity of esters towards alkali metal amides in liquid ammonia also suggests protonation of the leaving group[17].

Diamines and aminohydrazines may be acylated by esters but there is no evidence for general acid or general base catalysis[18]. Of these compounds, 3-dimethylaminopropylhydrazine shows considerable deviation from a Brønsted plot for aminolysis, which suggests that intramolecular general base catalysis is occurring. This may be of some significance in enzyme mechanisms[18].

Certain amines are very readily acylated by esters. Testa[19] noted that 3-hydroxymethyl-3-phenylazetidine (3) is N-acetylated on

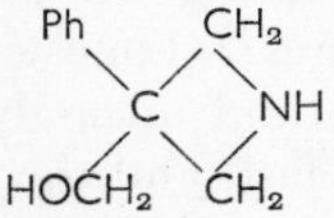

(3) 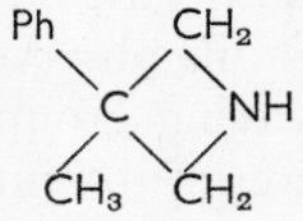　　　　　　　(4)

extraction from ethyl acetate. This was thought to arise from the presence of the hydroxyl group, but Bruice, Meinwald and coworkers [20] have shown that compounds without the OH group, 3-methyl-3-phenylazetidine (4) and aziridine, also show enhanced reactivity towards phenyl acetate. The most probable explanation is that, as the two CH_2 groups are 'pinned back', there is less crowding in the tetrahedral intermediate, a situation which will have greater effect upon the attack at an ester carbonyl bond than upon the sterically less demanding protonation.

Thiol esters are better acylating agents than oxygen esters and such reactions are of some biological significance. Studies by Bruice and his coworkers [16, 21, 22] have indicated that the pattern of thiolester aminolysis is similar to that of oxygen esters. The most significant outcome of this work, as mentioned previously, has been the demonstration of a tetrahedral intermediate by a detailed analysis of the kinetics of the reaction between hydroxylamine and n-butyl thiolacetate [22].

B. Acid Halides

The reaction between primary and secondary amines and acid halides is one of the most convenient methods of preparing amides. With primary amines a second acyl group may be introduced. An equilibrium is set up (equation 4) and with halides which are poor

$$RNH_2 + RCOCl \rightleftharpoons RNHCOR + HCl \tag{4}$$

acylating agents, or with weakly nucleophilic amines, it is usually necessary to remove the hydrochloric acid produced to obtain a reasonable yield of amide. Aqueous alkali is used in the Schotten–Baumann procedure [23, 24], but in most cases the reaction will proceed, sometimes violently, in an inert solvent without the addition of base.

Most of the available evidence points to a rate-controlling substitution by the amine on the acyl halide, like that for acylation by esters, although other mechanisms involving ionisation to give an acyl ion which then attacks the amine, are, in principle, possible. However, since ionisation of acid halides occurs most readily in acid media, conditions under which the amine would be protonated and therefore unreactive, any mechanism involving this process as the precursor to reaction seems unlikely. Also, slow ionisation of the acyl halide implies that the rate of reaction should be independent of the amine and

10*

extensive work by Hinshelwood[25-29] has shown that this is not the case. For any particular halide, the rate increases proportionately with amine basicity. In the acylation of aniline, changes in rate produced by nuclear substitution are reflected almost completely in changes in the activation energy, whereas the entropy of activation appears to be constant. For instance, electron-repelling groups in the acid chloride increase the activation energy, whereas the opposite applies to the amine. Litvinenko[30] has listed Hammett rho factors for acylation by substituted benzoyl chlorides ranging from -2.6 to -4.79 indicating a large charge separation in the transition state[29]. Furthermore, tracer studies using ^{18}O in the hydrolysis of benzoyl chlorides under neutral conditions indicate that carbonyl addition leading to a tetrahedral intermediate occurs[31]. It seems reasonable to extend this to the aminolysis of acid halides.

All this evidence points to a mechanism like that for acylation by esters, involving preequilibrium formation of a tetrahedral intermediate followed by decomposition to products (Scheme 3). Probably

$$\underset{\substack{\| \\ O}}{RC}-X + R_2NH \underset{k_2}{\overset{k_1}{\rightleftharpoons}} \underset{\substack{| \\ +NHR_2}}{R-\overset{O^-}{\underset{|}{C}}-X} \xrightarrow{k_3} \underset{\substack{\| \\ O}}{RC}-NR_2 + HX$$

SCHEME 3. Aminolysis of acid chlorides.

the halides are such good leaving groups that base catalysis of the decomposition step is never observed. Bender and Jones[32] have studied the effect of varying the halide upon the rate of reaction. Their results are summarised in Table 2. The figures clearly show a strong

TABLE 2.　Reaction between benzoyl halides and morpholine.

Benzoyl compound	k_{obs}(l/mole sec)
Fluoride	1.06
Chloride	2.64×10^3
Bromide	6.66×10^4
Iodide	2.39×10^5

dependence of the rate upon the halide moiety and suggest that the slow step is the breakdown of the tetrahedral intermediate, which involves the breaking of a carbon–halide bond. The removal of the proton attached to the nitrogen in the tetrahedral intermediate cannot be the rate-determining process, as there is no change in reaction rate between aniline and benzoyl chloride upon isotopic substitution on the amine nitrogen[33].

C. Anhydrides

Anhydrides are less reactive than acid halides and are therefore used in the preparation of amides where reaction with acyl halides is too vigorous (equation 5). By boiling or heating under pressure a

$$(RCO)_2O + RNH_2 \longrightarrow RNHCOR + RCO_2H \qquad (5)$$

second acyl group may be introduced with primary amines[34-37] (equation 5a). As expected, the ease of acylation of arylamines by

$$RNHCOR + (RCO)_2O \longrightarrow RN(COR)_2 + RCO_2H \qquad (5a)$$

anhydrides is decreased by electron-attracting substituents[34], and special procedures are required to render these reactions synthetically useful. Thus di-o-substituted arylamines, e.g. sym-tribromoaniline, are particularly difficult to acylate but strong acids are effective catalysts[38]. Normally, strong acids would be expected to reduce the rate of amine acylation owing to the formation of unreactive amine salts, but the inductive effect of two $-I$ groups in ortho positions must lessen this tendency. More important, it is probable that the acid reacts with the anhydride to form an acyl salt (equation 6) which is a

$$(RCO)_2O + HX \longrightarrow RCOX + RCO_2H \qquad (6)$$

more effective acylating agent. In aqueous solution, where the anhydride also undergoes hydrolysis, aminolysis is still the predominant reaction[39], providing the amine is a sufficiently powerful nucleophile to compete effectively with water. Acylation of amines also occurs in the solid phase[40] and this is said to have some advantages over the reaction in solution.

Acylation of amines by anhydrides has not been subjected to extensive mechanistic investigation but it seems highly probable that the mechanism is similar to that of anhydride hydrolysis, which brings the reaction in line with the mechanism of amine acylation by other reagents. Since acyl halides hydrolyse by an S_N2 mechanism[41], acyl acetate (acetic anhydride) should be more inclined to react in this way. Thus acetic anhydride would be expected to react with amines, stronger nucleophiles than water, via an S_N2 reaction also. More direct evidence in support of an S_N2 mechanism comes from the work of Remmers[34], who found that the ease of acylation of substituted aromatic amines by an anhydride is dependent upon the amine, which would not be the case if ionisation were the rate-determining step.

The reaction may be more complex, however, than a simple S_N2 process. Following oxygen-18 studies, Denney and Greenbaum[42] have suggested the following scheme (Scheme 4) for the reaction between

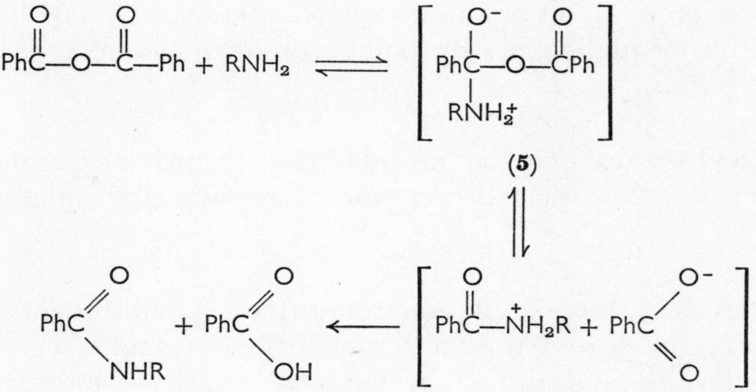

$$\text{PhC(=O)-O-C(=O)-Ph} + \text{RNH}_2 \rightleftharpoons \left[\begin{array}{c} \text{O}^- \quad \text{O} \\ \text{PhC-O-CPh} \\ \text{RNH}_2^+ \end{array} \right]$$

(5)

$$\left[\begin{array}{c} \text{O} \\ \text{PhC-}\overset{+}{\text{N}}\text{H}_2\text{R} + \text{PhC}(\text{O}^-)(=\text{O}) \end{array} \right] \longrightarrow \text{PhC(=O)-NHR} + \text{PhC(=O)-OH}$$

(6)

SCHEME 4. Anhydride aminolysis.

amines and benzoic anhydride. The tetrahedral intermediate (5) is expected, the second equilibrium is not, but is necessitated by the experimental results. If a single carbonyl oxygen in the anhydride is labelled with ^{18}O, then, on reaction with aniline (R = Ph), the proportion of isotopic oxygen in the products is benzanilide (33%) and benzoic acid (67%). This can only occur if all three oxygens in the anhydride are equivalent at some stage, e.g. if the ions in 6 are in equilibrium with unreacted amine and anhydride. With ammonia (R = H), the products, benzamide and benzoic acid, each contain 50% of the label, showing that in this case equilibration has not occurred.

Acylation of amines by anhydrides may show autocatalytic behaviour. Thus the reaction of various amines with anhydrides in benzene is catalysed by the monomeric form* of the acid produced[43]. Litvinenko[44] has suggested that the catalytic effect is due to complex formation (7 or 8) and the catalytic power of the carboxylic acid increases with acid strength until protonation of the amine occurs. Similar complexes have been proposed for the acid-catalysed acylation

* In benzene solution carboxylic acids exist in a dimer–monomer equilibrium and this situation leads to complex kinetics.

of amines by benzoyl chloride[45]. With picryl chloride, added acids have no catalytic effect[46], indicating the need for a carboxyl group. The evidence in favour of the formation of complexes of this type is tenuous and without significant precedent. Reactions in benzene often prove difficult to elucidate and further study of this solvent is required. Recent studies by Bruckenstein and Saito[47] indicate extensive ion-pair formation between acids and amines in benzene, which may be a factor influencing the rate of acylation.

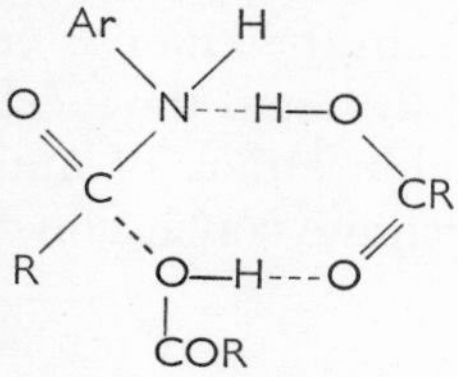

$$
\begin{array}{cc}
\text{structure (7)} & \text{structure (8)}
\end{array}
$$

Ar, H, O, N---H—O, C, CR, R, O—H---O, COR (7)

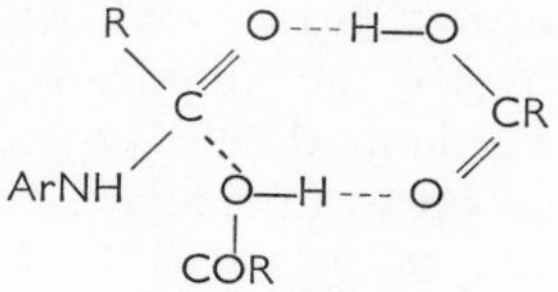

R, O---H—O, C, CR, ArNH, O—H---O, COR (8)

In mixtures of acetic anhydride and acetic acid, acylation of various nitroanilines reaches a maximum at 45% anhydride[48], which is also the mixture with maximum electrolytic conductivity[49]. This suggests that ions participate in the reaction. Nitroanilines are unprotonated in this solvent, so the ions must be acetylium ions (equation 7).

$$CH_3C(O)OC(O)CH_3 \longrightarrow CH_3\overset{+}{C}O + CH_3CO_2^- \tag{7}$$

As the rate of reaction depends upon the structure of the amine, ionisation cannot be the rate-determining step, which must be attack of the amine by the acetylium ion.

The reaction of amines with mixed anhydrides has been investigated by Gold[50-52]. Rationalising the products formed is a fairly complex matter and most of the considerations are irrelevant to the present discussion. However, in the acylation of aniline and 2,4-dichloroaniline by acetic chloroacetic anhydride, chloroacetylation predominates, as expected on electronic grounds, and the ratio of acetylation to chloroacetylation is independent of the amine despite marked changes in rate. The effect of further substitution in the anhydride moiety upon the rate of amine acylation is complicated by steric factors but is, in general, in agreement with an S_N2 mechanism.

D. Carboxylic Acids and Amides

The most usual method of preparing acetamide is by heating ammonium acetate (equation 8). The system is in equilibrium but may

$$CH_3CO_2NH_4 \rightleftharpoons CH_3CONH_2 + H_2O \tag{8}$$

be forced to the right by the removal of water. The reaction is less facile on increased substitution in either the carboxylic acid or the amine[53]. The reaction probably involves dissociation into acetic acid and ammonia followed by nucleophilic attack of the ammonia on the acetic acid. Morawetz and Otaki[54] have shown that the reactive species in amide formation are the free amine and the carboxylic acid anion (equation 9). The mechanism must involve an intramolecu-

$$RCO_2^- + RNH_2 \rightleftharpoons RCONHR + OH^- \tag{9}$$

lar proton transfer as expulsion of the O^{2-} ion from the anion is very unlikely. However, Bruice and Benkovic[55] have suggested the mechanism may involve the kinetically indistinguishable attack of amine anion on the unionised acid (equation 10). No definite conclusion can be reached on present evidence.

$$\underset{\substack{\| \\ RC-OH}}{\overset{O}{}} + RNH^- \longrightarrow \underset{\substack{\| \\ R-C-NHR}}{\overset{O}{}} + OH^- \tag{10}$$

Thiolacetic acid reacts more readily than acetic acid. This is not surprising in view of the greater ease of removal of volatile hydrogen sulphide.

The acylation of amines by amides is not of preparative significance, although the reaction does occur. Amides are obtained by the reaction between a primary amine and diacetamide or dibenzamide in toluene[56]. The reaction proceeds under neutral conditions, which can be advantageous in certain circumstances.

E. Miscellaneous Acylating Agents

Numerous other reagents acylate amines but few of these reactions have been subjected to mechanistic investigation. Only the more important ones are mentioned here.

According to Satchell[57] ketens readily acylate amines in inert media. Other work[58] has indicated that addition of water or acetic acid is necessary for the reaction to proceed. The obvious explanation is that acetic anhydride is formed (equation 11) which then reacts

$$CH_2{=}C{=}O + CH_3CO_2H \longrightarrow (CH_3CO)_2O \tag{11}$$

with the amine. The only puzzling feature of this interpretation is that keten appears to react more rapidly than acetic anhydride itself under the same conditions. Svetkin[59] has suggested that this is due to the initial formation of the enol form of acetic anhydride (equation 12)

$$CH_2{=}C{=}O + CH_3CO_2H \longrightarrow CH_2{=}\underset{\substack{| \\ OH}}{C}{-}O{-}COCH_3 \tag{12}$$

which should be more reactive than the keto form. Such an explanation is not unreasonable, but more direct evidence is necessary before it can be considered valid.

Methacryloyl chloride reacts more rapidly with basic amines than with water and is, therefore, a useful reagent in aqueous solution. Excess amine is required to remove the liberated hydrogen chloride[60]. Isopropenyl acetate is a mild acylating agent[61] and will convert amines into amides[62]. Acylations are also readily performed using α-ketonitriles, the most usual being benzoyl nitrile[63], while α-acetoxy-acrylonitrile selectively acylates the amino group, but does not react with an alcoholic or phenolic hydroxyl group[64]. N-Acetyl- and N-benzoyllactams[65], as well as 2- and 3-acetoxypyridine[66], acylate amines, but these reactions are not of preparative use. O-Acylhydroxylamines are weak acylating agents, like esters[67]. For preparative purposes O-acylketamines are the most useful, being stable to heat and unaffected by alcohols and water (equation 13). The reaction is

$$R_2NOCOR + RNH_2 \longrightarrow RNHCOR + R_2NOH \qquad (13)$$

catalysed by acids. The strength of these catalysts is quite critical as very weak acids are ineffective but strong acids reduce the reaction rate by protonating the amine.

Recently tetraacyloxysilanes have been demonstrated to be excellent acylating agents[68], giving 60–90% yields of amide (equation 14).

$$(RCO_2)_4Si + 4 Et_2NH \longrightarrow 4 RCONEt_2 + Si(OH)_4 \qquad (14)$$

The reaction has been extended to the acylation of di-t-butylamine with silico anhydrides of acetic, butyric and caproic acids. Ethylamine also reacts readily with these reagents.

F. Formylation

Formylation of amines, to give formamides, is a special case of acylation and warrants separate comment. Most studies of this reaction have been preparative rather than mechanistic, and although many procedures have been described, the range of applicability seems to be limited.

The most usual method is to transformylate an amine with a more reactive formamide (equation 15). Formamide itself will react with

$$HCONH_2 + PhNH_2 \rightleftharpoons HCONHPh + NH_3 \qquad (15)$$

aniline derivatives and amines such as benzylamine[69], but not with simple aliphatic amines. Glacial acetic acid is often used as a solvent[70] and the reaction also proceeds with the amine hydrochloride[71]. It seems probable that the reactive species is, in fact, the free amine rather than the hydrochloride, and that the equilibrium (equation 15) is displaced to the right by precipitation of ammonium chloride. Aromatic amines are also readily formylated by heating with dimethylformamide and sodium methoxide (equation 16): dimethylamine is

$$HCONMe_2 + PhNH_2 \longrightarrow PhNHCHO + NHMe_2 \qquad (16)$$

evolved and the product is obtained by aqueous decomposition of an intermediate[72]. The exact role of the methoxide ion is not clear.

Esters are also good reagents[73] and ethyl formate is particularly useful for formylating t-butylamine and t-octylamine under conditions where there is no acylation by methyl acetate[74].

Chloral reacts with many amines to furnish formamido compounds and chloroform[75] but complexes often result from this reaction[76]. Formic acid alone will formylate certain anilines such as 3,4-dichloroaniline[77]. The addition of acetic anhydride improves reactivity[78], presumably by the formation of acetic formic anhydride, but acetylation may then occur in preference to formylation. Huffman[79] has improved this method by preparing formic acetic anhydride before adding the amine[80]. Oláh[81] has described the use of formyl fluoride as a formylating agent.

III. ALKYLATION AND ARYLATION

Alkylation and arylation have been closely studied, and the results are particularly interesting because they contribute significantly to our understanding of substitution at an amino nitrogen in general. Usually the reaction occurs readily between most kinds of amine and the appropriate alkyl or aryl halide, although steric interference from either reagent retards the rate. Alkyl derivatives of other strong acids, such as alkyl sulphates and benzenesulphonates are also effective reagents, although the halides are by far the most useful. A rather different, but also common method, is reductive alkylation, a process in which the amine is first condensed with a carbonyl compound and subsequently reduced, either by catalytic hydrogenation or by formic acid or one of its derivatives.

A. Alkyl Halides

Ammonia and simple aliphatic amines react readily with alkyl halides. With reactive reagents, such as methyl iodide, a mixture of

primary, secondary and tertiary amines is obtained, together with
quaternary ammonium salts (Scheme 5). With branched-chain

$$NH_3 + MeI \longrightarrow NH_2Me + HI$$
$$NH_2Me + MeI \longrightarrow NHMe_2 + HI$$
$$NHMe_2 + MeI \longrightarrow NMe_3 + HI$$
$$NMe_3 + MeI \longrightarrow NMe_4I$$

SCHEME 5. Reaction between ammonia and methyl iodide.

alkyl halides the reaction is less facile and it is almost essential to use
the more reactive iodide in these cases. Tertiary alkylating reagents,
such as t-butyl halides, are ineffective, and elimination of hydrogen
halide occurs in preference to nitrogen substitution, indicating steric
interference, which also accounts for the lower reactivity of branched-
chain secondary amines compared to primary amines, despite their
higher basicity and, therefore, greater nucleophilicity.

Arylamines are less reactive, because of their lower basicity, and
therefore require higher temperatures. The reaction is not difficult
with small, reactive alkylating reagents and proceeds to the quaternary ammonium salt. Diarylamines are, as expected, even less reactive,
although diphenylamine can be methylated with methyl iodide at 40°
in dimethylformamide[82]. Generally speaking, however, diarylamines
are more easily alkylated via their sodium or magnesium salts[83].
Triarylamines do not react with the usual alkylating reagents.

Nearly all these reactions appear to proceed via an uncomplicated
S_N2 process[84], and usually follow a second-order rate law (equation
17). The rates depend upon the nucleophilicity of the amine[85] and the

$$\text{Rate} = k[\text{amine}][RX] \qquad (17)$$

leaving characteristics of the displaced group (X)[86], and are faster in
polar solvents[87] such as alcohol, consistent with the formation of
charge in the transition state. More direct evidence is the inversion
of configuration reported for the alkylation of trimethylamine with
D-sec-butyl chloride[88]. The steric requirements noted previously are
also consistent with an S_N2 process. For further discussion of these
reactions the reader is referred elsewhere[89, 90].

As with other nucleophilic substitutions, the alkylation of amines by
some reagents is accompanied by competing elimination of HX
(equation 18).

$$\begin{array}{c}\diagdown\\ \diagup\end{array}\!\!CH\!-\!CH_2X + R_3N \longrightarrow \begin{array}{c}\diagdown\\ \diagup\end{array}\!\!C\!=\!C\!\begin{array}{c}\diagup\\ \diagdown\end{array} + R_3NHX \qquad (18)$$

B. Other Alkylating Reagents

Alkylation may also be brought about by an isocyanate or isothiocyanate. With strongly basic amines addition occurs[91] (equation 19) but with less basic amines there is substitution[92] (equation 20). In

$$Ph_3CNCS + NHR_2 \rightleftharpoons Ph_3CNHCSNR_2 \qquad (19)$$

$$Ph_3CNCS + PhNH_2 \rightleftharpoons Ph_3CNHPh + HNCS \qquad (20)$$

solvents favouring carbonium ion formation, e.g. benzonitrile, the triphenylcarbinyl ion readily forms and reacts with even strongly basic amines to give substitution products[91] (equation 21).

$$Ph_3CNCS + NH_3 \rightleftharpoons Ph_3CNH_2 + HNCS \qquad (21)$$

Olefins containing strongly electron-withdrawing substituents, e.g. α,β-unsaturated nitriles, carbonyl compounds and nitroalkenes react with amines (equation 22)[93,94]. With sodium amide as a catalyst, amines will add to styrene[95] (equation 23).

$$RNH_2 + NO_2CH{=}CH_2 \longrightarrow NO_2CH_2CH_2NHR \qquad (22)$$

$$R_2NH + PhCH{=}CH_2 \longrightarrow PhCH_2CH_2NR_2 \qquad (23)$$

Terminal acetylenes react with amines in a 2:1 ratio in the presence of cuprous chloride to give substituted propargylamines[96]. The process occurs in two steps, the first being the production of an enamine (equation 24). This is followed by 1,2-addition of the acetylene across the enamine double bond (equation 25).

$$RC{\equiv}CH + R_2NH \longrightarrow RCH{=}CHNR_2 \qquad (24)$$

$$RCH{=}CHNR_2 + RC{\equiv}CH \longrightarrow \underset{\underset{C{\equiv}CR}{|}}{RCH_2CHNR_2} \qquad (25)$$

C. Aryl Halides

Amines replace halide attached to aromatic rings but only in fairly drastic conditions. Nucleophilic displacement is greatly enhanced, however, by the presence of nitro groups on the aromatic ring and such reactions have been extensively studied. It is generally assumed that less activated aromatic halides react in the same way, although there is little direct evidence for this. Most workers now agree on the general character of the reaction (Scheme 6). The formation of the tetrahedral intermediate (9) is reversible, and its decomposition to products occurs either spontaneously (step k_3) or in a base-catalysed step (k_4). Depending on the conditions and the reactants, either k_1,

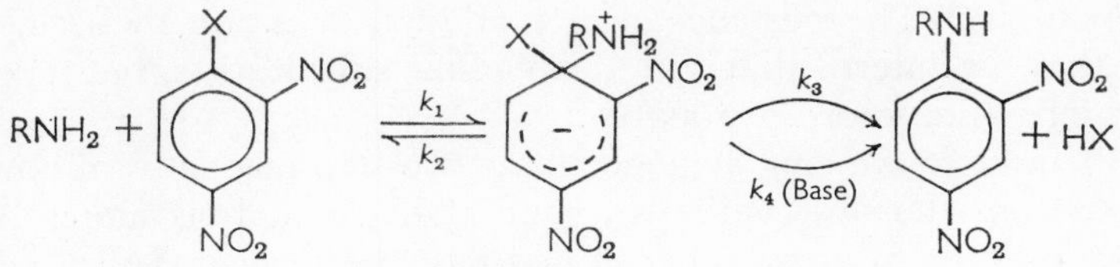

(9)

Scheme 6. Aminolysis of an aryl halide.

k_3 or k_4 can be rate controlling. A similar mechanism has already been discussed for ester aminolysis, but in the present case, evidence for a two-step process is more complete.

From a very elementary consideration of transition-state energies, one might reasonably conclude that the formation of a tetrahedral intermediate with aromatic halides will be favourable. For simple S_N2 displacement without intermediate formation, as in the case of alkylation, the transition state must involve the formation of partial bonds to both halide and amine moieties from the original covalent bond to the halide (equation 26). Partial-bond formation results in

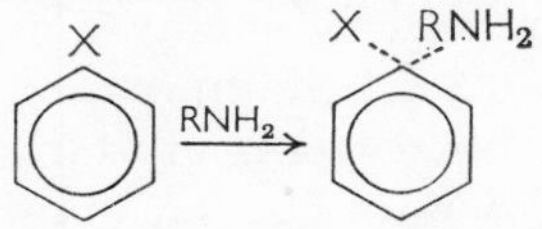

$$(26)$$

gross distortion of the bonds with loss of aromaticity. If aromaticity is lost, then the carbon atom at the seat of the reaction assumes sp^3 hybridisation and an intermediate with resonance stabilisation results (equation 27). Under certain conditions, these and similar inter-

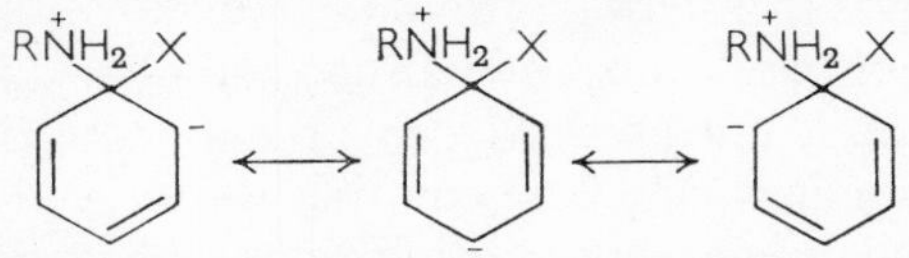

$$(27)$$

mediates have been isolated. The highly coloured adducts between alkoxide ions and polynitro compounds were first described by Meisenheimer[97] who suggested formulae of the type **10**. Similar coloured compounds result from mixing amines with polynitro

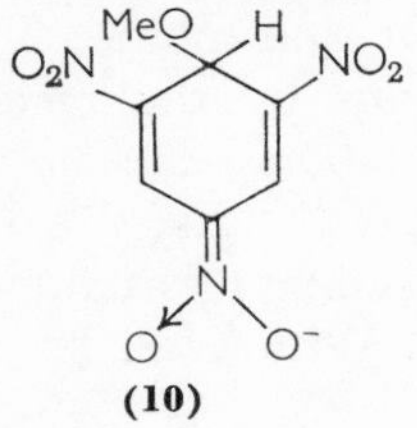

(10)

compounds[98]. The existence of such adducts does not, by itself, prove that they are intermediates in nucleophilic substitution, but it renders their formation more reasonable.

The most convincing evidence for a two-step mechanism, however, comes from the observation of general base catalysis either by the amine itself or by some other added base. Several groups of workers have reported the phenomenon. Brady and Cropper[99] found that triethylamine, while not reacting with 2,4-dinitrochlorobenzene, will catalyse the reaction with methylamine. The reaction between n-butylamine and 2,4-dinitrochlorobenzene (DNCB) in water–dioxan mixtures[100] follows the complex rate equation (28). Both the second

$$\text{Rate} = k_{\text{n}}[\text{BuNH}_2][\text{DNCB}] + k_{\text{gb}}[\text{BuNH}_2]^2[\text{DNCB}] +$$
$$k_{\text{OH}^-}[\text{BuNH}_2][\text{DNCB}][\text{OH}^-] \quad (28)$$

and third terms of equation (28) indicate general base catalysis; in the second term by another molecule of amine and in the third term by hydroxide ion. Bunnett and Randall[101] report similar kinetics for the reaction at 2,4-dinitrofluorobenzene with N-methylaniline. Piperidine, methanol, triethylenediamine and pyridine all catalyse the reaction of piperidine with 2,4-dinitrofluorobenzene and 2,4-dinitro-chlorobenzene[102]. The observation of third-order kinetics, is most significant because, in view of the low probability of termolecular collisions under the experimental conditions, it suggests that the amine and aryl halide combine initially to form an intermediate that has sufficient half-life to be attacked by a second molecule of base to give products. The observation of general base catalysis suggests that the second step involves transfer of a proton.

The incidence of base catalysis in aminolysis of aryl halides is by no means universal and, until recently, this remained a puzzling feature of the reaction. For instance, base catalysis is observed in the reaction between N-methylaniline and 2,4-dinitrofluorobenzene, but not with 2,4-dinitrochlorobenzene or 2,4-dinitrobromobenzene[101]. Again only the fluoro compound shows general base catalysis in the reaction with substituted pyridines[103]. These differences are readily understood if either formation (k_1) or decomposition (k_3 and k_4) of the tetrahedral intermediate can be rate controlling. That this is indeed the case has been elegantly shown by the recent studies of Bunnett and Bernasconi[104] of the reaction between piperidine and 2,4-dinitro-phenol ethers. This reaction, to give 2,4-dinitrophenylpiperidine, may be catalysed by bases including hydroxide ion, giving a rate expression shown in equation (29). With good leaving groups attached to the

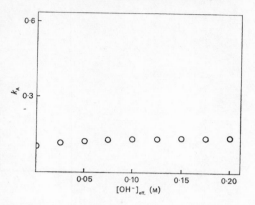

FIGURE 1. Reaction of 2,4-dinitrophenyl 4-nitrophenyl ether with piperidine, catalysed by NaOH.

$$\text{Rate} = k_n[\text{ether}][\text{piperidine}] + k_{OH^-}[\text{ether}][\text{piperidine}][OH^-] \quad (29)$$

benzene ring, such as 2,4-dinitrophenoxy, for which the rate-determining step is expected to be the formation of the tetrahedral intermediate, the reaction is not base catalysed (see Figure 1). With poor leaving groups, such as phenoxy, hydroxide ion catalysis is detected at low pH, consistent with a rate-determining breakdown of the intermediate catalysed by hydroxide. At higher pH, however, the rate of the hydroxide ion catalysed step increases until it is no longer rate

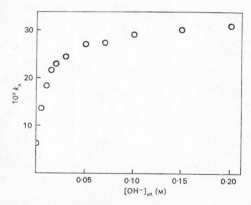

FIGURE 2. Reaction of 2,4-dinitrophenyl phenyl ether with piperidine, catalysed by NaOH.

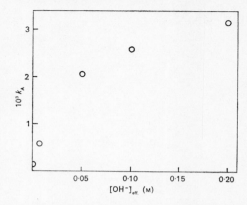

FIGURE 3. Reaction of 2,4-dinitroanisole with piperidine, catalysed by
NaOH.

determining and the rate of reaction becomes independent of hydroxide ion concentration (see Figure 2). The slow step is then formation of
the tetrahedral intermediate. With an extremely poor leaving group,
such as methoxy, the shift in rate-determining step occurs at much
higher hydroxide ion concentrations (see Figure 3).

Further confirmation for a two-step process comes from studies of
oxygen isotope effects in the arylation of piperidine with 2,4-dinitro-
phenyl phenyl ether catalysed by hydroxide ion[105] (Scheme 7). At

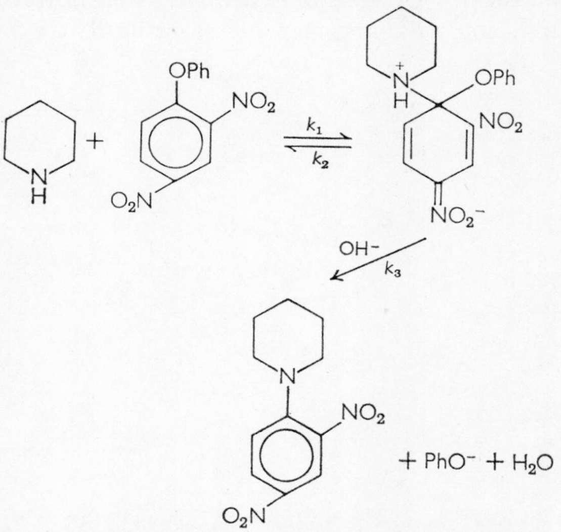

SCHEME 7. Reaction between piperidine and phenyl ethers.

low pH the rate of reaction is governed by the catalysed step k_3 and most of the intermediate formed breaks down to give reactants again. The rate-determining step, then, involves the breaking of a carbon–oxygen bond to liberate phenoxide ion, and the overall reaction displays an oxygen isotope effect, i.e. compounds containing oxygen-16 react slightly more rapidly than those containing oxygen-18. At high pH, where formation of the intermediate is the slow step and carbon–oxygen bond fission is not involved, the isotope effect is closer to unity. At intermediate pH's the isotope effect is intermediate in value. The experimental results are shown in Table 3.

TABLE 3. Reaction between piperidine and 2,4-dinitrophenyl phenyl ether.

[NaOH](M)	k^{16}/k^{18}
0·005	1·0109
0·033	1·0070
0·149	1·0024

In accord with the two-step process, it can be predicted that the rate of arylation will be largely independent of the displaced group in those cases where formation of the intermediate is rate determining. This situation has been realised for the reaction between piperidine and 1-substituted 2,4-dinitrobenzenes in methanol[106]. The data in Table 4 show that for the last six compounds, the rates are only slightly dependent on the leaving group, i.e. the 1-substituent. As expected, none of these reactions is subject to general base catalysis.

TABLE 4. Reaction between piperidine and 1-substituted 2,4-dinitrobenzenes.

1-Substituent	k_{obs}(l/mole min)
—F	90
—NO_2	24·2
—$OSO_2C_6H_4Me$	2·72
—SOPh	0·129
—Br	0·188
—Cl	0·117
—SO_2Ph	0·086
—$OC_6H_4NO_2$	0·081
—I	0·027

The steric bulk of the amine also influences the slow step. Thus no base catalysis is found in the reaction between 2,4-dinitrofluorobenzene (DNFB) and aniline[107]. However, with N-methylaniline general base catalysis can be detected by the presence of a third-order term in the rate expression (equation 30). This difference too may be explained

$$\text{Rate} = k_n[\text{PhNHMe}][\text{DNFB}] + k_{gb}[\text{PhNHMe}]^2[\text{DNFB}] \quad (30)$$

most satisfactorily in terms of a two-step process. Under pseudo-first-order conditions, i.e. excess amine, the observed rate constant for the catalysed process (k_{cat}) is given by equation (31). If k_4 [base] $\gg k_2$

$$k_{cat} = \frac{k_1 k_4[\text{base}]}{k_2 + k_4[\text{base}]} \quad (31)$$

(i.e. most of the intermediate breaks down to give products and little returns to reactants) then k_{cat} equals k_1 and no catalysis is observed. This applies in the case of unsubstituted aniline. However, for steric reasons, methyl substitution increases the size of k_2, making $k_2 > k_4$ [base] so that, in this instance, k_{cat} equals $k_1 k_4[\text{base}]/k_2$ and general base catalysis is observed. The small difference in the size of the amines reveals the delicate balance between rate-controlling formation and breakdown of the tetrahedral intermediate.

The exact nature of base catalysis is still not clear[108]. The most probable explanation is that it results from protonation of the leaving group by the conjugate acid of the base, formed in a rapid equilibrium step subsequent to formation of the tetrahedral intermediate (Scheme 8). Protonation will improve the leaving tendency of the displaced

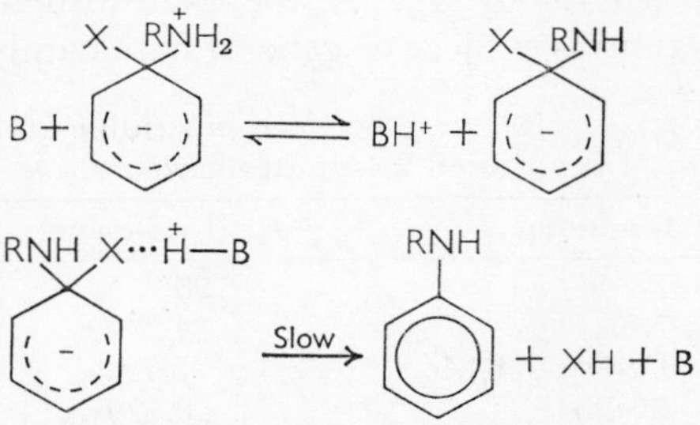

SCHEME 8.   Base catalysis in the arylation of amines.

group (X) and hence catalyse the reaction. A similar mechanism has been advanced for the reaction between esters and amines.

D. Reductive Alkylation

Ketones and aldehydes readily react with primary amines to give imines (equation 32). The first step must involve substitution on

$$RNH_2 + O{=}CR_2 \rightleftharpoons RNH{-}\underset{\underset{OH}{|}}{CR_2} \rightleftharpoons RN{=}CR_2 + H_2O \qquad (32)$$

$$(\mathbf{11})$$

nitrogen by the activated carbonyl compound to form a carbinol-amine (**11**), but dehydration occurs so readily that only with reactive aldehydes[109] is it possible to stop the reaction at this stage.

Imine formation is an equilibrium process[110] and removal of the water produced is necessary for the complete reaction. Addition of acid reduces the amount of product formed, by protonating the more basic amine and displacing the equilibrium to the left. On the other hand, acids catalyse the *rate* of imine formation by accelerating dehydration of the intermediate carbinolamine. This step is slow in neutral solution but, on raising the acidity, rapidly increases until formation of the intermediate carbinolamine becomes rate determining. This change in the rate-determining step results in a rate maximum at a pH of about four. Secondary amines also react, often spontaneously, with carbonyl compounds to form ternary immonium salts (equation 33). Net amine alkylation can be brought about by subsequent reduction of the imine (equation 34).

$$R_2\overset{+}{N}H_2ClO_4^- + R_2C{=}O \longrightarrow R_2\overset{+}{N}{=}CR_2ClO_4^- \qquad (33)$$

$$RCH{=}NH \xrightarrow{\;H_2\;} RCH_2NH_2 \qquad (34)$$

Primary amines cannot be usefully prepared, by reductive alkylation, from lower members of the aliphatic aldehydes and ammonia owing to side-reactions (for instance, formaldehyde and ammonia give hexamethylenetetramine). For aldehydes with more than five carbon atoms, however, conversion to the corresponding amine gives 60% yields[111]. Even better yields are reported for aromatic aldehydes, presumably because of the greater ease of imine formation[112]. With an excess of aromatic aldehyde the reaction with ammonia invariably leads to formation of the secondary amine[113], but with aliphatic compounds a mixture of primary and secondary amines is generally obtained. Alternatively, secondary amines can be prepared directly from primary amines (equation 35).

$$RNH_2 + RCHO \longrightarrow RN{=}CHR + H_2O$$

$$\Big\downarrow H_2 \qquad\qquad (35)$$

$$RNHCH_2R$$

There have been few detailed mechanistic studies of reductive alkylation, as the intermediacy of an imine is such an obvious route. However, Klebanskii and Vlesova[114] have compared the rate of amine production, by reduction with hydrogen over nickel, from an isolated imine with that from a mixture of amine and aldehyde. Rather surprisingly the latter was faster but imine formation during the reaction was detected by i.r. spectroscopy and by polarography. This unexpected result has been explained by a rate-determining absorption of the aldehyde and amine onto the heterogeneous catalyst, preceding the chemical transformations. In accord with this conclusion, when zinc and hydrochloric acid are used as the reducing agent, the imine reacts faster than a mixture of amine and aldehyde and formation of the imine is therefore rate controlling.

Alkylated amines may also be prepared by the reaction of an amine with alcohol in the presence of Raney nickel[115]. The initial step here seems to be dehydrogenation of the alcohol to an aldehyde, followed by reductive alkylation (equation 36)[116, 117]. The reaction of aniline

$$RCH_2OH \longrightarrow RCHO \xrightarrow{RNH_2} RCH{=}NR \xrightarrow{H_2} RCH_2NHR \qquad (36)$$

with benzyl alcohol in the presence of a nickel catalyst is facilitated by the presence of potassium benzylate[118]. The sequence of reactions is the same, except that reduction is brought about by benzyl alcohol and potassium benzylate (equation 37). Thorium oxide will also

$$PhCH{=}NPh + PhCH_2OH \xrightarrow{PhCH_2OK} PhCH_2NHPh + PhCHO \qquad (37)$$

catalyse the reaction between amines and alcohols in the gaseous phase but there are many side-reactions[119]. Tertiary amines are readily prepared from secondary amines by reaction with esters in the presence of lithium aluminium hydride[120]. An amide is formed by reaction between the ester and amine, which on reduction by lithium aluminium hydride, gives a tertiary amine.

E. Leuckart Reaction

The original synthesis discovered by Leuckart[121] was the reductive alkylation of benzaldehyde to form benzylamine using formamide or ammonium formate as the reducing agent. This was later extended by Wallach[122] and by the Eschweiler–Clarke procedure[123]. The scope of the Leuckart reaction has been reviewed by Moore[124].

Apart from the special case of formaldehyde, the reaction is most facile with aromatic aldehydes and water-insoluble ketones. Aliphatic aldehydes either do not react or produce tars[125]. Nuclear substitution in aromatic aldehydes tends to reduce their reactivity. In reactions involving aryl alkyl ketones the yields of alkylated amines are found to decrease as the size of the alkyl group increases[126].

A full discussion of the Leuckart and related reactions is outside the scope of this review and only those involving alkylation of an amine by a ketone in the presence of formic acid will be considered. The ease of imine formation suggests that condensation and elimination of water are the first steps (equation 38). The imine produced is then reduced

$$R_2C{=}O + RNH_2 \longrightarrow R_2C{=}NR \xrightarrow{HCO_2H} R_2CHNHR + CO_2 \qquad (38)$$

to an alkylated amine by the formic acid present (equation 38). Unfortunately there is no firm kinetic evidence to support this suggestion. However, increased yields are obtained by using the isolated imine[127]. The reducing action of formic acid probably occurs via the formation of an ester (equation 39) and Lukasiewicz[128] has produced

$$R_2C{=}NR + HCO_2H \longrightarrow R_2CNHR$$

$$\underset{H}{\overset{O\diagdown C{=}O}{|}} \qquad (39)$$

evidence to suggest that the decarboxylation of this ester to give the alkylated amine is a free-radical process.

The Mannich reaction between an aldehyde (usually formaldehyde) and a primary or secondary amine in the presence of a third substance having an active hydrogen, e.g. ketones, nitroalkanes, malonic esters, bears close similarity to the Leuckart reaction (equation 40). The net

$$R_2NH + HCHO + CH_3NO_2 \longrightarrow R_2NCH_2CH_2NO_2 + H_2O \qquad (40)$$

result is substitution of the active hydrogen by $R_2NCH_2{-}$. The reaction of dimethylamine and formaldehyde with ethylmalonic acid has been investigated mechanistically by Alexander and Underhill[129] (equation 41). This reaction is acid catalysed and is first order in

$$EtCH(CO_2H)_2 + HCHO + Me_2NH \longrightarrow Et\underset{CO_2H}{\overset{CO_2H}{|}}C{-}CH_2NMe_2 + H_2O \qquad (41)$$

each of the reactants. The pH-rate profile shows a maximum at pH 3·8. The fall off in rate at lower pH probably arises from a decrease in the concentration of reactive unprotonated amine; at higher pH the rate decreases owing to a reduction of acid catalysis in the rate-determining step. The first step in the reaction is substitution by formaldehyde on the amino nitrogen to give a carbinolamine (Scheme 9). This is followed by an acid-catalysed reaction between

$$Me_2NH + HCHO \rightleftharpoons Me_2NCH_2OH$$

$$Me_2NCH_2OH + HCEt \overset{CO_2H}{\underset{CO_2H}{\diagup}} \longrightarrow Me_2NCH_2CEt(CO_2H)_2 + H_2O$$

SCHEME 9. Mechanism of the Mannich reaction.

the carbinolamine and malonic acid, which is the rate-determining step. This scheme nicely explains the third-order kinetics. The exact nature of the acid-catalysed process is not clear, but probably arises from protonation of the intermediate carbinolamine.

The full scope of the Mannich reaction has been recently reviewed[130].

IV. HALOGENATION

Halogenation of an amine may occur either on carbon or on nitrogen, the former reaction being outside the scope of this review.

In many solvents N-halogenation by molecular halogen seems to be unimportant compared with C-halogenation; not because in principle it cannot occur, but because under the conditions of the reaction the N-halogen compound is converted back to halogen and amine, whereas C-halogenation is non-reversible.

However, hypochlorous acid and the hypochlorite ion are far better N- than C-chlorinating agents and react readily with primary, secondary and tertiary amines. The reason for this is not entirely clear. Chlorination of amines using sodium hypochlorite was discovered by Hofmann in 1881[131, 132] and the method was extended by Coleman[133, 134]. N,N-dichloroamine is obtained if excess hypochlorite is used. A solution of hypochlorous acid in ether reacts with aniline at −20° to give N,N-dichloroaniline[135]. Under different conditions, in the presence of hydrochloric acid at room temperature, only nuclear substitution occurs[136]. Ethyl hypochlorite[137], as well as t-butyl hypochlorite[138], have also been used for the N-chlorination of amines. N-Bromoamines are made in a corresponding way, using sodium hypobromite.

There have been no detailed investigations of the mechanism of amine halogenation by hypohalite. In chlorination the reactive species could be either hypochlorous acid or the hypochlorite ion, both of which are inefficient in C-chlorination. Mauger and Soper[139] have suggested that there is hydrogen bonding between the amine and the hypochlorite ion. Esters of hypochlorous acid, such as t-butyl hypochlorite, can react in a variety of ways, leading to both C- and N-substitution, but the conditions favouring N-chlorination have not been clearly defined. It seems unlikely that chlorination is brought about by free chlorine present in the hypochlorite solution as this should result in nuclear substitution. The active species in bromination must be the corresponding hypobromite ion.

A rather different method of preparation employs succinimides. N-Chloroamines result from the action of N-chlorosuccinimide in ether on an amine, and N-bromoamines are obtained in a corresponding manner from N-bromosuccinimide or N-bromophthalimide[140]. The action of N-bromo- and N-chlorosuccinimide is difficult to understand. A recent view[140a] of the action of N-bromosuccinimide suggests that it provides molecular bromine, and hence bromine atoms, at low concentration. It is not clear why this should bring about preferential N-bromination, but it is possible that an entirely different mechanism is operative in N-substitution. The action of N-chlorosuccinimides is so little understood that speculation about the course of N-chlorination is without significant value. Chloramine-T is an effective chlorinating reagent for amines[141]. Sulphuryl chloride reacts with p-aminophenol to give 4-hydroxy-2,3,5,6-tetrachlorophenyldichloroamine[142].

Rather special procedures are used in the preparation of fluoroamines. Anodic fluorination of methylamine in liquid hydrogen fluoride leads to N,N-difluorotrifluoromethylamine (equation 42)[143].

$$MeNH_2 \xrightarrow{\;F\;} CF_3NF_2 \tag{42}$$

Jet fluorination[144] of methylamine leads to a variety of products, the main one being the completely fluorinated compound, and a free-radical mechanism has been suggested. Reaction of alkyl iodides or azo compounds with perfluorohydrazine gives N,N-difluoroamines in which the alkyl group is unfluorinated[145] (equation 43). The mecha-

$$RI \xrightarrow{\;h\nu\;} R^\bullet \xrightarrow{\;F_2N^\bullet\;} RNF_2 \tag{43}$$

nism is almost certainly free radical. A small yield of difluoromethylamine is produced by the action of fluorine on an aqueous solution of N-methyl carbamate[146].

Aliphatic mono- and dichloroamines are unstable oils, insoluble in water; generally they are explosive, but a few may be distilled.

Aromatic N-chloroamines are unknown and the dichloro compounds are unstable and rapidly convert to ring-substituted products. N, N-Dichloroaniline explodes if removed from the freezing mixture. Electron-repelling substituents in the benzene ring increase the stability of dichloroamines: thus pentachloro-N, N-dichloroaniline is stable at room temperature in the absence of water, while 4-hydroxy-2,3,5,6-tetrachloro-N, N-dichloroaniline may be steam distilled. All the known N-bromo compounds are explosive, or at least, unstable. N,N-Difluorotrifluoromethylamine is an unstable liquid boiling at $-78°$ to $-75°$. The iodoamines are mostly solids, with colours ranging from sulphur-yellow for $(CH_3)_2NI$ and garnet-red for CH_3NI_2 to blue-black for $C_2H_5NI_2$.

N-Halogenation of tertiary amines occurs readily, but in aqueous solution one or two of the alkyl groups are eliminated as aldehydes[147]. The usual chlorinating agents are hypochlorous acid or calcium hypochlorite while N-bromination can be brought about by molecular bromine. The first product is an N-halo quaternary halide (equation 44). The chloro compounds have also been prepared by Böhme[148, 149]

$$Me_3N + Br_2 \longrightarrow Me_3\overset{+}{N}Br + Br^- \tag{44}$$

by the action of chlorine on an amine in carbon tetrachloride. Spectroscopic evidence for their production by the action of hypochlorous acid on an amine in aqueous solution was obtained by Ellis and Soper[150]. The subsequent course of the reaction is not certain but loss of a proton from the cation, followed by hydrolysis, is a reasonable way of explaining the production of aldehydes (Scheme 10). Addition

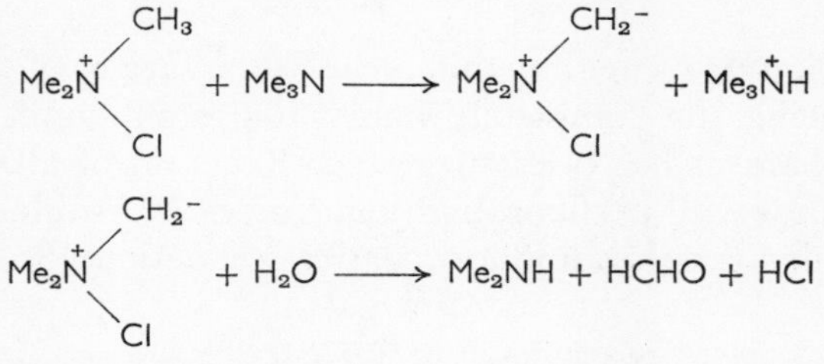

SCHEME 10. Chlorination of trimethylamine.

of chloride is thought to lower the rate of reaction by reducing the concentration of the reactive species $Me_3\overset{+}{N}Cl$. The secondary amine formed is then chlorinated to give N-chloroamine.

V. NITROSATION (DIAZOTISATION)

The various products obtained with nitrous acid can be regarded as intermediates along a common reaction path represented by equation (45). At low temperatures, for instance, both aromatic primary nitros-

$$RNH_2 \underset{(a)}{\overset{Slow}{\rightleftharpoons}} RNHNO \xrightarrow{(b)} RN_2{}^+ \xrightarrow{(c)} R^+ \xrightarrow{(d)} \text{Deamination products}$$

(45)

amines[151] and primary aliphatic diazonium ions[152] can be isolated from an ethereal solution of the amine and nitrosyl chloride. Except for diazotisation in very concentrated acids, as discussed later, the initial nitrosation reaction (step (a) in equation 45) appears to be rate controlling whatever subsequent reactions occur. Thus Kalatzis and Ridd[153] have shown recently that identical kinetic equations and remarkably similar rates apply to nitrosamine formation from N-methylaniline, as to the diazotisation of aniline, and many of the kinetic equations for diazotisation have exact counterparts in aliphatic deamination. This means that the initial task in understanding the mechanism for both primary and secondary amines is the common one of N-nitrosation.

Mechanistic studies of N-nitrosation in aqueous solutions are complicated, however, by the existence of several nitrous species in equilibrium with molecular nitrous acid, many of which are effective nitrosating reagents (a situation similar to that for aromatic nitration)[154]. Some of the reagents formed in aqueous solution are listed below in order of increasing activity. In the presence of basic

ON—NO$_2$	nitrous anhydride
ON—Hal	nitrosyl halide
ON—OH$_2$$^+$	nitrous acidium ion
ON$^+$	nitrosonium ion

anions (X^-) additional reagents with the general formula NO—X are also formed. Nitrous acid itself is notably absent for it appears that the molecular form of this acid does not react directly with amines.

To simplify the discussion, evidence for each reagent is treated separately, although under normal conditions reaction usually occurs simultaneously via more than one reagent. The effect of solvent acidity is also dealt with separately, and nitrosation by other reagents and the reactions of tertiary amines are also discussed briefly.

Early work on the nitrosation of amines forms an interesting part of the historical development of mechanistic organic chemistry, but the reader is referred elsewhere for these details[155]. Several recent reviews have dealt with the mechanism of N-nitrosation in aqueous solution[156], the effect of solvent acidity on diazotisation[157] and the nitrosation of secondary amines[158].

A. Nitrous Anhydride

Hammett[159] was the first to suggest this species as a reagent to explain the 'anomalous' second-order dependence on nitrous acid concentration reported earlier for diazotisation[160] and for the de-amination of both methylamine[161] and diethylamine[162] (equation 46). Hammett suggested that nitrous anhydride was formed in a

$$\text{Rate} = k[\text{HNO}_2]^2[\text{RNH}_2] \tag{46}$$

rapid preequilibrium step (equation 47) and then slowly attacked the

$$2\,\text{HNO}_2 \overset{\text{Fast}}{\rightleftharpoons} \text{N}_2\text{O}_3 + \text{H}_2\text{O} \tag{47}$$

unprotonated amine to form an ammonium ion intermediate as shown in Scheme (11). The elegant kinetic investigations of Hughes, Ingold

$$\text{RNH}_2 + \text{N}_2\text{O}_3 \xrightarrow{\text{Slow}} \text{R}\overset{+}{\text{N}}\text{H}_2\text{NO} + \text{NO}_2{}^-$$

$$\text{R}\overset{+}{\text{N}}\text{H}_2\text{NO} \xrightarrow{\text{Fast}} \text{RN}_2{}^+ \xrightarrow{\text{Fast}} \text{Deamination products}$$

SCHEME 11.

and Ridd[163, 164] confirmed Hammetts suggestion and simultaneously resolved earlier, apparently conflicting, observations. They found, like Schmid[160], that the rate of diazotisation of basic aromatic amines, such as aniline, in dilute acidic solutions (about 0·01 M) followed the overall third-order equation (46). On decreasing the acidity, however, the reaction rate progressively became independent of the amine concentration and eventually followed the second-order rate law given by equation (48)[163, 164]. Earlier workers[165] had assumed

$$\text{Rate} = k[\text{HNO}_2]^2 \tag{48}$$

incorrectly a first-order dependence each on amine and nitrous acid concentrations. Equation (48) refers to conditions where the formation of nitrous anhydride (equation 47) is rate controlling, because the increased concentration of highly nucleophilic amine base at lower acidities is sufficient to react with the nitrous anhydride as soon as it forms[163, 166]. Other evidence has accumulated to confirm the validity of this conclusion. Studies of the exchange of oxygen-18 between nitrous acid and water, which occurs mainly* via the hydrolysis and the formation of nitrous anhydride (equation 49), show

$$N_2O_3 + H_2{}^{18}O \rightleftharpoons HNO_2 + HNO^{18}O \qquad (49)$$

that it also follows equation (48) with a rate comparable to that for aniline diazotisation under similar experimental conditions[167,168]. Thus diazotisation and oxygen exchange must be controlled by the same rate-determining step—the formation of nitrous anhydride from two molecules of nitrous acid.

These results effectively exclude Kenner's[170] suggestion that third-order nitrosation corresponds to slow proton loss from a tetrahedral intermediate (12), with nitrite ion acting as a base (Scheme 12).

$$RNH_2 + H_2ONO^+ \xrightarrow{\text{Fast}} R\overset{+}{N}H_2NO + H_2O$$

$$(12)$$

$$R\overset{+}{N}H_2NO + NO_2{}^- \xrightarrow{\text{Slow}} RNHNO + HNO_2$$

$$RNHNO \xrightarrow{\text{Fast}} \text{Products}$$

SCHEME 12.

This mechanism is similar to that proposed for the acylation of amines and should be subject to general base catalysis, which is not the case[171].

Further studies have shown that nitrous anhydride is a very common reagent in diazotisation[172,173], deamination[174] and nitrosamine formation[153]. Ridd[156] has summarised the rate data for the reaction of several amines with nitrous anhydride and these are given with one or two additions in Table 5. As expected for a reaction in which the rate-determining step is nucleophilic attack by the amine base on nitrous anhydride, electron-withdrawing substituents decrease the diazotisation rate, and nitrous anhydride does not react with highly deactivated amines such as p-nitroaniline[176] and 2,4-dinitroaniline[176]. Table 5 shows that there is a fair correlation between reactivity and basicity within each class of amine[173], although aliphatic amines appear to be less reactive than their high basicity would suggest. The reason for this difference is not clearly understood. Also N-methyl substitution has a greater effect on the rate of nitrosation than on the basicity of aniline, suggesting that steric factors are important.

---

* One other mechanism[169] is known to become important when the nitrite concentration is very low.

11+C.A.G.

TABLE 5. Amine nitrosation by nitrous anhydride.

Amine	pK_A	$10^{-5}k$ (equation 46)[a]	Temperature (°c)	References
Aliphatic				
Dimethylamine	10·73	4·0	25	162, 175
Propylamine	10·69	2·8	25	162, 175
Methylamine	10·68	4·8	25	161, 175
(Ammonia)	9·24	0·04	25	161, 175
Substituted Anilines				
p-CH$_3$	5·08	58·5	25	173
m-CH$_3$	4·72	24·3	25	173
H	4·61	23·0	25	173, 175
o-CH$_3$	4·42	12·6	25	173
p-Cl	4·10	8·4	25	173
m-Cl	3·50	2·9	25	173
o-Cl	2·65	0·42	25	173
p-CH$_3$O	5·34	5·56	0	172
H	4·61	3·11	0	163
p-Cl	4·10	0·92	0	172
p-Me$_3$N$^+$	—	0·14	0	172
N-Methylaniline	4·85	12	0	153

[a] Rate coefficients from equation (46) in $l^2/mole^2$ sec.

B. Nitrosyl Halides

Many covalent nitrosyl compounds are effective nitrosating reagents in organic solvents. It is therefore not surprising to find that even low concentrations of nucleophilic anions promote nitrosation in aqueous solutions by forming nitrosyl compounds other than nitrous anhydride (nitrosyl nitrate). Most of the mechanistic evidence for these reagents comes from diazotisation studies either in halogen acids or catalysed by halide ions. We shall therefore confine our discussion to the reactions of nitrosyl halides.

Schmid[177] was the first to show that the expression for the rate of diazotisation of aniline in halo acids required another kinetic term, given by equation (50), as well as the term for reaction by nitrous

$$\text{Rate} = k[\text{ArNH}_2][\text{H}^+][\text{HNO}_2][\text{Hal}^-] \qquad (50)$$

anhydride. Once again it was Hammett[159], who noting that the concentration of nitrosyl halide was proportional to the product $[\text{H}^+][\text{HNO}_2][\text{Hal}^-]$, pointed out that equation (50) was consistent with a mechanism involving a rapid, preequilibrium formation of nitrosyl halide (NOHal), which then reacted with unprotonated amine in a slow step (Scheme 13). Hughes and Ridd[178] placed

$$HNO_2 + H^+ + Hal^- \underset{\text{Fast}}{\rightleftharpoons} NOHal + H_2O \qquad (51)$$

$$NOHal + ArNH_2 \xrightarrow{\text{Slow}} Ar\overset{+}{N}H_2NO + Hal^-$$

$$Ar\overset{+}{N}H_2NO \xrightarrow{\text{Fast}} ArN_2^+ + H_2O$$

SCHEME 13.

Hammett's suggestion goes beyond the realms of speculation by demonstrating that the formation of nitrosyl halide (equation 51) could be made rate controlling for bromide and iodide ion-catalysed diazotisation in perchloric acid. The rate under these conditions was found to be independent of the amine concentration and followed equation (52). They suggested[178] that the nitrosyl halide was formed

$$\text{Rate} = k[\text{H}^+]^2[\text{HNO}_2][\text{Hal}^-] \qquad (52)$$

by the reaction of halide ion with the nitrous acidium ion (H_2ONO^+)[166]. A comparable slow formation of nitrosyl chloride has not been realised in either deamination or diazotisation, but has been observed in the analogous nitrosation of the highly reactive azide ion in the presence of added chloride[179].

With low concentrations of nitrous acid, below 0·01 M, nitrosyl chloride becomes the dominant reagent for the diazotisation of aniline in aqueous hydrochloric acid stronger than 0·1 M[177b]. Since the concentration of nitrosyl chloride is proportional to the square of the hydrochloric acid concentration (see equation 52), the rate increases rapidly with acidity and should reach a maximum in 5 M hydrochloric acid when all the nitrous acid is converted to nitrosyl chloride[180]. This effect has been observed for the deamination of ammonia[181a] and of glycine[181b], but not for diazotisation itself.

Deamination and diazotisation on a preparative scale are frequently carried out in organic solvents with nitrosyl halides[182]. The mechanism of these reactions is probably similar to those in aqueous solutions, and recent studies in methanolic hydrochloric acid solvents show that diazotisation follows equation (50)[183,184]. The rates generally, however, are considerably slower than in aqueous solution[183] and this difference has been attributed to methanolysis lowering the concentration of nitrosyl chloride (equation 53)[184]. The simplest deamination

$$NOCl + MeOH \rightleftharpoons MeONO + HCl \qquad (53)$$

products from aliphatic amines are obtained with nitrosyl halides. At low temperatures, unstable primary nitrosamines are formed[152, 153], but the amino group is replaced by halide on raising the temperature[185]. A similar replacement occurs with aromatic amines in the Gatterman[186a] and Sandmeyer[186b] syntheses.

Nitrosyl halides, at least in aqueous solutions, appear to be more potent reagents than nitrous anhydride. Schmid and his coworkers have calculated *actual* rate coefficients for the reaction of unprotonated amine with both molecular nitrosyl chloride[175, 187, 188] and nitrosyl bromide[175, 189] (i.e. from the equation, Rate = $k[RNH_2][NOHal]$). These are listed in Table 6. The characteristic features of these rate

TABLE 6.   Amine nitrosation by nitrosyl halides at 25°C.

Amine	pK_A	$10^{-9}k^a$	Reference
Nitrosyl chloride			
Aliphatic			
Ammonia	9·24	0·05	175
Glycine	9·78	0·017	175
Substituted anilines			
p-CH$_3$	5·08	3·0	187
m-CH$_3$	4·72	2·70	187
H	4·61	2·60	187
o-CH$_3$	4·42	2·44	187
p-Cl	4·10	1·89	187
m-Cl	3·50	1·63	187
o-Cl	2·65	1·16	187
Nitrosyl bromide			
Aniline	4·61	3·2	189
Ammonia	9·24	0·032	175

^{a} Rate coefficients from the equation, Rate = $k[RNH_2][NOHal]$, in l/mole sec.

coefficients are firstly a remarkable independence from the amine basicity[187], and secondly, a magnitude in close proximity to the rate of molecular encounters in solution (about 10^{10} l/mole sec at 25°)[190]. Activation energies for these reactions are all of the order of 4 to 5 kcal/mole[188, 189], which is also about that expected for a diffusion-controlled reaction[190]. It therefore appears that nitrosyl halides are sufficiently powerful electrophiles to react on encounter with all but the most unreactive amino bases.

C. Nitrous Acidium Ion

Acid-catalysed nitrosation in the absence of halide ions may involve the nitrous acidium ion (H_2ONO^+). The diazotisation rate of aniline in aqueous perchloric or sulphuric acid at first decreases and then increases with rising acidity[157]. This implies a change in mechanism, because the reaction rate for nitrous anhydride itself should steadily decrease owing to increasing protonation of the amine.

In perchloric acid concentrations up to 0·5 M, unreactive aromatic amines diazotise in accordance with equation (54)[191]. Hughes,

$$\text{Rate} = k[ArNH_2][HNO_2][H^+] \tag{54}$$

Ingold and Ridd[166] were the first to interpret this result as evidence for rapid formation of the nitrous acidium ion, which then reacts with the unprotonated amine in a rate-determining step (Scheme 14).

$$HNO_2 + H^+ \xrightarrow{\text{Fast}} H_2ONO^+$$

$$ArNH_2 + H_2ONO^+ \xrightarrow{\text{Slow}} ArNH_2NO^+ + H_2O$$

$$ArNH_2NO^+ \xrightarrow{\text{Fast}} ArN_2^+ + H_2O$$

SCHEME 14.

Other interpretations were considered, but rejected[166], including the one involving slow nitrosation by the nitrosonium ion (NO^+), although this species is formed readily in strongly acidic solutions.

The nitrous acidium ion must be a very reactive species, for it readily reacts with nitroanilines[176] whereas nitrous anhydride does not. Larkworthy[176] has determined the rate coefficients from equation (54) for a few substituted anilines and these are listed in Table 7.

TABLE 7.   Aniline nitrosation by nitrous acidium ion at 0°c.

Substituted aniline	pK_A	k (equation 54)[a]	Reference
o-Cl	2·65	175	176
p-NO$_2$	1·00	161	176
o-NO$_2$	−0·26	145	176
2,4-(NO$_2$)$_2$	−4·53	3·7	176

[a] Rate coefficients from equation (54) in $l^2/mole^2$ sec.

Actual rate coefficients for the reaction of the nitrous acidium ion with the unprotonated amine cannot be computed, because the required equilibrium constants are unknown. However, nitrosyl chloride must react with many aromatic amines on encounter (Section V.B) and the nitrous acidium ion should be even more reactive. This may account for the similar diazotisation rates of the most basic amines recorded in Table 7. Aliphatic deamination by the nitrous acidium ion has not been reported, presumably because most aliphatic amines are completely protonated and therefore unreactive under the acidic conditions necessary for the formation of this reagent.

D. Acid-catalysed Nitrosation

Increasing solvent acidity favours the formation of the nitrosonium ion (NO^+)[192], which should be a particularly powerful reagent by analogy with nitration by the nitronium ion (NO_2^+). The other important effect of higher solvent acidity is to reduce the concentration of reactive, unprotonated amine. We noted above that with aliphatic amines protonation is so complete that deamination is not observed at acidities above 0.1 M. With aromatic amines, the situation is different, partly because of their lower basicity, but also from the intervention of a new mechanism for N-nitrosation, apparently involving the protonated amine. The importance of this new mechanism depends on the basicity of the aromatic amine.

The diazotisation rates of weakly basic amines, such as p-nitroaniline, increase with perchloric acid concentration as is evident from the data in Table 8[193]. This catalysis has been associated with a salt

TABLE 8. Effect of perchloric acid on the rate of diazotisation at $0°$C.

	$\bar{k}$ (l/mole sec)[a]						
Perchloric Acid (M)	0·5	1·0	2·0	3·0	4·0	5·0	6·0
Amine							
p-Nitroaniline[193]	16·1	24	51·5	92·5	—	—	—
Aniline[197, 198]	0·025	0·081	0·49	2·60	12·3[b]	57·5[b]	263[b]
p-Toluidine[197, 198]	0·063	0·285	2·45	15·0	123[b]	575[b]	—

[a] Rate coefficient calculated from the stoichiometric reactant concentrations present, i.e. Rate = $\bar{k}$ [Amine][Nitrous Acid].
[b] At a temperature of $2°$C.

effect rather than an increase in the hydronium ion activity, because neutral perchlorate salts produce an almost equivalent increase[194]. An alternative suggestion that reaction occurs via nitrosyl perchlorate[173, 195] seems most unlikely, as this compound is ionic and not covalent under the experimental conditions[192]. Also, the perchlorate ion catalysis is of an exponential character[194]. The salt effect of perchloric acid can be nullified if the ionic strength is kept constant by the addition of sodium perchlorate, and the diazotisation of p-nitro-aniline then follows equation (55), where h_0 refers to the Hammett

$$\text{Rate} = k[\text{ArNH}_2][\text{HNO}_2]h_0 \qquad (55)$$

acidity function[193]. The net rate of reaction is essentially independent of the acidity, because any increase in h_0 is offset by a corresponding decrease in the free-amine concentration. Despite current difficulties in understanding the exact significance of acidity-function correlations in general[196], equation (55) has been regarded as a logical extension of equation (54) to concentrated acids[193]. This implies an unchanged mechanism from that operating at lower acidities, which involves a rate-determining attack by the nitrous acidium on the unprotonated amine. However, the precise character of the perchlorate ion catalysis has still to be resolved. It may be a secondary effect increasing the concentration of nitrous acidium ion, and the observation that neutral perchlorates similarly catalyse the *rate* of formation of nitrous anhydride is consistent with this interpretation[194]; but a formation of the potentially more reactive nitrosonium ion cannot be excluded with certainty.

With very basic amines, the catalytic effect of perchloric acid is much stronger[173, 195, 197] than that for p-nitroaniline. For example, the rate coefficients for aniline and p-toluidine listed in Table 8 increase by a factor of 10^4 in going from 0·5 M to 6 M perchloric acid[198]. Part of this increase can be associated with the ionic strength effect noted above for feebly basic amines, but the bulk arises from the incursion of a new kinetic path, which follows equation (56)[197, 198], in addition to the reaction following equation (55).

$$\text{Rate} = k[\text{ArNH}_3{}^+][\text{HNO}_2]h_0 \qquad (56)$$

The details of this new mechanism have been examined only recently. It is certain that the initial N-nitrosation is again rate determining, because equation (56) is also observed in nitrosamine formation from N-methylaniline under similar experimental conditions[153]. Equations (55) and (56) differ only in their dependence on the amine moiety; the nitrosating reagent is probably the same positively charged species in both cases, but for equation (55) the rate-determining process is attack on the free amine, for equation (56) on the protonated amine[197,198]. The relative contributions from the two paths to the overall rate depends on the solvent acidity and the amine basicity. Reaction occurs much more readily through the free amine[198], but if the amine is almost entirely protonated, the interaction of the nitrosating agent with the protonated amine becomes significant.

Strong evidence for this unexpected reaction between two positively charged ions comes from recent studies of substituent effects[198]. These are summarised in Table 9 in which values of k (equation 56)

TABLE 9. Substituent effects for aniline nitrosation in 3 M perchloric acid at 0°C (from ref. 198).

Substituted aniline	pK_A	10^2k (equation 56)[a]
p-CH$_3$O	5·34	153
p-CH$_3$	5·08	119
m-CH$_3$	4·69	109
H	4·61	16·1
m-CH$_3$O	4·23	300
p-Cl	4·00	3·3
m-Cl	3·52	6·6
N-*Methylanilines*		
p-CH$_3$	5·36	25·5
H	4·85	4·44
p-Cl	4·25	0·905
m-Cl	3·77	2·34

[a] Rate coefficients from equation (56) in l/mole sec.

are tabulated for several primary and secondary arylamines. The most striking feature is a well-defined dependence on the structure but not on the basicity of the amine. The data for nitrosyl halides (Table 6) and the nitrous acidium ion (Table 7) show that these reactive reagents do not discriminate much between different amines and a similar result would therefore be expected for the more powerful reagents potentially available in concentrated acids. Generally speaking, however, when due allowance is made for differences in the free-amine concentrations, the effect of any substituent is in the opposite direction to that expected for a reaction of the unprotonated base[198]. Although all the mechanistic details are not entirely clear, the substituent effects in Table 9 can be accounted for by a reaction involving the rapid formation of a π complex (13) between the nitrosating reagent (X^+) and the protonated amine, followed by a slow rearrangement simultaneous to proton removal from nitrogen, as shown in Scheme (15) [197, 198].

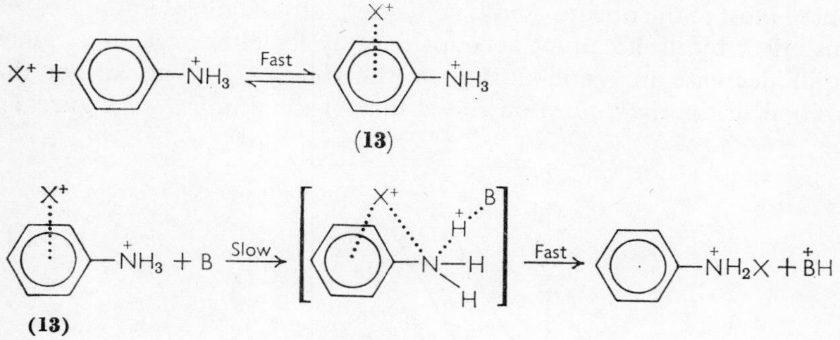

(13)

(13)

SCHEME 15.

Complexes are apparently formed between nitrous species and several aromatic compounds in concentrated acids [199], and the formation of a similar precursor in N-nitrosation may well explain why no comparable reaction has been observed with protonated aliphatic amines. Stedman and his coworkers [200], however, have obtained very convincing evidence for an analogous degradative nitrosation of protonated hydroxylamine ($NH_3{}^+OH$), and in this case the oxygen atom is an obvious basic site at which interaction can occur. Apart from this exception, N-nitrosation via the conjugate acid appears to be confined to the most basic aromatic amines.

The exact nature of the nitrosating agent in these strongly acidic solutions is also not clear; either the nitrous acidium ion or the nitrosonium ion (or even both) may be involved. In this connection, it is interesting to note that the nitrous acid deamination of benzamide in concentrated sulphuric acid has been interpreted as a reaction of the nitrosonium ion [201], although precisely the opposite conclusion has been reached for the nitrosation of phenol in concentrated perchloric acid [202].

E. Diazotisation at High Acidities

The diazotisation rate of aniline and similarly basic amines increases rapidly with perchloric acid concentration up to about 6 M [198, 203]. Further increases in acidity cause the rate to pass through a maximum and then to decrease rapidly [157, 204]. This effect is shown in Figure 4 for aniline, p-toluidine and p-chloroaniline. Throughout the region of decreasing rate, the reaction follows equation (57) [204], and

$$\text{Rate} = k[\text{ArNH}_3^+][\text{HNO}_2]h_0{}^{-2} \tag{57}$$

11*

a closely similar kinetic relationship is found in concentrated sulphuric acid (above 60%)[204]. Although the rate maxima occur at an acidity where most of the nitrous acid is converted into the nitrosonium ion[192], this effect by itself cannot account entirely for either the subsequent rapid decrease in reaction rate or the obvious dependence of the position of the maximum on the nature of the amine (see Figure 4).

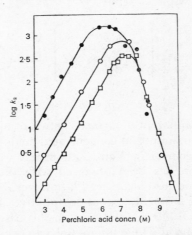

FIGURE 4. Diazotisation of aniline (○), p-chloroaniline (●) and p-toluidine (□) in concentrated perchloric acid.

Consequently the change in the acidity profile has been linked to a fundamental change in the rate-determining step. Thus unlike the situation at lower acidities, where the initial N-nitrosation is always rate controlling, the slow step at very high acidities may be proton loss from the intermediate nitrosamine ion (**14**) as shown in Scheme (16)[204].

$$NO^+ + ArNH_3^+ \xrightleftharpoons{Fast} ArNH_2NO^+ + H^+ \tag{58}$$

$$ArNH_2NO^+ \xrightarrow{Slow} ArNHNO + H^+ \tag{59}$$
$$(\mathbf{14})$$

$$ArNHNO \xrightarrow{Fast} ArN_2^+ + OH^-$$
$$\text{SCHEME 16.}$$

Most of the available experimental observations are consistent with this deduction; the reactions are almost insensitive to substituent effects (see Figure 4), and the diazotisation of aniline in concentrated sulphuric acid shows a large isotope effect for deuterium ($k_H/k_D = 10$), indicating a proton transfer in the rate-determining step[204]. Two factors probably contribute to the change in the rate-determining step as the acidity increases. One is a decrease in the rate of proton transfer to the solvent from nitrogen, the other is an increased rate of nitrosonium ion loss from the protonated nitrosamine (14), to give the reactants (equation 58). Direct studies show the rate of proton transfer from ammonium[205] and anilinium ions[206] to the solvent become very slow in concentrated acids, and nitrosamine formation is easily reversible under the same conditions[153].

Thus diazotisation in concentrated acids, like arylation under very different conditions, appears to involve slow decomposition of a tetrahedral intermediate (equation 59). The proton becomes a poor leaving group, mainly because the solvent is reluctant to accept it. A similar situation arises in the N-nitration of anilines in concentrated acids[207].

F. Miscellaneous Reagents

The diazotisation of aniline in aqueous nitric acid has also been studied. At low acidities (below 0·2 M), nitrous anhydride is the most important reagent[208]. At higher acidities the kinetics of the reaction are similar to those for equivalent concentrations of perchloric acid and the rate increases rapidly with acidity[209]. This catalysis has been attributed to nitrosation by dinitrogen tetroxide (nitrosyl nitrate)[209], but other studies of catalysis by neutral nitrate salts indicate the contribution to the overall rate of nitrosation by dinitrogen tetroxide is slight[194]. It seems, therefore, that the reagents at high acidity are the nitrous acidium and nitrosonium ions, as for perchloric acid.

Covalent acyl nitrites formed from nitrous acid and various carboxylic acids react readily with aromatic amines, and they are used often for nitrosating water-insoluble amines[210]. The mechanism must be similar to that for nitrous anhydride. In aqueous buffer solutions of carboxylic acids, however, the bulk of the reaction is brought about by nitrous anhydride itself. Acyl nitrites are formed, but these react more readily with the very basic nitrite ion, always present at pH > 3·5, than with the amine directly[164, 211].

Alkyl nitrites are useful mild reagents for diazotisation and deamination in organic solvents. In benzene, aniline is readily deaminated at elevated temperatures by these reagents and the product reacts with the solvent to form biphenyl in high yield[212]. Nitrosonium tetrafluoroborate ($NO^+BF_4^-$) also reacts readily with both aliphatic and aromatic amines in dry organic solvents[213].

Interest recently has focussed on the direct interaction between nitric oxide (NO) and various amines. Secondary and tertiary alkyl-amines undergo N-substitution at elevated temperatures to form secondary nitrosamines (equation 60). These reactions may be

$$R_2NH + NO \Big\} \longrightarrow R_2NNO \qquad (60)$$
$$R_3N + NO $$

catalysed by copper salts and have been the subject of innumerable patents[214]. Formally they can be described by a disproportionation of nitric oxide (NO) into NO^- and NO^+ (the NO^+ then nitrosates the amine in the usual way), but the mechanism is more complex and with secondary amines involves two concurrent paths[215]. Apparently the combination of cupric ion and nitric oxide is a powerful nitrosating reagent in alkaline media, and in one path (Scheme 17), this complex formed in a preequilibrium step (equation 61) reacts directly with the

$$Cu^{2+} + NO \rightleftharpoons Cu^{2+}NO \qquad (61)$$
$$Cu^{2+}NO + R_2NH \longrightarrow Cu^+ + H^+ + R_2NNO$$
$$\textbf{(15)}$$

SCHEME 17.

secondary amine to give cuprous ion and the nitrosamine $(\textbf{15})$[215]. In the other concurrent path (Scheme 18), copper acts as both a

$$R_2NH + 2NO \rightleftharpoons R_2NN_2O_2^- + H^+ \qquad (62)$$
$$R_2NN_2O_2^- + Cu^+ \longrightarrow R_2N\overset{.}{N}O^- + NO^- + Cu^{2+}$$
$$R_2N\overset{.}{N}O^- + Cu^{2+} \longrightarrow R_2NNO + Cu^+$$

SCHEME 18.

reducing and an oxidising agent[215]. The related nitrosation and de-alkylation of tertiary amines (equation 60) has not been investigated in detail, but presumably involves similar reagents. The anion $(R_2NN_2O_2{}^-)$ formed in the preequilibrium step (equation 62) has been isolated at low temperature by Drago and his coworkers[216]. They have also shown that this anion is formed in a multistage process involving an initial slow attack of nitric oxide on the amine base (Scheme 19)[217]. Secondary N-chloroamines react similarly with

$$R_2NH + NO \overset{Slow}{\rightleftharpoons} R_2NHNO$$
$$R_2NHNO + NO \xrightarrow{Fast} R_2NHN_2O_2 \xrightarrow{Fast} R_2NN_2O_2^- + H^+$$

SCHEME 19.

nitric oxide to produce secondary nitrosamines in high yield[218]. Also, Rigauldy and Vernières[219] have reported that diazonium salts precipitate immediately when nitric oxide is bubbled into an ethereal solution of either 2,4,6-triphenyl- or 2,4,6-tri-*t*-butylaniline. Thus the direct interaction of nitric oxide with amines may be a fairly general nitrosative method.

G. Tertiary Amines

Tertiary amines are generally regarded as being inert to nitrosation, and diagnostic tests are based upon this assumption. However, reaction does occur at elevated temperatures in weakly acidic media (pH > 3) with cleavage of an alkyl group to give a secondary nitrosamine (equation 63)[220]. Methyl and benzyl groups appear to be

$$R_2NCHR_2^1 \xrightarrow{HNO_2} R_2NNO + R_2^1C{=}O + H_2O \tag{63}$$

displaced more readily than other bulkier ones. With mixed tertiary alkyl aryl amines, nuclear substitution also occurs[221], probably via direct attack on the aromatic ring (see below). The mechanism of the dealkylation is not entirely understood. As the reaction does not occur at high acidities, it must involve electrophilic attack by the nitrosating reagent on the unprotonated amine, probably to form a nitroso-ammonium ion (**16**). Smith and Pars[220] have cited evidence to support a 1,2-elimination by the nitrosoammonium ion to form a dialkylimmonium ion intermediate (**17**), which then hydrolyses and reacts with another molecule of the nitrosating reagent to form the products (Scheme 20). An alternative mechanism involving homolytic

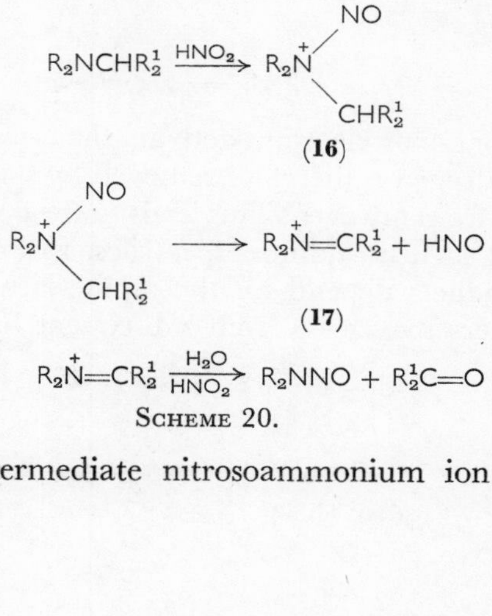

$$R_2NCHR_2^1 \xrightarrow{HNO_2} R_2\overset{+}{N}\overset{\displaystyle NO}{\underset{\displaystyle CHR_2^1}{}}$$

$$(\mathbf{16})$$

$$R_2\overset{+}{N}\overset{\displaystyle NO}{\underset{\displaystyle CHR_2^1}{}} \longrightarrow R_2\overset{+}{N}{=}CR_2^1 + HNO$$

$$(\mathbf{17})$$

$$R_2\overset{+}{N}{=}CR_2^1 \xrightarrow[HNO_2]{H_2O} R_2NNO + R_2^1C{=}O$$

SCHEME 20.

cleavage of the intermediate nitrosoammonium ion (**16**) has also been proposed[222].

At much lower temperatures (about 10°c) and in weakly acidic conditions, tertiary alkyl aryl amines appear to undergo nuclear rather than N-substitution. Thus N,N-dialkylanilines first form the p-nitroso derivative, which then reacts further to produce a p-diazonium ion (equation 64)[223]. Further discussion of these reactions is

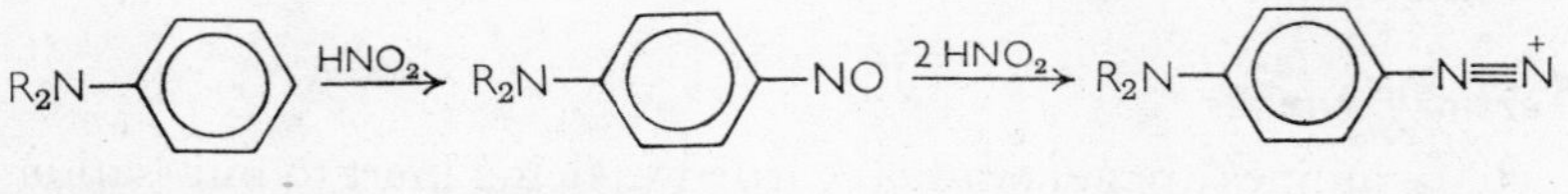

$$R_2N\text{—}\langle\bigcirc\rangle \xrightarrow{HNO_2} R_2N\text{—}\langle\bigcirc\rangle\text{—NO} \xrightarrow{2\,HNO_2} R_2N\text{—}\langle\bigcirc\rangle\text{—}N\equiv\overset{+}{N} \quad (64)$$

outside the scope of this chapter.

VI. OXIDATION

The oxidation of amines is complicated by the operation of several reaction paths. Not all of these involve substitution on nitrogen, for both hydrogen abstraction from either carbon or nitrogen and addition of oxygen to carbon may occur. The predominant path seems to depend as much on the reagent as on the structure of the amine.

Oxidative substitution of alkylamines results in the formation of either hydroxylamines or amine oxides (equations 65 to 67). The hydroxylamines can oxidise further to nitroso and nitro derivatives.

Primary alkylamines

$$RNH_2 \longrightarrow RNHOH \longrightarrow RNO \longrightarrow RNO_2 \quad (65)$$

Secondary alkylamines

$$R_2NH \longrightarrow R_2NOH \quad (66)$$

Tertiary alkylamines

$$R_3N \longrightarrow R_3\overset{+}{N}\text{—}\overset{-}{O} \quad (67)$$

However, the lone-pair electrons activate the α-hydrogens of aliphatic amines, and hydrogen is therefore removed by some reagents as easily from this site as from nitrogen. This leads to the formation of unsaturated compounds such as aldimines, nitriles, ketimines and enamines. The actual products depend on the amine structure (equations 68 to 71) and others may arise from subsequent hydrolysis, oxidation,

$$RCH_2NH_2 \longrightarrow RCH{=}NH \longrightarrow RC{\equiv}N \quad (68)$$

$$R_2CHNH_2 \longrightarrow R_2C{=}NH \quad (69)$$

$$(R_2CH_2)_2NH \longrightarrow R_2CHN{=}CR_2 \quad (70)$$

$$R_2CH_2CH_2NR_2 \longrightarrow R_2CH{=}CHNR_2 \quad (71)$$

condensation and dealkylation. The details of these reactions will not be pursued further.

With aromatic amines, substitution on nitrogen leads to products analogous to those obtained from aliphatic amines. An additional important reaction is hydrogen or hydride ion abstraction from nitrogen to form either a radical or a radical-cation (equations 72 and 73).

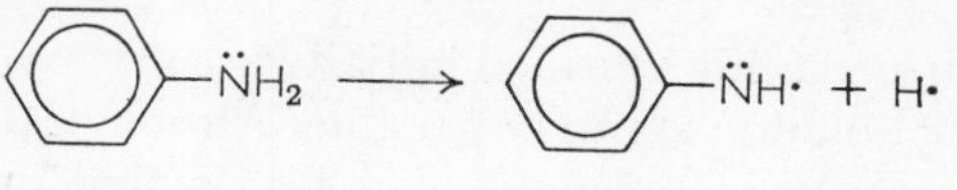

$$\langle\bigcirc\rangle\text{--}\ddot{N}H_2 \longrightarrow \langle\bigcirc\rangle\text{--}\ddot{N}H\cdot + H\cdot \tag{72}$$

$$\langle\bigcirc\rangle\text{--}\ddot{N}H_2 \longrightarrow \langle\bigcirc\rangle\text{--}\ddot{N}H^+ + H^- \tag{73}$$

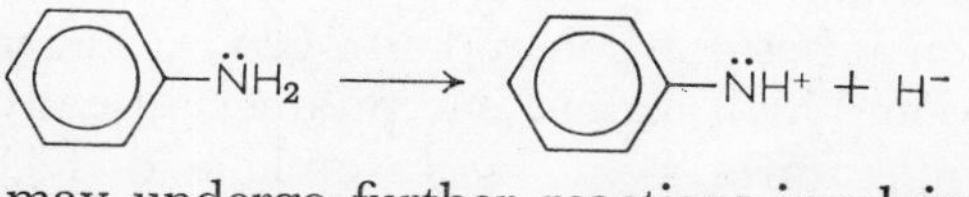

Both radicals may undergo further reactions involving nuclear substitution, condensation or quinone formation. These, too, will not be considered further.

As far as reagents are concerned, substitution on nitrogen generally occurs with peroxides, and we shall mainly consider these reactions. Non-peroxidic reagents, such as permanganate or chlorine dioxide, tend to abstract hydrogen. There are many exceptions, but these are most conveniently discussed as they arise in the text. Despite the wealth of products formed by peroxidic reagents, most of their reactions can be understood in terms of an initial attack by the electrophilic peroxidic oxygen on the nitrogen lone pair. Since this leads to a quaternisation of the nitrogen atom, it is not surprising to find that steric congestion as well as electron withdrawal in the amine moiety inhibits the reaction rate.

A. Hydroperoxides—Peroxyacids and Hydrogen Peroxide
I. Primary aliphatic amines

Early work revealed the ease with which primary aliphatic amines react in aqueous solution with many simple peroxides including Caro's acid (H_2SO_5) and hydrogen peroxide[224]. The products depend almost entirely on the structure of the alkyl group attached to nitrogen and are complicated by many subsequent reactions. The initial steps resulting from electrophilic attack by the peroxidic oxygen are shown in Scheme (21). For amines with either primary or secondary alkyl

Primary alkyl groups

$$RCH_2NH_2 \longrightarrow (RCH_2NHOH) \longrightarrow RCH=NOH \longrightarrow$$
$$R\text{--}\underset{\underset{\displaystyle O}{\|}}{C}\text{--}NHOH + \text{Other products}$$

Secondary alkyl groups

$$R_2CHNH_2 \longrightarrow (R_2CHNHOH) \longrightarrow R_2C{=}NOH \longrightarrow \text{Various products}$$

Tertiary alkyl groups

$$R_3CNH_2 \longrightarrow R_3CNHOH \longrightarrow R_3CNO \longrightarrow R_3CNO_2$$

SCHEME 21.

groups, the hydroxylamines formed initially are also powerful reducing agents and their isolation is rarely possible. Although oximes, the next product in this chain of reactions, can be isolated in fair yields[224], further oxidation at both carbon and nitrogen leading to the formation of nitroso, nitro and hydroxamic acid products, usually occurs readily at even low temperatures. Some hydroxamic acid $(RC({=}O)NHOH)$ is always formed from amines with primary alkyl groups[225]: this can be detected with ferric chloride (to give a wine-red colouration), and the reaction is often used as a test for the presence of RCH_2NH_2[225].

Recent work has concentrated on the adjustment of conditions to give a single product, primarily by using aqueous hydrogen peroxide at low temperatures (5° to 15°C) in the presence of catalytic concentrations of soluble tungsten, molybdenum and uranium oxides[226]. Oxidations with tungsten catalysts have been reviewed[227]. In this way, ketoximes can be obtained in high yields from secondary-alkyl primary amines (equation 74)[226b], but the rate is slow with bulky

$$R_2CHNH_2 + H_2O_2 \xrightarrow[\text{etc.}]{Na_2WO_4} R_2C{=}NOH \tag{74}$$

groups attached to nitrogen[226c]. The yield of aldoxime from primary-alkyl primary amines is much lower, however, due again to competing formation of hydroxamic acid[226b]. At temperatures below 0°C, it is apparently possible to isolate the initially formed hydroxylamine from these reactions[228]. Controlled oxidation of aliphatic amines to their corresponding oximes is also possible with molecular oxygen in the presence of catalysts or with ultraviolet light[229].

Peracetic acid is a good reagent for oxidising *t*-carbinamines and dialkylamines directly to their nitro derivatives[230]. The mechanism has not been studied, but this reaction, too, must proceed via the hydroxylamine[231] and the nitroso intermediate (equation 75). The

$$R_2CHNH_2 \xrightarrow{CH_3COO_2H} (R_2CHNHOH) \longrightarrow (R_2CHNO) \longrightarrow R_2CHNO_2 \tag{75}$$

advantage of using acidic peroxides is that the possibility of condensation reactions is diminished, because protonation reduces the free-amine concentration. Rather surprisingly, trifluoroperacetic acid does not oxidise primary aliphatic amines, and acylation takes place under neutral conditions (equation 76)[232]. Electron withdrawal by the CF_3

$$RCH_2NH_2 + CF_3\overset{\overset{\textstyle O}{\|}}{C}OOH \longrightarrow RCH_2\overset{\overset{\textstyle H}{|}}{N}COCF_3 + H_2O_2 \qquad (76)$$

group must make the carbonyl carbon more electrophilic than the peroxidic oxygen.

2. Primary aromatic amines

Caro's acid[233], perbenzoic acid[234], perphthalic acid[235], performic acid[236], peracetic acid[230, 237], trifluoroperacetic acid[232, 238] and permaleic acid[239] are all used to oxidise aromatic amines to their nitroso and nitro derivatives. Reactivity decreases along the series: trifluoroperacetic > permaleic > peracetic acid[239]. Trifluoroperacetic acid is a very powerful reagent capable of oxidising weakly basic amines[238a], including picramide[238b], to their nitro derivatives. Peracetic acid will not react with deactivated amines: it is most useful with very basic amines (such as anisidine)[237b] as trifluoroperacetic acid attacks the aromatic ring and not the amino nitrogen[238a].

As with aliphatic amines, oxidation by peroxidic reagents proceeds through the hydroxylamine intermediate (see below), but this rapidly reacts further to a nitroso compound and cannot be isolated*. Under slightly more vigorous conditions, the nitroso group is oxidised to nitro (equation 77)[237, 240].

$$ArNH_2 \xrightarrow{RCOO_2H} (ArNHOH) \longrightarrow ArNO \longrightarrow ArNO_2 \qquad (77)$$

Bimolecular condensation reactions between the products and un-reacted amine frequently interfere and lead to the formation of azo and azoxy compounds[242]. These side-reactions should be least extensive for amines containing electron-releasing substituents, which facilitiate oxidation but not condensation with nitrosobenzene[243]. Experimental techniques have also been developed to minimise the formation of condensation products. In particular, the reaction is carried out quickly with excess oxidiser under acidic conditions, where the concentration of unprotonated amine (which undergoes condensation) is small. Trifluoroperacetic acid is particularly useful in this respect, because of its strongly acidic properties. However, protonation

* Phenylhydroxylamine can be prepared, however, by oxidising the magnesium salt of aniline (PhNHMgBr) with hydrogen peroxide at −25°[241].

of the amine also retards the *rate* of oxidation, and it is therefore necessary to use an excess of the reagent.

Ibne–Rasa and Edwards[244] have recently studied the peracetic acid
oxidation of substituted anilines and some of their conclusions bear on
the mechanism of peroxidic oxidations of amines in general. The
reaction rates have a first-order dependence on the concentrations of
unprotonated amine and peracetic acid (equation 78) and the Ham-

$$\text{Rate} = k[\text{ArNH}_2][\text{CH}_3\text{COO}_2\text{H}] \tag{78}$$

mett *rho* value of $-1\cdot86$ implies that electron-attracting substituents
retard the rate. Other workers[245] have shown that similar substituents
in the peroxidic reagent have the opposite effect. Polar solvents also
increase the reaction rate indicating that charge is developed in the
transition state. Radical traps, such as methyl methacrylate have no
effect, and earlier oxygen-18 experiments[246] also exclude the possibility of hydroxyl-radical formation in amine oxidations with Caro's
acid. All these observations are most consistent with a mechanism in
which the electrophilic peroxidic oxygen attacks the amine lone pair
to form a tetrahedral intermediate. Since oxidation of both phenylhydroxylamine[244] and nitrosobenzene[240b] under similar conditions is
faster than that of aniline itself, the first steps in Scheme (22) must be

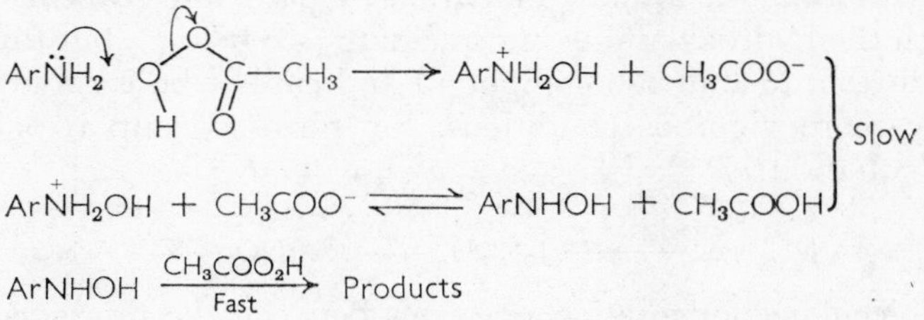

SCHEME 22.

rate controlling. In order to account for the highly negative activation
entropies in both water and ethanol ($\Delta S^{\ddagger} = -26$ cal/mole deg),
Ibne-Rasa and Edwards[244] argue that the transition state in hydroxylic
solvents probably incorporates a solvent molecule as in **18**. The

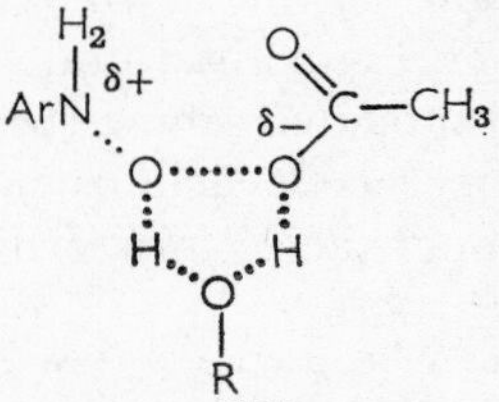

(18)

25-fold decrease in the rate of oxidation of *p*-nitroaniline in going from water to ethanol is then ascribed to differences in solvation arising from the latter's lower acidity rather than lower polarity.

A heterolytic process for oxidation by peroxides also implies that the rate depends on the leaving tendency of the displaced anion. In accordance with this conclusion, it is found that both hydrogen peroxide[244] and alkyl peroxides are less effective in amine oxidations than acyl peroxides. For the latter, an intramolecular proton transfer to the basic carbonyl oxygen may also facilitate the reaction.

3. Secondary amines

Secondary alkylamines are readily oxidised to hydroxylamines (equation 79). The reagent commonly used is a slight excess of hydro-

$$R_2NH + H_2O_2 \longrightarrow R_2NOH + H_2O \tag{79}$$

gen peroxide in aqueous solution[247]. By analogy with primary amines, the mechanism must involve attack by the electrophilic peroxidic oxygen on the unprotonated amine.

The reaction is facile with all secondary alkylamines except the simplest, dimethylamine, but yields of hydroxylamine are low owing to further oxidation. Thus *N*, *N*-dialkylhydroxylamines are prepared by indirect methods[247b], or as with primary amines, tungsten catalysts are used with hydrogen peroxide at low temperature[248]. As well as further oxidation[249], other side-reactions, including the formation of dialkylformamides, may occur[250]. These can be accounted for by the initial formation of a hydroperoxide (**19**), which concertedly undergoes Baeyer–Villiger rearrangement and attack by another molecule of secondary amine to give an intermediate **20**, followed by elimination of R^1OH and hydrolysis to give *N*, *N*-dialkylformamide (Scheme 23)[250].

$$\begin{array}{c} R \\ \diagdown \\ NH + H_2O_2 \longrightarrow RNHCHR^1 + H_2 \\ \diagup \qquad\qquad\qquad | \\ R^1CH_2 \qquad\qquad\qquad OOH \\ \qquad\qquad\qquad\qquad (\mathbf{19}) \end{array}$$

$$\begin{array}{c} RNHCHR^1 \\ | \qquad \longrightarrow [R\overset{+}{N}H{=}CHOR^1 \rightleftharpoons RN{=}CHOR^1 + H^+] \\ OOH \end{array}$$

$$\begin{array}{c} RN{=}CHOR^1 \xrightarrow{R_2NH} RN{-}CHOR^1 \\ \qquad\qquad\qquad\qquad | \quad | \\ \qquad\qquad\qquad\qquad H \quad NR_2 \\ \qquad\qquad\qquad\qquad (\mathbf{20}) \end{array}$$

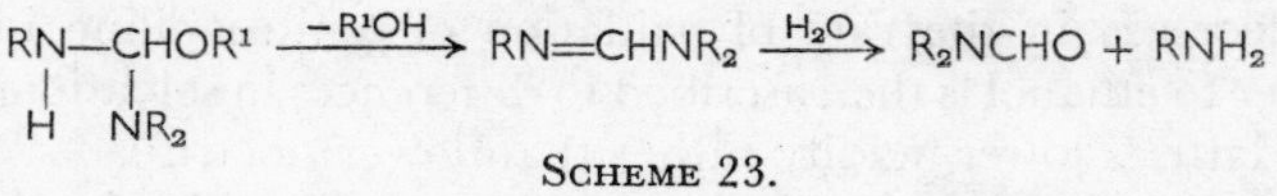

$$RN-CHOR^1 \xrightarrow{-R^1OH} RN=CHNR_2 \xrightarrow{H_2O} R_2NCHO + RNH_2$$
$$\underset{H \quad NR_2}{|\quad|}$$

SCHEME 23.

Secondary arylamines also form hydroxylamines initially, but these invariably oxidise further to the radicals $Ar_2\dot{N}$ and $Ar_2\dot{N}(O)$ [251]. The diphenylimino radical $(Ar_2\dot{N})$ also results from oxidation by non-peroxidic reagents [252], and it exists in equilibrium with its dimer, tetraphenylhydrazine (equation 80) [252]. Diaryl nitroxide radicals

$$Ar_2NH \xrightarrow{[O]} Ar_2\dot{N} \rightleftharpoons Ar_2N-NAr_2 \tag{80}$$

$(Ar_2\dot{N}(O))$ are formed in 70% yield from oxidations of diarylamines with perbenzoic acid in ether at 0°C [253]. Since the same radical is formed from diphenylhydroxylamine [253, 254], the intermediate formation of the latter seems very likely. Diaryl nitroxide radicals are reasonably stable, but on standing at room temperature undergo spontaneous disproportionation and rearrangement to a variety of products [253, 255].

With mixed aliphatic–aromatic secondary amines, reaction with either hydrogen peroxide or Caro's acid first results in cleavage of the alkyl group [249]. The residue is then oxidised to nitroso- and nitro-benzene as is normal for primary amines [249].

4. Tertiary amines

Most simple acyl- and hydroperoxides readily add oxygen to any tertiary amine giving amine oxide $(R_3N^+O^-)$. These facile reactions are usually carried out at room temperature in water, alcohol or benzene solvent with dilute solutions (even as low as 3%) of an organic peracid. Hydrogen peroxide and Caro's acid are less effective, presumably because of the poorer leaving characteristics of the bisulphate and hydroxide anions. The synthetic utility of the various reagents has been reviewed [256].

The mechanism of tertiary amine oxide formation has not been studied in detail, but by analogy with primary amines (see above), the reaction must involve attack by the electrophilic peroxidic oxygen on the amine lone pair, followed by anion elimination and proton loss (Scheme 24). In accord with this conclusion, the reactions are first

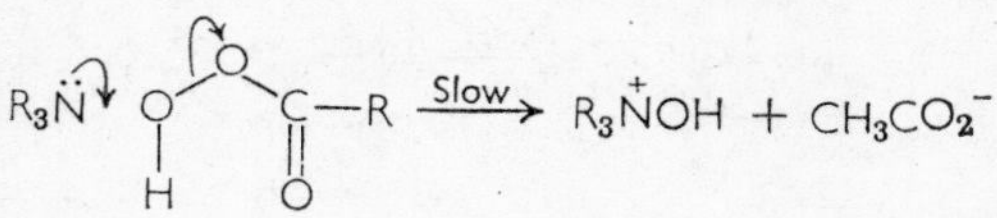

$$R_3\dot{N} \; O \underset{\underset{H}{|}}{\overset{\overset{O}{\frown}}{}} C-R \xrightarrow{Slow} R_3\overset{+}{N}OH + CH_3CO_2^-$$

$$R_3\overset{+}{N}OH \xrightarrow{\text{Fast}} R_3\overset{+}{N}O^- + H^+$$

Scheme 24.

order with respect to each reactant[257, 258] and the protonated amine is unreactive[257]. Ogata and Tabushi[258] have found that electron-withdrawing substituents retard the rate of oxidation of N, N-dimethyl-aniline by Caro's acid, and this result suggests that the initial attack by the oxidising agent is rate controlling. At low temperatures and with high concentrations of hydrogen peroxide, hydrogen-bonded per-oxide adducts ($R_3N.H_2O_2$), reminiscent of hydrated amine oxides, can be isolated[259]. It is very unlikely, however, that amine oxide formation normally proceeds via these adducts.

B. Diacyl Peroxides

The most extensive studies of oxidation by diacyl peroxides have been carried out on secondary amines, and because many of the con-clusions are applicable to other types of amines, these reactions will be considered first.

I. Secondary amines

The reaction of secondary amines with benzoyl peroxide has en-gendered considerable interest. Except in the case of arylamines (see below), this reagent leads to the formation of O-benzoylhydroxyl-amines from which the hydroxylamine can be obtained by the addition of base[260] (Scheme 25). For a long time it was thought that these

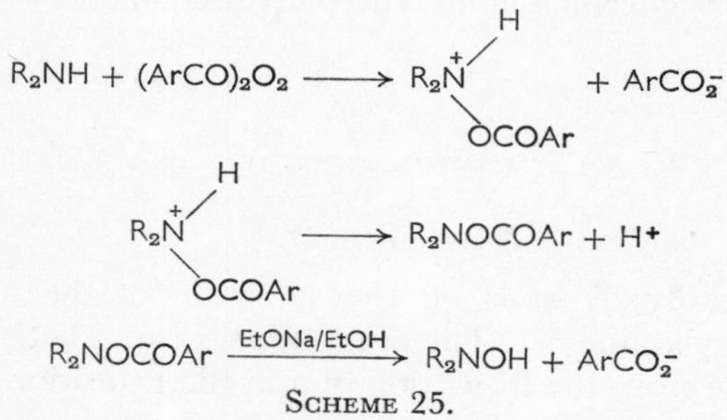

$$R_2NH + (ArCO)_2O_2 \longrightarrow R_2\overset{+}{N}\overset{H}{\diagup}_{OCOAr} + ArCO_2^-$$

$$R_2\overset{+}{N}\overset{H}{\diagup}_{OCOAr} \longrightarrow R_2NOCOAr + H^+$$

$$R_2NOCOAr \xrightarrow{\text{EtONa/EtOH}} R_2NOH + ArCO_2^-$$

Scheme 25.

reactions were free-radical processes, but the arguments against this interpretation have been cogently marshalled by Walling[261]. His conclusions have now been substantiated by showing that if the oxida-tion of dibenzylamine is carried out with benzoyl peroxide labelled with oxygen-18 in only the carbonyl group, the intermediate O-benzoylhydroxylamine (**21**) contains 50% of the isotopic oxygen, whereas the product dibenzylhydroxylamine (**22**), obtained after the addition of base, contains none (Scheme 26)[262]. Hence homolysis of

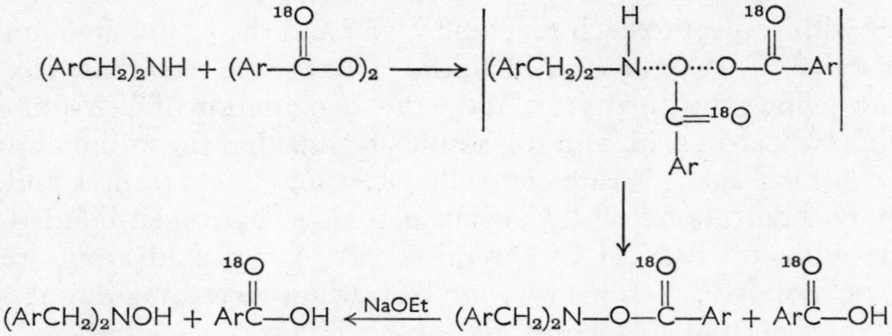

$$(ArCH_2)_2NH + (Ar\overset{\overset{\displaystyle ^{18}O}{\|}}{C}-O)_2 \longrightarrow \left| (ArCH_2)_2-\overset{\overset{\displaystyle H}{|}}{N}\cdots O\cdots O-\overset{\overset{\displaystyle ^{18}O}{\|}}{C}-Ar \right|$$

$$(ArCH_2)_2NOH + Ar\overset{\overset{\displaystyle ^{18}O}{\|}}{C}-OH \xleftarrow{\ NaOEt\ } (ArCH_2)_2N-O-\overset{\overset{\displaystyle ^{18}O}{\|}}{C}-Ar + Ar\overset{\overset{\displaystyle ^{18}O}{\|}}{C}-OH$$

(**22**)   (**21**)

SCHEME 26.

the benzoyl peroxide, depicted by equation (81), which would lead

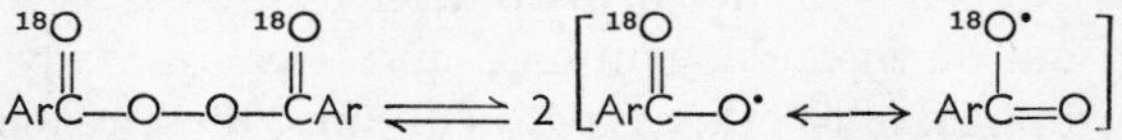

$$Ar\overset{\overset{\displaystyle ^{18}O}{\|}}{C}-O-O-\overset{\overset{\displaystyle ^{18}O}{\|}}{C}Ar \rightleftharpoons 2 \left[Ar\overset{\overset{\displaystyle ^{18}O}{\|}}{C}-O^{\bullet} \longleftrightarrow Ar\overset{\overset{\displaystyle ^{18}O^{\bullet}}{\|}}{C}=O \right] \tag{81}$$

to equilibration of the oxygen-18, cannot occur. Also Huisgen and Bayerlein[263] have found that the reaction between benzoyl peroxide and diethylamine does not initiate polymerisation of styrene and they conclude that oxidations of secondary alkylamines with diacyl-peroxides in general proceed exclusively by a heterolytic path. Other evidence[264, 265], however, mainly the evolution of carbon dioxide from disproportionation of the acetoxy radical ($CH_3COO^{\bullet} \rightarrow CO_2 + CH_3^{\bullet}$) suggests that up to 10% of the oxidation of amines with diacetyl peroxide may occur via a homolytic path (Scheme 27)[264]. The actual

$$R_2NH + (CH_3CO)_2O_2 \longrightarrow R_2\overset{\bullet}{N}H^+ + \overset{\bullet}{O}-CO-CH_3 + \overset{-}{O}-CO-CH_3$$
(23)

$$\longrightarrow R_2\overset{\bullet}{N} + \overset{\bullet}{O}-CO-CH_3 + HO-CO-CH_3 \longrightarrow R_2N-O-CO-CH_3$$
(**24**)   + CH_3CO_2H

SCHEME 27.

amount probably depends on the structure of the amine, and is greatest for arylamines[264]. The predominance of a heterolytic process may well arise from the formation of a cyclic transition state (**25**), in

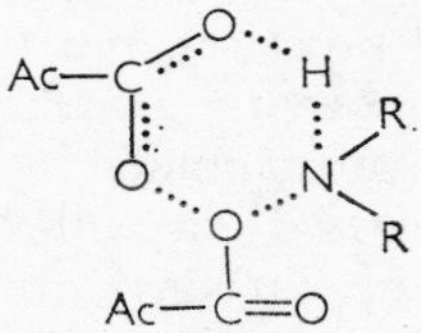

(25)

which the neighbouring acyl (or aroyl) oxygen acts as a base and facilitates proton abstraction from the amine in a concerted process[266]. Since a similar interaction would also account for the higher reactivity of acyl peroxides compared to alkyl peroxides, cyclic transition states may be a feature of amine oxidations by all types of acyl and aroyl peroxides, at least in hydroxylic solvents.

Reaction of secondary aryl and mixed alkyl aryl amines with benzoyl peroxide results in the formation of some *o*-hydroxybenzanilides as well as the expected *O*-benzoyl derivative. In some cases the yield of *o*-hydroxyl derivative may be as high as 40%[267]. Several mechanisms have been proposed[268], but most of the recent evidence points to a reaction (Scheme 28) involving the initial heterolytic forma-

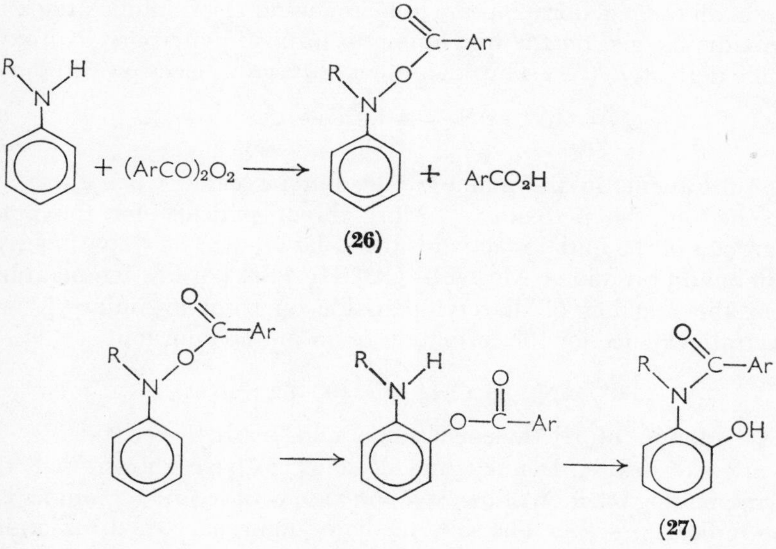

(26)

(27)

Scheme 28.

tion of the *N*-alkyl-substituted *O*-benzoyl phenylhydroxylamine (**26**), which then rearranges to give the ring-substituted product (**27**). For example, the initial formation of **26** has been clearly demonstrated by reacting diphenylhydroxylamine with benzoyl chloride, from which *N*-phenyl-*o*-hydroxybenzanilide is obtained in high yield[263]. The exact nature of the rearrangement process is not entirely understood, but the predominance of a heterolytic process seems most likely. The related rearrangement of arylhydroxylamines to aminophenols by an ionic mechanism is well known[269]. Also, Huisgen and Bayerlein[263] have reported that the reaction of N-ethylaniline with benzoyl peroxide, which leads to ring-substituted products, does not initiate the polymerisation of styrene. However, studies of the oxidation of diphenylamine with oxygen-18 labelled benzoyl peroxide[262, 270], and of carbon dioxide evolution in oxidation with diacetyl peroxide[264] (see p. 328), suggest that a small fraction of the hydroxylamine formed initially, may, in fact, rearrange to ring-substituted products via a homolytic path.

2. Tertiary amines

The products from diacyl and benzoyl peroxide oxidation depend very much on the structure of the tertiary amine[271, 272]. The initial step in all the reactions appears to be the usual electrophilic attack by peroxidic oxygen on the free amine to form a quaternary hydroxylamine derivative (**28**). Thus the rate is reduced by electron-withdraw-

$$R_3N + (ArCO)_2O_2 \longrightarrow R_3\overset{+}{N}-O-COAr + ArCO_2^- \qquad (82)$$
$$(28)$$

ing substituents in the amine[273, 274], but increased when these are attached to the peroxide[275]. More direct evidence for the polar character of the initial reaction is the isolation of N-acetoxytrimethylammonium bromide [$(Me_3NO-COCH_3)^+Br^-$] at low temperatures from the reaction of diacetyl peroxide on trimethylamine[276], and spectral evidence for the formation of an ammonium ion

$$[N(CH_2CH_2)_3\overset{+}{N}-O-COPh]$$

in a solution of triethylenediamine and benzoyl peroxide[271]. As mentioned before, Denney and Denney[262] have shown that the corresponding reaction of benzoyl peroxide with secondary amines is a non-radical process. The intermediate quaternary hydroxylamine derivative (**28**) can undergo subsequent reaction in a number of ways, depending on the structure of the tertiary amine.

Thus tertiary alkylamines containing an alkyl group beta to nitrogen (i.e. containing the structure $R_2N-CH_2-CH_2-$) first eliminate benzoic acid to form a quaternary imine (**29**), which subsequently loses a proton to give an enamine (Scheme 29)[271]. The enamine may

$$\begin{array}{ccc} R_2N + (ArCO)_2O_2 & \longrightarrow & R_2\overset{+}{N}-O-CO-Ar + ArCO_2^- \\ | & & | \\ CH_2 & & CH_2 \\ | & & | \\ CH_3 & & CH_3 \end{array}$$

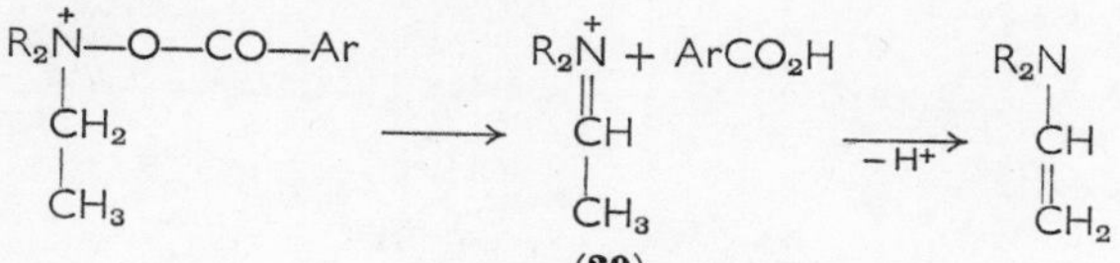

$$\begin{array}{ccc}
\overset{+}{\underset{|}{R_2N}}\text{—O—CO—Ar} & \overset{+}{\underset{\parallel}{R_2N}} + ArCO_2H & R_2N \\
\underset{|}{CH_2} & \underset{|}{CH} & \underset{\parallel}{CH} \\
CH_3 & CH_3 & CH_2
\end{array}$$

$$\longrightarrow \qquad \xrightarrow{-H^+}$$

(29)

SCHEME 29.

undergo hydrolysis to form a secondary amine and an aldehyde: this reaction is useful for demethylating tertiary alkylamines[275]. If, however, one of the alkyl groups is benzyl, the quaternary imine (**30**) is hydrolysed to a secondary amine and benzaldehyde directly (Scheme 30)[271]. Reagents other than benzoyl peroxide, such as perphthalic

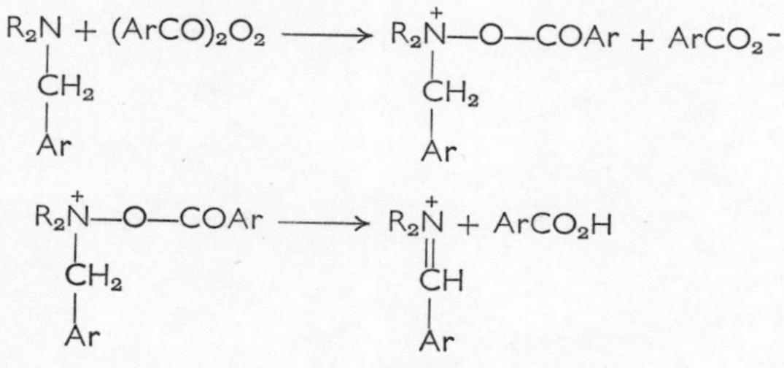

$$\begin{array}{cc}
\underset{|}{R_2N} + (ArCO)_2O_2 & \longrightarrow \quad \overset{+}{\underset{|}{R_2N}}\text{—O—COAr} + ArCO_2{}^- \\
\underset{|}{CH_2} & \underset{|}{CH_2} \\
Ar & Ar
\end{array}$$

$$\begin{array}{cc}
\overset{+}{\underset{|}{R_2N}}\text{—O—COAr} & \longrightarrow \quad \overset{+}{\underset{\parallel}{R_2N}} + ArCO_2H \\
\underset{|}{CH_2} & \underset{|}{CH} \\
Ar & Ar
\end{array}$$

(30)

$$\begin{array}{c}
\overset{+}{\underset{\parallel}{R_2N}} \xrightarrow{H_2O} R_2NH + ArCHO + H^+ \\
\underset{|}{CH} \\
Ar
\end{array}$$

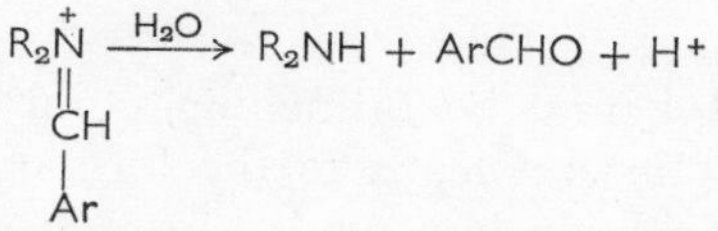

(30)

SCHEME 30.

acid[271, 247b], produce similar transformations in tertiary alkylamines, but it is not clear whether these are always heterolytic processes.

Tertiary aryl alkyl amines, e.g. *N,N*-dimethylaniline, generally undergo dealkylation with benzoyl and other diacyl peroxides[272]: primary alkyl groups other than methyl are cleaved more readily than secondary, and secondary more readily than methyl[275]. These reactions have been subjected to extensive investigation, but many important details remain unclear. Very recent work has again emphasised the complexities[277] and a full discussion of these is outside the scope of this review. Walling and Indictor[278] have suggested that the rate-controlling step is the formation of the quaternary hydroxylamine derivative (**31**). This intermediate probably decomposes rapidly, as in Scheme (31), via two competing paths, although there is some

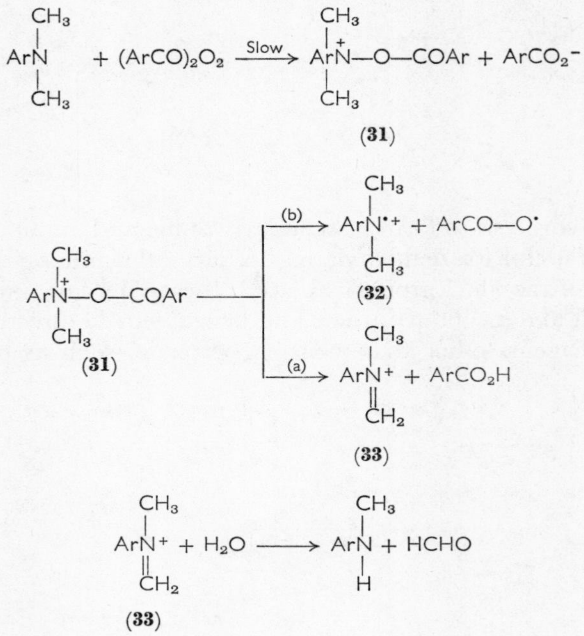

$$\underset{\overset{|}{CH_3}}{\overset{\overset{CH_3}{|}}{ArN}} + (ArCO)_2O_2 \xrightarrow{\text{Slow}} \underset{\overset{|}{CH_3}}{\overset{\overset{CH_3}{\overset{|}{+}}}{ArN}}-O-COAr + ArCO_2^-$$

(31)

$$\underset{\overset{|}{CH_3}}{\overset{\overset{CH_3}{\overset{|}{+}}}{ArN}}-O-COAr \longrightarrow \begin{array}{l} \xrightarrow{\text{(b)}} \underset{\overset{|}{CH_3}}{\overset{\overset{CH_3}{|}}{ArN^{\bullet+}}} + ArCO-O^{\bullet} \\[4pt] \quad\quad\quad (32) \\[10pt] \xrightarrow{\text{(a)}} \underset{\overset{||}{CH_2}}{\overset{\overset{CH_3}{|}}{ArN^+}} + ArCO_2H \\[4pt] \quad\quad\quad (33) \end{array}$$

(31)

$$\underset{\overset{||}{CH_2}}{\overset{\overset{CH_3}{|}}{ArN^+}} + H_2O \longrightarrow \underset{\overset{|}{H}}{\overset{\overset{CH_3}{|}}{ArN}} + HCHO$$

(33)

SCHEME 31.

evidence against **31** being the branching point[279]. One competing path (a), is heterolytic and leads to the formation of a quaternary imine (**33**), which on hydrolysis leads to the formation of the major products—dealkylated amine and formaldehyde[280]. The isolation of other minor products also requires the formation of **33**[276, 281]. The other competing path (b) involves disproportionation to give a radical-cation (**32**). There is also spectral evidence for the formation of radical intermediates[282], and the formation of N, N, N', N'-tetramethylbenzidine in fairly high yield in some solvents[276, 283] seems to imply the intervention of a homolytic path. Controversy still exists over the relative importance of the radical and ionic paths. The system's low efficiency as a polymerisation initiator[277, 284], and Horner's[285] related studies of oxidation with diacetyl peroxide, suggest that homolytic decomposition is relatively small. On the other hand, recent kinetic studies have been interpreted in terms of a radical rather than an ionic mechanism[283, 286]. In all probability, the contribution from each path depends on the experimental conditions and the reactants involved.

3. Primary amines

Primary amines react less readily with diacyl peroxides than either secondary or tertiary amines. The reaction results in a complex mixture of products[287]. From aniline and benzoyl peroxide, for instance, benzoic acid, benzanilide, azoxybenzene and a little *o*-hydroxybenzanilide can be isolated[287b]. Their formation can be rationalised from an initial attack by benzoyl peroxide to give both *O*-benzoylphenylhydroxylamine (**34**) plus benzoic acid and benzanilide (**35**) plus perbenzoic acid (Scheme 32). Subsequent hydrolysis,

$$ArNH_2 + (ArCO)_2O_2 \longrightarrow$$

$$\longrightarrow ArNH-O-COAr + ArCO_2H$$
$$(34)$$

$$\longrightarrow ArNH-COAr + ArCOO_2H$$
$$(35)$$

SCHEME 32.

oxidation, condensation and rearrangement, like those discussed previously for secondary and tertiary amines, would account for the other products, but the details at present are speculative.

From related studies with diacetyl peroxide, Horner and Anders[264] conclude that oxidation of primary aromatic amines is an entirely heterolytic process. They find, for instance, that no carbon dioxide is evolved as would be expected if acetoxy radicals were formed in the oxidation process $(CH_3COO \cdot \rightarrow CO_2 + CH_3 \cdot)$. Since the products are analogous to those obtained with benzoyl peroxide, the same conclusion may also apply to the latter reagent. More recent studies[282], however, claim to detect the presence of free radicals in the oxidation of aniline with benzoyl peroxide in solvent benzene.

C. tert-Butyl Hydroperoxide

At somewhat higher temperatures (40° to 100°c) than is usual for peroxidic oxidations, *t*-butyl hydroperoxide reacts with amines in an entirely different way from other hydroperoxides. The overall result is not substitution on nitrogen (although this may be the initial step), but the formation of various products arising from either hydrogen abstraction or dealkylation. In this sense, oxidation by *t*-butyl hydroperoxide at high temperatures is reminiscent of non-peroxidic reagents.

Primary alkylamines possessing α-hydrogen atoms form imines in good yield (60%)[288]. The overall reaction is illustrated by equation (83). The imine may then either hydrolyse or oxidise further to a

$$RCH_2NH_2 + t\text{-BuOOH} \longrightarrow RCH{=}NH + t\text{-BuOH} + H_2O \qquad (83)$$

nitrile. Primary alkylamines without α-hydrogens (such as *t*-octyl-amine) react less readily to give the corresponding nitroalkane as the principal isolable product[288]. This reaction is therefore typical of other hydroperoxides (see section A).

Secondary alkylamines also form imines with *t*-butyl hydroperoxide (Scheme 33), which then undergo reversible hydrolysis to a primary

$$(RCH_2)_2NH + t\text{-BuOOH} \longrightarrow RCH{=}N{-}CH_2R + t\text{-BuOH} + H_2O$$

$$RCH{=}N{-}CH_2R + H_2O \rightleftharpoons RCHO + RCH_2NH_2$$

SCHEME 33.

amine. For example, diisopropylamine yields isopropylamine almost quantitatively[288].

Tertiary amines with *t*-butyl hydroperoxide at high temperature (about 70°C) dealkylate in an extremely rapid reaction (equation 84)[288, 289]. The resultant secondary amines may then react further[289].

$$(RCH_2)_3N + t\text{-BuOOH} \longrightarrow (RCH_2)_2NH + RCHO + t\text{-BuOH} \tag{84}$$

At lower temperatures in the presence of catalytic amounts of V_2O_5, dealkylation does not occur and an amine oxide, the usual product obtained with hydroperoxides, is formed[290].

The mechanism of these oxidations is by no means clear-cut, but inhibition by free-radical traps indicates a homolytic process probably involving *t*-butoxy radicals (*t*-BuO·). De la Mare[288] originally suggested that his observations could be explained by the initial formation of a radical-cation (equation 85), which then underwent a complex

$$(RCH_2)_2NH + t\text{-BuOOH} \longrightarrow \left[\begin{array}{c} RCH_2\dot{N}{-}CH_2R \\ | \\ H \end{array} \right]^+ + t\text{-BuO}^{\cdot} + OH^- \tag{85}$$

series of chain-transfer reactions to give the products. More recent electron spin resonance studies, however, reveal the formation of a relatively stable *N*-oxide radical $\left[\begin{array}{c} RCH_2{-}\ddot{N}{-}CH_2R \\ | \\ \underset{\cdot}{O} \end{array} \right]$ in these reactions[289, 291]. This can be explained by a mechanism in which the initial reaction is the usual *N*-substitution by peroxidic oxygen to form the hydroxylamine, which then reacts with another molecule of *t*-butyl hydroperoxide to give the *N*-oxide radical (equation 86)[288, 291].

$$(RCH_2)_2NH + t\text{-BuOOH} \longrightarrow (RCH_2)_2NOH + t\text{-BuOH}$$

$$(RCH_2)_2NOH + t\text{-BuOOH} \longrightarrow (RCH_2)_2N\text{—}O\cdot + t\text{-BuO}^{\cdot} + H_2O \quad (86)$$

$$(RCH_2)_2N\text{—}O\cdot + RCH_2\underset{\underset{H}{|}}{N}CH_2R \longrightarrow R\overset{\cdot}{C}H\text{—}\underset{\underset{H}{|}}{N}\text{—}CH_2R + RCH_2\underset{\underset{OH}{|}}{N}CH_2R$$

$$R\overset{\cdot}{C}H\text{—}\underset{\underset{H}{|}}{N}\text{—}CH_2R + t\text{-BuO}\cdot \longrightarrow RCH\!=\!N\text{—}CH_2R + t\text{-BuOH}$$

SCHEME 34.

The latter reacts further as shown in Scheme (34) to form a ketimine from the secondary amine.

The *t*-butoxy radical itself can abstract α-hydrogens directly. This is evident from studies with di-*t*-butyl peroxide, which decomposes homolytically at temperatures above 125°c ($(t\text{-Bu})_2O_2 \rightarrow 2\ t\text{-BuO}\cdot$). Primary and secondary alkylamines react to form imines[292]. The mechanism has been viewed as a radical chain sequence following abstraction of the α-hydrogen[292] (Scheme 35). Tertiary alkylamines

$$(t\text{-Bu})_2O_2 \longrightarrow 2\ t\text{-BuO}\cdot$$

$$RCH_2NH_2 + t\text{-BuO}\cdot \longrightarrow R_2\overset{\cdot}{C}\text{—}NH_2 + t\text{-BuOH}$$

$$R_2\overset{\cdot}{C}\text{—}NH_2 + (t\text{-Bu})_2O_2 \longrightarrow R_2C\!=\!NH + t\text{-BuOH} + t\text{-BuO}\cdot$$

SCHEME 35.

apparently do not react[292], but Henbest and Patton[293] have shown that tertiary aryl alkyl amines dimerize to form ethylenediamine derivatives. For example, dimethylaniline is converted to N,N'-dimethyl-N,N'-diphenylethylenediamine in 28% yield (Scheme 36).

$$(t\text{-Bu})_2O_2 \longrightarrow 2\ t\text{-BuO}\cdot$$

$$C_6H_5N(CH_3)_2 + t\text{-BuO}\cdot \longrightarrow \underset{\underset{CH_3}{|}}{\overset{\overset{C_6H_5}{|}}{N}}\text{—}\overset{\cdot}{C}H_2 + t\text{-BuOH}$$

$$2\ \underset{\underset{CH_3}{|}}{\overset{\overset{C_6H_5}{|}}{N}}\text{—}\overset{\cdot}{C}H_2 \longrightarrow \underset{\underset{CH_3}{|}}{\overset{\overset{C_6H_5}{|}}{N}}\text{—}CH_2\text{—}CH_2\text{—}\underset{\underset{CH_3}{|}}{\overset{\overset{C_6H_5}{|}}{N}}$$

SCHEME 36.

D. Ozone (O₃)

Primary and secondary aliphatic amines are decomposed by ozone[294]. Tertiary aliphatic amines are converted readily to the corresponding amine oxides ($R_3\overset{+}{N} \rightarrow \overset{-}{O}$). This reaction is of synthetic value and has been reviewed from this standpoint[256a, 295]. Tertiary aromatic amines do not normally give well-defined products and carbon–nitrogen bond fission often occurs[294, 296]. Kolsaker and Meth-Cohn[296], however, have recently identified N-formyl-N-methyl-p-nitroaniline (36), N-methyl-p-nitroaniline (37) and the amino peroxide (38) as products from the ozonolysis of N,N-dimethyl-p-nitroaniline (equation 87).

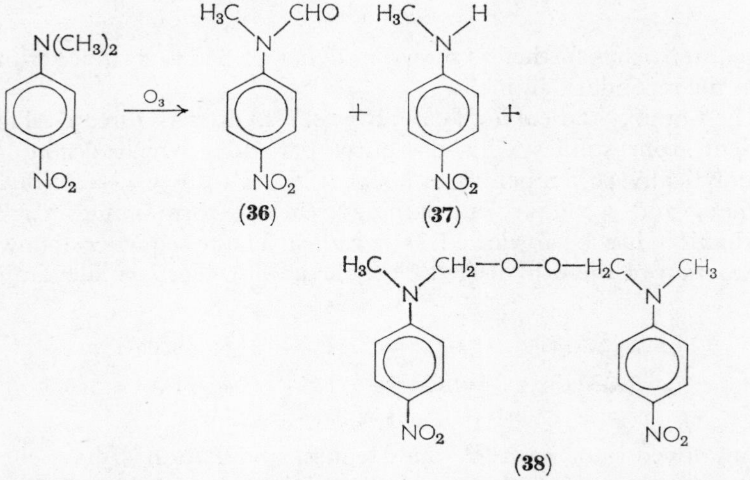

(87)

The ozonolysis of tertiary aliphatic amines is by far the most important reaction: it is exceedingly facile and may reach explosive proportions unless carried out at low temperature. Several groups of workers[294c, 297, 298] have recently described the preparation of amine oxides, via this reaction, in high yield. The solvent appears to be very important, and the highest yields of N-oxide are obtained in chloroform, carbon tetrachloride and methanol[297]. In petroleum solvents, amine oxide formation is inhibited and appreciable carbon–nitrogen bond fission occurs[294, 297]. Henbest and Stratford[297] have made the most detailed study of the products obtained under various conditions. In chloroform or methanol, they find that tributylamine, for example, is converted to about 60% of the N-oxide: other products include small amounts of dibutylamine, dibutylformamide and dibutylbutyramide. In hydrocarbon solvents, the yield of N-oxide is negligible, dibutyl amine being the major product at $-45°$, dibutylformamide at 15°c. These findings are explained by suggesting that the amine can form either the N-oxide directly or a carbinolamine (Scheme 37). Decomposition of the carbinolamine via various paths leads to the dibutyl

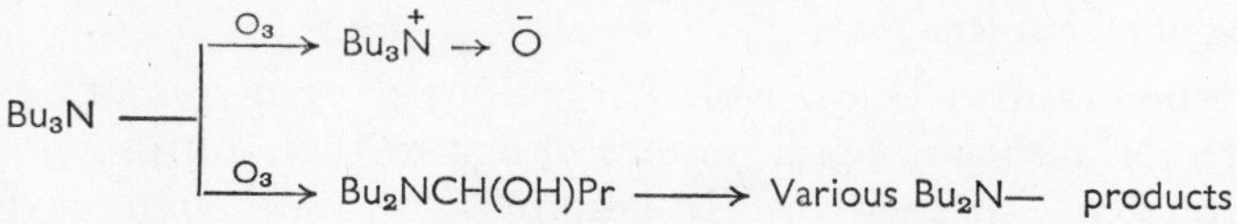

$$Bu_3N \quad \Biggl[\begin{array}{l} \xrightarrow{O_3} Bu_3\overset{+}{N} \to \overset{-}{O} \\[1em] \xrightarrow{O_3} Bu_2NCH(OH)Pr \longrightarrow \text{Various } Bu_2N\text{— products} \end{array}$$

SCHEME 37.

products. The function of the solvent is not clear. It has been suggested that either hydrogen bonding[297], or the formation of charge-transfer complexes[298], between the tertiary amine and polar solvents facilitates substitution on nitrogen.

The mechanism of amine oxide formation has not been studied closely. The reaction has a first-order dependence on both amine and ozone concentrations[298] and almost certainly involves a slow electrophilic attack by the terminal oxygen on the amine lone-pair electrons (equation 88) as the conjugate acid of the amine is unreactive[294c].

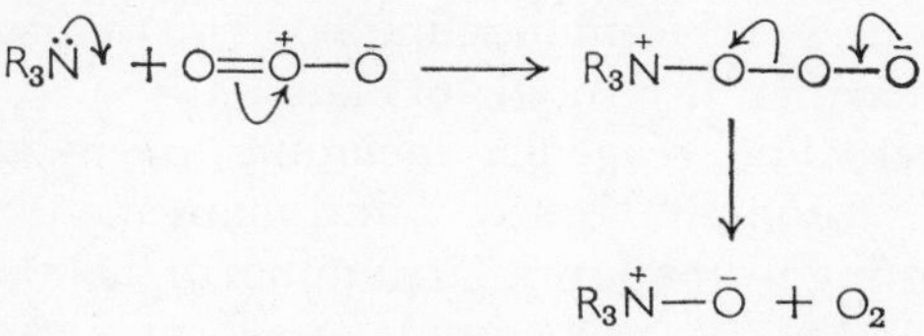

$$R_3\overset{\frown}{\ddot{N}} + O{=}\overset{+}{\underset{\smile}{\ddot{O}}}{-}\overset{-}{O} \longrightarrow R_3\overset{+}{N}{-}\overset{\frown}{\overset{-}{O}}{-}O{-}\overset{\frown}{\overset{-}{O}}$$

$$\downarrow$$

$$R_3\overset{+}{N}{-}\overset{-}{O} + O_2 \tag{88}$$

E. Potassium Permanganate

Neutral or alkaline solutions of permanganate usually abstract α-hydrogens rather than substitute on nitrogen[299]. The resultant imine (equation 89) reacts further to give a multiplicity of products.

$$RCH_2NH_2 \xrightarrow{KMnO_4} RCH{=}NH \tag{89}$$

For example, benzylamine forms benzaldehyde[299a], or condensation products[299b,c], and diethylamine is oxidised to a mixture of acetic acid, ammonia, ethanol and acetohydroxamic acid[300].

When the alkyl group is tertiary, as in t-butylamine, there is no α-hydrogen to abstract and the reaction takes a different course. Substitution then occurs readily on nitrogen to give a nitroalkane in yields of up to 80% (equation 90). Kornblum[301] has recently reviewed

$$(CH_3)_3CNH_2 \xrightarrow{KMnO_4} (CH_3)_3CNO_2 \tag{90}$$

this reaction. The formation of the nitroalkane presumably proceeds via the hydroxylamine and nitroso intermediates, as with peroxidic reagents.

F. Miscellaneous Reagents

At temperatures below $-42°C$, gaseous oxygen difluoride (OF_2) converts primary aliphatic amines either to their nitroso or to the isomeric oxime derivatives (equation 91)[302]. Although oxygen di-

$$3 \, RCH_2NH_2 + OF_2 \longrightarrow RCH_2N{=}O + 2 \, RCH_2\overset{+}{N}H_3\overset{-}{F}$$

$$\big\updownarrow$$

$$RCH{=}NOH$$

(91)

fluoride must be handled with great care, the reagent appears to be most useful as the only one available for oxidising alicyclic amines. Cyclopropylamine, for example, yields 40% of nitrosocyclopropane. The mechanism has been viewed as a direct attack on the amine lone pair by oxygen, made strongly electrophilic by the presence of two fluorine atoms. Aromatic amines, however, react slowly even at ambient temperatures and form indefinite polymeric products, typical of homolytic rather than heterolytic oxidation.

Several non-peroxidic reagents, including phenyliodoacetate[303], lead tetraacetate[304], sodium borate[305] and manganese dioxide[306] are effective in converting aromatic primary amines to their azo derivatives. It is unlikely that the azo compounds are formed, as in peroxidic oxidations, via condensation of unreacted amine with the nitroso intermediate. Instead, the mechanism is probably homolytic involving initial hydrogen abstraction from nitrogen to give an anilino radical, followed by dimerisation to hydrazobenzene and further oxidation of the latter to the azo product (Scheme 38). The very facile auto-

$$ArNH_2 \xrightarrow{\,'O'\,} ArNH{\cdot} + H{\cdot}$$

$$2 \, ArNH{\cdot} \longrightarrow ArNH{-}NHAr$$

$$ArNH{-}NHAr \xrightarrow{\,'O'\,} ArN{=}NAr + 2 \, H{\cdot}$$

SCHEME 38.

oxidation of aromatic amines to azo compounds occurring in strongly basic solutions[307], and in pyridine catalysed by cuprous chloride[308], may also proceed via a similar homolytic pathway and not require the formation of peroxidic intermediates.

VII. REFERENCES

1. H. K. Hall, *J. Am. Chem. Soc.*, **79**, 5441 (1957).
2. H. K. Hall, *J. Org. Chem.*, **29**, 3539 (1964).
3. C. G. Swain and C. B. Scott, *J. Am. Chem. Soc.*, **75**, 141 (1953).

4. H. C. Brown, *Record Chem. Progr.* (*Kresge-Hooker Sci. Lib.*), **14**, 83 (1953).
5. R. P. Bell, *The Proton in Chemistry*, Cornell Univ. Press, Ithaca, New York, 1959, p. 175 ff.
6. L. L. Fellinger and L. F. Audrieth, *J. Am. Chem. Soc.*, **60**, 579 (1938).
7. P. K. Glasoe, J. Kleinberg and L. F. Audrieth, *J. Am. Chem. Soc.*, **61**, 2387 (1939).
8. R. L. Betts and L. P. Hammett, *J. Am. Chem. Soc.*, **59**, 1568 (1937).
9. W. H. Watanabe and L. R. DeFonso, *J. Am. Chem. Soc.*, **78**, 4542 (1956).
10. J. F. Bunnett and G. T. Davis, *J. Am. Chem. Soc.*, **82**, 665 (1960).
11. W. P. Jencks and J. Carriuolo, *J. Am. Chem. Soc.*, **82**, 675 (1960).
12. T. C. Bruice and M. F. Mayahi, *J. Am. Chem. Soc.*, **82**, 3067 (1960).
13. T. C. Bruice and S. J. Benkovic, *J. Am. Chem. Soc.*, **86**, 418 (1964).
14. W. P. Jencks and M. Gilchrist, *J. Am. Chem. Soc.*, **88**, 104 (1966).
15. M. L. Bender, *J. Am. Chem. Soc.*, **73**, 1626 (1951).
16. T. C. Bruice and L. R. Fedor, *J. Am. Chem. Soc.*, **86**, 4886 (1964).
17. R. Levine and W. C. Fernelius, *Chem. Rev.*, **54**, 523 (1954).
18. T. C. Bruice and R. G. Willis, *J. Am. Chem. Soc.*, **87**, 531 (1965).
19. E. Testa, L. Fontanella and M. Bovara, *Ann. Chem.*, **671**, 97 (1964).
20. L. R. Fedor, T. C. Bruice, K. L. Kirk and J. Meinwald, *J. Am. Chem. Soc.*, **88**, 108 (1966).
21. T. C. Bruice, J. J. Bruno and W-S. Chou, *J. Am. Chem. Soc.*, **85**, 1659 (1963).
22. L. R. Fedor and T. C. Bruice, *J. Am. Chem. Soc.*, **86**, 4117 (1964).
23. C. Schotten, *Chem. Ber.*, **21**, 2235 (1888).
24. E. Baumann, *Chem. Ber.*, **19**, 3218 (1886).
25. E. G. Williams and C. N. Hinshelwood, *J. Chem. Soc.*, 1079 (1934).
26. F. J. Stubbs and C. N. Hinshelwood, *J. Chem. Soc.*, S71 (1949).
27. A. N. Bose and C. N. Hinshelwood, *J. Chem. Soc.*, 4085 (1958).
28. H. S. Venkataraman and C. N. Hinshelwood, *J. Chem. Soc.*, 4977 (1960).
29. H. S. Venkataraman and C. N. Hinshelwood, *J. Chem. Soc.*, 4986 (1960).
30. L. M. Litvinenko, *Izv. Akad. Nauk SSSR*, 1737 (1962).
31. C. A. Bunton, T. A. Lewis and D. R. Llewellyn, *Chem. Ind.* (*London*), 1154 (1954).
32. M. L. Bender and J. M. Jones, *J. Org. Chem.*, **27**, 3771 (1962).
33. J. J. Eliott and S. F. Mason, *Chem. Ind.* (*London*), 488 (1959).
34. L. Remmers, *Chem. Ber.*, **7**, 346 (1874).
35. F. Ulffers and A. von Janson, *Chem. Ber.*, **27**, 93 (1894).
36. J. J. Sudborough, *J. Chem. Soc.*, **79**, 533 (1901).
37. K. J. P. Orton, *J. Chem. Soc.*, **81**, 495 (1902).
38. A. E. Smith and K. J. P. Orton, *J. Chem. Soc.*, **93**, 1242 (1908).
39. J. Koskikallio, *Suomen Kemistilehti*, **32**, 133 (1959).
40. M. Kh. Gluzman and R. S. Mil'ner, *Izv. Vysshikh. Uchebn. Zavedenii, Khim. i Khim. Tekhnol.*, **3**, 684 (1960); *Chem. Abstr.*, **55**, 2557c (1961).
41. B. L. Archer and R. F. Hudson, *J. Chem. Soc.*, 3259 (1950).
42. D. B. Denney and M. A. Greenbaum, *J. Am. Chem. Soc.*, **78**, 877 (1956).
43. M. H. Loucheux and A. Banderet, *Bull. Soc. Chim. France*, 2242 (1961).
44. L. M. Litvinenko, D. M. Aleksandrova and A. A. Zhilinskaya, *Ukr. Khim. Zh.*, **26**, 476 (1960); *Chem. Abstr.*, **55**, 10022h (1961).
45. L. M. Litvinenko and N. M. Oleinik, *Zh. Obshch. Khim.*, **33**, 2287 (1963).

46. L. M. Litvinenko and D. M. Aleksandrova, *Ukr. Khim. Zh.*, **27**, 212 (1961); *Chem. Abstr.*, **55**, 23389h (1961).
47. S. Bruckenstein and A. Saito, *J. Am. Chem. Soc.*, **87**, 698 (1965).
48. V. E. Bel'skii and M. I. Vinnik, *Izv. Akad. Nauk SSSR, Ser. Khim.*, 40 (1964); *Chem. Abstr.*, **60**, 14343d (1964).
49. V. E. Bel'skii and M. I. Vinnik, *Izv. Akad. Nauk SSSR, Ser. Khim.*, 2132 (1963); *Chem. Abstr.*, **60**, 10491b (1964).
50. A. R. Emery and V. Gold, *J. Chem. Soc.*, 1443 (1950).
51. A. R. Emery and V. Gold, *J. Chem. Soc.*, 1447 (1950).
52. A. R. Emery and V. Gold, *J. Chem. Soc.*, 1455 (1950).
53. H. Goldschmidt and C. Wachs, *Z. Physik. Chem.*, 353 (1897).
54. H. Morawetz and P. S. Otaki, *J. Am. Chem. Soc.*, **85**, 463 (1963).
55. T. C. Bruice and S. J. Benkovic, *Bioorganic Mechanisms*, Vol. I, W. A. Benjamin, Inc., New York, 1966, p. 91.
56. S. Meyer, M. Kócor and E. Taschner, *Roczniki Chem.*, **32**, 277 (1958).
57. D. P. N. Satchell, *Quart. Rev. (London)*, **17**, 60 (1963).
58. Yu. V. Svetkin, *Zh. Obshch. Khim.*, **26**, 1216 (1956).
59. Yu. V. Svetkin, *Zh. Obshch. Khim.*, **27**, 1676 (1957).
60. T. A. Sokolova and L. A. Ovsyannikova, *Zh. Obshch. Khim.*, **28**, 779 (1958).
61. E. A. Jeffery and D. P. N. Satchell, *J. Chem. Soc.*, 1876 (1962).
62. H. J. Hagemeyer and D. C. Hull, *Ind. Eng. Chem.*, **41**, 2920 (1949).
63. A. Dornow and H. Theidel, *Angew. Chem.*, **66**, 605 (1954).
64. J. Pokorny, *Chem. Zvesti*, **18**, 218 (1964).
65. N. Tani, N. Oguni and T. Araki, *Bull. Chem. Soc. Japan*, **37**, 1245 (1964).
66. Y. Ueno, T. Takaya and E. Imoto, *Bull. Chem. Soc. Japan*, **37**, 864 (1964).
67. Yu. V. Mitin, T. Yu. Stolyarova and G. P. Vlasova, *Zh. Obshch. Khim.*, **33**, 3628 (1963).
68. Yu. K. Yur'ev and Z. V. Belyakova, *Zh. Obshch. Khim.*, **28**, 1 (1958).
69. M. Sekiya, *J. Pharm. Soc. Japan*, **70**, 553 (1950).
70. H. R. Hirst and J. B. Cohen, *J. Chem. Soc.*, **67**, 830 (1895).
71. A. Galat and G. Elion, *J. Am. Chem. Soc.*, **65**, 1566 (1943).
72. G. P. Pettit and E. G. Thomas, *J. Org. Chem.*, **24**, 895 (1959).
73. J. P. E. Human and J. A. Mills, *J. Chem. Soc.*, 1457 (1948).
74. E. McC. Arnett, J. G. Miller and A. R. Day, *J. Am. Chem. Soc.*, **72**, 5635 (1950).
75. F. F. Blicke and C. Lu, *J. Am. Chem. Soc.*, **74**, 3933 (1952).
76. L. Schmid and B. Becker, *Monatsh. Chem.*, **46**, 675 (1926).
77. D. J. Beaver, D. P. Roman and P. J. Stoffel, *J. Am. Chem. Soc.*, **79**, 1236 (1957).
78. C. E. Dalgliesch, *J. Chem. Soc.*, 137 (1952).
79. C. W. Huffman, *J. Org. Chem.*, **23**, 727 (1958).
80. C. D. Hurd and A. S. Roe, *J. Am. Chem. Soc.*, **61**, 3355 (1939).
81. G. Oláh and S. Kuhn, *Chem. Ber.*, **89**, 2211 (1956).
82. S. Patai and S. Weiss, *J. Chem. Soc.*, 1035 (1959).
83. F. E. King, T. J. King and I. H. M. Muir, *J. Chem. Soc.*, 5 (1946).
84. A. Streitwieser, *Chem. Rev.*, **56**, 602 (1956).
85. K. J. Laidler, *J. Chem. Soc.*, 1786 (1938); E. Hertel and H. Lührmann, *Z. Elektrochem.*, **45**, 405 (1939); D. P. Evans, H. B. Watson and R. Williams, *J. Chem. Soc.*, 1345 (1939).

86. J. W. Baker, *J. Chem. Soc.*, 2631 (1932).
87. N. Menschutkin, *Z. Physik. Chem.*, **6**, 41 (1890).
88. Cited in P. A. S. Smith, *Open-chain Nitrogen Compounds*, Vol. I, W. A. Benjamin, Inc., New York, 1965, p. 23.
89. A. Streitwieser, *Solvolytic Displacement Reactions*, McGraw-Hill, New York, 1962.
90. C. A. Bunton, *Nucleophilic Substitution at a Saturated Carbon Atom*, Elsevier, New York, 1963.
91. R. G. R. Bacon and J. Köchling, *J. Chem. Soc.*, 5366 (1965).
92. A. Illiceto, A. Fava and U. Mazzuccato, *J. Org. Chem.*, **25**, 1445 (1960).
93. F. Müller, *Methoden der Organischen Chemie* (Houben-Weyl), Vol. 11/1 (Ed. E. Müller), Georg Thieme Verlag, Stuttgart, 1957, p. 267.
94. W. J. Dale and G. Buell, *J. Org. Chem.*, **21**, 45 (1956).
95. R. Wegler and G. Pieper, *Chem. Ber.*, **83**, 1 (1950).
96. J. A. Gervasi, M. Brown and L. A. Bigelow, *J. Am. Chem. Soc.*, **78**, 1679 (1956).
97. J. Meisenheimer, *Ann. Chem.*, **323**, 205 (1902).
98. G. N. Lewis and G. T. Seaborg, *J. Am. Chem. Soc.*, **62**, 2122 (1940).
99. O. L. Brady and F. R. Cropper, *J. Chem. Soc.*, 507 (1950).
100. S. D. Ross, *J. Am. Chem. Soc.*, **80**, 5319 (1958).
101. J. F. Bunnett and J. J. Randall, *J. Am. Chem. Soc.*, **80**, 6020 (1958).
102. C. Bernasconi and H. Zollinger, *Tetrahedron Letters*, 1083 (1965).
103. F. Pietra, *Tetrahedron Letters*, 745 (1965).
104. J. F. Bunnett and C. Bernasconi, *J. Am. Chem. Soc.*, **87**, 5209 (1965).
105. C. R. Hart and A. N. Bourns, *Tetrahedron Letters*, 2995 (1966).
106. J. F. Bunnett, E. W. Garbisch and K. M. Pruitt, *J. Am. Chem. Soc.*, **79**, 385 (1957).
107. J. F. Bunnett and R. H. Garst, *J. Am. Chem. Soc.*, **87**, 3875 (1965).
108. A. J. Kirby and W. P. Jencks, *J. Am. Chem. Soc.*, **87**, 3217 (1965).
109. A. Lowry and E. H. Balz, *J. Am. Chem. Soc.*, **43**, 344 (1921).
110. E. H. Cordes and W. P. Jencks, *J. Am. Chem. Soc.*, **84**, 826, 832 (1962).
111. G. Mignonac, *Compt. Rend.*, **172**, 223 (1921).
112. K. Baur, *U.S. Pat.*, 1,966,478 (1934).
113. C. F. Winano, *J. Am. Chem. Soc.*, **61**, 3566 (1939).
114. A. L. Klebanskii and M. S. Vlesova, *Zh. Obshch. Khim.*, **28**, 1767 (1958).
115. G. N. Kao, B. D. Tilak and K. Venkataraman, *J. Sci. Ind. Res. (India)*, **14B**, 624 (1955).
116. R. G. Rice and E. J. Kohn, *J. Am. Chem. Soc.*, **77**, 4052 (1955).
117. C. Ainsworth, *J. Am. Chem. Soc.*, **78**, 1635 (1956).
118. E. F. Pratt and E. J. Frazza, *J. Am. Chem. Soc.*, **76**, 6174 (1954).
119. M. Darrien and J. C. Jungers, *Bull. Soc. Chim. France*, 2164 (1962).
120. W. B. Wright, *J. Org. Chem.*, **25**, 1033 (1960); **27**, 1042 (1962).
121. R. Leuckart, *Chem. Ber.*, **18**, 2341 (1885).
122. O. Wallach, *Chem. Ber.*, **24**, 3992 (1891).
123. H. T. Clarke, H. B. Gillespie and S. Z. Weisshaus, *J. Am. Chem. Soc.*, **55**, 4571 (1933).
124. M. L. Moore, *Organic Reactions*, Vol. V (Ed. R. Adams), John Wiley and Sons, New York, 1949, p. 301.

125. W. S. Emerson, F. W. Neumann and T. P. Moundres, *J. Am. Chem. Soc.*, **63**, 972 (1941).
126. M. Metáyer, P. Mastagli and A. Bricard, *Bull. Soc. Chim. France*, 1054 (1950).
127. E. Staple and E. C. Wagner, *J. Org. Chem.*, **14**, 559 (1949).
128. A. Lukasiewicz, *Tetrahedron*, **19**, 1789 (1963).
129. E. R. Alexander and E. J. Underhill, *J. Am. Chem. Soc.*, **71**, 4014 (1949).
130. B. Reichert, *Die Mannich-Reaktion*, Springer Verlag, Berlin 1959.
131. A. W. Hofmann, *Chem. Ber.*, **12**, 984 (1879).
132. A. W. Hofmann, *Chem. Ber.*, **16**, 558 (1883).
133. G. H. Coleman, G. Nichols and T. F. Martens, *Organic Syntheses*, Vol. 25 (Ed. W. E. Bachmann), John Wiley and Sons, New York, 1945, p. 14.
134. G. H. Coleman, *J. Am. Chem. Soc.*, **55**, 3001 (1933).
135. S. Goldschmidt, *Chem. Ber.*, **46**, 2728 (1913).
136. A. Seyewetz and E. Chaix, *Bull. Soc. Chim. France*, **41**, 196 (1927).
137. S. Goldschmidt, R. Endres and R. Dirsch, *Chem. Ber.*, **58**, 573 (1925).
138. H. E. Baumgarten and J. M. Petersen, *J. Am. Chem. Soc.*, **82**, 459 (1960).
139. R. P. Mauger and F. G. Soper, *J. Chem. Soc.*, 71 (1946).
140. A. Schönberg, R. Moubasher and M. Z. Barakat, *J. Chem. Soc.*, 2504 (1951).
140 (a) P. S. Skell, D. L. Tuleen and P. D. Readio, *J. Am. Chem. Soc.*, **85**, 2850 (1963).
141. S. W. Fox and M. W. Bullock, *J. Am. Chem. Soc.*, **73**, 2754 (1951).
142. W. Eller and L. Klemm, *Chem. Ber.*, **55**, 217 (1922).
143. C. J. Hoffman and R. G. Neville, *Chem. Rev.*, **62**, 1 (1962).
144. P. Robson, V. C. R. McLoughlin, J. B. Hynes and L. A. Bigelow, *J. Am. Chem. Soc.*, **83**, 5010 (1961).
145. R. C. Petry and J. P. Freeman, *J. Am. Chem. Soc.*, **83**, 3912 (1961).
146. R. E. Banks, R. N. Hazeldine and J. P. Lalu, *Chem. Ind. (London)*, 1803 (1964).
147. C. W. Crane, J. Forrest, O. Stephenson and W. A. Waters, *J. Chem. Soc.*, 827 (1946).
148. H. Böhme and W. Krause, *Chem. Ber.*, **84**, 170 (1951).
149. H. Böhme, E. Mundlos and O-E. Herboth, *Chem. Ber.*, **90**, 2003 (1957).
150. A. J. Ellis and F. G. Soper, *J. Chem. Soc.*, 1750 (1954).
151. E. Müller and H. Hais, *Chem. Ber.*, **96**, 570 (1963).
152. E. Müller, H. Hais and W. Rundel, *Chem. Ber.*, **93**, 1541 (1960).
153. E. Kalatzis and J. H. Ridd, *J. Chem. Soc. (B)*, 529 (1966).
154. For a summary see P. B. D. de la Mare and J. H. Ridd, *Aromatic Substitution*, Butterworths, London, 1959.
155. K. H. Saunders, *The Aromatic Diazo-Compounds*, E. Arnold and Co., London, 1949; H. Zollinger, *Diazo and Azo Chemistry*, Interscience Publ. Inc., New York, 1961.
156. J. H. Ridd, *Quart. Rev. (London)*, **15**, 418 (1961); H. Schmid, *Chemiker Ztg.*, **86**, 809 (1962).
157. J. H. Ridd, *J. Soc. Dyers Colourists*, **81**, 355 (1965).
158. H. Leotte, *Rev. Port. Quim.*, **6**, 108 (1964).
159. L. P. Hammett, *Physical Organic Chemistry*, McGraw-Hill Inc., New York, 1940, p. 294.

160. H. Schmid, *Z. Elektrochem.*, **42**, 579 (1936).
161. T. W. J. Taylor, *J. Chem. Soc.*, 1099 (1928).
162. T. W. J. Taylor and L. S. Price, *J. Chem. Soc.*, 2052 (1929).
163. E. D. Hughes, C. K. Ingold and J. H. Ridd, *J. Chem. Soc.*, 65 (1958).
164. E. D. Hughes and J. H. Ridd, *J. Chem. Soc.*, 70 (1958).
165. A. Hantzsch and M. Schümann, *Chem. Ber.*, **32**, 1691 (1899).
166. E. D. Hughes, C. K. Ingold and J. H. Ridd, *J. Chem. Soc.*, 83 (1958).
167. E. A. Bunton, D. R. Llewellyn and G. Stedman, *Chem. Soc. (London)*, *Spec. Publ.*, (10), 113 (1957); C. A. Bunton, D. R. Llewellyn and G. Stedman, *J. Chem. Soc.*, 568 (1959).
168. C. A. Bunton, J. E. Burch, B. C. Challis and J. H. Ridd, unpublished results.
169. C. A. Bunton and G. Stedman, *J. Chem. Soc.*, 3466 (1959).
170. J. Kenner, *Chem. Ind. (London)*, **19**, 443 (1941).
171. A. T. Austin, *Ph.D. Thesis*, London, 1950.
172. L. F. Larkworthy, *J. Chem. Soc.*, 3116 (1959).
173. H. Schmid and C. Essler, *Monatsh. Chem.*, **91**, 484 (1960).
174. (a) A. T. Austin, *Nature*, **188**, 1086 (1960).
 (b) G. J. Ewing and N. Bauer, *J. Phys. Chem.*, **62**, 1449 (1958).
 (c) A. T. Austin, E. D. Hughes, C. K. Ingold and J. H. Ridd, *J. Am. Chem. Soc.*, **74**, 555 (1952).
175. H. Schmid, *Monatsh, Chem.*, **85**, 424 (1954).
176. L. F. Larkworthy, *J. Chem. Soc.*, 3304 (1959).
177. (a) H. Schmid, *Z. Elektrochem.*, **43**, 626 (1937).
 (b) H. Schmid and G. Muhr, *Chem. Ber.*, **70**, 421 (1937).
178. E. D. Hughes and J. H. Ridd, *J. Chem. Soc.*, 82 (1958).
179. G. Stedman, *J. Chem. Soc.*, 2949 (1959).
180. H. Schmid and A. Maschka, *Z. Physik. Chem.* (Part B), **49**, 171 (1941).
181. (a) H. Schmid and R. Pfeifer, *Monatsh. Chem.*, **84**, 829 (1953).
 (b) H. Schmid and R. Pfeifer, *Monatsh. Chem.*, **84**, 842 (1953).
182. H. Söll in *Methoden der Organischen Chemie* (Houben-Weyl), Vol. 11, part 2 (Ed. E. Müller), Georg Thieme Verlag, Stuttgart, 1958, p. 133.
183. H. Schmid and G. Muhr, *Monatsh. Chem.*, **91**, 1198 (1960).
184 (a) H. Schmid and G. Muhr, *Monatsh. Chem.*, **93**, 102 (1962).
 (b) H. Schmid and G. Morawetz, *Ber. Bunsen. Physik. Chem.*, **67**, 797 (1963).
185. (a) P. A. S. Smith, D. R. Baer and S. N. Eğe, *J. Am. Chem. Soc.*, **76**, 4564 (1954).
 (b) H. Felkin, *Compt. Rend.*, **236**, 298 (1953).
186. H. Krauch and W. Kunz, *Organic Name Reactions*, J. Wiley and Sons, Inc., New York, 1964, (a) p. 186; (b) p. 404.
187. H. Schmid and C. Essler, *Monatsh. Chem.*, **88**, 1110 (1957).
188. H. Schmid and E. Hallaba, *Monatsh. Chem.*, **87**, 560 (1956).
189. H. Schmid and M. G. Fouad, *Monatsh. Chem.*, **88**, 631 (1957).
190. S. W. Benson, *The Foundations of Chemical Kinetics*, McGraw-Hill Co. Inc., New York, 1960, p. 497.
191. E. D. Hughes and J. H. Ridd, *J. Chem. Soc.*, 3466 (1959).
192. (a) K. Singer and P. A. Vamplew, *J. Chem. Soc.*, 3971 (1956).
 (b) N. S. Bayliss, R. Dingle, D. W. Watts and R. J. Wilkie, *Australian J. Chem.*, **16**, 933 (1963).

193. B. C. Challis, L. F. Larkworthy and J. H. Ridd, *J. Chem. Soc.*, 5203 (1962).
194. B. C. Challis and J. H. Ridd, *J. Chem. Soc.*, 5197 (1962).
195. H. Schmid and C. Essler, *Monatsh. Chem.*, **90**, 222 (1959).
196. B. C. Challis, *Ann. Reports*, **62**, 249 (1965).
197. B. C. Challis and J. H. Ridd, *J. Chem. Soc.*, 5208 (1962).
198. E. C. R. de Fabrizio, E. Kalatzis and J. H. Ridd, *J. Chem. Soc.* (B), 533 (1966).
199. (a) D. Dimitrov and F. Fratev, *Compt. Rend. Acad. Bulgare Sci.*, **16**, 825 (1963).
    (b) Z. J. Allen, J. Podstata, D. Snobe and J. Jarkovsky, *Tetrahedron Letters*, 3565 (1965).
200. (a) M. N. Hughes, T. D. B. Morgan and G. Stedman, *Chem. Commun.*, 241 (1966).
    (b) M. N. Hughes and G. Stedman, *J. Chem. Soc.*, 2824 (1963).
201. H. Ladenheim and M. L. Bender, *J. Am. Chem. Soc.*, **82**, 1895 (1960).
202. D. A. Morrison and T. A. Turney, *J. Chem. Soc.*, 4827 (1960).
203. B. C. Challis, unpublished results.
204. B. C. Challis and J. H. Ridd, *Proc. Chem. Soc.*, 245 (1960).
205. Cf. E. F. Caldin, *Fast Reactions in Solution*, J. Wiley and Sons, Inc., New York, 1964, p. 244.
206. J. H. Ridd, personal communication.
207. E. A. Halevi, A. Ron and S. Speiser, *J. Chem. Soc.*, 2560 (1965).
208. H. Schmid and A. Woppman, *Monatsh. Chem.*, **83**, 346 (1952).
209. H. Schmid and A. F. Sami, *Monatsh. Chem.*, **86**, 904 (1955).
210. V. V. Kozlov and B. I. Belov, *Zh. Obshch. Khim.*, **33**, 1951 (1963).
211. F. Seel and W. Hufnagel, *Z. Physik. Chem. (Frankfurt)*, **26**, 269 (1960).
212. (a) Shu Huang, *Hua Hsüeh Hsüeh Pao.*, **25**, 171 (1959); *Chem. Abstr.*, **54**, 4489 (1960).
    (b) J. I. G. Cadogan, *J. Chem. Soc.*, 4257 (1962).
213. (a) U. Wannagut and G. Hohlstein, *Chem. Ber.*, **88**, 1839 (1955).
    (b) G. Oláh, L. Nosko, S. Kuhn and M. Szelke, *Chem. Ber.*, **89**, 2374 (1956).
    (c) G. G. Yakobsen, A. I. D'yachenko and F. A. Bel'chikova, *Zh. Obshch. Khim.*, **32**, 849 (1962).
    (d) G. Oláh, N. A. Overchuk and J. C. Lapierre, *J. Am. Chem. Soc.*, **87**, 5785 (1965).
214. E. L. Reilly (to E. I. du Pont de Nemours and Co.), *Ger. Pat.* 1,085,166 (1960); E. L. Reilly (to E. I. de Pont de Nemours and Co.), *U.S. Pat.* 3,153,094 (1964); J. F. Haller (to Olin Mathieson Chemical Corp), *U.S. Pat.* 3,065,270 (1962); D. R. Lavering and L. G. Maury (to Hercules Powder Co.), *U.S. Pat.* 3,090,786 (1963).
215. W. Brackman and P. J. Smit, *Rec. Trav. Chim.*, **84**, 357 (1965).
216. R. S. Drago and F. E. Paulick, *J. Am. Chem. Soc.*, **83**, 1819 (1961).
217. R. S. Drago, R. O. Ragsdale and D. P. Eyman, *J. Am. Chem. Soc.*, **83**, 4337 (1961).
218. F. Minisci and R. Galli, *Chem. Ind. (London)*, **46**, 173 (1964).
219. J. Rigaudy and J. C. Vernières, *Compt. Rend.*, **261**, 5516 (1965).
220. P. A. S. Smith and H. G. Pars, *J. Org. Chem.*, **24**, 1325 (1959).
221. S. Ghosal and B. Mukherjee, *Indian J. Chem.*, **4**, 30 (1966).

222. J. Glazer, E. D. Hughes, C. K. Ingold, A. T. James, G. T. Jones and E. Roberts, *J. Chem. Soc.*, 2671 (1950).
223. (a) H. P. Patel and J. M. Tedder, *J. Chem. Soc.*, 4889 (1963).
     (b) J. M. Tedder, *J. Chem. Soc.*, 4003 (1957).
224. E. Bamberger and R. Seligman, *Chem. Ber.*, **36**, 701 (1903).
225. E. Bamberger, *Chem. Ber.*, **36**, 710 (1903).
226. (a) V. A. Cherkasova, *Zh. Obshch. Khim.*, **29**, 2804 (1959).
     (b) K. Kahr and C. Berther, *Chem. Ber.*, **93**, 132 (1960).
     (c) O. L. Lebedev and S. N. Karzarnovskii, *Zh. Obshch. Khim.*, **30**, 1631 (1960).
     (d) P. Burckard, J-P. Fleury and F. Weiss, *Bull. Chim. Soc. France*, 2730 (1965).
     (e) O. L. Lebedev and S. N. Karzarnovskii, *Zh. Obshch. Khim.*, **30**, 3105 (1960).
227. C. H. Kline and V. Kollonitsch, *Ind. Eng. Chem.*, **57**, 73 (1965).
228. Synthese Chemie G.m.b.H., *Ger. Pat.* 951,933 (1953).
229. Badische Anilin and Soda Fabrik A.G., *Ger. Pat.* 1,021,358 (1955).
230. W. D. Emmons, *J. Am. Chem. Soc.*, **79**, 5528 (1957).
231. W. D. Emmons and A. S. Pagano, *J. Am. Chem. Soc.*, **77**, 4557 (1955).
232. W. D. Emmons, *J. Am. Chem. Soc.*, **76**, 3468 (1954).
233. C. K. Ingold, *J. Chem. Soc.*, **125**, 87 (1924).
234. (a) A. Baeyer and V. Villiger, *Chem. Ber.*, **33**, 1569 (1900).
     (b) N. Prileschajew, *Chem. Ber.*, **42**, 4811 (1909).
235. A. Baeyer and V. Villiger, *Chem. Ber.*, **34**, 762, (1901).
236. J. A. Castellano, J. Green and J. M. Kauffman, *J. Org. Chem.*, **31**, 821 (1966).
237. (a) J. D'Ans and A. Kneip, *Chem. Ber.*, **48**, 1136 (1915).
     (b) R. R. Holmes and R. P. Bayer, *J. Am. Chem. Soc.*, **82**, 3454 (1960).
238. L. I. Khmel'nitskii, T. S. Novikova and S. S. Novikov, *Izv. Akad. Nauk SSSR Otd. Khim. Nauk*, 516 (1962).
239. R. W. White and W. D. Emmons, *Tetrahedron*, **17**, 31 (1962).
240. (a) J. H. Boyer and S. E. Ellzey, *J. Org. Chem.*, **24**, 2038 (1959).
     (b) K. M. Ibne-Rasa, C. G. Lauro and J. O. Edwards, *J. Am. Chem. Soc.*, **85**, 1165 (1963).
241. (a) B. Oddo and R. Binaghi, *Gazz. Chim. Ital.*, **54**, 193 (1924).
     (b) J. F. Durand and R. Navas, *Compt. Rend.*, **180**, 521 (1925).
242. P. Ruggli and J. Rohner, *Helv. Chim. Acta*, **25**, 1553 (1925).
243. Y. Ogata and Y. Takaji, *J. Am. Chem. Soc.*, **80**, 3591 (1958).
244. K. M. Ibne-Rasa and J. O. Edwards, *J. Am. Chem. Soc.*, **84**, 763 (1962).
245. P. Robson, *J. Chem. Soc.*, 5173 (1964).
246. I. P. Gragerov and A. F. Levit, *Zh. Obshch. Khim.*, **30**, 3726 (1960).
247. (a) L. Mamlock and R. Wolffenstein, *Chem. Ber.*, **33**, 159 (1900).
     (b) M. A. T. Rodgers, *J. Chem. Soc.*, 769 (1955).
248. H. Q. Smith (to Pennsalt Chem. Corp.), *Belg. Pat.* 615,736 (1962).
249. (a) E. Bamberger and M. Vuk, *Chem. Ber.*, **35**, 703 (1902).
     (b) R. Hubner, *Chem. Ber.*, **35**, 731 (1902).
250. A. A. R. Sayigh and U. Ulrich, *J. Chem. Soc.*, 3144 (1963).
251. A. T. Koritskii and A. F. Lukornikov, *Dokl. Akad. Nauk SSSR*, **147**, 1126 (1962).

252. H. Wieland and S. Gambarjan, *Chem. Ber.*, **39**, 1499 (1906).
253. K. Tokumara, H. Sakuragi and O. Simamura, *Tetrahedron Letters*, 3945 (1964).
254. (a) H. Wieland and M. Offenbaecher, *Chem. Ber.*, **47**, 2111 (1914).
     (b) H. Wieland and K. Roth, *Chem. Ber.*, **53**, 210 (1920).
255. S. Gambarajan, *Chem. Ber.*, **42**, 4003 (1909).
256. (a) H. Freytag in *Methoden der Organischen Chemie* (Houben-Weyl), Vol. 11, part 2 (Ed. E. Müller), George Thieme, Stuttgart, 1958, p. 190.
     (b) D. Swern, *Chem. Rev.*, **45**, 34 (1949).
257. S. D. Ross, *J. Am. Chem. Soc.*, **68**, 1484 (1946).
258. Y. Ogata and I. Tabushi, *Bull. Chem. Soc. Japan*, **31**, 969 (1958).
259. A. A. Oswald and D. L. Guertin, *J. Org. Chem.*, **28**, 651 (1958).
260. (a) C. W. Capp and E. G. E. Hawkins, *J. Chem. Soc.*, 4106 (1953).
     (b) S. Gambarajan, *Chem. Ber.*, **60A**, 390 (1927).
     (c) S. Gambarajan, *Chem. Ber.*, **58B**, 1775 (1925).
261. C. Walling, *Free Radicals in Solution*, J. Wiley and Sons Inc., New York, 1957, pp. 590–595.
262. D. B. Denney and D. Z. Denney, *J. Am. Chem. Soc.*, **82**, 1389 (1960).
263. R. Huisgen and F. Bayerlein, *Ann. Chem.*, **630**, 138 (1960).
264. L. Horner and B. Anders, *Chem. Ber.*, **95**, 2470 (1962).
265. O. A. Chaltikyan, E. N. Atanasyan, A. A. Saskisyan, G. A. Marmaryan and D. S. Gaibakyan, *Zh. Fiz. Khim.*, **32**, 2601 (1958).
266. Ya. K. Syrkin and I. I. Moiseev, *Uspekki Khim.* (Engl. Trans.), **29**, 209 (1960).
267. (a) J. T. Edwards and S. A. Samad, *Can. J. Chem.*, **41**, 1027 (1963).
     (b) J. T. Edwards, *J. Chem. Soc.*, 1464 (1954).
268. A summary is given in ref. 262.
269. C. K. Ingold, *Structure and Mechanism in Organic Chemistry*, Cornell Univ. Press, Ithaca, New York, 1953, pp. 621–624.
270. R. I. Milyutinskaya and K. S. Bagdasaryn, *Zh. Fiz. Khim.*, **34**, 405 (1960).
271. D. Buckley, S. Dunstan and H. B. Henbest, *J. Chem. Soc.*, 4901 (1957).
272. Ref. 261, p. 590.
273. L. Horner and K. Scherf, *Ann. Chem.*, **573**, 35 (1951).
274. G. Favini, *Gazz. Chim. Ital.*, **89**, 2121 (1959).
275. L. Horner and W. Kirmse, *Ann. Chem.*, **597**, 48 (1955).
276. W. B. Geiger, *J. Org. Chem.*, **23**, 298 (1958).
277. R. B. Roy and G. A. Swan, *Chem. Commun.*, 427 (1966).
278. C. Walling and N. Indictor, *J. Am. Chem. Soc.*, **80**, 5815 (1958).
279. R. Huisgen, F. Beyerlein and F. Heydkamp, *Chem. Ber.*, **93**, 363 (1960).
280. L. Horner and C. Betzel, *Ann. Chem.*, **579**, 175 (1953).
281. J. M. Fayadh, D. W. Jessop and G. A. Swan, *J. Chem. Soc.* (C), 1605 (1966).
282. S. D. Stavrova, G. V. Peregudov and M. F. Margaritova, *Dokl. Akad. Nauk SSSR*, **157**, 636 (1964).
283. D. M. Graham and R. B. Mesrobian, *Can. J. Chem.*, **41**, 2938 (1963).
284. M. Imoto, T. Otsu and T. Ota, *Makromol. Chem.*, **16**, 10 (1955).
285. L. Horner, H. Brüggerman and K. H. Knapp, *Ann. Chem.*, **626**, 1 (1959).
286. F. Hrabák and M. Vacek, *Collection Czech. Chem. Commun.*, **30**, 573 (1965).
287. (a) S. Gambarajan and L. Kazarian, *J. Gen. Chem. USSR*, **3**, 222 (1933).
     (b) L. Horner and W. Kirmse, *Ann. Chem.*, **597**, 66 (1955).

288. H. E. de la Mare, *J. Org. Chem.*, **25**, 2114 (1960).
289. L. A. Harris and J. S. Olcott, *J. Am. Oil Chemists' Soc.*, **43**, 11 (1966).
290. L. Kuhnen, *Chem. Ber.*, **99**, 3384 (1966).
291. G. M. Coppinger and J. D. Swalen, *J. Am. Chem. Soc.*, **83**, 4900 (1961).
292. E. S. Huyser, C. J. Bredeweg and R. M. Vanscoy, *J. Am. Chem. Soc.*, **86**, 4148 (1964).
293. H. B. Henbest and R. Patton, *J. Chem. Soc.*, 3557 (1960).
294. (a) W. Strecker and H. Thienemann, *Chem. Ber.*, **53**, 2096 (1920).
     (b) W. Strecker and M. Baltes, *Chem. Ber.*, **54**, 2693 (1921).
     (c) L. Horner, H. Schaefer and W. Ludwig, *Chem. Ber.*, **91**, 75 (1958).
295. P. S. Bailey, *Chem. Rev.*, **58**, 925 (1958).
296. P. Kolsaker and O. Meth-Cohn, *Chem. Commun.*, 423 (1965).
297. H. B. Henbest and M. J. W. Stratford, *J. Chem. Soc.*, 711 (1964).
298. G. P. Shulman, *Can. J. Chem.*, **43**, 3069 (1965).
299. (a) M. M. Wei and R. Stewart, *J. Am. Chem. Soc.*, **88**, 1974 (1966).
     (b) H. Schechter, S. S. Rawalay and M. Tubis, *J. Am. Chem. Soc.*, **86**, 1701 (1964).
     (c) H. Schechter and S. S. Rawalay, *J. Am. Chem. Soc.*, **86**, 1706 (1964).
300. V. S. Smirnov and E. A. Shklyarak, *J. Gen. Chem. USSR*, **16**, 1443, 1687 (1946).
301. N. Kornblum in *Organic Reactions*, Vol. 12 (Ed. A. C. Cope), J. Wiley and Sons Inc., New York, 1962, p. 116; cf. N. Kornblum and W. J. Jones, *Org. Syn.*, **43**, 87 (1963).
302. R. F. Merritt and J. F. Ruff, *J. Am. Chem. Soc.*, **86**, 1392 (1964).
303. (a) G. B. Barlin, K. H. Pausacker and N. V. Riggs, *J. Chem. Soc.*, 3122 (1954).
     (b) K. H. Pausacker, *J. Chem. Soc.*, 1989 (1953).
304. K. H. Pausacker and J. G. Scroggie, *J. Chem. Soc.*, 4003 (1954).
305. S. M. Mehta and M. V. Vakihnala, *J. Am. Chem. Soc.*, **74**, 563 (1952).
306. O. H. Wheeler and G. Gonzalez, *Tetrahedron*, **20**, 189 (1964).
307. (a) T. Kawabata, S. Tanimoto and R. Odda, *Kogyo Kagaku Zasshi*, **67**, 1151 (1964); *Chem. Abstr.*, **61**, 14597 (1964).
     (b) L. Horner and J. Dehnert, *Chem. Ber.*, **96**, 786 (1963).
     (c) G. A. Russell, *J. Am. Chem. Soc.*, **84**, 2652 (1962).
     (d) K. H. Pausacker and J. G. Scroggie, *Australian J. Chem.*, **12**, 430 (1959).
308. (a) K. Kinoshita, *Bull. Chem. Soc. Japan*, **32**, 777 (1959).
     (b) K. Kinoshita and T. Hitomi, *Kogyo Kagaku Zasshi*, **62**, 1387 (1959); *Chem. Abstr.*, **57**, 8467 (1962).
     (c) J. T. Yoke, *Inorg. Chem.*, **2**, 1210 (1963).

CHAPTER 7

Carbon–nitrogen and nitrogen–nitrogen double bond condensation reactions

Paula Y. Sollenberger and R. Bruce Martin

University of Virginia, Charlottesville, Virginia, U.S.A.

I. INTRODUCTION

The formation of $>$C$=$N— and —N$=$N— double bonds may take place by several different types of reactions. In order to keep the subject to manageable size, this chapter is concerned only with the formation of such bonds by condensation reactions. By describing a reaction as a condensation we mean to imply that after the reactants have come together some substance such as water, alcohol, or ammonia is eliminated. The emphasis of the chapter is on the mechanism and catalysis of these reactions. Preparative methods have not been emphasized since they have recently been treated elsewhere[1, 2]. The bibliography does not represent an exhaustive survey of the field and is weighted in favor of recent publications.

II. THE AZOMETHINE LINKAGE, $>$C$=$N—

A. Formation in Condensation Reactions

Aldehydes and ketones undergo reversible condensation reactions with primary amines to give products containing an azomethine, $>$C$=$N—, linkage and water. The reaction of carbonyl compounds with secondary and tertiary amines generally does not result in the formation of an azomethine linkage. Carbonyl compounds having an α-hydrogen atom react with secondary amines to give enamines. When, however, the perchlorate salts of the secondary amines are employed, ternary iminium salts are obtained[3]. The reaction of certain cyclic ketones with tertiary amines at low temperatures results in unstable complex formation[4].

Imines are generally not obtained from the reaction of carbonyl compounds with ammonia as the carbon–nitrogen double bond has no particular stabilization and tends to enter into polymerization reactions[5]. The product of the reaction of formaldehyde and ammonia is hexamine (1)[6, 7], and reactions of other aliphatic aldehydes

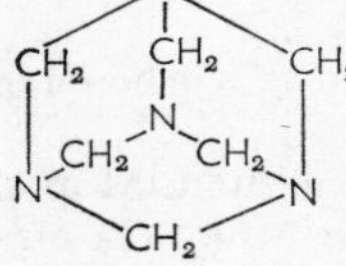

(1) 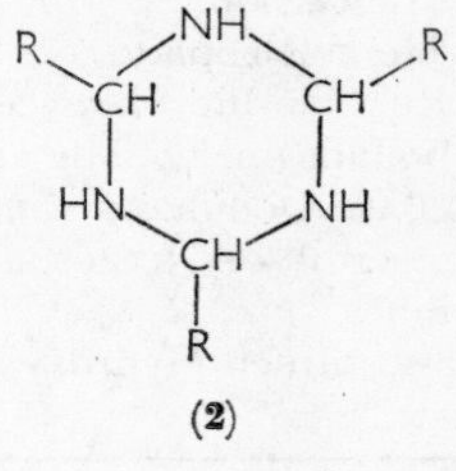   (2)

with ammonia result in the formation of hexahydrotriazine compounds (**2**) or α-amino alcohols[8, 9]. Aromatic aldehydes react with aqueous or alcoholic ammonia at room temperature to give hydroamides[10, 11] (equation 1) although the reaction has been found to stop at the imine when carried out in a very dilute aldehyde solution[12, 13].

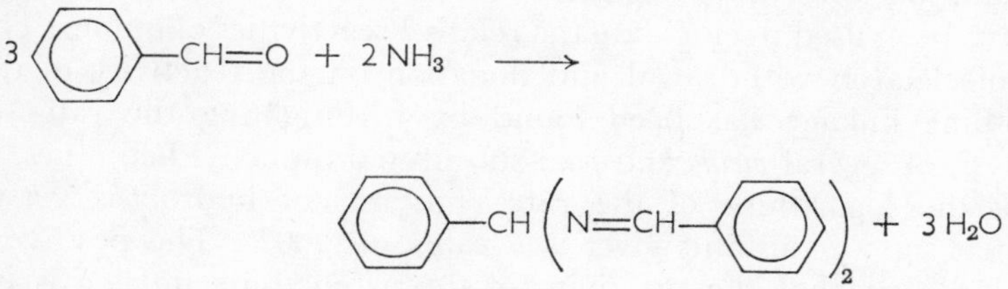

$$3\ C_6H_5\text{—CH}{=}O + 2\ NH_3 \longrightarrow$$

$$C_6H_5\text{—CH}\left(N{=}CH\text{—}C_6H_5\right)_2 + 3\ H_2O \qquad (1)$$

Diaryl ketimines are more stable than alkyl aryl ketimines which in turn are more stable than the purely aliphatic ketimines since conjugation increases the thermodynamic stability of the azomethine linkage[14]. Benzophenone imine (**3**), with a melting point of 48°c is quite stable[15]. Although the monoimine of *p*-benzoquinone is very unstable, the diimine (**4**) is sufficiently stable to be studied spectroscopically[16].

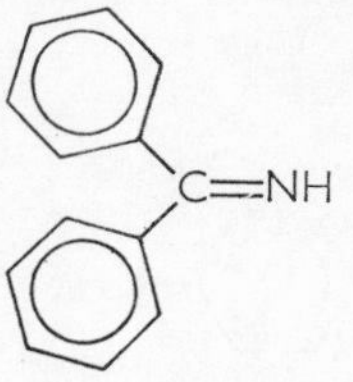

(**3**)

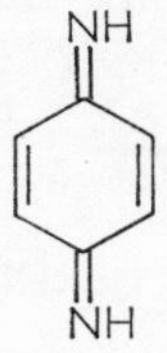

(**4**)

N-Substitution considerably increases the thermodynamic stability of imines. Thus Schiff bases, as *N*-substituted imines are commonly called, are generally obtained as the condensation products of the reaction of primary amines with carbonyl compounds (equation 2).

$$>\!C{=}O + RNH_2 \rightleftharpoons\ >\!C{=}NR + H_2O \qquad (2)$$

The reaction is reversible and reaches equilibrium noticeably short of completion[17]. When the reaction is carried out with an amine that has an electronegative atom containing at least one pair of unshared electrons adjacent to the attacking nitrogen atom, such as hydroxylamine, semicarbazide or hydrazine, the equilibrium is essentially complete and the condensation product may be readily isolated.

When aromatic amines are condensed with n-butyraldehyde in the absence of even trace amounts of acid, Schiff bases are not the exclusive condensation products. Dimers have been isolated from the reaction medium and investigation has shown them to have an enamine structure[18].

In addition to the above equilibrium considerations, kinetic studies also have been used to elucidate the relative reactivities of imines. The effect of electron withdrawal and donation on the reactivity of the azomethine linkage has been found by investigating the rates of hydrolysis of several *meta-* and *para*-substituted diphenyl ketimines. A plot of the logarithms of the rate constants of hydrolysis versus Hammett sigma constants gives a ρ value of $2 \cdot 00$[19]. The positive ρ value indicates that the reactivity of the azomethine linkage is increased by electron-withdrawing substituents while substituents capable of electron donation decrease its reactivity.

Recently, a case was reported where the reaction of an aromatic aldehyde with an aromatic amine does not lead to a Schiff base[20]. When *o*-nitroaniline is heated with an excess of benzaldehyde, the expected Schiff base, *N*-benzylidene-*o*-nitroaniline (**5**), is obtained as the major product. When, however, the same reaction is carried out with excess *o*-nitroaniline, **6** is obtained as the sole product. The

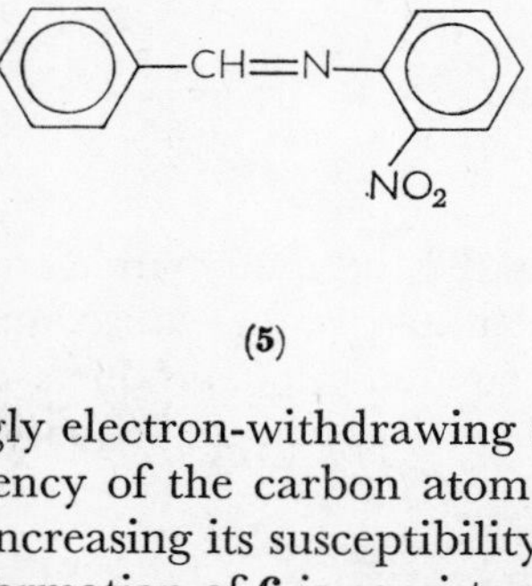

(5) 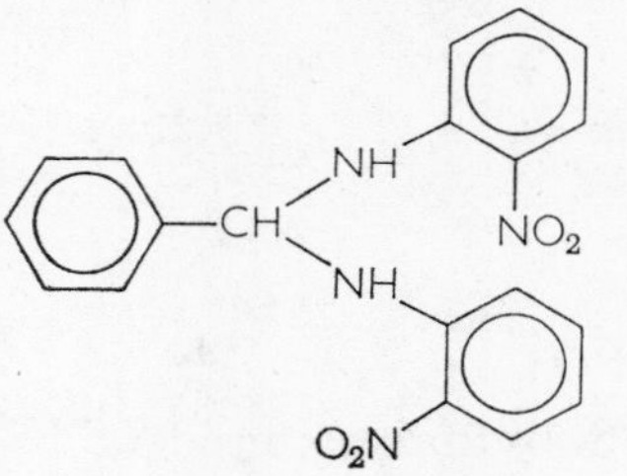   (6)

strongly electron-withdrawing nitro substituent enhances the electron deficiency of the carbon atom of the carbon–nitrogen double bond, thus increasing its susceptibility to attack by another amine molecule. The formation of **6** is consistent with the positive ρ value obtained in the Hammett plot. One other stable compound similar to **6** has been reported, *N,N*-trichloroethylidene-di(*o*-nitroaniline) obtained by the condensation of *o*-nitroaniline with chloral[21].

Several factors other than conjugation and inductive effects have been found to affect the reactivity of the azomethine linkage. *Ortho* and *para* hydroxy-substituted diaryl ketimines are unusually stable toward hydrolysis[22]. This inertness could be due to a phenol–imine keto–amine tautomerism[19] as in (3). Extensive existence of *ortho* and

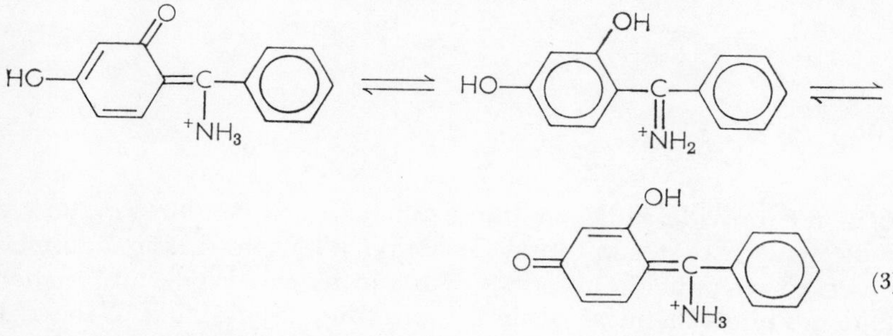

$$(3)$$

para hydroxy-substituted ketimines in the keto–amine form might account for the slow rate of hydrolysis since the keto–amine form should not be subject to hydrolysis. *Ortho* and *para* methoxy-substituted diaryl ketimines also show comparatively slow rates of hydrolysis. These compounds are not capable of tautomerism but resonance contributions from the amine forms could be responsible for their resistance to hydrolysis (equation 4).

$$(4)$$

The nuclear magnetic resonance spectra of Schiff bases formed from primary amines and *ortho*-hydroxy aldehydes and ketones show that the Schiff bases derived from 1-hydroxy-2-acetonaphthone and from 2-hydroxy-1-naphthaldehyde exist as keto amines (**7a**) although their formation involves loss of most of the resonance energy of one of the aromatic rings[23-25]. When R is a phenyl group, the phenol–imine tautomer (**7b**) predominates[24]. Schiff bases derived from *ortho*-hydroxy aldehydes and ketones have the phenol–imine structure (8)[23, 26]. Evidently, in such compounds the keto–amine tautomer

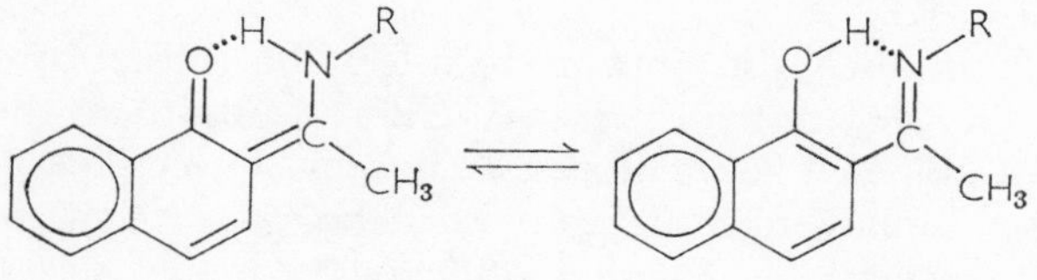

(7a)   (7b)

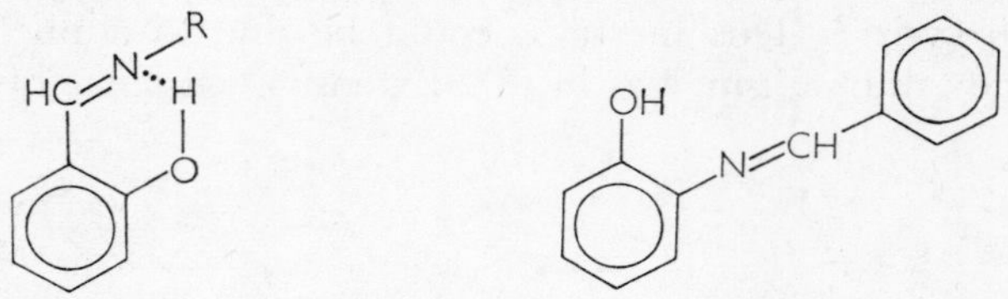

(8)   (9)

does not provide sufficient stabilization to compensate for the resonance energy that is lost in its formation. The keto–amine tautomer is
observed when acidic solutions of **8** and its *para* isomer are made to
undergo photochemical isomerizations, but its half-life is of the order
of 1 msec[27]. The infrared spectra of Schiff bases of amino acids with
3-hydroxypyridine-4-aldehyde, 3-hydroxypyridine-2-aldehyde and
salicylaldehyde show the presence of two tautomeric species although
the compounds have been found to exist predominantly in the keto–
amine form[28]. Spectroscopic studies show the existence of strong
intramolecular hydrogen bonding in Schiff bases such as **8**[29-34], but
hydrogen bonding is not observed in the isomeric compound **9** because of unfavorable steric conditions.

Nuclear magnetic resonance studies were also made of Schiff bases
obtained from amines and aliphatic β-diketones. These compounds
can exist in any of three tautomeric forms, the keto–imine (**10**), the
keto–enamine (**11**) and the enol–imine (**12**)[35]. It was found that in

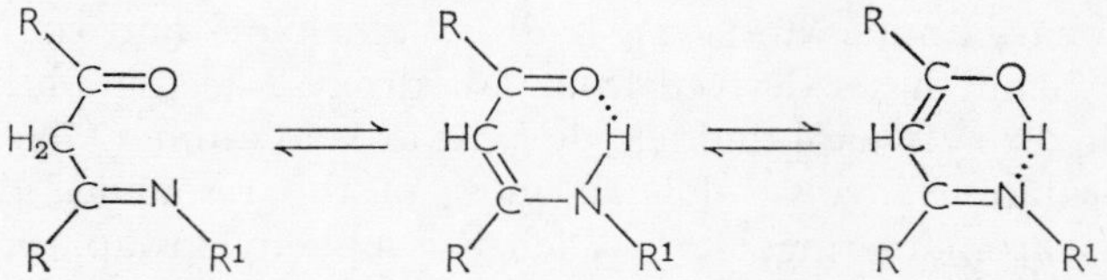

(10)   (11)   (12)

solution these bases exist predominantly in the keto–enamine form[36, 37].
The position of the equilibrium is not altered by variation of the
solvent or of R^1 [38].

Steric hindrance also appears to affect the resistance of the azomethine linkage to hydrolysis. A chloro substituent in the *ortho*
position of diphenyl ketimine causes a much slower rate of hydrolysis
than does a similar substituent in either the *meta* or *para* position. The
rates of hydrolysis of dimethyl diphenyl ketimines decrease in the
order shown. Several hours of refluxing in aqueous solution are re-

quired for the 2,6-disubstituted compound to show any hydrolysis at all[19].

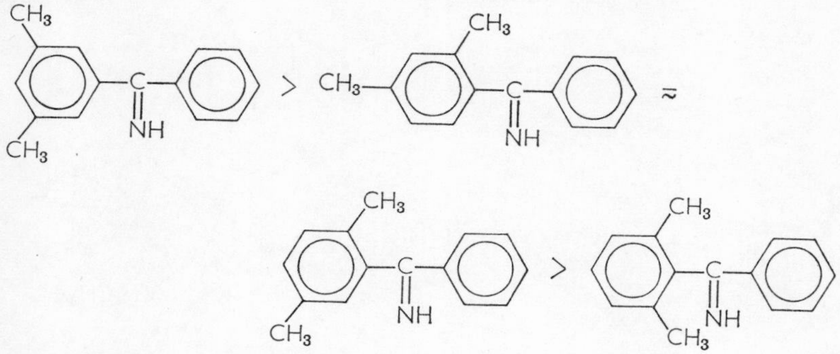

B. Metal Ion Promoted Condensation Reactions

The condensation reactions of carbonyl compounds with primary amines are influenced by the presence of metal ions. Metal ions bring together reactants in a structure suitable for complex formation so that products are obtained in their presence that are obtained sparingly or not at all in their absence.

Metal ions in condensation reactions may trap a reaction intermediate which would otherwise proceed to form a different final product. The condensation of methylamine with α-diketones in the presence of a metal ion results in Schiff base formation (equation 5).

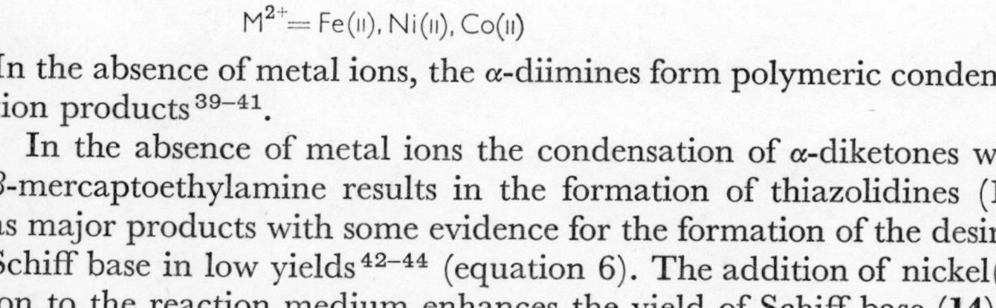

$$M^{2+} = Fe(II), Ni(II), Co(II)$$

In the absence of metal ions, the α-diimines form polymeric condensation products[39–41].

In the absence of metal ions the condensation of α-diketones with β-mercaptoethylamine results in the formation of thiazolidines (13) as major products with some evidence for the formation of the desired Schiff base in low yields[42–44] (equation 6). The addition of nickel(II) ion to the reaction medium enhances the yield of Schiff base (14) to greater than 70% (equation 7)[45]. Nickel(II) ion combines with the reactants and with the thiazolidine as well as with the Schiff

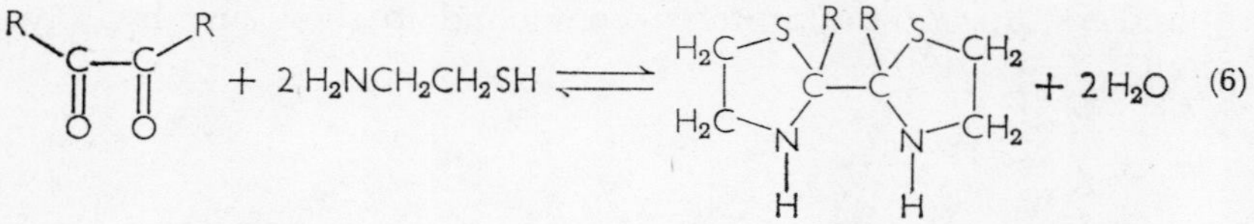

$$(6)$$

(13)

$$(7)$$

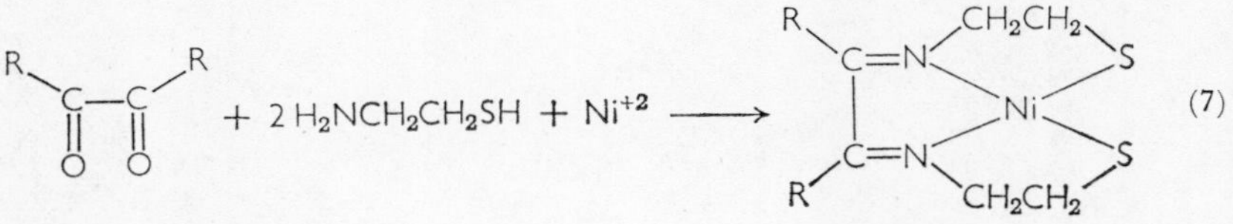

(14)

base, but the equilibria are such that metal ion complex formation with the Schiff base is favored. The Schiff base complex can also be obtained from the thiazolidine by heating it in a solution containing the metal ion[46]. Reactions in which metal ions can extract a particular compound from either reactants or products have been referred to as equilibrium template reactions[45].

A metal ion may also serve as a template to organize the course of a stereochemically selective multistep reaction. The coordination sphere of the metal ion or metal–chelate compound induces the ligand molecules to orient themselves in a manner suitable for condensation and chelation, the succession of reaction steps being determined by the metal ion. Such reactions, known as kinetic template reactions, are useful in the synthesis of large organic ring structures[45].

In the absence of metal ions, the self-condensation of o-amino-benzaldehyde results in a trimer (15)[47, 48]. When the condensation is carried out in the presence of metal ions, a closed tetradentate macrocyclic ligand (16) is formed. This reaction has been established

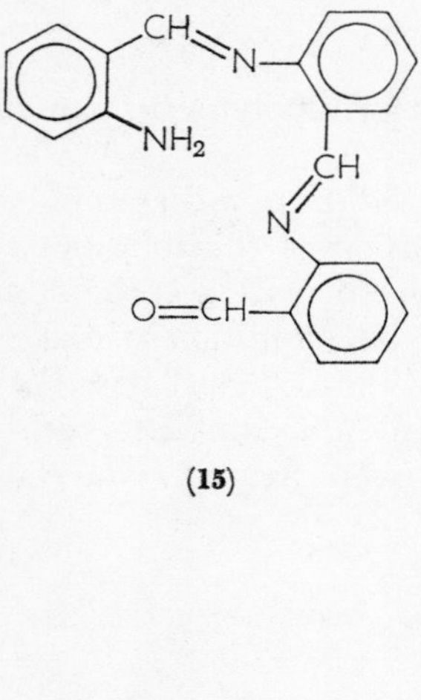

(15)

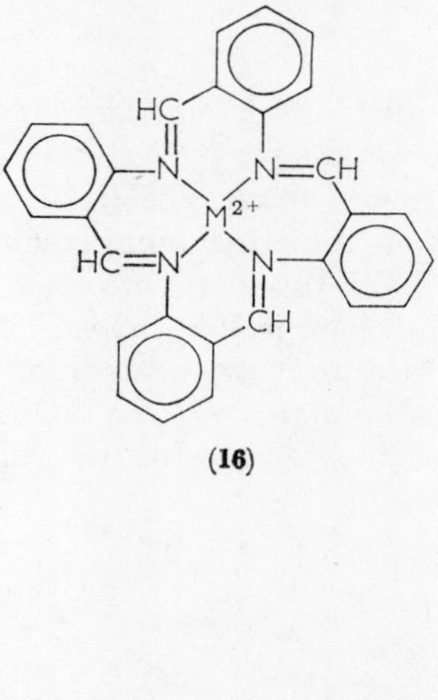

(16)

for the three metal ions, Ni(II), Zn(II) and Co(II)[49]. When Ni(II) is used as the metal ion, a second product, a closed tridentate macrocyclic ligand (17), is isolated from the condensation. This com-

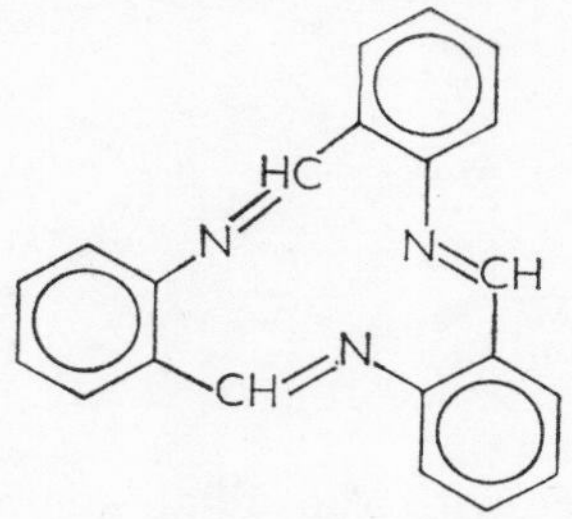

(17)

pound does not enclose the metal ion in the usual sense but is coordinated on one face of the pseudo-octahedral environment about the Ni(II)[50].

Compounds such as 18 and 19 have been produced by the amine-catalyzed condensation of bis-ethylenediaminenickel(II) chloride with β-diketones, β-ketoimines and with substituted salicylaldehydes and

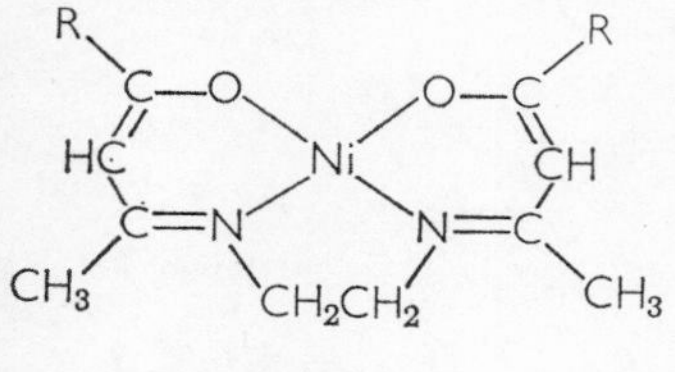

(18)

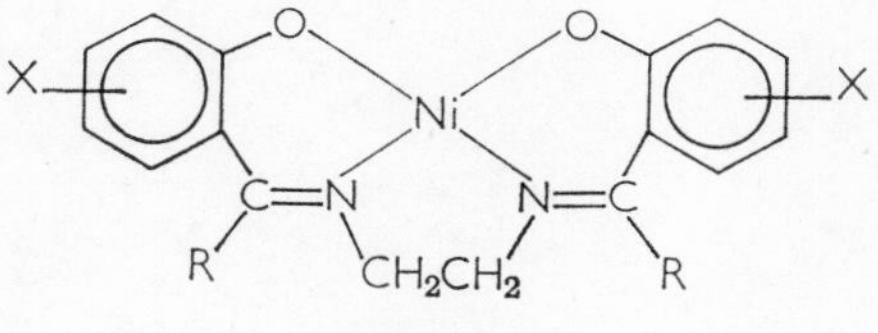

(19)

o-hydroxyacetophenones[51]. It has been suggested that these condensations are also examples of kinetic template reactions.

C. Miscellaneous Condensation Reactions

In addition to the formation of compounds containing a $>$C$=$N— linkage by the condensation of primary amines with carbonyl compounds, several other condensation reactions leading to azomethine linkages are known.

When diethyl ketals are refluxed with primary amines, Schiff bases are obtained (equation 8)[52]. Aromatic amines give considerably better yields than aliphatic amines.

Schiff bases have been obtained from the condensation of primary amines with gem-dichloro compounds (equation 9)[53].

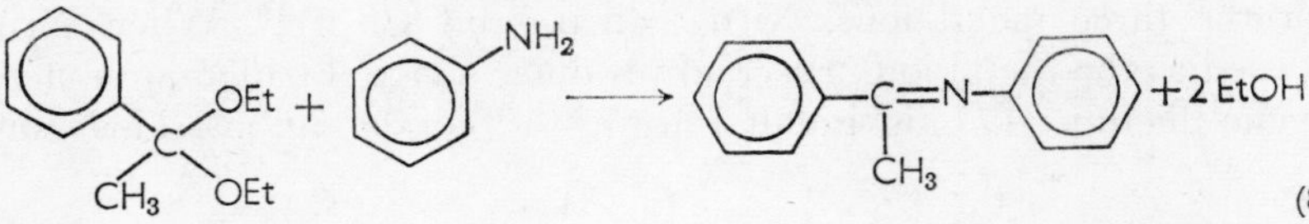

$$+ 2 \text{ EtOH} \tag{8}$$

$$+ 2 \text{ HCl} \tag{9}$$

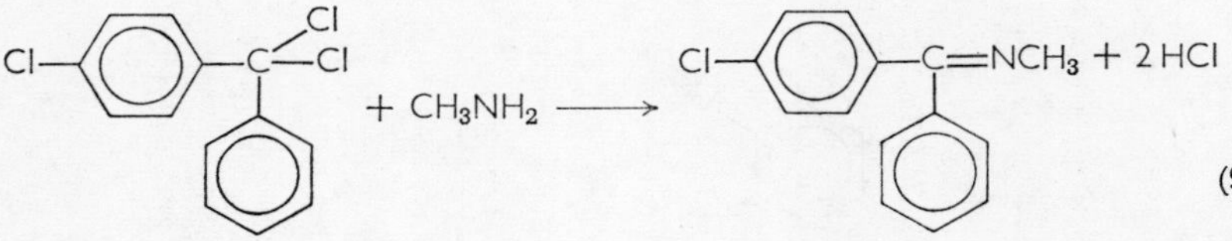

Aromatic ketones react with the alkali metal or calcium salt of primary amines to give Schiff bases (equation 10)[54].

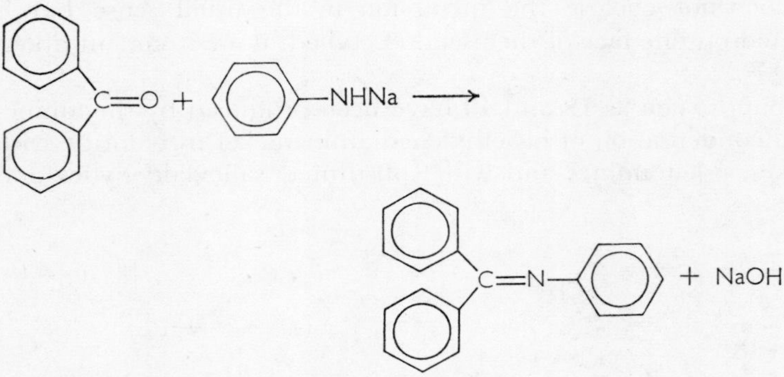

$$+ \text{ NaOH} \tag{10}$$

Schiff bases have been obtained by the reaction of phenyl azide with thio ketones (equation 11)[55] or with carbonyl compounds

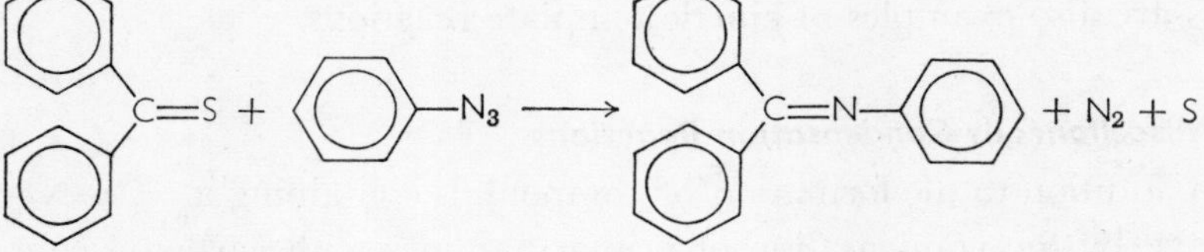

$$+ N_2 + S \tag{11}$$

(equation (12)[56].

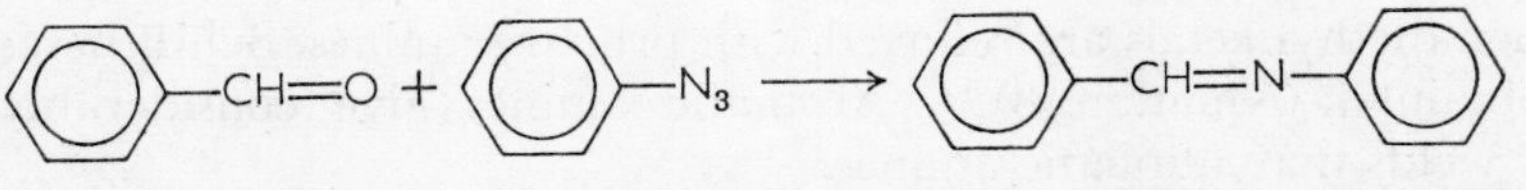

$$\tag{12}$$

Cyclohexanone when refluxed with urea in a basic medium gives cyclohexylidene 2-carbamylcyclohex-1-enylamine in 28% yield[57].

Urea splits into ammonia and isocyanic acid. The reaction is believed to involve an enamine intermediate (equation 13).

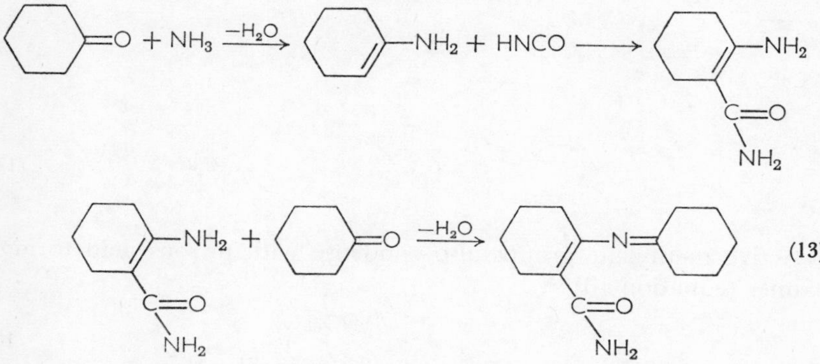

Phenylisocyanate with p-dimethylaminobenzaldehyde[58] as well as with thioketones yields Schiff bases (equations 14 and 15).

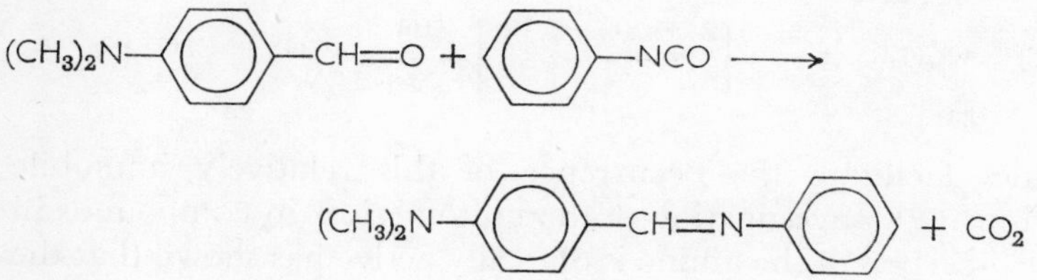

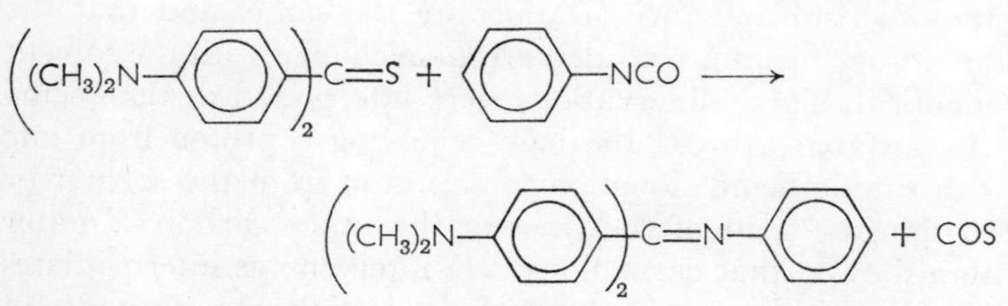

Nitrosobenzene reacts with alkylidenetriphenylphosphoranes to give Schiff bases (equation 16)[59].

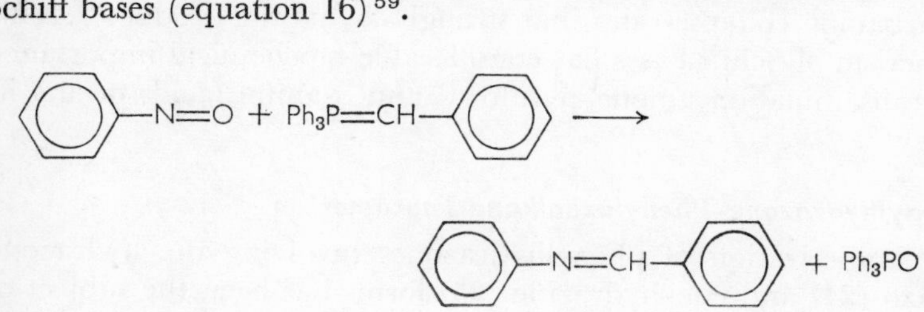

Nitroso compounds also give Schiff bases by condensing with compounds that have active methylene groups (equation 17)[60].

$$\text{C}_6\text{H}_5-\text{N}{=}\text{O} + \text{H}_2\text{C(CR)}_2 \longrightarrow$$

$$\text{C}_6\text{H}_5-\text{N}{=}\text{C(CR)}_2 + \text{C}_6\text{H}_5-\overset{\downarrow}{\text{N}}{=}\text{C(CR)}_2 \qquad (17)$$

Active methylene groups also condense with nitrous acid to give oximes (equation 18)[61].

$$\text{RCH}_2\overset{\text{O}}{\text{C}}\text{R}^1 + \text{HONO} \longrightarrow \text{R}\overset{\text{NOH}}{\text{C}}{-}{-}\overset{\text{O}}{\text{C}}{-}\text{R}^1 + \text{H}_2\text{O} \qquad (18)$$

D. Base-catalyzed Tautomerism of Schiff Bases

Imines are capable of the following type of tautomerism. Strong

$$\text{RCH}{=}\text{NCH}_2\text{R}^1 \underset{\text{HB}}{\overset{\text{B}}{\rightleftharpoons}} \left[\text{RCH}{\cdots}\text{N}{\cdots}\text{CHR}^1 \right]^- \underset{\text{B}}{\overset{\text{HB}}{\rightleftharpoons}} \text{RCH}_2\text{N}{=}\text{CHR}^1$$

$$(20) \qquad\qquad\qquad (21) \qquad\qquad\qquad (22)$$

nucleophiles facilitate the occurrence of this relatively immobile tautomerism[62-64]. Investigation of the tautomerism in compounds in which the α-carbon of the amine is optically active has shown that the rates of tautomerism and racemization are the same, and that the initial rates of racemization and deuterium exchange with the solvent are also identical. These observations were interpreted as the occurrence of a concerted process, the base removing a proton from one carbon with simultaneous donation of a proton from the solvent or from the conjugate acid of the base to the other carbon[65]. Later investigation showed that carbanions (21) intervene as intermediates in these tautomerisms. The identity of the initial rate constants of tautomerism, racemization and deuterium exchange are attributed to a carbanion collapse-ratio that strongly favors the product[66]. This tautomerism of Schiff bases has considerable biochemical importance since transamination among pyridoxal and α-amino acids occurs in this way[67].

E. Phenylhydrazone–Phenylazoalkane Tautomerism

The isomerization of phenylhydrazones involving the hydrazone (23), azo (24) and ene–hydrazine (25) forms has been the subject of

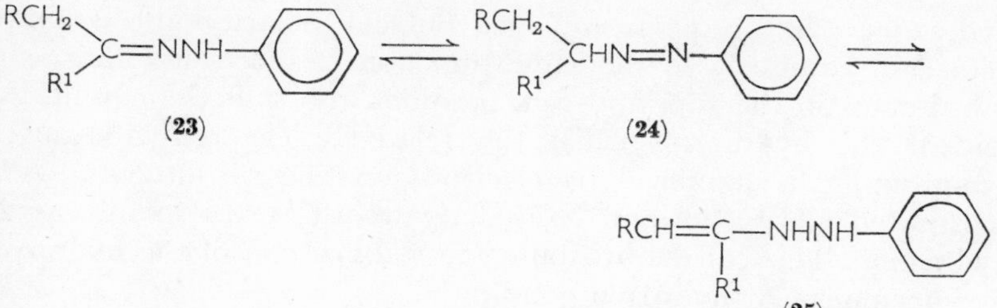

$$\begin{matrix} RCH_2 \\ \diagdown \\ R^1 \diagup \end{matrix} C{=}NNH{-}\bigcirc \ \rightleftharpoons \ \begin{matrix} RCH_2 \\ \diagdown \\ R^1 \diagup \end{matrix} CHN{=}N{-}\bigcirc \ \rightleftharpoons$$

(23) (24)

$$RCH{=}\underset{\underset{R^1}{|}}{C}{-}NHNH{-}\bigcirc$$

(25)

considerable investigation[68]. On the basis of spectroscopic evidence, phenylhydrazones had been reported to tautomerize readily in neutral solutions to the corresponding phenylazoalkanes[69,70]. Further investigation showed that the spectroscopic changes observed were due not to tautomerism but to autoxidation of the phenylhydrazones to 1-hydroperoxy-1-phenylazoalkanes[71–73] (equation 19). Spectral evi-

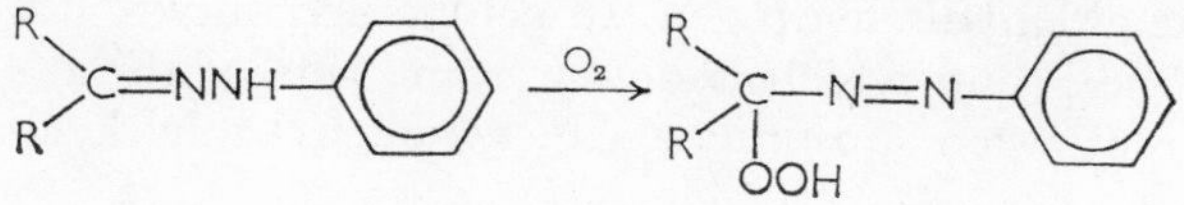

$$\begin{matrix} R \\ \diagdown \\ R \diagup \end{matrix} C{=}NNH{-}\bigcirc \ \xrightarrow{O_2} \ \begin{matrix} R \\ \diagdown \\ R \diagup \end{matrix} \underset{\underset{OOH}{|}}{C}{-}N{=}N{-}\bigcirc \qquad (19)$$

(26)

dence confirmed the structure of **26**[74]. Recent infrared and nuclear magnetic resonance studies indicate that in non-polar solvents or as pure liquids, phenylhydrazones exist in the hydrazone form[72,75,76] while the presence of a carbon–carbon double bond in their infrared spectra in methanol has been reported[77]. This difference indicates a tautomeric shift to the ene–hydrazine structure in polar solvents. The polarographic curves of twenty-four phenyl- and methylphenyl-hydrazones in aqueous methanol show the existence of all three tautomers[78].

Some phenylhydrazones show hydrazone–azo tautomerism in the pure state as well as in non-polar solvents. For instance, 4-arylazo-1-naphthols (**27**) are capable of keto–enol as well as hydrazone–azo tautomerism[79–83]. The position of equilibrium moves toward the

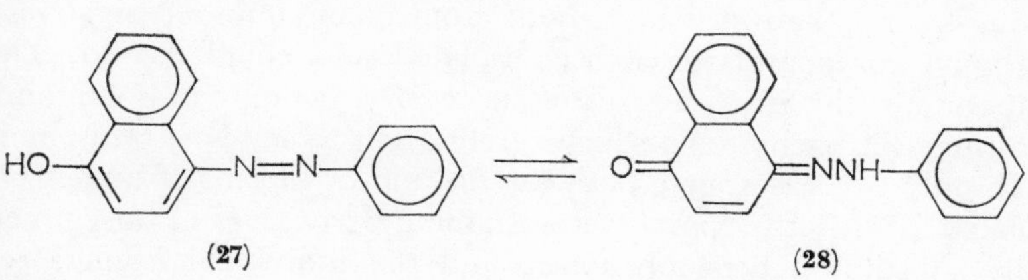

$$HO{-}\bigcirc\hspace{-4pt}\bigcirc{-}N{=}N{-}\bigcirc \ \rightleftharpoons \ O{=}\bigcirc\hspace{-4pt}\bigcirc{=}NNH{-}\bigcirc$$

(27) (28)

hydrazone (28) in polar solvents. Solvent polarity affects simple keto–enol equilibria in the same direction[84]. Electron-withdrawing substituents in the *meta* and *para* positions also shift the equilibrium toward the hydrazone while the relatively electron-withdrawing azo-grouping is stabilized by electron-donating substituents. *Ortho*-substituted CH_3O, Cl and NO_2 derivatives all exist mainly as the hydrazone. This can be attributed to stabilization of the hydrazone by intramolecular hydrogen bonding[85].

F. Geometrical Isomerism

A $>C=N-$ linkage may play the same part in geometrical isomerism as a $>C=C<$ linkage. The nomenclature is different, the terms *cis* and *trans* used in alkene chemistry being replaced by *syn* and *anti*. In aldehyde derivatives, *syn* refers to the isomer in which the substituent on the nitrogen atom is on the same side of the double bond as the aldehydic hydrogen. In ketone derivatives, it is necessary to specify the group which is on the same side of the double bond (*syn*) as the nitrogen substituent, e.g. *syn*-phenyl tolyl ketoxime (29).

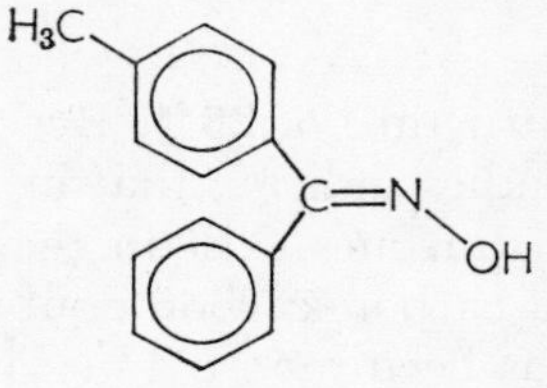

(29)

Rotation about the azomethine linkage resulting in interconversion of stereoisomers occurs considerably more easily than rotation about a doubly bonded carbon. This tendency may be accounted for by the greater electronegativity of nitrogen compared to that of carbon causing a lowering of the double-bond character of the azomethine linkage by polarization, $>C=N- \leftrightarrow >\overset{+}{C}-\overset{-}{N}-$. Nevertheless, many examples are known where both isomers of amine derivatives of carbonyl compounds have been isolated and characterized. These compounds, therefore, are distinguished from compounds containing saturated nitrogen atoms which rarely occur as stable stereoisomers.

In only two cases have both the *syn* and *anti* forms of imines been isolated[53, 86]. Both reports were of amine derivatives of unsymmetrically substituted benzophenones, and the individual isomers were found to equilibrate rapidly in solution at room temperature. When, however, there is an electronegative group adjacent to the nitrogen of the imino group, the separate geometric isomers are considerably more stable and, therefore, more readily isolated. Several reports have been made of the isolation of the isomeric forms of phenyl-hydrazones[87–91], semicarbazones[92,93], oximes[94–97] and N-halo-imines[98,99]. The presence of the electronegative group increases the resistance to rotation about the double bond[100] by decreasing the normal polarization. This decrease in polarization might be attributed

$$\begin{array}{c}\diagdown \\ \diagup\end{array} C{=}N{-}\overset{\delta-}{Z}{-} \quad\longleftrightarrow\quad \begin{array}{c}\diagdown \\ \diagup\end{array} \overset{+}{C}{-}\overset{-}{N}{-}\overset{\delta-}{Z}{-}$$

to electrostatic repulsion of adjacent negative charges that gives the imino group more double-bond character and allows the separation of geometrical isomers.

The *syn* and *anti* isomers of amine derivatives of carbonyl compounds can be distinguished by their proton magnetic resonance spectra. The chemical shift of the proton depends on whether it is *syn* or *anti* to the

$$\begin{array}{c}\diagdown \\ {}^{\alpha}_{}CH \\ \diagdown \\ \qquad\qquad C{=}N\,{\sim}\,Z \\ \diagup \\ H_{(1)}\end{array}$$

anisotropic group, Z, and on the nature of Z. When Z is NHY, as in semicarbazones and phenylhydrazones, $H_{(1)}$ comes into resonance at lower magnetic fields (deshielded) when it is *syn* than when it is *anti* to the anisotropic group, while α-hydrogens *syn* to the anisotropic group resonate at higher magnetic fields (shielded) than the cor-responding *anti* hydrogens[75,101–103]. Although $H_{(1)}$ is always de-shielded when it is *syn* to Z, isomeric assignments from $H_{(\alpha)}$ must be made with care since changes in R or in the solvent may reverse the chemical shifts of the *syn* and *anti* α-hydrogens. For example, α-methyl and α-methylene hydrogens are shielded when they are *syn* to Z with the difference in chemical shifts being smaller for the latter, but α-methine hydrogens behave as aldehydic hydrogens and are de-shielded[104]. The α-hydrogens of semicarbazones and 2,4-dinitro-phenylhydrazones are deshielded in solvents such as acetone and dimethyl sulfoxide[105]. When Z is OH or OR, both $H_{(1)}$ and $H_{(\alpha)}$ are deshielded when they are *syn* to the anisotropic group[106,107]. When the peaks in a spectrum have been assigned, the *syn*:*anti* ratio can be determined from integration of the peak areas. Such ratios are re-ported to be accurate to $\pm 5\%$[75].

Syn–anti assignments can also be made from solvent effects. In aromatic solvents both *cis* and *trans* hydrogens come into resonance at higher fields than they do in aliphatic solvents. The degree of upfield shift varies depending upon whether the hydrogens are *syn* or *anti*. Thus $\Delta\nu$ values ($\Delta\nu = \nu$ in aromatic solvent $-\nu$ in aliphatic solvent) can be used to determine whether a given compound is the *syn* or the *anti* isomer. For example, when Z is NHY, $\Delta\nu_{syn} > \Delta\nu_{anti}$, whereas when Z is OH or OR, $\Delta\nu_{syn} < \Delta\nu_{anti}$[105].

The information obtained from these proton magnetic resonance studies has been used to determine relative stabilities of *syn* and *anti* isomers. Steric effects largely control the stability of the two isomers. In general, the isomer *syn* to the smaller group is favored. For example, for a series of aliphatic ketone semicarbazones in solution the percentage of the *syn*-methyl isomer is always greater than that of the *anti*-methyl isomer and decreases in the order methyl *t*-butyl > methyl isopropyl > methyl n-propyl > methyl ethyl[101]. In addition to steric considerations, solvents also have an effect on the stereoisomeric composition. Those solvents capable of hydrogen bonding with the N—H in compounds with the group $\diagup$C=NNHY stabilize the *syn* isomer with respect to the *anti*[102].

III. THE AZO LINKAGE, —N=N—

A. Formation in Condensation Reactions

Compounds containing an azo linkage are most commonly produced by diazo coupling reactions (equation 20) and by oxidation of hydrazines (equation 21) or primary aromatic amines (equation 22).

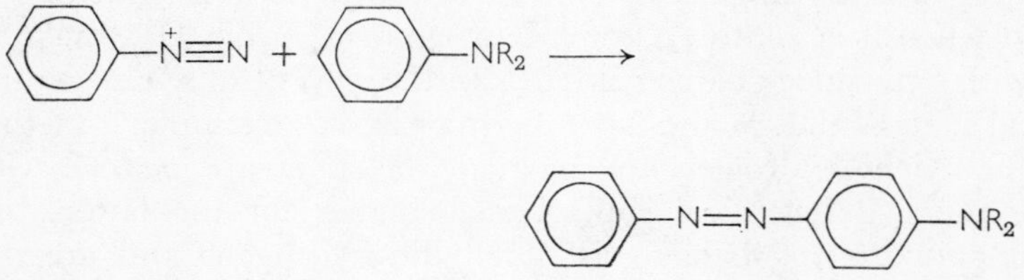

$$\text{RNHNHR} \xrightarrow[\text{H}_2\text{SO}_4]{\text{Na}_2\text{Cr}_2\text{O}_7} \text{RN=NR} \tag{21}$$

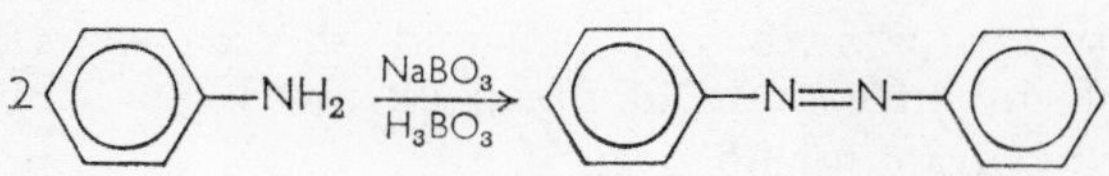

Formation of azo compounds by condensation reactions is a less common means of synthesis but can occur by the reaction of primary aromatic amines with nitrosobenzenes (equation 23)—the Mills reaction[108,109]. Aliphatic azo compounds have not been prepared in this manner.

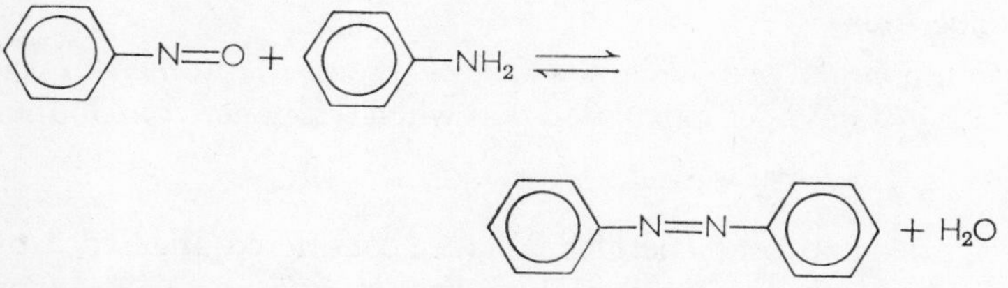

(23)

The sodium derivatives of pyridylamines condense with p-nitrosodimethylaniline (equation 24) but the amines themselves do not[110].

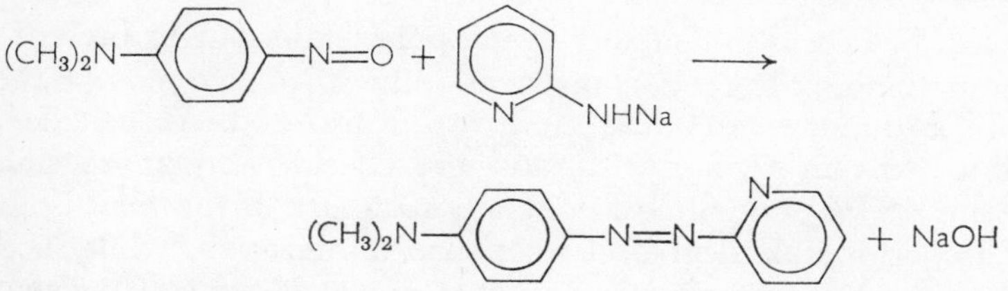

(24)

The condensation of N-alkyl- and N-arylhydroxylamines with nitroso compounds leads to the formation of azoxy compounds[111] (equation 25).

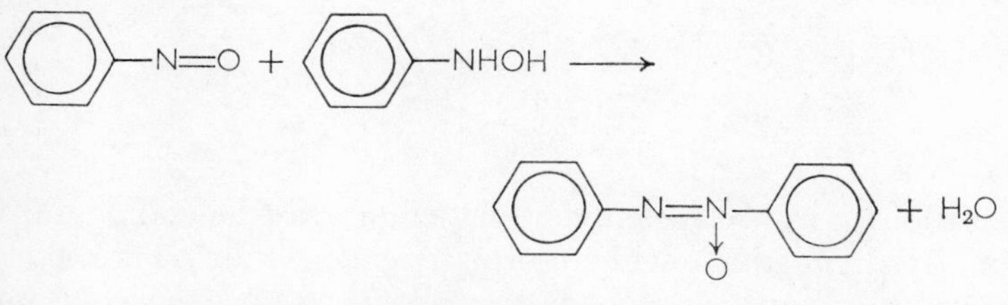

(25)

The aromatic azo compounds are resonance stabilized (equation 26).

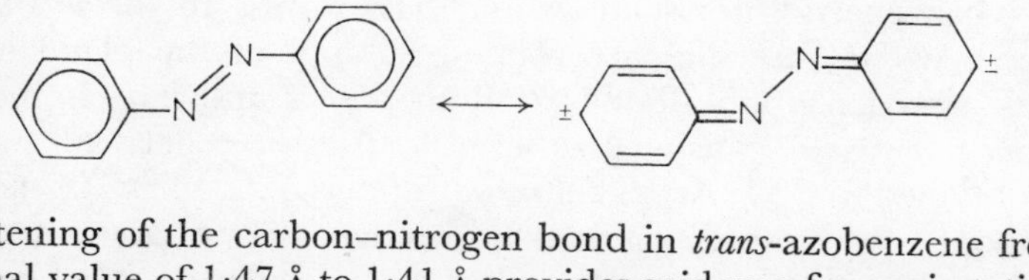

(26)

Shortening of the carbon–nitrogen bond in *trans*-azobenzene from its normal value of 1·47 Å to 1·41 Å provides evidence for conjugation[112].

Aliphatic azo compounds of the type R—N=NH decompose readily into nitrogen and a hydrocarbon. In organic solvents this decomposition occurs by both a base-catalyzed anionic elimination reaction and a homolytic reaction [113].

B. Tautomerism

Azo compounds are capable of undergoing tautomerism (see section II.E). It has been calculated that when resonance contributions

$$-CH_2-N=N- \quad \rightleftharpoons \quad -CH=N-NH-$$

are disregarded and bond energies alone are considered, the hydrazone form is favored over the azo form by 9 kcal/mole [75].

C. Geometrical Isomerism

The azo linkage is not subject to the polarization forces present in the azomethine linkage. Geometrical isomers of azo compounds, therefore, are more stable and more readily isolated. Azobenzene is known to exist in both the *cis* and the considerably more stable *trans* forms [114, 115]. The almost coplanar structure of the *trans* isomer allows a considerable degree of resonance stabilization, while in the *cis* isomer the benzene rings are rotated about 50° out of the plane containing the nitrogen atoms [116]. In 2-hydroxyazobenzenes hydrogen bonding further stabilizes the *trans* isomer (**30**) [85, 117, 118]. *Cis–trans*

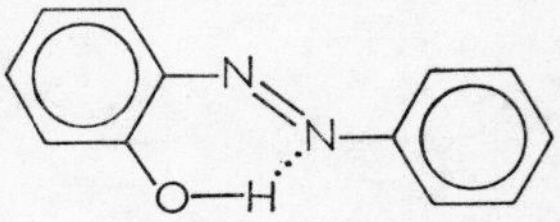

(30)

isomerization has only recently been demonstrated in aliphatic azo compounds. Azomethane which normally exists as the *trans* isomer [119] upon irradiation at − 196°c has been converted to the *cis* isomer, which has been isolated [120].

Azoxybenzenes also exist in *cis* and *trans* forms. In the less stable *cis*-azoxybenzene the rings are rotated 56° out of the plane of the nitrogen–nitrogen double bond [121]. Both the *cis* and *trans* isomers of azoxybenzene itself [122] as well as several substituted derivatives have been obtained [123, 124]. Azoxybenzenes undergo *cis–trans* isomerizations about fifty times faster than azobenzenes [125]. This greater rate is attributed to weakening of the nitrogen–nitrogen double bond by the electron-withdrawing oxygen atom. Likewise, azoxybenzenes with substituents capable of electron donation undergo isomerization more slowly than the unsubstituted compounds.

IV. CONDENSATION OF CARBONYL COMPOUNDS WITH AMINO GROUPS

A. General Two-step Mechanism

The kinetics of the formation of semicarbazones[126,127], Schiff bases[128,129], oximes[126], and phenylhydrazones[130,131] in aqueous solution have been extensively studied. All four reaction systems have been found to be kinetically second order[132–134]; first order in carbonyl compound and in nitrogen base, subject to general acid catalysis by each acid species present in accord with the Brønsted theory of general acid catalysis[135–138], and not subject to kinetic salt effects[139]. These condensation reactions have been shown to proceed by a two-step mechanism involving a carbinolamine addition intermediate (equation 27). The rate-determining step of the reaction is pH

$$\diagdown_{\diagup}C{=}O + RNH_2 \; \rightleftharpoons \; \diagdown C \diagup^{OH}_{NHR} \; \rightleftharpoons \; \diagdown_{\diagup}C{=}NR + H_2O \qquad (27)$$

dependent. In neutral solutions the rate-determining step is the dehydration of the tetrahedral carbon addition intermediate. At low pH, however, the dehydration becomes very fast, and, as a result of the decrease in concentration of free amine, the attack of the amine upon the carbonyl compound becomes the rate-limiting step. This transition in rate-determining step takes place at higher pH's with more basic amines[126]. A similar two-step mechanism has been proposed for the reaction of imido esters with amines to give amidines. In amidine formation, however, the attack of the amine on the protonated imido ester to give the tetrahedral carbon addition intermediate is rate determining on the basic side of the maximum in the pH–rate profile, while decomposition of the intermediate is rate determining on the acidic side[140].

Evidence for the two-step mechanism proposed above was obtained by observing the ultraviolet absorption of a carbonyl compound before and immediately after addition of a nitrogen base. The addition of hydroxylamine, methoxyamine or semicarbazide at neutral pH decreases the absorption of furfural to about one-third its original value within a minute[126]. This decrease cannot be attributed to the formation of the condensation product, which has a larger extinction coefficient than furfural at the wavelength studied. A similar observation had been made earlier in phenylhydrazone formation[141]. This initial decrease in absorption was credited to rapid formation of the non-absorbing carbinolamine intermediate. The slow reappearance of the peak is due to formation of the condensation product in a slow dehydration step. That a carbinolamine is an intermediate has been supported by the isolation of stable carbinolamines[90,142-145].

Early kinetic work on the reaction of carbonyl compounds with nitrogen bases showed that these reactions all exhibit maxima in their pH–rate profiles (Figure 1)[126,136] when pseudo first-order rate constants are plotted against pH[126,135,136,146-148]. Such maxima require

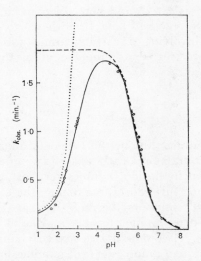

FIGURE 1. The observed first-order rate constants for the reaction of hydroxylamine with acetone as a function of pH[126]. (Dashed line: calculated for rate-limiting dehydration of carbinolamine. Dotted line: calculated for rate-limiting addition of free hydroxylamine. The solid line may be reproduced by considering the change in rate-limiting step to occur at about pH 2·7[163].)

a mechanism of at least two steps for their interpretation. At neutral pH a build up of the carbinolamine intermediate is demonstrated by the spectroscopic evidence discussed above. In solutions of low buffer concentrations rate-determining dehydration of this intermediate proceeds by an hydronium ion-catalyzed reaction pathway at a rate described by equation (28). An increase, therefore, in hydrogen ion

$$\text{Rate} = k_2 \left[\begin{array}{c} \text{OH} \\ \diagup \\ \text{C} \\ \diagdown \\ \text{NHR} \end{array} \right] [\text{H}_3\text{O}^+] \qquad (28)$$

concentration from neutrality to the pK_a of the protonated amine increases the rate of the acid-catalyzed dehydration. As the acidity is increased further, its rate-accelerating effect begins to be countered by extensive transformation of the amine into its inert conjugate acid. As the acidity is increased still further, practically every molecule of carbinolamine formed undergoes immediate dehydration, but the concentration of free base is so low that its addition to the carbonyl compound becomes rate limiting, and the overall reaction rate is decreased, giving rise to the bell-shaped pH–rate profile.

Predominance of acid catalysis on one side of the pH maximum constitutes further evidence for a change in rate-determining step. In oxime formation, the dehydration step, on the basic side of the maximum, is catalyzed by general acids, but the addition step, on the acidic side of the maximum, shows much less sensitivity to such catalysis[126]. In semicarbazone formation, on the other hand, the addition step appears much more sensitive to catalysis by general acids than does the dehydration step[126]. Because of the difference in sensitivity to catalysis shown by the two steps, if the reaction is studied at a pH just below the one at which the change in rate-determining step occurs, this change can be shown by increasing the catalyst concentration.

Figure 2.A[127] shows that at pH 3·27, the rate of semicarbazone formation increases linearly with increasing concentration of general acid catalyst. When the reaction is studied at pH 4·10, close to the pH at which the change in rate-limiting step occurs, the linear relationship between rate and catalyst concentration is no longer observed (Figure 2.B)[127]. At this pH the addition step is still rate determining at low buffer concentrations. As the catalyst concentration is increased, the rate of the addition step increases linearly with increasing amount of catalyst until it reaches the rate of the dehydration step. At this point the dehydration step becomes rate limiting and, since this step is not as sensitive to general acid catalysis, the overall rate of semicarbazone formation levels off with increasing concentration of general acid and becomes essentially independent of catalyst concentration.

Providing that complexing of the catalyst to itself or to substrate is ruled out, leveling off of a rate versus buffer concentration curve is presumptive evidence for a change in rate-limiting step.

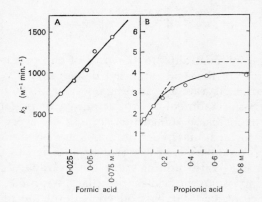

FIGURE 2. A. Second-order rate constants for *p*-nitrobenzaldehyde semi-carbazone formation as a function of formic acid concentration at pH 3·27. B. Second-order rate constants for acetophenone semicarbazone formation as a function of propionic acid concentration at pH 4·10. (The dashed line is the rate of carbinolamine dehydration at pH 4·10[127].)

Further evidence in support of this proposed two-step mechanism was obtained by studying the reactions of a series of *para*-substituted benzaldehydes with semicarbazide to determine the electronic effects upon the individual steps of the reaction[149]. The equilibrium constants for the formation of semicarbazide addition compounds show a linear logarithmic correlation with Hammett's substituent constants, with a ρ value of $+1\cdot81$. The aldehydes are therefore destabilized, relative to the addition intermediate, by electron-withdrawing substituents. The rates of acid-catalyzed dehydration show an almost equal and opposite ρ value of $-1\cdot74$, indicating that dehydration is aided by substituents capable of electron donation (Figure 3)[149].

Investigation of the effect of substituents on the overall rate of benzaldehyde semicarbazone formation shows that there is a change in rate-determining step with a change in hydrogen ion concentration. At neutrality, where the overall rate should depend upon both the equilibrium constant for addition compound formation and the rate constant for its dehydration, the two substituent effects effectively cancel each other and $\rho = +0.07$. At a pH of 1·75, where the overall rate should depend upon the rate of addition of the semicarbazide to the carbonyl compound, the ρ value is $+0.91$, indicating that the overall rate is increased by electron-withdrawing substituents[149].

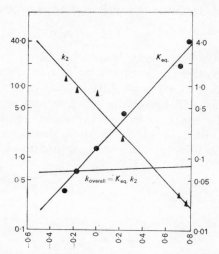

FIGURE 3. Logarithmic plot of the effect of substituents on benzaldehyde semicarbazone formation at neutral pH in 25% ethanol. (Equilibrium constant for carbinolamine formation, K_{eq}; rate constant for carbinolamine dehydration, k_2; rate constant for the overall reaction, $k_{overall}$[149].)

In studying condensation reactions of carbonyl compounds and nitrogen bases, several investigators have reported non-linear Hammett correlations[128, 150–155]. Since much of the work had been done at the intermediate pH where the maximum overall rates had been observed, it was suggested that at the intermediate pH a transition in the rate-determining step may be observed at a constant pH but with varying substituents due to a shift in the ρ value as the rate-limiting step of the reaction changes from addition at low σ to dehydration at high σ. A linear Hammett correlation will be obtained for a multistep reaction only if the rate-determining step is preceded by rapidly established equilibria without the accumulation of intermediates. For example, if k_1, k_{-1} and k_2 in equation (29) all obey Hammett

$$\text{ArCH=O} + \text{RNH}_2 \underset{k_{-1}}{\overset{k_1}{\rightleftarrows}} \text{ArCH} \overset{\text{OH}}{\underset{\text{NHR}}{\diagup}} \overset{k_2}{\longrightarrow} \text{ArCH=NR} + \text{H}_2\text{O} \quad (29)$$

13+c.a.g.

equations and the second step is rate determining, then the overall relationship

$$\log k = \log \frac{k_1 k_2}{k_{-1}} = (\rho_1 + \rho_2 - \rho_{-1})\sigma$$

will be obtained. Defining the ratio k_2/k_{-1} as k_r and $\rho_2 - \rho_{-1}$ as ρ_r,

$$\log k = (\rho_1 + \rho_r)\sigma$$

Linear Hammett correlations, however, will not be obtained if the rates of the two steps are comparable so that no single step is rate determining, or if there is a shift in rate-determining step since different ρ values will be obtained before and after the shift. This was found to be the case in semicarbazone formation. When the reaction was studied at pH 3·9 a sharp break occurred at $\sigma = 0$ (Figure 4)[149]. By using a steady-state approximation and solving for

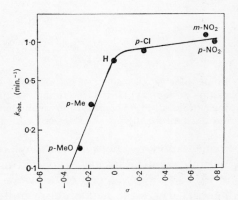

FIGURE 4. Logarithmic plot of the observed rate constants for semicarbazone formation from substituted benzaldehydes at pH 3·9 in 25% ethanol[149].

the sigma corresponding to the maximum rate, it has been shown[156] that if $\rho_r < 0 < \rho_1$, and $|\rho_r| > \rho_1$, the maximum rate as a function of sigma will occur when

$$\frac{k_2}{k_{-1}} = \left|\frac{\rho_r}{\rho_1}\right| - 1$$

as the rate-determining step shifts from addition at low sigma to dehydration at high sigma.

Although some non-linear Hammett correlations may be adequately explained in terms of a shift in rate-determining step with a change in σ, not all such deviations can be so justified. There is, however, another reason for the curvature in plots observed in condensation reactions involving aromatic carbonyl compounds. Such carbonyl compounds are stabilized by *para* substituents capable of electron donation by resonance due to stabilization of canonical form **32**. The σ^+ constants were derived to indicate the effect of

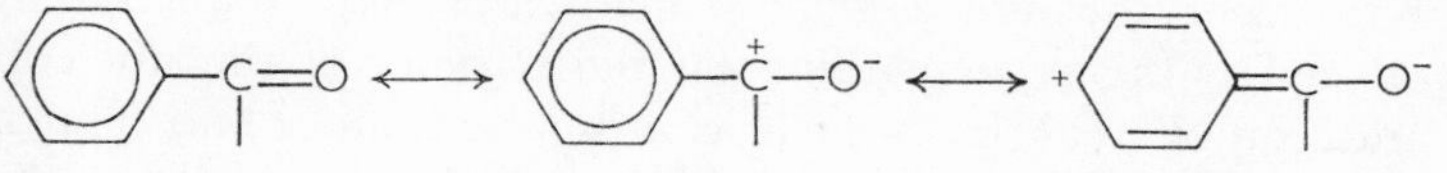

(31)   (32)

substituents on the conjugation represented by structures **33** and **34**. Since **33** and **34** are analogous to **31** and **32**, reactions involving

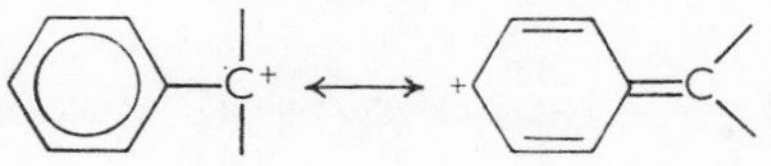

(33)   (34)

aromatic carbonyl compounds should be better correlated by σ^+ substituent constants than by the ordinary σ constants[157]. For example, the base strengths of *para*-substituted benzylidene anilines[158] and the protonation of aromatic aldehydes[159] have both been found to be correlated by σ^+ substituent constants. The carbinolamine intermediate is not subject to such conjugation effects. Dehydration, therefore, of the intermediate is correlated by σ constants. In dilute solutions, at a pH which would make dehydration rate limiting, the overall rate of reaction is the resultant of the equilibrium constant for addition compound formation and the rate constant for its dehydration. If both steps followed σ, a linear Hammett correlation would result. If, however, as is the case, one step of the reaction is correlated by σ^+ while the other step is correlated by σ, a non-linear correlation is obtained. In other words, electron donation by resonance is more important in stabilizing the carbonyl compound than in promoting dehydration of the intermediate. Substituents such as p-OH and p-OCH$_3$, therefore, hinder the addition step more than they accelerate the dehydration step. They cause a decrease in the observed overall rate, and negative deviations in Hammett plots result. The data obtained for the reaction of substituted benzaldehydes with n-butylamine has been correlated by the two-parameter equation

$$\log \frac{k}{k_{\mathrm{H}}} = -1{\cdot}355\sigma^{\circ} + 0{\cdot}80\sigma^{+}\, {}^{160}.$$

Similar two-parameter equations have been used to correlate oxime formation from substituted benzophenones[161].

B. Catalysis

I. Catalysis of the addition step

The addition of nitrogen bases to carbonyl compounds has been found to occur through a general acid-catalyzed or an apparently uncatalyzed pathway depending on the basicity of the nucleophile. The addition of weak bases such as aromatic amines and semicarbazide to carbonyl compounds exhibits general acid catalysis. A substantial body of kinetic evidence and chemical intuition now indicates that the addition reaction proceeds via a transition state (equation 30) where HB represents a general acid and B a general base.

$$\underset{\mathrm{H}}{\overset{\mathrm{H}}{\mathrm{RN}}}{:} + \underset{/}{\overset{\backslash}{\mathrm{C}}}{=}\mathrm{O} + \mathrm{HB} \rightleftharpoons \left[\underset{\mathrm{H}}{\overset{\mathrm{H}}{\mathrm{RN}}} \cdots \overset{\mid}{\underset{\mid}{\mathrm{C}}} \cdots \mathrm{O} \cdots \mathrm{H} \cdots \mathrm{B} \right] \rightleftharpoons \underset{/}{\overset{\backslash}{\mathrm{C}}} \overset{\mathrm{OH}}{\underset{\underset{+}{\mathrm{NH_2R}}}{}} + \mathrm{B} \quad (30)$$

(Throughout this chapter catalysts will be written without regard to charges. The bonds made and broken in this transition state as well as in the ones which follow are not colinear as represented.) Equation (30) represents general acid-catalyzed assistance of the attack of a free amine on the carbonyl compound. The kinetically equivalent general base-catalyzed attack of a protonated amine upon the carbonyl compound may be ruled out by the chemical inertness of an amine without a free electron pair.

In all rate-limiting steps such as equation (30) and others to be discussed, simple proton transfer alone is considered insufficient for rate limitation unless concerted with bond making and breaking. In general, proton transfers involving atoms with free electron pairs such as oxygen, nitrogen or sulfur, occur much too readily to be by themselves rate limiting in the kinds of reactions considered in this chapter. Exceptions could occur, however, in the case of proton transfers from relatively weak acids present in low concentration.

A plot of the logarithms of the catalytic constants for various catalyzing acids versus the ionization constants of the acids yields a straight line as described by the Brønsted catalysis equation (31)[162]. The value of the slope of the line, α, commonly referred to as the Brønsted

$$\log k_{\mathrm{c}} = \alpha \mathrm{p}K_{\mathrm{a}} + C \qquad (31)$$

exponent, depends on the degree to which a proton has been transferred from its initial state to the activated complex in the transition state.

Brønsted α_1 values for the addition of five nitrogen bases to benzaldehydes are presented in the third column of Table 1 [131, 163]. The

TABLE 1.   Brønsted α values for the addition and dehydration steps [131, 163].

Nucleophile	pK_a	α_1	α_2
Semicarbazide	3·6	0·25	1·0
Aniline	4·6	0·25	1·0
Phenylhydrazine	5·2	0·20	> 0·9
Hydroxylamine	6·0	0·1	> 0·8
t-Butylamine	10·4	0·0	0·73

weaker bases, semicarbazide and aniline, exhibit α_1 values of 0·25, a quantitative expression of the general acid-catalyzed addition mentioned above. The stronger bases hydroxylamine and t-butylamine exhibit α_1 values approaching zero. When values of α approach zero, water is the dominant general acid catalyst and swamps the effects of other general acids so that the reaction is often described as uncatalyzed. The so-called uncatalyzed additions of hydroxylamine and t-butylamine are more suitably decribed as general acid catalyzed with water as the general acid catalyst. Only when the addition of hydroxylamine is carried out in strongly acid solutions has general acid catalysis by hydronium ion been demonstrated.

Consideration of the role of solvent components in general acid–base catalysis is not simply an idle exercise, but is important for describing reaction pathways and in writing transition states. For example, the reverse of the general acid-catalyzed reaction given in equation (30) is general base-catalyzed removal of a proton from the oxygen atom of the protonated tetrahedral addition intermediate. When the general base is water, the transition state possesses a positive charge, when the general base is hydroxide ion the transition state is neutral, but in both cases the protonated addition intermediate is involved and never the dipolar ion form.

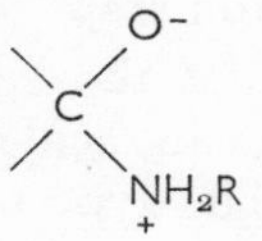

A representation of the addition reaction kinetically equivalent to equation (30), involving the same atoms in the transition state, can be formulated from general base-catalyzed removal of a proton from a free amine nucleophile attacking an already protonated carbonyl compound to yield the transition state shown in equation (32). Both reactions combine proton transfers with bond making and breaking and the general base-specific acid-catalyzed reaction (equation 32) is

$$
B + \overset{R}{\underset{H}{HN:}} + \overset{+}{\underset{/}{\diagdown}}C{=}\overset{+}{O}H \rightleftharpoons \left[B\cdots H\cdots \overset{R}{\underset{H}{N}}\cdots \overset{|}{\underset{|}{C}}\cdots OH \right] \rightleftharpoons HB + \overset{\diagdown}{\underset{/}{C}}\overset{OH}{\underset{NHR}{\diagup}} \tag{32}
$$

formally indistinguishable from the general acid-catalyzed reaction (equation 30).

Two arguments in favor of the transition state in equation (30) have been presented. First, the calculated second-order rate constant for attack of free semicarbazide on protonated p-nitrobenzaldehyde[127] according to equation (32) exceeds the maximum rate for a diffusion-controlled reaction[164]. Second, application of the solvation and reacting bond rules appears to favor equation (30) over equation (32)[165].

The solvation rule states that a proton being transferred from one oxygen or nitrogen atom to another lies closer to the more basic oxygen or nitrogen atom in the transition state and moves away as substituents are changed to make the atom less basic[166]. The reacting bond rule predicts that as the bascicity of the amine is increased the nitrogen–carbon bond in the transition state becomes longer and the carbon–oxygen bond becomes shorter[167]. Shortening of the carbon–oxygen bond decreases the basicity of the carbonyl oxygen. By a combination of the solvation and reacting bond rules it can be predicted that in a series of reactions with different amines, as the amine becomes more basic, the proton moves further from the carbonyl oxygen in the transition state given in equation (30) and the value of α decreases. On the other hand, if equation (32) is correct, as the amine becomes more basic the proton moves closer to the nitrogen and α increases. Table 1 shows that as the amines become more basic, α_1 does decrease indicating that the observed general acid catalysis is best described by equation (30).

2. Catalysis of the dehydration step

Rate-limiting dehydration of the carbinolamine addition intermediate may take place by an acid-catalyzed, base-catalyzed or uncatalyzed reaction path. The type of catalysis observed depends upon the pH region in which the reaction is carried out and the amine component of the carbinolamine. Carbinolamine intermediates derived from strongly basic amines may dehydrate by expelling an hydroxide ion without the aid of catalysts. Intermediates derived from more weakly basic amines require acid or base catalysts for dehydration to occur. These conclusions are often inferred from hydrolysis studies on the imine compound.

Mechanism (33), involving addition of a proton from a general acid to the leaving hydroxyl group, appears to be the most satisfactory for the acid-catalyzed dehydration step. This mechanism would, from a

$$\text{HN}\!-\!\overset{\displaystyle R}{\underset{\displaystyle |}{\text{C}}}\!-\!\text{OH} + \text{HB} \rightleftharpoons \left[\text{H}\!-\!\overset{\displaystyle R}{\underset{\displaystyle |}{\text{N}}}\!\cdots\!\overset{}{\underset{\displaystyle \text{H}}{\text{C}}}\!\cdots\!\text{O}\!\cdots\!\text{H}\!\cdots\!\text{B}\right] \rightleftharpoons \text{R}\overset{+}{\text{N}}\text{H}\!=\!\text{C}\!\diagup^{\diagdown} + \text{H}_2\text{O} + \text{B}$$

(33)

consideration of the solvation and reacting bond rules discussed above, predict the observed decrease in the Brønsted exponent, α_2, with an increase in the basicity of the amine (Table 1). Such trends in the values of Brønsted exponents are useful in elucidating mechanisms but interpretation of the absolute values of Brønsted exponents in terms of mechanism appears difficult when α values refer to reactions where proton transfers are concerted with bond making and breaking.

The large values of α_2 for the reaction of phenylhydrazine, hydroxylamine and t-butylamine with carbonyl compounds reflect the important role of catalysis played by the solvated proton in this mechanism. Values of α_2 of unity for the dehydration of carbinolamines derived from semicarbazide and aniline indicate that such dehydrations appear to be only specific acid catalyzed, except in fairly basic solutions where the concentration of hydronium ion is low[163, 168].

Equation (33) is analogous to equation (30) in that catalysis occurs in both at the oxygen atom. A mechanism analogous to equation (32) involving general base-catalyzed removal of a nitrogen-bound proton in the protonated carbinolamine intermediate to yield water and unprotonated Schiff base may be eliminated. For such a mechanism to be operative, the imino nitrogen atom must possess a free electron pair to act as a proton acceptor from a general acid in the reverse of the carbinolamine dehydration reaction. No free electron pair exists on the fully coordinated nitrogen atom in the cationic Schiff base benzhydrylidenedimethylammonium ion, yet the rate behavior can be accounted for in a way similar to that for its less methylated analogs[168]. The ploy of exhaustive N-methylation to clarify the role of nitrogen-bound hydrogens in the hydrolysis of carbon–nitrogen double bonds was also used in a study of N-methylthiazoline perchlorate hydrolysis[169]. These thiazoline reactions exhibit many of the general base-catalyzed characteristics of the hydrolysis of Schiff bases formed from aliphatic amines.

The uncatalyzed dehydration of carbinolamines has been observed in basic solutions. The unshared pair of electrons on the nitrogen atom provides the driving force for expulsion of the hydroxide ion (equation 34). When the amine component of the addition intermediate is

$$\underset{\underset{NHR}{\big|}}{\overset{\overset{OH}{\big|}}{C}} \; \rightleftharpoons \; C{=}\overset{+}{N}HR + OH^- \tag{34}$$

strongly basic, uncatalyzed dehydration is the predominant reaction pathway in basic solutions.

In the reverse reaction, the hydrolysis of Schiff bases composed of substituted benzaldehydes and t-butylamine has been found to be independent of hydrogen ion concentration and not subject to general catalysis from pH 9 to 14 (Figure 5)[129]. Three different lines of reasoning

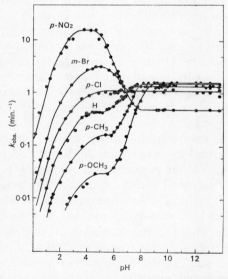

FIGURE 5. The rate of hydrolysis of Schiff bases derived from substituted benzaldehydes and t-butylamine as a function of pH[129].

indicate that the pH independence is not due to attack of water on free Schiff base but rather to attack of hydroxide ion on protonated Schiff base. First, electron-withdrawing polar substituents in the benzaldehyde actually slightly retard the hydrolysis rate ($\rho^+ = -0.21$), reflecting the competing effects of rate acceleration of nucleophilic attack and retardation due to the necessity of a preequilibrium protonation[129]. A similar conclusion is reached by considering that electron-withdrawing polar substituents slow the hydrolysis rates of Schiff bases derived from substituted anilines[131]. Second, the rate calculated for hydroxide ion attack on protonated benzylidene-1,1-dimethylethylamine exceeds that estimated for the general base-catalyzed region by means of the Brønsted equation[163]. This increase in rate is typical of direct nucleophilic attack rather than general base catalysis by hydroxide ion. Third, the observed second-order rate constant for attack by hydroxide ion on the necessarily cationic benzhydrylidenedimethylammonium ion is closely similar to that calculated for attack of hydroxide ion on the conjugate acid of benzhydrylidenemethylamine[168]. The hydrolysis of Schiff bases composed of substituted benzaldehydes and aniline shows a pH-independent reaction from pH 9 to pH 12 also ascribed to the reverse of equation (34). Above pH 12, however, the rate is no longer independent of pH but increases with increasing pH presumably due to hydroxide ion attack on unprotonated Schiff base[128,163,170,171].

Base-catalyzed dehydration has been observed in semicarbazone[149], oxime[135,172,173] and hydrazone[174] formation at pH values of 9 or greater. It is not known whether the reactions are specific or general base catalyzed. The results may be united in a mechanism involving general base-catalyzed expulsion of hydroxide ion in the dehydration step and nucleophilic attack by hydroxide ion concerted with general acid catalysis by the solvent water in the reverse reaction. This

$$B + HN\underset{|}{\overset{R \quad |}{-C}}-OH \rightleftharpoons \left[B\cdots H\cdots N\underset{|}{\overset{R}{\cdots}}C\cdots OH \right] \rightleftharpoons BH + RN{=}C{\overset{/}{\underset{\backslash}{}}} + OH^- \quad (35)$$

mechanism also has the merit of not invoking anionic nitrogen groups in aqueous solutions at pH 9. The base-catalyzed oximation of benzaldehyde occurs 1·4 times more rapidly with OD^- than with OH^-. This result does not seem to be consistent with general base catalysis (equation 35) but is accommodated by preequilibrium proton transfer followed by rate-limiting expulsion of hydroxide ion from the conjugate base of the carbinolamine[173] (equation 36).

$$RN\underset{|}{\overset{H \quad |}{-C}}-OH \rightleftharpoons R\bar{N}\underset{|}{-C}-OH \rightleftharpoons \left[RN\cdots\underset{|}{C}\cdots OH \right]^- \rightleftharpoons RN{=}C{\overset{/}{\underset{\backslash}{}}} + OH^-$$

$$(36)$$

13*

Basic conditions have been used in the synthesis of amine derivatives of carbonyl compounds in aqueous ethanolic solutions[149]. This preparative method is useful when the carbonyl group bears strongly electron-withdrawing groups. Carbinolamine intermediates of such compounds undergo dehydration slowly under the acid conditions generally employed in syntheses since the depletion of electron density at the reaction center decreases the tendency both for general acid catalysis at the oxygen atom and for the expulsion of water (equation 33)[175].

C. Simplified Mechanisms

Equations (30) and (33) when combined with the results recorded in Table 1 reduce to one of two simplified mechanisms, depending upon the basicity of the amine nucleophile. Mechanism (37) can be used to describe condensation reactions in which the amine is a relatively weak base. This mechanism accounts for the general acid

$$
\begin{aligned}
&\underset{/}{\overset{\backslash}{}}C{=}O + RNH_2 + HB \underset{k_{-1}}{\overset{k_1}{\rightleftharpoons}} \underset{/}{\overset{\backslash}{}}C\!\!\begin{array}{l}\text{OH}\\[-2pt]\\[-2pt]\overset{+}{N}H_2R\end{array} + B
\end{aligned}
$$

$$
\underset{/}{\overset{\backslash}{}}C\!\!\begin{array}{l}\text{OH}\\[-2pt]\\[-2pt]\overset{+}{N}H_2R\end{array} \underset{k_{-2}}{\overset{k_2}{\rightleftharpoons}} \underset{/}{\overset{\backslash}{}}C{=}\overset{+}{N}HR + H_2O
$$

(37)

catalysis on the acidic side and the specific hydrogen ion catalysis on the basic side of the pH–rate profile that is observed in semicarbazone (pK_a semicarbazide = 3·6) formation, in Schiff base formation from aromatic amines (pK_a aniline = 4·6) and in phenylhydrazone formation (pK_a phenylhydrazine = 5·2).

Reactions of carbonyl compounds with strong bases have been found to be apparently uncatalyzed on the acidic side of the pH–rate profile and general acid catalyzed on the basic side. Oxime (pK_a hydroxylamine = 6.0) formation and Schiff base formation from aliphatic amines ($pK_a \sim 10$) are such reactions and can be described by mechanism (38).

$$
\underset{/}{\overset{\backslash}{}}C{=}O + RNH_2 \underset{k_{-1}}{\overset{k_1}{\rightleftharpoons}} \underset{/}{\overset{\backslash}{}}C\!\!\begin{array}{l}\text{OH}\\[-2pt]\\[-2pt]NHR\end{array}
$$

$$
\underset{/}{\overset{\backslash}{}}C\!\!\begin{array}{l}\text{OH}\\[-2pt]\\[-2pt]NHR\end{array} + HB \underset{k_{-2}}{\overset{k_2}{\rightleftharpoons}} \underset{/}{\overset{\backslash}{}}C{=}\overset{+}{N}HR + H_2O + B
$$

(38)

As the basicity of the amine is allowed to increase in a series of condensation reactions, there will be a gradual transition from mechanism (37) to mechanism (38). Reactions using amines of intermediate basicity are described by a combination of the two mechanisms. Also, a reaction that can be described by one of the mechanisms may require a combination of the two mechanisms for a description of its behavior under particular conditions. For example, mechanism (38) completely describes the course of oxime formation except when the reaction is carried out in strong acids. Under these conditions carbinolamine formation is acid catalyzed and a small contribution from mechanism (37) must be included in the description. It is of interest to note that the mechanism used to describe Schiff base formation depends on whether the amine used is aliphatic (mechanism 38) or aromatic (mechanism 37)[163].

The difference between the hydrolysis of aliphatic and aromatic Schiff bases is dramatically illustrated in strongly acid solutions where the activity of water is a variable and where decomposition of the carbinolamine intermediate to give carbonyl compound and amine is rate limiting[163]. The logarithms of the observed first-order rate constants for hydrolysis of the aromatic Schiff base N-p-chlorobenzylideneaniline in strong H_2SO_4 solutions when plotted against the logarithm of the water activity yield a straight line with slope of $3 \cdot 1$[128] not inconsistent with water acting as a proton-transfer agent in the rate-limiting step as indicated in mechanism (37)[176]. A similar plot for the hydrolysis of the aliphatic Schiff base p-nitrobenzylidene-1,1-dimethylethylamine yields a curve with an initial slope in dilute H_2SO_4 of 11 or greater[163]. When replotted according to the prescription developed for reactions exhibiting inhibition by additional acid[177,178], a straight line of less than unit slope is obtained[163], a result indicating no involvement of water in the rate-limiting step[176]. This result is inconsistent with mechanism (37) but is accommodated by mechanism (38)[163].

Though not synthesized by a carbon–nitrogen double bond condensation reaction, 2-phenyliminotetrahydrofuran undergoes hydrolysis by the reverse of this route in aqueous solutions. In addition to providing interesting comparisons with the hydrolysis of Schiff bases, the nature of the products in the iminolactone hydrolysis is dependent upon the pH and catalyst employed. Butyrolactone and aniline are obtained in acidic solutions and γ-hydroxybutyramide in basic solutions[179]. Bifunctional buffer reagents such as $H_2PO_4^-$ and HCO_3^- with both acidic and basic groups yield the former pair of products almost exclusively even in regions of pH where the latter product is predominant in the absence of such buffers[180]. This strong dependence of reaction pathway on buffer composition in aqueous solutions is a most significant development that deserves detailed study in this and other systems.

A special catalytic effect appears to be indicated for phosphate buffers in other systems discussed in this review. When the results from phosphate buffer containing solutions are employed to calculate the Brønsted exponent α_2 for carbinolamine dehydration in phenyl-hydrazone and oxime formation, the low values of less than 0·7 obtained (compare Table 1) suggest a special catalytic effect. Indeed, oxime formation might be more appropriately classed in mechanism (37) rather than mechanism (38) if the results from bifunctional buffer solutions are excluded.

D. Nucleophilic Catalysis

The anilinium ion and its ring-substituted derivatives have been found to be considerably more effective as catalysts in semicarbazone and oxime formation than their pK_a values would lead one to predict. Investigation showed that they are not functioning as classical general acid catalysts but as nucleophilic catalysts, through the formation of reactive Schiff base intermediates[181]. Similar Schiff base intermediates have been proposed to occur in the amine-catalyzed enolization of acetone[182]. In reactions (39 and 40) the overall rate of

$$\tag{39}$$

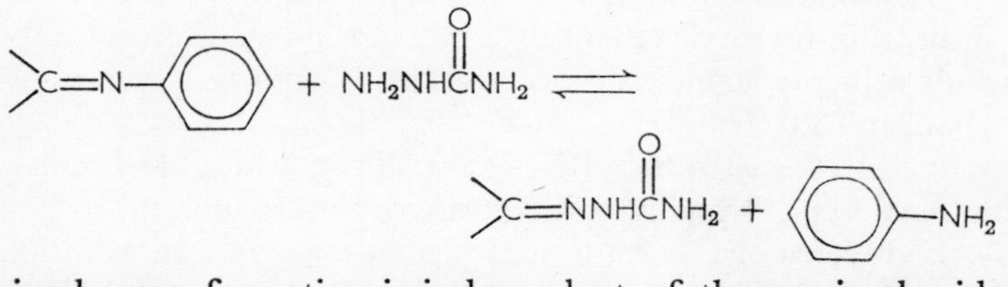

$$\tag{40}$$

semicarbazone formation is independent of the semicarbazide concentration if it is greater than 0·02 mole. The semicarbazone is quantitatively and instantaneously formed when N-p-chlorobenzyl-ideneaniline is added to an aqueous solution of semicarbazide. When hydroxylamine is used instead of semicarbazide, the rate of oxime formation is, within experimental error, equal to the rate of semicarbazone formation under the same conditions. These observations establish that equation (39) is rate limiting.

The rate enhancement caused by intermediate Schiff base formation might at first seem surprising since the $>$C=NR group is less reactive than the $>$C=O group. The greater basicity of $>$C=NR, however, permits the easy formation of the extremely reactive $>$C=$\overset{+}{\text{N}}$HR group. The amine catalyst is not limited to anilinium ions. Any primary amine can be used as long as it is more reactive towards the carbonyl compound than is the final acceptor amine. In addition, the equilibrium constant for the reaction with the acceptor amine must be larger than that for the reaction with the amine catalyst.

The study of the mechanism of Schiff base formation in aqueous solution has been approached by hydrolysis studies because of the unfavorable equilibrium constants of formation. The formation reaction can be studied directly in the presence of semicarbazide or hydroxylamine since these bases serve as a trap for the reactive Schiff base, and the rate of semicarbazone or oxime formation is identical to the rate of Schiff base formation. This technique has been used to study Schiff base formation from methylamine and acetone[173]. Nucleophilic catalysis is also useful in synthesis. For example, certain mesitylketoximes that have not been obtained from the ketones and hydroxylamine have been synthesized from the appropriate ketimines[183].

Equation (40) is known as a transimination or a 'trans-Schiffization' reaction. It probably proceeds via a two-step mechanism involving a *gem*-diamine intermediate, analogous to the carbinolamine intermediate observed in imine formation from carbonyl compounds. The transimination reactions of Schiff bases derived from either aliphatic or aromatic amines with semicarbazide, hydroxylamine or methoxyamine, of oximes with semicarbazide and of semicarbazones with hydroxylamine all appear general base catalyzed in terms of one protonated and one free base reactant or general acid catalyzed in terms of two reactants of free base. These results may be accounted for by a nearly symmetrical mechanism for transimination (equations 41 and 42). Simple proton transfers are considered to be rapid and

$$\underset{/}{\overset{\backslash}{\text{C}}}\text{=}\overset{+}{\text{N}}\text{HR} + \text{R}^1\text{NH}_2 + \text{B} \rightleftharpoons \underset{/}{\overset{\backslash}{\text{C}}}\overset{\text{NHR}}{\underset{\text{NHR}^1}{}} + \text{HB} \qquad (41)$$

$$\underset{\diagup}{\overset{\diagdown}{C}}\overset{\displaystyle NHR}{\underset{\displaystyle NHR^1}{}} + HB \rightleftharpoons \underset{\diagup}{\overset{\diagdown}{C}}{=}\overset{+}{N}HR^1 + RNH_2 + B \qquad (42)$$

the protonic charge is distributed at any time in favor of the stronger base.

Regardless of whether the first or second step in the mechanism proposed for transimination is rate limiting, the corresponding rate law describes the general catalytic alternatives observed experimentally and described in the previous paragraph. Therefore a choice as to which step is rate limiting cannot be made on the simple observation of general catalysis though a decision might be made from an evaluation of the Brønsted exponents for each step since they may exhibit a numerical difference. This difference is apt to be quite small as is any difference between Hammett ρ values for the two possible rate-limiting steps. Ultimately the nature of the rate-limiting reaction is determined by the relative rates of general acid-catalyzed amine expulsion from the approximately symmetrical tetrahedral carbon addition intermediate to yield the reverse of equation (41) or the forward reaction (42).

Transimination reactions are useful in synthetic work. Distillation of an imine in the presence of a high-boiling amine will cause the imine to exchange the lower for the higher-boiling amine[64]. In the same manner ketones can react with ketimines to give the exchange products[184, 185]. A considerable body of evidence suggests that transimination reactions are involved in the catalytic process of pyridoxal-dependent enzymes (see section V).

E. Structure and Reactivity Correlations

I. The carbonyl compound

The rates of reaction of carbonyl compounds with nitrogen bases are dependent upon the structure of the carbonyl compound, being affected by inductive, resonance and steric effects. It must be remembered that factors governing reactivity are distinct from those governing stability. For example, the rate of cyclohexanone semicarbazone formation is about fifty times greater than the rate of furfural semicarbazone formation, although the equilibrium constant for the latter is almost three hundred times as favorable[137].

Both the rate and equilibrium constants for the addition of nitrogen bases to carbonyl compounds will be affected by inductive effects.

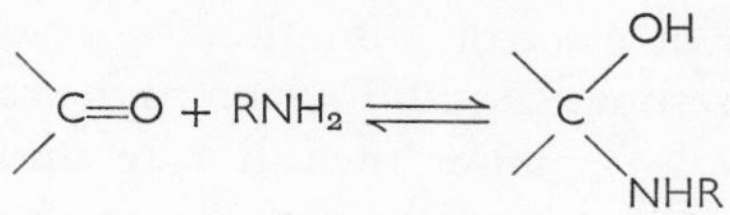

$$\overset{\diagdown}{\underset{\diagup}{C}}=O + RNH_2 \rightleftharpoons \overset{\diagdown}{\underset{\diagup}{C}}\overset{\displaystyle OH}{\underset{\displaystyle NHR}{}}$$

Electron-withdrawing substituents will favor the addition by increasing the positive character of the carbonyl carbon atom, thereby making it more susceptible to nucleophilic attack. Positive ρ (or ρ^+) values obtained in Hammett plots for equilibrium and rate constants of the addition step verify the destabilization of the carbonyl compound relative to the carbinolamine by electron-withdrawing substituents. The ρ value for the equilibrium constants of the addition step in semicarbazone formation from substituted benzaldehydes is $+1\cdot81$ [186], while that for Schiff base formation from toluidine and substituted benzaldehydes, which follows σ^+ more closely than σ, has a ρ^+ value of $+0\cdot66$ [187]. The ρ^+ value for the rate constants of the addition step in semicarbazone formation from substituted benzaldehydes is $+0\cdot94$ [127], while that for the addition step leading to benzylidene-aniline formation is $+0\cdot39$ [128]. The apparently uncatalyzed addition of nitrogen bases to carbonyl compounds is more susceptible to inductive effects than is the acid-catalyzed addition. In the case of semicarbazide attack on substituted benzaldehydes $\rho^+ = +0\cdot94$ for the water-catalyzed and $+0\cdot71$ for the hydronium ion-catalyzed addition step [127].

The second step of the reaction, the dehydration step, is affected very differently by inductive effects. Electron donation to the reaction center will aid the dehydration both by aiding the protonation of the

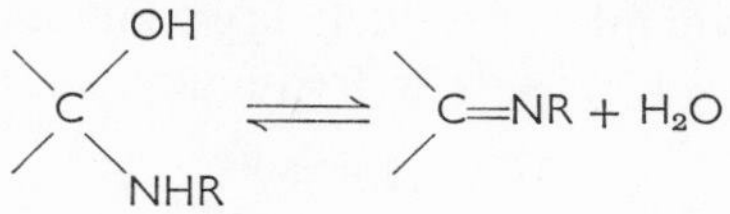

$$\overset{\diagdown}{\underset{\diagup}{C}}\overset{\displaystyle OH}{\underset{\displaystyle NHR}{}} \rightleftharpoons \overset{\diagdown}{\underset{\diagup}{C}}=NR + H_2O$$

intermediate and by aiding the expulsion of a water molecule. The rate constants for the dehydration step in semicarbazone formation show an expected negative ρ value of $-1\cdot74$. The equilibrium constants for the above reaction give a ρ value of $-0\cdot17$ [186]. Since the addition step is aided by electron withdrawal and the dehydration step is aided by electron donation, the overall rate of dehydration is little affected by substituent effects. The overall rate of dehydration in semicarbazone formation has been found to have a ρ value of $+0\cdot07$.

The effect of resonance on the reactivity of the carbonyl group has been discussed above (see section IV.A) where it was pointed out that substituents capable of electron donation by resonance give a considerable amount of resonance stabilization to the carbonyl compound. Such substituents cause smaller overall rate and equilibrium constants as well as smaller equilibrium constants for carbinolamine formation than would be predicted from Hammett σ values[186].

The rates of carbonyl addition reactions are strongly dependent on the steric requirements of the carbonyl compound. Bulky groups near the reaction center will usually stabilize the carbonyl compound relative to the carbinolamine, thereby decreasing the equilibrium constant for addition compound formation and giving smaller overall rates of product formation. Several workers have investigated the steric effect of ketones in oxime formation[188-191]. Methyl ketones are found to be more reactive than ethyl ketones, which in turn are more reactive than propyl ketones. A further increase in chain length beyond propyl has no effect on the rate. Cyclic ketones react more rapidly than straight-chain ketones with the same number of carbon atoms and a centrally located carbonyl group. A methyl group on an α carbon causes a greater decrease in reaction rate than a similar substituent on a β carbon while γ substitution appears to have no effect on the rate[192]. Investigation of steric effects in aromatic ketones showed that an o-methyl group decreases the rate of oxime[193] and thiosemicarbazone[194] formation from butyrophenone by a factor of twenty, while a m-methyl group has no significant effect on the rate of reaction[161]. Introduction of a second *ortho* substituent into the ketone often prevents these condensation reactions from occurring at all[195-198].

Every rate is in principle a product of two factors, as expressed in the Arrhenius equation (43). A is a frequency, or entropy factor and

$$k = A \, e^{-E/RT} \qquad (43)$$

$e^{-E/RT}$ is an energy factor, dependent on the energy of activation E. Changes in the degrees of freedom of the reactants and solvent determine the entropy of activation, while the energy of activation is determined by polar, resonance and steric effects. Early work on semicarbazone formation had suggested that the entropies of activation could be correlated with the rigidity of the carbonyl compound[199]. Significant deviations from this correlation were found in oxime formation[200], thiosemicarbazone formation[201], and guanylhydrazone formation[134] leading to the conclusion that the relationship between the entropy of activation and molecular rigidity is entirely random in nature. Most of the attempts to date to separate the variations in rate into differences in enthalpies and entropies of activation have been done under conditions where dehydration is probably rate determining. These thermodynamic differences, therefore, are based on the overall rate of reaction and correlations are difficult since it cannot be known whether a given enthalpy change reflects primarily an enthalpy change in the equilibrium constant for addition compound formation or in the rate constant for its dehydration. Any correlations that might be made, therefore, must await determination of enthalpy and entropy changes for the individual steps of the reaction.

Unexpectedly high *ortho*:*para* rate ratios have been reported for oxime, semicarbazone, phenylhydrazone and Schiff base formation from methoxy- and hydroxy-substituted benzaldehydes[151, 201–204]. In order to find an explanation for the rate-accelerating effect of these *ortho* substituents, an investigation was made of the effects of several *ortho* substituents on the individual steps of benzaldehyde semicarbazone formation[186]. The *o*:*p* ratios for the overall reaction decrease in the order $CH_3O > HO \sim Cl > CH_3 > H > NO_2$. In each case, except for the nitro-substituted derivatives, the *para* isomer reacts more slowly than unsubstituted benzaldehyde, while the *ortho* isomer reacts as fast or faster than the unsubstituted compound. The differences in the reactivity of the *ortho* and *para* isomers are due primarily to differences in the equilibrium constants for carbinolamine formation, with only slight differences in the rates of dehydration for each pair of isomers. The results in the above series headed by the methoxy group rule out hydrogen bonding as the sole explanation for the rate acceleration in the *ortho*-hydroxy derivative.

Rate acceleration by *ortho* substituents can be best explained by considering the quinoid forms of the isomers (**35**) and (**36**). Studies

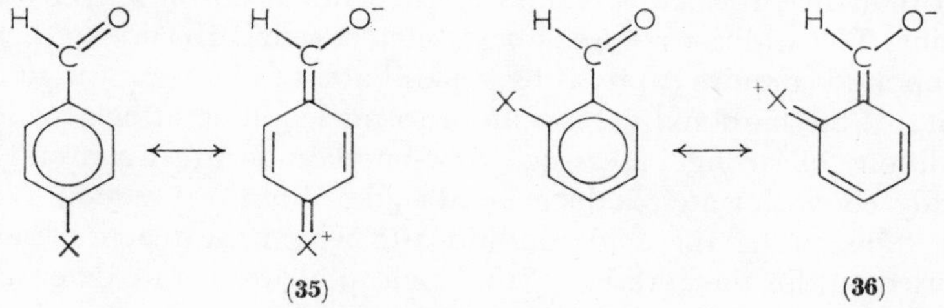

(35)   (36)

of electrophilic substitution reactions have shown that electron donation by resonance is more effective from the *para* than from the *ortho* position[205]. The *ortho*-substituted benzaldehydes, therefore, are more susceptible to nucleophilic attack. Thus the equilibrium constant for addition compound formation and the overall rate of semicarbazone formation are greater for the *ortho* isomer than for the *para* when the substituent is capable of electron donation by resonance.

2. The nucleophile

The effect of the amine on the overall equilibrium constants of condensation reactions of carbonyl compounds and amines has been discussed above (see section II.A) where it was pointed out that the thermodynamic stability of the $>$C$=$N— linkage increases with the type of amine used, in the order NH_3 < aliphatic amines < aromatic amines < amines containing an adjacent electronegative atom with a free electron pair. In contrast to the overall equilibrium constants, the rate and equilibrium constants for addition compound formation appear to be dependent on the basicity of the amine. In studies in which different amines have been reacted with the same carbonyl compound under the same conditions, the following observations have been made. The equilibrium constants for addition compound formation with *p*-chlorobenzaldehyde were found[149] to be 21·7, 9·11, 4·14 l/mole for hydroxylamine (pK_a = 6·0)[126], methoxyamine (pK_a = 4·6)[206], and semicarbazide (pK_a = 3·6)[127], respectively. Aniline (pK_a = 4·6)[128] and its ring-substituted derivatives react more readily than semicarbazide with *p*-chlorobenzaldehyde[181]. Under conditions in which addition compound formation is probably rate determining, the reaction of butyrophenone with semicarbazide is faster than its reaction with thiosemicarbazide (pK_a = 2·3)[194].

Two factors make quantitative correlations of base strength and reactivity rather difficult to obtain. The first difficulty arises from the fact that the Brønsted α_1 values are dependent on the basicity of the amine. The additions of weak bases such as semicarbazide to carbonyl compounds require catalysis by general acids to increase the susceptibility of the carbonyl carbon atom to nucleophilic attack, while the additions of strong bases such as *t*-butylamine are apparently uncatalyzed, with water as the general acid catalyst (see section IV.B.1). Secondly, under the acid conditions in which the reaction must be carried out for the addition of the nucleophile to be rate determining, the nucleophile will exist as its conjugate acid to some extent. Thus the rate of the addition reaction will be dependent on the pK_a of the nucleophile, not as a measure of its nucleophilic reactivity but as a measure of the extent to which it exists as the free base in the reaction medium. Two reactions, however, have been found to give good correlations between rates of the addition reactions and strengths of the nucleophiles. A ρ value of $-1\cdot10$ was obtained for the acid-catalyzed addition of substituted benzamides to formaldehyde[207], and a ρ value of $-2\cdot00$ has been reported for the acid-catalyzed condensation of substituted anilines with benzaldehydes under conditions in which addition of the base is rate determining[208].

The rate constants for the uncatalyzed reaction of piperonal with a series of primary aliphatic amines were found to parallel not the basicity of the amine but rather the free energies of dissociation of the corresponding addition compounds, $RNH_2 \cdot B(CH_3)_3$, thus establishing the inhibitory effect of branching on the reactivity of the amine[209]. The enthalpy–entropy plot of the reaction shows that branching in the alpha position increases the energy of activation of the reaction by approximately $0\cdot75(n-1)$ kcal/mole where n is the number of methyl groups attached to the α carbon of the amine (Figure 6)[160].

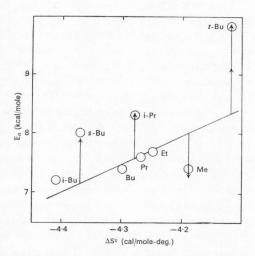

FIGURE 6. Enthalpy–entropy plot for the reaction of piperonal with primary aliphatic amines[160]. (The vertical arrows indicate activation energies calculated from the term $0\cdot75(n-1)$ kcal/mole.)

3. Relationships between reactivities of carbonyl compound, nucleophile, and catalyst

There is a relationship between the magnitude of the Brønsted α value (see section IV.B) and the reactivity of the carbonyl compound[127, 210, 211]. The Hammett and Brønsted equations for i carbonyl compounds and j general acid catalysts are

$$\log k_{ij} = \log k_{oj} + \sigma_i \rho_j \tag{44}$$

and

$$\log k_{ij} = \log G_i - \alpha_i pK_{a_j} \tag{45}$$

where k's are rate constants, σ and ρ are respectively the Hammett substituent and reaction constants, G is a constant, and α is the Brønsted exponent. Assuming the validity of these equations in describing the systems under study and twice differentiating (the order of differentiation making no difference), we obtain, for a constant nucleophile, equation (46),

$$\frac{\partial \rho_j}{\partial pK_{a_j}} = \frac{\partial^2 \log k_{ij}}{\partial pK_{a_j} \partial \sigma_i} = -\frac{\partial \alpha_i}{\partial \sigma_i} \tag{46}$$

the first equality arising from equation (44) and the second from equation (45)[212, 213]. Theoretically it is expected, and experimentally it is observed, that the middle term in equation (46) is a positive number less than unity. The left hand side of equation (46) is dependent only upon the j catalysts and indicates that the sensitivity of the rate of the reaction to substituent effects increases as the strength of the catalyzing acid decreases. The right hand side is dependent only upon the i carbonyl compounds and implies that the value of α is decreased by the presence of electron-withdrawing substituents in the carbonyl compound[214–216]. Thus the two sides of equation (46) are independently variable, and for a constant nucleophile, a plot of a ρ function versus pK_a for a series of catalysts ought to yield a straight line with a positive slope equal to the negative of that obtained from a plot of α versus σ values for a series of carbonyl compounds. An equation similar to equation (46) may be derived for general base catalysis employing the Brønsted β value and yielding a negative middle term. In this case reaction with hydroxide ion should be avoided in comparisons unless it is certain that the reaction represents an example of general base catalysis and not direct nucleophilic attack.

The Brønsted α (or β) value may also be related to the reactivity of the nucleophile by employing the Brønsted and Swain–Scott[217] equations for j catalysts and k nucleophiles (equations 47 and 48)

$$\log k_{jk} = \log G_k - \alpha_k pK_{a_j} \tag{47}$$

$$\log k_{jk} = k_{oj} + s_j n_k \tag{48}$$

where s measures the susceptibility of a reagent to attack by a nucleo-phile of nucleophilicity n. Twice differentiating this pair of equations, we obtain, for a constant carbonyl compound, equation (49).

$$\frac{\partial s_j}{\partial pK_{a_j}} = \frac{\partial^2 \log k_{jk}}{\partial n_k\, \partial pK_{a_j}} = -\frac{\partial \alpha_k}{\partial n_k} \tag{49}$$

The middle term of equation (49) is also positive. The left hand side of equation (49) is again dependent only upon the j catalysts and denotes that the sensitivity of the reaction to the reactivity of the nucleophile increases as the strength of the acid catalyst decreases. The right hand side, being dependent only upon the k nucleophiles, indicates that the value of α decreases as the reactivity of the nucleo-phile increases. (Section IV.B.1 recorded that as the basicity of the nucleophile increases, α decreases.) Again the two sides of the equation are independently variable, and a plot of s versus pK_a for a series of catalysts should yield a straight line with a positive slope equal to the negative of that obtained by plotting α versus n for a series of nucleophiles.

Finally by considering the Hammett and Swain–Scott equations for i carbonyl compounds and k nucleophiles, we obtain, for a con-stant catalyst, equation (50).

$$\frac{\partial \rho_k}{\partial n_k} = \frac{\partial^2 \log k_{ik}}{\partial \sigma_i\, \partial n_k} = \frac{\partial s_i}{\partial \sigma_i} \tag{50}$$

In this case the middle term of equation (50) is negative (if applied to electrophilic reactions it is positive). Once again there are two independent equations. The left hand side of equation (50) depends only upon the k nucleophiles and shows that the sensitivity of the reaction to substituent effects is decreased by an increase in the reactivity of the nucleophile. The right hand side, dependent only upon the i carbonyl compounds, denotes that the sensitivity of the rate of the reaction to the reactivity of the nucleophile is decreased by the presence of electron-withdrawing substituents in the carbonyl compound. A plot of ρ versus n for a series of nucleophiles should yield a straight line with a negative slope identical to that obtained by plotting s versus σ for a series of carbonyl compounds.

Sufficient results are lacking to quantitatively test equations (46), (49), and (50). Procurement of appropriate data, therefore, is de-sirable. The right hand side of equation (49) has been tested by plotting the variation of α against the variation of pK_a of the conjugate acid of the nucleophile (taken as a measure of nucleophilic reactivity) to obtain a nearly straight line of appropriate slope[131]. The lack of straight-line plots or the non-identity of slopes where indicated above would indicate that the Hammett, Brønsted, or Swain–Scott equations do not describe adequately the systems under study. Straight-line plots of the requisite slopes are not, however, sufficient evidence for the applicability of the equations employed.

V. SCHIFF BASE OCCURRENCE IN ENZYMES AND THEIR MODELS

Much of the investigation of the mechanism and catalysis of $>$C$=$N— forming condensation reactions has been designed to gain insight into the catalytic activity of enzymes which require pyridoxal phosphate (**37**), for their activity and are involved in the metabolism of amino acids.

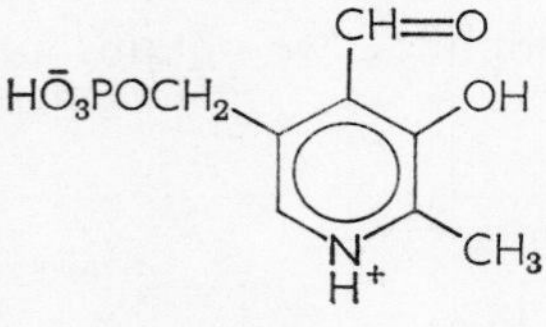

(**37**) 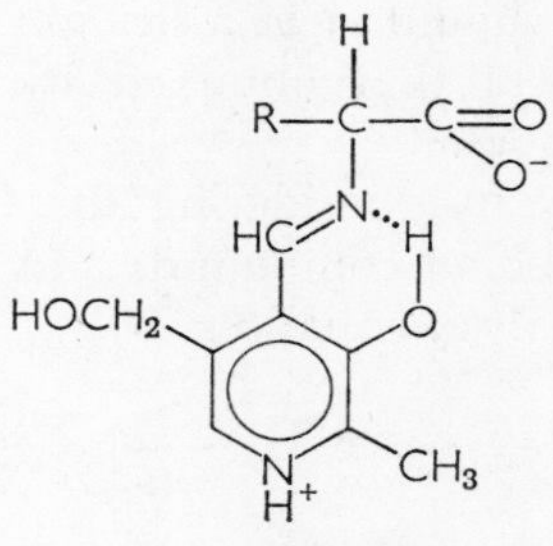   (**38**)

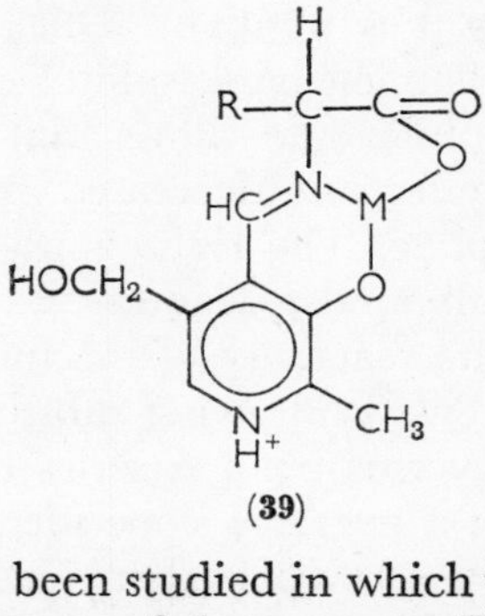

(**39**)

Model systems have been studied in which pyridoxal reacts with an amino acid in the absence of the enzyme. The results of these non-enzymatic studies show that a Schiff base (**38**) is formed between the pyridoxal and the amino acid[218–221]. The various seemingly unrelated transformations of amino acids that follow are all made possible by the fact that the strongly electron-withdrawing pyridinium nitrogen atom causes a weakening of the three bonds to the α-carbon atom of the amino acid residue. The amino acid is thus activated for any of a large number of subsequent reactions such as transamination, decarboxylation, racemization, elimination, etc[222]. Which of these reactions actually occurs is determined by the structure of the amino acid, the reaction conditions and the presence of catalysts[223-226]. In enzymatic systems the nature of the enzyme protein is the primary determining factor.

Non-enzymatic pyridoxal-catalyzed transamination reactions (equation 51) take place slowly in aqueous solutions. The addition of

$$\underset{\substack{|\\O}}{\overset{\overset{\displaystyle NH_2}{|}}{RCHCOOH}} + R^1CCOOH \rightleftharpoons RCCOOH + \underset{\substack{|\\O}}{\overset{\overset{\displaystyle NH_2}{|}}{R^1CHCOOH}} \qquad (51)$$

appropriate metal ions (Fe^{3+}, Cu^{2+}, Al^{3+}) enhances the reaction rates although they are still considerably slower than the rates of the comparable enzymatic reactions[227]. The mechanism proposed for these metal ion-catalyzed reactions involves formation of a chelate complex (**39**) of the metal ion with the Schiff base, formed from pyridoxal and the amino acid[219,228]. The primary catalytic role of the metal ion is not clear. Several possibilities have been suggested[229,230] such as promotion of Schiff base formation, maintenance of the planarity of the conjugated system through chelate ring formation, or promotion of the requisite electron displacements. If the amino acid is esterified, rapid non-enzymatic transamination occurs without metal ion catalysts[231]. This finding suggests that a function of the metal ion lies in masking the carboxyl group.

The requirement for metal ions in these non-enzymatic transamination reactions suggests that such ions may be involved as catalysts in the enzymatic reactions. However, most of the pyridoxal phosphate dependent enzymes that have been purified to date do not contain metal ions, and addition of the latter to the reaction medium does not increase the rate of enzymatic reactions[230,232-235]. Metal ions, therefore, appear to fulfill some of the roles in non-enzymatic reactions played by the protein in enzymatic reactions, but the protein is apparently a much more efficient catalyst.

Spectral studies show that in the acid pH range, pyridoxal phosphate is bound to the protein as a Schiff base (**40**) and infrared evidence suggests that there is a strong hydrogen bond between the phenolic hydroxyl group and the imino nitrogen atom[236]. At higher pH values

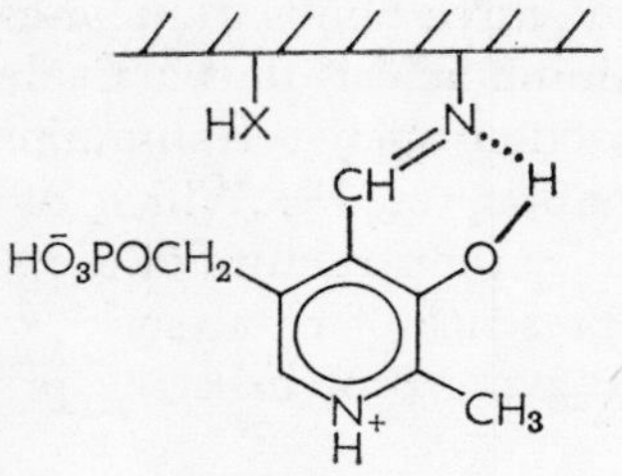

(40)

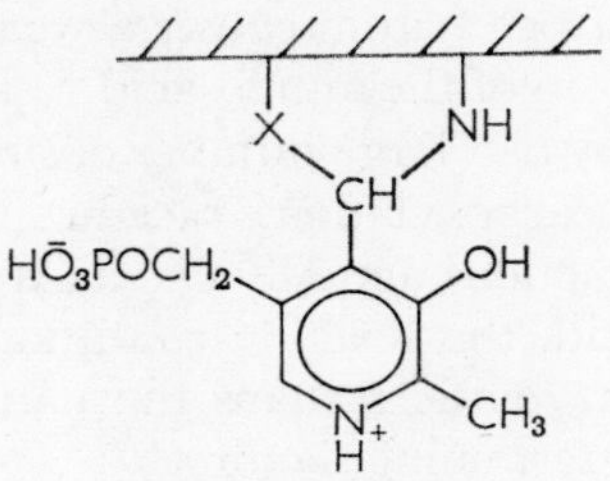

(41)

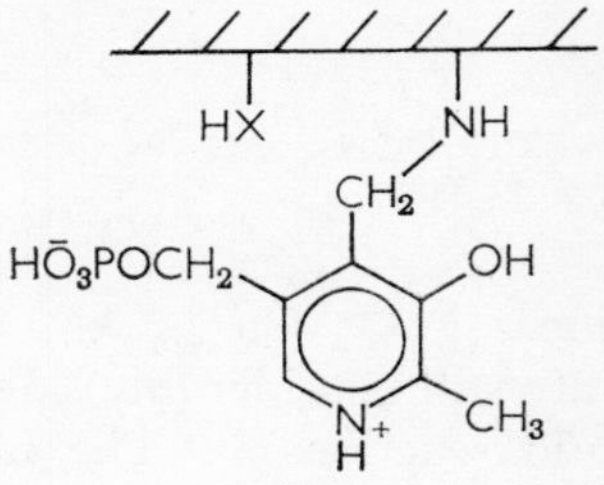

(42)

an additional group of the enzyme may add reversibly across the Schiff base linkage to give a substituted aldamine derivative (41)[237-239]. (The hatched lines indicate the protein.) Whenever the place of attachment of pyridoxal phosphate to the enzyme has been investigated, pyridoxal phosphate has been found to be linked to an ε-amino group of a lysine residue[240-246]. In the few cases where the point has been examined, pyridoxal phosphate-dependent enzymes have been found to require free sulfhydryl groups for their activity[247, 248]. This requirement does not imply that a sulfhydryl group is directly involved in the binding of the pyridoxal phosphate. Its function may be in catalysis of the enzymatic reaction or in maintenance of the required conformation of the active site.

Preexistence of a Schiff base between pyridoxal phosphate and the enzyme may account for the greatly enhanced rates of enzymatic reactions as compared to the rates of the corresponding non-enzymatic reactions. Subsequent reactions of the enzyme with amino acids must involve Schiff base formation via a fast transimination step (see section IV.D). Once the new Schiff base is formed, the ε-amino group of the lysine residue that was originally bound to pyridoxal phosphate is free and is in a favorable position to act as a catalyst in subsequent steps of the enzymatic reaction.

At pH values where the enzyme is bound to pyridoxal as a Schiff base (**40**), reaction with sodium borohydride destroys the activity of the enzyme by reducing the Schiff base to a pyridoxylamine (**42**). This observation gives strength to the hypothesis that the subsequent reaction of enzyme with an amino acid involves transamination. Skeletal muscle phosphorylase is an exception[240], indicating that this particular enzyme does not require a Schiff base linkage for its activity. Further research is required to determine whether the role of pyridoxal phosphate in this enzyme can be attributed simply to maintenance of the active site of the enzyme in the required conformation or whether a new set of catalytic properties must be ascribed to the vitamin B_6 aldehyde.

Details on biological transamination and related processes are described in chapter 9 of this book.

VI. CONDENSATION OF NITROSOBENZENES WITH AMINO GROUPS

A. Condensation of Nitrosobenzenes with Anilines

The condensation of nitrosobenzenes with anilines has not been as extensively studied kinetically as the condensation of benzaldehydes with anilines. Those results that have been reported, however, show a not unexpected similarity in the two reactions.

A study of the condensation in acetate-buffered ethanolic solutions showed it to be a second-order reaction; first order each in aniline and in nitrosobenzene. When the reaction is carried out in an unbuffered medium, the rate equation becomes more complex. Under such conditions, the second-order equation is obeyed only when the ratio of aniline to nitrosobenzene is greater than two[249]. The reaction is not subject to kinetic salt effects and is evidently catalyzed by general acids. Strong inorganic acids, however, are usually unsatisfactory catalysts because of the formation of tarry materials. Because of the decomposition of nitrosobenzene, the reaction cannot be studied in alkaline solutions.

The condensation probably proceeds via an intermediate hydroxyhydrazine (**43**) analogous to the carbinolamine intermediate encountered in the condensation of amines with carbonyl groups. Though hydroxyhydrazines have never been isolated, evidence for their existence can be found in the occasional appearance of azoxy compounds as by-products. These are presumably formed when oxidation of the hydroxyhydrazine (by excess nitrosobenzene) occurs fast enough to compete with dehydration[250].

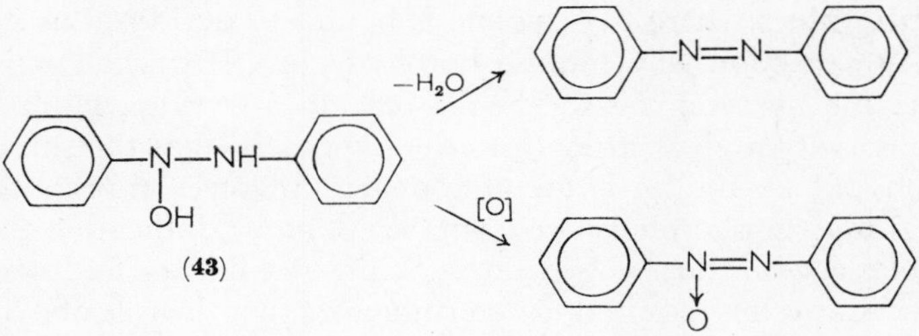

(43)

Satisfactory Hammett correlations are obtained for the overall rate constants determined in acetate-buffered solutions. The condensation of substituted anilines with nitrosobenzene gives a ρ value of $-2\cdot14$, showing increased reactivity as the basicity of the amine is increased. The condensation of aniline with substituted nitrosobenzenes gives a ρ value of $+1\cdot22$, indicating that the reaction is aided by electron-withdrawing groups on the nitrosobenzene[249]. Similar substituent effects have been observed in earlier studies of this reaction[251].

The marked positive ρ value corresponds to the sign obtained when nitrogen bases react with a series of carbonyl compounds under conditions in which carbinolamine formation is rate limiting and contrasts with the near zero value of ρ obtained from the observed rate constants when carbinolamine dehydration is rate limiting. Thus we conclude that hydroxyhydrazine formation is rate limiting, and since it is evidently general acid catalyzed, it can be described by mechanism (52). This mechanism is analogous to mechanism (37) proposed

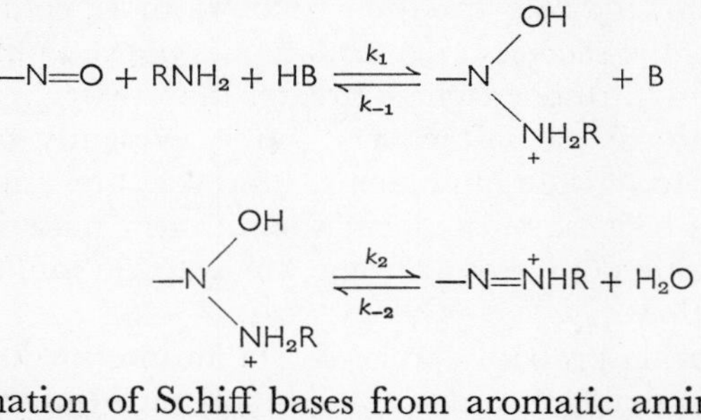

(52)

for the formation of Schiff bases from aromatic amines (see section IV.C). Under the conditions employed in the nitrosobenzene study, however, hydroxyhydrazine dehydration was never rate limiting.

B. Condensation of Nitrosobenzenes with N-Substituted Hydroxylamines

The condensation of nitrosobenzene with N-phenylhydroxylamine, known as the azoxy coupling reaction, results in the formation of azoxybenzene (equation 53). When the starting materials have

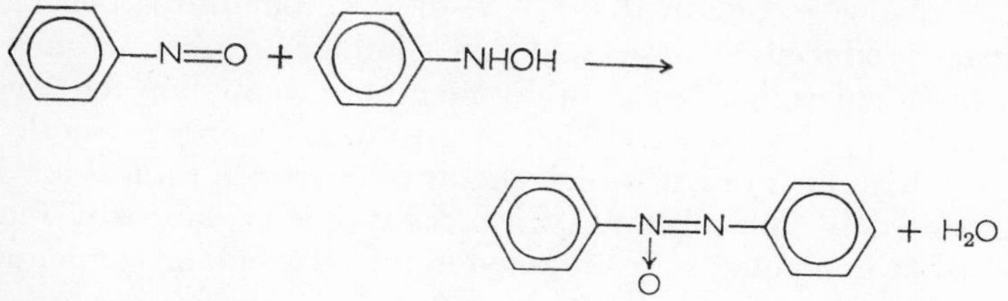

$$\text{(53)}$$

differently substituted phenyl groups, not only are both possible asymmetrical azoxy compounds formed, but the symmetrical ones as well[252–255]. Isotopic studies were undertaken in order to gain insight into the mechanism of the reaction. When N-phenylhydroxylamine reacts with ^{15}N-labeled nitrosobenzene, the ^{15}N is distributed equally in both nitrogens of the azoxybenzene[111]. The condensation of ^{18}O-labeled N-phenylhydroxylamine with nitrosobenzene results in azoxybenzene, whose [18]O content is almost exactly half that of the N-phenylhydroxylamine[256,257]. These results are interpreted as evidence for the reversible formation of an intermediate N,N'-diol analogous to the C,N-diol known to occur in the reactions of hydroxylamines with carbonyl-containing compounds (see equation (27) in the case where R is OH).

Electron spin resonance studies show that nitrosobenzene radical-ions are produced throughout the course of the base-catalyzed azoxy coupling reaction[258,259]. This observation has led to the proposal of

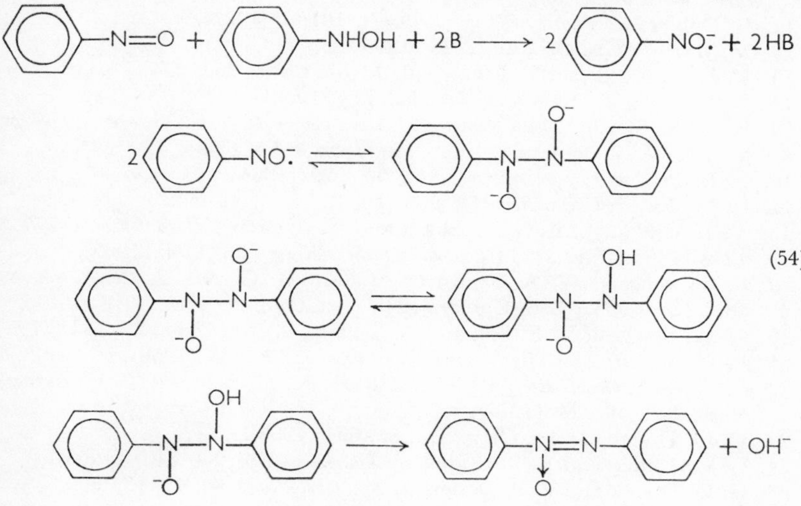

$$\text{(54)}$$

mechanism (54) to account for the results of the isotopic studies without involving an intermediate N,N'-diol[260]. The nitrosobenzene radical-ions produced in basic solutions combine to give an intermediate bis-anion which apparently undergoes protonation followed by irreversible loss of hydroxide ion to give azoxybenzene. Another possible mechanism is one in which the nitrosobenzene radical-ion is first protonated by the solvent. Both mechanisms explain why four different azoxy compounds are obtained in mixed aromatic condensation reactions.

VII. REFERENCES

1. P. A. S. Smith, *The Chemistry of Open-Chain Organic Nitrogen Compounds*, Vol. 1, W. A. Benjamin, Inc., New York, 1965.
2. P. A. S. Smith, *The Chemistry of Open-Chain Organic Nitrogen Compounds*, Vol. 2, W. A. Benjamin, Inc., New York, 1966.
3. N. J. Leonard and J. V. Paukstelis, *J. Org. Chem.*, **28**, 3021 (1963).
4. O. H. Wheeler and E. M. Levy, *Can. J. Chem.*, **37**, 1727 (1959).
5. C. Moureu and G. Mignonac, *Compt. Rend.*, **156**, 1801 (1913); **169**, 237 (1919); **170**, 936 (1920); *Ann. Chim.*, **14**, 332 (1920).
6. A. Butlerow, *Ann. Chem.*, **115**, 322 (1860).
7. E. Baur and W. Rüetschi, *Helv. Chim. Acta*, **24**, 754 (1941); H. H. Richmond, G. S. Myers and G. F. Wright, *J. Am. Chem. Soc.*, **70**, 3659 (1948); Y. Ogata and A. Kawasaki, *Bull. Chem. Soc. Japan*, **37**, 514 (1964).
8. M. M. Sprung, *Chem. Rev.*, **26**, 297 (1940).
9. Y. Ogata and A. Kawasaki, *Tetrahedron*, **20**, 855 (1964).
10. M. A. Laurent, *Ann. Chem.*, **21**, 130 (1837); F. Sachs and P. Steinert, *Chem. Ber.*, **37**, 1733 (1904).
11. F. Dobler, *Z. Physik. Chem.* (*Leipzig*), **101**, 1 (1922).
12 M. Busch, *Chem. Ber.*, **29**, 2143 (1896).
13. R. K. McLeod and T. I. Crowell, *J. Org. Chem.*, **26**, 1094 (1961).
14. J. B. Cloke, *J. Am. Chem. Soc.*, **62**, 117 (1940).
15. P. A. S. Smith, *The Chemistry of Open-Chain Organic Nitrogen Compounds*, Vol. 1, W. A. Benjamin, Inc., New York, 1965, Chap. 7.
16. F. E. Prichard, *Spectrochim. Acta*, **20**, 925 (1964).
17. R. W. Layer, *Chem. Rev.*, **63**, 489 (1963).
18. V. I. Minkin, L. E. Nivorozhkin and V. A. Bren, *J. Gen. Chem. USSR*, **35**, 1276 (1965) translated from *Zh. Obshch. Khim.*, **35**, 1270 (1965).
19. L. do Amaral, W. A. Sandstrom and E. H. Cordes, *J. Am. Chem. Soc.*, **88**, 2225 (1966); J. B. Culbertson, *J. Am. Chem. Soc.*, **73**, 4819 (1951).
20. G. W. Stacy, B. V. Ettling and A. J. Papa, *J. Org. Chem.*, **29**, 1537 (1964).
21. P. Grammaticakis, *Bull. Soc. Chim. France*, **17**, 158 (1950).
22. K. Hoesch, *Chem. Ber.*, **48**, 1122 (1915); K. Hoesch and T. V. Zarzecki, *Chem. Ber.*, **50**, 462 (1917).
23. G. O. Dudek, *J. Am. Chem. Soc.*, **85**, 694 (1963).
24. G. O. Dudek and E. P. Dudek, *J. Am. Chem. Soc.*, **88**, 2407 (1966).
25. G. O. Dudek and E. P. Dudek, *J. Am. Chem. Soc.*, **86**, 4283 (1964).

26. O. A. Osipov, V. I. Minkin and V. A. Kogan, *Zh. Fiz. Khim.*, **37**, 1492 (1963); *Chem. Abstr.*, **59**, 11218 (1963).
27. D. G. Anderson and G. Wettermark, *J. Am. Chem. Soc.*, **87**, 1433 (1965).
28. D. Heinert and A. E. Martell, *J. Am. Chem. Soc.*, **84**, 3257 (1962); **85**, 183 (1963); **85**, 188 (1963).
29. E. R. Blout and R. M. Gofstein, *J. Am. Chem. Soc.*, **67**, 13 (1945).
30. F. C. McIntire, *J. Am. Chem. Soc.*, **69**, 1377 (1947).
31. E. D. Bergmann, Y. Hirshberg and S. Pinchas, *J. Chem. Soc.*, 2351 (1950).
32. L. N. Ferguson and I. Kelly, *J. Am. Chem. Soc.*, **73**, 3707 (1951).
33. O. Gerngross and A. Olcay, *Chem. Ber.*, **96**, 2550 (1963).
34. F. E. Prichard, *Spectrochim. Acta*, **20**, 925 (1964).
35. H. P. Schad, *Helv. Chim. Acta*, **38**, 1117 (1955); J. Dabrowski and U. Dabrowska, *Roczniki Chem.*, **32**, 821 (1958); *Chem. Abstr.*, **53**, 4896 (1959); J. Dabrowski, *Bull. Acad. Polon. Sci., Ser. Sci., Chim.*, **7**, 93 (1959); *Chem. Abstr.*, **54**, 18342 (1960); J. Dabrowski, *Spectrochim. Acta*, **19**, 475 (1963).
36. V. M. Potapov, F. A. Trofimov and A. P. Tetrent'ev, *Dokl. Akad. Nauk SSSR*, **134**, 609 (1960); *Chem. Abstr.*, **55**, 6818 (1961); V. M. Potapov, F. A. Trofimov and A. P. Tetrent'ev, *Zh. Obshch. Khim.*, **31**, 3344 (1961); *Chem. Abstr.*, **57**, 3352 (1962).
37. G. O. Dudek and R. H. Holm, *J. Am. Chem. Soc.*, **83**, 3914 (1961); **84**, 2691 (1962); G. O. Dudek and G. P. Volpp, *J. Am. Chem. Soc.*, **85**, 2697 (1963).
38. G. O. Dudek and R. H. Holm, *J. Am. Chem. Soc.*, **83**, 2099 (1961).
39. P. Krumholtz, *J. Am. Chem. Soc.*, **75**, 2163 (1953).
40. P. E. Figgins and D. H. Busch, *J. Am. Chem. Soc.*, **82**, 820 (1960).
41. P. E. Figgins and D. H. Busch, *J. Phys. Chem.*, **65**, 2236 (1961).
42. M. P. Schubert, *J. Biol. Chem.*, **114**, 341 (1936).
43. G. Hesse and G. Ludwig, *Ann. Chem.*, **632**, 158 (1960).
44. M. C. Thompson and D. H. Busch, *J. Am. Chem. Soc.*, **84**, 1762 (1962).
45. M. C. Thompson and D. H. Busch, *J. Am. Chem. Soc.*, **86**, 213 (1964).
46. D. H. Busch, *Record Chem. Progr. (Kresge-Hooker Sci. Lib.)*, **25**, 107 (1964).
47. E. Bamberger, *Chem. Ber.*, **60**, 314 (1927).
48. F. Seidel and W. Dick, *Chem. Ber.*, **60**, 2018 (1927).
49. G. A. Melson and D. H. Busch, *J. Am. Chem. Soc.*, **86**, 4834 (1964).
50. G. A. Melson and D. H. Busch, *J. Am. Chem. Soc.*, **87**, 1706 (1965).
51. E. J. Olszewski, L. J. Boucher, R. W. Oehmke, J. C. Bailar, Jr. and D. F. Martin, *Inorg. Chem.*, **2**, 661 (1963); E. J. Olszewski and D. F. Martin, *J. Inorg. Nucl. Chem.*, **26**, 1577 (1964).
52. L. Claisen, *Chem. Ber.*, **29**, 2931 (1896).
53. D. Y. Curtin and J. W. Haussu, *J. Am. Chem. Soc.*, **83**, 3474 (1961).
54. E. C. Britton and F. Bryner, *U.S. Pat.* 1,938,890; *Chem. Abstr.*, **28**, 1715 (1934).
55. A. Schönberg and W. Urban, *J. Chem. Soc.*, 530 (1935).
56. L. A. Neiman and V. I. Maimind, *Izv. Akad. Nauk SSSR, Ser. Khim.*, 1831 (1964); *Chem. Abstr.*, **62**, 2724 (1965).
57. A. F. McKay, E. J. Tarlton and C. Podesva, *J. Org. Chem.*, **26**, 76 (1961).
58. H. Staudinger and R. Endle, *Chem. Ber.*, 1042 (1917).
59. U. Schöllkopf, *Angew. Chem.*, **71**, 260 (1959).
60. R. C. Azzam, *Proc. Egypt. Acad. Sci.*, **9**, 89 (1953); *Chem. Abstr.*, **50**, 16685

400   Paula Y. Sollenberger and R. Bruce Martin

(1956); F. Krohnke, H. Leister and J. Vogt, *Chem. Ber.*, **90**, 2792 (1957).
61. E. F. Degering, C. Bordenca and B. H. Gwynn, *An Outline of Organic Nitrogen Compds.*, John S. Swift Co., Inc., Cincinnati, Ohio, 1942, Chap. 8.
62. C. K. Ingold and C. W. Shoppe, *J. Chem. Soc.*, 1199 (1929).
63. R. Perez-Ossorio and E. D. Hughes, *J. Chem. Soc.*, 1225 (1931); 696 (1932); J. W. Baker, W. S. Nathan and C. W. Shoppe, *J. Chem. Soc.*, 1847 (1935).
64. R. W. Layer, *Chem. Rev.*, **63**, 489 (1963).
65. S. K. Hsü, C. K. Ingold and C. L. Wilson, *J. Chem. Soc.*, 1778 (1935); R. Perez-Ossorio and E. D. Hughes, *J. Chem. Soc.*, 426 (1952); R. Perez-Ossorio, F. G. Herrera and R. M. Utrilla, *Nature*, **179**, 40 (1957).
66. D. J. Cram and R. D. Guthrie, *J. Am. Chem. Soc.*, **87**, 397 (1965).
67. B. Witkop and T. W. Beiler, *J. Am. Chem. Soc.*, **76**, 5589 (1954).
68. P. C. Free, *J. Prakt. Chem.*, **47**, 238 (1893); *Ann. Chem.*, **283**, 391 (1894); E. Fisher, *Chem. Ber.*, **29**, 793 (1896); J. Thiele and K. Heuser, *Ann. Chem.*, **290**, 1 (1896); E. C. C. Baly and W. B. Tuck, *J. Chem. Soc.*, **89**, 982 (1906); K. V. Auwers and H. Wunderling, *Chem. Ber.*, **64**, 2748 (1931); T. Uemura and Y. Inamura, *Bull. Chem. Soc. Japan*, **10**, 169 (1935); P. Ramart-Lucas, J. Hoch and M. Martynoff, *Bull. Soc. Chim. France*, **4**, 481 (1937); P. Grammaticakis, *Bull. Soc. Chim. France*, **14**, 438 (1947); A. E. Arbuzov and Y. P. Kitaev, *Dokl. Akad. Nauk SSSR*, **113**, 577 (1957); *Chem. Abstr.*, **51**, 14605 (1957).
69. R. O'Connor, *J. Org. Chem.*, **26**, 4375 (1961).
70. R. O'Connor and G. Henderson, *Chem. Ind.* (*London*), 850 (1965).
71. A. J. Bellamy and R. D. Guthrie, *Chem. Ind.* (*London*), 1575 (1964).
72. A. J. Bellamy and R. D. Guthrie, *J. Chem. Soc.*, 2788 (1965).
73. A. J. Bellamy and R. D. Guthrie, *J. Chem. Soc.*, 3528 (1965).
74. H. C. Yao and P. Resnick, *J. Org. Chem.*, **30**, 2832 (1965).
75. G. J. Karabatsos and R. A. Taller, *J. Am. Chem. Soc.*, **85**, 3625 (1963).
76. R. R. Shagidullin, F. K. Sattarova, T. V. Troepol'skaya and Y. P. Kitaev, *Izv. Akad. Nauk SSSR, Otd. Khim. Nauk*, 473 (1963); *Chem. Abstr.*, **59**, 7347 (1963); R. R. Shagidullin, F. K. Sattarova, N. V. Semenova, T. V. Troepol'skaya and Y. P. Kitaev, *Izv. Akad. Nauk SSSR, Otd. Khim. Nauk*, 663 (1963); *Chem. Abstr.*, **59**, 7347 (1963).
77. R. R. Shagidullin, F. K. Sattarova, T. V. Troepol'skaya and Y. P. Kitaev, *Izv. Akad. Nauk SSSR, Otd. Khim. Nauk*, 385 (1963); *Chem. Abstr.*, **58**, 13761 (1963).
78. Y. P. Kitaev and T. V. Troepol'skaya, *Izv. Akad. Nauk SSSR, Otd. Khim. Nauk*, 454 (1963); *Chem. Abstr.*, **59**, 7347 (1963).
79. R. Kuhn and R. Bär, *Ann. Chem.*, **516**, 143 (1935).
80. A. Burawoy and A. R. Thompson, *J. Chem. Soc.*, 1443 (1953).
81. E. Sawicki, *J. Org. Chem.*, **22**, 743 (1957).
82. E. Fischer and Y. F. Frei, *J. Chem. Soc.*, 3159 (1959).
83. Y. Yagi, *Bull. Chem. Soc. Japan*, **37**, 1875 (1964).
84. C. K. Ingold, *Structure and Mechanism in Organic Chemistry*, Cornell University Press, Ithaca, New York, 1953, p. 380.
85. K. J. Morgan, *J. Chem. Soc.*, 2151 (1961).
86. S. C. Bell, G. L. Conklin and S. J. Childress, *J. Am. Chem. Soc.*, **85**, 2868 (1963).

87. R. Kuhn and W. Münzing, *Chem. Ber.*, **86**, 858 (1953).
88. F. Ramirez and A. F. Kirby, *J. Am. Chem. Soc.*, **75**, 6026 (1953); **76**, 1037 (1954); H. Van Duin, *Rec. Trav. Chim.*, **73**, 78 (1954); F. A. Isherwood and R. L. Jones, *Nature*, **175**, 419 (1955); P. de Mayo and A. Stoessl, *Can. J. Chem.*, **38**, 950 (1960).
89. D. Schulte-Frohlinde, *Ann. Chem.*, **622**, 47 (1959).
90. E. J. Poziomek, D. N. Kramer, B. W. Fromm and W. A. Mosher, *J. Org. Chem.*, **26**, 432 (1961).
91. L. Tschetter, *Proc. S. Dakota Acad. Sci.*, **43**, 165 (1964); *Chem. Abstr.*, **63**, 8165 (1965).
92. G. W. Wheland, *Advanced Organic Chemistry*, John Wiley and Sons, Inc., New York, 1960, p. 452.
93. I. V. Hopper, *J. Chem. Soc.*, **127**, 1282 (1925).
94. E. Odernheimer, *Chem. Ber.*, **16**, 2988 (1883); H. Goldschmidt, *Chem. Ber.*, **16**, 2176 (1883); E. Beckmann, *Chem. Ber.*, **22**, 1531 (1889).
95. O. L. Brady and R. F. Goldstein, *J. Chem. Soc.*, 1959 (1927).
96. S. Ginsburg and T. B. Wilson, *J. Am. Chem. Soc.*, **79**, 481 (1957).
97. W. D. Phillips, *Ann. N.Y. Acad. Sci.*, **70**, 817 (1958).
98. J. Stieglitz, *Am. Chem. J.*, **40**, 36 (1908); P. P. Peterson, *Am. Chem. J.*, **46**, 325 (1911).
99. W. Theilacher and K. Fauser, *Ann. Chem.*, **539**, 103 (1939).
100. N. V. Khromov-Borisov, *Zh. Obshch. Khim.*, **25**, 136 (1955); *Chem. Abstr.*, **49**, 8257 (1955).
101. G. J. Karabatsos, J. D. Graham and F. M. Vane, *J. Am. Chem. Soc.*, **84**, 753 (1962).
102. G. J. Karabatsos, B. L. Shapiro, F. M. Vane, J. S. Fleming and J. S. Ratka, *J. Am. Chem. Soc.*, **85**, 2784 (1963).
103. G. J. Karabatsos, R. A. Taller and F. M. Vane, *Tetrahedron Letters*, **18**, 1081 (1964).
104. G. J. Karabatsos, R. A. Taller and F. M. Vane, *J. Am. Chem. Soc.*, **85**, 2327 (1963).
105. G. J. Karabatsos, R. A. Taller and F. M. Vane, *J. Am. Chem. Soc.*, **85**, 2326 (1963).
106. E. Lustig, *J. Am. Chem. Soc.*, **65**, 491 (1961).
107. G. J. Karabatsos, J. D. Graham and F. M. Vane, *J. Am. Chem. Soc.*, **84**, 37 (1962).
108. C. Mills, *J. Chem. Soc.*, **67**, 925 (1895); R. Clauser, *Chem. Ber.*, **34**, 889 (1901); R. Clauser and G. Schweitzer, *Chem. Ber.*, **35**, 4280 (1902).
109. P. Ruggli and J. Rohner, *Helv. Chim. Acta*, **25**, 1533 (1942).
110. R. W. Faessinger and E. V. Brown, *J. Am. Chem. Soc.*, **73**, 4606 (1951).
111. M. M. Shemyakin, V. I. Maimind and B. K. Vaichunaite, *Izv. Akad. Nauk SSSR, Otd. Khim. Nauk.*, 1260 (1957); *Chem. Abstr.*, **52**, 6231 (1958).
112. J. J. de Lange, J. M. Robertson and I. Woodward, *Proc. Roy. Soc. (London)*, **A171**, 398 (1939).
113. D. J. Cram and J. S. Bradshaw, *J. Am. Chem. Soc.*, **85**, 1108 (1963).
114. J. P. Freeman, *J. Org. Chem.*, **28**, 2508 (1963).
115. G. S. Hartley, *Nature*, **140**, 281 (1937); G. S. Hartley, *J. Chem. Soc.*, 633 (1938).
116. J. H. Collins and H. H. Jaffe, *J. Am. Chem. Soc.*, **84**, 4708 (1962).

117. S. B. Hendricks, O. R. Wulf, G. E. Hilbert and U. Liddel, *J. Am. Chem. Soc.*, **58**, 1991 (1936).
118. G. M. Wyman, *Chem. Rev.*, **55**, 648 (1955).
119. H. Boersch, *Monatsh. Chem.*, **65**, 327 (1935); W. West and R. B. Killingsworth, *J. Chem. Phys.*, **6**, 1 (1938).
120. R. F. Hutton and C. Steel, *J. Am. Chem. Soc.*, **86**, 745 (1964).
121. D. L. Webb and H. H. Jaffe, *J. Am. Chem. Soc.*, **86**, 2419 (1964).
122. E. Müller and W. Kreutzmann, *Ann. Chem.*, **495**, 132 (1932); K. A. Gehrckens and E. Müller, *Ann. Chem.*, **500**, 296 (1933).
123. V. M. Dziomko and K. A. Dunawaskaya, *Zh. Obshch. Khim.*, **31**, 68 (1961); *Chem. Abstr.*, **55**, 23394 (1961).
124. C. S. Hahn, B. H. Yun and H. K. Lee, *J. Korean Chem. Soc.*, **7**, 197 (1963).
125. P. Luner and C. A. Winkler, *Can. J. Chem.*, **30**, 679 (1952).
126. W. P. Jencks, *J. Am. Chem. Soc.*, **81**, 475 (1959).
127. E. H. Cordes and W. P. Jencks, *J. Am. Chem. Soc.*, **84**, 4319 (1962).
128. E. H. Cordes and W. P. Jencks, *J. Am. Chem. Soc.*, **84**, 832 (1962).
129. E. H. Cordes and W. P. Jencks, *J. Am. Chem. Soc.*, **85**, 2843 (1963).
130. J. C. Powers and F. H. Westheimer, *J. Am. Chem. Soc.*, **82**, 5431 (1960).
131. L. do Amaral, W. A. Sandstrom and E. H. Cordes, *J. Am. Chem. Soc.*, **88**, 2225 (1966).
132. S. F. Acree and J. M. Johnson, *Am. Chem. J.*, **38**, 258 (1907).
133. E. G. R. Ardagh, B. Kellam, F. C. Rutherford and H. T. Walstaff, *J. Am. Chem. Soc.*, **54**, 721 (1932).
134. D. W. Brooks and J. D. Gettler, *J. Org. Chem.*, **27**, 4469 (1962).
135. L. Barrett and A. Lapworth, *J. Chem. Soc.*, **93**, 85 (1908); A. Olander, *Z. Physik. Chem. (Leipzig)*, **129**, 1 (1927).
136. W. P. Jencks, 'Mechanism and Catalysis of Simple Carbonyl Group Reactions' in *Progress in Physical Organic Chemistry*, Vol. 2 (Eds. S. G. Cohen, A. Streitwieser, Jr. and R. W. Taft), Interscience Publishers, New York, 1963, p. 70.
137. J. B. Conant and P. D. Bartlett, *J. Am. Chem. Soc.*, **54**, 2881 (1932).
138. L. P. Hammett, *Physical Organic Chemistry*, McGraw-Hill Book Co., Inc., New York, 1940, p. 329.
139. G. H. Stempel, Jr. and G. S. Schaffel, *J. Am. Chem. Soc.*, **66**, 1158 (1944)
140. E. S. Hand and W. P. Jencks, *J. Am. Chem. Soc.*, **84**, 3505 (1962); R. B. Martin, A. Parcell and R. I. Hedrick, *J. Am. Chem. Soc.*, **86**, 2406 (1964).
141. S. Bodforss, *Z. Physik. Chem. (Leipzig)*, **109**, 233 (1924).
142. A. Hantzsch, *Chem. Ber.*, **25**, 701 (1892).
143. A. Kling, *Compt. Rend.*, **148**, 569 (1909); G. Knopfer, *Monatsh. Chem.*, **32**, 768 (1911).
144. P. K. Chang and T. L. V. Ulbricht, *J. Am. Chem. Soc.*, **80**, 976 (1958).
145. R. Cantarel and J. Guenzet, *Bull. Soc. Chim. France*, 1285 (1961).
146. E. G. R. Ardagh and J. G. Williams, *J. Am. Chem. Soc.*, **47**, 2976 (1925).
147. E. G. R. Ardagh, B. Kellam, F. C. Rutherford and W. T. Walstaff, *J. Am. Chem. Soc.*, **54**, 721 (1932).
148. F. H. Westheimer, *J. Am. Chem. Soc.*, **56**, 1962 (1934).
149. B. M. Anderson and W. P. Jencks, *J. Am. Chem. Soc.*, **82**, 1773 (1960).
150. B. Oddo and F. Toguacchini, *Gazz. Chim. Ital.*, **52**, 347 (1922).
151. G. Vavon and P. Montheard, *Bull. Soc. Chim. France*, **7**, 551 (1940).

152. A. V. Willi and R. E. Robertson, *Can. J. Chem.*, **31**, 361 (1953).
153. D. S. Noyce, A. T. Bottini and S. G. Smith, *J. Org. Chem.*, **23**, 752 (1958).
154. G. M. Santerre, C. J. Hansrote, Jr. and T. I. Crowell, *J. Am. Chem. Soc.*, **80**, 1254 (1958).
155. J. D. Dickinson and C. Eaborn, *J. Chem. Soc.*, 3641 (1959).
156. T. I. Crowell, in *The Chemistry of the Alkenes* (Ed. S. Patai), Interscience Publishers, New York, 1964, p. 265.
157. Y. Yukawa and Y. Tsuno, *Bull. Chem. Soc. Japan*, **32**, 971 (1959).
158. J. Weinstein and E. McIninch, *J. Am. Chem. Soc.*, **82**, 6064 (1960).
159. K. Yates and R. Stewart, *Can. J. Chem.*, **37**, 664 (1959).
160. T. I. Crowell, C. E. Bell, Jr. and D. H. O'Brien, *J. Am. Chem. Soc.*, **86**, 4973 (1964).
161. J. D. Dickinson and C. Eaborn, *J. Chem. Soc.*, 3036 (1959).
162. J. N. Brønsted and K. Pedersen, *Z. Physik. Chem. (Leipzig)*, **108**, 185 (1924).
163. R. B. Martin, *J. Phys. Chem.*, **68**, 1369 (1964).
164. M. Eigen and L. DeMaeyer, in *The Structure of Electrolytic Solutions* (Ed. W. J. Hammer), John Wiley and Sons, New York, 1959, p. 64.
165. C. G. Swain and J. C. Woroxz, *Tetrahedron Letters*, **36**, 3199 (1965).
166. C. G. Swain, D. A. Kuhn and R. L. Schowen, *J. Am. Chem. Soc.*, **87**, 1553 (1965).
167. C. G. Swain and E. R. Thornton, *J. Am. Chem. Soc.*, **84**, 817 (1962).
168. K. Koehler, W. Sandstrom and E. H. Cordes, *J. Am. Chem. Soc.*, **86**, 2413 (1964).
169. R. B. Martin and A. Parcell, *J. Am. Chem. Soc.*, **83**, 4830 (1961).
170. G. Kresze and H. Manthey, *Z. Elektrochem.*, **58**, 118 (1954).
171. B. Kastening, L. Hollock and G. A. Melkonian, *Z. Elektrochem.*, **60**, 130 (1956).
172. G. Baddeley and R. M. Topping, *Chem. Ind. (London)*, 1693 (1958).
173. A. Williams and M. L. Bender, *J. Am. Chem. Soc.*, **88**, 2508 (1966).
174. D. H. R. Barton, R. E. O'Brien and S. Sternhell, *J. Chem. Soc.*, 470 (1962).
175. J. H. Simons, W. T. Black and R. F. Clark, *J. Am. Chem. Soc.*, **75**, 5621 (1953).
176. J. F. Bunnett, *J. Am. Chem. Soc.*, **83**, 4956, 4968 (1961).
177. R. B. Martin, *J. Am. Chem. Soc.*, **84**, 4130 (1962); R. B. Martin, *J. Am. Chem. Soc.*, **86**, 5709 (1964).
178. R. B. Martin, *J. Org. Chem.*, **29**, 3197 (1964).
179. G. L. Schmir and B. A. Cunningham, *J. Am. Chem. Soc.*, **87**, 5692 (1965).
180. B. A. Cunningham and G. L. Schmir, *J. Am. Chem. Soc.*, **88**, 551 (1966).
181. E. H. Cordes and W. P. Jencks, *J. Am. Chem. Soc.*, **84**, 826 (1962).
182. M. L. Bender and A. Williams, *J. Am. Chem. Soc.*, **88**, 2502 (1966).
183. C. R. Hauser and D. S. Hoffenberg, *J. Am. Chem. Soc.*, **77**, 4885 (1955).
184. J. A. Gautier, J. Renault and C. Fauran, *Bull. Soc. Chim. France*, 2738 (1963).
185. V. E. Haury, *U.S. Pat.* 2,692,284; *Chem. Abstr.*, **49**, 15946 (1955).
186. R. Wolfenden and W. P. Jencks, *J. Am. Chem. Soc.*, **83**, 2763 (1961).
187. O. Bloch-Chaude, *Compt. Rend.*, **239**, 804 (1954).
188. P. Petrenko-Kritschenko and S. Lordkipanidze, *Chem. Ber.*, **34**, 1702 (1901).

189. P. Petrenko-Kritschenko and W. Kantscheff, *Chem. Ber.*, **39**, 1452 (1906).
190. L. Ruzicka and J. B. Buijs, *Helv. Chim. Acta.*, **15**, 8 (1932).
191. A. R. Poggi, *Gazz. Chim. Ital.*, **77**, 536 (1947).
192. P. G. Kletzke, *J. Org. Chem.*, **29**, 1363 (1964).
193. E. C. Suratt, J. R. Proffitt, Jr. and C. T. Lester, *J. Am. Chem. Soc.*, **72**, 1561 (1950).
194. J. L. Maxwell, M. J. Brownlee and M. P. Holden, *J. Am. Chem. Soc.*, **83**, 589 (1961).
195. A. Claus and C. Foecking, *Chem. Ber.*, **20**, 3097 (1887).
196. E. Feith and S. H. Davies, *Chem. Ber.*, **24**, 3546 (1891).
197. F. Baum, *Chem. Ber.*, **28**, 3207 (1895).
198. R. G. Kadesch, *J. Am. Chem. Soc.*, **66**, 1207 (1944).
199. F. P. Price, Jr. and L. P. Hammett, *J. Am. Chem. Soc.*, **63**, 2387 (1941).
200. F. W. Fitzpatrick and J. D. Gettler, *J. Am. Chem. Soc.*, **78**, 530 (1956).
201. I. D. Fiarman and J. D. Gettler, *J. Am. Chem. Soc.*, **84**, 961 (1962).
202. G. Vavon and P. Anziani, *Bull. Soc. Chim. France*, **4**, 2026 (1937).
203. G. Vavon and P. Montheard, *Bull. Soc. Chim. France*, **7**, 560 (1940).
204. D. G. Knorre and N. M. Emanuel, *Dokl. Akad. Nauk SSSR*, **91**, 1163 (1953); *Chem. Abstr.*, **49**, 12936 (1955).
205. M. J. S. Dewar, *J. Chem. Soc.*, 463 (1949); P. B. D. de la Mare and J. H. Ridd, *Aromatic Substitution*, Academic Press, Inc., New York, 1959, p. 82.
206. T. C. Bissot, R. W. Parry and D. H. Campbell, *J. Am. Chem. Soc.*, **79**, 796 (1957).
207. M. Imoto and M. Kobayashi, *Bull. Chem. Soc. Japan*, **33**, 1651 (1960).
208. E. F. Pratt and M. J. Kamlet, *J. Org. Chem.*, **26**, 4029 (1961).
209. R. L. Hill and T. I. Crowell, *J. Am. Chem. Soc.*, **78**, 2284, 6425 (1956).
210. S. I. Miller, *J. Am. Chem. Soc.*, **81**, 101 (1959).
211. W. P. Jencks, 'Mechanism and Catalysis of Simple Carbonyl Group Reactions' in *Progress in Physical Organic Chemistry*, Vol. 2 (Eds. S. G. Cohen, A. Streitwieser, Jr. and R. W. Taft), Interscience Publishers, New York, 1963, p. 63.
212. J. Hine, *J. Am. Chem. Soc.*, **81**, 1126 (1959).
213. C. D. Ritchie, J. D. Saltiel and E. S. Lewis, *J. Am. Chem. Soc.*, **83**, 4601 (1961).
214. J. E. Leffler, *Science*, **117**, 340 (1953).
215. G. S. Hammond, *J. Am. Chem. Soc.*, **77**, 334 (1955).
216. C. G. Swain and E. R. Thornton, *J. Am. Chem. Soc.*, **84**, 817 (1962).
217. C. G. Swain and C. B. Scott, *J. Am. Chem. Soc.*, **75**, 141 (1953).
218. A. E. Braunstein and M. M. Shemyakin, *Biochemistry (USSR) (English Transl.)*, **18**, 393 (1953); *Chem. Abstr.*, **48**, 4603 (1954).
219. D. E. Metzler, M. Ikawa and E. E. Snell, *J. Am. Chem. Soc.*, **76**, 648 (1954); D. E. Metzler, *J. Am. Chem. Soc.*, **79**, 485 (1957).
220. T. C. Bruice and S. J. Benkovic, *Biorganic Mechanisms*, Vol. 2, W. A. Benjamin, Inc., New York, 1966.
221. A. E. Braunstein, in *The Enzymes*, Vol. 2, 2nd edn. (Eds. P. D. Boyer, H. Lardy and K. Myrbäck), Academic Press, New York, 1960, p. 113.
222. E. E. Snell, in *Proceedings of the Symposium on Chemical and Biological Aspects of Pyridoxal Catalysts, Rome, 1962* (Eds. E. E. Snell, P. M. Fasella, A. Braunstein and A. Rossi-Fanelli), Pergamon Press, New York, 1963, p. 1.

223. J. Olivard, D. E. Metzler and E. E. Snell, *J. Biol. Chem.*, **199**, 669 (1952).
224. M. Ikawa and E. E. Snell, *J. Am. Chem. Soc.*, **76**, 653 (1954).
225. M. Ikawa and E. E. Snell, *J. Am. Chem. Soc.*, **76**, 4900 (1954).
226. F. Binkley and M. Boyd, *J. Biol. Chem.*, **217**, 67 (1955).
227. D. E. Metzler and E. E. Snell, *J. Am. Chem. Soc.*, **74**, 979 (1952).
228. P. Fasella, H. Lis, N. Siliprandi and C. Baglionic, *Biochim. Biophys. Acta*, **23**, 417 (1957).
229. E. E. Snell, in *Vitamins and Hormones* (Eds. R. S. Harris, G. F. Marrian and K. V. Thimann), Academic Press, New York, 1958, p. 78.
230. E. E. Snell and W. T. Jenkins, *J. Cellular Comp. Physiol.* (Suppl. 1), **54**, 161 (1959).
231. C. Cennamo, *Biochim. Biophys. Acta*, **93**, 323 (1964).
232. Y. Matsuo and D. M. Greenberg, *J. Biol. Chem.*, **230**, 561 (1958).
233. W. T. Jenkins, D. A. Yphantis and I. W. Sizer, *J. Biol. Chem.*, **234**, 51 (1959).
234. P. Fasella, G. G. Hammes and B. L. Vallee, *Biochem. Biophys. Acta*, **65**, 142 (1962).
235. W. B. Dempsey and E. E. Snell, *Biochemistry*, **2**, 1414 (1963).
236. D. Heinert and A. E. Martell, *J. Am. Chem. Soc.*, **84**, 3257 (1962).
237. H. N. Christensen, *J. Am. Chem. Soc.*, **80**, 99 (1958).
238. A. B. Kent, E. G. Krebs and E. H. Fischer, *J. Biol. Chem.*, **232**, 549 (1958).
239. E. M. Wilson and H. L. Kornberg, *Biochem. J.*, **88**, 578 (1963).
240. E. H. Fischer, A. B. Kent, E. R. Snyder and E. G. Krebs, *J. Am. Chem. Soc.*, **80**, 2906 (1958).
241. R. C. Hughes, W. T. Jenkins and E. H. Fischer, *Proc. Nat. Acad. Sci. U.S.*, **48**, 1615 (1962).
242. F. H. Westheimer, *Proc. Chem. Soc.*, 253 (1963).
243. O. L. Polyanovskii and B. A. Keil, *Biochemistry (USSR) (English Transl.)*, **28**, 306 (1963).
244. L. G. Schirch and M. Mason, *J. Biol. Chem.*, **238**, 1032 (1963).
245. M. Fugioka and E. E. Snell, *J. Biol. Chem.*, **240**, 3044 (1965).
246. J. A. Anderson and H. W. Chang, *Arch. Biochem. Biophys.*, **110**, 346 (1965).
247. C. Turano, A. Giartosio and P. Fasella, *Arch. Biochem. Biophys.*, **104**, 524 (1964).
248. M. Fujioka and E. E. Snell, *J. Biol. Chem.*, **240**, 3050 (1965).
249. Y. Ogata and Y. Takagi, *J. Am. Chem. Soc.*, **80**, 3591 (1958).
250. P. A. S. Smith, *The Chemistry of Open-Chain Organic Nitrogen Compounds*, Vol. 2, W. A. Benjamin, Inc. New York, 1966, p. 365.
251. K. Ueno and S. Akiyoshi, *J. Am. Chem. Soc.*, **76**, 3671 (1954).
252. E. Bamberger and E. Renauld, *Chem. Ber.*, **30**, 2278 (1897).
253. E. Bamberger and A. Rising, *Ann. Chem.*, **316**, 257 (1901).
254. V. O. Lukashevich, *Compt. Rend. Acad. Sci. URSS*, **21**, 376 (1938); *Chem. Abstr.*, **33**, 3769 (1939).
255. Y. Ogata, M. Tsuchida and Y. Takagi, *J. Am. Chem. Soc.*, **79**, 3397 (1957).
256. S. Oae, T. Fukumoto and M. Yamagami, *Bull. Chem. Soc. Japan*, **36**, 728 (1963).
257. L. A. Neiman, V. I. Maimind and M. M. Shemyakin, *Tetrahedron Letters*, **35**, 3157 (1965).

258. G. A. Russell, E. G. Janzen and E. T. Strom, *J. Am. Chem. Soc.*, **86**, 1807 (1964).
259. G. A. Russell, E. G. Janzen, H. D. Becker and F. J. Smentowski, *J. Am. Chem. Soc.*, **84**, 2652 (1962).
260. G. A. Russell and E. J. Geels, *J. Am. Chem. Soc.*, **87**, 122 (1965).

CHAPTER **8**

Cleavage of the carbon–nitrogen bond

EMIL H. WHITE and DAVID J. WOODCOCK

Department of Chemistry, The Johns Hopkins University, Baltimore, Maryland 21218

I. INTRODUCTION

Reactions in which a bond from carbon to nitrogen is broken are important in several areas of organic chemistry. The Hofmann and Cope eliminations are widely used in synthetic chemistry, and the deamination of aliphatic amines (via the nitrous acid, nitrosoamide and triazene methods) has been widely studied, usually with a primary emphasis on establishing the mechanism of the reaction. Since there has been much research activity in the latter area recently, a major fraction of the chapter is devoted to an interpretive and critical review of the deamination reactions.

A wide variety of methods for the cleavage of carbon–nitrogen bonds is now available (equation 1). The chemistry of these reactions is dominated by two fundamental properties of the nitrogen atom: (1) the ability to form a positively charged tetracovalent group that can, on the one hand, labilize the adjacent carbon–hydrogen bonds for chemical attack and, on the other, serve as a good 'leaving group'; and (2) the capability of combining with another nitrogen atom to

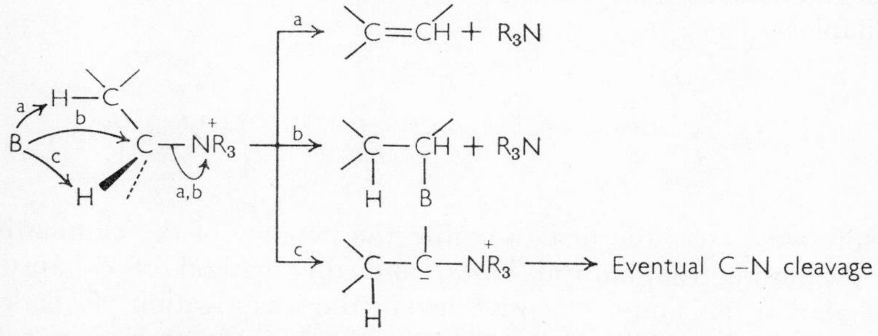

$$
\begin{aligned}
&\xrightarrow{a}\quad \text{\Large$>$}C{=}CH + R_3N\\[4pt]
&\xrightarrow{b}\quad \text{\Large$>$}C{-}CH\underset{\;H\;\;B}{}+ R_3N \qquad (1)\\[4pt]
&\xrightarrow{c}\quad \text{\Large$>$}C{-}C{-}\overset{+}{N}R_3 \longrightarrow \text{Eventual C–N cleavage}
\end{aligned}
$$

form the very stable N_2 molecule. Reactions of the second type have unique characteristics, as described in the section on deamination. In the following catalog of methods, some attempt is made to evaluate the synthetic usefulness of the methods, and to outline the mechanisms of the reactions.

II. QUATERNARY SALTS

A. Hofmann Elimination

The thermal decomposition of a quaternary ammonium hydroxide to alkene, tertiary amine and water was first observed by Hofmann in 1851 (equation 2)[1a]. From his work with unsymmetrical amines,

$$(C_2H_5)_4\overset{+}{N}\ \overset{-}{O}H \xrightarrow{\;\Delta\;} (C_2H_5)_3N + CH_2{=}CH_2 + H_2O \qquad (2)$$

Hofmann suggested that a quaternary salt containing an ethyl group always eliminated ethylene in preference to other alkenes. This rule has been generalized to state that the alkyl group eliminated is that one having the most β-hydrogens. A general exception to this rule is found when the quaternary salt contains a β-substituent that can stabilize a negative charge; in this case the alkene eliminated contains the β-substituent (equation 3). The Hofmann rule has also been stated

$$C_6H_5CH_2CH_2\overset{+}{\underset{\underset{C_2H_5}{|}}{N}}(CH_3)_2\overset{-}{O}H \longrightarrow C_6H_5CH{=}CH_2 + \underset{\underset{C_2H_5}{|}}{N}(CH_3)_2 + H_2O \qquad (3)$$

to the effect that the product of elimination from a quaternary hydroxide is the least-substituted alkene[2]. However, this formulation does not apply to eliminations from quaternary salts containing a β-substituted electron-withdrawing group (as above) and in many cases does not apply for quaternary salts containing α-substituted alkyl groups, as in the case of dimethylethylisopropylammonium hydroxide (equation 4). Thus no concise and correct statement of the Hofmann rule is available[1c].

$$C_2H_5-\overset{\overset{\displaystyle CH_3}{|}}{\underset{\underset{\displaystyle CH_3}{|}}{N}}\overset{+}{}{-}\overset{\overset{\displaystyle CH_3}{|}}{CH}{-}CH_3 \xrightarrow{\quad\Delta\quad} \underset{60\%}{CH_3{=}CHCH_3} + C_2H_5N(CH_3)_2 \qquad (4)$$

Hofmann[1b] was the first to realize the potency of the elimination for examining the natural bases, and the method of exhaustive methylation has found extensive use in the investigation of alkaloid structures[3]. Essentially, exhaustive methylation is carried out by preparing the methiodide of the tertiary base, treating this with silver oxide, and concentrating the solution of the quaternary hydroxide under reduced pressure until decomposition takes place. The exclusion of carbon dioxide is essential for formation of good yields of elimination product[4]. In the examination of a natural product, the procedure is usually repeated until the nitrogen is liberated as trimethylamine. The degraded natural product now contains olefinic double bonds for further degradation. In addition, the number of exhaustive methylation procedures carried out to free the nitrogen determines the number of rings to which the nitrogen was attached in the natural product. An example is given by the degradation of quinolizidine (equation 5). The nitrogen is attached to two rings and it requires three cycles of the exhaustive methylation procedure to eliminate the

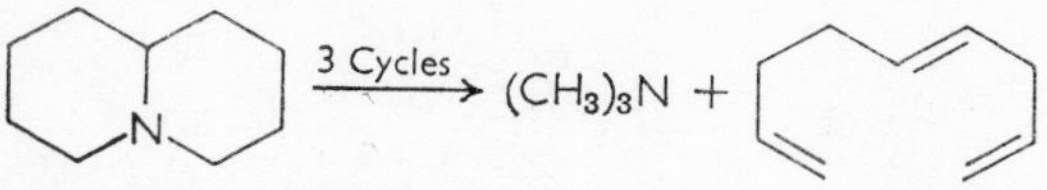

$$\text{[structure]} \xrightarrow{3 \text{ Cycles}} (CH_3)_3N + \text{[structure]} \tag{5}$$

nitrogen. Reported[3] yields for the degradation reaction are usually good to very good (60–100%). A survey of products and yields in the Hofmann elimination has appeared[3].

The Hofmann elimination has also been used in the synthesis of alkenes not readily available by other elimination reactions. A recent example is the synthesis of trimethylenecyclopropane (equation 6)[5a].

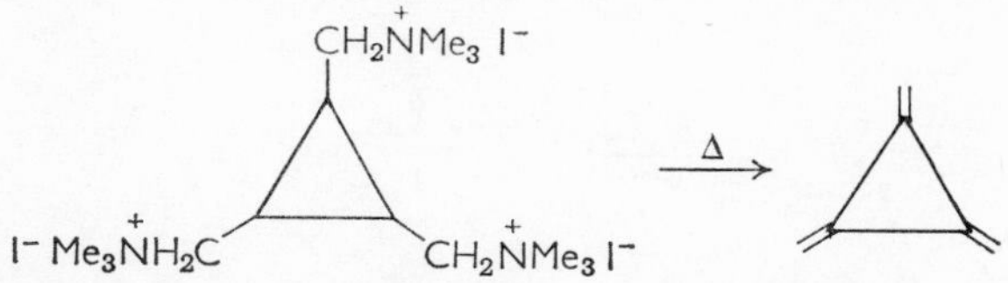

$$\text{[structure]} \xrightarrow{\Delta} \text{[structure]} \tag{6}$$

1,6-Elimination has been observed for a special case and used for the synthesis of a paracyclophane (equation 7)[5b].

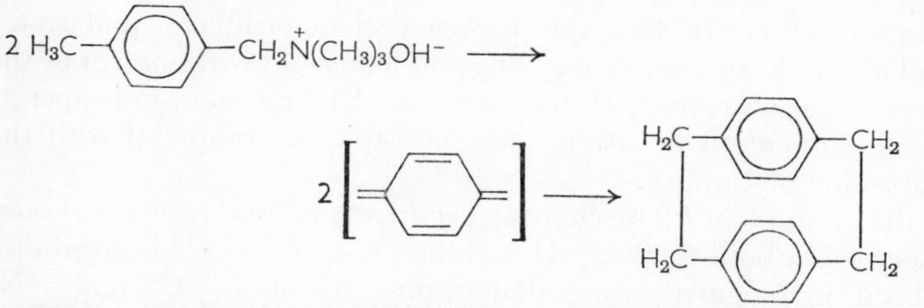

$$2\ H_3C\!-\!\!\text{[ring]}\!-\!CH_2\overset{+}{N}(CH_3)_3OH^- \longrightarrow$$

$$2\left[\text{[structure]}\right] \longrightarrow \text{[structure]} \tag{7}$$

The mechanism of the Hofmann elimination has received considerable attention in the last two decades and a number of reviews on the mechanism of this and other elimination reactions have appeared[3,6]. Two facets of the reaction have been studied in particular, the degree of concertedness of the reaction and the orientation of the elimination.

Ingold examined the elimination[4a,7] and found that it was first order in both hydroxide ion and quaternary ammonium ion[7a]. The reaction was described as having an $E2$ (elimination, bimolecular) mechanism, though Ingold noted that a second mechanism, namely $E1_{cb}$ (elimination, unimolecular, from the conjugate base (3) of the substrate), would also fit the kinetic data. Attempts have been made to distinguish between these two mechanistic paths. Since the $E1_{cb}$ mechanism involves an equilibration between the quaternary ion (1) and the zwitterion intermediate (3) it should be possible to observe an

14*

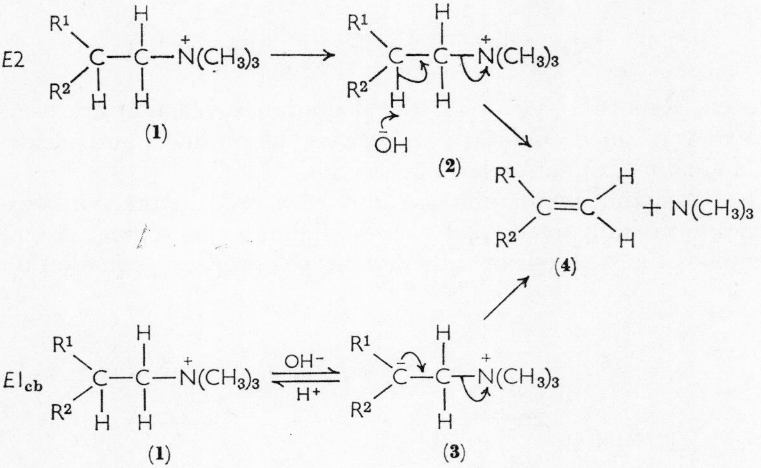

$$(8)$$

exchange of the β-hydrogens with deuterium from the solvent, for example, providing that the elimination to alkene (4) is slower than the equilibration. In fact, for the reaction of ordinary quaternary salts of aliphatic amines, no exchange of deuterium with solvent at the β-position was observed[8]. If, therefore, the El_{cb} mechanism is operating, the elimination to alkene must be rapid as compared with the equilibration of 1 and 3.

In the event of an $E2$ mechanism occurring, an isotope effect should be observed at both the β-C—H and the α-C—N sites. This possibility has been investigated using deuterium-, tritium-, ^{14}C- and ^{15}N-substituted compounds. An isotope effect (k_H/k_D) of 4 for elimination from the β,β,β-trideuteroethyltrimethylammonium ion[8a] and one of 3 for elimination from the β,β-dideutero-β-phenethyltrimethylammonium ion[8b] have been observed. These indicate that β-C—H bond cleavage is involved to some extent in the transition state. The nitrogen isotope effect in the elimination reaction of phenethyltrimethylammonium ion was found to be 30% of the theoretical maximum, a result indicating that for this reaction, the α-C—N bond is partially cleaved in the transition state[9]. A study of the tritium isotope effect has been made using n-propyltrimethylammonium salts[10a]. The isotope effects (k_H/k_T) observed by substitution of tritium at the α-, β- and γ-carbons of the n-propyl group were 1·1, 2·86 and 1·1 respectively. The effect of β-substituents on the tritium isotope effect in substituted β-tritioethyltrimethylammonium ions has been studied[10b] (Table 1). These results seem to indicate little C—H bond

TABLE 1. Tritium isotope effect in Hofmann elimination for $(CH_3)_3\overset{+}{N}CH_2CHTR$ at 60°.

R	k_H/k_T
H	3·0
CH_3	2·9
C_6H_5	~4·2
p-$NO_2C_6H_4$	8

cleavage in the transition state, except, perhaps, in the case of the p-nitrophenethyl group. The smaller isotope effect observed for the monotritiated derivatives as opposed to that observed for the fully deuterated compounds is attributed to a secondary isotope effect due to the additional deuterium atoms present in the deuterated molecules studied[10]. The carbon-14 isotope effect has been studied with the compounds $R\overset{+}{N}(CH_3)_3$[10a] (Table 2). The results shown in

TABLE 2. Carbon-14 isotope effect for $R\overset{+}{N}(CH_3)_3$ in the Hofmann elimination.

R	k_{12C}/k_{14C}	Temperature (°C)
$^{14}CH_2CH_3$	1·060	40
$^{14}CH_2CH_2CH_3$	1·067	50
$CH_2{}^{14}CH_2CH_3$	1·032	57
$^{14}C(CH_3)_3$	1·045	91
$^{14}CH_2CH_2C_6H_4NO_2$-p	1·026	100

the table are about a half to a third of the maximum theoretical value ($k_{12C}/k_{14C} = 1·15$ at 40°) showing participation of C—N bond cleavage to this degree in the transition state. The low ^{14}C isotope effect for the p-nitrophenethyl group suggests that the increased acidity of the β-proton has aided a shift in the elimination mechanism from $E2$ towards $E1_{cb}$.

Cram[11] has studied the elimination from threo- (equation 9) and erythro-1,2-diphenyl-1-propyltrimethylammonium iodides (equation 10). Since the $E2$ mechanism for elimination requires a trans-coplanar arrangement of the β-proton, the carbon atoms of the incipient double bond and the ammonium ion in the transition state, these isomers should lead to different alkene isomers if the reaction follows the $E2$ path. With ethoxide ion as base in ethanol, trans elimination was observed for both isomers and a rate ratio ($k_{threo}/k_{erythro}$) of 57 was calculated. This is in accord with the steric factors which hinder the *erythro* form from attaining the *trans*-coplanar configuration. It appears that with ethoxide ion as base an *E2* mechanism for the elimination occurs. However, upon changing the base to *t*-butoxide ion in

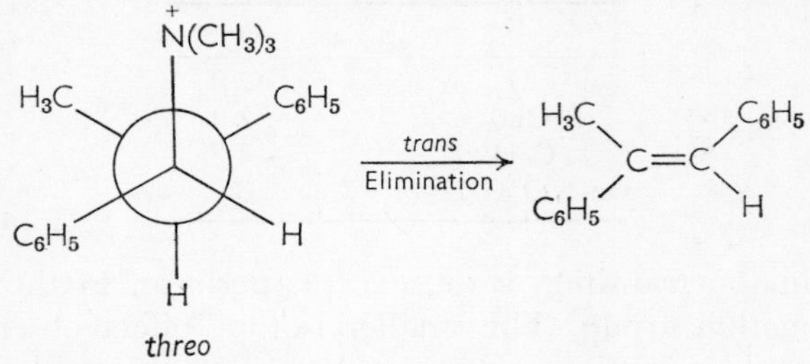

$$(9)$$

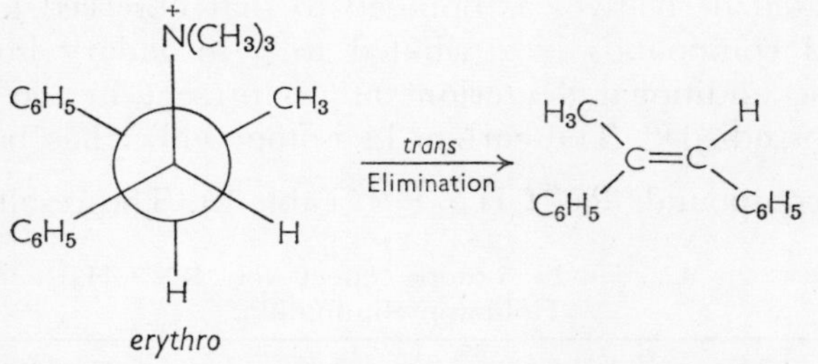

$$(10)$$

t-butanol, both isomers were found to produce the same alkene at the same rate. The alkene was that obtained by *trans* elimination from the *threo* derivative, and therefore, in effect, by *cis* elimination from the *erythro* derivative. This equality of the rates of elimination suggests that there is a common intermediate—the carbanion—giving the same product from both isomers. That is, in the case of *t*-butoxide ion as base, the elimination occurs via an $E1_{cb}$ mechanism. It is probable then that the Hofmann elimination can pass through a whole range of transition states[12] extending from $E1$ (cleavage of C—N bond before β-C—H bond) to $E1_{cb}$ (cleavage of β-C—H bond before C—N bond) with most of the normal Hofmann eliminations falling at a mid-point with near synchronous $E2$ mechanisms[13].

A third mechanism for the Hofmann elimination was suggested by Wittig[14]. This involves the formation of an ylide (**5**) by attack of base on an α'-hydrogen, followed by attack of the α'-carbanion on the β-hydrogen (equation 11). This mechanism requires that a β-proton be transferred to the tertiary amine (**6**) eliminated. Deuterium studies have shown[15], however, that in the normal Hofmann elimination the ylide mechanism is not an important one. On the other hand, evidence for an ylide mechanism has been obtained in the case of a highly branched alkylammonium ion, which for steric requirements cannot attain the *trans*-coplanar configuration required for $E2$ elimination[16], and in the case when very strong base (n-butyllithium) is used to bring about the reaction[15b].

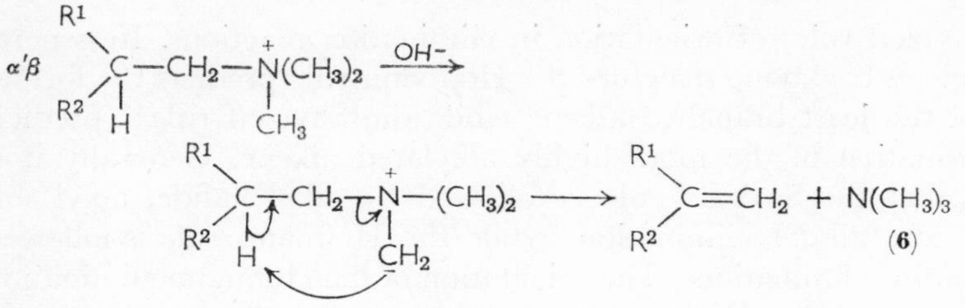

cis Elimination has been observed in the Hofmann elimination for certain special cases. In the case of sterically hindered molecules which cannot easily attain a configuration with a *trans* β-hydrogen, *cis* elimination can occur, perhaps by the ylide mechanism, as above[16,17]. The presence of an acidic β-hydrogen *cis* to the nitrogen can also induce *cis* elimination in preference to *trans*. The elimination from the 2-phenylcyclohexyltrimethylammonium ion[18], illustrates the latter case (equation 12). The *trans* form of the 2-phenylcyclohexyltrimethylammonium ion has a *trans* β-hydrogen at the 6-position and an acidic *cis* β-hydrogen at the 2-position, and upon elimination, formation of 1-phenylcyclohexene by elimination of the *cis* hydrogen at the 2-position predominates. Two mechanisms are available for

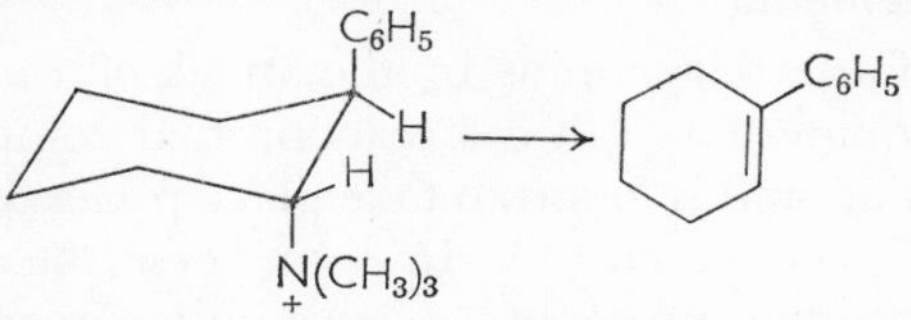

such a *cis* elimination, $E1_{cb}$ and $\alpha'\beta$ (Wittig), although a *cis* $E2$ mechanism could be considered. Examination of the *trans*-2-phenyl-2-d_1-cyclohexylamine derivative showed that elimination occurred without transfer of deuterium to the trimethylamine thereby excluding the ylide mechanism. Also since the *trans*-ammonium ion was shown not to isomerize to the *cis*-ammonium ion under the reaction conditions there could have been no formation of anion required by the $E1_{cb}$ mechanism, unless the anion when formed had eliminated tertiary amine at a rate much faster than equilibration of the anion and solvent.

Research into the mechanism of the Hofmann elimination has been stimulated by the controversy on the reasons for the Hofmann and the Saytzeff rules of orientation in elimination reactions. In general for a given backbone structure the Hofmann rule predicts the formation of the least branched alkene while the Saytzeff rule[19] predicts the formation of the more highly alkylated alkene. Generally it is found that the Saytzeff rule is followed by alkyl halide, alkyl sulfonate and all $E1$ eliminations while the Hofmann rule is followed by 'onium' eliminations. The orientation of the Hofmann elimination was suggested by Hughes and Ingold[20] to be due to the strong inductive effect caused by the positive charge of the 'onium' group which made the β-hydrogens acidic. Since β-substituted alkyl groups tend to reduce the acidity of the β-hydrogens, these hydrogens tend not to be lost. Brown[21], however, attributed the orientation entirely to steric factors, a direct result of the bulkiness of the 'onium' group which forced elimination to the least alkylated olefin. Although Brown used rather large alkyl groups, the theory was to account for all eliminations down to the simplest. It is of interest to note that the nitro group, which is smaller than iodine, should give Saytzeff products on the basis of Brown's theory and Hofmann products on the basis of Ingold and Hughes' theory. In fact, Hofmann products were observed[23]. Although this problem has not been completely solved it seems likely that the steric effect plays only a small role in the orientating effect unless the alkyl groups are bulky[22, 24].

B. Amine Displacements

Displacement of a tertiary amine by the attack of a nucleophile on the α-carbon of a quaternary salt is a reaction that competes with the Hofmann elimination and is found to take place preferably with weak bases (carbonate, acetate, etc.)[4a]. In some cases steric influences rather than basicity may promote the substitution reaction in favor of elimination. For example, the reaction of propyltrimethylammonium ion with hydroxide ion yields 19% methanol and 81% propylene, whereas the reaction with phenoxide ion yields 65% anisole and 15% propylene (equation 13)[4a]. Phenyl groups attached

$$\text{n-C}_3\text{H}_7\overset{+}{\text{N}}(\text{CH}_3)_3\text{X}^- \quad \underset{\overset{\displaystyle\longrightarrow}{\text{Displacement}}}{\overset{\text{Elimination}}{\longrightarrow}} \quad \begin{array}{l} \text{CH}_3\text{CH}{=}\text{CH}_2 + \text{N}(\text{CH}_3)_3 \\ \text{n-C}_3\text{H}_7\text{N}(\text{CH}_3)_2 + \text{CH}_3\text{X} \end{array} \qquad (13)$$

$$(\text{X} = \text{OH or OC}_6\text{H}_5)$$

to nitrogen apparently favor the displacement reaction[26], presumably by making the amine a better 'leaving group'. An S_N2 mechanism is indicated for the substitution reaction since the reaction of 1-phenylethyltrimethylammonium ion with acetate ion yielded 1-phenylethyl-acetate with 98–100% overall inversion of configuration[25]. When no β-hydrogen is present in the quaternary salt, nucleophilic displacement becomes the exclusive mode of decomposition (equation 14)[27].

$$C_6H_5CH_2\overset{+}{N}(CH_3)_3\overset{-}{O}H \begin{cases} \longrightarrow C_6H_5CH_2OH + N(CH_3)_3 & (65\%) \\ \longrightarrow CH_3OH + C_6H_5CH_2N(CH_3)_2 & (35\%) \end{cases} \quad (14)$$

A useful synthetic alkylation reaction occurs when the nucleophilic reagent is a carbanion (equation 15)[28]. Other nucleophiles have been

$$(CH_3)_4\overset{+}{N} + CH_2(COOC_2H_5)_2 \xrightarrow{\overset{-}{O}C_2H_5} CH_3CH(COOC_2H_5)_2 + (CH_3)_3N \quad \text{(Ref. 28d)} \quad (15)$$

observed to displace tertiary amines from quaternary salts[28a]; examples include amines and halide, cyanide, sulphide and mercaptide ions.

In special cases, displacement of the amine by a γ-carbon of the quaternary salt has been observed to take place in preference to β-elimination (equation 16). Strong base and an acidic γ-hydrogen are required to promote this type of displacement[29]. A similar reaction

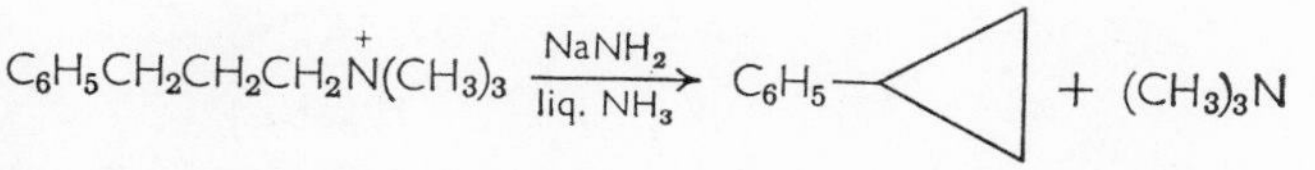

$$C_6H_5CH_2CH_2CH_2\overset{+}{N}(CH_3)_3 \xrightarrow[\text{liq. NH}_3]{\text{NaNH}_2} C_6H_5 -\!\!\triangleleft + (CH_3)_3N \quad (16)$$

occurs when β-hydroxyammonium salts are subjected to Hofmann elimination conditions (equation 17). In this case, displacement of the tertiary amine by oxygen occurs to form an epoxide[30].

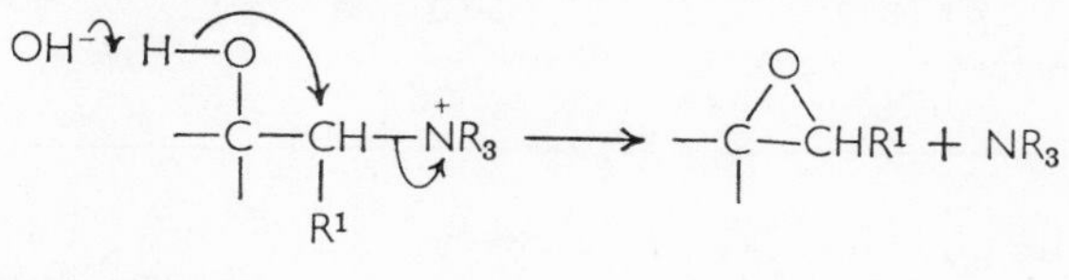

$$\overset{OH\curvearrowright H\!-\!O}{\underset{\underset{R^1}{|}}{\overset{|}{-}}\!\!\!\overset{|}{C}\!-\!\overset{|}{CH}\!-\!\overset{+}{N}R_3} \longrightarrow \overset{O}{\underset{|}{-}\!\!\overset{\triangle}{C}\!-\!CHR^1} + NR_3 \quad (17)$$

C. Rearrangements

I. The Stevens rearrangement

The Stevens rearrangement[31, 32] is a base-promoted 1,2-migration of an alkyl group from quaternary nitrogen to carbon (equations 18 and 19). The migration terminus must contain an acidic hydrogen

$$\underset{\underset{CH_2C_6H_5}{|}}{C_6H_5COCH_2\overset{+}{N}(CH_3)_2} \xrightarrow{OH^-} \underset{\underset{CH_2C_6H_5}{|}}{C_6H_5COCH\!-\!N(CH_3)_2} \quad (18)$$

$$C_6H_5CH_2\overset{+}{N}(CH_3)_3 \xrightarrow[\text{ether}]{C_6H_5Li,} \underset{\underset{CH_3}{|}}{C_6H_5CH\!-\!N(CH_3)_2} \quad (19)$$

for the rearrangement to occur; the weaker the acidity of this hydrogen, the stronger the base required for the reaction to proceed[31b]. The rearrangement of two isomeric brominated benzyldimethylphenacyl-ammonium bromides in the same reaction mixture gave no product of a 'cross' reaction showing the rearrangement to be intramolecular[32a]. A more rigorous experiment using ^{14}C-labeled compounds also gave no 'cross' products (equation 20)[32d]. The effect of substituents on the

$$
p\text{-BrC}_6\text{H}_4\text{COCH}_2\overset{+}{\text{N}}(\text{CH}_3)_2 \quad \text{C}_6\text{H}_5\text{COCH}_2\overset{+}{\text{N}}(\text{CH}_3)_2
$$
$$
\underset{^{14}\text{CH}_2\text{C}_6\text{H}_5}{|} \quad + \quad \underset{\text{CH}_2\text{C}_6\text{H}_5}{|}
$$

$$
\downarrow
$$

$$
p\text{-BrC}_6\text{H}_4\text{COCH—N}(\text{CH}_3)_2 \quad \text{C}_6\text{H}_5\text{COCH—N}(\text{CH}_3)_2
$$
$$
\underset{^{14}\text{CH}_2\text{C}_6\text{H}_5}{|} \quad + \quad \underset{\text{CH}_2\text{C}_6\text{H}_5}{|}
$$

(20)

benzene ring of the migrating group of benzyldimethylphenacyl-ammonium ion was such that electron-withdrawing substituents increased the rate of reaction (equation 21) (Table 3)[32b]. For sub-

$$
\text{C}_6\text{H}_5\text{COCH}_2\overset{+}{\text{N}}(\text{CH}_3)_2 \xrightarrow{\text{OH}^-} \text{C}_6\text{H}_5\text{COCHN}(\text{CH}_3)_2
$$
$$
\underset{\text{CH}_2\text{C}_6\text{H}_4\text{X}}{|} \qquad \qquad \underset{\text{CH}_2\text{C}_6\text{H}_4\text{X}}{|}
$$

(21)

TABLE 3. Rates of Stevens rearrangement of benzyldimethyl-phenacylammonium ion substituted in the benzene ring of the migrating group.

X	p-CH$_3$O	H	p-CH$_3$	p-Cl	p-NO$_2$
k/k_0	0·76	1·0	1·06	2·65	73

stituents in the phenacyl group the reverse was found, p-CH$_3$O mildly increasing the rate and m-NO$_2$ mildly decreasing it[32e]. Stevens interpreted these results on the basis of the formation of an ylide intermediate followed by migration involving dissociation into a tightly bound ion pair (equation 22).

$$
\text{C}_6\text{H}_5\text{COCH}_2\overset{+}{\text{N}}(\text{CH}_3)_2 \xrightarrow{\text{OH}^-} \text{C}_6\text{H}_5\text{CO}\overset{\frown}{\text{CH}}\overset{+}{\text{—N}}(\text{CH}_3)_2 \longrightarrow
$$
$$
\underset{\text{CH}_2\text{C}_6\text{H}_5}{|} \qquad \qquad \underset{\text{CH}_2\text{C}_6\text{H}_5}{\overset{\frown}{|}}
$$

$$
\text{C}_6\text{H}_5\text{COCH}=\overset{+}{\text{N}}(\text{CH}_3)_2 \longrightarrow \text{C}_6\text{H}_5\text{COCHN}(\text{CH}_3)_2 \quad (22)
$$
$$
\underset{\overset{-}{\text{CH}}_2\text{C}_6\text{H}_5}{|} \qquad \qquad \underset{\text{CH}_2\text{C}_6\text{H}_5}{|}
$$

Retention of optical activity for the migrating group to an extent of 97% was observed[33] and Hauser[34] suggested that to account for this the reaction mechanism must involve a cyclic transition state. However, the intimate ion-pair mechanism of Stevens is not ruled out by this evidence, since the migration of an ion with retention of configuration is feasible. Jenny and Druey[35a] observed that re-arrangement of optically active allyldimethyl(1-phenylethyl)am-monium bromide gave both 1,2- and 1,4-rearrangement products with retention of optical activity (equation 23). In benzene at 80°,

$$CH_2{=}CH{-}CH_2{-}\overset{+}{N}(CH_3)_2 \xrightarrow[\substack{\text{or} \\ \text{liq, NH}_3}]{\underset{C_6H_6}{NaNH_2}} CH_2{=}CH{-}CH{-}N(CH_3)_2$$

with the $H{-}\underset{\underset{C_6H_5}{|}}{\overset{|}{C}}{-}CH_3$ group below.

$$
\begin{array}{c}
CH_2{-}CH{=}CH{-}N(CH_3)_2\\
+\ H{-}\underset{\underset{C_6H_5}{|}}{\overset{|}{C}}{-}CH_3
\end{array} \tag{23}
$$

1,2-migration was favored over 1,4-migration by a factor of 1·4 and the product of 1,4-migration was found with 82 ± 10% retained configuration. In liquid ammonia at −33°, 1,2-migration was much more favored and the product of 1,4-migration was found with 72 ± 10% retained configuration. The authors claimed that this evidence is in favor of the discrete but tightly bound ion-pair mecha-nism of Stevens, since a 1,4 concerted cyclic mechanism would in-volve an inversion of configuration. The mechanism is probably right, but molecular models indicate that a cyclic mechanism proceeding with retention of configuration is also possible[35b].

The transfer of asymmetry from nitrogen to carbon has been observed in the Stevens rearrangement[37]; a transition state 7 in which the phenyl and ethylene groups are *cis* to each other is ap-parently involved (equation 24).

$$
\begin{array}{c}
\underset{CH_3}{\overset{CH_2C_6H_5}{|}}\\
C_6H_5{-}\overset{|}{\underset{|}{N^*}}{-}CH_2{-}CH{=}CH_2
\end{array}
\longrightarrow
\text{(7)}
\longrightarrow
\begin{array}{c}
\underset{CH{=}CH_2}{\overset{C_6H_5}{|}}\\
C_6H_5N{-}\overset{|}{\underset{|}{C^*}}{-}H
\end{array}
\tag{24}
$$

$[\alpha]_D = +39\cdot6$ $[\alpha]_D = -18\cdot4°$

Certain tertiary amines also undergo this type of reaction (equation 25)[36].

$$C_6H_5COCH_2N{-}C_6H_5 \xrightarrow[150°]{KOH} C_6H_5COCHNHC_6H_5 \tag{25}$$
with $CH_2C_6H_5$ substituents below each structure.

2. The Sommelet–Hauser rearrangement

The Sommelet–Hauser rearrangement, first observed by Sommelet[38] in 1937 but developed by Hauser[39] since 1951, involves the migration of an alkyl group from a quaternary ion bearing a benzyl group to the *ortho* position of that benzyl group (equation 26). The rearrange-

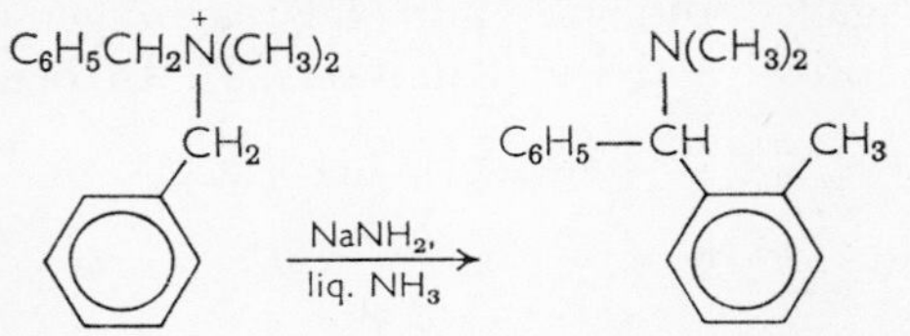
$C_6H_5CH_2\overset{+}{N}(CH_3)_2$
CH_2
$NaNH_2$, liq. NH_3
$N(CH_3)_2$
C_6H_5—CH
CH_3

(26)

ment is similar to the Stevens rearrangement and the two reactions can occur simultaneously, in which case it is generally found that the Sommelet–Hauser rearrangement predominates at low temperatures and the Stevens rearrangement at high temperatures (equation 27)[41].

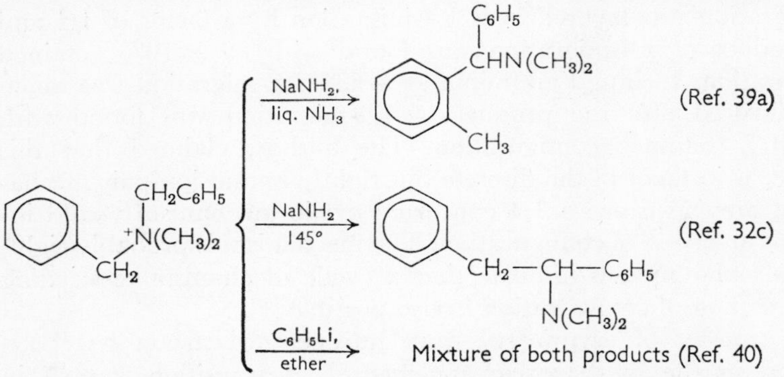
$CH_2C_6H_5$
$^+N(CH_3)_2$
CH_2
$NaNH_2$, liq. NH_3
C_6H_5
$CHN(CH_3)_2$
CH_3
(Ref. 39a)
$NaNH_2$
145°
CH_2—CH—C_6H_5
$N(CH_3)_2$
(Ref. 32c)
C_6H_5Li, ether
Mixture of both products (Ref. 40)

(27)

Hauser found that benzyltrimethylammonium ion was rearranged by sodamide in liquid ammonia to give dimethyl-2-methylbenzylamine in 97% yield[39a]. Quaternization of this base followed by a second reaction with sodamide resulted in the substitution of a second methyl on the benzene ring, and this procedure could be continued until the ring was fully methylated (equation 28). Hauser proposed that the mechanism of the reaction was initial proton abstraction by base at a benzylic site, and equilibration to a methyl anion, followed by anionic substitution at the ring *ortho* position[39a]. The resulting *exo*-methylene derivative would aromatize under the reaction conditions (equation 29). Support for this mechanism came when Hauser isolated the intermediate **8** from the reaction of trimethyl-(2,4,6-trimethylbenzyl)ammonium ion with sodamide in liquid ammonia (equation 30)[39b]. The ammonium ion contains an *ortho*

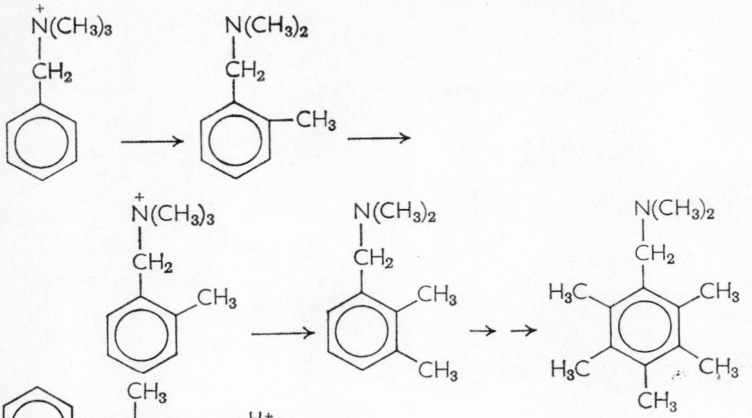

$$(28)$$

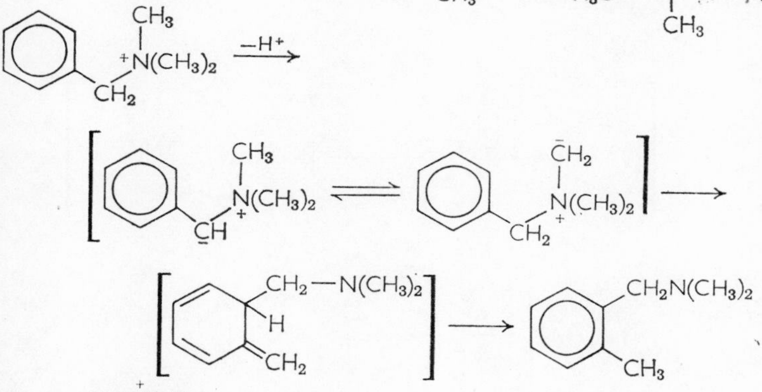

$$(29)$$

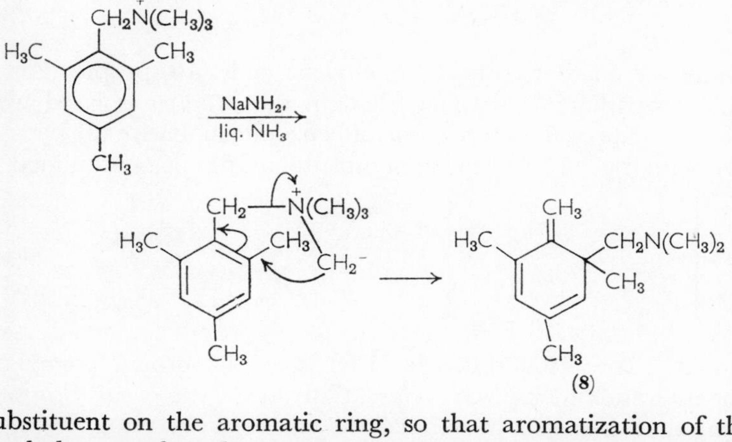

$$(30)$$

$$(8)$$

substituent on the aromatic ring, so that aromatization of the *exo*-methylene product does not occur under the conditions of the experiment. The *exo*-methylene derivative, formed in a 70% yield, could, however, be aromatized by action of heat or acids (equation 31).

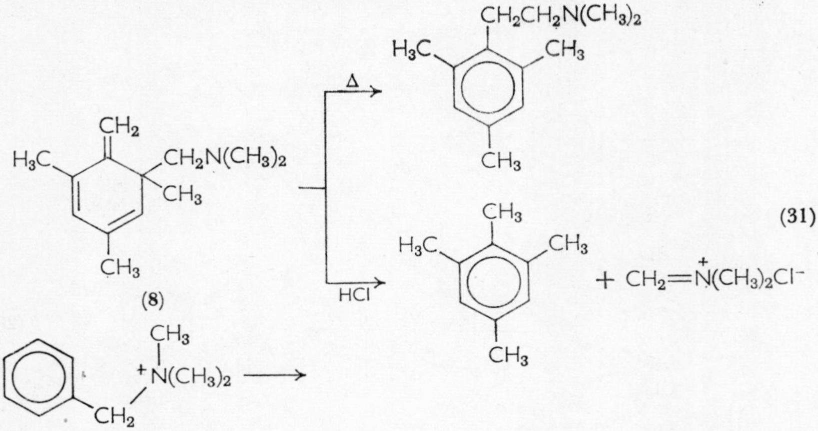

(31)

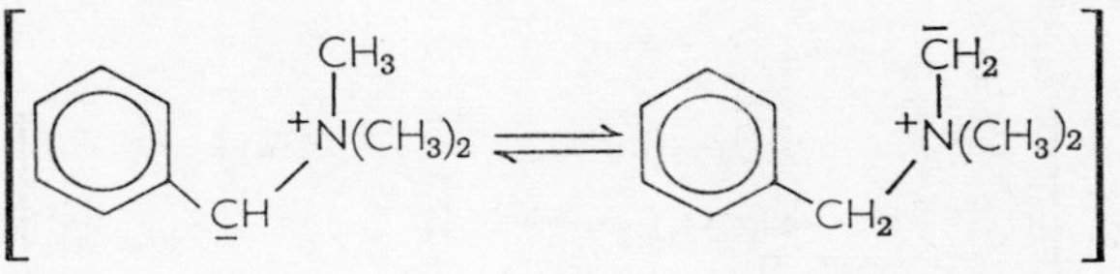

(32)

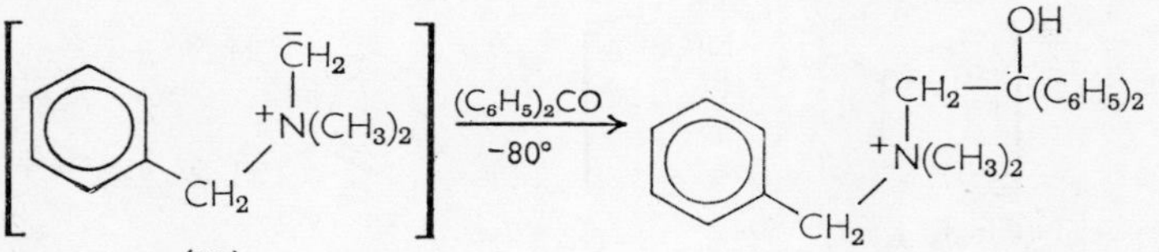

(33)

An attempt was made to trap the methylene ylide (**10**) proposed as a reaction intermediate[39c]. In fact, though, the product isolated was the derivative expected from reaction of benzophenone with the benzyl ylide (**9**) (equation 34). The effect of substituents has been examined[42]

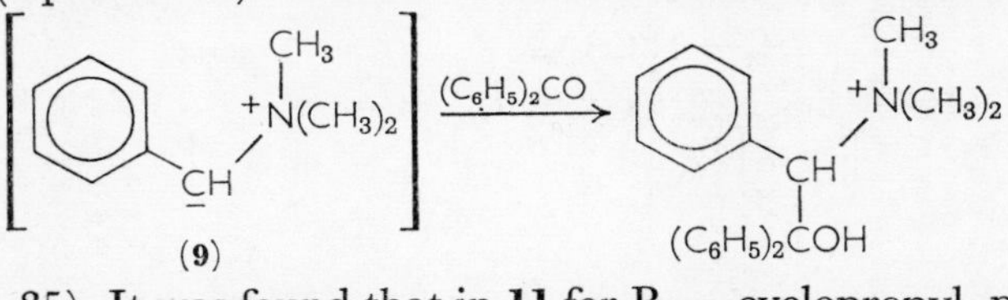

(34)

(equation 35). It was found that in **11** for R = cyclopropyl, phenyl or allyl, the migration took place preferentially at the site of substitution, indicating that these groups stabilize the anion **12**.

Bumgardener[42,43] has suggested as an alternative mechanism (equation 36): the formation and collapse of an ion-pair intermediate similar to that proposed for the Stevens rearrangement (equation 22).

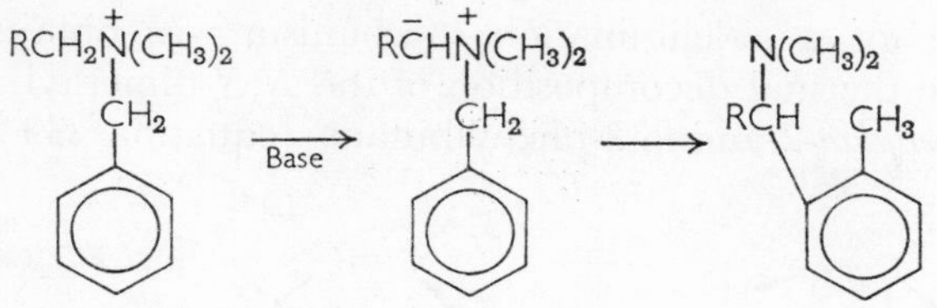

$$\text{(35)}$$

(11)   (12)

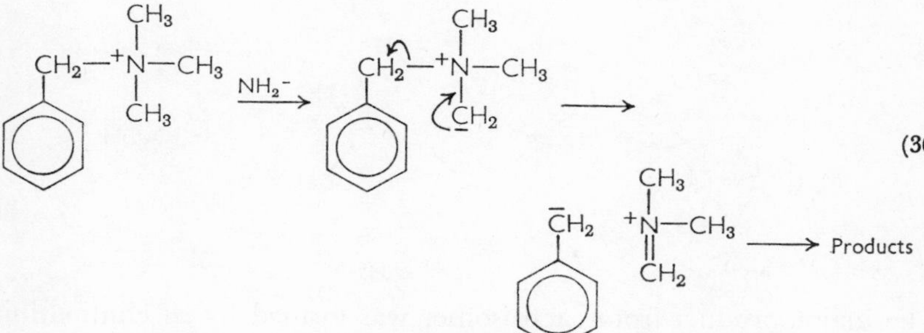

$$\text{(36)}$$

Products

There is as yet no evidence to support this mechanism in distinction to the Hauser mechanism.

III. AMINE OXIDES

A. The Cope Elimination

Although first observed in 1898[45], the thermal decomposition of tertiary amine oxides to give an alkene and hydroxylamine has received serious attention only since 1949 when the first of a series of papers by Cope and coworkers appeared[44, 46]. The reaction has found use in the synthesis of alkenes (equation 37)[47], where high

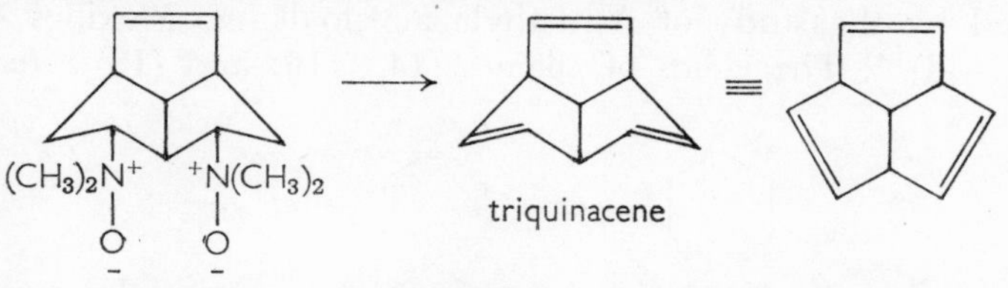

$$\text{(37)}$$

triquinacene

yields are obtained[44], and of dialkylhydroxylamines (equation 38)[48].

$$R_2\overset{+}{-}NCH_2CH_2COOC_2H_5 \longrightarrow R_2NOH + CH_2\!\!=\!\!CHCOOC_2H_5 \qquad \text{(38)}$$

In the example of equation (38) the elimination proceeds in the required direction due to the presence of the acidic hydrogens α to the carbonyl group.

Evidence for a *cis*-elimination mechanism was obtained by the study of the thermal decomposition of the *N,N*-dimethyl *N*-oxides of *threo*- and *erythro*-2-amino-3-phenylbutane (equations 39 and 40)[49].

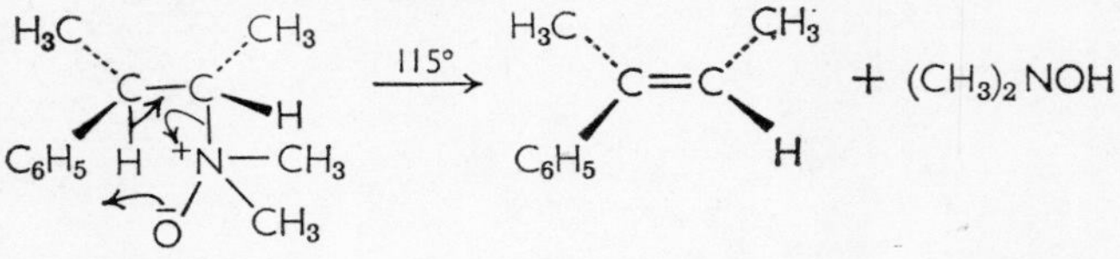

(39)

threo   *cis*

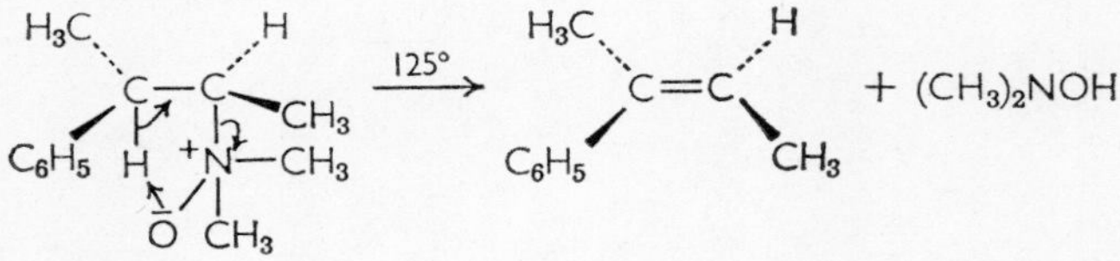

(40)

erythro   *trans*

The major product from each isomer was formed by *cis* elimination. In the case of the *threo* isomer the predominance of *cis* to *trans* alkene was greater than 400 to 1, and in the case of the *erythro* isomer the predominance of *trans* over *cis* alkene was greater than 20 to 1. Substituted-alkyl groups in general appear to have no effect on the course of the elimination[50], the product ratio depending only upon the number of available β-hydrogens. An exception to this generalization is the *t*-butyl group which causes elimination of isobutylene more readily than of ethylene (by a factor of 2, after correcting for the relative numbers of β-hydrogens) probably due to steric factors[50]. Evidence for a cyclic, five-membered, planar transition state was obtained by the study of *N*-methylazacycloalkane *N*-oxides (equations 41–43)[51]. The yields of alkenes (**14**), (**16**) and (**18**) reflect the

Yield

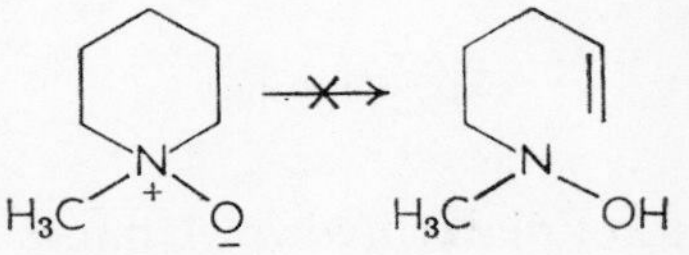

0%   (41)

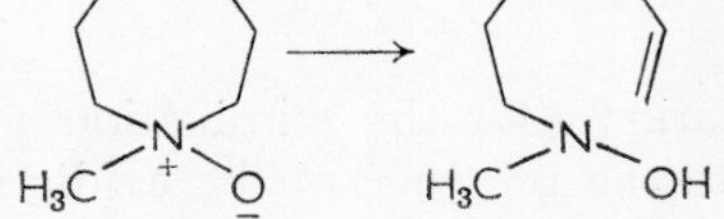

(**13**)   (**14**)

57%   (42)

(**15**)   (**16**)

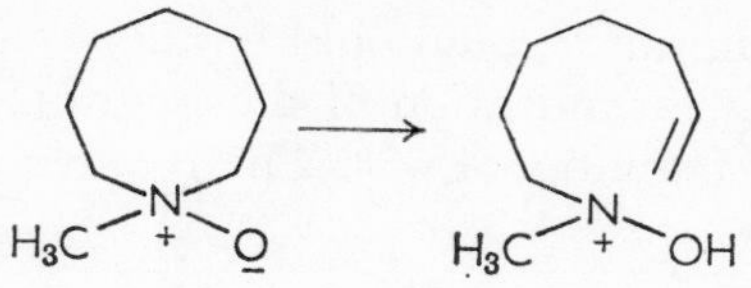

$$78\% \qquad (43)$$

(17)   (18)

impossibility of forming the transition state in the case of the six-membered ring (13), and the increasing ease of forming the transition state in the case of the seven-membered ring (15) and the eight-membered ring (17). A study of the elimination from 2-butyl-N,N-dimethyl-3-phenylamine oxide has shown that the reaction is unimolecular over at least 85–90% of the reaction in a number of solvents[52]. Hydrogen-bonding solvents were found to strongly retard the reaction, the interaction presumably reducing the basicity of the oxygen.

B. The Meisenheimer Rearrangement

Tertiary amine oxides which contain a benzylic or an allylic group attached to nitrogen rearrange with migration of this group from nitrogen to oxygen (equation 44)[53a,b]; alkyl groups also migrate

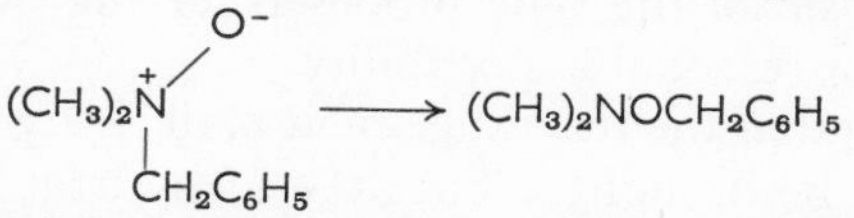

$$(CH_3)_2\overset{+}{N} \longrightarrow (CH_3)_2NOCH_2C_6H_5 \qquad (44)$$

provided that β-hydrogens are absent (otherwise the Cope elimination predominates)[53c]. The reaction was shown to be predominantly intramolecular and the rate was shown to be independent of added base[54]. Further, substitution in the ring of the migrating benzyl group has been examined and it was found that electron-withdrawing groups facilitate the rearrangement (Table 4)[55]. Cope originally

TABLE 4. Rearrangement rates.

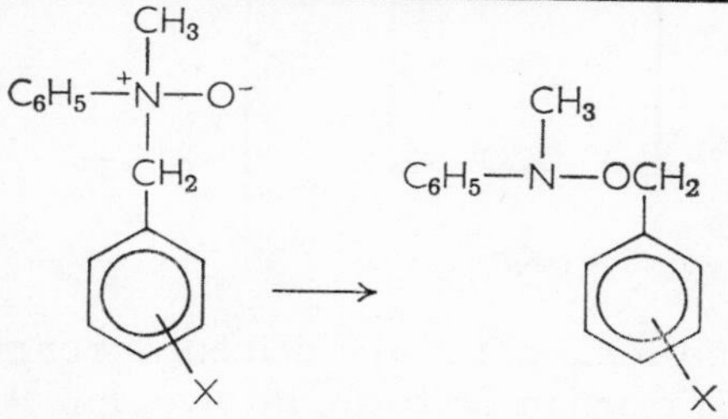

X	p-NO$_2$	p-C$_6$H$_5$	m-Cl	H	p-CH$_3$	p-CH$_3$O
k/k_0	14·8	1·95	1·1	1	0·32	0·24

proposed an S_Ni mechanism[53b], but doubt has been cast on this by the observation that strong racemization of the migrating groups occurs during rearrangement (equation 45)[55]. The relative rates of Table 4

$$
\underset{\substack{| \\ D}}{\overset{\substack{H \quad CH_3 \\ | \quad + |}}{C_6H_5C-N-CH_3}} \underset{O^-}{\overset{}{}} \longrightarrow \underset{\substack{| \\ D}}{\overset{\substack{H \\ |}}{C_6H_5C-ON}}\overset{\substack{CH_3 \\ }}{\underset{CH_3}{}} \tag{45}
$$

product 60–80 % racemized

rule out a carbonium ion mechanism, thus leaving a carbanion and a free-radical mechanism for consideration. Evidence in favor of a radical-pair intermediate was obtained by the observation of a nitroxide radical by e.s.r. spectroscopy in the rearrangement of benzylmethylphenylamine oxide (equation 46), although no proof

$$
\underset{\substack{| \\ CH_2C_6H_5}}{\overset{\substack{CH_3 \\ + |}}{C_6H_5-N-O^-}} \longrightarrow \underset{\substack{\cdot \\ CH_2C_6H_5}}{\overset{\substack{CH_3 \\ |}}{C_6H_5-N=O}} \longrightarrow \overset{\substack{CH_3 \\ |}}{C_6H_5-N-OCH_2C_6H_5} \tag{46}
$$

that this was the sole intermediate in the reaction was given[55]. An ion-pair mechanism of the type proposed for the Stevens rearrangement also seems a reasonable possibility.

It is interesting that the rearrangement of dimethyl-(1-phenylethyl)-amine oxide can be brought about at $-5°$ by irradiation with ultraviolet light[55].

C. Cleavage with Acylating Agents

The cleavage of amine oxides with acylating agents gives aldehydes and acylated secondary amines in good yield (equation 47)[58]. The

$$
\underset{R^2}{\overset{R^1}{}}\!\!\diagdown\!\!\underset{O^-}{\overset{+}{N}}\!\!-CH_3 \xrightarrow{(CH_3CO)_2O} \left[\underset{R^2}{\overset{R^1}{}}\!\!\diagdown\!\!\underset{OCOCH_3}{\overset{+}{N}}\!\!-CH_3\right] \longrightarrow \left[\underset{R^2}{\overset{R^1}{}}\!\!\diagdown\!\!\underset{\substack{O=C-O \\ | \\ CH_3}}{N-CH_2}\right] \longrightarrow \underset{R^2}{\overset{R^1}{}}\!\!\diagdown\!\!\underset{O=CCH_3}{N} \tag{47}
$$

CH₃COO⁻   + CH₃COOH   + CH₂O

reaction has found some use in the demethylation of alkaloids. The mechanism of the reaction probably involves initial acylation of the amine oxide followed by a rearrangement. An ion-pair intermediate[59] seems a likely possibility, although a radical-pair intermediate has also been suggested for the reaction[60]. Related rearrangements are those of N,N-dimethylaniline oxides to give *ortho* ring-substituted products in addition to the acylated secondary amine (equation 48)[59].

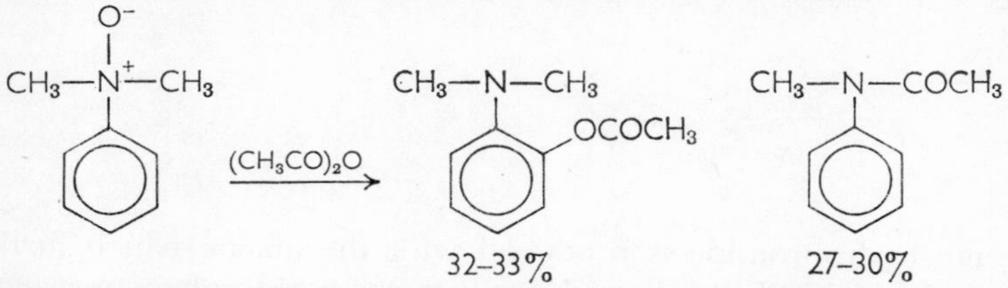

$$32\text{–}33\% \qquad\qquad 27\text{–}30\% \tag{48}$$

Pyridine N-oxide also rearranges to a ring-substituted product (equation 49)[56], while α-picoline N-oxide gives 2-pyridylmethyl

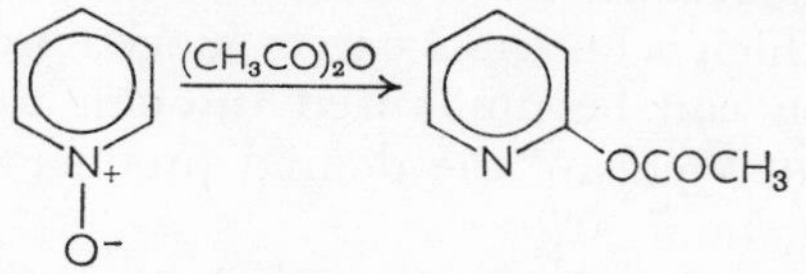

$$\tag{49}$$

acetate (equation 50)[57]. The cleavage has also been achieved with

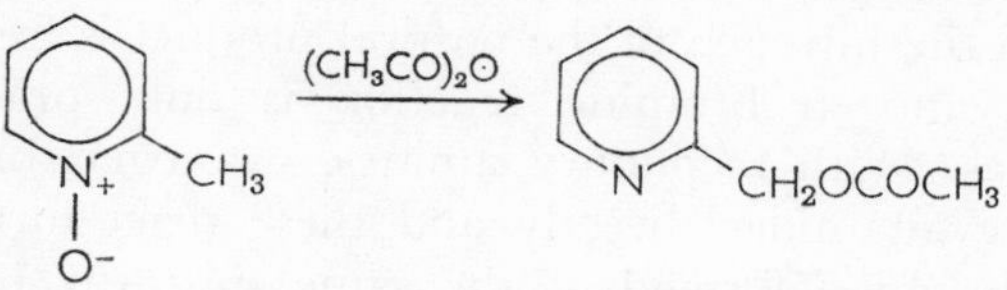

$$\tag{50}$$

ferric ions[61]; in this case the reaction is thought to proceed through two 1-electron steps[61b].

IV. TERTIARY AMINES

A. *The von Braun Cyanogen Bromide Reaction*

The cleavage of tertiary amines by cyanogen bromide (equation 51) was first observed by von Braun[62] and Scholl and Nör[63] in 1900. The reaction was investigated extensively by von Braun; it has recently been reviewed by Hageman[64].

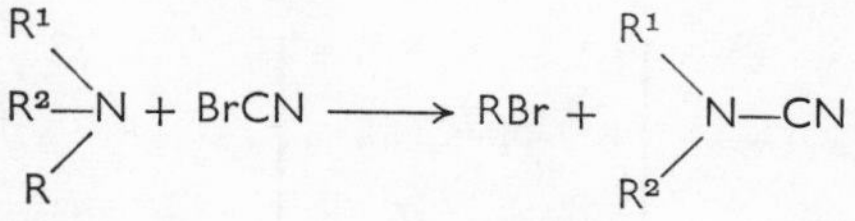

$$\tag{51}$$

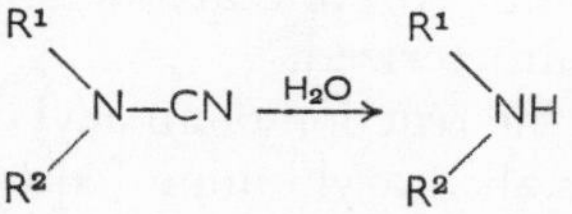

$$\tag{51a}$$

In some cases the formation of alkene occurs as a side-reaction, and this is especially prevalent when the tertiary amine contains a secondary or tertiary carbon attached to the nitrogen (equation 52). A mole

$$
2 \quad
\begin{array}{c} R^1 \\ \diagdown \\ R \diagup \end{array} C
\begin{array}{c} CH_3 \\ \diagup \\ \diagdown \\ R^2 \\ N \diagdown \\ R^3 \end{array}
\xrightarrow{\text{BrCN}}
\begin{array}{c} R^1 \\ \diagdown \\ R \diagup \end{array} C = CH_2 +
\begin{array}{c} R^2 \\ \diagdown \\ N - CN + \\ R^3 \diagup \end{array}
\begin{array}{c} R^1 \\ \diagdown \\ R \diagup \end{array} C
\begin{array}{c} CH_3 \\ \diagup \\ \diagdown \\ R^2 \\ \overset{+}{N} \diagdown \\ | \quad R^3 \\ H \end{array} \quad Br^- \quad (52)
$$

of amine hydrobromide is produced with the alkene which further reduces the yield of the desired product, since the salt is unreactive towards cyanogen bromide. A second side-reaction can occur if the alkyl bromide (20) reacts with the starting amine (19) to give a quaternary salt, which is unreactive to cyanogen bromide. In practice this latter reaction can be minimized through use of an excess of cyanogen bromide. Yields of the desired product are thus variable, but usually greater than 50%.

The reaction has found use in the degradation of alkaloids[64]. Unlike the Hofmann exhaustive methylation procedure which may be repeated until the nitrogen of the natural product is freed as tertiary amine, the cyanogen bromide reaction is only practicable with tertiary amines. With secondary amines, cyanogen bromide yields disubstituted cyanamides directly and these react further with the amine hydrobromide formed at the same time to give guanidines (equation 53)[65].

$$
\begin{array}{c} R \\ \diagdown \\ N-H \\ R^1 \diagup \end{array}
\xrightarrow{\text{BrCN}}
\begin{array}{c} R \\ \diagdown \\ N-CN + \\ R^1 \diagup \end{array}
\begin{array}{c} R \\ \diagdown \\ \overset{+}{N}H_2 \ Br^- \\ R^1 \diagup \end{array}
\longrightarrow
\begin{array}{c} RR^1N-CNRR^1 \\ \| \\ NH_2Br^- \\ + \end{array} \quad (53)
$$

The mechanism of the reaction seems to involve the formation of a quaternary salt, observed as an initial transient precipitate[62, 66], followed by a nucleophilic displacement by bromide ion (equation 54). The group most susceptible to nucleophilic attack is generally

$$
\begin{array}{c} R^1 \\ \diagdown \\ R^2 - N + BrCN \\ R \diagup \end{array}
\longrightarrow
\left[
\begin{array}{c} R^1 \\ \diagdown \\ R^2 - \overset{+}{N} - CN \ Br^- \\ R \diagup \end{array}
\right]
\longrightarrow
\begin{array}{c} R^1 \\ \diagdown \\ N-CN + RBr \\ R^2 \diagup \end{array} \quad (54)
$$

displaced. The intermediate salt is stable at low temperatures but it has not been isolated and characterized.

Similar cleavages occur in the reaction of an acyl chloride, bromide or anhydride with a tertiary alkyl aryl amine (equation 55)[64, 67].

$$C_6H_5N(CH_3)_2 \xrightarrow{CH_3COBr} C_6H_5NCOCH_3 + C_6H_5\overset{+}{N}(CH_3)_3Br^- \qquad (55)$$
$$\underset{CH_3}{\overset{|}{}}$$

B. Cleavage by Nitrous Acid

The cleavage of tertiary amines by nitrous acid (equation 56) was

$$R_2NCH_2R^1 \xrightarrow{HONO} R_2N\!-\!NO + R^1CHO \qquad (56)$$

first recorded in 1864[68]. The reaction has found only limited use in alkaloid degradation[69], possibly because of side-reactions and conflicting reports on its use in the literature; the fact that tertiary amines are stable to nitrous acid at low pH, where the amine is present as the inert ion, accounts for the early conflicting observations. The mechanism probably involves initial nitrosation of the tertiary amine to form a nitrosammonium ion; decomposition via an ylide-type intermediate can then lead to the observed products (equation 57)[70].

$$
\begin{array}{c}
R_2\overset{+}{N}CH_2R^1 \longrightarrow R_2\overset{+}{N}{=}CHR^1 \quad + \quad HNO \\
\underset{NO}{\overset{|}{}} \qquad\qquad \downarrow H_2O \qquad\qquad\qquad \downarrow \tfrac{1}{2}H_2N_2O_2 \\
\Big[\!\begin{array}{c} R_2\overset{+}{N}{-}\overset{-}{C}HR^1 \\ \underset{NO}{\overset{|}{}} \end{array}\!\Big] \underset{-NO^-}{\longrightarrow} R_2NH + R^1CHO \qquad \downarrow \tfrac{1}{2}N_2O + \tfrac{1}{2}H_2O \\
\qquad\qquad\qquad \downarrow HONO \\
\qquad\qquad\qquad R_2N\!-\!NO
\end{array}
\qquad (57)
$$

Smith[70b] has obtained nitrous oxide in approximately the correct yield predicted by this mechanism. A free-radical pathway for the decomposition of the ylide intermediate has also been proposed[71].

C. Oxidative Cleavage

The cleavage of a tertiary amine to a secondary amine and an aldehyde or ketone has been observed for a number of oxidizing agents[72]. The mechanisms of the various reactions are not fully known, but they probably involve initial attack at an α-carbon, possibly after formation of 4-coordinated nitrogen (equation 58).

$$
\begin{array}{c}
(R_2CH)NR^1_2 \xrightarrow{Path\,(a)} R_2CH\overset{-}{\overset{+}{N}}R^1_2 \xrightarrow{-H^+} R_2\overset{-}{C}\overset{+}{N}R^1_2 \xrightarrow{-X^-} R_2C{=}\overset{+}{N}R^1_2 \\
\underset{X}{\overset{|}{}} \qquad\qquad \underset{X}{\overset{|}{}} \\
\underset{-1e^-}{\overset{Path\,(b)}{\longrightarrow}} (R_2CH)\overset{+}{N}R^1_2 \xrightarrow{-1e^-,\,-H^+}
\end{array}
\qquad (58)
$$

$$R_2C{=}\overset{+}{N}R^1_2 \xrightarrow{H_2O} R_2CO + R^1_2NH \qquad (59)$$

The dichromate oxidation of tertiary alkyl aryl amines in a sulfate–bisulfate buffer was found[73a] to give secondary amines in good yields. Trialkylamines did not react under the conditions used, suggesting that path (b) in equation (58) is operating; that is, an aryl ring may be necessary to stabilize the radical intermediate. This reaction has been adapted[73] to degrade primary or secondary amines by first converting them to the corresponding 2,4-dinitroaniline derivative, and subsequently oxidizing this amine (equation 60).

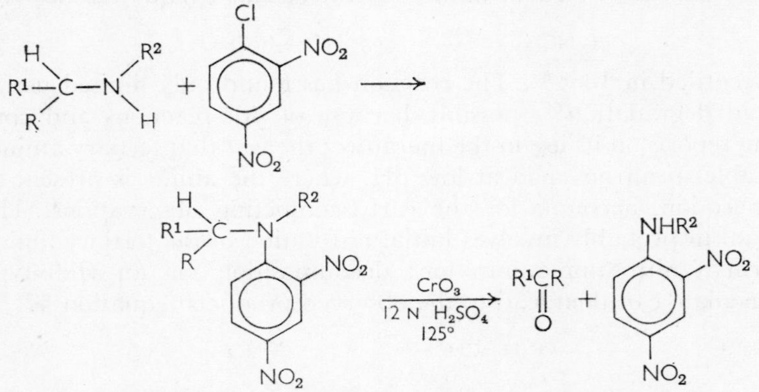

Yield of ketone: 94% with $R = R^1 = CH_3$, $R^2 = H$
21% with $R = CH_3$, $R^1 = n\text{-}C_6H_{13}$, $R^2 = H$

Yield of ketone: 94% with $R = R^1 = CH_3$, $R^2 = H$
21% with $R = CH_3$, $R^1 = n\text{-}C_6H_{13}$, $R^2 = H$

Dibenzoyl peroxide oxidizes tertiary amines to secondary amines and aldehydes (equation 61)[72]. The effect of substituents has been

$$(RCH_2)_3N + (C_6H_5COO)_2 \xrightarrow{H_2O} (RCH_2)_2NH + RCHO + 2 C_6H_5COOH \qquad (61)$$

examined quantitatively[74]; for alkyl groups the order of elimination was found to be $RCH_2— > CH_3— > R_2CH—$. A phenyl group substituted either α or β to the nitrogen seemed to have no directive influence on the group eliminated.

The reaction of tertiary amines with N-bromosuccinimide can give good yields of secondary amines[75], although tertiary amines containing N-aryl groups were found to give only ring-brominated tertiary amines[76].

Lead tetraacetate oxidizes tertiary alkyl aryl amines to acetyl derivatives of secondary amines (equation 62)[76]. Tertiary alkyl-amines are not effected.

$$C_6H_5N(CH_3)_2 \xrightarrow{Pb(OCOCH_3)_4} C_6H_5N—CH_3 + HCHO \qquad (62)$$
$$\underset{COCH_3}{|}$$

83% yield

Chlorine dioxide oxidizes triethylamine to the secondary amine and acetaldehyde (equation 63); a free-radical mechanism was proposed[77].

$$2\,ClO_2 + H_2O + (C_2H_5)_3N \longrightarrow CH_3CHO + (C_2H_5)_2NH + 2\,H^+ + 2\,ClO_2^- \quad (63)$$

Manganese dioxide oxidation of tertiary acyclic amines[78] has been found to give N,N-dialkylformamides and secondary amines as major products (equation 64). Similarly, N,N-dimethylcyclohexylamine on

$$(n\text{-}C_4H_9)_3N \xrightarrow[\text{cyclohexane}]{\underset{19°}{MnO_2,}} (n\text{-}C_4H_9)_2NCHO + (n\text{-}C_4H_9)_2NH \quad (64)$$
$$\qquad\qquad\qquad\qquad\qquad 51\%\text{ yield} \qquad 33\%\text{ yield}$$

oxidation with manganese dioxide gave cyclohexanone in 50% yield[78c].

Acidic potassium permanganate is reported[79] to convert tertiary amines containing benzyl groups to benzaldehydes or benzoic acids. Neutral permanganate oxidizes tributylamine to a mixture of dibutylamine, butyraldehyde, N,N-dibutylbutyramide and butyric acid[80]. Tribenzylamine reacted with neutral permanganate to give benzaldehyde and benzoic acid[80]; no dibenzylamine was isolated.

The oxidation of tributylamine by ozone has been studied[101]. In polar solvents ($CHCl_3$, CH_3OH) tributylamine N-oxide was formed (61%) together with small amounts of dibutylamine and dibutylformamide. In hydrocarbon solvents at -78 to $-45°$, dibutylamine (49–63%) was the major product, but at 15°, dibutylformamide was the major product (44%). N,N-Dibutylbutyramide was a minor product under all conditions. The suggested mechanism is given in equation (65)[101].

$$Bu_3N \begin{cases} \overset{+}{Bu_3N} - \overset{-}{O} \\ Bu_2NCHPr \\ \quad| \\ \quad OH \end{cases}$$
$$\begin{array}{l} \overset{O}{\underset{\|}{Bu_2NCPr}} \\ Bu_2NCH{=}CHEt \to Bu_2NCHO + (EtCHO) \\ Bu_2NH + PrCHO \end{array} \quad (65)$$

D. Amine Displacement

The displacement of amine from tertiary amine salts requires rather drastic conditions[81], although good yields are often obtained (equation 66). The cleavage of methyl groups from tertiary amines by HI at

$$(CH_3)_3\overset{+}{N}HCl^- \xrightarrow[300°]{\Delta} (CH_3)_2NH + CH_3Cl \quad (66)$$

300–360° has been used in a quantitative estimation of N-methyl groups[82]. The optimum conditions for displacement of alkyl group from tertiary and secondary aromatic amines have been studied[81d]. Treatment of the salt of a tertiary amine with hydrogen bromide at 150° gives good yields of the secondary aromatic amine (equation 67).

$$(CH_3)_2\overset{+}{N}HC_6H_5Br^- \xrightarrow{\text{HBr}}_{150°} CH_3Br + C_6H_5\overset{+}{N}H_2CH_3Br^- \qquad (67)$$

Ease of elimination of alkyl groups is in the order methyl > n-propyl > ethyl, and electron-withdrawing groups substituted in the aromatic ring increase the rate of displacement.

A study of the thermal elimination from hindered tertiary aliphatic amine salts has produced results consistent with an $E1$ mechanism[82a]. The alkene eliminated appeared to be determined mainly by the stability of the carbonium ion (equation 68).

$$
\begin{array}{c}
(CH_3)_2C\overset{+}{}NHC(CH_3)_3 \\
\underset{R}{|}\ \underset{CH_3}{|} \\
Cl^-
\end{array}
\quad
\begin{array}{l}
CH_3NHC(CH_3)_3 + (CH_3)_2\overset{+}{C}R \\
\qquad\qquad \rightarrow \text{Alkene} + H^+ \\[2mm]
CH_3NHC(CH_3)_2 + (CH_3)_3C^+ \\
\quad\ \underset{R}{|} \qquad \rightarrow (CH_3)_2C{=}CH_2 + H^+
\end{array}
\qquad (68)
$$

Displacement of alkyl groups by silicon tetrabromide in tertiary aromatic amines has been observed (equation 69)[81f]. The proposed

$$C_6H_5NR_2 + HBr \rightarrow C_6H_5NHR + RBr \xrightarrow{\text{SiBr}_4} C_6H_5NRSiBr_3 + HBr \qquad (69)$$

mechanism involves displacement of the alkyl bromide and subsequent reaction of silicon tetrabromide with the secondary amine.

E. Amine Exchange in Mannich Bases

Salts of Mannich bases undergo facile amine exchange[28a, 83] and the reaction has been used to synthesize Mannich bases which are unobtainable by the direct Mannich reaction (equation 70). Two

$$
CH_3COCH_2CH_2\overset{+}{N}(C_2H_5)_2Cl^- + C_6H_5NH_2 \longrightarrow
$$
$$
\underset{H}{|}
$$
$$
CH_3COCH_2CH_2NHC_6H_5 + (C_2H_5)_2\overset{+}{N}H_2Cl^- \qquad (70)
$$
$$
(21) \qquad\qquad\qquad 86\% \text{ yield}
$$

mechanisms are possible for this transamination reaction: a direct nucleophilic displacement (equation 71) and an elimination–addition mechanism (equations 72 and 73). The generally accepted mechanism

$$
RCOCH_2CH_2\overset{\overset{+}{\frown}}{\underset{\underset{H\ddot{N}R_2^2}{\nearrow}}{N}R_2^1}{-}H \longrightarrow RCOCH_2CH_2\overset{+}{N}R_2^2 + HNR_2^1 \qquad (71)
$$
$$
\qquad\qquad\qquad\qquad\qquad\qquad\qquad \underset{H}{|}
$$

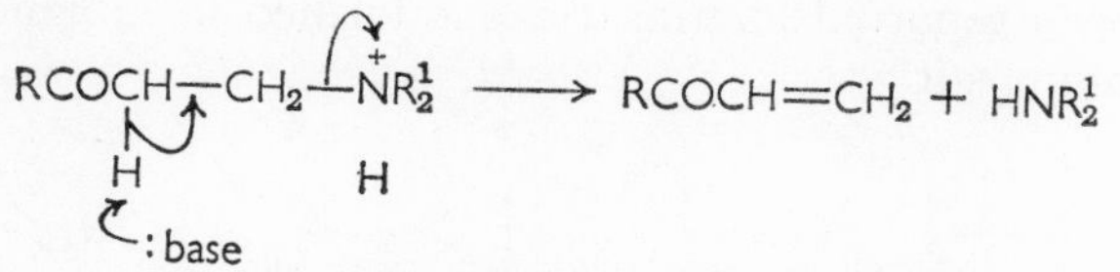

$$RCOCH{-}CH_2{-}\overset{+}{N}R_2^1 \longrightarrow RCOCH{=}CH_2 + HNR_2^1 \qquad (72)$$

$$RCOCH{=}CH_2 + HNR_2^2 \longrightarrow RCOCH^2CH_2NR_2^2 \qquad (73)$$

is elimination–addition. Craig, however, has studied proton exchange with solvent during the exchange reactions of aniline and N-methylaniline with the Mannich base (21)[83b]. His results are compatible with a mixed mechanism of about 90% substitution and 10% elimination–addition for exchange with aniline and one of about 65% substitution and 35% elimination–addition for exchange with N-methylaniline. An optically active Mannich base (22) was found[84] to racemize during exchange (equation 74). This was shown not to be due to enolization, and therefore supports the elimination–addition mechanism. Further support for this mechanism was provided by the isolation of the unsaturated ketone (23) using exchange conditions, and by the addition of amine to this ketone to give the Mannich base (24) (equation 75).

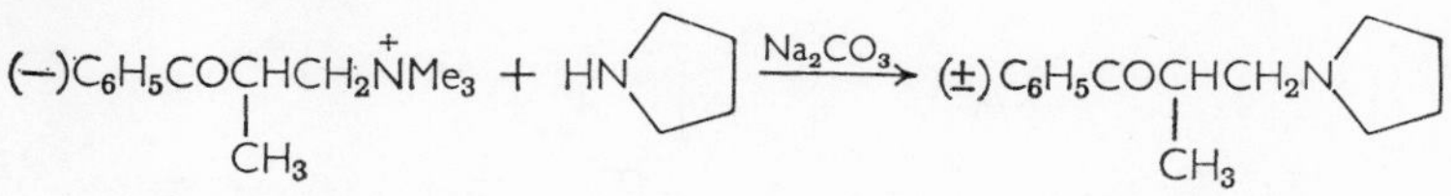

$$(-)C_6H_5COCHCH_2\overset{+}{N}Me_3 + HN{\left\langle\;\right\rangle} \xrightarrow{Na_2CO_3} (\pm)C_6H_5COCHCH_2N{\left\langle\;\right\rangle} \qquad (74)$$
$$\underset{CH_3}{|} \qquad\qquad\qquad\qquad\qquad \underset{CH_3}{|}$$

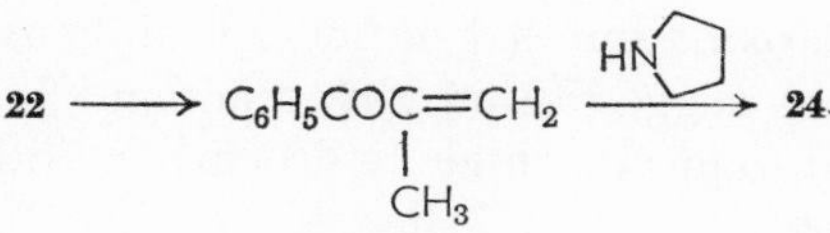

(22)   (24)

$$22 \longrightarrow C_6H_5COC{=}CH_2 \xrightarrow{HN{\left\langle\;\right\rangle}} 24 \qquad (75)$$
$$\underset{CH_3}{|}$$

(23)

V. SECONDARY AMINES

A. Aziridines

Treatment of an aziridine with nitrosyl chloride, 3-nitro-N-nitrosocarbazole or methyl nitrite leads to the formation of an alkene and nitrous oxide (equation 76)[85a–d]; an intermediate N-nitroso derivative was isolated at low temperatures. The deamination of cis- and $trans$-2,3-dimethylaziridine gave the corresponding cis- and $trans$-2-butenes with complete stereospecificity in each case. The reaction was found to be first order with respect to the N-nitroso intermediate[85b,c] and the transition state was suggested to involve simultaneous cleavage of both C—N bonds. A similar decomposition has also been reported for the triazene formed from aziridine and a tolydiazonium salt[85e].

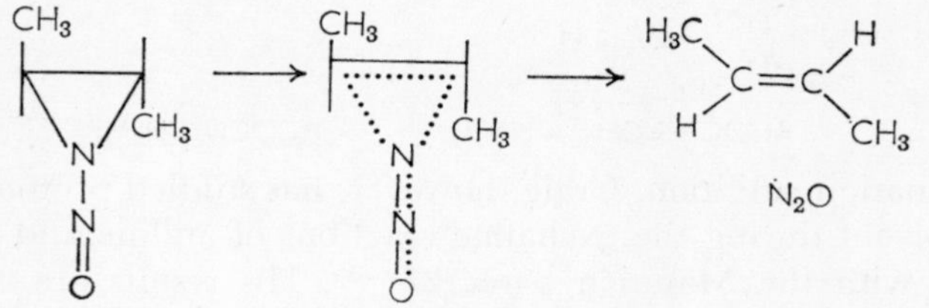

$$ (76) $$

B. Benzylic and Allylic Secondary Amines

N-Nitrosodibenzylamines give good yields of hydrocarbon products when treated with sodium hydrosulfite in base, or lithium in liquid ammonia (equations 77 and 78)[97]. A diazene intermediate previously

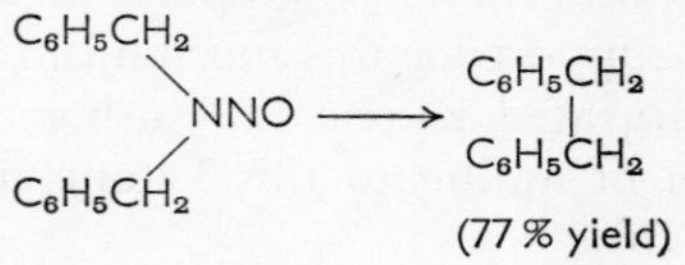

$$ (77) $$

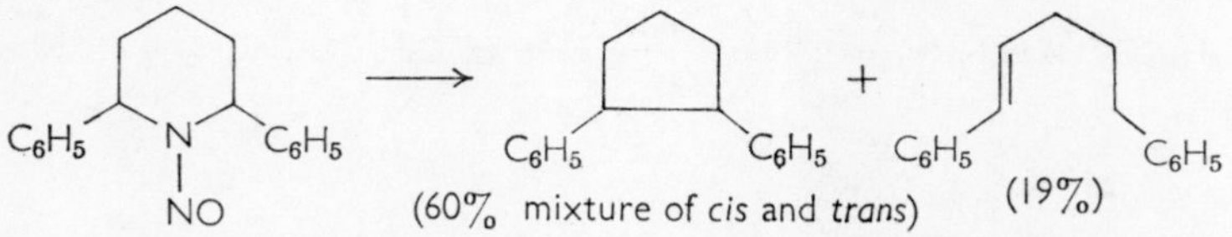

(77% yield)

$$ (78) $$

(60% mixture of *cis* and *trans*)   (19%)

formulated for the oxidation of 1-amino-2,6-diphenylpiperidines with mercuric oxide was suggested for this reaction[97d]. Mixed dibenzylamines were found to form a single hydrocarbon with no formation of

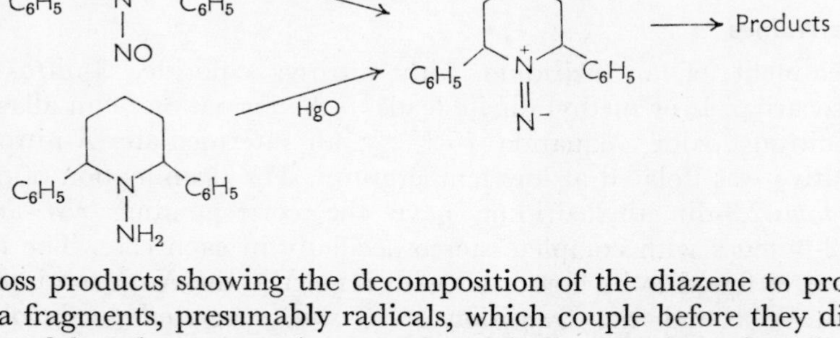

$$ (79) $$

cross products showing the decomposition of the diazene to proceed via fragments, presumably radicals, which couple before they diffuse out of the solvent cage. As examples of similar reactions, Angeli's salt,

Na_2ONNO_2 has been shown to react with dibenzylamine to give dibenzyl in 70% yield (equation 80)[100a], and with 3-pyrrolines to give

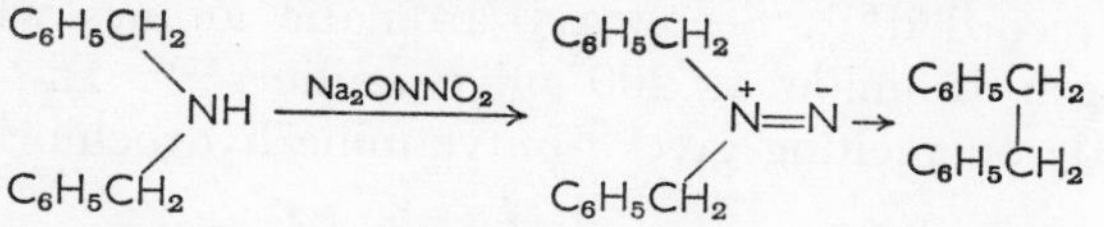
C6H5CH2
NH
Na2ONNO2
C6H5CH2
N=N
C6H5CH2
C6H5CH2

(80)

dienes with complete stereospecificity (equations 81–83)[100b]. Diazene intermediates were also suggested for these reactions.

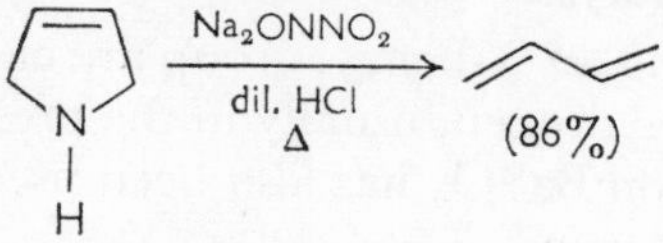
Na2ONNO2
dil. HCl
Δ
(86%)
N
H

(81)

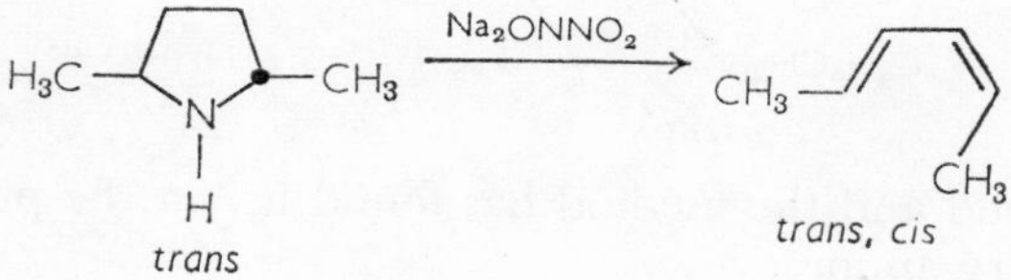
Na2ONNO2
H3C
CH3
N
H
trans
CH3
CH3
trans, cis

(82)

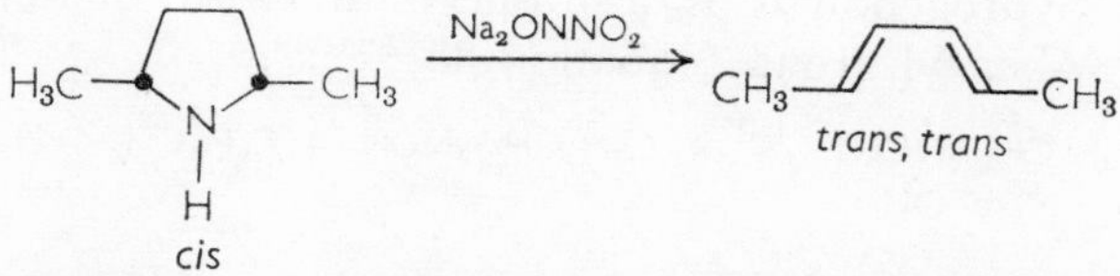
Na2ONNO2
H3C
CH3
N
H
cis
CH3
CH3
trans, trans

(83)

Certain secondary amines react with difluoramine to give hydrocarbons by pathways which probably involve a diazene intermediate (equations 84 and 85)[95]. The same diazene intermediate is

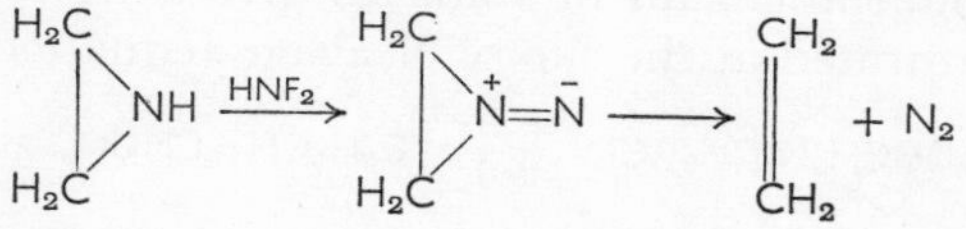
H2C
NH
H2C
HNF2
H2C
N=N
H2C
CH2
CH2
+ N2

(84)

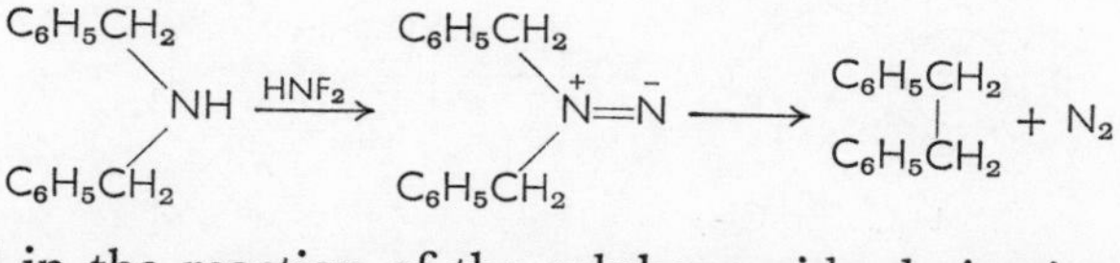
C6H5CH2
NH
C6H5CH2
HNF2
C6H5CH2
N=N
C6H5CH2
C6H5CH2
C6H5CH2
+ N2

(85)

postulated in the reaction of the sulphonamide derivative (**25**) with base (equation 86)[97].

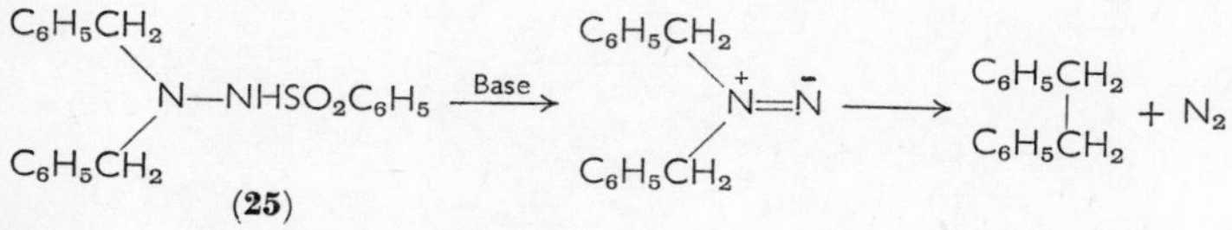
C6H5CH2
N—NHSO2C6H5
C6H5CH2
Base
C6H5CH2
N=N
C6H5CH2
C6H5CH2
C6H5CH2
+ N2

(86)

(**25**)

C. Thermal Dealkylation

A few examples of the thermal dealkylation of secondary amines have been recorded[81b]. Secondary aromatic amines when treated with hydrogen bromide at 200° give anilines[81d]. Di-t-butylamine hydrochloride on melting gives t-butylamine hydrochloride and iso-butylene[82a].

VI. REDUCTIVE CLEAVAGE

A. Catalytic Hydrogenolysis

Tertiary amines having an N-benzyl group are cleaved to secondary amines and toluene by hydrogen, usually in the presence of palladium oxide, though 10% Pd on $BaSO_4$ has also been used (equation 87)[86].

$$C_6H_5CH_2N\begin{smallmatrix}R\\\\R^1\end{smallmatrix} \xrightarrow{[H]} C_6H_5CH_3 + NHRR^1 \qquad (87)$$

Yields are good and the method has found use in the preparation of pure secondary amines.

Catalytic hydrogenolysis is also successful with quaternary salts having an N-benzyl group (equation 88)[86a, 88].

$$(C_6H_5CH_2)_3\overset{+}{N}CH_3 \underset{OH^-}{} \xrightarrow{[H]} (C_6H_5CH_2)_2NCH_3 + C_6H_5CH_3 + H_2O \qquad (88)$$

B. Dissolving Metals

Emde[89a, b] observed that quaternary ammonium halides containing an unsaturated group either attached to, or β to the nitrogen were cleaved by sodium amalgam in water to give a tertiary amine and hydrocarbon saturated at the site of cleavage (equations 89 and 90).

$$C_6H_5CH{=}CHCH_2\overset{+}{N}(CH_3)_3Cl^- \xrightarrow[{[H]}]{Na/Hg} C_6H_5CH{=}CHCH_3 + N(CH_3)_3 \qquad (89)$$

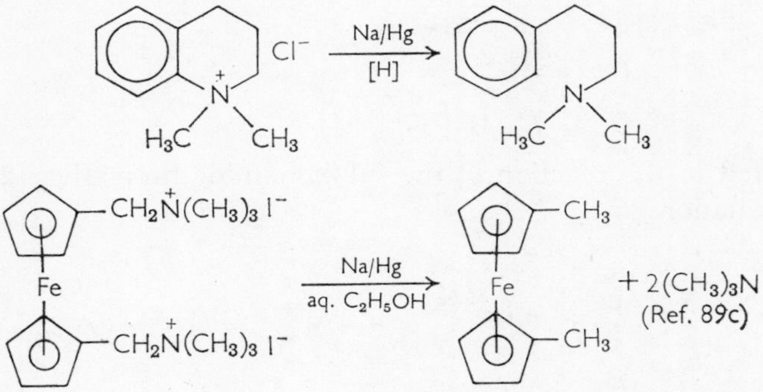

(90)

Under similar conditions fully saturated quaternary salts usually do not cleave.

However, tetraalkylammonium salts can be cleaved by sodium in dioxane[90a] or by sodium or potassium in liquid ammonia[90b]. The mechanism for reduction in liquid ammonia is thought to involve the direct reaction of the quaternary salt with electrons. Both free-radical and carbanion intermediates have been proposed to explain the observed differences in relative rates of cleavage from alkyltrimethylammonium ions[91]. Secondary and tertiary alkyl groups were suggested to cleave by a one-electron process giving free-radical intermediates while primary alkyl groups were suggested to cleave by a two-electron process giving carbanion intermediates (equations 91 and 92).

$$R_4\overset{+}{N} + e^- \longrightarrow R_3N + R^\bullet \tag{91}$$

$$R_4\overset{+}{N} + 2e^- \longrightarrow R_3N + R^-: \tag{92}$$

Lithium in tetrahydrofuran has been found to cleave tertiary aromatic amines (equation 93)[102a]. N-Phenylcarbazole was cleaved at 30°

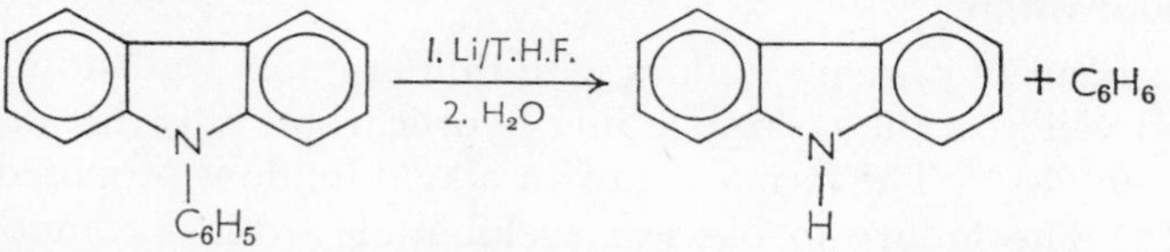

$$\tag{93}$$

over 3 hours to give carbazole, and N-benzylcarbazole was cleaved over 3 hours at reflux to give carbazole. Only a small yield of diphenylamine was obtained from triphenylamine[102].

C. Electrolytic Cleavage

The electrolytic reductive cleavage of quaternary salts containing an N-benzyl group has been observed (equation 94)[92]. The reaction

$$C_6H_5CH_2\overset{+}{N}(CH_3)_2NO_3^- \xrightarrow{\text{Aprotic solvent}} (CH_3)_2NC_6H_5 + (C_6H_5CH_2)_2 \ (35\% \ \text{yield})$$
$$\underset{C_6H_5}{\big|} \xrightarrow{\text{Protic solvent}} (CH_3)_2NC_6H_5 + C_6H_5CH_3 \tag{94}$$

appears to involve a one-electron transfer to the ammonium ion and cleavage to a free radical (equation 95). A tertiary amine and the

$$R_4\overset{+}{N} + e^- \longrightarrow [R_4N] \longrightarrow R_3N + R^\bullet \longrightarrow \text{Products} \tag{95}$$

product of dimerization of the free radical are obtained in aprotic solvents.

The electrolytic reduction of α-aminoketones with a tertiary amino group at a lead cathode in 30% sulfuric acid was observed to give secondary amines (equation 96)[93]. This reaction has been used to

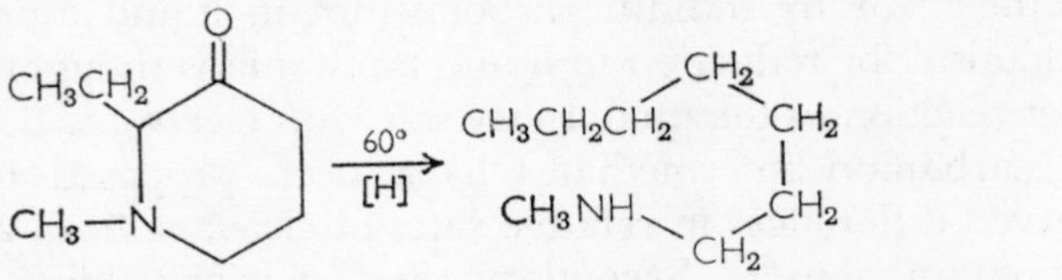

$$\text{(96)}$$

prepare medium-sized ring compounds containing nitrogen (equation 97)[94].

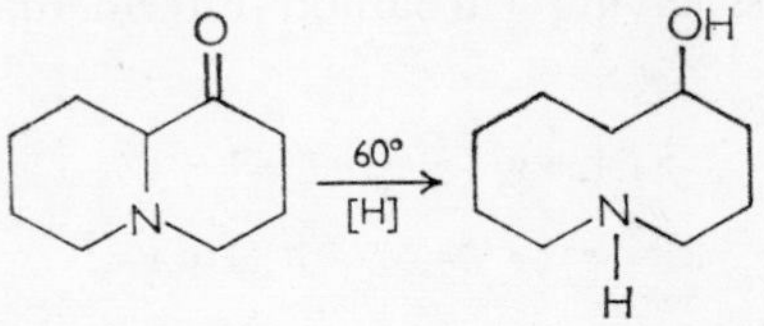

$$\text{(97)}$$

D. Miscellaneous Methods

I. Difluoramine

Treatment of primary amines with difluoramine leads to moderate to good yields of the corresponding hydrocarbon plus the amine salt (equation 98)[95]. The formation of an alkyldiimide is proposed (equation 99). The failure to observe cyclobutane from decomposition of

$$3\ RNH_2 + HNF_2 \longrightarrow 2\ R\overset{+}{N}H_3F + N_2 + RH$$

$$\text{(98)}$$

Yields obtained: R = n-Bu, 61 %; R = s-Bu, 40 %;
R = cyclopropyl, 77 %; R = phenyl, 20 %

$$HNF_2 \longrightarrow NF + HF$$

$$RNH_2 + NF \xrightarrow[-HF]{} [RN{=}NH] \longrightarrow RH + N_2$$

$$\text{(99)}$$

cyclopropylcarbinamine rules out a cleavage of the diimide to a carbonium ion, and the absence of methylcyclopropane in the products from the same reaction argues against a homolytic cleavage[95a]. The same diimide intermediate has been proposed as an intermediate in the reaction of a primary amine sulfonamide with hydroxylamine-O-sulfonic acid[96].

2. Hydroxylamine-O-sulfonic acid

A sulfonamide derived from a primary amine, treated with hydroxylamine-O-sulfonic acid or chloramine in alkaline solution reacts to give a hydrocarbon in good yield (equation 100)[98]. Primary,

$$RNH_2 \xrightarrow{ArSO_2Cl} RNHSO_2Ar \xrightarrow[NH_2X]{OH^-} ArSO_2H + N_2 + RH \qquad (100)$$

$$(X = OSO_3H \text{ or } Cl)$$

secondary and tertiary carbinamines (including benzylic and bridge-head carbinamines) are reactive.

The mechanism of the reaction is thought to involve an alkyldi-imide as in the reaction of primary amines with difluoramine (equation 101)[96, 99]. Other reactions thought to involve the same

$$RNHSO_2Ar \xrightarrow{NH_2X} \begin{bmatrix} RN-SO_2Ar \\ | \\ NH_2 \end{bmatrix} \xrightarrow{OH^-} [R-N=N-H] + ArSO_2^- \qquad (101)$$

$$\downarrow$$

$$RH + N_2$$

$$(X = Cl \text{ or } OSO_3H)$$

alkyldiimide intermediate are the oxidation of hydrazines and the treatment of the p-toluenesulfonyl derivative (**26**) with base (equation 102)[99]. The results obtained by use of an optically active

$$\left. \begin{array}{l} RNHNH_2 \xrightarrow{[O]} \\ RNHNHSO_2C_6H_4CH_3 \xrightarrow{B^-} \end{array} \right] \longrightarrow R-N=N-H \longrightarrow RH \qquad (102)$$

$$(\textbf{26})$$

alkyl group in these deamination reactions suggest that the alkyldi-imide intermediate decomposes by both a base-catalyzed anionic elimination (with some stereospecificity) and a homolytic reaction (non-stereospecific)[99].

3. Catalytic cleavage with hydrazine

The reaction of secondary and tertiary benzylamines with hydrazine in the presence of palladium catalyst (5% on carbon) was found to give toluene in yields greater than 85%[87].

4. Photochemical cleavage

Hlavka and Bitha have reported a photochemical reduction of a tertiary amine derivative of a tetracyclene (equation 103)[102b].

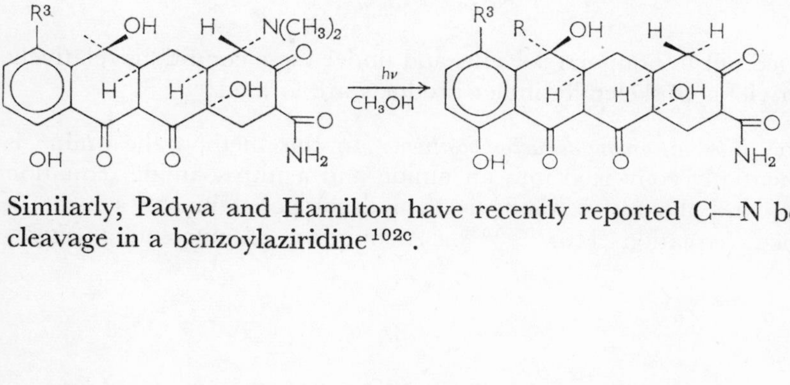

$$(103)$$

Similarly, Padwa and Hamilton have recently reported C—N bond cleavage in a benzoylaziridine[102c].

VII. PRIMARY AMINES

A. Aliphatic Amines

I. Reactions in which nitrogen is the leaving group

These reactions are related methods for the conversion of C—N bonds into C—O bonds (equations 104–107). Despite the apparent

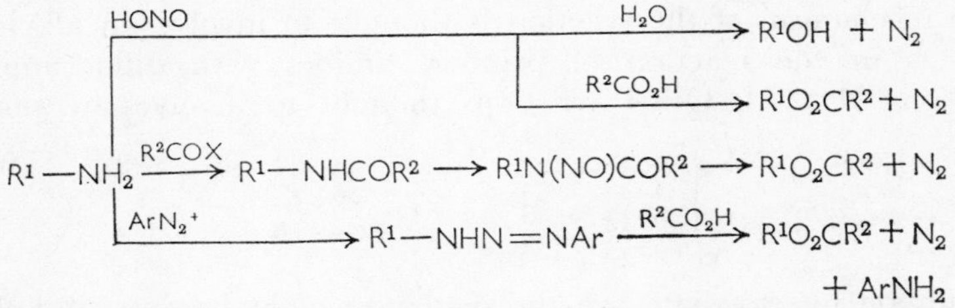

$$R^1\text{—NH}_2 \xrightarrow{\text{HONO}} \begin{cases} \xrightarrow{H_2O} R^1OH + N_2 & (104) \\ \xrightarrow{R^2CO_2H} R^1O_2CR^2 + N_2 & (105) \end{cases}$$

$$R^1\text{—NH}_2 \xrightarrow{R^2COX} R^1\text{—NHCOR}^2 \longrightarrow R^1N(NO)COR^2 \longrightarrow R^1O_2CR^2 + N_2 \quad (106)$$

$$\xrightarrow{ArN_2{}^+} R^1\text{—NHN}=\text{NAr} \xrightarrow{R^2CO_2H} R^1O_2CR^2 + N_2 \quad (107) \\ + ArNH_2$$

differences in reagents, the three reactions proceed through similar intermediates that, for the moment, can be designated as aliphatic diazonium salts, $RN_2{}^+X^-$.

In gross outline, the reactions proceed through carbonium ions, and they yield the manifold products characteristic of carbonium ion reactions (equation 108). In water and other polar solvents,

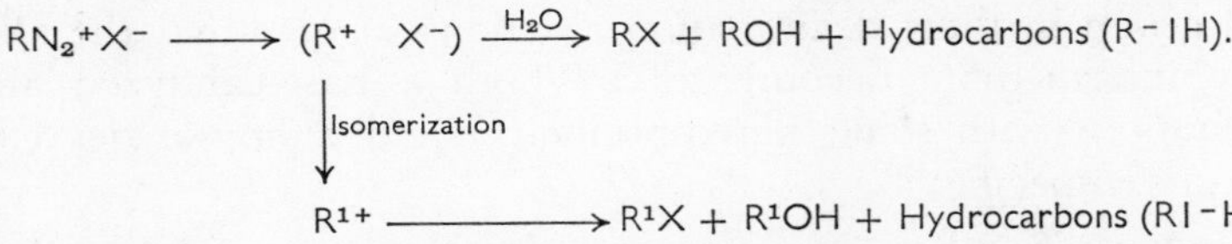

$$RN_2{}^+X^- \longrightarrow (R^+ \quad X^-) \xrightarrow{H_2O} RX + ROH + \text{Hydrocarbons } (R^-IH).$$

$$\downarrow \text{Isomerization} \qquad (108)$$

$$R^1{}^+ \longrightarrow R^1X + R^1OH + \text{Hydrocarbons } (RI^-H).$$

serious skeletal rearrangements are the rule, and under these conditions the conversion of amines into alcohols is not successful[103]. On the other hand, skeletal rearrangements, as in the pinacolic deamination, are often so complete as to be useful in synthesis (equation 109)[104]. The nitrosoamide and triazene reactions can be readily

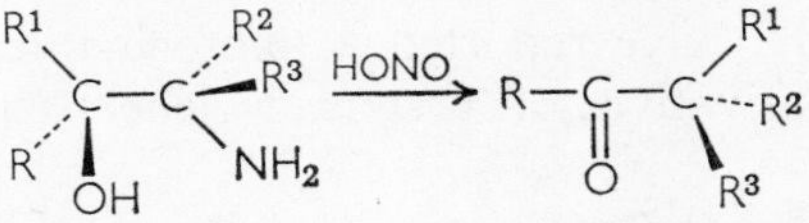

$$(109)$$

carried out in nonpolar solvents and under these conditions relatively high yields of skeletally intact products are formed[103,105].

(a) *The nitrosoamide decomposition.* In this method the amine is sequentially converted into an amide and a nitrosoamide (equation 106), and the nitrosoamide is then decomposed in the solvent of choice (equation 110)[106,103b]. A number of nitrosating agents

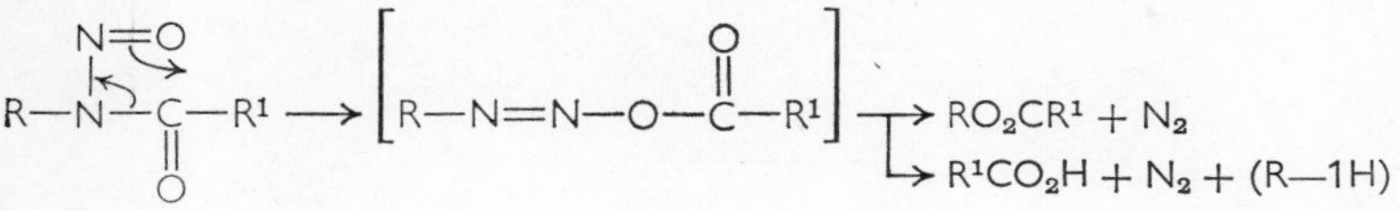

$$R-\overset{\displaystyle N=O}{\underset{\displaystyle O}{N}}-\overset{}{\underset{}{C}}-R^1 \longrightarrow \left[R-N=N-O-\overset{\displaystyle O}{\overset{\|}{C}}-R^1 \right] \begin{array}{l} \longrightarrow RO_2CR^1 + N_2 \\ \longrightarrow R^1CO_2H + N_2 + (R-1H) \end{array} \quad (110)$$

(27)

have been used, but dinitrogen tetroxide ($N_2O_4 \rightleftharpoons 2\,NO_2$) is the reagent of choice[106b]. In the author's opinion, the nitrosoamide decomposition is the best of the three methods of deamination from the standpoint of synthetic utility, stereochemical correlations and adaptability to mechanistic studies.

The yields of esters are high for primary carbinamines, and moderate for secondary and tertiary carbinamines (Table 5)[103b, 107, 108]. In

TABLE 5.   Yields of esters from the decomposition of the nitroso-amides (equation 110).

R	RO_2CR^1
n-Butyl	78–83
Isobutyl	61–66
s-Butyl	23–26
Cyclohexyl	50
1-Phenylethyl	32–38
Benzhydryl	90–100
2-Phenylbutyl	14–24

non-polar solvents, the amount of skeletal rearrangement is small; for example, the decomposition of N-isobutyl-N-nitrosoacetamide in carbon tetrachloride yielded isobutyl acetate, s-butyl acetate and t-butyl acetate in the ratio 63/3/1[103b]. The reaction of isobutylamine with nitrous acid in water, in contrast, yielded isobutyl alcohol, s-butyl alcohol and t-butyl alcohol in the ratio 0·14/0·27/1·0. As a last advantage, except in rare instances, the reaction proceeds with retention of configuration[106c, 107, 109].

The nitrosoamide reaction is a clean one, and only nitrogen, carboxylic acids and hydrocarbons are formed as by-products (equation 110); these compounds are usually readily removed from the substitution products. Two examples from the literature will illustrate the use to which the reaction has been put. Alvarez has used the method recently in his conversion of pregnenolone acetate (28) into epitestosterone (29) (equation 111)[110]. Based on our own stereochemical results[106c, 107, 109], it is certain that considerable testosterone acetate was formed in this reaction as well. Also in the steroid

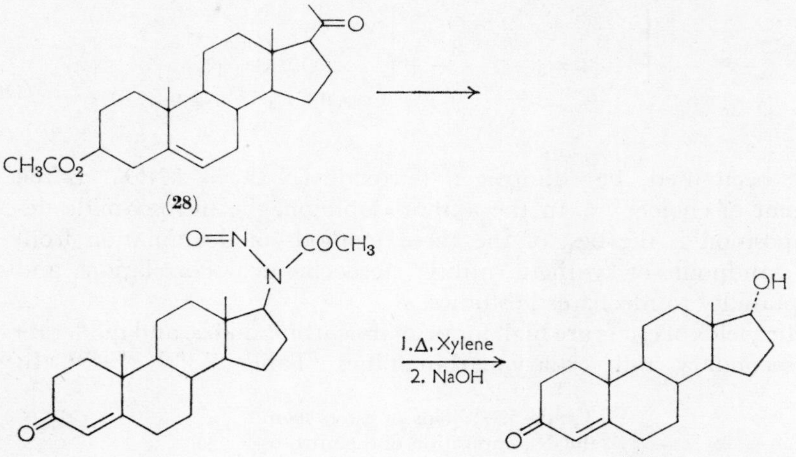

(111)

field, Sato and Latham have used the method to interrelate tomatidine and dihydroneotigogenin (**30**) (equation 112); similar reactions were

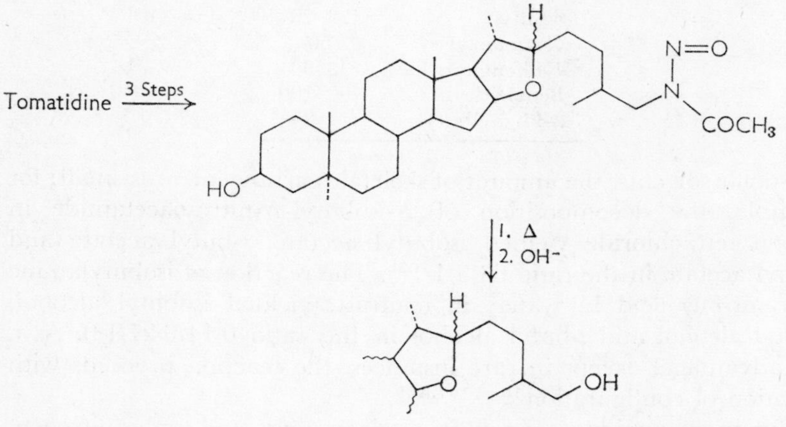

(112)

used to interrelate solasidine and dihydrotigogenin [111].

The corresponding *N*-nitroamides (**31**) decompose under similar conditions to yield similar products (equation 113) [112]. The nitro-amides are useful in that the critical intermediate **32** can be prepared by an independent path—the reaction of a nitroamine salt (**33**) with an acyl halide, a reaction that can be studied at much lower temperatures than are feasible for the decomposition of the nitroamides

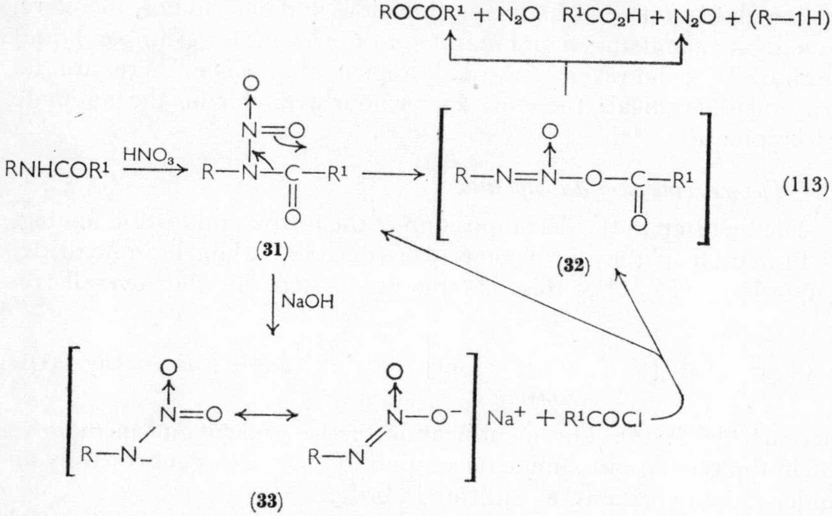

$$ROCOR^1 + N_2O \qquad R^1CO_2H + N_2O + (R\text{---}1H)$$

$$RNHCOR^1 \xrightarrow{HNO_3} \underset{(31)}{R\text{---}N\text{---}C\text{---}R^1} \longrightarrow \left[R\text{---}N\text{---}N\text{---}O\text{---}C\text{---}R^1 \right] \qquad (113)$$

$$\underset{(33)}{\left[R\text{---}N^- \overset{N=O}{} \longleftrightarrow R\text{---}N \overset{N\text{---}O^-}{} \right]} Na^+ + R^1COCl$$

(**31**) themselves [107, 112, 113]. The 'salt' reaction and the thermal decomposition of the nitroamides (**31**) have been shown to yield the same products [107, 112]. In addition to these derivatives, nitroso and nitrocarbamates [112, 114] have been examined, as well as nitroso-sulfamides [103b, 115] and nitrosohydroxylamine derivatives [107, 112, 116].

(1) *The role of free radicals*

The known free-radical decomposition of aryl nitrosoamides $(ArN(NO)COR^1)$ [117] and the report that nitrosoamides of O-alkyl-hydroxylamines [116] decompose by a free-radical pathway indicate that free-radical processes might occur in the normal nitrosoamide decomposition. In fact, the aliphatic nitrosoamides have been used as initiators at elevated temperatures for the polymerization of styrene and other olefins [118]. At, or near, room temperature, however, it appears that free radicals are *not* formed in the nitrosoamide decomposition. It has been found, for example, that (1) CO_2 (a product of the decarboxylation of carboxyl radicals) [119] is not formed in the decomposition [106c]; (2) the scavenger nitric oxide [120], has no effect on the reaction [106c]; (3) normal products and no polymer are formed in the decomposition of N-(s-butyl)-N-nitrosobenzamide [106c] and N-nitroso-N-(1-phenylethyl)acetamide [121, 122] in styrene; and (4) no difference in acetate yields is observed when N-nitroso-N-(1-phenyl-ethyl)acetamide is decomposed [121] in benzene in the presence or absence of $0\cdot1$ M I_2 [123]. Styrene and 1-phenylethyl acetate react with

15*

iodine and bromine at high concentrations and one cannot, therefore, investigate geminate recombination of radicals in, for example, liquid bromine[124a]; however, it would appear that since there are no scavengable radicals, there are few or none available for the geminate recombination[124b].

(2) The mechanism of decomposition

The first step in the decomposition of the nitrosoamides and analogs is formation of the diazo ester (**34**) (diazoic carboxylic anhydride) (equation 114)[106c]; this is the slow step in the overall re-

$$27 \xrightarrow{\text{Slow}} \left[R—N{=}N—O—\overset{\overset{\textstyle O}{\|}}{C}—R^1 \right] \xrightarrow{\text{Fast}} \text{Products of equation (110)} \quad (114)$$

$$(34)$$

action[106c, 114a, 125]. The identification of the critical intermediate as **32** in the related nitroamide decomposition has been confirmed by an independent synthesis, as indicated above.

The subsequent fast decomposition of the diazo esters is of special interest in that it relates to the step in which nitrogen is lost in the nitrous acid deamination of aliphatic amines. The gross mode of decomposition of the diazo ester (**34**) is a function of the nature of the alkyl group R, and in general terms, substitution, β-elimination or α-elimination occurs (equation 115).

$$34 \longrightarrow \left[\underset{\underset{\textstyle H}{|}}{\overset{|}{C}}—\underset{\underset{\textstyle H}{|}}{\overset{|}{C}}—N{=}N^+\bar{O}_2CR^1 \right] \longrightarrow \begin{cases} \underset{\underset{\textstyle H}{|}}{\overset{|}{C}}—\underset{\underset{\textstyle H}{|}}{\overset{|}{C}}—O_2CR^1 + N_2 \\[2em] —C{=}C—H + R^1CO_2H + N_2 \\[2em] —\overset{|}{C}—\underset{\underset{\textstyle H}{|}}{C}{=}\overset{+}{N}{=}\bar{N} + R^1CO_2H \end{cases}$$

$$(35)$$

$$(115)$$

(a) β-Elimination and hydrocarbon formation. Olefins are formed in the decomposition of all nitrosoamides with R groups bearing a β-hydrogen. The olefin may possibly stem from some intramolecular pathway, but in polar solvents it is more likely to result from attack of the solvent on the β-hydrogen (equation 116). The nitrosoamides of 1- and 3-phenylpropylamine[121] and 2-methyl-2-butylamine[126] give only olefins in the hydrocarbon fraction on decomposition. The nitro-carbamate of 2-butylamine, however, yielded methyclyclopropane (2%) in addition to the butenes (47%)[126, 127]. The cyclopropane

$$35 \longrightarrow \left[\begin{array}{c} \overset{(a)}{\underset{B}{\overset{}{\underset{}{\text{—C—C}}}}} + \overset{}{\underset{}{\text{O}_2\text{CR}^1}} \\ \end{array} \right] \quad \begin{array}{l} \overset{(a)}{\longrightarrow} \text{Ester} \\ \\ \overset{(b)}{\longrightarrow} \text{R}^1\text{CO}_2\text{H} + \overset{}{\underset{}{\text{—C=C—H}}} \end{array} \tag{116}$$

(36)

formation step is probably closely related to the process leading to olefins (equation 117), although the proximity of the anion may favor the intramolecular path.

$$36 \longrightarrow \quad \triangle + \text{R}^1\text{CO}_2\text{H} \tag{117}$$

(b) *α-Elimination.* When R^+, the potential carbonium ion from **34** is relatively unstable (from primary alkyl groups, groups that bear α-carbonyl groups, etc.), an α-elimination occurs to give the corresponding diazo compound and carboxylic acid (equation 118). When R bears an α-carbonyl group, e.g. R = carbomethoxymethyl, the

$$-\overset{|}{\underset{\underset{R}{\overset{}{\underset{}{\text{H}}}}}{\text{C}}}\text{—N=N} \quad \overset{}{\underset{\text{O}_2\text{CR}^1}{}} \longrightarrow \overset{}{\underset{}{\text{C}=\overset{+}{\text{N}}=\bar{\text{N}}}} + \text{R}^1\text{CO}_2\text{H} \tag{118}$$

diazo compound can be easily isolated[128], whereas if R = primary alkyl, the diazoalkane must be 'trapped'[129] since the formation of esters from diazoalkanes and acids is very fast; indirect methods have also been used to show the existence of this pathway[130]. That is, for nitrosoamides of primary carbinamines, esters are still the main product, but they are formed via diazoalkanes. The stability of the diazoalkane is apparently another factor to consider, since we have recently found that the decomposition of the nitrosobenzamide of benzhydrylamine in benzene yields about 4% of diphenyldiazo-methane[108b]. The only sure way of avoiding this pathway is to carry out the decomposition in polar solvents. We have found, for example, that nitrosoamides of 1-phenylethylamine[107], 3-cholestanylamine[131] and benzhydrylamine[108b] in *O*-deuteroacetic acid yield the corresponding esters containing no detectable deuterium.

(c) *Displacement.* In non-polar solvents, nitrosoamides of simple secondary carbinamines show a displacement component. The decomposition of 0·04 M *N-s*-butyl-*N*-nitrosobenzamide in pentane yields *s*-butyl benzoate with 45% retention of configuration and 55% inversion (this is equivalent to 10% inversion + 90% racemization; since we treat the reaction in terms of competing pathways, the former scheme will be used throughout this article). When the decomposition was carried out in the presence of 0·1 M benzoic acid, the ester was formed with 58% inversion, and when the reaction was carried out in the presence of 0·4 M acetic acid, the acetate ester was formed with 67% inversion of configuration while the benzoate was formed with 48% inversion (equation 119)[106c]. The displacement reaction is

$$R^1 - CO_2H \longrightarrow \underset{(37)}{\overset{\displaystyle}{C - \overset{+}{N_2}}} \quad \bar{O}_2CR^1 \longrightarrow R^1 - CO_2C \quad \text{(inv.)} \tag{119}$$

not an important one here, and in the decomposition of the nitroso-amide of 1-phenylethylamine in which a benzylic cation is formed, the displacement could not be detected, either with the free acid[106c] or the carboxylate ion[132]. An extreme case of the lack of the displacement reaction is given by the finding that the neighboring group effect is not very important for the nitrosoamide decomposition (*vide infra*). Judging from the effect of substitution on nucleophilic displacement reactions of alkyl halides, however, this mode of decomposition could be important for the substitution portion of the decomposition of nitrosoamides of primary carbinamines. Since path (b) (*α*-elimination) occurs extensively in non-polar solvents, one should look for the displacement mode in polar media

(d) *Retention of configuration.* In polar solvents, the nitrosoamide decomposition of acyclic and monocyclic amines proceeds invariably with retention of configuration[133] (Table 6); the reaction can thus be used in stereochemical correlation studies. Experiments employing ^{18}O (and others to be discussed later) suggest the mechanism shown in equation (120)

$$(37) \longrightarrow \left[\overset{\displaystyle N_2}{\underset{\displaystyle C^+}{\big|}} \quad \bar{O}_2CR^1 \right] \longrightarrow C - O_2CR^1 \tag{120}$$

(e) *Front-side exchange.* The decomposition of a nitrosoamide in the presence of a carboxylic acid (the 'foreign acid') different from that included in the nitrosoamide leads to the formation of two esters (equation 121). In pentane, the normal ester (RO_2CR^1; R = *s*-butyl)

$$\mathbf{27} + R^2CO_2H \longrightarrow N_2 + RO_2CR^1 + RO_2CR^2 \tag{121}$$

TABLE 6. Retention of configuration in the nitrosoamide decomposition.

R	Solvent	Retention (%)	Reference
$C_2H_5CHCH_3$	CH_3CO_2H	68	106c
3-β-Cholestanyl	CH_3CO_2H (D)	82	131
$C_6H_5CHCH_3$	CH_3CO_2H (D)	81	107
$C_6H_5CHC_6H_4Cl$	CH_3CO_2H (D)	68	108b
$C_6H_5C(CH_3)C_2H_5$	Ether–Methanol (1:2)	97	109

was formed with overall retention of configuration and the foreign ester (RO_2CR^2) with overall inversion[106c], reflecting the occurrence of the displacement mode of reaction. In dioxane, on the other hand, it was observed that both esters were formed with retention of configuration and with the same degree of retention[106c]. The dioxane, present in large excess, presumably interacts with the back side of the *s*-butyl group to the exclusion of the 'foreign' dinitrobenzoic acid (equation 122). It is felt that actual bonding to the solvent occurs

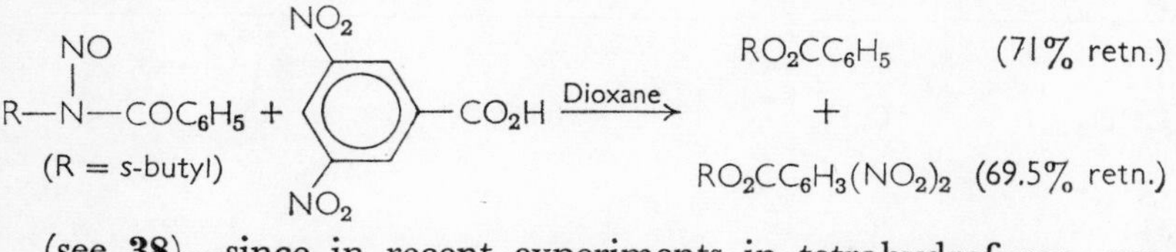
NO
NO_2
R—N—COC_6H_5 +
CO_2H
Dioxane
(R = s-butyl)
NO_2
$RO_2CC_6H_5$ (71% retn.)
+
$RO_2CC_6H_3(NO_2)_2$ (69.5% retn.)

(122)

(see **38**)—since in recent experiments in tetrahydrofuran, products

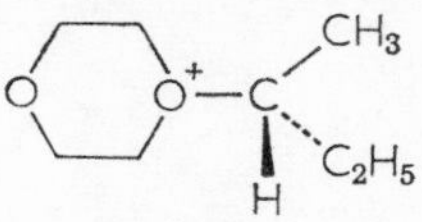
O
O
C
CH_3
C_2H_5
H

(**38**)

containing the elements of the solvent have been obtained (equation 123)[134].

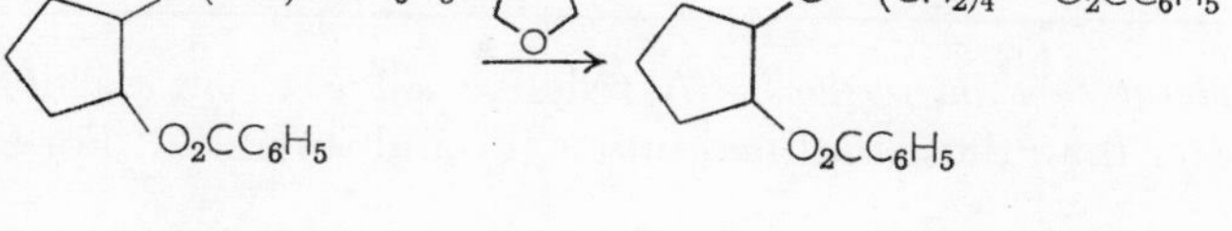
$N(NO)COC_6H_5$
$O_2CC_6H_5$
O
O—$(CH_2)_4$—$O_2CC_6H_5$
$O_2CC_6H_5$

(123)

The entry of the foreign acid to the front side of the reacting species is visualized as in equation (124). Complexes of carboxylic acids and

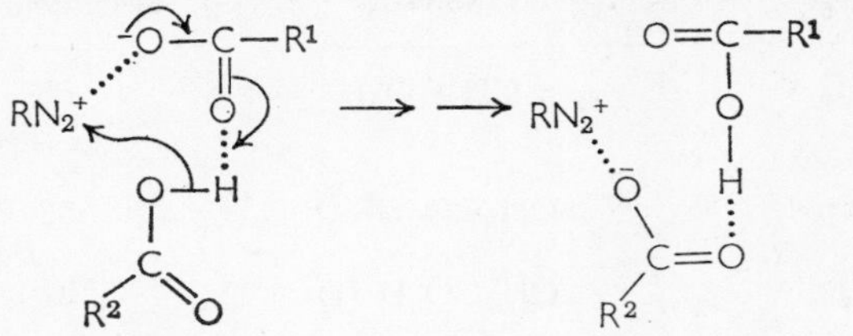

$$\text{(124)}$$

(39)

their anions similar to the anion depicted in **39** are well known[135]. It was noted in the dioxane work that in the reverse situation, the reaction of the nitroso-3,5-dinitrobenzamide in the presence of benzoic acid, the exchange reaction did not occur. 3,5-Dinitrobenzoic is stronger than benzoic acid by a factor of 27; it appears, therefore, that the exchange can be thought of in terms of an acid–base reaction. A similar 'front-side' exchange probably occurs in acetic acid as solvent, since alkyl acetates, the 'solvolysis' products, are often formed with partial retention of configuration (Table 7).

TABLE 7.   Intramolecular and intermolecular products in the nitrosoamide decomposition in acetic acid.

R	R¹	RO_2CR^1		RO_2CCH_3		Reference
		Yield	% Retention	Yield	% Retention	
H | $C_2H_5CCH_3$ |	C_6H_5	12	68	5	43	106c
H | $C_6H_5CCH_3$ |	$C_{10}H_7$	32	81	17	56	107
H | $C_6H_5CC_6H_4Cl$-p |	$C_{10}H_7$	46	68	53	54	108b
3-β-Cholestanyl (equatorial)	$C_{10}H_7$	62	82	10	75	131

(f) *Solvent cage interactions.* A reactive solvent can participate by paths other than those outlined under (c) and (e) above. For example, the decomposition of N-nitroso-N-(1-phenylethyl)naphthamide in pure acetic acid yields the naphthoate ester (32% yield; 81% retention of configuration), and also the acetate ester (17% yield; 56% retention of configuration)[107]. Part of the acetate is certainly formed via path (e); the low retention of configuration observed must then stem from some type of attack on the back side of the cation (path (g) could account for only a small part of the inversion). Since displacement reactions (section c) are not detected in the 1-phenylethyl system, we assume that the inverted acetate stems largely from the collapse of the solvent cage with capture of the cation, **41**, by an acetic acid molecule on the unhindered back side (equation 125).

$$
\text{(125)}
$$

Some type of irreversible process is envisaged, as in the interaction with tetrahydrofuran and dioxane (section e), but the intermediacy of species such as $CH_3CO_2 \cdots \overset{|}{\underset{}{C}} \cdots O_2CC_{10}H_7$ may be possible. For further examples of solvent interaction, see Table 7.

(g) *Intramolecular inversion*

(i) *Deuterium results.* The decomposition of nitrosoamides in O-deuteroacetic acid yields both foreign (RO_2CCH_3) and normal esters (RO_2CR^1) (equation 126), neither of which contains appre-

$$
RN(NO)COR^1 \xrightarrow{CH_3CO_2D} RO_2CR^1 + RO_2CCH_3 \qquad \text{(126)}
$$

ciable amounts of deuterium[107, 131]. Under these conditions the normal ester could not stem from diazoalkane intermediates, or from displacement reactions, since any 'normal' acid liberated would be swamped by the excess of deuteroacetic acid and deuterated, or only acetate esters would be obtained. A further check with ^{18}O-labeled carboxylic acid showed that little of the free acid ended up as ester[107]. Thus the formation of the normal ester is an intramolecular process; and yet, a part of the ester is formed with inversion of configuration (Table 6). Nitrosoamides of tertiary carbinamines decompose at $\sim -30°$, and it was not possible in this case to use pure acetic acid as the solvent. However, methanol has been used as the reactive solvent in the reaction of 2-phenyl-2-butylamine and some inversion (3%) was observed (Table 6). In addition, a run in tetrahydrofuran with an excess of diazomethane to act as a scavenger for acid formed in the elimination process (equation 110) yielded the normal ester with 95% retention of configuration and 5% inversion[109]. We are forced to the conclusion, therefore, that an intramolecular path exists for inversion of configuration[136]. Studies with ^{18}O-labeled compounds confirm this view.

(ii) [18]O *Results.* The decomposition of nitrosoamides of secondary and tertiary carbinamines labeled in the carbonyl group with [18]O yields esters with most of the [18]O still in the carbonyl group (equation 127) (Table 8). It was shown, further, that nitroamides gave approxi-

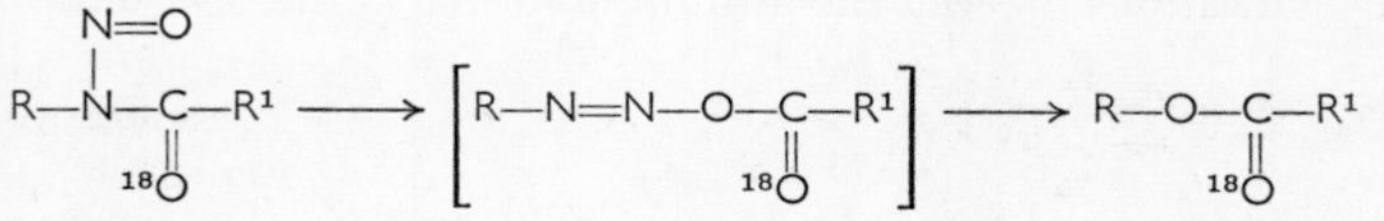

$$
\underset{\substack{| \\ {}^{18}\text{O}}}{\overset{\substack{\text{N}=\text{O} \\ |}}{\text{R}-\text{N}-\text{C}-\text{R}^1}} \longrightarrow \left[\underset{\substack{\| \\ {}^{18}\text{O}}}{\text{R}-\text{N}=\text{N}-\text{O}-\text{C}-\text{R}^1} \right] \longrightarrow \underset{\substack{\| \\ {}^{18}\text{O}}}{\text{R}-\text{O}-\text{C}-\text{R}^1} \quad (127)
$$

TABLE 8.   [18]O Distribution in the nitrosoamide reaction.

R	Solvent	[18]O in carbonyl group of ester (%)	Reference
Cyclohexyl	CH_3CO_2H	65	107
$C_6H_5-\overset{H}{\underset{|}{\overset{|}{C}}}-CH_3$	CH_3CO_2H	69	107
$C_6H_5-\overset{H}{\underset{|}{\overset{|}{C}}}-C_6H_5$	CH_3CO_2H	65	108b
$C_6H_5-\overset{CH_3}{\underset{|}{\overset{|}{C}}}-C_2H_5$	Ether–Methanol (1:2)	65	109
$C_6H_5-\overset{H}{\underset{|}{\overset{|}{C}}}-CH_3$	Dioxane + CH_2N_2	55	107
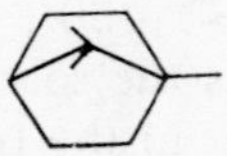	CH_3CO_2H, CH_2Cl_2	50[a]	107

[a] Now known to be a free-radical reaction.

mately the same results, and as expected the nitrous oxide from this reaction contained only the normal abundance of ^{18}O (equation 113)[107].

The results of Table 8 suggest (1) that the ions must be free for a finite period of time since some mixing of the ^{18}O was observed; (2) that the steps leading to the ester are extremely fast ones since they compete successfully with fast processes leading to randomization of the oxygens (rotation of the anion, translation, etc); and (3) that with the exception of the extreme case—the bridgehead one—the ^{18}O results for runs in a common solvent are remarkably independent of the nature of the alkyl group (**R**)

In one further experiment, the ^{18}O distribution was determined for both the L and D esters obtained from the decomposition of optically pure N-nitroso-N-(L-1-phenylethyl)naphthamide labeled in the carbonyl group with ^{18}O (Table 9). In acetic acid (Table 6),

TABLE 9. ^{18}O Distribution in L and DL forms of 1-phenylethanol from the nitrosoamide decomposition in acetic acid[a].

	C₆H₅CH(CH₃)OH			R¹CH₂OH
	81% L, 19% D	100% L	DL	
Excess ^{18}O	0·37	0·38	0·38	0·84
(Atom %)	1·82	1·83	1·81	3·64

[a] Similar results were obtained in dioxane and in a solid-state reaction.

81% of the ester was formed with retention of configuration and 19% with inversion. This ester was cleaved with lithium aluminum hydride and the 1-phenylethanol was resolved into L and DL fractions and analyzed (Table 9). It was found that the L and DL fractions had the

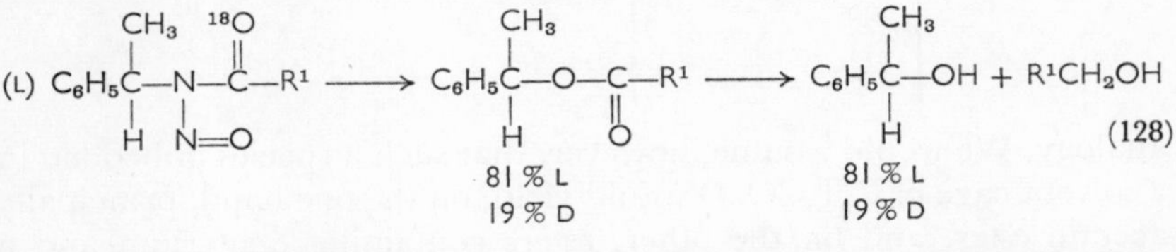

same content of ^{18}O; thus the L and D portions must also have had the same ^{18}O-content, and it follows that the ^{18}O distribution in both the L and D esters must be the same, namely 69% is present in the carbonyl position and 31% in the ether position (Table 8). That is, we not only find an example of intramolecular inversion, but also a mode of inversion that has certain points in common with the reaction mode that leads to retention of configuration. As controls for the reaction, we have shown that recovered nitrosoamide at the half-life is not racemized and it contains all the ^{18}O in the carbonyl group; furthermore, the product ester shows no ^{18}O scrambling or racemization under the conditions of the experiment and no racemization occurs during the resolution of the alcohols.

(3) The loss of nitrogen and ester formation

We now discuss the stage at which the product ester is formed (equation 129).

$$R-N=N-O-\overset{\overset{\displaystyle O}{\|}}{C}-R^1 \longrightarrow R-O-\overset{\overset{\displaystyle O}{\|}}{C}-R^1 \qquad (129)$$
$$(34)$$

Several mechanisms will be examined in an effort to account for the results just outlined. The 1-phenylethyl compounds will be used as examples since most of our results have been obtained with these compounds; where comparisons have been made with analogs, e.g. Table 6, the results have been similar. A reaction via free diazoalkanes is clearly ruled out. Some type of close complex between the diazoalkane and carboxylic acid involving a rotation of the partners about one another is conceivable (equation 130), although we know of no

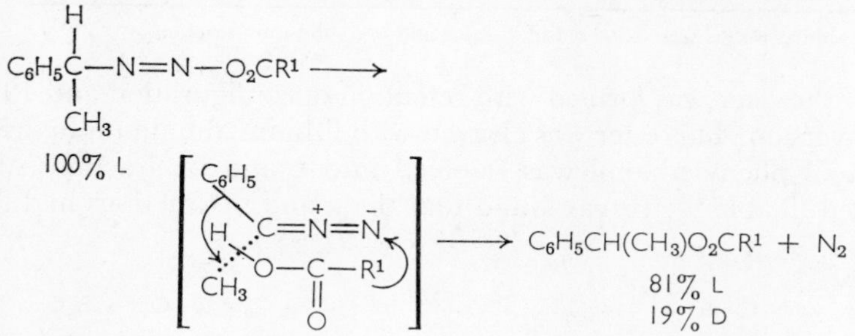

$$\underset{\overset{\displaystyle |}{CH_3}}{\overset{\overset{\displaystyle H}{|}}{C_6H_5C}}-N=N-O_2CR^1 \longrightarrow$$

100% L

$$\left[\begin{array}{c} C_6H_5 \\ \underset{CH_3}{\overset{H}{\diagdown}}C=\overset{+}{N}=\overset{-}{N} \\ O-C-R^1 \\ \| \\ O \end{array} \right] \longrightarrow C_6H_5CH(CH_3)O_2CR^1 + N_2 \qquad (130)$$

81% L
19% D

analogy. We would assume, however, that such a species imbedded in a solvent cage of CH_3CO_2D would yield, on the one hand, principally acetate ester, and on the other, esters containing deuterium and a completely randomized ^{18}O distribution, since proton transfers can be extremely rapid[137]. Furthermore, such a mechanism cannot operate in the tertiary carbinamine case, where some inversion is still occurring (7% in CH_2Cl_2 [109]; Table 6).

A second possibility involves a concerted loss of nitrogen, where, as the R—N bond is stretched, the R—O smoothly approaches the bond distance in the ester (equation 131). In addition to a lack of

$$\textbf{34} \longrightarrow R\cdots\overset{\overset{\displaystyle N\equiv N}{\cdots\cdots}}{\cdots\cdots}O\text{—COR}^1 \longrightarrow N_2 + ROCOR^1 \qquad (131)$$

analogy for a process of this type, difficulty would be experienced in accounting for the stereochemical results (Table 6), the ^{18}O results (Table 8), the distribution in the enantiomers (Table 9) and the extensive skeletal rearrangement observed in the nitrosoamide decomposition of isobutylamine in polar solvents[103b], (which implies a high degree of carbonium ion character during the reaction). It is also worth pointing out that the decomposition of 1-phenylethylazomethane (**40**) is thought to involve sequential C—N bond cleavage

$$\underset{\underset{H}{|}}{\overset{\overset{CH_3}{|}}{C_6H_5\text{—C—N}}}=N\text{—CH}_3 \longrightarrow \underset{\underset{H}{|}}{\overset{\overset{CH_3}{|}}{C_6H_5\text{—C}^\bullet}} + {}^\bullet N{=}N\text{—CH}_3 \longrightarrow$$

$$\textbf{(40)}$$

$$\underset{\underset{H}{|}}{\overset{\overset{CH_3}{|}}{C_6H_5\text{—C}^\bullet}} + N_2 + CH_3{}^\bullet \qquad (132)$$

rather than a concerted cleavage[138]. This does not help the analysis, however, since it could be argued that the ionic character of the transition state for the nitrosoamide decomposition is such that coulombic attraction would favor a concerted cleavage. The case thus rests on the

$$\overset{\overset{\displaystyle N\equiv N}{\cdots\cdots}}{R^+\cdots\cdots O^-\text{—COR}^1}$$

$$\textbf{(41)}$$

other arguments in this section and those of the following paragraphs.

The third possible mechanism[107], we believe, accounts for all our observations, and in addition, it is generally applicable to the triazene and nitrous acid methods of deamination. Based on the structures of aromatic diazonium salts, the formation of ion pair **42** seems to be a logical first step in the reaction. By analogy to other ion pairs, the N—O bond in **42** could have some covalent character and the nitrogen could be bonded to one of the oxygens in particular[139]. However, for the decomposition in a solvent as polar as acetic acid (Table 8), it appears that the C—N and N—O bonds (**34**) are either broken simultaneously or in rapid succession in view of the independence of the ^{18}O scrambling with respect to the carbonium ion character of R. In the free-radical field the simultaneous cleavage of two bonds has been reported for the decomposition of certain peresters[140].

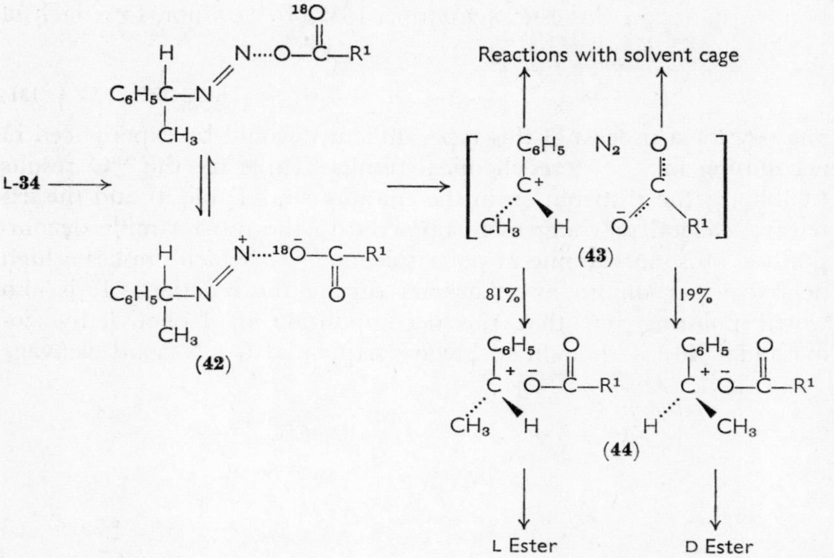

(133)

In any event, the key step involves the loss of nitrogen. The loss of nitrogen from an aliphatic diazonium ion is probably exothermic[141]; the transition state should resemble the reactant molecule[142], and no specific bonding to solvent would be expected. Furthermore, by virtue of the poor solvating ability of nitrogen, the loss of nitrogen is a practically irreversible process; we propose that the loss of nitrogen from an aliphatic diazonium ion forms a cation and an anion at a separation greater than that in the corresponding ground-state intimate ion pair 44, and that fast follow-up reactions of the cation occur with the anion and with molecules in the solvent cage (species 43 is discussed in greater detail in section c on the nitrous acid deamination). An analogy to the formation of 43 and subsequent reactions would be the formation of two radicals in a solvent cage from the thermal decomposition of an azoalkane (RN=NR), and by fast follow-up reactions of the radicals so formed (geminate recombination, reaction with scavengers, etc.). A fraction of the cations formed on the loss of nitrogen in the nitrosoamide reaction will react with the solvent cage, but most will react with the anion, because the coulombic attraction of the ions will make this a very fast process[143]. Most of the anions will bond on the front side of the cation to give L-44, and subsequently the ester with retention of configuration. However, in a fraction of the cases, in the time interval between the loss of nitrogen and the formation of the ester, and probably as a consequence of the loss of nitrogen, the cation rotates so far that bond formation with the anion can occur on what was the back side of the ion. Rotation of the cation about an axis through the center of mass, approximated in this case by the bond from the central carbon to the phenyl group, need not appreciably alter the position of the electron-deficient carbon atom with respect to the two oxygens, and the subsequent approach of the ions could give D and L molecules having the same or very nearly the same distribution of ^{18}O. The non-labeled oxygen atom in **42** and **43** is the closer of the two to the carbon undergoing substitution and presumably it interacts more strongly electronically with the cation; it seems reasonable that the ester will be formed with a major amount of the ^{18}O still in the carbonyl group. A number of other ion-pair reactions involving the carboxylate group (carbonyl ^{18}O) have been shown to proceed with retention of ^{18}O in the carbonyl group[144]. It is felt that most of the ^{18}O scrambling occurs at this point in the overall reaction. As seen in Table 8 and reference 107, less scrambling of the ^{18}O occurs in protic solvents; without a doubt, in these solvents the anion must be hydrogen bonded. In acetic acid, for example, the complex ion **45** discussed in section (2) (e) would be

$$\overset{^{18}O}{\underset{(45)}{R^1C-\overset{-}{O}\cdots HO_2CCH_3}}$$

formed—in which a specific barrier is imposed on the rotation of the carboxylate ion. The ^{18}O results show that the four oxygen atoms of the complex anion do not bear the same relationship to the cation, since as shown in Tables 7 and 8, the normal ester is the chief product of the reaction and most of the ^{18}O is retained in the carbonyl group.

The mechanism outlined predicts that large cations should show a greater retention of configuration than smaller cations, with a large difference expected between secondary and tertiary ions because of the convenient avenue of rotation past a hydrogen atom. The data of Table 6, although limited, are consistent with these ideas. The mechanism of equation (133) thus accounts for the steric effects, the stereochemical results, the ^{18}O results and the negligible influence of electronic effects on the ^{18}O and stereochemical[132] results.

It might be argued that the inversion of configuration and the ^{18}O mixing could occur during the lifetime of the ground-state intimate ion pair **44**. Such reactions might be possible for ion pairs of relatively stable carbonium ions (e.g. R = triphenylmethyl); it is unlikely, however, that reactions other than rapid collapse (to yield the ester and possibly the olefin) are possible for the ion pairs derived from simple alkylamines. For such ion pairs, the R—O bond in **44** (to the strongly nucleophilic counter ion) would be relatively strong with some covalent character[139]; it is not apparent how racemization of the naphthoate ion pair could occur at this stage and why molecules which might have succeeded in inverting their configuration should have the same [18]O distribution as those formed with retention of configuration. Nor is it apparent on this basis why the observed 'racemization' should be so little affected by the solvent or why two R[+] groups as different as cyclohexyl and 1-phenylethyl should give almost the same [18]O mixing. In support of these qualitative statements is the finding by Goering and Levy[145] that in the solvolysis of *p*-chlorobenzhydryl *p*-nitrobenzoate in 80% aqueous acetone, internal return (from the intimate ion pair[146]) is completely stereospecific (retention) and only external ion-pair return[146] results in partial or complete racemization.

We believe, in addition, that solvent-separated ion pairs (134)[146]

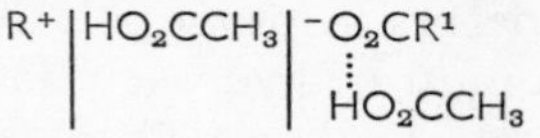

$$\text{R}^+ \; | \; \text{HO}_2\text{CCH}_3 \; | \; {}^-\text{O}_2\text{CR}^1 \qquad\qquad (134)$$
$$\text{HO}_2\text{CCH}_3$$

are not involved in the formation of the intramolecular products of the nitrosamide decomposition (RO_2CR^1, Table 7), since complete [18]O scrambling should surely occur in such species, and since a simple proton transfer would give $\text{R}^+ \; {}^-\text{O}_2\text{CCH}_3$ and thus the foreign ester (RO_2CCH_3); note that the principle product is the normal ester RO_2CR^1 (Table 7) and that the [18]O and most of the stereochemical results that have been discussed pertain to this product. The solvolysis product obtained with net *retention* of configuration in the decomposition of *N*-(2-butyl-2-phenyl)-*N*-nitrosobenzamide in methanol (equation 135) may have resulted, however, from some type

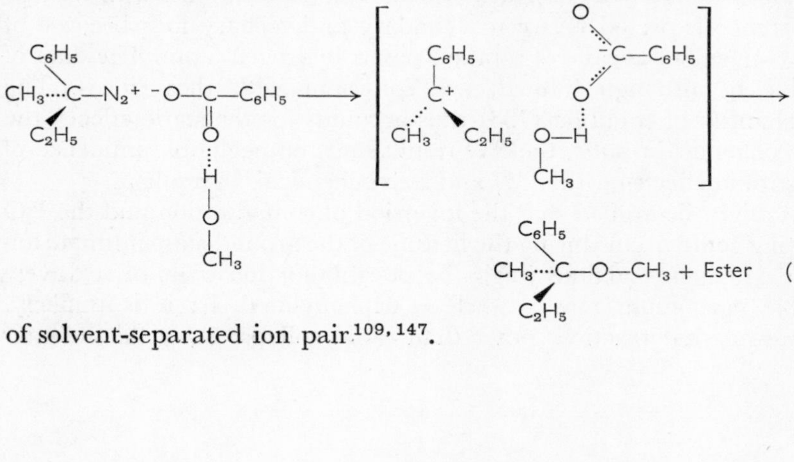

$$(135)$$

of solvent-separated ion pair[109, 147].

Finally, the mechanism outlined in equation (133) can account for other observations on the 'non-concertedness' of the nitrosoamide decomposition and related reactions. For example, although the solvolysis of *trans*-2-bromocyclopentyl acetate yields only *trans*-diacetate, presumably via **46** formed by a neighboring group interaction (equation 136)[148], the nitrosoamide decomposition on an

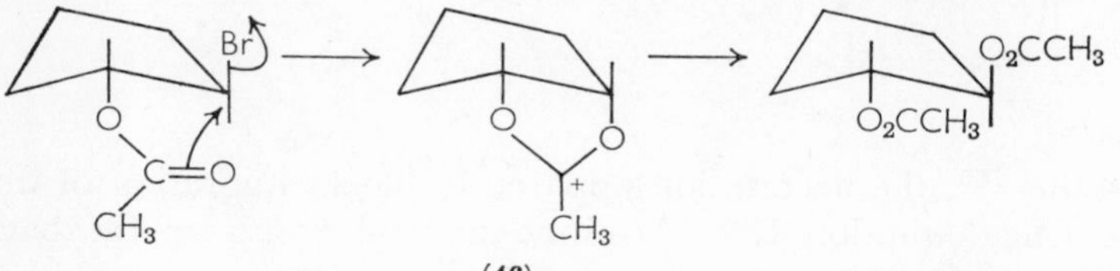

$$(136)$$

$$(46)$$

analog in methylene dichloride yielded a large number of products, but pertinent to the discussion, a *trans*-dibenzoate fraction (13% yield) that was only about 50% racemic[134], and 16% of the *cis* isomer. We conclude that the neighboring group interaction is not an important one for the nitrosoamide decomposition, presumably because the formation of ester (equation 137) is sufficiently fast to

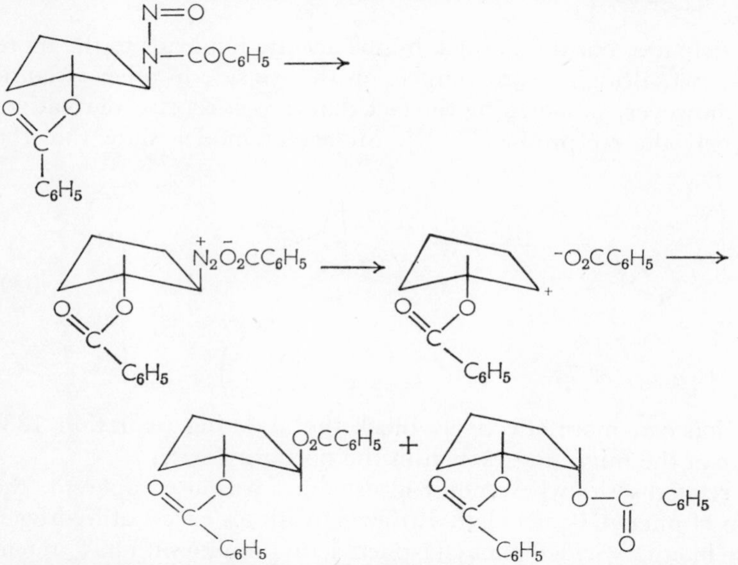

$$(137)$$

compete with the rotation about single bonds that would be required to give species related to **46**[149]. On the other hand, where a neighboring aryl group is fixed in a position permitting immediate access to the back side (equation 138), complete interaction occurs[150]. Interestingly,

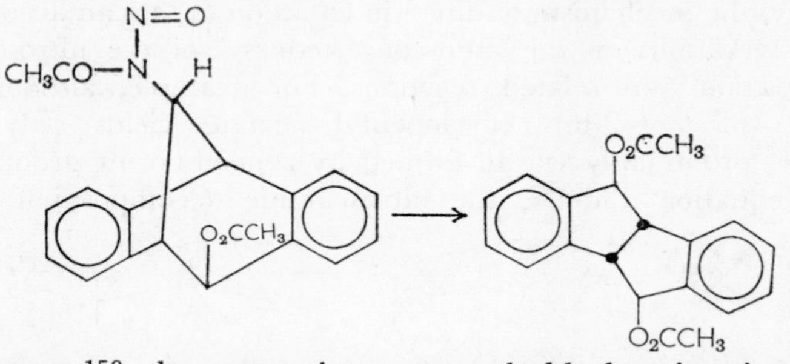

$$\text{(138)}$$

in the isomer[150], the acetate ion apparently blocks migration of this aromatic ring (equation 139). Possibly the aryl group on the back

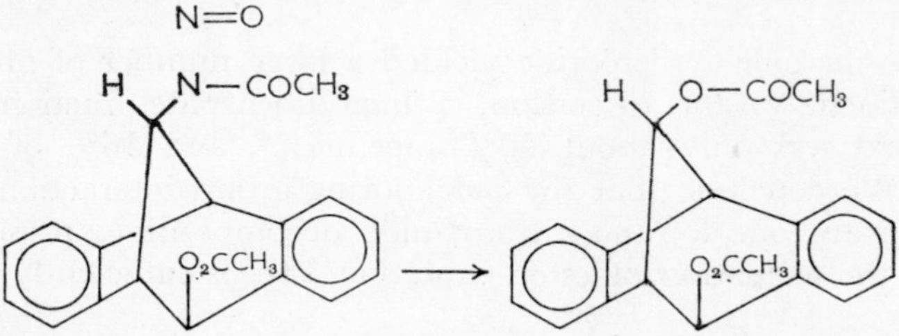

$$\text{(139)}$$

side participates, but if so, attack by the acetate ion leads to the more stable isomer. Bridging can compete on the *syn* side in other circumstances, however, as shown by the fact that *endo*-norbornyl derivatives give largely the *exo* product[126, 151]. Molecular models show that the

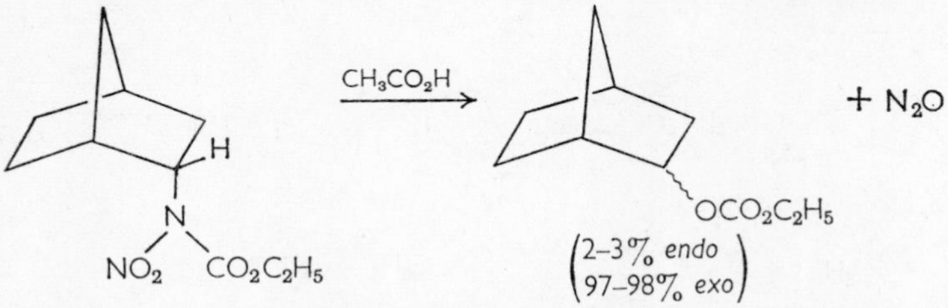

$$\left(\begin{array}{l}\text{2–3\% } endo \\ \text{97–98\% } exo\end{array}\right)$$

$$\text{(140)}$$

acetate ion can more effectively block the aryl ring (equation 138) than it can the migrating carbon in the norbornyl case.

In a related series where non-concertedness was also apparent, the reaction of phenyl(1-phenylethyl)triazene with an excess of hydrogen chloride in ether yielded phenyl(1-phenylethyl)amine with 89% retention of configuration, as well as optically active ring-alkylated products and active 1-phenylethyl chloride[152]. No reasonable concerted process for forming these compounds comes to mind. Lastly, the isomerizations and skeletal rearrangements characteristic of the nitrosoamide de-

$$C_6H_5N = NNHCH(CH_3)C_6H_5 \xrightarrow{HCl} C_6H_5NH - N = N - CH(CH_3)C_6H_5 \xrightarrow{HCl}$$

$$\begin{bmatrix} & N_2 & \\ C_6H_5NH_2 & {}^+CHCH_3C_6H_5 \\ & Cl^- & \end{bmatrix} \begin{array}{l} \longrightarrow C_6H_5NHCH(CH_3)C_6H_5 \\ \\ \longrightarrow ClCH(CH_3)C_6H_5 \end{array} \qquad (141)$$

composition in polar media[103b, 153] are reasonable in terms of the relatively large electron deficiency of the cation (equation 133) which is formed largely without benefit of solvent or neighboring group interactions.

b. *The triazene reaction.* In this method, the aliphatic amine is allowed to react with a diazonium salt, and the triazene formed is treated with an acid in the solvent of choice (equation 142). Since

$$RNH_2 \longrightarrow RNH-N=N-Ar \xrightarrow{HX} \begin{array}{l} RX + N_2 + ArNH_2 \\ \\ \longrightarrow Olefins + N_2 + ArNH_2 \end{array} \qquad (142)$$

$$(47)$$

the arylamine and other by-products can be easily separated from the ester RX, the method is a useful one. One or the other of the two steps had appeared in the early literature[154], but the overall reaction was first applied to deamination in 1961[155]. The triazenes are in effect alkylating agents, and they have been used successfully to alkylate acids directly[155], and alcohols, phenols and mercaptans in the presence of aluminum catalysts[156]. The triazenes are thus stable, crystalline substitutes for diazomethane, diazoethane, etc.[155]

As in the nitrosoamide decomposition, the yields are satisfactory for primary carbinamines and moderate for secondary carbinamines (Table 10). The decrease in yields in the latter case is largely the

TABLE 10. Yields for the triazene method of deamination[155].

R	Aryl	Solvent	Acid HX	RX (%)	Olefin (%)
Methyl	p-Tolyl	Ether	3,5-DNB[a]	95	
n-Butyl	p-Chlorophenyl	Ether	3,5-DNB or HBr	63	
Isobutyl	p-Tolyl	Ether	3,5-DNB	56	
Cyclohexyl	p-Tolyl	CH$_2$Cl$_2$	Acetic	38	30
1-Phenylethyl	p-Tolyl	Hexane	Acetic	44	28
1-Phenylethyl	p-Tolyl	CH$_3$OH	Acetic	40[b]	

[a] 3,5-Dinitrobenzoic acid.

[b] Little acetate was formed; the value pertains to the solvolysis product ROCH$_3$ (formed with predominant inversion of configuration).

result of the formation of olefins and secondary amines (*vide infra*). Skeletal isomerization in non-polar solvents is minimal. No isomers were detected in the deamination of n-butylamine, and less than 2% of the iso and *tert* isomers was formed in the deamination of isobutylamine[155]. As noted also in the nitrosoamide decomposition (previous section), however, isomerization is quite extensive in polar solvents[153].

Applied to 1-phenylethylamine, the triazene reaction proceeds with overall retention of configuration, although the stereospecificity is less than in the nitrosoamide reaction (Table 11).

TABLE 11.   Stereochemistry of the triazene reaction of 1-phenylethylamine
$(C_6H_5CH(CH_3)NHN_2C_6H_4R\text{-}p)^a$.

Run	R	Solvent	HX (equation 142)	% Retention 1-Phenylethyl-X
1	CH_3[155]	Hexane	Acetic acid	54
2	CH_3[155]	Ether	Acetic acid	55
3	CH_3[155]	Acetic acid	Acetic acid	57
4	Cl[152]	Acetic acid	3,5-Dinitrobenzoic acid	61[b]
5	Cl[152]	Ether[c]	HCl	66

[a] At 25°.
[b] 1-Phenylethyl acetate was also formed (57% Retn.).
[c] No ethylbenzene or ethyl phenylethyl ether was formed.

In the triazene deamination of secondary carbinamines, *N*-alkylarylamines (**48**) and the ring-alkylated isomers are prominent by-products (equation 143). In the deamination of 1-phenylethylamine

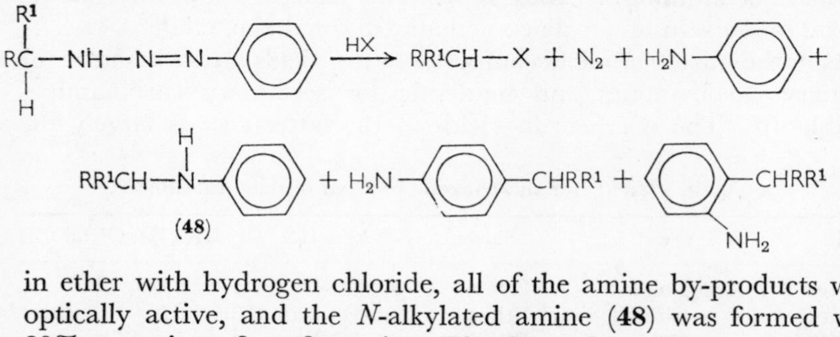

(143)

in ether with hydrogen chloride, all of the amine by-products were optically active, and the *N*-alkylated amine (**48**) was formed with 89% retention of configuration. The formation of these products is suggestive evidence that the reaction is not a concerted one, and that carbonium ions are involved (free-radical products were not obtained in either the presence or absence of thiophenol)[152]. The following mechanism, related to that proposed for the nitrosoamide decomposition, accounts for our observations. As in the nitrosoamide decomposi-

$$47 \underset{HX}{\rightleftharpoons} \begin{array}{c} H \\ | \\ N-Ar \\ R \end{array} N{=}N \xrightarrow{HX} \left[\begin{array}{c} H \\ | \\ N-Ar \\ \overset{\delta+}{N}{\equiv}N \cdots H \\ R \qquad X^{\delta-} \end{array} \right] \longrightarrow$$

$$\left[\begin{array}{c} N_2 \qquad H \\ | \\ N-Ar \\ R^+ \qquad | \\ H \\ -X \end{array} \right] \begin{array}{l} \longrightarrow R^+X^- \longrightarrow RX \\[8pt] \quad\quad H \\ \quad\quad | \\ \longrightarrow RN-Ar \\[6pt] \longrightarrow Olefins \end{array}$$

(144)

tion, it is believed that the loss of nitrogen leaves R^+, X^- and $ArNH_2$ in a solvent cage at separations greater than those in the corresponding ground-state ion pair and solvated ion. The various products, including those from attack on the solvent cage, are formed in the rapid approach of these species to the final covalent bonding distances. The rather low extent of retention (Table 11) relative to the values obtained for the nitrosoamide decomposition (Table 6) may result from the different positions of the anion X^- with respect to the cation R^+ in the two reactions. The role of diazoalkanes in the racemization observed in the triazene reaction is not known at the present time.

c. *The deamination of amines with nitrous acid.* The reaction of nitrous acid with aliphatic amines is the oldest method of deamination[157] and the simplest to effect, but it is the least useful, at least as far as substitution products are concerned. A number of reviews of this reaction have appeared recently[158], and therefore the present treatment will be selective rather than exhaustive.

The reaction is usually carried out in water with nitrous acid generated from silver or alkali metal nitrites, although in much recent work acetic acid has been used as the solvent. The rate-determining step in the overall reaction is apparently the formation of a nitrosoamine (equation 145)[159]. Isomerization to the tautomer **50** yields a

$$RNH_2 \longrightarrow \begin{array}{c} N{=}O \\ | \\ RNH \\ (49) \end{array} \longrightarrow \left[\begin{array}{c} N-OH \\ \diagup\!\!\diagdown \\ R-N \end{array} \right] \longrightarrow Products \qquad (145)$$

$$(50)$$

diazonium species, the rapid decomposition of which should parallel the decomposition of related intermediates in the nitrosoamide and triazene methods of deamination.

The deamination has been oversimplified in many studies, by consideration of only a single diazonium ion and a single free carbonium ion, with no attention paid to the counter ion. This view is manifestly wrong in deaminations in acetic acid and partly wrong in the aqueous systems as shown by recent results[109, 131, 160]. This practice probably arose by an over emphasis on the alcohol product. In these studies, the total product was usually hydrolyzed or treated with lithium aluminum hydride to achieve the maximum yield of alcohol. Conclusions based on this pooled alcohol product are of limited value since there are a large number of products formed in the reaction (ROH, RONO, $RONO_2$ and $ROCOCH_3$ in runs in acetic acid)[109], each with its own stereochemistry, particular mode of formation, etc.

The occurrence of several competing modes of reaction is to be expected in view of the low free energy of activation for the loss of nitrogen from the diazonium ion[141, 161], and for subsequent reactions of the cation. The activation energy scale for the nitrous acid deamination ($E_{act} \sim$ 3–5 kcal) is a low-energy analog of that for the solvolysis reaction ($E_{act} \sim$ 25–30 kcal)[162]. Since the barrier is lower in the deamination, differences in the E_{act} between competing modes of reaction are smaller and the rates of competing reactions become more similar. Some caution is needed in applying this concept; the possible role of the entropy of activation in selecting among several possible reactions with similar energies of activation has been pointed out recently[163a].

The nitrous acid deamination appears to be as complex as the nitrosoamide reaction, and in a common solvent it appears that similar mechanisms are followed. For example, the nitrosonaphthamide of 1-phenylethylamine in acetic acid, and the reaction of 1-phenylethylamine with nitrous acid in the same solvent yield 1-phenylethyl naphthoate (Table 7) and 1-phenylethanol[132], respectively, both with 79–81% retention of configuration. These products are the intramolecular products from the respective reactions, and the stereochemical results suggest that they are formed by a common path. Similar results were also found in the deaminations of 3-cholestanylamine[131], 2-phenyl-2-butylamine[109] and 1,2,2-triphenylethylamine[133]. Also pertinent, is the finding that the acetates from the deamination of n-propylamine in acetic acid, by the two methods, contained similar amounts of the isomeric product, isopropyl acetate; the amount of isomerization observed was far greater than in the solvolysis of n-propyl tosylate in the same solvent[141a].

Because of these results and because of structural similarities in the reaction intermediates, it seems a reasonable conclusion that the two deamination reactions have similar mechanisms. The following modes of reaction (also outlined in the section on nitrosoamides) are therefore considered for the nitrous acid deamination:

Path
(a) β-Elimination and hydrocarbon formation
(b) α-Elimination
(c) Displacement
(d) Retention of configuration
(e) Front-side exchange
(f) Solvent cage interactions
(g) Intramolecular inversion

In equation (146), species **51** could be considered as a solvent-separated ion pair[146], **52** and **55** as ground-state intimate ion pairs[139, 146] and **53** and **54** as descriptions of the point of maximum separation of the cation and anion (*vide infra*). As indicated in the section on nitroso-amides, **51** and **52** may have very short lifetimes in polar solvents.

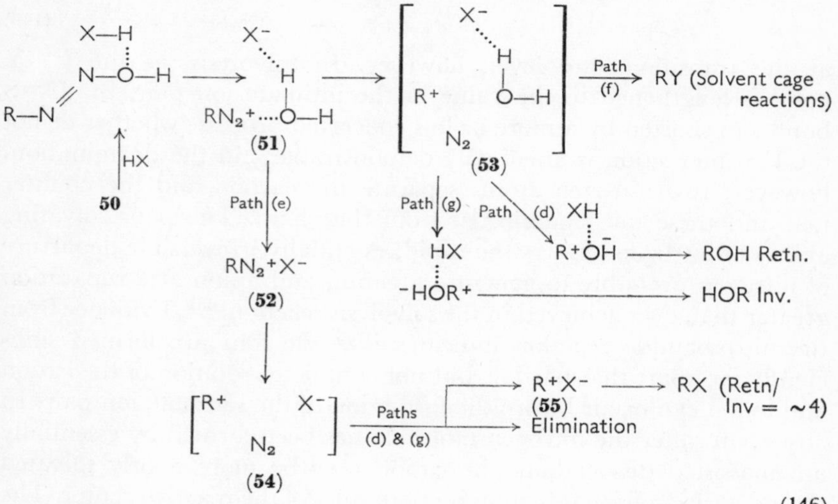

$$(146)$$

(1) *The counter-ion hypothesis*

In contrast to earlier treatments of the nitrous acid deamination, the present hypothesis gives an explicit and important role to the counter ion. The experimental results, outlined in later sections, justify this view.

The nitrous acid deamination will be discussed in comparative terms with the solvolysis of the corresponding halide or sulfonate in the same solvent. Application of the Hammond Postulate[142] to the

$$\text{>C-N=N-X} \longrightarrow \left[\text{HS} \overset{a}{\cdots} \text{>C} \overset{b}{\cdots} \text{N=N}^+ {}^-\text{X} \right] \longrightarrow \text{Intermediate} \qquad (147)$$

$$\text{>C-Y} \longrightarrow \left[\text{HS} \overset{c}{\cdots} \text{>C} \overset{d}{\cdots} \text{Y} \right] \longrightarrow \text{Intermediate} \qquad (148)$$

reactions leads to the view that the transition state for the deamination should resemble the starting diazonium ion pair (bonds to solvent should be weak and long), and that for solvolysis should resemble the first intermediate in the reaction (bonds to solvent should be relatively strong). As a simple argument, the total interaction with the central carbon, a + b and c + d, could be equal; the net electron deficiency of carbon in the transition state might even be less in deamination than in solvolysis (equation 149). One important difference is seen

$$\text{C-N=N}^+\text{X}^- \longleftrightarrow \text{C-}\overset{+}{\text{N}}\equiv\text{N X}^- \longleftrightarrow \text{C}^+ \overset{-}{\text{N}}=\text{N-X} \qquad (149)$$

at the transition-state level, however. In solvolysis, as the C—Y bond is lengthened to the value in the intimate ion pair, the C—S bond can shorten by a more or less concerted process (whether or not the HS interaction is kinetically demonstrable); in the deamination, however, two nitrogen atoms separate the carbon and the counter ion, and these must be ejected from the chain. The poor solvating ability of nitrogen ensures the rapid, essentially irreversible departure of nitrogen probably to generate a cation and anion at a separation greater than ever achieved in the solvolysis reaction[163b]. Evidence from the nitrosoamide reaction indicates that the ions are formed sufficiently far apart that partial (but not complete) rotation of the cation and anion can occur before the formation of the intimate ion pair. In any event, after the nitrogen molecule has been formed by essentially an unassisted dissociation, the cation may be more poorly solvated than the cation from solvolysis on both sides of the reactive center. The resulting greater electron deficiency on carbon in deamination can then account for the skeletal rearrangements[164, 165] and hydride shifts[166, 167a] that occur in deamination, usually to a greater extent than in solvolysis[141a, 165, 171]. The relative 'bareness' of the electron-deficient carbon could also account for the high rate at which follow-up reactions occur, for example reaction with a neighboring phenyl group is faster than rotation about a single bond[149], and reaction with a carboxylate ion (labeled with ^{18}O, a nitrosoamide experiment) occurs faster than rotation of the ion. This view of the reaction also accounts for the formation of the final bond to the counter ion predominantly on the front side of the cation, and bond formation with molecules in the solvent cage predominantly on the back side (in acetic acid and solvents of lower polarity).

Even at a separation of several Å, the cation and counter ion will experience a considerable attractive force, and if the p orbital of carbon is in a line with the anion, a bond can be considered to exist. This species can be considered a vibrationally excited intimate ion pair (Figure 1). This portion of the deamination differs from solvolysis

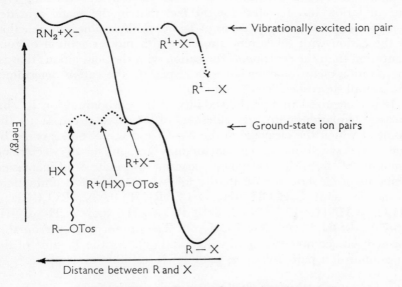

FIGURE 1. Energy diagram for the deamination.

in that the intimate ion pair is formed from a different and higher-energy transition state (Figure 1). Since the ground-state intimate ion pair cannot account for the results obtained in the nitrosoamide decomposition (section a (3)) (and by analogy in the nitrous acid deamination), it is presumed that the characteristic reactions occur from the higher vibrational levels before the ion pair is collisionally deactivated, and that very low-energy barriers separate the vibrationally excited ion pair from intermediates formed by the loss of a proton, shift of an alkyl or phenyl group, or a hydride ion, etc. The photo-isomerization of cycloheptatriene to toluene in the gas phase has been interpreted recently as occurring from a high vibrational level of the ground electronic state of cycloheptatriene[167b]. It seems possible that intramolecular reactions of an electron-deficient species could compete with vibrational relaxation, even in the liquid phase.

If, after the nitrogen molecule has been ejected, the p orbital of the cation and the counter ion are not aligned for orbital overlap, some rotation of the cation will be required for bonding. In some cases, sufficient rotation will occur so that bond formation can occur on what was the back side of the cation; thus the intramolecular inversion observed in the nitrosoamide decomposition and the nitrous acid deamination can be accounted for at this stage in the reaction. This interpretation also accounts for the fact that in the acetic acid de-composition of the tolyltriazene of 1-phenylethylamine, the reaction of the cation with tolylamine proceeds with more retention of con-figuration than the reaction of the cation with the acid anion; that is, the position of the counter ion with respect to the cation determines the overall stereospecificity (section b).

It is recognized that as the stability of the carbonium ion R^+ in-creases (equation 146), the lifetime of **51** may decrease to the point at which the cleavage of the N—O and C—N bonds could be considered synchronous. This factor may account for the decreasing amounts of 'front-side exchange' (leading to acetate) and increasing amounts of alcohol formed in the following series of deaminations in acetic acid: $C_6H_5CH_2NH_2$ (2·1% ROH, 97·9% RO_2CCH_3); $(C_6H_5)_2CHNH_2$ (12% ROH, 88% RO_2CCH_3) and $(C_6H_5)_3CNH_2$ (>95% ROH, <5% RO_2CCH_3)[168]. The products are presumably formed via competitive paths (d) and (e). As the lifetime of the diazonium ion pair decreases, path (e) is less able to compete.

(2) *Alternative mechanisms for the nitrous acid deamination*

Reaction mechanisms involving concerted or synchronous proces-ses[162] for the deamination are unlikely, since application of the Hammond postulate[142] to an exothermic reaction leads to the view that the transition state should resemble the reactant, and that bonds to solvent or neighboring groups should be long and weak[162]. If the developing carbonium ion receives negligible aid from these inter-actions, it would appear to be preferable to consider the reaction as involving essentially an unassisted loss of nitrogen to form a highly electron-deficient species, followed by fast product-forming steps.

That is, the characteristic feature is the formation of the electron-deficient ion, not the weak initial participation.

Considering the experimental data, furthermore, no reasonable concerted process for the formation of alcohol with retention of configuration (section (4) d) seems possible, other than the formation of a pyramidal species, a process for which there is no evidence in the

$$\begin{array}{c} \diagdown \quad \diagup \! {}^{N_2^+} \\ C \\ \diagup \, \diagdown \! _{OH_2^+} \end{array}$$

(56)

chemistry of carbon compounds[169]. The migration of a phenyl group with retention of configuration at the reaction terminus[149] is no less difficult to explain on the basis of a concerted reaction (equation 150). Also, the occurrence of transannular hydride shifts in the

$$\begin{array}{ccc} C_6H_5 & NH_2 & C_6H_5 \\ \diagdown & \diagup & \diagdown \\ HO\!-\!C\!-\!CH\!-\!\!-\!\!- \longrightarrow C_6H_5^*CO\!-\!C\!-\!\!-\!H & (150) \\ \diagup \quad \diagdown & \diagdown \\ C_6H_5^* \quad CH_3 & CH_3 \end{array}$$

* = [14]C labeled

deamination of cyclodecylamine[167a], and the formation of an ethyl ether in the deamination of a bridgehead amine in diethyl ether[161] are more reasonable in terms of reactions of electron-deficient carbon atoms than in terms of concerted reactions. Furthermore, in the related nitrosoamide deamination, the intramolecular inversion and predominant retention of configuration observed (section a) are not interpretable in terms of concerted reactions. Lastly, in the triazene decomposition, the alkylations of aniline on carbon and on nitrogen (section b) are similarly difficult to account for by concerted processes.

Reaction mechanisms involving the formation of a large fraction of open, free carbonium ions as intermediates are unlikely for the nitrous acid deamination in view of the following facts (outlined in greater detail in later sections): (1) alcohols are formed in deaminations in pure acetic acid, (2) they are usually formed with retention of configuration, (3) the solvolysis product (alkyl acetate) is often formed with overall retention of configuration, (4) alkyl nitrates are often formed in appreciable yield despite the low nucleophilicity of nitrate ion and the low concentrations of nitric acid in the medium and (5) axial amines lead to extensive elimination and to substitution with prominent inversion of configuration, whereas equatorial amines lead to little elimination and predominant retention of configuration;

16+C.A.G.

if an open carbonium ion were formed, the same ion would be formed from both isomers (equation 159). Similarly, axial and equatorial isomers usually lead to different types of rearrangements (equation 159).

Hot[170], non-solvated[171] and vibrationally excited[172] carbonium ions have also been proposed for the nitrous acid deamination to account for the differences between deamination and solvolysis. The carbonium ion from solvolysis is presumably 'normal' by right of primogeniture. The concept of a non-solvated carbonium ion as a 'little-solvated, high-energy, very short-lived, open carbonium ion'[171] is close to our view of the *cation* in the counter-ion hypothesis. The hotness of a 'hot' ion has not been defined explicitly, but in the sense of hot ions having more energy than ions obtained in solvolysis, the ions in the counter-ion hypothesis are hot. Attention has previously been drawn to vibrationally excited species, but in the treatment of Corey and coworkers, the vibrational excitation of the cation alone was considered[172].

(3) *Application of the counter-ion hypothesis*

Path (d) retention of configuration. Alcohols are formed in low yield in the nitrous acid deamination of amines in acetic acid, and they have been isolated in only a few instances. The alcohol represents the 'intramolecular' product of the deamination and with the exception of the norbornanols, which will be discussed later, they are formed from secondary and tertiary carbinamines with overall retention of configuration (Table 12). The results are so striking and so similar to those of the nitrosoamide decomposition (Tables 6 and 7) that we feel a common reaction mechanism must be involved. Paths (d) and (g) (equation 146) involving predominant collapse on the front side of the cation are proposed for the formation of the alcohol[109, 131, 160].

In runs in acetic acid, the oxygen atom of the alcohol probably originates in the nitrite ion through species **49** and paths (d) and (g) (equation 146). Under certain conditions, a portion may also arise through path (e) involving the formation of a nitrite ester followed by hydrolysis; this path, however, does not account for the alcohol–ester ratios given by the series of amines mentioned in the last part of section (1). Dilution of the acetic acid solvent by water (up to 8 mole % of water) had no effect on the stereochemical results in the deamination of 2-phenyl-2-butylamine[108a], and dilution up to 50 mole % water had only a small effect in other deaminations[131, 160]

TABLE 12. Stereochemistry of alcohols and acetates obtained from the nitrous acid deamination of aliphatic amines in acetic acid.

$$RNH_2 \longrightarrow ROH + RO_2CCH_3$$

R	ROH		RO_2CCH_3		
	% Yield	% Retention[a]	% Yield	% Retention[a]	Reference
CH₃ | C₆H₅C— | H	2	79	35	54	132
CH₃ | C₆H₅C— | C₂H₅	15	75	41	61	109
α-3-Aminocholestane (axial)[c]	9	89	15	40	131
trans-trans-2-Decalyl (axial)[d]	7·5	94	18·5	45	160
trans-cis-2-Decalyl (equatorial)	—	96	—	75	173
exo-Norbornylamine	10[b]	99[e]	87[b]	98[e]	174

[a] Retn. + Inv. = 100.
[b] Based on recovered substitution product.
[c] Plus 38% olefin.
[d] Plus 33% olefin.
[e] exo Product.

indicating that similar mechanisms for alcohol formation (i.e. equation 146) are followed in the presence and absence of moderate amounts of water. In pure water, however, an ^{18}O study on the deamination of cyclohexylamine has shown that most of the oxygen atoms of the alcohol are derived from the solvent[175]. It is not known whether this result stems from capture of a carbonium ion by the solvent or whether exchange with the solvent water molecules is occurring at some stage such as 50 or 51.

Path (g) intramolecular inversion. The formation of alcohol with inversion of configuration in acetic acid (Table 12; 100%–% retn.) by a process that is not sensitive to the presence of molar quantities of water (previous section) is most easily visualized in terms of the intramolecular inversion process outlined in the section on nitroso-amides. That is, a cation is formed which can rotate before final bond formation to oxygen occurs. The ratio of intramolecular retention to inversion was found to be about 4 for the nitrosoamide decomposition of 1-phenylethylamine and a similar ratio is probable for the nitrous acid deamination.

Path (e) front-side exchange.   Anions of acids present in the system are often found in the products of the nitrous acid deamination despite the fact that S_N2 reactions are not detectible in the deamination of secondary carbinamines in water and acetic acid (path (c), section (3)). By analogy to the nitrosoamide decomposition, path (e) is proposed for a large fraction of the foreign ions incorporated (the remainder occurring through path (f)). Path (e) accounts for the fact that the acetate esters (the 'solvolysis' product) obtained from the nitrous acid deamination of optically active 1-phenylethylamine[132] and 2-phenyl-2-butylamine[109] in acetic acid are formed with net *retention* of configuration.

Front-side exchange (path (e)) partakes of the character of an acid–base reaction in the nitrosoamide decomposition (section a (2) (e)), and in view of the basicity of the incipient 'hydroxide' ion, one would expect extensive exchange of the counter ion in the nitrous acid deamination (equation 146, path (e)). In fact, acetate esters are the chief products from deaminations in acetic acid (ref. of Table 12), and the alcohol component of the reaction has only rarely been reported (Table 12). In acetic acid runs, the acids present in the medium (HX, equation 146) are HO_2CCH_3, HONO and $HONO_2$ (the latter from the disproportionation of nitrous acid). Acetate esters have invariably been reported as products in deamination and in a few instances nitrite[176,177] and nitrate esters[162,174,178] have been isolated.

Path (c) displacement.   The concentration of acetate ions in the nitrous acid deaminations of the 1- and 3-methylallylamines in acetic acid was shown to have no effect on either the product distribution or the stereochemistry of the alkyl acetates obtained[179]. Similarly, acetate ion had no effect on the deamination of *trans-trans*-2-decalyl-amine[160,173]. These results rule out the occurrence of a displacement reaction (in the sense of an S_N2 process) by acetate ion and by the conjugate acid (acetic) on a reaction intermediate (equation 151).

$$CH_3CO_2^-, \ CH_3CO_2H \ \overset{\backslash}{\underset{/}{C}} \! - \! N \! = \! N \cdots X \tag{151}$$

On the other hand, such a process may occur in the deamination of primary carbinamines, since it has been shown that optically active

1-deutero-1-butylamine in acetic acid yields 1-deutero-1-butanol with 84% inversion of configuration (+16% retn.)[162]. The longer life-time of a primary alkyl diazonium compound and the relative lack of steric interactions makes this difference between the primary and secondary carbinamine cases understandable. It is true that the alkyl acetate, as a 'solvolysis' product, is a product of several reaction paths, but a displacement path must be one of these, since predomi-nant inversion occurred (as mentioned in the previous section, the corresponding acetate in the decomposition of secondary carbin-amines can be formed with overall retention of configuration).

Path (f) solvent cage reactions. From the stereochemical results discussed, it is apparent that the deamination of secondary carbin-amines does not involve, to any extent, free carbonium ions, or the attack of solvent on a reaction intermediate. Front-side exchange accounts for only a part of the solvent incorporation in acetic acid. By analogy to the nitrosoamide reaction, and from results such as those shown in equation (152)[161], it is proposed that the remainder of

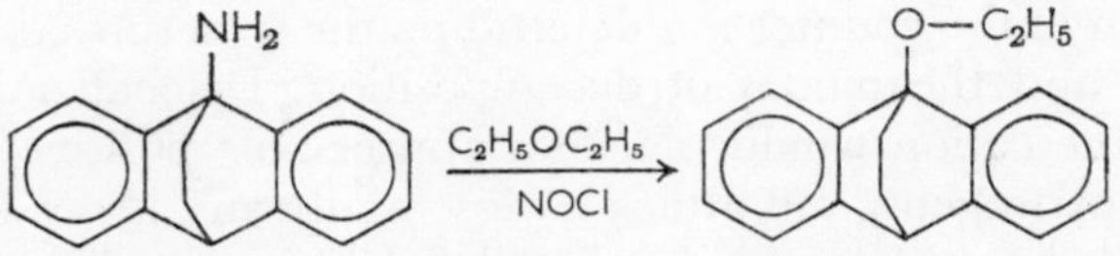

$$\text{(152)}$$

the solvolysis product is formed by attack of the cation on the solvent cage—in competition with collapse on the counter ion X (equation 153). For deamination in acetic acid and less polar solvents in which

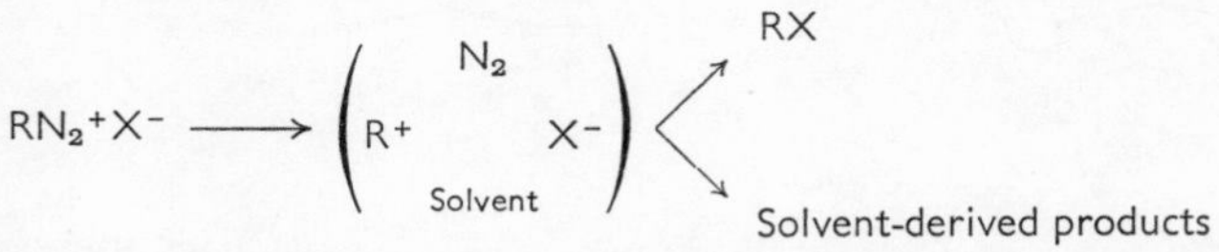

$$\text{(153)}$$

symmetrical ions are not formed, the solvent cage interaction is expected to occur largely, but not exclusively, on the back side of the cation.

Path (a) elimination and hydrocarbon formation. Olefins are usually formed in deamination reactions. With alicyclic amines, it has often been observed that axial amines yield far larger amounts of olefins than equatorial amines[180–182]. The *trans*-coplanar arrangement of the reacting atoms in the axial case is almost certainly responsible for this fact[183]. Cyclic processes involving the counter ion appear not to be responsible purely from the standpoint of their location. The unlikeli-ness of concerted processes has been commented on earlier (section

(2) and section (3), path (c)). Free carbonium ions cannot be involved since isomers such as *cis*- and *trans*-4-*t*-butylcyclohexyl-amine would then yield the same ion (and products) (equation 154).

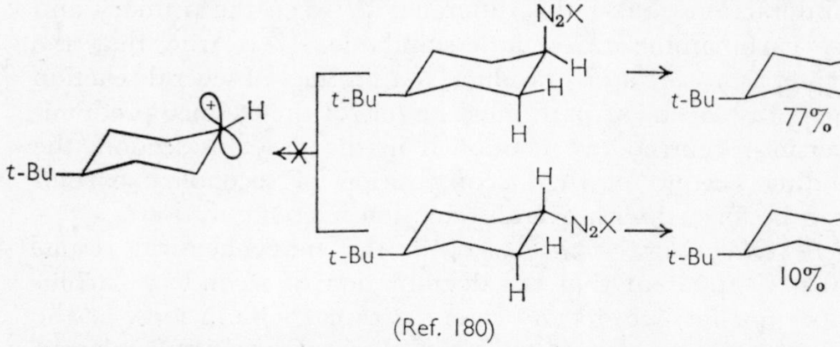

$$(154)$$

(Ref. 180)

Our interpretation of the results is that in the time interval between the loss of nitrogen and the attack of solvent molecules on the β-hydrogen, the counter ion maintains a difference between the species; that is, olefin is formed from a vibrationally excited ion pair (section (1)). The position of the counter ion determines the structures of the reacting species and the modes of decomposition. Delocalization of the charge on the cation would place an appreciable positive charge on the β-axial hydrogens, and attack by essentially any Lewis base in the solvent cage should be sufficient for olefin formation (equation 155).

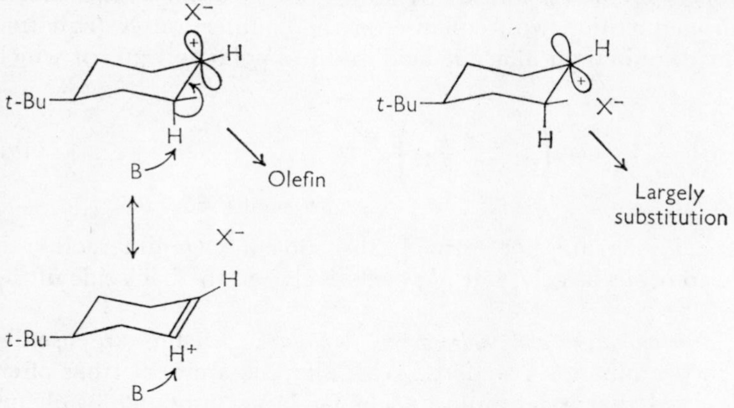

$$(155)$$

(57)

Implicit in these arguments is the notion that the activation energies are low and the products are determined largely by the distribution of conformers in the starting compounds[171,190]. The olefin distribution in the deamination of 2-butylamine in water[106c,162] can also be accounted for by a consideration of the conformations of the reacting molecules[162]. The role of hydrogen-bridged ions in the elimination reaction is not known at the present time; however, unless the counter ion is considered, hydrogen-bridged ions alone cannot account for the stereospecificity of the reactions.

Cyclopropanes are very often formed in the nitrous acid deamination[127, 184]. It appears that protonated cyclopropanes formed from intramolecular reactions of the cation are intermediates in their formation[184c, 185].

Path (b) α-elimination. When loss of nitrogen would lead to a high-energy carbonium ion, α-elimination becomes a competing mode of decomposition (equation 156). Amino ketones may also react by this

$$H_2NCH_2CO_2C_2H_5 \longrightarrow X^- {}^+N_2 {\underset{\underset{B \nearrow}{\overset{\displaystyle|}{\underset{H}{}}}}{-}} CHCO_2C_2H_5 \longrightarrow \bar{N}{=}\overset{+}{N}{=}\underset{\underset{H}{\overset{\displaystyle|}{}}}{C}CO_2C_2H_5 \qquad (156)$$

(Ref. 186)

pathway[187]. Even primary alkyl carbinamines lead to diazoalkanes under aprotic deamination conditions[188]. However, through the use of deuterated derivatives and solvents, it has been shown that diazoalkanes are not intermediates in the nitrous acid deamination of primary carbinamines in polar solvents such as water[189] and acetic acid[162, 179].

(4) Carbon rearrangements

The nitrous acid deamination of n-propylamine in acetic acid yields an acetate fraction containing far more isopropyl acetate (32%) than the acetate from the solvolysis of n-propyl tosylate in the same solvent (2·8%)[141a]. Also, the deamination of n-butylamine in acetic acid yields a butanol fraction containing 35% of 2-butanol, whereas the solvolysis of n-butyl *p*-nitrobenzenesulfonate in the same solvent was reported to give exclusively n-butyl acetate[162]. These results are reasonable in terms of the larger positive charge on the cation formed in deamination, relative to the charge on the more solvated ion formed in solvolysis.

The unassisted loss of nitrogen to form a relatively 'bare' carbonium ion with fast follow-up reactions accounts for the lack of neighboring group interactions in the nitrous acid deamination[170a, 171], the high rate of phenyl migration—faster than rotation about a carbon–carbon single bond (equation 157)[149], the lack of selectivity in the migration

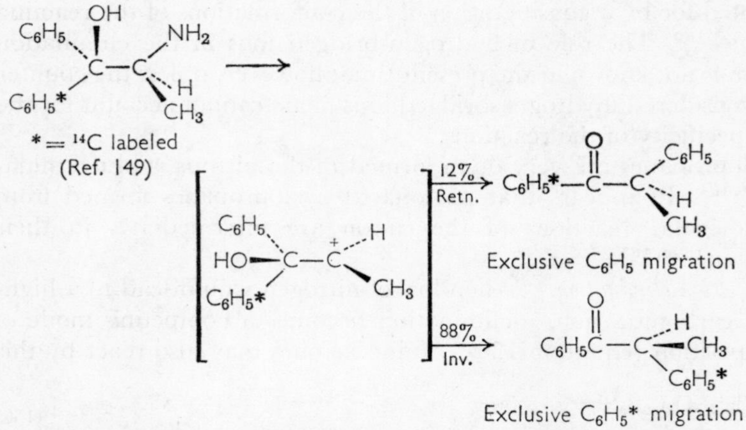

$$* = {}^{14}C \text{ labeled}$$
(Ref. 149)

(157)

Exclusive C₆H₅ migration

Exclusive C₆H₅* migration

of phenyl groups (equation 142)[170a], and the ground-state control of the group that migrates[171, 190]. For the reaction in which water is the leaving group (equation 158), the migration ratio of anisyl to phenyl

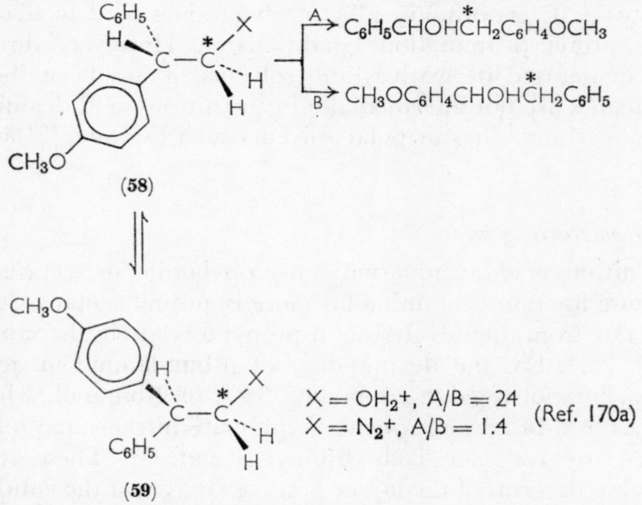

(158)

$X = OH_2^+, A/B = 24$
$X = N_2^+, A/B = 1.4$ (Ref. 170a)

is 24, indicating participation by the aromatic rings in the transition state. For the reaction in which nitrogen is the leaving group, the ratio is 1·4, indicating a very low degree of participation in the transition state for the loss of nitrogen. The low ratio is consistent with the view that the transition-state energy is low. Under these conditions, the migration ratio is determined largely by the distribution of isomers in the ground state[171, 190]; in the example of equation (158), with

$X = N_2{}^+$, conformations **58** and **59** should be approximately equally populated.

The highly stereospecific reactions of cyclohexylamines (e.g. equation 159)[191], are accounted for by the maintenance of gross stereochemistry by the counter ion (similar to the interpretation given in section (3)—path (a); see section (1)). By this means, the *trans*-coplanar arrangement of groups favorable for rearrangement on stereoelectronic grounds[192] is obtained.

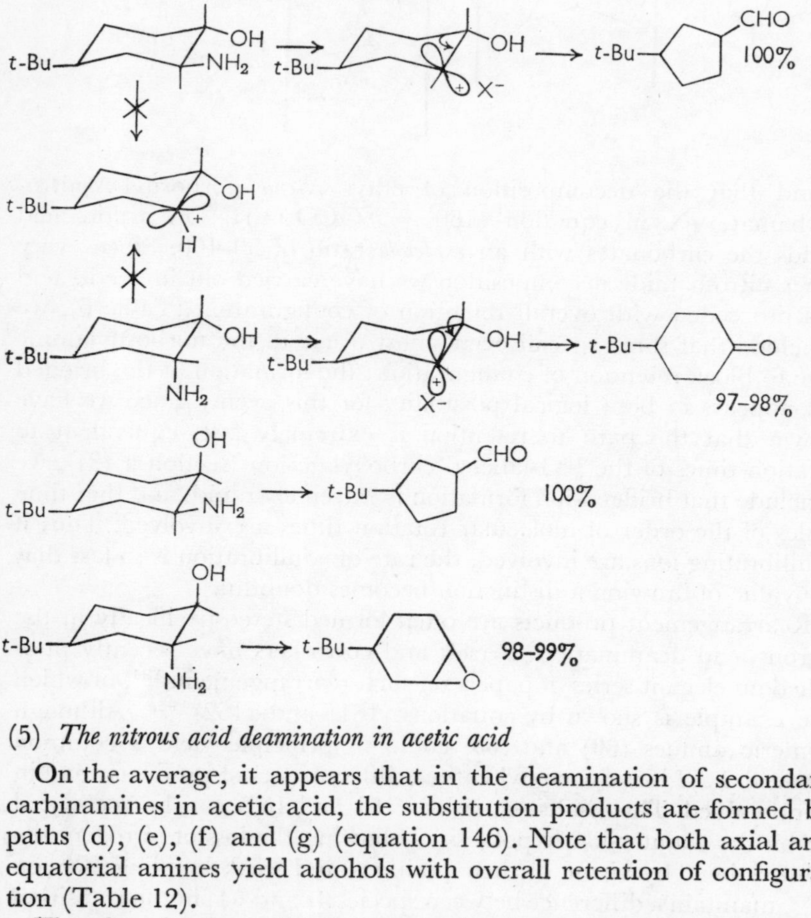

(159)

(5) *The nitrous acid deamination in acetic acid*

On the average, it appears that in the deamination of secondary carbinamines in acetic acid, the substitution products are formed by paths (d), (e), (f) and (g) (equation 146). Note that both axial and equatorial amines yield alcohols with overall retention of configuration (Table 12).

The deamination of *endo*-norbornylamine yields 87% norbornyl acetates (~99% *exo*), 10% norbornanols (98% *exo*) and 2% of the norbornyl nitrates[174]. The *exo* acetate was 89% racemic; this fact

16*

and the large *exo/endo* ratio indicates the involvement of the bridged ion (or of rapidly equilibrating ions) (equation 160)[193]. We have

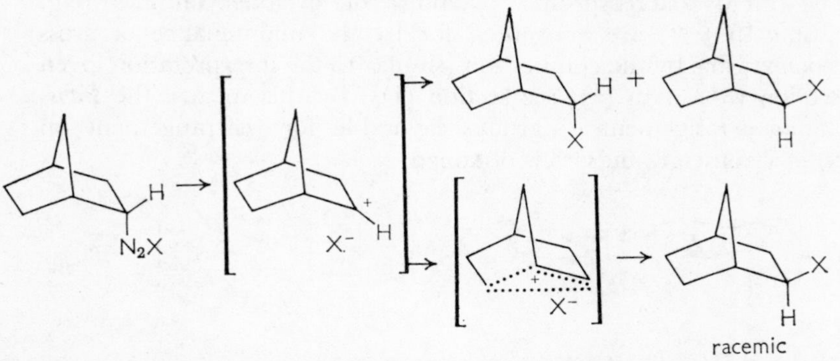

(160)

racemic

found that the decomposition of ethyl *N-endo*-norbornyl-*N*-nitro-carbamate (X in equation (160) = $^-OCO_2C_2H_5$) in acetic acid yields the carbonates with an *exo/endo* ratio of 50–100. Since every other nitrosoamide decomposition we have carried out in acetic acid has proceeded with overall retention of configuration (Table 6), we conclude that some special event must occur in the norbornylamine case to block retention of configuration; the formation of the bridged ion appears to be a logical possibility for this event. Since we have shown that the path to retention is extremely fast, equivalent to rotation times of the ^{18}O-labeled carboxylate ion (section a (3)), we conclude that bridged-ion formation is extremely rapid, and that time scales of the order of molecular rotation times are involved. Thus if equilibrating ions are involved, the rate of equilibration is so fast that the value of drawing a distinction becomes doubtful.

Rearrangement products are often formed stereospecifically in the nitrous acid deamination. Berson and coworkers have recently published an elegant series of papers on such rearrangements[194], of which one example is shown by equations (161) and (162)[195a]. Although isomeric amines (**60**) and (**63**) could in principle yield a common classical 2,2,2-bicyclooctenyl cation, different sets of products were in fact obtained. The results (equations 161 and 162) can be accounted for by the counter-ion hypothesis. Following the loss of nitrogen, the counter ions, by virtue of their positions and weak bonding (section (1)), maintain a difference between species **61** and **64**; in the transition states for rearrangement, the interaction with a neighboring carbon occurs on the opposite side of the electron-deficient carbon relative to the anion (see also sections (3) path (a) and (4)). Skeletal rearrangements can be fast relative to reactions between ions (nor-bornylamine and nitrosoamide results and also equation 159), and most of the product probably comes from bridged ions **62** and **65**. The deamination of **60** also led to the formation of a fair amount (~30%) of the product set obtained from amine **63**, this crossover probably occurs on the way to **61** by a process related to the intra-molecular inversion observed in simpler systems (path g).

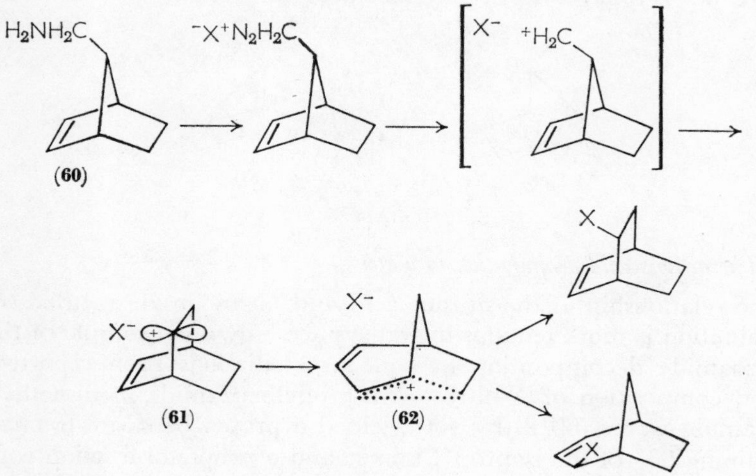

(161)

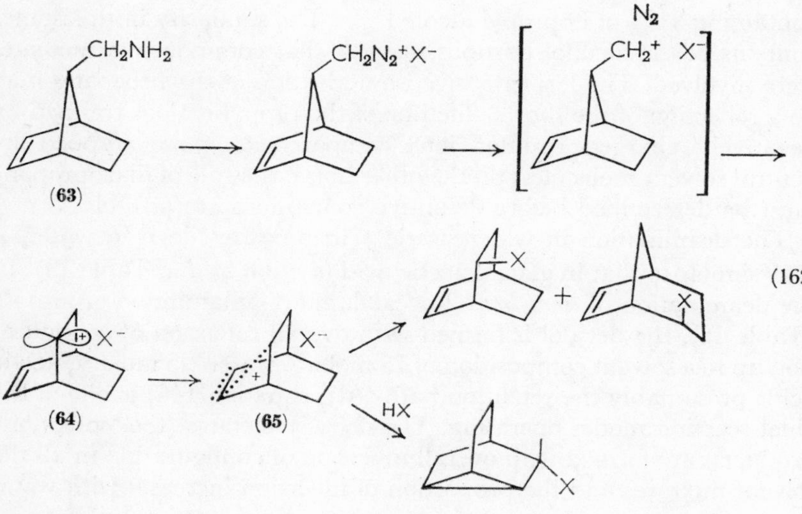

(162)

Berson and Gajewski[195a] have given an explanation of their experimental results in which the products of the reaction are determined by the formation of the 'deformationally isomeric' ions **66** and **67**. As an alternative, they propose that the different results in the two systems may be 'attributed to transitory differences in the local environment of the cations rather than to two structurally different cations', an explanation that encompasses the counter-ion mechanism given above. The reader is referred to the original work for details.

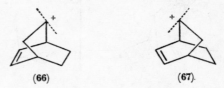

(6) *The nitrous acid deamination in water*

The relationship of the nitrous acid and nitrosoamide methods of deamination is more tenuous in water since only one example of the nitrosoamide decomposition in aqueous media has been reported. The decomposition of N-nitroso-N-n-propylbenzamide in dimethylformamide–water (60/40 by vol) yielded a propyl benzoate fraction containing 9% of the isopropyl isomer and a propanol fraction containing 33% of isopropyl alcohol[141a]. The nitrous acid deamination of n-propylamine under the same conditions yielded an alcohol fraction containing 31% of isopropyl alcohol[141a]. The similarity in the isomer contents of the alcohol portions suggests that common intermediates were involved. The less extensive isomerization of the benzoates may be a reflection of the shorter lifetime of the propyl cations trapped by negatively charged counter ions compared to those trapped by neutral solvent molecules; on the other hand, the role of diazopropane must be determined before definitive conclusions are possible.

The deamination in water–acetic acid mixtures, lean in water, is very similar to that in glacial acetic acid (section 5) and Table 13). In the deamination of *trans-trans*-2-decalylamine (axial amino group)[160] (Table 13), the decalol is formed with overall retention of configuration up to a solvent composition of 75 mole % water–25 mole % acetic acid; presumably the retention path ((d), equation 146) is one of the chief reaction modes operating. The decalyl acetates (the 'solvolysis' products) are formed with overall inversion of configuration in all the solvent mixtures and the proportion of inversion increases with water

TABLE 13.   Deamination of the 9,10-*trans*-2-decalylamines.

Mole % acetic acid (+ water)	100	75	50	25	3·4
A. *trans-trans*[a]-2-Decalylamine		Yields of products			
trans-2-Decalyl acetate	32	29	25	17	7
cis-2-Decalyl acetate	39	35	29	24	26
trans-2-Decalol	27	31	33	32	16
cis-2-Decalol	2	5	13	27	51
B. *trans-cis*[a]-2-Decalylamine					
trans-2-Decalyl acetate	18	15	11	6	3
cis-2-Decalyl acetate	55	48	46	37	26
trans-2-Decalol	1	3	3	4	3
cis-2-Decalol	26	36	40	53	68

[a] Refers to the hydrogen atom in the 2-position.

content. Note that the solvolysis product from nitrosoamide reactions in acetic acid is always formed with more inversion of configuration than the intramolecular product (Table 7). In these cases, the net extra inversion is probably the result of capture of the cation by the solvent cage with predominant, but not complete, inversion of configuration (path (f)).

In the water-rich 3·4 mole % acetic acid mixture, (1) the alcohol is formed with net inversion, (2) 77% of the substitution products are formed with inversion of configuration and (3) most of the product (67%) is now alcohol. In highly polar media, the force between the cation and counter ion (or solvated counter ion) may be so weakened, that collapse to the counter-ion product (paths (d) and (g)) is not inevitable and reactions with molecules of the solvent cage can lead to 'cooler' solvated ion pairs **68** and solvent-separated ion pairs **69** of

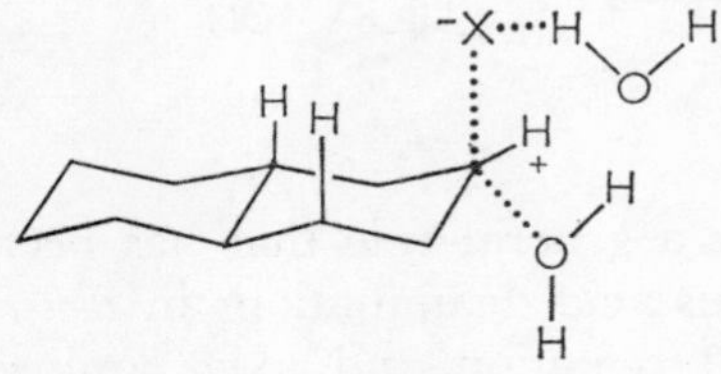

(68) 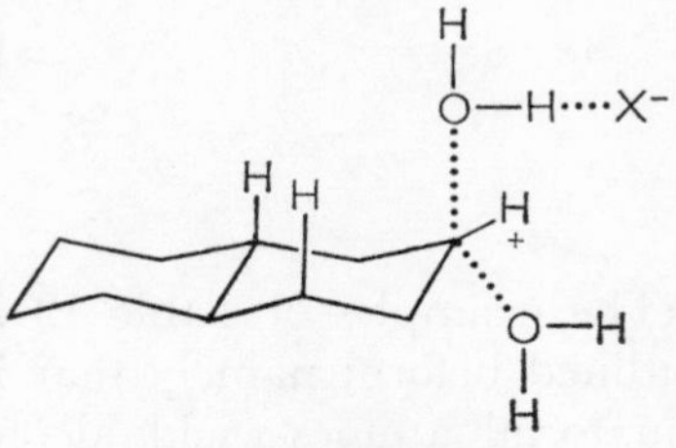   (69)

the type proposed for solvolytic reactions[195b]. Under these circumstances, the competition between inversion and retention modes of decomposition will lead to a dominance of inversion, since non-bonded interactions of the solvent with the axial hydrogens at positions

3 and 5 and the ring itself will raise the energy of bonding from the axial direction over that from the equatorial direction[182].

The deamination of *trans-cis*-2-decalylamine[173] (equatorial amino group) leads to approximately the same ratio of acetate to alcohol as observed in the axial case (Table 13). Over the entire range of water concentrations, in contrast, both the acetate and alcohol are formed with overall retention of configuration—with the proportion of alcohol increasing with the water content of the medium. Presumably in water-lean mixtures, paths (d), (e), (f) and (g) are dominant, and in water-rich mixtures, the mechanism outlined in the axial case is operating, except that the 3,5-axial interactions now lead to overall retention of configuration. It appears that as the solvent becomes richer in water, the alcohol distributions for the *trans* and *cis* amines are approaching a common value, suggesting that a common intermediate is formed which is sufficiently long lived to reach an equilibrium distribution of the solvating molecules.

The substitution reaction is still a fast process in water since it has been found in the deamination of a monocyclic carbinamine that substitution is faster than chair–chair interconversions. It was shown that the deamination of *cis*-2-deuterocyclohexylamine yielded over 94% of *cis*-2-deuterocyclohexanol (equation 163)[196].

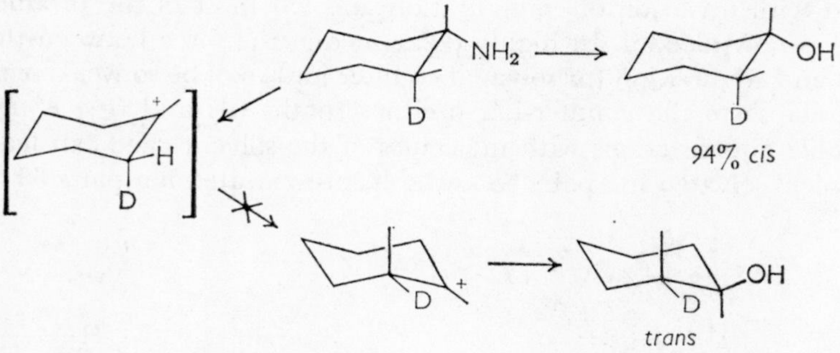

$$94\% \; cis \qquad (163)$$

trans

The examples of Table 13 illustrate a general rule that has been outlined before; namely that in nitrous acid deaminations in *water*, equatorial amines yield alcohols with retention and axial amines yield alcohols with predominant inversion of configuration[181, 182]. The interpretation of this rule by Dauben and coworkers[182] is related to that outlined above. Recent work indicates that equatorial amines also yield some alcohol with *inversion* of configuration[131, 180], but the rule still holds for the *predominant* mode of decomposition (e → e; a → e).

(7) *Related methods*

Nitrosyl chloride has occasionally been used in the deamination of aliphatic amines[197]; the chloride is often a chief product of the reaction, although alcohols have also been isolated (equation 164). The formation of the nitrosoamine has been demonstrated in the deamination of methylamine at low temperatures[159].

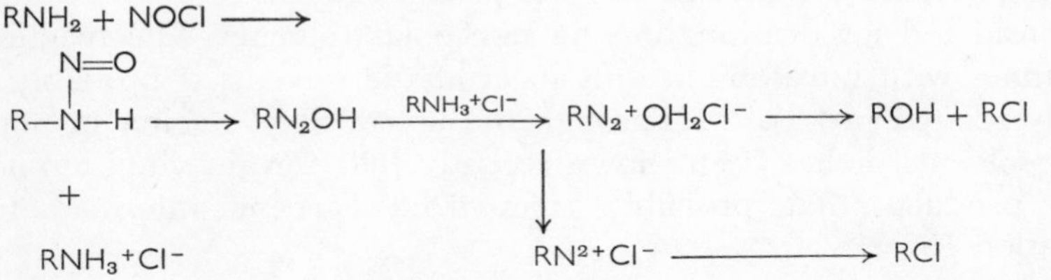

$$2\ RNH_2 + NOCl \longrightarrow$$

$$\underset{\underset{R-N-H}{\overset{N=O}{|}}}{} \longrightarrow RN_2OH \xrightarrow{RNH_3^+Cl^-} RN_2^+OH_2Cl^- \longrightarrow ROH + RCl \quad (164)$$

$$+$$

$$RNH_3^+Cl^- \qquad\qquad\qquad RN^{2+}Cl^- \longrightarrow RCl$$

The reaction of nitrosyl chloride with n-butylamine in methylene dichloride yields n-butyl chloride free of isomers[198] and the reaction with 1-phenylethylamine in dioxane yields 1-phenylethyl chloride with overall retention of configuration[199]. These results are similar to those obtained in the nitrosoamide decomposition in non-polar solvents, and a similarity in mechanisms seems certain. The reaction of nitrosyl chloride with various amines at $-75°$ in methylene dichloride was shown to yield both the alcohol and the chloride (Table 14). 1-Phenylethylamine yielded a larger fraction of alcohol than did

TABLE 14. Products from the reaction of aliphatic amines with nitrosyl chloride at $-75°$[200].

Amines (RNH₂)	Products (%)		
R	ROH	RCl	Olefins
n-Butyl	7	42	
1-Phenylethyl	20	30	
t-Butyl	—	—	64

butylamine; possibly with a decrease in the lifetime of the diazonium hydroxide (50, 51; equation 146), the exchange reaction (path (e)) has less of a chance to compete with the intramolecular modes of decomposition (paths (d) and (g)). Under the conditions of the experiment, a number of side-reactions are occurring since only about one-third of the theoretical amount of nitrogen is evolved at $-75°$, with another third being obtained in warming the reaction mixture to room temperature. An intractable mixture of nitrogen-containing compounds is also obtained*, presumably formed in part via symmetrical triazenes (RN=N—NHR).

The deamination of aliphatic amines is occasionally effected through reaction of the amine with an alkyl nitrite in a non-polar solvent[188, 201, 202]. Primary carbinamines apparently react in this system via diazoalkanes[188b] (see section (3), path (b)). Triazenes (RN$_2$NHR) and free radicals are also possible intermediates that must be considered for deaminations in non-polar solvents. The reaction of amines with nitrosonium salts in aromatic solvents is probably a closely related process[203]. Through the use of alkyl nitrites in non-polar solvents, it has been shown recently that certain vinyl amines yield products that probably stem from carbene intermediates (equation 165)[202].

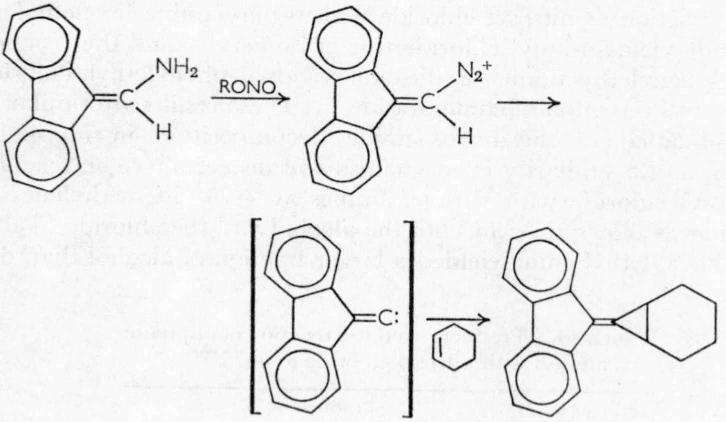

(165)

Lastly, an interesting method of deamination under basic conditions has been reported recently by Moss[204] (equation 166).

---

* Preliminary evidence indicates the following structures for compounds obtained from n-butylamine and tritylamine:

$$C_4H_9N(NO)CH(C_3H_7)N(NO)C_4H_9, \quad Ph_3CNHCPh_3,$$

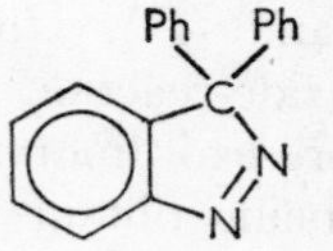

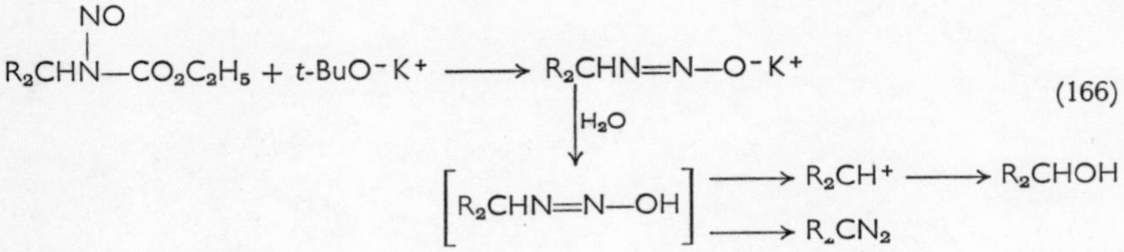

$$R_2CHN\overset{\displaystyle NO}{|}{-}CO_2C_2H_5 + t\text{-}BuO^-K^+ \longrightarrow R_2CHN{=}N{-}O^-K^+ \tag{166}$$

$$\downarrow H_2O$$

$$\left[R_2CHN{=}N{-}OH\right] \begin{array}{l} \longrightarrow R_2CH^+ \longrightarrow R_2CHOH \\ \longrightarrow R_*CN_2 \end{array}$$

2. Reactions in which nitrogen is not the leaving group

a. *Von Braun reaction of amides.* The replacement of the amino group of primary aliphatic amines by a halogen may be brought about in good yield by the von Braun reaction of the N-substituted benzamide derivatives[205, 206]. Treatment of the benzamide with phosphorus pentahalide gives the halide and benzonitrile (equation 167). Thionyl chloride has been employed successfully in the place of phosphorus pentahalide[207].

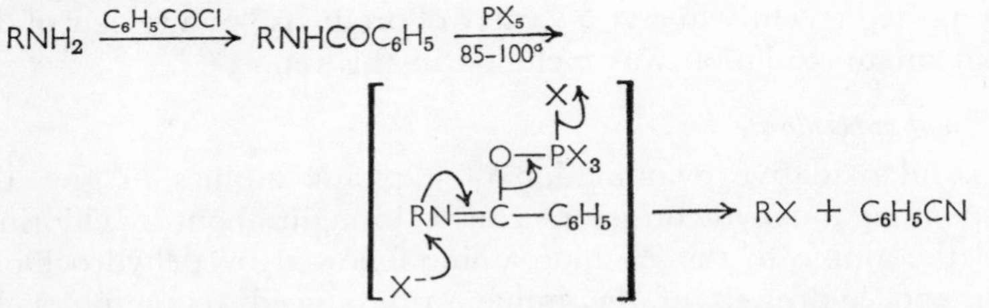

$$RNH_2 \xrightarrow{C_6H_5COCl} RNHCOC_6H_5 \xrightarrow[85-100°]{PX_5}$$

$$\left[\begin{array}{c} X \\ \overset{\displaystyle\vert}{N} \\ O{-}PX_3 \\ RN{=}C{-}C_6H_5 \\ X^- \end{array} \right] \longrightarrow RX + C_6H_5CN \tag{167}$$

b. *Pyrolysis of amides.* The pyrolysis of N-alkylamides can lead to C—N bond cleavage with formation of an alkene and the parent

$$RCH_2CH_2NHCOCH_3 \xrightarrow{\Delta} RCH_2{=}CH_2 + CH_3CONH_2 \tag{168}$$

amide (equation 168)[208]. N-t-Alkylamides and N-alkylanilides are found to pyrolyze with the same ease as many esters, though in general a higher temperature (> 500°) is required for pyrolysis and low yields (with charring) are obtained. The acid-catalyzed pyrolysis of amides has been observed[209] and N-alkyl acetamides form olefins when boiled with 15% hydrochloric acid[210].

c. *Oxidation methods.*

(1) *Oxidation with permanganate*

The oxidation of primary carbinamines with basic potassium permanganate yields the corresponding carboxylic acid (equation 169).

$$RCH_2NH_2 \longrightarrow RCO_2H + NH_3 \tag{169}$$

The reaction has been used in a degradation scheme for the location of ^{14}C in cyclic compounds[211]. At the 1,5-diaminopentane stage, a

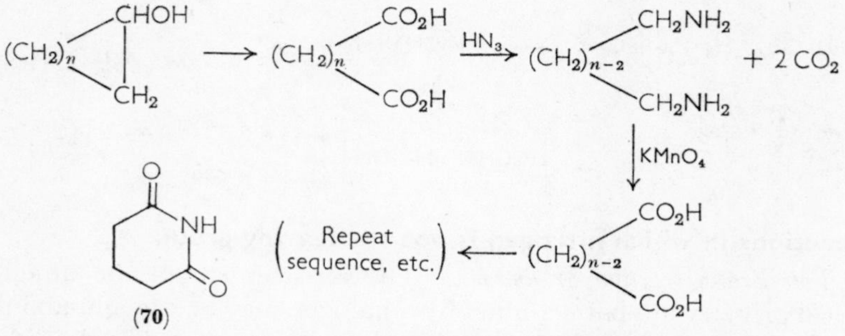

$$(CH_2)_n \overset{CHOH}{\underset{CH_2}{\Big\langle}} \longrightarrow (CH_2)_n \overset{CO_2H}{\underset{CO_2H}{\Big\langle}} \xrightarrow{HN_3} (CH_2)_{n-2} \overset{CH_2NH_2}{\underset{CH_2NH_2}{\Big\langle}} + 2\,CO_2$$

$$\Big\downarrow KMnO_4 \qquad\qquad (170)$$

(70) $\left(\begin{array}{c}\text{Repeat}\\\text{sequence, etc.}\end{array}\right) \longleftarrow (CH_2)_{n-2} \overset{CO_2H}{\underset{CO_2H}{\Big\langle}}$

neighboring group interaction occurred to give glutarimide (70)[211]; hydrolysis in the presence of nitrous acid yielded glutaric acid which was then treated with hydrazoic acid to continue the degradation. A similar scheme has also been used for radioactive 1,3-diaminocyclo-pentane[140b]. The oxidation of amines with neutral permanganate has been reported recently to give a variety of products[212]; a review of the permanganate oxidation was included in this report[212].

(2) t-Butyl hypochlorite

A useful oxidative deamination of aliphatic amines to give the corresponding aldehyde or ketone can be brought about by chlorination of the amine to the N-chloroamine followed by dehydrochlorination and hydrolysis of the imine so produced (equation 171).

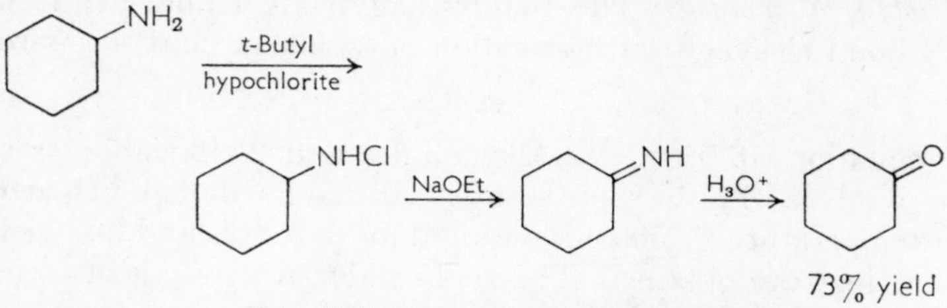

$$\text{cyclohexyl-}NH_2 \xrightarrow[\text{hypochlorite}]{t\text{-Butyl}}$$

$$\text{cyclohexyl-}NHCl \xrightarrow{NaOEt} \text{cyclohexyl=}NH \xrightarrow{H_3O^+} \text{cyclohexanone} \qquad (171)$$

73% yield

$$C_6H_5CH_2NH_2 \longrightarrow C_6H_5CHO \quad 80\% \text{ yield} \qquad (172)$$

Hypochlorous acid has been used to effect the chlorination[213], but t-butyl hypochlorite has been found to be a more convenient reagent[214].

(3) Photochemical oxidation

The deamination of cyclohexylamine has been brought about photochemically using benzophenone as sensitizer[215]. Cyclohexanone was produced in 80–90% yields (equation 173). The reaction is also successful with secondary amines.

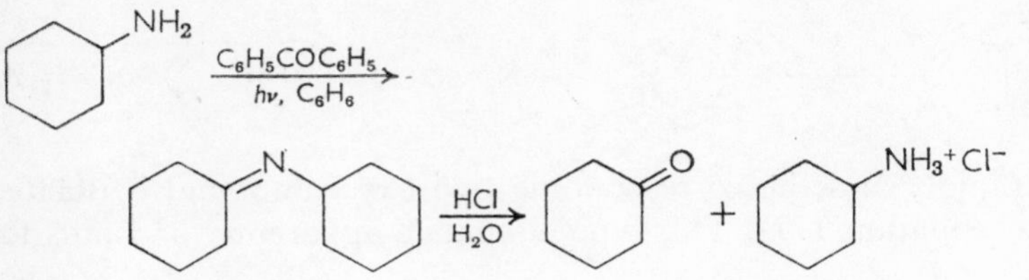

$$(173)$$

(4) Transamination

Transamination represents a potentially useful method for the synthesis of amines—and also for the cleavage of C—N bonds with a transfer of asymmetry (equation 174). Snell has reported the synthesis

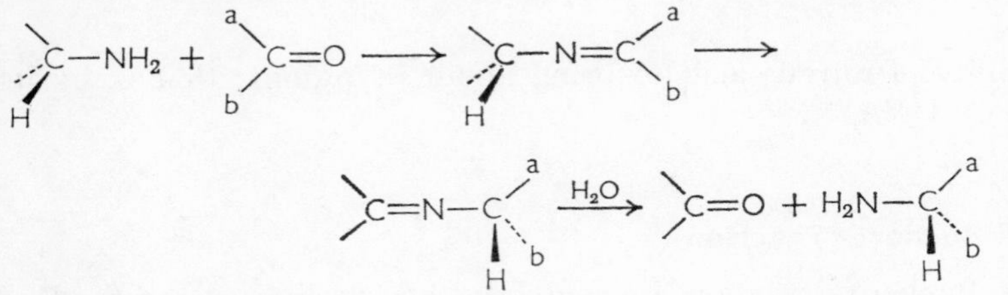

$$(174)$$

of glutamic acid containing an excess of the L isomer from the reaction of ketoglutaric acid with L-alanine, catalyzed by the copper salt of pyridoxal[216]. A discussion of conformational factors in this reaction has appeared recently[217]. The biological applications of transamination are covered in chapter 9 of this volume.

B. Aromatic and Heterocyclic Amines

1. Reactions involving diazonium salts

Treatment of a primary amine with aqueous nitrous acid almost invariably gives solutions of more or less stable diazonium salts (equation 175)[218]. The decomposition of the diazonium salt with loss

$$ArNH_2 \xrightarrow[HCl]{NaNO_2} ArN_2^+Cl^- \qquad (175)$$

of nitrogen, which can be brought about thermally or catalytically in the presence of cuprous salts or finely divided copper, gives high yields of substituted aromatic derivatives and the reaction has found much use in organic synthesis. This subject will be covered in detail in another volume of this series.

Heterocyclic aromatic amines also react to form diazonium salts (equation 176). With many heterocyclic amines, the substitution

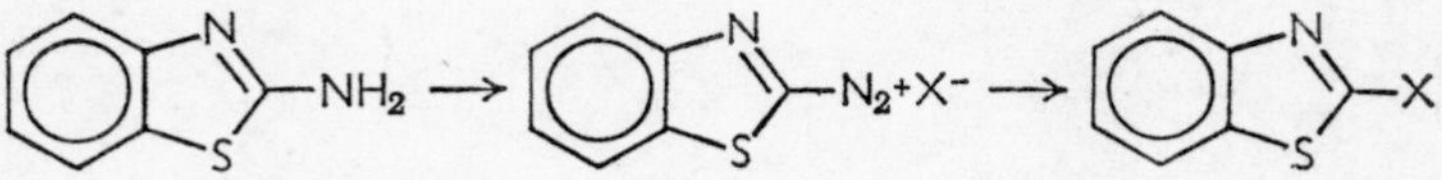

$$\text{(176)}$$

occurs rapidly and the corresponding hydroxy compound is isolated directly (equation 177). This type of change apparently accounts for

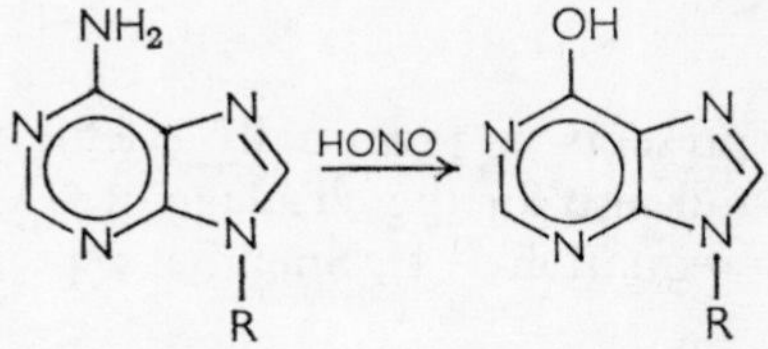

$$\text{(177)}$$

the ability of nitrous acid to bring about mutations; that is, by its action on DNA[219].

2. The Bucherer reaction

The Bucherer reaction is a useful, though limited, method of deamination[220]. As an example, treatment of naphthylamines with aqueous sodium bisulfite at 160° followed by sodium hydroxide at 100° gives naphthols (equation 178)[220]. The reaction is reversible

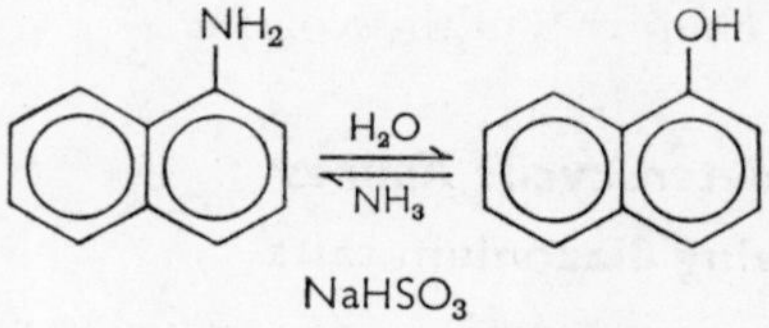

$$\text{(178)}$$

and limited in its application to naphthalene derivatives, resorcinol and compounds of this type. Secondary and tertiary naphthylamines also react to give naphthols.

The mechanism of the reaction has been investigated recently by Seeboth[221] who has isolated 1-tetralone-3-sulfonic acid by the reaction of both 1-naphthylamine and 1-naphthol with excess sodium bisulfite (equation 179); **72** was stable in aqueous media and could be chlorinated or brominated in the 2-position before hydrolysis to the corresponding naphthol; **71** was isolated only from nonaqueous media. Treatment of the 1-tetralone-3-sulfonic acid with arylamines gave good yields of arylnaphthylamines (equation 180). The reaction is

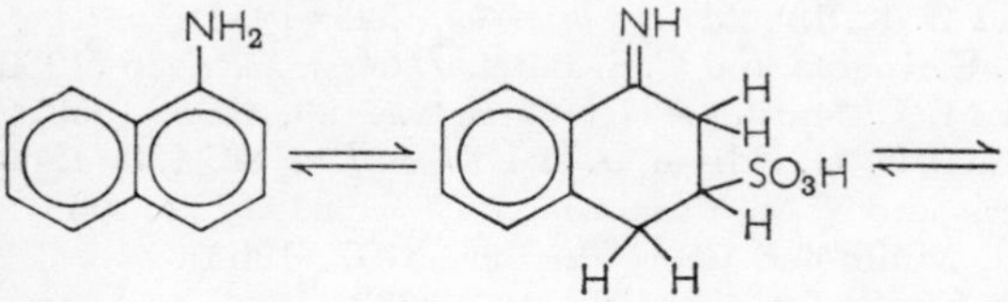

(71)

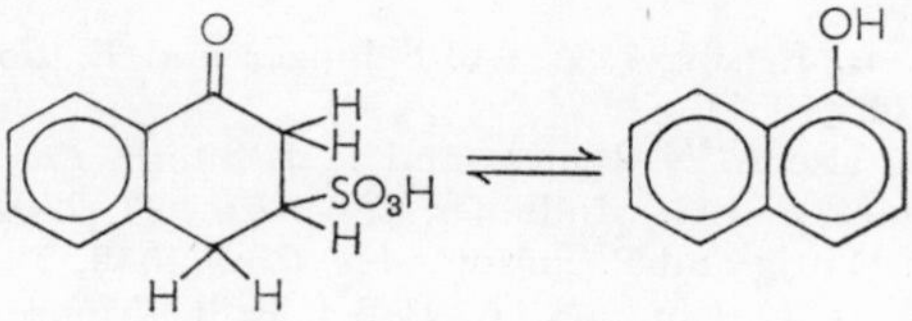

(179)

(72)

not successful with phenol itself, presumably because of the loss of resonance energy in the addition of bisulfite ion.

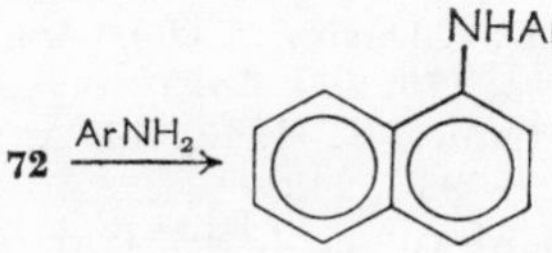

VIII. REFERENCES

1. (a) A. W. Hofmann, *Ann. Chem.*, **78**, 253 (1851); *Ann. Chem.*, **79**, 11 (1851).
   (b) A. W. Hofmann, *Chem. Ber.*, **14**, 494, 695 (1881).
   (c) L. D. Freedman, *J. Chem. Ed.*, **43**, 662 (1966).
2. See for example, D. J. Cram and G. S. Hammond, *Organic Chemistry*, 2nd ed., McGraw-Hill Book Co., Inc., New York, 1964, p. 334.
3. For a review of the method of exhaustive methylation and numerous examples of its use in alkaloid degradation, see A. C. Cope and E. R. Trumbull, *Org. Reactions*, **11**, 317 (1960).
4. (a) W. Hanart and C. K. Ingold, *J. Chem. Soc.*, 997 (1927).
   (b) J. von Braun, W. Teuffert and K. Weissbach, *Ann. Chem.*, **472**, 121 (1929).
5. (a) P. A. Waitkus, L. I. Peterson and G. W. Griffin, *J. Am. Chem. Soc.*, **88**, 181 (1966).
   (b) H. E. Winberg and F. S. Fawcett, *Org. Syn.*, **42**, 83 (1962).
6. (a) C. K. Ingold, *Proc. Chem. Soc.*, 265 (1962).
   (b) J. F. Bunnett, *Angew. Chem., Intern. Ed., Engl.*, **1**, 225 (1962).
   (c) D. V. Banthorpe, 'Elimination Reactions', in *Reaction Mechanisms in Organic Chemistry*, Vol. 2, (Ed. E. D. Hughes), Elsevier Publishing Co., 1963.
   (d) W. H. Saunders, in *The Chemistry of the Alkenes* (Ed. S. Patai), Interscience Publishers, 1963, p. 149.

7. (a) E. D. Hughes and C. K. Ingold, *J. Chem. Soc.*, 523 (1933).
   (b) E. D. Hughes, C. K. Ingold and C. S. Patel, *J. Chem. Soc.*, 526 (1933).
8. (a) V. J. Shiner and M. L. Smith, *J. Am. Chem. Soc.*, **80**, 4095 (1958).
   (b) W. H. Saunders and D. H. Edison, *J. Am. Chem. Soc.*, **82**, 138 (1960).
9. G. Ayrey, A. N. Bourns and V. A. Vyas, *Can. J. Chem.*, **41**, 1759 (1963).
10. (a) H. Simon and G. Müllhofer, *Chem. Ber.*, **96**, 3167 (1963).
    (b) H. Simon and G. Müllhofer, *Chem. Ber.*, **97**, 2202 (1964).
11. D. J. Cram, F. D. Greene and C. H. DePuy, *J. Am. Chem. Soc.*, **78**, 790 (1956).
12. R. L. Letsinger, A. W. Schnizer and E. Bobko, *J. Am. Chem. Soc.*, **73**, 5708 (1951).
13. See also A. N. Bourns and P. J. Smith, *Proc. Chem. Soc.*, 366 (1964); H. Simon and G. Müllhofer, *Chem. Ber.*, **97**, 2202 (1964).
14. G. Wittig and P. Polster, *Ann. Chem.*, **559**, 1 (1956).
15. (a) A. C. Cope, N. A. LeBel, P. T. Moore and W. R. Moore, *J. Am. Chem. Soc.*, **83**, 3861 (1961).
    (b) F. Weygand, H. Daniel and H. Simon, *Ann. Chem.*, **654**, 111 (1962).
16. A. C. Cope and A. S. Mehta, *J. Am. Chem. Soc.*, **85**, 1949 (1963).
17. See ref. 24a.
18. (a) S. J. Cristol and F. R. Stermitz, *J. Am. Chem. Soc.*, **82**, 4692 (1960).
    (b) S. J. Cristol and D. I. Davies, *J. Org. Chem.*, **27**, 293 (1962).
19. A. Saytzeff, *Ann. Chem.*, **179**, 296 (1875).
20. M. L. Dhar, E. D. Hughes, C. K. Ingold, A. M. M. Mandour and L. I. Woolfe, *J. Chem. Soc.*, 2093 (1948).
21. H. C. Brown and I. Moritani, *J. Am. Chem. Soc.*, **78**, 2203 (1956), and preceding papers.
22. D. V. Banthorpe, E. D. Hughes and C. K. Ingold, *J. Chem. Soc.*, 4054 (1960).
23. W. H. Jones, *Science*, **118**, 387 (1953).
24. (a) A. C. Cope and D. L. Ross, *J. Am. Chem. Soc.*, **83**, 3854 (1961).
    (b) P. A. S. Smith and S. Frank, *J. Am. Chem. Soc.*, **74**, 509 (1952).
    (c) E. D. Hughes and J. Wiley, *J. Chem. Soc.*, 4094 (1960); T. H. Brownlee and W. H. Saunders, *Proc. Chem. Soc.*, 314 (1961); R. Ledger, A. J. Smith and J. McKenna, *Tetrahedron*, **20**, 2413 (1964); J. McKenna and J. B. Slinger, *J. Chem. Soc.*, 2759 (1958).
25. H. R. Snyder and J. H. Brewster, *J. Am. Chem. Soc.*, **71**, 291 (1949).
26. D. A. Archer and H. Booth, *J. Chem. Soc.*, 322 (1963).
27. E. D. Hughes and C. K. Ingold, *J. Chem. Soc.*, 69 (1933).
28. (a) J. H. Brewster and E. L. Eliel, *Org. Reactions*, **7**, 99 (1953).
    (b) H. Helman, *Angew Chem.*, **65**, 473 (1953).
    (c) H. Helman and G. Opitz, *α-Aminoalkyleriung*, Verlag Chemie, Weinheim, 1960.
    (d) G. Wittig, M. Heintzler and M.-H. Wetterling, *Ann. Chem.*, **557**, 201 (1947).
29. (a) C. L. Bumgardener, *J. Am. Chem. Soc.*, **83**, 4420 (1961).
    (b) H. Rinderknecht and C. Niemann, *J. Am. Chem. Soc.*, **73**, 4259 (1951).
    (c) J. Weinstock, *J. Org. Chem.*, **21**, 540 (1956).
    (d) M. A. T. Rogers, *J. Org. Chem.*, **22**, 350 (1957).
30. P. Rabe and J. Hellensleben, *Chem. Ber.*, **43**, 884 (1910).

31. (a) T. S. Stevens, E. M. Creighton, A. B. Gordon and M. MacNicol, *J. Chem. Soc.*, 3193 (1928).
    (b) G. Wittig, R. Mangold and G. Felletschin, *Ann. Chem.*, **560**, 117 (1948).
32. (a) T. S. Stevens, *J. Chem. Soc.*, 2107 (1930).
    (b) T. Thomas and T. S. Stevens, *J. Chem. Soc.*, 55 (1932).
    (c) T. Thomas and T. S. Stevens, *J. Chem. Soc.*, 1932 (1932).
    (d) R. A. W. Johnson and T. S. Stevens, *J. Chem. Soc.*, 4487 (1955).
    (e) J. L. Dunn and T. S. Stevens, *J. Chem. Soc.*, 1926 (1932).
33. (a) A. Campbell, A. H. J. Houston and J. Kenyon, *J. Chem. Soc.*, 93 (1947).
    (b) J. H. Brewster and M. W. Kline, *J. Am. Chem. Soc.*, **74**, 5179 (1952).
34. C. R. Hauser and S. W. Kantor, *J. Am. Chem. Soc.*, **73**, 1437 (1951).
35. (a) E. E. Jenny and J. Druey, *Angew. Chem. Intern. Ed. Engl.*, **1**, 155 (1962).
    (b) See also reference 53 (c) for evidence for free radicals.
36. (a) W. F. Cockburn, R. A. W. Johnson and T. S. Stevens, *J. Chem. Soc.*, 3340 (1960).
    (b) R. A. W. Johnson and T. S. Stevens, *J. Chem. Soc.*, 3346 (1960).
37. R. K. Hill and T. H. Chan, *J. Am. Chem. Soc.*, **88**, 866 (1966).
38. M. Sommelet, *Compt. Rend.*, **205**, 56 (1937).
39. (a) S. W. Kantor and C. R. Hauser, *J. Am. Chem. Soc.*, **73**, 4122 (1951).
    (b) C. R. Hauser and D. N. van Eenam, *J. Am. Chem. Soc.*, **79**, 5513 (1957).
    (c) W. H. Puterbauch and C. R. Hauser, *J. Am. Chem. Soc.*, **86**, 1105 (1964).
    (d) C. R. Hauser and D. N. van Eenam, *J. Org. Chem.*, **23**, 865 (1958).
    (e) W. Q. Beard and C. R. Hauser, *J. Org. Chem.*, **25**, 334 (1961); **26**, 371 (1961).
40. G. Wittig, H. Tenhaeff, W. Schoch and G. Koenig, *Ann. Chem.*, **572**, 1 (1951).
41. See H. E. Zimmerman in *Molecular Rearrangements*, Vol. 1 (Ed. P. de Mayo), John Wiley and Sons, Inc., New York, 1963, p. 387.
42. C. L. Bumgardener, *J. Am. Chem. Soc.*, **85**, 73 (1963).
43. See D. J. Cram, *Fundamentals of Carbanion Chemistry*, Academic Press, New York, 1965, p. 229.
44. The pyrolysis of amine oxides is reviewed in ref. 3.
45. W. Wernick and R. Wolffenstein, *Chem. Ber.*, **31**, 1553 (1898).
46. A. C. Cope, T. T. Foster and P. H. Towle, *J. Am. Chem. Soc.*, **71**, 3929 (1949).
47. R. B. Woodward, T. Fukunaga and R. C. Kelly, *J. Am. Chem. Soc.*, **86**, 3162 (1964).
48. M. A. T. Rogers, *J. Chem. Soc.*, 769 (1955).
49. D. J. Cram and J. E. McCarty, *J. Am. Chem. Soc.*, **76**, 5740 (1954).
50. A. C. Cope, N. A. LeBel, H.-H. Lee and W. R. Moore, *J. Am. Chem. Soc.*, **79**, 4720 (1957).
51. A. C. Cope and N. A. LeBel, *J. Am. Chem. Soc.*, **82**, 4656 (1960).
52. (a) D. J. Cram, M. R. V. Sahyun and G. R. Knox, *J. Am. Chem. Soc.*, **84**, 1735 (1962).
    (b) M. R. V. Sahyun and D. J. Cram, *J. Am. Chem. Soc.*, **85**, 1263 (1963).
53. (a) J. Meisenheimer, *Chem. Ber.*, **52**, 1617 (1919).

    (b) A. C. Cope and P. H. Towle, *J. Am. Chem. Soc.*, **71**, 3423 (1949).

    (c) J. I. Brauman and W. A. Sanderson, *Tetrahedron*, **23**, 37 (1967).

54. A. H. Wragg, T. S. Stevens and D. M. Ostle, *J. Chem. Soc.*, 4057 (1958).

55. (a) U. Schoellkopf, M. Patsch and H. Schaefer, *Tetrahedron Letters*, 2515 (1964).

    (b) R. B. Woodward and R. Hoffmann, *J. Am. Chem. Soc.*, **87**, 2511 (1965).

56. (a) V. J. Traynelis, A. I. Gallagher and R. F. Martello, *J. Org. Chem.*, **26**, 4365 (1961).

    (b) M. Katada, *J. Pharm. Soc. Japan*, **67**, 51 (1947); *Chem. Abstr.*, **45**, 9536 (1951).

57. (a) V. J. Traynelis and A. I. Gallagher, *J. Am. Chem. Soc.*, **87**, 5710 (1965), and preceding papers.

    (b) S. Oae, T. Kitao and Y. Kitaoka, *J. Am. Chem. Soc.*, **84**, 3362 (1962).

    (c) J. H. Margraf, H. B. Brown, S. C. Mohr and R. G. Peterson, *J. Am. Chem. Soc.*, **85**, 958 (1963).

58. M. Polonovski and M. Polonovski, *Bull. Soc. Chim. France*, **41**, 1190 (1927).

59. R. Huisgen, F. Bayerlein and W. Heykamp, *Chem. Ber.*, **92**, 3223 (1959).

60. (a) J. C. Craig, N. Y. Mary and L. Wolf, *J. Org. Chem.*, **29**, 2868 (1964).

    (b) V. Boekleheide and D. L. Harrington, *Chem. Ind. (London)*, 1423 (1955).

61. (a) M. S. Fish, N. M. Johnson and E. C. Harding, *J. Am. Chem. Soc.*, **76**, 3668 (1956).

    (b) J. C. Craig, F. P. Dwyer, A. N. Glazer and E. C. Horning, *J. Am. Chem. Soc.*, **83**, 1871 (1961).

62. J. von Braun, *Chem. Ber.*, **33**, 1438 (1900).

63. R. Scholl and W. Nör, *Chem. Ber.*, **33**, 1550 (1900).

64. H. A. Hageman, *Org. Reactions*, **7**, 198 (1953).

65. J. von Braun, *Chem. Ber.*, **42**, 2035 (1909).

66. J. von Braun, *Chem. Ber.*, **40**, 3914 (1907).

67. (a) W. Staedel, *Chem. Ber.*, **19**, 1947 (1886).

    (b) M. Tiffeneau and K. Fuhrer, *Bull. Soc. Chim. France*, **15**, 162 (1914).

    (c) G. Drefahl, H. Lückert and W. Köhler, *J. Prakt. Chem.*, **11**, 341 (1960).

68. G. E. Hein, *J. Chem. Educ.*, **40**, 181 (1963).

69. (a) E. Speyer and L. Walther, *Chem. Ber.*, **63**, 852 (1930).

    (b) E. Ochiai, *J. Pharm. Soc. Japan*, **49**, 91 (1929), through ref. 68.

    (c) N. K. Abubakirov and S. V. Yunosov, *J. Gen. Chem.*, USSR, **24**, 733 (1954).

    (d) R. C. Cookson and M. E. Trevitt, *J. Chem. Soc.*, 2689 (1956).

70. (a) P. A. S. Smith and H. G. Pars, *J. Org. Chem.*, **24**, 1325 (1959).

    (b) P. A. S. Smith, *Open-chain Nitrogen Compounds*, Vol. 1, W. A. Benjamin, Inc., New York, 1965, p. 33; P. A. S. Smith and R. N. Loepky, *J. Am. Chem. Soc.*, **89**, 1147 (1967).

71. J. Glazer, E. D. Hughes, C. K. Ingold, A. T. James, G. T. Jones and E. Roberts, *J. Chem. Soc.*, 2671 (1950).

72. E. Müller, Ed., *Houben-Weyl's Methods of Organic Chemistry*, Vol. XI/I, Georg Thieme Publishing Co., Stuttgart, 1957, pp. 976–984.
73. (a) F. W. Neumann and C. W. Gould, *Anal. Chem.*, **25**, 751 (1953).
   (b) A. T. Bottini and R. E. Olsen, *J. Org. Chem.*, **27**, 452 (1962).
74. L. Horner and W. Kirmse, *Ann. Chem.*, **597**, 48 (1956) and preceding papers.
75. S. Dunstan and H. B. Henbest, *J. Chem. Soc.*, 4905 (1957).
76. L. Horner, E. Winkelmann, K. H. Knapp and W. Ludwig, *Chem. Ber.*, **92**, 288 (1959).
77. D. H. Rosenblatt, A. J. Hayes, B. L. Harrison, R. A. Streaty and K. A. Moore, *J. Org. Chem.*, **28**, 2790 (1963); D. H. Rosenblatt, *J. Am. Chem. Soc.*, **89**, 1158, 1163 (1967).
78. (a) H. B. Henbest and M. J. Stratford, *J. Chem. Soc.*, (C), 995 (1966).
   (b) H. B. Henbest and A. Thomas, *J. Chem. Soc.*, 3032 (1957).
   (c) E. F. Curragh, H. B. Henbest and A. Thomas, *J. Chem. Soc.*, 3559 (1960).
79. R. A. Labriola, M. Ishii and I. Mariani, *Anales. Asoc. Quim. Arg.*, **33**, 156 (1945).
80. H. Shechter and S. S. Rawalay, *J. Am. Chem. Soc.*, **86**, 1706 (1964).
81. (a) A. W. Hofmann, *Proc. Roy. Soc.*, **10**, 595 (1860).
   (b) C. D. Hurd, *The Pyrolysis of Carbon Compounds*, A. C. S. Monograph 50, Chemical Catalog Co., N.Y., N.Y., 1929, pp. 310–329.
   (c) W. J. Hickinbottom, *Reactions of Organic Compounds*, 3rd ed., Longmans, Green and Co., N.Y., N.Y., 1957, p. 416.
   (d) R. A. Chambers and D. E. Pearson, *J. Org. Chem.*, **28**, 3144 (1963).
   (e) E. S. Gould, *Mechanism and Structure in Organic Chemistry*, Henry Holt and Co., N.Y., 1959, p. 652.
   (f) H. Breederveld, *Rec. Trav. Chim.*, **78**, 589 (1959).
82. J. Herzig and H. Meyer, *Chem. Ber.*, **27**, 319 (1894); *Monatsh. Chem.*, **18**, 379 (1897).
82. (a) C. Ainsworth and N. R. Easton, *J. Org. Chem.*, **27**, 4118 (1962).
   (b) M. Gorman, N. Neuss and K. Biemann, *J. Am. Chem. Soc.*, **84**, 1058 (1962).
83. (a) J. C. Craig, S. R. Johns and M. Moyle, *J. Org. Chem.*, **28**, 2779 (1963).
   (b) J. C. Craig, M. Moyle and L. F. Johnson, *J. Org. Chem.*, **29**, 410 (1964).
84. A. F. Casy and J. L. Myers, *J. Chem. Soc.*, 4639 (1964).
85. (a) C. L. Bumgardener, K. S. McCallum and J. P. Freeman, *J. Am. Chem. Soc.*, **83**, 4417 (1961).
   (b) W. Rundel and E. Müller, *Chem. Ber.*, **96**, 2528 (1963).
   (c) R. D. Clark and G. H. Helmkamp, *J. Org. Chem.*, **29**, 1316 (1964).
   (d) G. Drefall, K. Ponsold and B. Schönecker, *Chem. Ber.*, **97**, 2014 (1964).
   (e) C. S. Rondestvedt and S. J. Davis, *J. Org. Chem.*, **22**, 200 (1957).
86. (a) W. H. Hartung and R. Simonoff, *Org. Reactions*, **7**, 263 (1953).
   (b) H. J. Haas, *Chem. Ber.*, **94**, 2442 (1961).
   (c) B. Marchand, *Chem. Ber.*, **95**, 557 (1962).

87. F. Walls and J. Caballero, *Bol. Inst. Quim. Univ. Nal. Auton. Mex.*, **15**, 74 (1963); *Chem. Abstr.*, **61**, 8215 (1964).

88. (a) E. E. Van Tamelen, T. A. Spencer, D. S. Allen and R. L. Orvis, *Tetrahedron*, **14**, 8 (1961).
   (b) K. Bernauer, H. Schmid and P. Karrer, *Helv. Chim. Acta*, **40**, 731 (1957).

89. (a) H. Emde, *Arch. Pharm.*, **244**, 289 (1906); *Chem. Ber.*, **42**, 2590 (1909).
   (b) H. Emde and H. Kull, *Arch. Pharm.*, **272**, 469 (1934).
   (c) P. L. Pauson, M. A. Sandhu and W. E. Watts, *J. Chem. Soc.*, (C), 251 (1966).

90. (a) E. Grovenstein, E. P. Blanchard, D. A. Gorden and R. Stevenson, *J. Am. Chem. Soc.*, **81**, 4842 (1959).
   (b) E. Grovenstein and R. Stevenson, *J. Am. Chem. Soc.*, **81**, 4850 (1959).

91. (a) E. Grovenstein, S. Chandra, C. E. Collum and W. E. Davis, *J. Am. Chem. Soc.*, **88**, 1275 (1966).
   (b) E. Grovenstein and L. C. Rogers, *J. Am. Chem. Soc.*, **86**, 854 (1964).

92. (a) B. Emmert, *Chem. Ber.*, **42**, 1507 (1909).
   (b) B. C. Southwood, R. Osteryoung, K. D. Fleischer and F. C. Nachod, *Anal. Chem.*, **33**, 208 (1961).
   (c) J. S. Mayell and A. J. Bard, *J. Am. Chem. Soc.*, **85**, 421 (1963).
   (d) S. D. Ross, M. Finkelstein and R. C. Peterson, *J. Am. Chem. Soc.*, **82**, 1582 (1960).

93. N. J. Leonard, S. Swann and H. L. Dryden, *J. Am. Chem. Soc.*, **74**, 2871 (1952).

94. (a) N. J. Leonard, S. Swann and J. Figueras, *J. Am. Chem. Soc.*, **74**, 4620 (1952).
   (b) N. J. Leonard, S. Swann and E. H. Mottus, *J. Am. Chem. Soc.*, **74**, 6251 (1952).
   (c) N. J. Leonard, S. Swann and G. Fuller, *J. Am. Chem. Soc.*, **76**, 3193 (1954).

95. (a) C. L. Bumgardener, K. J. Martin and J. P. Freeman, *J. Am. Chem. Soc.*, **85**, 97 (1963).
   (b) C. L. Bumgardener and J. P. Freeman, *J. Am. Chem. Soc.*, **86**, 2233 (1964).

96. D. J. Cram, J. S. Bradshaw, W. Lwowski and G. R. Know, *J. Am. Chem. Soc.*, **84**, 2832 (1962).

97. (a) C. G. Overberger, J. G. Lombardino and R. G. Hiskey, *J. Am. Chem. Soc.*, **80**, 3009 (1958).
   (b) C. G. Overberger, N. P. Marullo and R. G. Hiskey, *J. Am. Chem. Soc.*, **83**, 1372 (1961).
   (c) C. G. Overberger and N. P. Marullo, *J. Am. Chem. Soc.*, **83**, 1378 (1961).
   (d) C. G. Overberger, J. G. Lombardino and R. G. Hiskey, *J. Am. Chem. Soc.*, **79**, 6430 (1957).

98. (a) A. Nickon and A. Sinz, *J. Am. Chem. Soc.*, **82**, 753 (1960).
   (b) A. Nickon and A. S. Hill, *J. Am. Chem. Soc.*, **86**, 1152 (1964).

99. D. J. Cram and J. S. Bradshaw, *J. Am. Chem. Soc.*, **85**, 1108 (1963).

100. (a) D. M. Lemal and T. W. Rave, *J. Am. Chem. Soc.*, **87**, 393 (1965).
    (b) D. M. Lemal and S. D. McGregor, *J. Am. Chem. Soc.*, **88**, 1335 (1966).

101. H. B. Henbest and M. J. W. Stratford, *J. Chem. Soc.*, 711 (1964).
102. (a) H. Gilman and J. J. Dietrich, *J. Am. Chem. Soc.*, **80**, 380 (1958).
    (b) J. J. Hlavka and P. Bitha, *Tetrahedron Letters*, 3843 (1966).
    (c) A. Padwa and L. Hamilton, *J. Am. Chem. Soc.*, **87**, 1821 (1965).
103. (a) F. C. Whitmore and D. P. Langlois, *J. Am. Chem. Soc.*, **54**, 3441 (1932).
    (b) E. H. White, *J. Am. Chem. Soc.*, **77**, 6011 (1955).
    (c) L. G. Cannell and R. W. Taft, Jr., *J. Am. Chem. Soc.*, **78**, 5812 (1956).
104. G. W. Wheland, *Advanced Organic Chemistry*, 3rd ed., John Wiley and Sons, Inc., New York, 1960; Y. Pocker in *Molecular Rearrangements*, Vol. I, (Ed. P. de Mayo), Interscience Publishers, New York, 1963.
105. The deamination of n-butylamine with nitrosyl chloride in the non-polar solvent ethyl ether apparently yielded n-butyl chloride free of isomers (P. A. S. Smith, D. R. Baer and S. N. Ege, *J. Am. Chem. Soc.*, **76**, 4564 (1954)).
106. (a) E. H. White, *J. Am. Chem. Soc.*, **76**, 4497 (1954).
    (b) **77**, 6008 (1955).
    (c) **77**, 6014 (1955).
107. E. H. White and C. A. Aufdermarsh, Jr., *J. Am. Chem. Soc.*, **83**, 1179 (1961).
108. (a) J. E. Stuber, unpublished work.
    (b) C. A. Elliger, unpublished work.
109. E. H. White and J. E. Stuber, *J. Am. Chem. Soc.*, **85**, 2168 (1963).
110. F. Alvarez, *Steroids*, **2**, 393 (1963).
111. Y. Sato and G. H. Latham, Jr., *J. Org. Chem.*, **22**, 981 (1957).
112. E. H. White and D. W. Grisley, Jr., *J. Am. Chem. Soc.*, **83**, 1191 (1961).
113. E. H. White and C. A. Aufdermarsh, Jr., *J. Am. Chem. Soc.*, **83**, 1174 (1961).
114. (a) E. H. White and L. A. Dolak, *J. Am. Chem. Soc.*, **88**, 3790 (1966).
    (b) *J. Org. Chem.*, **31**, 3038 (1966).
115. Th. J. de Boer and H. J. Backer, *Koninkl. Ned. Akad. Wetenschap. Proc.*, **55B**, 44 (1954).
116. J. H. Cooley, P. T. Jacobs, M. A. Kahn, L. Heasley and W. D. Goodman, *J. Org. Chem.*, **30**, 3062 (1965); T. Koenig and M. Deinzer, *J. Am. Chem. Soc.*, **88**, 4518 (1965).
117. W. A. Pryor, *Free Radicals*, McGraw-Hill Book Co., New York, 1966.
118. H. Craubner and A. Hrubesch, *Macromol. Chem.*, **72**, 19 (1964).
119. M. S. Kharasch, H. N. Friedlander and W. H. Urry, *J. Org. Chem.*, **14**, 91 (1949); F. R. Edwards and F. R. Mayo, *J. Am. Chem. Soc.*, **72**, 1265 (1950).
120. B. A. Gingras and W. A. Waters, *J. Chem. Soc.*, 1920 (1954).
121. A. C. Watterson, Jr., unpublished work.
122. R. Huisgen and H. Reimlinger (*Ann. Chem.*, **599**, 183 (1956)) have reported similar negative findings.
123. E. L. Eliel and V. G. Saha, *J. Am. Chem. Soc.*, **86**, 3581 (1964).
124. (a) H. P. Waits and G. S. Hammond, *J. Am. Chem. Soc.*, **86**, 1911 (1964).
    (b) Similar conclusions are generally applicable to the nitrous acid deamination; however, free radicals have been reported recently in the

deamination of aminoperchlorohomocubane (K. V. Scherer, Jr. and R. S. Lunt III, *J. Am. Chem. Soc.*, **88**, 2860 (1966).

125. R. Huisgen and H. Reimlinger, *Ann. Chem.*, **599**, 161 (1956); see also K. Heyns and W. Bedenburg, *Ann. Chem.*, **595**, 55 (1955).

126. L. A. Dolak, unpublished results.

127. Cyclopropanes have been reported from nitrous acid deaminations (P. S. Skell and I. Starer, *J. Am. Chem. Soc.*, **82**, 2971 (1960); M. S. Silver, *J. Am. Chem. Soc.*, **82**, 2971 (1960); **83**, 3482 (1961)), and from the deoxidation of alcohols (P. S. Skell and I. Starer, *loc. cit.*).

128. E. H. White and R. J. Baumgarten, *J. Org. Chem.*, **29**, 2070 (1964).

129. E. H. White and C. A. Aufdermarsh, Jr., *J. Am. Chem. Soc.*, **83**, 1174 (1961).

130. A. Streitwieser and W. D. Schaeffer, *J. Am. Chem. Soc.*, **79**, 2893 (1957).

131. E. H. White and F. W. Bachelor, *Tetrahedron Letters*, 77 (1965).

132. R. Huisgen and C. Rüchardt, *Ann. Chem.*, **601**, 21 (1956).

133. C. J. Collins, J. B. Christie and V. F. Raaen, *J. Am. Chem. Soc.*, **83**, 4267 (1961).

134. M. Billig, unpublished work.

135. G. M. Barrow and E. A. Yerger, *J. Am. Chem. Soc.*, **76**, 5211 (1954); **77**, 4474, 6206 (1955); E. Erlenmeyer, *Chem. Ber.*, **42**, 516 (1909).

136. A similar phenomenon has been reported in carbanion chemistry, see D. J. Cram and L. Gosser, *J. Am. Chem. Soc.*, **86**, 2950 (1964).

137. E. Grunwald in *Progress in Physical Organic Chemistry*, Vol. 3, Interscience Publishers, New York, 1965, p. 317.

138. S. Seltzer and F. T. Dunne, *J. Am. Chem. Soc.*, **87**, 2628 (1965).

139. S. Winstein and G. C. Robinson, *J. Am. Chem. Soc.*, **80**, 169 (1958).

140. P. D. Bartlett and C. Rüchardt, *J. Am. Chem. Soc.*, **82**, 1756 (1960).

141. (a) R. Huisgen and C. Rüchardt, *Ann. Chem.*, **601**, 1 (1956).
     (b) A. Streitweiser, Jr., *J. Org. Chem.*, **22**, 861 (1957).

142. G. S. Hammond, *J. Am. Chem. Soc.*, **77**, 334 (1955).

143. G. N. Lewis, *J. Franklin Inst.* **226**, 293 (1938); G. B. Kistiakowsky and R. Williams, *J. Chem. Phys.* **23**, 334 (1955).

144. D. B. Denney and D. G. Denney, *J. Am. Chem. Soc.*, **79**, 4806 (1957); D. B. Denney and B. Goldstein, *J. Am. Chem. Soc.*, **79**, 4948 (1957); ref. 139. It should be pointed out, however, that the problem of concomitant inversion of configuration was not a factor in these cases. See also H. L. Goering and R. W. Thies, *Abstr. 152nd National Meeting Am. Chem. Soc.*, New York, Sept. 1966, paper 38S.

145. H. L. Goering and J. F. Levy, *J. Am. Chem. Soc.*, **86**, 120 (1964).

146. S. Winstein, P. E. Klinedinst, Jr. and E. Clippinger, *J. Am. Chem. Soc.*, **83**, 4986 (1961), and earlier papers.

147. Retention of configuration has been observed recently in a solvolysis reaction (ref. 145).

148. S. Winstein and R. M. Roberts, *J. Am. Chem. Soc.*, **75**, 2297 (1953).

149. For similar conclusions in the nitrous acid deamination, see B. M. Benjamin, H. J. Schaeffer and C. J. Collins, *J. Am. Chem. Soc.*, **79**, 6160 (1957).

150. S. J. Cristol, J. P. Mohrig, F. P. Parungo, D. E. Plorde and K. Schwarzenbach, *J. Am. Chem. Soc.*, **85**, 2675 (1963).

151. See ref. 141a for a related case of phenyl participation; however, it is not clear here whether or not diazoalkanes were intermediates.

152. M. Schroeder and E. H. White, *Abstr. 152nd Meeting Am. Chem. Soc.*, New York, Sept. 1966, paper 54S.

153. H. Maskill, R. M. Southam and M. C. Whiting, *Chem. Communs.* 496 (1965).

154. H. Goldschmidt and J. Holm, *Chem. Ber.*, **21**, 1016 (1888); O. Dimroth, *Chem. Ber.*, **38**, 670 (1905).

155. E. H. White and H. Scherrer, *Tetrahedron Letters*, **21**, 758 (1961).

156. V. Y. Pochinok and L. P. Limarenko, *Ukr. Khim. Zh.*, **21**, 496, 628 (1955); V. Y. Pochinok and V. A. Portayagina, *Ukr. Khim. Zh.*, **18**, 631 (1952).

157. R. Piria, *Ann. Chem.*, **68**, 348 (1848).

158. (a) J. H. Ridd, *Quart. Rev. (London)*, **15**, 418 (1961).
(b) A. Streitwieser, *J. Org. Chem.*, **22**, 861 (1957).
(c) R. J. Baumgarten, *J. Chem. Educ.*, **43**, 398 (1966).
(d) E. Müller, Ed. *Methoden der Organischen Chemie*, Georg Thieme, Stuttgart, 1958, X1/2, pp. 133–181.

159. See reference 158a. Methylnitrosoamine has been prepared at low temperatures by E. Müller, H. Haiss and W. Rundel, *Chem. Ber.*, **93**, 1541 (1960).

160. T. Cohen and E. Jankowski, *J. Am. Chem. Soc.*, **86**, 4217 (1964).

161. M. Wilhelm and D. Y. Curtin, *Helv. Chim. Acta.*, **40**, 2129 (1957).

162. A. Streitwieser, Jr. and W. D. Schaffer, *J. Am. Chem. Soc.*, **79**, 2888 (1957).

163. (a) J. C. Martin and W. G. Bentrude, *J. Org. Chem.*, **20**, 1902 (1959).
(b) E. H. White and C. A. Elliger, *J. Am. Chem. Soc.*, **89**, 165 (1967).

164. M. S. Silver, M. C. Caserio, H. E. Rice and J. D. Roberts, *J. Am. Chem. Soc.*, **83**, 3671 (1961).

165. P. A. S. Smith and D. R. Baer, *Org. Reactions*, Vol. 11, (Ed. A. C. Cope), John Wiley and Sons, Inc., New York, 1960, pp. 157–188.

166. W. Nagata, T. Sugasawa, Y. Hayase and K. Sasakura, *Proc. Chem. Soc.*, 241 (1964); L. G. Cannell and R. W. Taft, *J. Am. Chem. Soc.*, **78**, 5812 (1956).

167. (a) V. Prelog, H. J. Urech, A. A. Bothner-By and J. Wursch, *Helv. Chim. Acta.*, **38**, 1095 (1955).
(b) R. Srinivasan, *J. Am. Chem. Soc.*, **84**, 3432 (1962); See also E. F. Ullman and Wm. A. Henderson, *J. Am. Chem. Soc.*, **88**, 4942 (1966).

168. S. J. Cristol and J. R. Mohrig, *Abstr. 18th National Organic Chem. Symp.*, Columbus, Ohio, 1963.

169. J. Hine, *Physical Organic Chemistry*, 2nd ed., McGraw-Hill Book Co. Inc., New York, 1962, pp. 127–128.

170. (a) L. S. Ciereszko and J. G. Burr, Jr., *J. Am. Chem. Soc.*, **74**, 5431 (1952).
(b) J. D. Roberts, C. C. Lee and W. H. Saunders, Jr., *J. Am. Chem. Soc.*, **76**, 4501 (1954).

171. D. J. Cram and J. E. McCarty, *J. Am. Chem. Soc.*, **79**, 2866 (1957).

172. E. J. Corey, J. Casanova, Jr., P. A. Vatakencherry and R. Winter, *J. Am. Chem. Soc.*, **85**, 169 (1963).

173. T. Cohen and E. Jankowski, private communication.

174. J. A. Berson and A. Remanick, *J. Am. Chem. Soc.*, **86**, 1749 (1964).

175. D. L. Boutle and C. A. Bunton, *J. Chem. Soc.*, 761 (1961).

176. A. Brodhag and C. R. Hauser, *J. Am. Chem. Soc.*, **77**, 3024 (1955).

177. Nitrite esters have apparently been isolated in an impure state by F. C. Whitmore and D. P. Langlois, ref. 103a.

178. The carbonyl compounds often isolated[168,170b] may originate in the elimination of nitrite ion from the alkyl nitrate, or nitroxide from the nitrite.

179. D. Semenow, C. Shih and W. G. Young, *J. Am. Chem. Soc.*, **80**, 5472 (1958).

180. W. Hückel and K. Heyder, *Chem. Ber.*, **96**, 220 (1963).

181. J. A. Mills, *J. Chem. Soc.*, 260 (1953); A. K. Bose, *Experientia*, **9**, 256 (1953).

182. W. G. Dauben, R. C. Tweit and C. Mannerskantz, *J. Am. Chem. Soc.*, **76**, 4420 (1954).

183. S. Winstein, D. Pressman and W. Young, *J. Am. Chem. Soc.*, **61**, 1645 (1939).

184. (a) W. G. Dauben and P. Lang, *Tetrahedron Letters*, 453 (1962).

(b) J. Hora, V. Černý and F. Šorm, *Tetrahedron Letters*, 501 (1962).

(c) G. J. Karabatsos, C. E. Orzech, Jr. and S. Meyerson, *J. Am. Chem. Soc.*, **87**, 4394 (1965).

185. A. A. Aboderin and R. L. Baird, *J. Am. Chem. Soc.*, **86**, 2300 (1964).

186. N. E. Searle, *Organic Synthesis*, Vol. 36, John Wiley and Sons, Inc., New York, 1956, p. 25.

187. O. E. Edwards and M. Lesage, *J. Org. Chem.*, **24**, 2071 (1959).

188. (a) J. H. Bayless, F. D. Mendicino and L. Friedman, *J. Am. Chem. Soc.*, **87**, 5790 (1965).

(b) L. Friedman, private communication.

189. J. D. Roberts and J. A. Yancey, *J. Am. Chem. Soc.*, **74**, 5943 (1952).

190. D. Y. Curtin and M. C. Crew, *J. Am. Chem. Soc.*, **77**, 354 (1955).

191. M. Chérest, H. Felkin, J. Sicher, F. Šipoš and M. Tichý, *J. Chem. Soc.*, 2513 (1965).

192. D. H. R. Barton and R. C. Cookson, *Quart. Rev. (London)*, **10**, 65 (1956).

193. G. D. Sargent, *Quart. Rev. (London)*, **20**, 301 (1966).

194. J. A. Berson and P. Reynolds-Warnhoff, *J. Am. Chem. Soc.*, **84**, 682 (1962); **86**, 595 (1964); J. A. Berson and D. Willner, *J. Am. Chem. Soc.*, **84**, 675 (1962); **86**, 609 (1964); J. A. Berson and M. S. Poonian, *J. Am. Chem. Soc.*, **88**, 170 (1966).

195. (a) J. A. Berson and J. J. Gajewski, *J. Am. Chem. Soc.*, **86**, 5020 (1964).

(b) W. E. Doering and H. H. Zeiss, *J. Am. Chem. Soc.*, **75**, 4733 (1953).

196. A. Streitwieser and C. E. Coverdale, *J. Am. Chem. Soc.*, **81**, 4275 (1959).

197. P. A. Levine and R. E. Marker, *J. Biol. Chem.*, **103**, 373 (1933); P. D. Bartlett and L. H. Knox, *J. Am. Chem. Soc.*, **61**, 3184 (1939).

198. P. A. S. Smith, D. R. Baer and S. N. Ege, *J. Am. Chem. Soc.*, **76**, 4564 (1954).

199. H. Felkin, *Compt. Rend.*, **236**, 298 (1953).

200. B. E. Weller, unpublished work.

201. S. Smirnov, *J. Russ. Phys. Chem. Soc.*, **43**, 1 (1912); A. T. Jurewicz, J. H. Bayless and L. Friedman, *J. Am. Chem. Soc.*, **87**, 5788 (1965).

202. D. Y. Curtin, J. A. Kampmeier and B. R. O'Connor, *J. Am. Chem. Soc.*, **87**, 863 (1965).

203. G. A. Olah, N. A. Overchuk and J. C. Lapierre, *J. Am. Chem. Soc.*, **87**, 5785 (1965).
204. R. A. Moss, *J. Org. Chem.*, **31**, 1082 (1966).
205. J. von Braun and F. Jostes, *Chem. Ber.*, **59**, 1091 (1926).
206. J. von Braun and G. Lemke, *Chem. Ber.*, **55**, 3526 (1922).
207. W. R. Vaughan and R. D. Carlson, *J. Am. Chem. Soc.*, **84**, 769 (1962).
208. W. J. Bailey and C. N. Bird, *J. Org. Chem.*, **23**, 996 (1958); H. E. Baumgarten, F. A. Bower, R. A. Setterquist and R. E. Allen, *J. Am. Chem. Soc.*, **80**, 4588 (1958).
209. J. W. Cook, G. T. Dickenson, D. Ellis and J. D. Loudon, *J. Chem. Soc.*, 1074 (1949).
210. J. J. Ritter and P. P. Minieri, *J. Am. Chem. Soc.*, **70**, 4045 (1948).
211. V. Prelog, H. H. Kägi and E. H. White, *Helv. Chim. Acta*, **45**, 1658 (1962).
212. H. Shechter, S. S. Rawalay and M. Tubis, *J. Am. Chem. Soc.*, **86**, 1701 (1964); H. Shechter and S. S. Rawalay, *J. Am. Chem. Soc.*, **86**, 1706 (1964).
213. A. Berg, *Am. Chim. Phys.*, **7**, 289 (1894).
214. W. E. Bachmann, M. P. Cava and A. S. Dreiding, *J. Am. Chem. Soc.*, **76**, 5554 (1954).
215. R. J. Baumgarten and S. G. Cohen, *J. Am. Chem. Soc.*, **87**, 2996 (1965); ref. 158c.
216. J. B. Longenecker and E. E. Snell, *Proc. Natl. Acad. Sci. U.S.*, **42**, 221 (1956).
217. H. C. Dunathan, *Proc. Natl. Acad. Sci. U.S.*, **55**, 712 (1966).
218. (a) K. H. Saunders, *The Aromatic Diazo Compounds*, Richard Clay and Sons, Ltd., Suffolk, 1936.
    (b) H. Zollinger, *Azo and Diazo Chemistry*, Interscience Publishers, Inc., New York, 1961.
    (c) J. H. Ridd, *Quart. Rev. (London)*, **15**, 418 (1961).
219. D. M. Bonner and S. E. Mills, *Heredity*, 2nd ed., Prentice-Hall, Inc., Englewood Cliffs, N.J., 1964, pp. 58–59.
220. N. L. Drake, *Org. Reactions*, **1**, 105 (1942).
221. (a) H. Seeboth, *Monatsber. Deut. Akad. Wiss. Berlin*, **6**, 268 (1964).
    (b) H. Seeboth, *Monatsber. Deut. Akad. Wiss. Berlin*, **3**, 43 (1961).
    (c) A. Rieche and H. Seeboth, *Angew. Chem.*, **70**, 52, 312 (1958).
    (d) A. Rieche and H. Seeboth, *Ann. Chem.*, **638**, 43, 57, 66, 76, 81, 92, 101 (1960).

CHAPTER **9**

Biological formation and reactions of the amino group

Barbara E. C. Banks

Physiology Department, University College, London

I. INTRODUCTION

The elements of amino groups obviously originate in the atmosphere but only a few of the simplest, unicellular living organisms are capable of using or fixing atmospheric nitrogen directly as a precursor of amino compounds. Higher plants, as well as those microorganisms which cannot fix atmospheric nitrogen, can use either ammonia or nitrate as a nitrogen source. Animals, which are more complex in structure, function and degree of specialisation, require organic nitrogen compounds, some of which already contain amino groups. The sequence of reactions by which either free or combined nitrogen is converted to naturally occurring amino compounds is not yet completely worked out, but it is clear that the steps in each sequence are all enzyme catalysed. The ability of different life forms to use free or combined nitrogen is governed, to some extent, by the presence of particular enzymes. Since all enzymes are proteins, and therefore derivatives of the most important naturally occurring amino compounds, the α-L-amino acids, any argument concerned with the purely biological formation of amino compounds is essentially circular. The implication is, then, that at some very early stage, before the evolution of life on this planet, amino acids must have been formed directly from the elements of the atmosphere.

Although Miller[1, 2] has shown that the passage of an electric discharge through mixtures of methane, ammonia and water vapour (possible constituents of a primeval atmosphere), results in the production of some α-amino acids, these always occur as racemic mixtures of the L and D forms. In biological systems, both structural and functional proteins are built almost exclusively from L-amino acids. It is clearly easier to construct a well-ordered macromolecule from residues having the same stereochemistry, but the original selection of one-handed residue over the other is as yet unexplained.

Such speculations as to the origin of α-L-amino acids may seem out of place, but they are designed to emphasise the importance of these particular amino compounds in biological systems. Not only must amino acids have been among the earliest formed amino compounds, on the evolutionary scale, since they are the precursors of the natural catalysts of all metabolic processes, but they also appear to be the first-formed products following ingestion of inorganic nitrogen compounds by either plants or microorganisms. All other amino compounds (e.g. purines and pyrimidines, amino sugars, vitamins and chemical transmitters) must, therefore, be derived from amino acids. Although some of the biosynthetic routes are incompletely understood, many are known to involve the transfer of amino groups from either amino acids or from amides of amino acids to receptors such as carbohydrate residues or heterocyclic ring systems.

Between the various life forms now extant, there is a remarkable conservation of so-called fixed nitrogen. Plants, animals and microorganisms enjoy a symbiotic relationship, with microorganisms holding the balance between fixed and atmospheric nitrogen. Apart from the nitrogen-fixing microorganisms, there exist a number of nitrifying bacteria which can convert ammonia, or more importantly organic nitrogen compounds of plant or animal origin, into nitrate. The nitrate may then be assimilated by plants and converted to organic nitrogen compounds. Of these, certain amino acids, readily synthesised by plants, are necessary in the diet of herbivorous animals while carnivores obtain the same essential amino acids by consuming herbivores. When animals and unconsumed or inedible plants die, the organic amino and nitrogen compounds may be reconverted to nitrate, by nitrifying bacteria, and so the principal cycle of fixed-nitrogen conservation is completed. Alternatively, the nitrogen of nitrate may be returned to the atmosphere by the action of denitrifying bacteria.

Not all animals obtain the essential amino acids from plants. Some (the ruminants) can by-pass the main route by direct assimilation of the amino acids produced by symbiotic bacteria in the gut. These bacteria can break down cellulose, the main structural carbohydrate of plants, and can also synthesise all the amino acids, required for their growth and development, from a simple fixed-nitrogen source. The ruminant obtains its more complex nitrogen requirements by digestion of dead bacteria. (Goats, for example, can survive on a diet of hay and ammonia.) This may appear to be a somewhat one-sided symbiosis, with the bacteria carrying out all the important functions but, in return, the bacteria obtain from the animal a copious supply of carbohydrate and are maintained in a favourable environment with respect to temperature and medium composition.

It is fortunate, not least from the point of view of this chapter, that the fundamental differences in the dietary nitrogen requirements of different forms of life are not followed by an equivalent diversity in the principal routes to amino compounds and their derivatives. Figure 1 shows, in outline, the most important reactions and these are the ones to which the present discussion will be limited. The reactions shown with bold arrows are common to most bacteria and to some cells of multicellular organisms. The reactions shown with light arrows are no less important but they only occur in particular life forms. For example, the decarboxylation of α-L-amino acids to form amines occurs only in bacteria and (to a lesser extent) in plants. Although excess amino acids in animals may be removed by decarboxylation, it is the intestinal bacteria which carry out the reaction and not cells of the animal itself. The production of alkaloids is also marked with a light arrow. Alkaloids, which are formed almost exclusively in plants, with the exception of the poisonous skin secretions of certain toads and salamanders, have profound physiological effects when administered to higher animals, but their function in plants remains quite unknown. Because the function of the product is

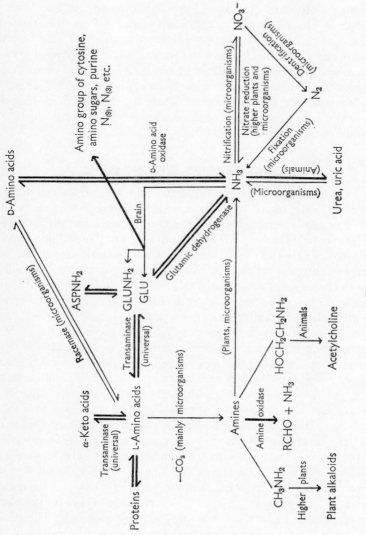

FIGURE 1.

unknown, the formation of plant alkaloids is considered as a *secondary* rather than a *primary* metabolic process.

The first two sections of this chapter are concerned with the general problems involved in elucidating metabolic pathways and with the properties of the necessary catalysts. In later sections, the particular reactions given in Figure 1 are considered separately. The emphasis given in each case reflects more the present state of knowledge of the 'mechanism' of each reaction than the biological importance. Although the formation of proteins from amino acids and of nucleic acids from purine and pyrimidine bases are included in Figure 1, discussion of the biosynthesis of these vital macromolecules is beyond the scope of this chapter.

II. PROPERTIES OF ENZYMES

As stated in the opening paragraph, the basic reason for the profound differences in the nitrogen requirements of different organisms lies in the presence or absence of specific enzymes. Since the great majority of metabolic processes are enzyme catalysed, so each reaction involved in the synthesis and metabolism of amino compounds may be discussed in terms of the properties of individual protein catalysts. In each cell of multicellular organisms and in every unicellular organism, many reactions occur, some only during particular phases of cell development and others throughout the existence of the cell. This fact implies the presence of many different catalysts. Metabolic studies in whole animals, plants and microorganisms, or in tissue slices or homogenates of particular parts of higher organisms, have led to the partial identification of many hundreds of enzymes but relatively few of these have been isolated in forms which satisfy the criteria of purity applied to proteins. (It should be remembered that these criteria are still far less stringent than those applied to conventional organic molecules). However, certain generalisations about some of the properties of enzymes can be made with reasonable safety. For example, all enzymes appear to be proteins (molecular weight, 10^4–10^6) which may or may not require a non-protein adjunct (cofactor) for catalytic activity. The cofactor may be a carbohydrate or polynucleotide fragment, a metal or a relatively small organic molecule. The proteins are all built from some or all of the twenty known, naturally occurring amino acids (Table 1) and therefore the specificity of each enzyme as a catalyst for a particular reaction (or a set of very closely related reactions) must in part be a function of the sequence in which the

TABLE 1.   Common α-L-amino acids.

Common name	Chemical structure
Glycine	NH_2CH_2COOH
Alanine	$NH_2CH(CH_3)COOH$
Valine	$CH(CH_3)_2$ $\mid$ $NH_2CHCOOH$
Leucine	$CH_2CH(CH_3)_2$ $\mid$ $NH_2CHCOOH$
Isoleucine	$CH_3CHC_2H_5$ $\mid$ $NH_2CHCOOH$
Serine	CH_2OH $\mid$ $NH_2CHCOOH$
Threonine	$HC(OH)CH_3$ $\mid$ $NH_2CHCOOH$
Methionine	$CH_2CH_2SCH_3$ $\mid$ $NH_2CHCOOH$
Phenylalanine	$CH_2C_6H_5$ $\mid$ $NH_2CHCOOH$
Tyrosine	$CH_2C_6H_4OH\text{-}p$ $\mid$ $NH_2CHCOOH$
Tryptophane	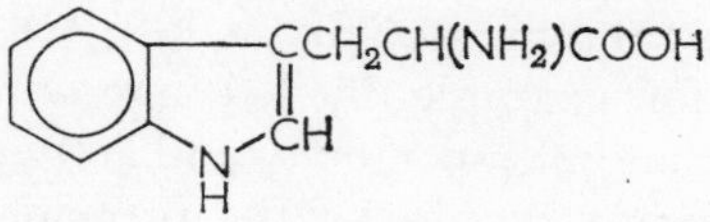
Proline	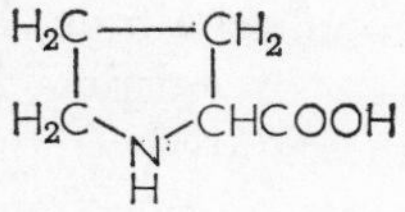
Hydroxyproline	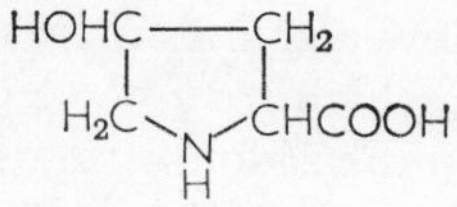
Aspartic acid	CH_2COOH $\mid$ $NH_2CHCOOH$

TABLE 1. *(continued)*

Common name	Chemical structure
Asparagine	CH_2CONH_2 $\mid$ $NH_2CHCOOH$
Glutamic acid	CH_2CH_2COOH $\mid$ $NH_2CHCOOH$
Glutamine	$CH_2CH_2CONH_2$ $\mid$ $NH_2CHCOOH$
Lysine	$CH_2(CH_2)_3NH_2$ $\mid$ $NH_2CHCOOH$
Arginine	$CH_2CH_2CH_2NHC{-}NH_2$ $\mid\qquad\qquad\;\;\parallel$ $NH_2CHCOOH\quad NH$
Histidine	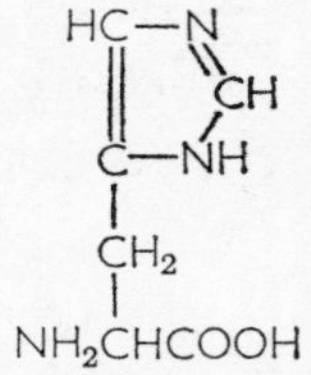 $HC{-}N$ $\parallel\quad\diagdown CH$ $C{-}NH$ $\mid$ CH_2 $\mid$ $NH_2CHCOOH$
Cystine	$CH_2{-\!-}S{-\!-}S{-\!-}CH_2$ $\mid\qquad\qquad\qquad\mid$ $NH_2CHCOOH\qquad NH_2CHCOOH$

amino acids are joined. The relative ease with which, for many enzymes, catalytic activity is lost, without any peptide bonds being broken, for example by heating or by change of pH or solvent, suggests that both specificity and activity are functions of the secondary and tertiary structure of the protein, that is, the three-dimensional arrangement of the protein chain or chains.

In spite of the mass of existing literature concerned with the 'mechanisms' of enzyme catalysis (more, perhaps, than in the field of physical organic chemistry), in no single case can an answer be given to the question of how enzymes increase the rates of known chemical reactions by factors which may be as high as 10^8 or 10^9. In one case, the complete three-dimensional structure of a small enzyme (lysozyme (egg-white) mol. wt. $\sim 14{,}600$) is now known, as a result of collaborative efforts of protein chemists and crystallographers. Unfortunately, the chemistry of the substrate of this particular enzyme (complex polysaccharide molecules containing *N*-acetylglucosamine and *N*-acetylmuramic acid present in some bacterial cell walls—see section V.E) is not well understood, and so even for this enzyme, theories of the mechanism of catalysis, although plausible, are largely speculative[3].

Although it is not known, in detail, how enzymes catalyse metabolic reactions, it is still useful to discuss which enzymes are present in living cells, particularly in relation to cell function and degree of specialisation. It should at once be pointed out that the presence of a particular enzyme in a cell is a necessary but not a sufficient condition for the reaction catalysed by that enzyme to occur. The standard free-energy changes of metabolic processes are all relatively low (about ± 12 kcal/mole), hence the reactant concentrations are as important as the presence of the necessary catalyst in that they determine the direction in which reaction proceeds. An example will serve to illustrate this important factor. It has been emphasised that plant and animal cells do not contain the enzymes, present in some bacteria, which are responsible for converting atmospheric nitrogen to ammonia. Multicellular organisms must be supplied with fixed nitrogen, but the next step in the production of amino compounds from ammonia is most probably that given in equation (1). The reaction is catalysed by an enzyme (glutamic dehydrogenase, see section IV.B) which occurs in bacteria, plants and also animals. (It is not implied that enzymes catalysing the same reaction but present in different species are identical proteins.) In principle, the reaction (equation 1) could be, in all living matter, the first step in the production of amino compounds, as it appears to be in bacteria and probably also in plants.

$$
\begin{array}{c}
\text{COOH} \\
| \\
\text{(CH}_2)_2 \\
| \\
\text{CO} \\
| \\
\text{COOH}
\end{array}
+ \text{NH}_4{}^+ + \text{NADH*} \rightleftharpoons
\begin{array}{c}
\text{COOH} \\
| \\
\text{(CH}_2)_2 \\
| \\
\text{CHNH}_2 \\
| \\
\text{COOH}
\end{array}
+ \text{NAD}^+ + \text{H}_2\text{O} \qquad (1)
$$

α-ketoglutaric acid   glutamic acid

However, it is well known that animals, with the exception of ruminants, cannot survive on ammonia as a source of nitrogen. The reason for this apparent anomaly is that although glutamic dehydrogenase is present in animals (predominantly in the liver), the level of ammonia is too low to drive the reaction to the formation of glutamate.

* See Figure 2a and section IV.A.

17*

The concentration of ammonium ions is kept at a low level by other enzymes, also active in the liver, notably those concerned with the conversion of free ammonia to the normal waste products (urea or uric acid) and also those concerned with the synthesis of glutamine, subsequently used in a variety of synthetic reactions. Ammonia is toxic in the central nervous system of higher animals and one function of the liver is to keep the concentration in the bloodstream low. It is alleged that the main function of liver glutamic dehydrogenase is in its involvement in the series of reactions by which excess L-amino acids are deaminated. The reactions are summarised in equations (2–5). (See section VI.B). Liver glutamic dehydrogenase is, therefore,

$$
\underset{\substack{\text{α-L-amino acid}}}{\underset{\substack{|\\ NH_2}}{RCHCOOH}} + \underset{\substack{\text{α-ketoglutaric acid}}}{\underset{\substack{|\\ (CH_2)_2\\ |\\ COOH}}{\overset{\substack{COOH\\ |}}{CO}}} \underset{\xrightarrow{\hspace{1cm}}}{\overset{\text{Transaminase}}{\rightleftharpoons}} \underset{\text{α-keto acid}}{RCOCOOH} + \underset{\text{glutamic acid}}{\underset{\substack{|\\ (CH_2)_2\\ |\\ COOH}}{\overset{\substack{COOH\\ |}}{CHNH_2}}} \tag{2}
$$

$$
\underset{\substack{|\\ (CH_2)_2\\ |\\ COOH}}{\overset{\substack{COOH\\ |}}{CHNH_2}} + NAD^+ + H_2O \rightleftharpoons \underset{\substack{|\\ (CH_2)_2\\ |\\ COOH}}{\overset{\substack{COOH\\ |}}{CO}} + NADH + NH_4^+ \tag{3}
$$

$$
NH_4^+ + CO_2 + 2\,ATP \rightleftharpoons \underset{\text{carbamyl phosphate}}{NH_2COOPO_3H_2} + 2\,ADP + H_3PO_4 \tag{4}
$$

$$
\text{Carbamyl phosphate} \xrightarrow{\text{Krebs cycle}} CO_2 + NH_3 + H_3PO_4 \tag{5}
$$

more concerned with the production of ammonia than of glutamate. The ammonia is fed directly into the cycle of reactions by which excess nitrogen compounds are converted into waste products and so cannot gain access to other parts of the body where it might be toxic. Glutamic dehydrogenase may be involved in a similar series of reactions in plants and bacteria, where the balance between the production of necessary amino acids (reversal of reactions 2 and 3) and the removal of unwanted amino acids (reactions 2–5) is delicate.

It is clear, then, that the earlier statement that dietary nitrogen requirements depend on the presence of particular enzymes is incomplete. It is not possible to deduce, merely from the presence of an enzyme, whether a particular reaction will occur in a given direction. Not only is the location of the enzyme important in multicellular organisms, but also the tolerance of the whole organism for high concentrations of the appropriate reactants. This is particularly true for those reactions for which the standard free-energy change is low.

III. METHODS OF STUDYING METABOLIC PATHWAYS

The details of the methods used in elucidating metabolic pathways inevitably depend on the particular organisms studied, but the most common approach involves administration of isotopically labelled compounds and detection of the label in metabolic intermediates. Work on whole animals and plants is subject to certain obvious limitations. For example:

(a) A large dilution factor is inevitable since relatively small quantities of labelled compounds are introduced into a large mass of material. In studying the metabolism of amino compounds, this problem is less severe in plants than in animals because the total protein concentration is lower. In animals, proteins are the main structural material whereas plants are built primarily of polysaccharides.

(b) It is difficult to isolate the few possible intermediates from the large number of irrelevant compounds present initially.

(c) Certain intermediates in a reaction sequence may only be present at very low, stationary concentrations. Apart from the problem of identifying such intermediates, there is the additional problem of characterising the enzymes catalysing their interconversion. It is interesting to note that, to date, the best characterised enzymes are those for which at least one substrate is present, *in vivo*, at a significant concentration level, e.g. enzymes of the glycolysis cycle and the Krebs–Henseleit cycle (section VI).

(d) The complex control devices, by means of which life is maintained, may easily be upset. This imposes a limit on the quantity of isotope introduced, the form in which it is administered and the actual method of administration (e.g. by mouth or by injection). Isotope studies on isolated, functional parts of animals or plants (e.g. heart or liver, leaves or seeds) are also subject to these limitations. Indeed, the problem of maintaining an approximation to normal functioning is even more severe in functional parts than in the whole organism. Work on tissue slices or homogenates does not suffer from the same disadvantages but the results obtained are more difficult to interpret unambiguously. In destroying both the inter- and intracellular structure, the distribution of enzymes between and within cells is altered and the spatial arrangement, whereby the product of one enzyme-catalysed reaction passes directly as a substrate for another reaction, breaks down. Intermediates detected under these conditions may not be formed to any significant extent in whole cells. As pointed out previously, the mere presence of an enzyme in tissue is not indicative of its importance *in vivo*.

Studies of metabolic pathways in microorganisms are not subject to the same limitations as those considered above in relation to plants and animals. There are a number of advantages in working with unicellular organisms. For example:

(a) It is generally easier to keep microorganisms alive under conditions used for the introduction of isotopically labelled compounds.

(b) The dilution of any isotopically labelled compound is less than in multicellular organisms because of the lower mass of material in each microorganism.

(c) The high rate of reproduction, and hence the increased frequency of mutation of microorganisms, permits the culture of mutant strains in which particular metabolic pathways may be blocked. Intermediates present at low stationary concentrations in the wild strain may well accumulate in mutants, and so be more easily identified.

(d) The considerable adaptability shown by many microorganisms, with respect to nutrients in the growth medium, greatly facilitates the production of mutants with desired characteristics. For example, it is possible, in some cases, to produce mutants in which a particular metabolic path is blocked, simply by culturing the organism in the presence of the end-product of that particular path.

Among the disadvantages are:

(a) The difficulty of culturing microorganisms on a large scale.

(b) All the reactions necessary for growth, development and reproduction must occur within each unicellular organism. In multicellular organisms, on the other hand, particular cells are specialised to carry out functions for the organism as a whole and within these specialised cells, the total number of enzyme catalysed reactions may be reduced.

(c) Because of the size of microorganisms, the only feasible method of introducing isotopically labelled compounds is by assimilation. Since transport of material across cell walls is limited to certain chemical compounds, the labelled compounds which can be introduced are similarly limited.

In studying the metabolism of amino compounds, use is commonly made of the stable isotope ^{15}N, the abundance of which may be determined mass spectrometrically. The radioactive isotope of nitrogen, ^{13}N, has too short a half-life (~ 10 min) to be of much value in this field, though some experiments in the limited field of nitrogen fixation have been reported[4].

IV. FORMATION OF α-L-AMINO ACIDS

There are two main reasons for considering the formation of α-L-amino acids separately from other amino compounds. The first is that the α-L-amino acids are the basic components of enzymes and the second is that those organisms which can use inorganic nitrogen compounds as the sole source of nitrogen appear to form all amino compounds from amino acid precursors. The reactions, included in Figure 1, which are primarily involved in amino acid formation are:

(a) *Ammonia formation*, by nitrogen fixation and nitrate reduction.

(b) *Glutamate formation* by incorporation of ammonia into an existing carbon skeleton (α-ketoglutarate), catalysed by the enzyme glutamic dehydrogenase.

(c) *Transamination* between (i) glutamate and some α-keto acids to form α-L-amino acids with the regeneration of α-ketoglutarate; (ii) aspartate, (formed by (c)(i) from oxaloacetate (or possibly by direct incorporation of ammonia into fumaric acid) and some α-keto acids to form α-L-amino acids with the regeneration of oxaloacetate.

(d) *Miscellaneous* reactions of amino acids formed by reactions (a)–(c).

A. Ammonia Formation

I. Nitrogen fixation

The ability of certain microorganisms to take up and use atmospheric nitrogen was recognised late in the nineteenth century. Since 1890, an increasing number of microorganisms with this ability have been isolated in pure strain and the somewhat surprising fact to emerge is that nitrogen-fixing microorganisms have little else in common. It is conventional to classify microorganisms broadly according to diet, (e.g. autotrophs require only inorganic foodstuffs, heterotrophs require some organic material), energy source (e.g. chemosynthetic microorganisms require a chemical 'energy source', photosynthetic microorganisms can obtain energy directly from sunlight) and type of oxidation reactions (e.g. in anaerobes, oxidation of foodstuffs occurs in the absence of molecular oxygen). Inorganic oxidation–reduction systems such as nitrate–nitrite or sulphate–sulphide, may be involved

$$NO_3^- \rightleftharpoons NO_2^-$$
$$SO_4^{2-} \rightleftharpoons H_2S$$

or, in the case of fermenting microorganisms, carbohydrate products such as pyruvate may undergo the type of reaction

$$2\ CH_3COCOOH \rightleftharpoons CH_3CH(OH)COOH + CH_3COOH + CO_2$$

Aerobic microorganisms use molecular oxygen in the terminal stages of oxidation of foodstuffs.

Nitrogen-fixing microorganisms have representatives in all these classes. For example, among the free living bacteria, *Clostridium pasteurianum* is an anaerobic chemosynthetic heterotroph while *Azotobacter agilis* is an aerobic heterotroph; blue–green algae (nitrogen-fixers which may be found living in symbiosis with higher plants and lichens) and certain photosynthetic bacteria are autotrophs.

There have been a number of claims for nitrogen fixation by higher organisms, including chick embryos[5] and some fungi[6,7] but the most controversial issue is concerned with nitrogen fixation by plants. The requirement of higher plants for an adequate supply of fixed nitrogen (as nitrate or ammonium salts) has been recognised since the late eighteenth century, while crop rotation has been practised since Roman times. In particular, some leguminous plants have a beneficial effect on soil for subsequent crops, the effect being most marked when the plants are turned into the soil as green manure. The possibility that these leguminous plants in some way increased the concentration of fixed nitrogen in the soil led to the proposal that the plants themselves fixed atmospheric nitrogen. The situation was confused by the observation, late in the nineteenth century, that legumes grown in sterile soil do not fix atmospheric nitrogen but examination of the plants grown under sterile and non-sterile conditions showed that only those grown under non-sterile conditions had nodulated roots. Root nodules, characteristic of both legumes and other species capable of fixing nitrogen, were shown to contain microorganisms. These (many bacteria of the species *Rhizobium*, blue–green algae, etc.) live in true symbiosis with the plants, being responsible for fixing atmospheric nitrogen and supplying nitrogenous material to the plant which, in turn, provides the symbiotes with carbohydrate foodstuffs. Isolated bacteria from root nodules cannot themselves fix atmospheric nitrogen.

In the following period of fifty years little progress was made towards elucidating the mechanism of the process. Many attempts were made to obtain cell-free preparations containing the enzymes presumably involved in nitrogen fixation, but these were unsuccessful until 1960. The intracellular organisation of enzyme systems is known to be complex, groups of enzymes involved in sequential reactions frequently being localised in or on subcellular particles. One would predict that nitrogen fixation would involve a complex series of reactions so it is not surprising that the difficulty experienced in obtaining active, cell-free preparations is associated with the conditions under which the cells are disrupted. Successful methods of cell disruption include sonic, mechanical and enzymic disruption in an atmosphere of hydrogen, but the method of choice depends on the organism studied. Activity is judged by incorporation of $^{15}N^8$ or ^{13}N into ammonia and organic nitrogen compounds by the action of the supernatant liquid obtained after high speed centrifugation of the cell extract (25,000–50,000 g), the supernatant liquid having been shown to be cell-free. Knowledge of the mechanism of nitrogen fixation is still in a primitive state but some interesting facts have emerged[9-11]. Firstly, there appears to be a close but non-stoichiometric connection between nitrogen fixation and carbohydrate metabolism. In extracts of *Clostridium pasteurianum*, *Bacillus polymyxa*, *Rhodospirillum rubrum* and *Chromatium* the particular intermediate in carbohydrate metabolism required is pyruvate, believed to be metabolised to acetyl phosphate (equation 6). Working with extracts of *Cl. pasteurianum*, Carnahan and

$$CH_3COCOOH + H_3PO_4 \rightleftharpoons CH_3COOPO_3H_2 + CO_2 + H_2 \qquad (6)$$

coworkers have shown that there is a large molar excess (~ 100-fold) of pyruvate metabolised over amount of nitrogen fixed, and also an initial lag period in which nitrogen fixation cannot be detected. Had there been a stoichiometric connection between pyruvate metabolism and nitrogen fixation, the obvious overall reaction would be as shown in equation (7). However, as frequently occurs in biological systems,

$$3\,CH_3COCOOH + 3\,H_3PO_4 + N_2 \rightleftharpoons 3\,CH_3COOPO_3H_2 + 3\,CO_2 + 2\,NH_3 \qquad (7)$$

the connection between the oxidation (of pyruvate) and reduction (of nitrogen) appears to involve an electron-transfer system rather than hydrogen transfer. Since reference will be made, in later sections, to those oxidation–reduction reactions involved in the metabolism of amino compounds, a brief account follows of the currently accepted position with regard to natural redox systems.

The enzymes catalysing reactions involving oxidation and reduction (oxido-reductases) are of two types:

(a) those requiring a readily dissociable, nicotine–adenine nucleotide (NADP) cofactor (Figure 2a, 2b) catalysing reactions of the type

$$MH_2 + NAD^+ \rightleftharpoons M + NADH + H^+ \qquad (8)$$
$$(MH_2 = \text{reduced metabolite})$$

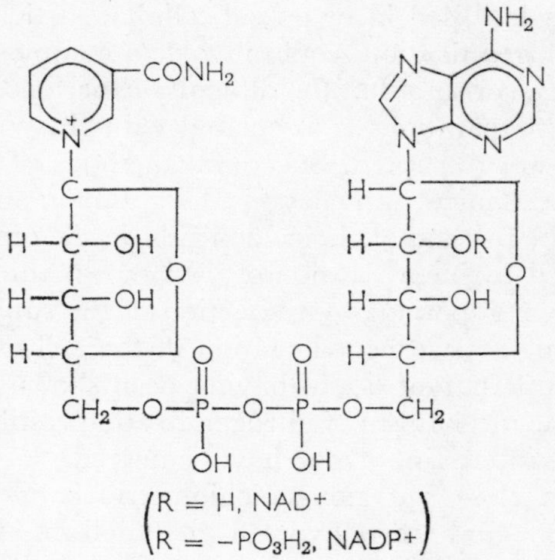

FIGURE 2a.   Nicotinamide—adenine nucleotide cofactors (oxidised).

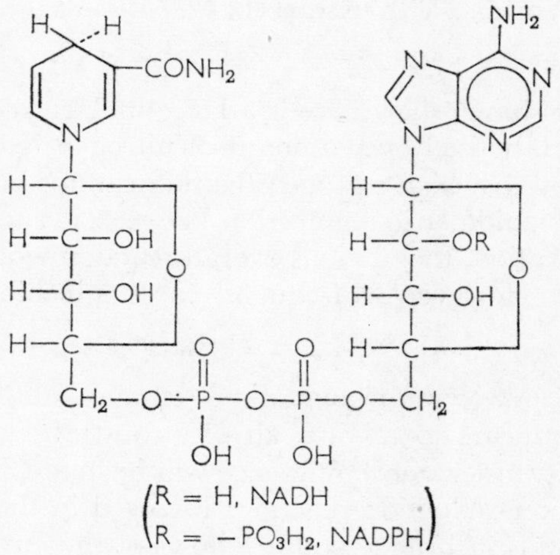

FIGURE 2b.   Nicotinamide—adenine nucleotide cofactors (reduced).

(b) those containing a tightly bound flavine cofactor (Figure 2c, 2d) and catalysing reactions of the type

$$MH_2 + FP \rightleftharpoons M + FPH_2 \tag{9}$$

Reoxidation of the reduced cofactors occurs by one of the reactions shown in Figure 3. The abbreviation 'cyt-system' refers to the complex

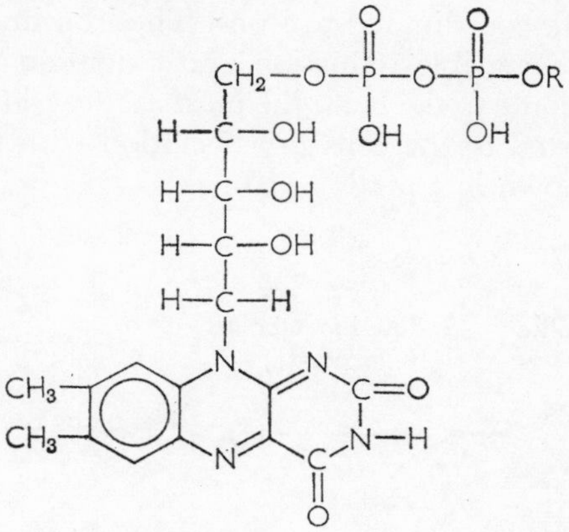

$$
\begin{pmatrix}
R = H,\ FMN \\
R = \text{adenosine (linked through } C'_{(5)} \text{ of} \\
\text{D-ribose), FAD}
\end{pmatrix}
$$

FIGURE 2c.   Flavin cofactors (oxidised).

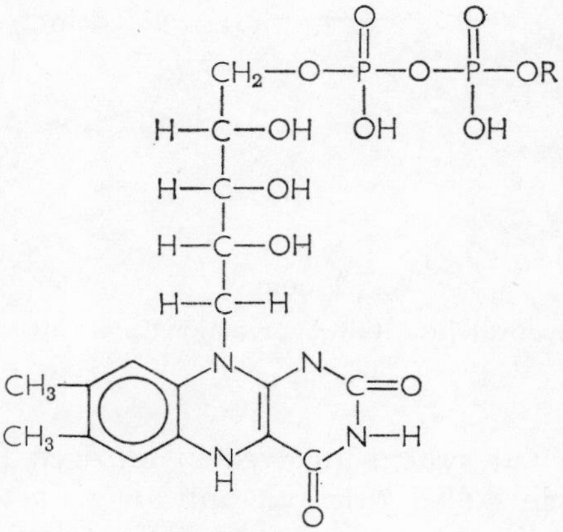

$$
\begin{pmatrix}
R = H,\ FMN \\
R = \text{adenosine, FAD}
\end{pmatrix}
$$

FIGURE 2d.   Flavin cofactors (reduced).

arrangement of flavoproteins, a quinone-type derivative and the so-called cytochromes which exist in particular subcellular particles called mitochondria. The cytochromes are protein-bound haem complexes (haem is a porphyrin ring system in which the four pyrrole ring nitrogens are coordinated to iron). The iron in these haemoproteins undergoes reversible reduction and oxidation as electrons pass from one cytochrome to the next, the ultimate reaction with molecular oxygen, transported to the cells in association with the haemoglobin of blood, is as shown in equation (10).

$$2 \text{ cyt a}_3 \text{ Fe}^{2+} + 2 \text{ H}^+ + \tfrac{1}{2} O_2 \rightleftharpoons 2 \text{ cyt a}_3 \text{ Fe}^{3+} + H_2O \tag{10}$$

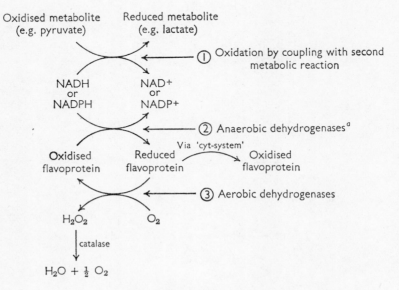

FIGURE 3.

[a] The reduced flavoproteins catalysing reaction ② are always reoxidised by the cytochrome system.

For the particular system involved in nitrogen fixation, an iron-containing protein called *ferredoxin* appears to act as an electron-carrying intermediary between pyruvate metabolism and nitrogen reduction. Ferredoxin was first isolated from *Cl. pasteurianum* in 1962[12]. The iron in ferredoxin is not associated with a porphyrin ring system nor is the protein associated with a flavin entity. The ferredoxin isolated from *Cl. pasteurianum* is a small protein (mol. wt. ~6,000), showing absorption maxima at 280, 300 and 390 mμ when oxidised (brown colour) but only the 280 mμ absorption remains when the protein is reduced. The soluble, nitrogen-fixing system, from *Cl. pasteurianum*, has been fractionated into two parts[13], one called the hydrogen-donating system (HDS) and the other the nitrogen-activating system (NAS). Both fractions are required for nitrogen fixation. The HDS catalyses the reaction between pyruvate and inorganic phosphate (equation 6) and also the production of molecular hydrogen from hydrogen ions. Both these activities of the HDS fraction are stimulated by ferredoxin. Figure 4 summarises the proposed connection between the nitrogen-activating system and the hydrogen-donating system.

Pyruvate etc.   +   Oxidised ferredoxin

HDS $\longrightarrow H_2$

Acetylphosphate + CO_2 + Reduced ferredoxin

NAS $\longrightarrow N_2$

Oxidised ferredoxin + NH_3

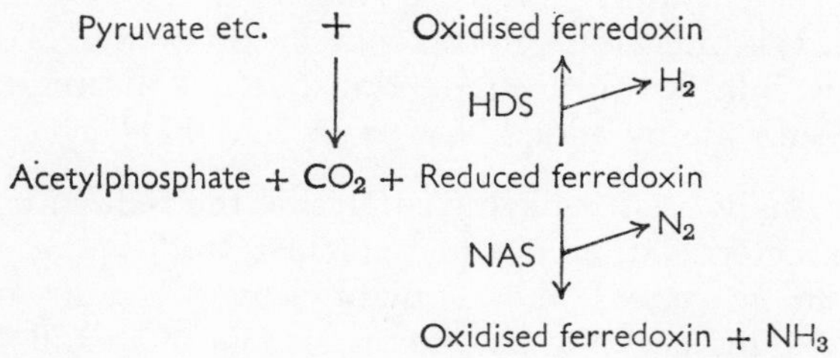

FIGURE 4.

The above account ignores many facets of the problem of nitrogen fixation. For example:

(a) The role of molybdenum, which is known to be an essential trace metal for nitrogen fixation in root nodules[14] and possibly in free living microorganisms[15].

(b) The long-standing argument as to the nature of the 'key-intermediate', that is, the compound immediately preceding incorporation of nitrogen into a carbon compound.

(c) The possibility of a route involving first oxidation and then reduction, for example via nitrous oxide, nitramide or hyponitrous acid to hydroxylamine, and thence to ammonia.

On balance, the evidence so far favours the theory that ammonia is the end-product of nitrogen fixation.

2. Nitrate reduction (in higher plants and some microorganisms)

The biological reduction of nitrate is a stepwise process. The intermediates have not all been characterised but enzymes catalysing some of the postulated steps have been identified in certain higher plants and microorganisms.

a. *Nitrate reductase.* An enzyme catalysing the reduction of nitrate to nitrite has been detected in some moulds[16], bacteria[17], plant leaves and yeasts. In all cases the enzyme activity is associated with a flavoprotein (FP), containing flavine adenine dinucleotide (Figure 2c, 2d) and requires two cofactors, one of which is either NADH or NADPH (Figure 2a, 2b), depending on the source, and the other is a specific metal, molybdenum, known to be essential, in trace amounts, for healthy plant growth. This particular oxidation–reduction system does not come into either of the separate categories given in section IV.A.1 but the possibility still exists that more than one enzyme may be involved. It is interesting, in this context, that removal of the metal, by chelation, inhibits nitrate reduction but not the reduction of the FAD of the enzyme by added NADH or NADPH[18].

b. *Nitrite reductase.* An enzyme catalysing the reduction of nitrite is less well characterised than nitrate reductase, but its presence has been demonstrated in yeasts[19] and plant leaves[19] and some micro-organisms[20]. This enzyme also appears to be a metalloflavoprotein requiring NADH or NADPH, but the heavy metal cofactor has not been positively identified. Likely candidates are iron, copper or manganese. The reduction product is uncertain. The enzyme may well be the same as that designated as hyponitrite reductase, detected in *Neurospora*[21].

c. *Hydroxylamine reductase.* The reduction of hydroxylamine to ammonia is catalysed by another metalloflavoprotein, present in some plant leaves[19], yeasts and microorganisms[20]. The enzyme requires NADH and the metal is most likely to be manganese.

Hydroxylamine is known to be highly toxic to plants so the possibility that it is an obligatory intermediate in nitrate reduction must be considered with caution.* Interest in hydroxylamine arises from the lengthy controversy, briefly mentioned here, concerning the relative merits of hydroxylamine or ammonia as the key intermediate in nitrogen fixation. The toxicity of hydroxylamine may well be due to its action as an enzyme inhibitor. However, in spite of its toxicity, there are still some claims[22] that hydroxylamine may be an intermediate in the formation of amino acids by oxime formation with such carbonyl compounds as glyoxilic acid (formed in photosynthesis), pyruvate, α-ketoglutarate or oxaloacetate (intermediates in carbo-

* In a sequence of reactions, occurring *in vivo*, if one step is blocked, by addition of a specific inhibitor, the resulting increase in the concentrations of intermediates preceding the block may well prove toxic.

hydrate metabolism) and subsequent reduction of the oximes to α-amino acids (equation 11). Virtanen and coworkers, the main

$$NH_2OH + RCOCOOH \rightleftharpoons \underset{\underset{COOH}{|}}{RC}{=}NOH \rightleftharpoons \underset{\underset{NH_2}{|}}{RCH}{-}COOH \qquad (11)$$

supporters of hydroxylamine as the key intermediate in nitrogen fixation, originally proposed that the main route to α-amino acids might be through oxime formation; however, this route is now generally believed to be only of minor importance.

The sequence of reactions alleged to occur in nitrate reduction are given in equation (12).

$$HNO_3 \longrightarrow HNO_2 \longrightarrow [HNO] \longrightarrow NH_2OH \longrightarrow NH_3 \qquad (12)$$

3. Nitrification and denitrification

The nitrifying microorganisms, which can convert ammonia and organic amino compounds to nitrate, and the denitrifying microorganisms which produce elementary nitrogen from nitrate and nitrite, are important in that they control the supply of fixed nitrogen to higher organisms. The mechanisms of the two processes are obscure. It is possible that, at least in *Nitrosomonas*, nitrification may occur by the reverse of the reactions of equation (12). Certainly, enzymes catalysing the production of nitrite from hydroxylamine have been detected[23]. In denitrification, cell-free preparations of *Pseudomonas stutzeri* have been shown to catalyse the reduction of nitrate and nitrite to nitrous and nitric oxides and to gaseous nitrogen. Among the proposed cofactors for the enzymes involved are NAD, NADP, FAD, FMN, iron and copper, but nothing is certain. It is alleged that some of the enzymes contain bound cytochromes, and in this respect, differ from the metalloflavoproteins involved in the reduction of nitrate to ammonia.

B. Glutamic Dehydrogenase

Glutamic dehydrogenase catalyses the formation of glutamate from α-ketoglutarate and ammonia. The cofactor for the enzyme may be NAD (as shown in equation 1), or NADP, depending on the source of the enzyme. In plants and in most animal tissues, the cofactor is NAD while some bacterial and yeast enzymes require NADP. The mammalian liver enzyme works with either cofactor. The equilibrium constant is such that, at the hydrogen ion and substrate concentrations in most regions containing the enzyme, the production of glutamate is favoured. It has been suggested[24] that the reaction may occur in two steps, in the first of which an imino acid is formed, and in the second, the imino acid is reduced by the cofactor (equations 13 and 14). The

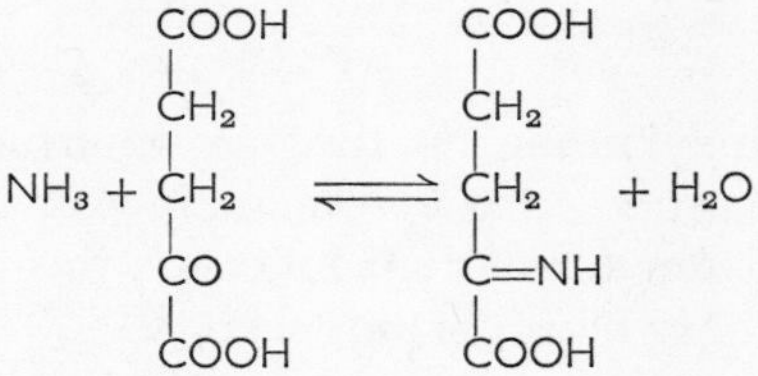

$$
NH_3 +
\begin{array}{c}
COOH \\
| \\
CH_2 \\
| \\
CH_2 \\
| \\
CO \\
| \\
COOH
\end{array}
\rightleftharpoons
\begin{array}{c}
COOH \\
| \\
CH_2 \\
| \\
CH_2 \\
| \\
C{=}NH \\
| \\
COOH
\end{array}
+ H_2O \tag{13}
$$

$$
\begin{array}{c}
COOH \\
| \\
CH_2 \\
| \\
CH_2 \\
| \\
C{=}NH \\
| \\
COOH
\end{array}
+ NADH + H^+ \rightleftharpoons
\begin{array}{c}
COOH \\
| \\
CH_2 \\
| \\
CH_2 \\
| \\
CHNH_2 \\
| \\
COOH
\end{array}
+ NAD^+ \tag{14}
$$

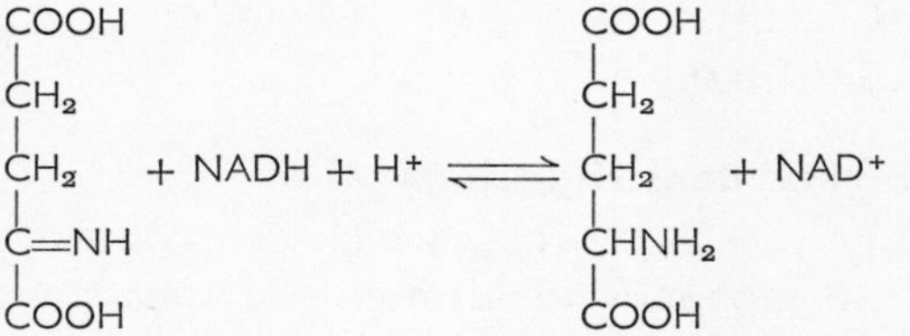

enzyme, which has been crystallised from beef liver, forms active protein aggregates (mol. wt. $\sim 10^6$) which can be dissociated (e.g. by dilution) to give four enzymically active subunits (mol. wt. $\sim 2\cdot5 \times 10^5$). The subunits can be further fragmented, by treatment with urea or dodecyl sulphate, to inactive units (mol. wt. $\sim 5 \times 10^4$)[25], suggesting that the enzymically active entities possess a number of polypeptide chains. The original aggregate may be held together by coordination to zinc[26]. The mechanism of catalysis remains obscure.

The biological importance of this enzyme has already been discussed (section II). Its role in producing glutamate, as the first organic amino compound, in bacteria and plants seems reasonably well established. In animals, which do not have the ability to produce all amino compounds from a simple nitrogen source, the enzyme seems to be concerned with the removal of excess amino compounds (equations 2–5) as well as with the production of glutamate for conversion to the acid amide (section V.B) or to take part in transamination reactions to form the non-essential amino compounds. Cellular control of the direction in which reaction occurs may well lie in the ratio of the concentrations of the oxidised and reduced forms of the cofactor. This ratio is not a fixed quantity but depends on the metabolic activity of the cell (NAD and NADP are cofactors for many oxidation–reduction reactions) as well as on the availability of molecular oxygen for the terminal step in respiration, by which the reduced cofactor is reoxidised by the cytochrome system (section IV.A.1).

C. Transamination

I. General

The enzyme-catalysed, reversible interchange of amino groups between carbon skeletons without the production of ammonia has been recognised since 1937 (equation 15). The actual number of

$$R^1CHCOOH + R^2COCOOH \rightleftharpoons R^1COCOOH + R^2CHCOOH \qquad (15)$$
$$\underset{NH_2}{|} \qquad\qquad\qquad\qquad\qquad \underset{NH_2}{|}$$

transaminases and the relevant specificities of these enzymes are still in dispute but indirect evidence suggests that many pairs of α-amino and α-keto acids can take part in transamination reactions. Only two transaminases have been purified to states approaching homogeneity and of these two, (catalysing reactions (16) and (17)) the glutamic–aspartic transaminase has been the subject of most mechanistic studies, partly because one of the substrates (oxaloacetic acid) can be relatively easily estimated. The enzymes catalysing reactions (16)

$$
\begin{array}{ccccc}
COOH & COOH & & COOH & COOH \\
| & | & & | & | \\
CH_2 & (CH_2)_2 & & CH_2 & (CH_2)_2 \\
| & + \; | & \rightleftharpoons & | & + \; | \\
CHNH_2 & CO & & CO & CHNH_2 \\
| & | & & | & | \\
COOH & COOH & & COOH & COOH \\
\text{aspartate} & \text{α-ketoglutarate} & & \text{oxaloacetate} & \text{glutamate}
\end{array}
\qquad (16)
$$

$$
\begin{array}{ccccc}
CH_3 & COOH & & CH_3 & COOH \\
| & | & & | & | \\
CHNH_2 & (CH_2)_2 & & CO & (CH_2)_2 \\
| & + \; | & \rightleftharpoons & | & + \; | \\
COOH & CO & & COOH & CHNH_2 \\
& | & & & | \\
\text{alanine} & COOH & & \text{pyruvate} & COOH
\end{array}
\qquad (17)
$$

and (17) do not show absolute specificity for these particular substrates. For example, in equation (16), oxaloacetate can be replaced by α-keto-malonate, aspartate by cysteic acid ($HSO_3CH_2CH(NH_2)COOH$) or glutamate by α-methylglutamate. However, the activities observed with substrates other than the principal ones is too low to account for the production of α-amino acids other than aspartate and alanine by transamination between glutamate and α-keto acids. Indirect evidence for the existence of other transaminases comes from studies of the nutritional requirements of whole animals and mutants of some microorganisms, and from the effects of vitamin B_6 deficiency. The dietary nitrogen requirements of higher animals include certain so-called essential amino acids (listed in Table 2 for man and rat), but

TABLE 2.   Essential amino acids.

Arginine[a]	Methionine
Histidine[a]	Phenylalanine
Isoleucine	Threonine
Leucine	Tryptophan
Lysine	Valine

[a] Only in rat.

it has been shown[27] that young rats develop equally well if supplied with the α-keto analogues of at least five of these amino acids (leu, ileu, val, phe and met) together with glutamic acid. Since development of the animal depends on protein biosynthesis which, in turn, depends on the presence simultaneously of all twenty-odd amino acids, the implication is that transamination between glutamate and the dietary α-keto acids must occur rapidly. Organisms requiring certain amino acids do not, then, appear to lack the enzymes catalysing the introduction of the amino group to form these amino acids, but enzymes involved in forming the carbohydrate skeletons into which the amino group is incorporated. The synthesis of particular transaminases can be either induced or repressed in certain microorganisms. For example, if *Escherichia coli* is grown in a medium deficient in valine, the ability of this organism to transaminate between alanine and α-ketoisovalerate increases[28]. If, on the other hand, the growth medium is supplemented with tyrosine, the ability of *E. coli* to form tyrosine and phenylalanine from glutamate and *p*-hydroxyphenylpyruvate or pyruvate, decreases[29].

As will be seen later, enzymes catalysing a number of reactions of α-amino acids, including transamination, require a derivative of vitamin B_6 as cofactor. Vitamin B_6 deficiency in both rats and microorganisms causes, among other things, a decrease in the level of transaminase activity. For example, *Streptococcus faecalis* can grow in the presence of the α-keto acid analogues of normally essential α-amino acids, but when grown in a vitamin B_6 deficient medium, the ability to transaminate between glutamate and α-keto acids is lost. Transaminase activity can be restored by addition of a sufficient quantity of the cofactor.

The transamination reaction represented by equation (15) involves the reversible transfer of α-amino groups. Although there are probably more enzymes catalysing this type of reaction, some transaminases catalysing ω-amino group transfer have also been partially identified. For example, the α-ketoglutarate—γ-aminobutyrate transaminase localised in the mitochondria of the grey matter of brain[30] and also present in some microorganisms[31], catalyses the reaction given in equation (18). Transamination between ornithine and α-keto acids

$$
\begin{array}{cccc}
\text{CH}_2\text{NH}_2 & \text{COOH} & \text{CHO} & \text{COOH} \\
| & | & | & | \\
\text{CH}_2 & (\text{CH}_2)_2 & \text{CH}_2 & (\text{CH}_2)_2 \\
| & + \; | & \rightleftharpoons \quad | & + \; | \\
\text{CH}_2 & \text{CO} & \text{CH}_2 & \text{CHNH}_2 \\
| & | & | & | \\
\text{COOH} & \text{COOH} & \text{COOH} & \text{COOH}
\end{array} \tag{18}
$$

γ-amino- α-ketoglutarate   succinic   glutamate
butyrate   semi-
                               aldehyde

(equation 19) has been demonstrated in liver and in *Neurospora crassa*, while an enzyme specific for the reaction between ornithine and α-ketoglutarate has been partially purified from rat liver mitochondria[32]. It may be noted that transamination reactions involving

$$
\begin{array}{cccc}
\text{CH}_2\text{NH}_2 & & \text{CHO} & \\
| & \text{R} & | & \text{R} \\
(\text{CH}_2)_2 & | & (\text{CH}_2)_2 & | \\
| & + \; \text{CO} & \rightleftharpoons \quad | & + \; \text{CHNH}_2 \\
\text{CHNH}_2 & | & \text{CHNH}_2 & | \\
| & \text{COOH} & | & \text{COOH} \\
\text{COOH} & & \text{COOH} &
\end{array} \tag{19}
$$

ornithine   glutamic-
                       γ-semialdehyde

glycine may be regarded as both α- and ω-amino group transfers. It is of interest that transamination between glyoxylate and α-amino acids occurs readily, even in the absence of enzyme catalysts[33].

In spite of the failure, to date, to isolate and adequately characterise more than two transaminases, both of which catalyse α-amino group transfer, it is now generally accepted that transamination plays an important part in amino acid metabolism.

2. Mechanism of transamination

The reactions of amino acids, catalysed by enzymes requiring a derivative of vitamin B_6 (Figure 5a) as cofactor, have been the subject of a number of mechanistic studies. Enzymes catalysing decarboxylation, racemisation, dehydration (of serine or threonine) or desulphydration (of cysteine) require pyridoxal-5'-phosphate as cofactor (Figure 5b), and are dealt with in later sections. Enzymes catalysing transamination, on the other hand, require either pyridoxal-5'-phosphate or pyridoxamine-5'-phosphate (Figure 5c). Snell and coworkers[33] showed that the reactions of amino acids normally

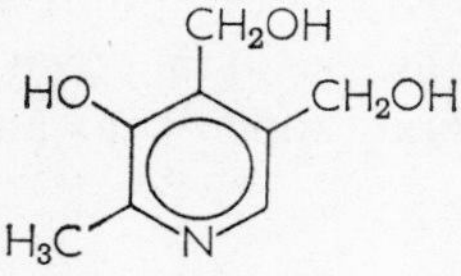

FIGURE 5a.   FIGURE 5b.   Pyridoxal-   FIGURE 5c.   Pyridoxamine-
Pyridoxine.   5′-phosphate.   5′-phosphate.

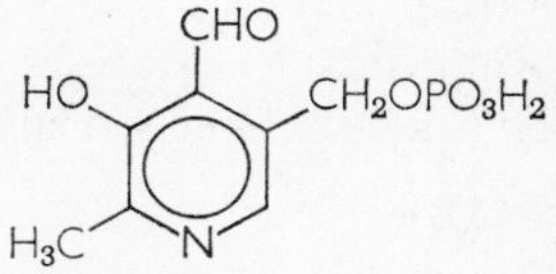

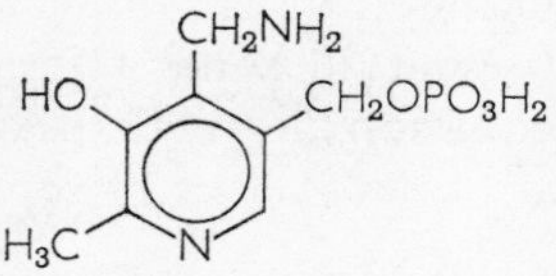

catalysed by pyridoxal phosphate-dependent enzymes occurred in aqueous solutions at 100° in the presence of the non-phosphorylated cofactor and metal ions (e.g. Al^{3+}), the relative extents of decarboxylation, racemisation, transamination etc., depending on pH. The general theory put forward by Snell to account for the reactions occurring in the absence of enzymes is summarised in Figure 6. Metzler[34] later confirmed that imines are formed between pyridoxal and amino acids in aqueous solutions. No detailed analysis of the subsequent reactions of the imines was possible because of the complexity of the product mixture.

It has since been shown[35] that reversible, non-enzymic transamination occurs between pyridoxal and alanine and pyridoxamine and pyruvate in neutral, aqueous solutions, at 25° and in the absence of metal ions. (No evidence has yet been obtained for the presence of metal ions in purified mammalian transaminases.) Under these conditions, no decarboxylation or racemisation of the amino acid could be detected. From the results of kinetic and spectrophotometric studies in this simple model system for transamination it has been shown that imines are formed rapidly and reversibly between the two pairs of reactants and that tautomerisation of the two Schiff's base intermediates is rate limiting in the overall transamination reaction. The reaction scheme is given in Figure 7.

A more complete investigation has recently been made of a model system more relevant to the reaction given in equation (16). (The enzyme catalysing this reaction has proved the most amenable to purification and characterisation). The reactions between pyridoxal-5′-phosphate and glutamate or aspartate and between pyridoxamine-5′-phosphate and oxaloacetate or α-ketoglutarate have been studied individually. Transamination in these systems again occurs by rapid imine formation and slow isomerisation of two imine intermediates. The complete system is best described by the reaction scheme given in Figures 8a and b. Values have been assigned to the six equilibrium

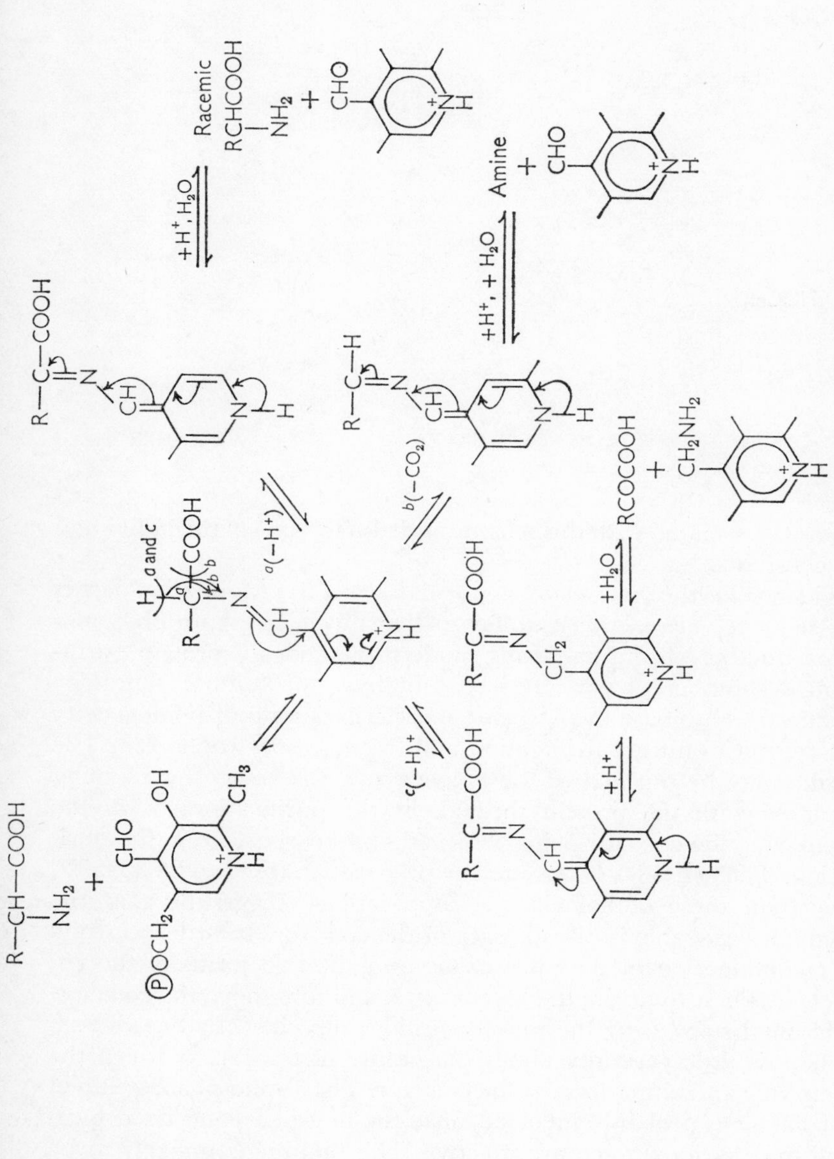

FIGURE 6.   Mechanism of reactions catalysed by three pyridoxal phosphate-dependent classes of enzymes.

[a] Racemisation, [b] Decarboxylation, [c] Transamination

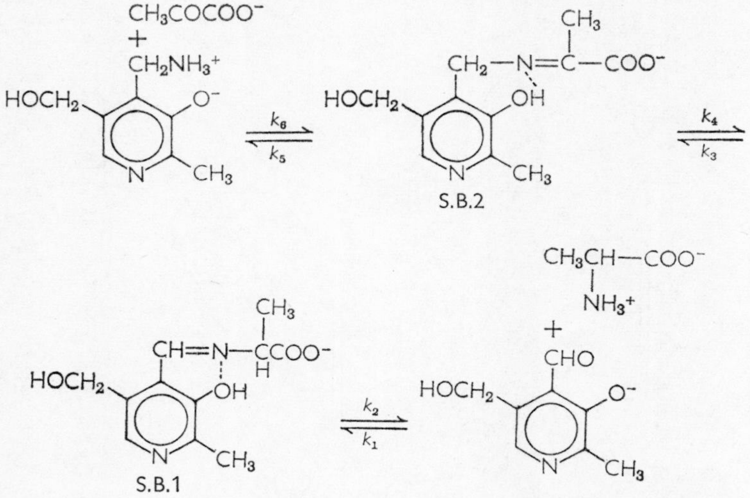

FIGURE 7.

constants associated with this scheme and also to some of the individual rate constants.

Evidence for the view that enzyme-catalysed transamination occurs be a sequence of reactions similar to that given in Figure 8b comes from a number of sources. Using modern methods for protein purification, a number of workers have obtained the soluble glutamic–aspartic transaminase of pig heart muscle in a state of homogeneity with respect to ultracentrifugation and free flow electrophoresis. The enzyme may be purified as the holoenzyme, the cofactor remaining associated with the protein throughout the purification, or as the apoenzyme, the cofactor being removed at an early stage in the purification. The activities of apoenzyme preparations tend to be 30–50% lower than those of holoenzyme preparations, suggesting that the enzyme is less stable in the absence of the cofactor. In order to study the recombination of the apoenzyme and the two forms of the cofactor, it is best to purify the holoenzyme and to remove the cofactor, as the final step, using the mild conditions described by Scardi[37].

There is little certainty about the nature of the link between the apoenzyme and either form of the cofactor. The 5′-phosphate group of the cofactor is probably involved since the non-phosphorylated pyridoxal and pyridoxamine are inactive. The binding is unlikely to be

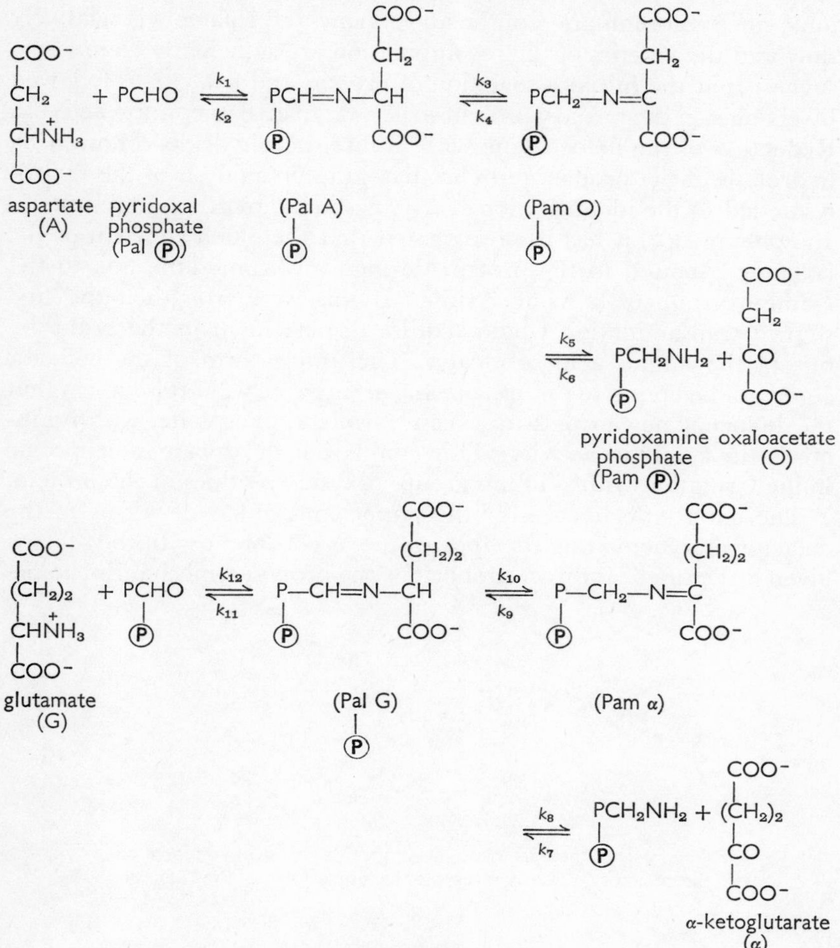

FIGURE 8a

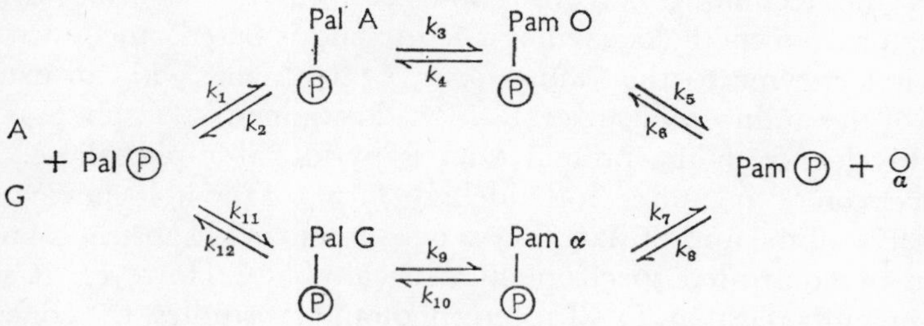

FIGURE 8b.

ionic since recombination of the apoenzyme and cofactors is relatively slow and the kinetics of the recombination are sufficiently complex to suggest that the initial association of protein and cofactor is followed by changes in the tertiary protein structure, to give maximum activity. Reduction of the holoenzyme with lithium borohydride, followed by hydrolysis of the protein and chromatographic analysis of the hydrolysate led to the identification of a lysine-bound pyridinium residue[36]. In consequence, it has been suggested that the aldehyde form of the cofactor is bound to the protein through an azomethine link to the ε-amino group of a lysine residue. If this were true then the first step in transamination would require displacement of the lysine by one of the amino acid substrates. The amine form of the cofactor could not be bound to the apoenzyme in the same way. It is known that the holoenzyme can pick up excess pyridoxal phosphate, with no increase in catalytic activity. This could well be due to non-specific imine formation with ε-amino groups of lysine residues in the protein.

The equal effectiveness of the amine and aldehyde forms of the cofactor, in reactivating the apoenzyme, is consistent with the scheme given in Figure 9. Indirect support for the proposed mechanism comes

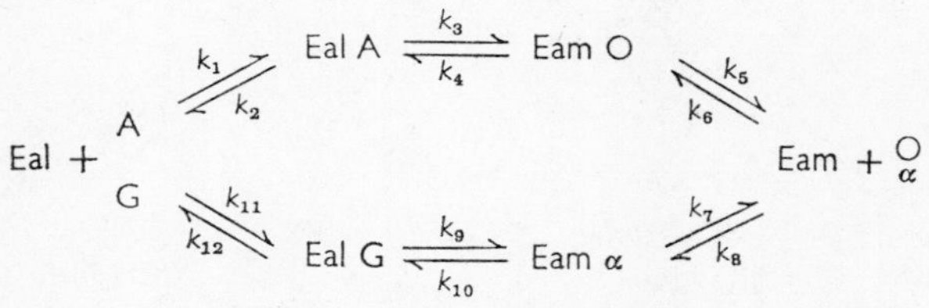

E represents the enzyme.   al represents the aldehyde form of the cofactor.   am represents the amine form of the cofactor

FIGURE 9.

from the technique of Scardi and coworkers[37] for removing the cofactor from the holoenzyme. The method is based on conversion of the holoenzyme to the amine form by treatment with an excess of one of the amino acid substrates. Pyridoxamine-5'-phosphate is more readily lost from the protein than is pyridoxal-5'-phosphate. Interconversion of the amine and aldehyde forms of the holoenzyme in the presence of the appropriate α-keto or α-amino acid substrates has now been demonstrated spectrophotometrically[38, 39]. However, it should be remembered that, in all spectrophotometric studies, the concentrations of holoenzyme required are vastly in excess of the catalytic quantities used in kinetic studies. Interpretation of data obtained using 10^{-3}M solutions of protein in terms of the mechanism of transamination in the presence of about 10^{-8}M protein solutions should be viewed with caution.

Attempts have been made to estimate the amounts of cofactor present in the catalytically active form of the enzyme. The main problem is that the methods available for such determinations usually require high protein concentrations. For example, in determining the molecular weight of a protein in an analytical ultracentrifuge, 0·3 to 1% protein solutions are required (i.e. 10^{-4}M for an average-sized protein). Many proteins (e.g. insulin and glutamic dehydrogenase) aggregate at this concentration. The literature value[40] for the molecular weight of glutamic–aspartic transaminase (110,000) is in error, the true value being $\sim 80,000$[42]. A recent, simple method of molecular weight assessment applicable to globular proteins, involves filtration through a molecular sieve (G-100 sephadex), and has been reviewed recently[41]. The virtue of this technique is that it can be used over a wide range of protein concentrations and so gives information about association phenomena. In the case of transaminase, the molecular weight, estimated by gel filtration, is $\sim 77,000$ and shows no variation in the concentration range 10^{-4}–10^{-8}M[42]. Repeated estimations of the amount of cofactor bound in the holoenzyme have given, consistently, one mole per 40,000 g protein[42].

It is well established that the glutamic–aspartic transaminase, while showing no concentration-dependent association phenomena, contains two moles of cofactor per mole of protein, molecular weight $\sim 80,000$. This may appear inconsistent with the proposed scheme, in which transamination is regarded as the sum of two independent half-reactions, since the presence of two cofactor molecules might seem to imply that a ternary mechanism (Figure 10) operates, the two half-reactions occurring simultaneously. However, evidence that the true equivalent weight of the transaminase is 40,000 rather than 80,000,

$$E + A \rightleftharpoons EA$$

$$EA + B \rightleftharpoons EAB \relbar\joinrel\dashrightleftharpoons ECD$$

$$ECD \rightleftharpoons EC + D$$

$$EC \rightleftharpoons E + C$$

Either A and C or B and D are the two amino acid substrates and the two keto acid substrates.

FIGURE 10

has recently been obtained by a method of quantitative estimation of the N-terminal amino acid of the protein.

A considerable number of kinetic investigations of the reaction catalysed by pig heart glutamic–aspartic transaminase have been reported, the majority of which are based on ambiguous methods of measurement of reaction rates. Unambiguous assay conditions have now been established[43] and used in a recent and complete investigation[44]. The results are consistent with the binary mechanism (Figure 9) rather than the ternary mechanism (Figure 10). The equilibrium constants associated with all six steps have been evaluated. From conventional studies, for this type of system, of the way in which initial velocities of forward and reverse reactions depend on substrate concentrations, it is normally possible to obtain eight independently variable parameters, each of which is a function of some of the twelve individual rate constants. However, in this particular system, additional information can be obtained from the effects, on measured initial velocities, of deuterium substitution in the α-position of the amino acid substrates. It is possible to predict which of the eight parameters will be affected by deuterium substitution. The scheme in Figure 11

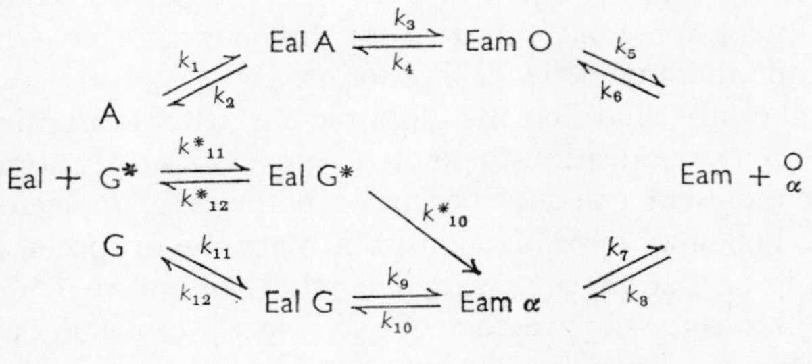

FIGURE 11.

illustrates the situation for transamination between oxaloacetate and α-deuteroglutamate. If, as is known to be the case in the model system, α-deuteration of the amino acids affects only the rate of the slowest steps in transamination, i.e. isomerisation of the Schiff's base intermediates, and not the fast steps, i.e. Schiff's base formation, only parameters containing k_{10} ($k_{10}{}^*$ for α-deuteroglutamate) will be affected in the presence of α-deuteroglutamate (and k_3 in the presence of α-deuteroaspartate). Using the results obtained with normal and α-deutero amino acid substrates, values have been assigned to k_3 and

k_9 (~ 400 sec^{-1}) and k_4 and k_{10} (~ 1000 sec^{-1}). Corresponding values in the model system are $k_3 = 1\cdot6 \times 10^{-6}$ sec^{-1}, $k_9 = 1\cdot75 \times 10^{-6}$ sec^{-1} and $k_{10} = 0\cdot82 \times 10^{-6}$ sec^{-1}. The presence of the enzyme therefore increases the rate of transamination by a factor of 10^8–10^9, which is some 10^3 times higher than the estimated figure quoted by Meister[24] in his recent review.

The above values of certain rate constants are at variance with those quoted by Hammes and Fasella[45]. These workers used the temperature-jump technique to determine the relaxation times associated with intermediates believed to be formed in the course of transamination, but the assignment of the observed phenomena to particular intermediates is somewhat arbitrary. The main source of error in applying this relatively new method to enzyme-catalysed reactions lies in the very high protein concentrations (greater than 10^{-5}M) required. It is quite possible that, under these conditions, enzyme–substrate complexes which do not lie on the normal reaction path may be present in significant amounts.

Enzymes catalysing transamination reactions are clearly of importance in controlling the distribution of amino groups in biological systems. From the mechanistic point of view, transamination is better understood than many other enzyme-catalysed reactions, largely because of the involvement of a well-defined cofactor. If, as seems reasonable, catalysis involves the cofactor directly, it is possible, as here, to design and study model systems from which the enzyme is absent. However, the mechanism of *catalysis* of transamination is as obscure as for other enzyme-catalysed reactions, but at least the magnitude of the problem is now well defined.

3. Formation of individual amino acids

It is clearly not possible to discuss here at any length, the metabolism of individual amino acids. In addition, the details of the biosynthesis and catabolism of amino acids, well reviewed in Volume II of Meister's recent book[46], are concerned more with the formation and breakdown of the carbon skeleton than with the introduction or loss of the amino group. Modifications of some of the twenty amino acids normally found in proteins have been detected in some protein hydrolysates, e.g. iodinated tyrosine, phosphoserine and hydroxylysine. In some cases the modification appears to be made before the amino acid is incorporated into protein (e.g. iodination of tyrosine) while in other cases modification is believed to occur when the amino acid is already present in proteins (e.g. hydroxylation of lysine, and in some cases, of proline). In addition to these less common amino acid residues in proteins, many non-protein amino acids are known to exist, but the functions of these are, in general, obscure. None of the less common amino acids are considered here but reviews are available[47,48]. The suspected origins of the amino groups of the common α-L-amino acids are summarised in Table 3. Transamination, commonly from gluta-

TABLE 3. Origins of amino groups of the common α-L-amino acids.

Amino acid	Principal origin of NH_2	Type of reaction
Glutamic acid	NH_4^+	Reductive amination of α-keto acid
Aspartic acid	(a) Glutamate	Transamination
	(b) NH_4^+ ?	Direct amination of fumarate?
Alanine	(a) Glutamate	Transamination
	(b) Aspartate	β-Decarboxylation (bacteria)
	(c) NH_4^+ ?	Reductive amination of pyruvate (bacteria)
Glycine	(a) Glutamate	Transamination
	(b) Serine	Dehydroxymethylation
Serine	(a) Alanine	Transamination
	(b) Glycine	Hydroxymethylation
Cysteine	Serine	(a) Trans-sulphuration from homo-cysteine
		(b) Direct sulphuration (bacteria)
Methionine	Cysteine	
	Homocysteine (NH₂ from asp.)	See 3.3 c (iii)
Valine	Glutamate?	Transamination
Leucine	Glutamate	Transamination
i-Leucine	Glutamate?	Transamination
Phenylalanine	Glutamate	Transamination
Tyrosine	Glutamate	Transamination
Threonine	Aspartate	Reduction and isomerisation
Histidine	Glutamate	Transamination
Tryptophane	Serine	Conjugation with indole
Arginine		
α-NH₂	Ornithine	Krebs–Hensleit cycle
ω-NH₂	Aspartate, NH_4^+	
Ornithine		
α-NH₂	Glutamate	Reduction of ω-COOH
ω-NH₂	Glutamate	Transamination
	Alanine	
Citrulline		
α-NH₂	Ornithine	
ω-NH₂	NH_3	Krebs–Hensleit cycle
Lysine		
α-NH₂	Aspartate or glutamate	
δ-NH₂	Glutamate	Transamination

mate or aspartate, to α-keto acids is the final step in many cases. Apart from reductive aminations of the type discussed in section IV.B, in which ammonia is incorporated into α-keto acids, the main exceptions to terminal transamination are listed below:

(i) The formation of histidine. The α-amino group of histidine is apparently introduced before the carboxylic acid group is formed. The proposed sequence of reactions, deduced from studies of mutants of *Escherichia coli* and *Neurospora crassa*, is illustrated in Figure 12. The enzyme catalysing transamination between glutamate and imidazolylacetol phosphate has been partly purified from *Neurospora crassa*[49] and shown not to be entirely specific for these substrates.

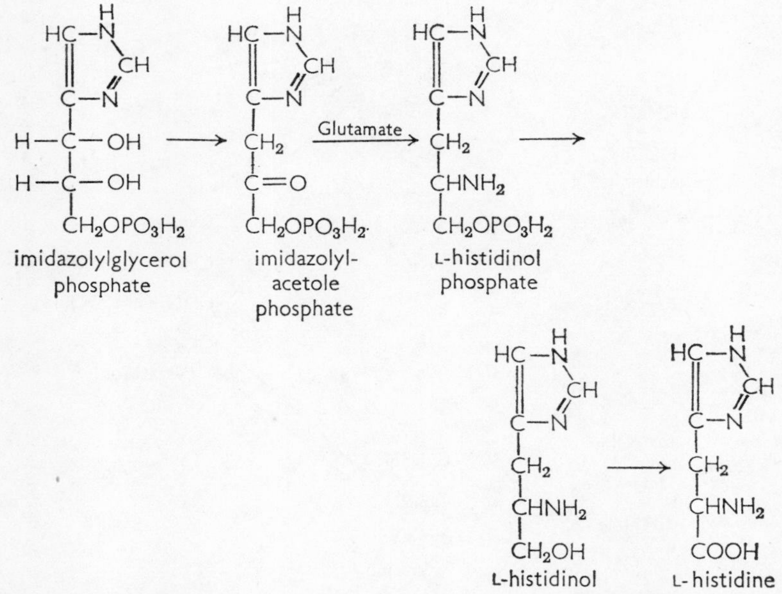

FIGURE 12.   Formation of histidine from imidazolylglycerol phosphate.

(ii) Formation of serine and glycine from each other by transhydroxymethylation, mediated by tetrahydrofolic acid (FH_4) (Figure 13). The overall reaction is represented by equation (20).

$$\begin{array}{l} CH_2OH \\ | \\ CHNH_2 + FH_4 \rightleftharpoons \quad \begin{array}{l} CH_2NH_2 + HOCH_2{-}FH_4 \\ | \\ COOH \end{array} \\ | \\ COOH \\ \text{serine} \qquad\qquad\qquad \text{glycine} \end{array} \qquad (20)$$

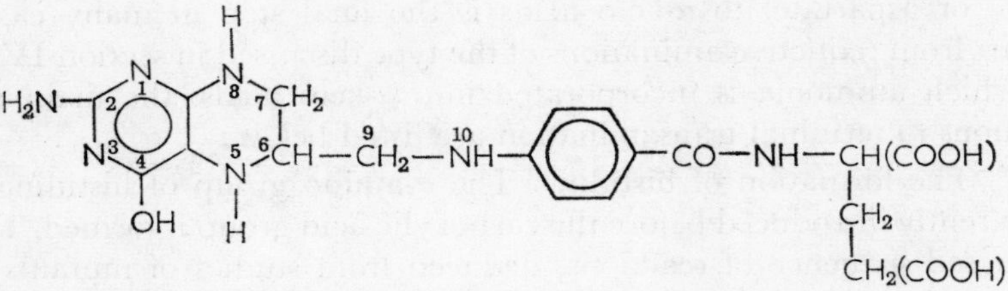

FIGURE 13. Tetrahydrofolic acid (FH$_4$).

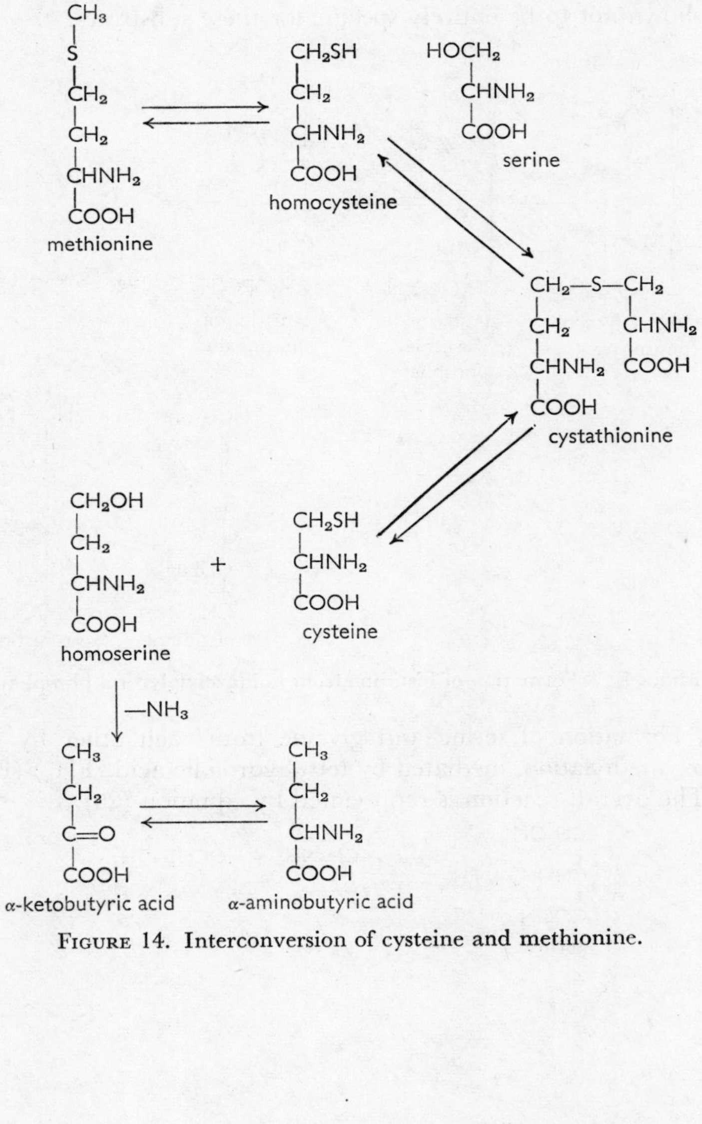

FIGURE 14. Interconversion of cysteine and methionine.

(iii) Formation of the sulphur-containing amino acids, cysteine and methionine, which are interconvertible by the reactions given in Figure 14.

(iv) Formation of tryptophane from serine and indole (or indolyl-3-glycerolphosphate) according to equation (21).

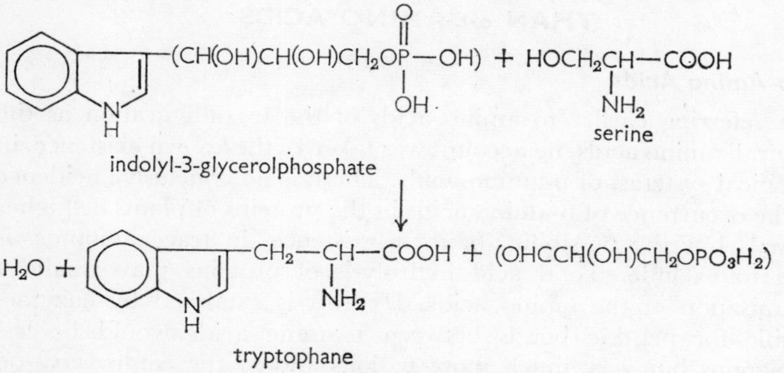

$$\text{indolyl-3-glycerolphosphate} \quad + \quad \text{serine} \tag{21}$$

tryptophane

(v) Formation of threonine from aspartate summarised in Figure 15.

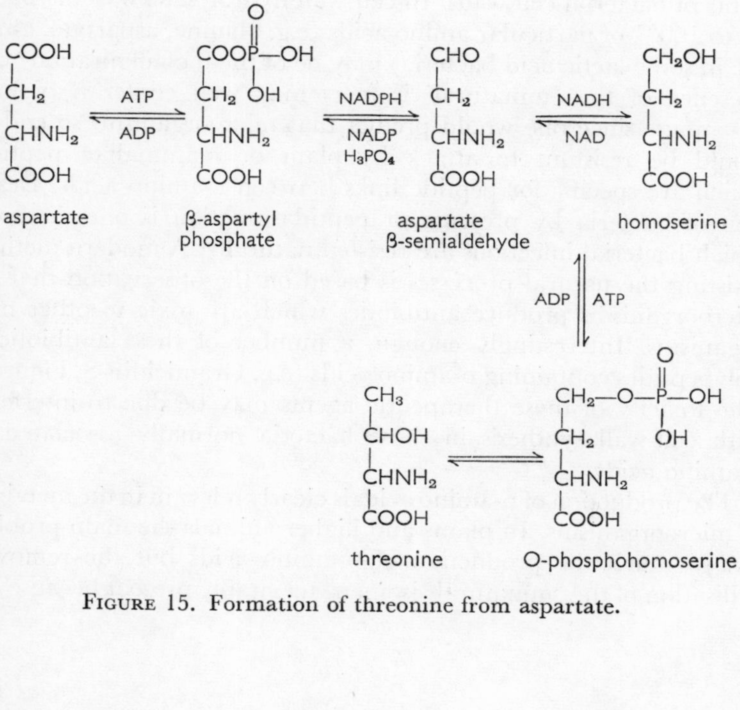

FIGURE 15. Formation of threonine from aspartate.

(iv) Formation of arginine, ornithine and citrulline, all of which are involved in the production of urea, which will be considered in section VI.B.

V. FORMATION OF AMINO COMPOUNDS OTHER THAN α-L-AMINO ACIDS

A. D-Amino Acids

In referring earlier to amino acids of the L configuration as the 'natural' amino acids, no account was taken of the known existence, in biological systems, of D-amino acids. There is no conclusive evidence for the occurrence of D-amino acids in the proteins of plants or higher animals but the possibility of their existence, in trace amounts, is hard to exclude. Total acid hydrolysis of proteins may result in racemisation of the amino acids. Hydrolysis catalysed by enzymes specific for peptide bonds between L-amino acids would be less ambiguous but very much more tedious and so the controversy on this subject remains unresolved[50,51]. It is now well established that peptide-linked, D-amino acids do occur, to a significant extent, in bacteria, particularly in association with the mucopeptide components of bacterial cell walls. In cell-wall hydrolysates of some bacteria 50 to 100% of particular amino acids (e.g. alanine, aspartate, glutamate in some lactic acid bacteria) may be of the D configuration[52]. The presence of the 'unnatural' isomers may well confer a biological advantage since one would predict that microorganisms so endowed would be resistant to attack by plant or mammalian peptidases which are specific for peptide links between L-amino acids. Destruction of bacteria by protease or peptidase action is one method by which bacterial infections are resisted naturally. A modern method of assisting the natural processes is based on the observation that some microorganisms produce antibiotics which are toxic to other microorganisms. Interestingly enough, a number of these antibiotics are polypeptides containing D-amino acids (e.g. Gramicidin-S, Figure 16). The toxicity of these therapeutic agents may be due to interference with cell-wall synthesis in those bacteria normally associated with D-amino acids.

The production of D-amino acids is clearly relevant in the metabolism of microorganisms. In plants and higher animals the main problem is not so much the production of D-amino acids but the removal or utilisation of the 'unnatural' isomers, inevitably present because of the ubiquity of microorganisms. It is known from nutritional studies that some of the essential L-amino acids can be replaced in the diet by the D-isomers (e.g. methionine and phenylalanine in man[53]) without ill effects, implying that a route exists for inversion of amino acid configuration. (The low concentration of free amino acids, of either configuration, in the waste products of animals also leads to the conclusion that animals have some mechanism for destroying D-amino acids).

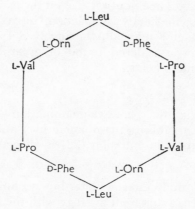

FIGURE 16.   Gramicidin—S.

The two types of reaction by which D-amino acids may be formed *in vivo* are dealt with in this section. The mechanisms by which D-amino acids are incorporated into cell-wall mucopeptides and bacterial antibiotics are not well understood but will be mentioned briefly in section V.E.

I. Racemisation

Enzymes catalysing the racemisation of some amino acids have been partly purified only from bacterial sources (e.g. alanine racemase from *Streptococcus faecalis*, methionine racemase from strains of *Pseudomonas*[54]). No direct evidence for their existence in animals is available.

Pyridoxal is known to catalyse racemisation of amino acids in the model systems studied by Snell and coworkers[33] but the evidence for involvement of pyridoxal-5'-phosphate in the enzyme-catalysed reaction is indirect. Nutritional studies on *S. faecalis*[55] have shown that this microorganism has a growth requirement for D-alanine, and that

neither L-alanine nor the α-keto analogue, pyruvate, can replace
D-alanine. The vitamin B_6 content of these microorganisms is low.
Purified glutamic acid racemase from *Lactobacillus arabinosus*[56] is
reported to be inhibited by hydroxylamine, consistent with involvement of an aldehyde group, but there are conflicting reports on activation of the enzyme by added pyridoxal-5′-phosphate[24]. In no case
has a racemase been shown, *in vitro*, to require pyridoxal-5′-phosphate
as a cofactor.

A possible mechanism of racemisation, in the model system, is
included in Figure 6. Some doubt is cast on the exact relevance of this
scheme, to the enzyme-catalysed reaction, by recent reports that a
reduced flavin cofactor may also be involved[57]. Enzyme-catalysed
racemisation is not considered to be of great importance in the
metabolism of D-amino acids.

2. D-Amino acid oxidase

Enzymes catalysing the oxidative deamination of D-amino acids are
widespread (microorganisms, mammalian liver and kidney etc.). The
overall reaction, given in equation (22) is alleged[24] to occur in two

$$RCH(NH_2)COOH + \tfrac{1}{2}O_2 \rightleftharpoons RCOCOOH + NH_3 \qquad (22)$$

steps, through an imine intermediate, but no imine has so far been
identified. There is an apparent analogy with the reverse of reductive
amination as catalysed by glutamic dehydrogenase (section IV.B) but
there is also an important difference. The amino acid oxidases are all
flavoproteins, the flavin cofactor being most commonly FAD (Figure
2c and d), and no pyridine nucleotide cofactor is involved. Reoxidation of the reduced flavoprotein (see Figure 3) by molecular oxygen
produces hydrogen peroxide which is, in turn, destroyed by the exceptionally efficient enzyme catalase. Equations (23–26) summarise

$$RCH(NH_2)COOH + FAD \rightleftharpoons RC(=NH)COOH + FADH_2 \qquad (23)$$

$$RC(=NH)COOH + H_2O \rightleftharpoons RC(=O)COOH + NH_3 \qquad (24)$$

$$FADH_2 + O_2 \rightleftharpoons FAD + H_2O_2 \qquad (25)$$

$$H_2O_2 \longrightarrow H_2O + \tfrac{1}{2}O_2 \qquad (26)$$

the reactions involved. In the absence of molecular oxygen, that is,
under anaerobic conditions, reactions (25) and (26) cannot occur[58].
The reduced flavoprotein resulting by reactions (23) and (24) from
one D-amino acid can then be used to catalyse reductive amination of
α-keto acids to form other D-amino acids.

There is no certain evidence for the existence of specific D-amino acid oxidases. The highly purified, crystalline enzyme[58, 59] from hog or sheep kidney oxidises most D-amino acids, attacking tyrosine, proline and methionine more rapidly than other substrates[60, 61].

The enzyme forms complexes with benzoic acid, which are more resistant to denaturation than is the native protein. Use is made of this fact in purifying the enzyme. This particular flavoprotein is relatively easily resolved (e.g. by dialysis[62] or dilution[61]) into apo-enzyme and FAD. Reported estimates of the size and composition of the active enzyme are at variance, partly because the protein tends to polymerise. The values given by Massey[58] (mol. wt. 182,000, 4 FAD/mole) and by Yagi[63] (mol. wt. 115,000, 2 FAD/mole) are based on ultracentrifuge data and spectrophotometric estimation of FAD concentrations. The apparent discrepancy is discussed by Charlwood[64] in terms of the effects, on sedimentation constants, of changes in shape as well as in size. The most recent value of the minimum molecular weight, quoted by Dixon[61], is 54,000 g/mole FAD. This estimate is based on a polarographic study of the resolution of the holoenzyme on dilution.

Inhibition of the enzyme by reagents combining with sulphydryl groups is consistent with the involvement of a sulphydryl group in the active site of the protein. The rate of inhibition by one of these re-agents (p-chloromercuribenzoate) has recently been shown[61] to be the same as the rate of loss of FAD from the holoenzyme. The susceptible sulphydryl group may, therefore, be concerned only with binding the cofactor to the protein and not with the mechanism of action of the enzyme. The kinetics of action of the enzyme and inhibition of activity by straight-chain fatty acids has led Dixon[61] to propose that substrates are bound to the holoenzyme by the carboxylic acid group and up to five carbons of the amino acid side-chain. Little can be concluded with safety from spectrophotometric studies because these require very high protein concentrations, (see section IV.C), but some such studies have been reported. The holoenzyme is normally yellow but addition of amino acid substrates, under anaerobic conditions, causes a shift in the absorption maximum, a red-coloured intermediate being formed.

No firm conclusions can be drawn so far concerning the mechanism of action of this enzyme, in spite of its availability in crystalline form and its apparent importance in the metabolism of D-amino acids. The reactions catalysed by D-amino acid oxidase probably account for the production of D-amino acids by microorganisms and also for the removal of D-amino acids from higher animals.

18*

B. L-Amino Acid Amides

This book is not concerned with the behaviour of acid amides but brief mention must be made in this section of two naturally occurring amides of L-amino acids, glutamine and asparagine, both of which occur in proteins. Glutamine is of particular importance because of its involvement in many metabolic processes of the type given in equation (27), where the carbon to which the hydroxyl group is

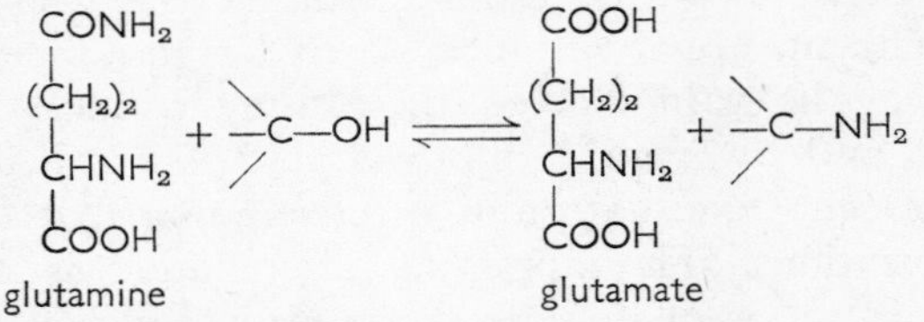

$$
\begin{array}{c}
\text{CONH}_2 \\
| \\
\text{(CH}_2)_2 \\
| \\
\text{CHNH}_2 \\
| \\
\text{COOH} \\
\text{glutamine}
\end{array}
+ \quad \diagdown\!\!\text{C—OH} \rightleftharpoons
\begin{array}{c}
\text{COOH} \\
| \\
\text{(CH}_2)_2 \\
| \\
\text{CHNH}_2 \\
| \\
\text{COOH} \\
\text{glutamate}
\end{array}
+ \quad \diagdown\!\!\text{C—NH}_2 \tag{27}
$$

attached may be part of the ring of a sugar, a purine or a pyrimidine. This is an over-simplification of the situation since a phosphate ester, adenosine triphosphate (ATP, Figure 17) is also commonly involved

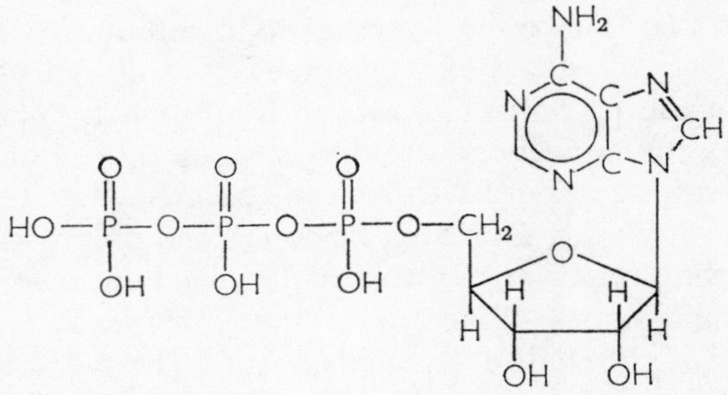

FIGURE 17. Adenosine triphosphate (ATP).

in this type of reaction. The same ester is also involved in the sequence of reactions leading to the formation of glutamine (equation 28).

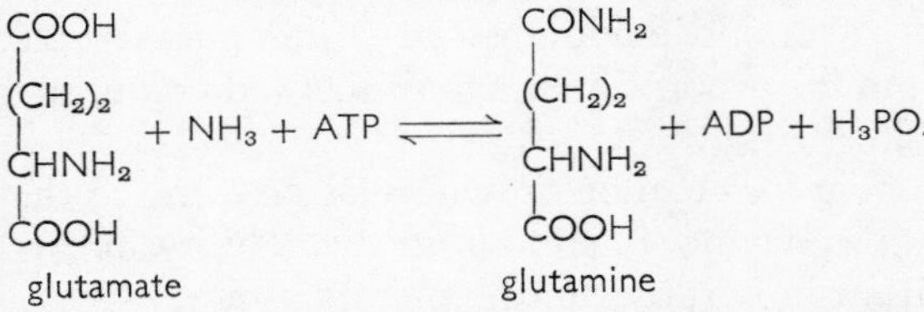

$$
\begin{array}{c}
\text{COOH} \\
| \\
\text{(CH}_2)_2 \\
| \\
\text{CHNH}_2 \\
| \\
\text{COOH} \\
\text{glutamate}
\end{array}
+ \text{NH}_3 + \text{ATP} \rightleftharpoons
\begin{array}{c}
\text{CONH}_2 \\
| \\
\text{(CH}_2)_2 \\
| \\
\text{CHNH}_2 \\
| \\
\text{COOH} \\
\text{glutamine}
\end{array}
+ \text{ADP} + \text{H}_3\text{PO}_4 \tag{28}
$$

Glutamine synthetase, the enzyme catalysing this reaction, has been purified from sheep brain[65]. The possible mechanism of its action has been reviewed by Meister[24], but none of the intermediate steps, proposed on the basis of isotope effects, substrate specificity or catalysis of possible intermediate reactions, have been confirmed. The apparent coupling of synthetic reactions with the formation of ADP and inorganic phosphate from ATP has long been recognised and gave rise to the curious myth that 'energy', derived from catabolic reactions, was somehow stored in ATP. Subsequent hydrolysis of ATP was then thought to release the stored 'energy' which could then be used to drive anabolic processes. It is now abundantly clear[66] that the one reaction which must not occur, if there is to be a link between degradative and synthetic processes, is the hydrolysis of ATP. Metabolic reactions involving ATP and other phosphate esters appear now to occur by transphosphorylation, and the coupling of metabolic processes is best discussed in terms of the presence of enzymes catalysing phosphate group transfer reactions. Enzymes catalysing the hydrolytic reactions are localised, within cells, in regions of low metabolic activity (e.g. in association with cell membranes, and the structural protein of muscle). Synthesis of glutamine may well involve the formation of a phosphorylated derivative of glutamate. The failure to identify such an intermediate may be explained if the phosphorylated intermediate is not released from the surface of the enzyme.

The presence of glutamine synthetase in mammalian brain is interesting in view of the account already given (section II) of differences between animals and microorganisms in their tolerance of ammonia. Since ammonia is toxic to higher animals, apparently acting on the central nervous system, the presence of glutamine synthetase in central nervous tissue confers a two-fold advantage. Firstly, given an adequate supply of glutamic acid, ammonia can be converted to and stored as glutamine, which is non-toxic, and secondly, the glutamine so formed can replace ammonia as a source of amino groups.

The function of asparagine is obscure. It can be identified in most living tissues and occurs at high concentrations in some higher plants in which it may replace glutamine as a non-toxic source of available nitrogen. Asparagine and glutamine are interconvertible by transamidation (equation 29). Direct formation of asparagine from ammonia

$$
\begin{array}{cc}
\begin{array}{c}
CONH_2 \\
| \\
(CH_2)_2 \\
| \\
CHNH_2 \\
| \\
COOH
\end{array}
\begin{array}{c}
COOH \\
| \\
CH_2 \\
| \\
CHNH_2 \\
| \\
COOH
\end{array}
& \rightleftharpoons &
\begin{array}{c}
COOH \\
| \\
(CH_2)_2 \\
| \\
CHNH_2 \\
| \\
COOH
\end{array}
\begin{array}{c}
CONH_2 \\
| \\
CH_2 \\
| \\
CHNH_2 \\
| \\
COOH
\end{array}
\end{array}
\qquad (29)
$$

glutamine aspartate   glutamate asparagine

and aspartate may occur by a reaction analogous to that given for glutamine (equation 28) but a recent report suggests that the products of ATP cleavage may in this case be AMP and inorganic pyrophosphate[67].

Glutamine (and probably asparagine) can be destroyed by two independent routes. The enzyme catalysing direct deamidation to glutamate is activated by phosphate, suggesting a mechanism related to that of glutamine synthetase. The second path involves transamination (equation 30) and subsequent deamidation (equation 31) of α-ketoglutaramic acid. The two reactions are catalysed by separate enzymes[68].

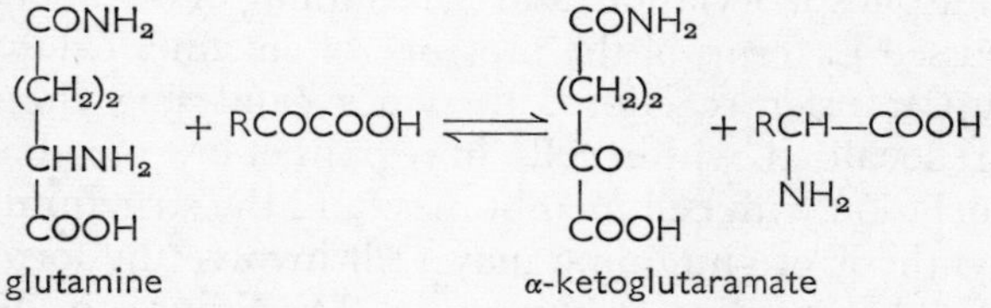

$$
\begin{array}{c}
CONH_2 \\
| \\
(CH_2)_2 \\
| \\
CHNH_2 \\
| \\
COOH \\
\text{glutamine}
\end{array}
+ RCOCOOH \rightleftharpoons
\begin{array}{c}
CONH_2 \\
| \\
(CH_2)_2 \\
| \\
CO \\
| \\
COOH \\
\text{α-ketoglutaramate}
\end{array}
+
\begin{array}{c}
RCH-COOH \\
| \\
NH_2
\end{array}
\tag{30}
$$

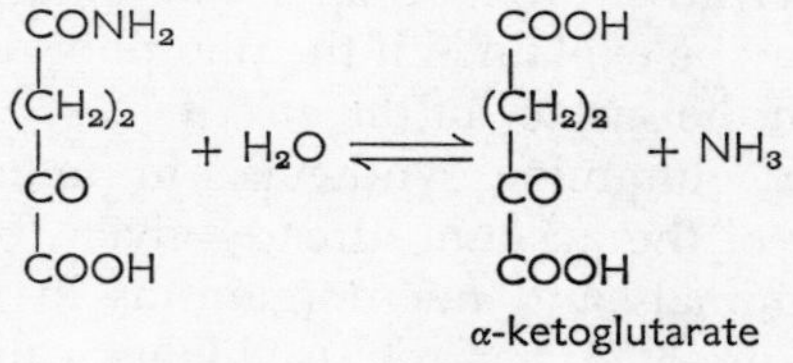

$$
\begin{array}{c}
CONH_2 \\
| \\
(CH_2)_2 \\
| \\
CO \\
| \\
COOH
\end{array}
+ H_2O \rightleftharpoons
\begin{array}{c}
COOH \\
| \\
(CH_2)_2 \\
| \\
CO \\
| \\
COOH \\
\text{α-ketoglutarate}
\end{array}
+ NH_3
\tag{31}
$$

C. Amines

I. General

The importance of amines in biological systems is undoubted, but they are the hardest to discuss under the heading of this chapter. One reason for this is that different life forms show remarkable differences in relation both to the formation and function of amines. In general, the words 'amine' and 'detoxication' are closely linked in the minds of most biochemists, implying that amines are undesirable metabolites which should be removed forthwith. Putrefaction of organic matter (spoilage of foodstuffs) is largely due to the formation of amines with familiar, offensive smells. The agents primarily responsible for this type of amine formation are microorganisms which contain a range of enzymes catalysing the decarboxylation of amino acids. The toxicity to mammals of amines produced by bacterial decay may be due to effects on blood pressure. Certain amines found normally in mammals (carried in the blood stream and so classified as *hormones*) are known, among other actions, to affect indirectly the size of certain blood vessels, causing them to constrict (vasopressors) or expand (vasodilators), so increasing or decreasing arterial blood pressure. The specific effects of particular amines depend on the site of action and on the level of concentration so no general statement can be made about their mode of action. In addition, in common with most biological control systems, other factors are involved in the overall control of blood pressure and all the types of control are interdependent.

It appears, from the preceeding paragraph, that in mammals as well as in bacteria there must be a mechanism for the production of at least some amines. The route is now known to be formally the same as that in bacteria, i.e. decarboxylation of amino acids, but a distinction must be made between the bacterial and mammalian enzymes, not only on the grounds of differences in the physiological function of the products, but also because of differences in the substrate specificities of enzymes from the two sources.

Plants can use urea and a variety of relatively simple amines as a source of nitrogen. The first step in the metabolism of amines is probably oxidation to the corresponding aldehyde and ammonia (see section V.A). The amines occurring naturally in plants include methylamine, isoamylamine and other volatile amines, which are found in some flowering species. The route to these amines is not thought to involve decarboxylation of amino acids, but little is yet known with certainty about the metabolism of amines in plants. Glutamic decarboxylase activity has been detected in a wide variety of higher plants, consistent with the widespread distribution of the product of decarboxylation, γ-aminobutyric acid.

2. Mechanism of α-decarboxylation

A number of enzymes catalysing decarboxylation of amino acids, and showing substrate specificity, have been isolated from bacteria and well characterised[69, 70]. In all cases, pyridoxal phosphate was found to be a cofactor. The mammalian decarboxylases are less well characterised but in common with the plant glutamic decarboxylase they also require pyridoxal phosphate for activity. The possible mechanism of decarboxylation is illustrated in Figure 18[71]. Data obtained by Mandeles and coworkers[72] on the decarboxylation of α-amino acids in D_2O are consistent with this mechanism, the product amine containing only one deuterium atom attached to the α-carbon of the substrate. An earlier proposal, that loss of hydrogen from the

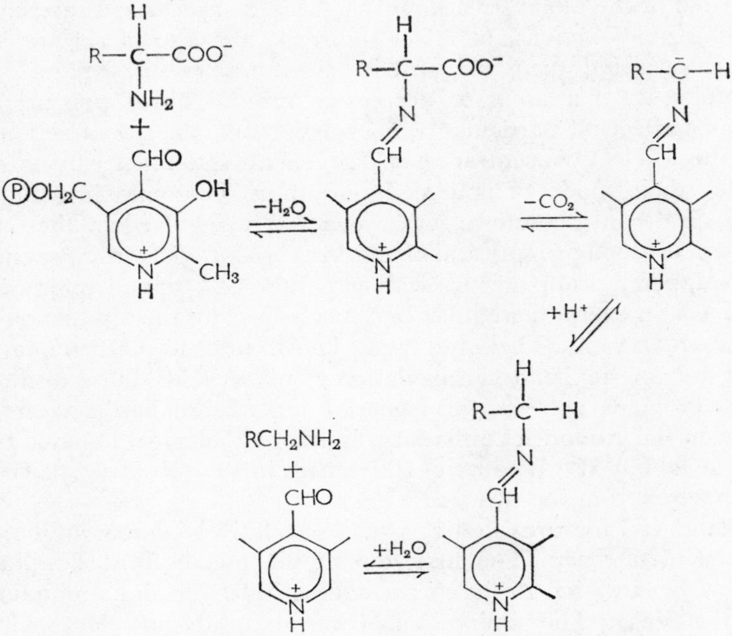

FIGURE 18. Mechanism of α-decarboxylation of amino acids.

α-carbon of the substrate preceeded decarboxylation, is disproved by this experiment, since such a mechanism would require that two deuterium atoms be incorporated into the amine from the solvent. The amine produced by enzyme-catalysed decarboxylation of L-glutamic acid in D_2O is asymmetric about the α-carbon atom (equation 32).

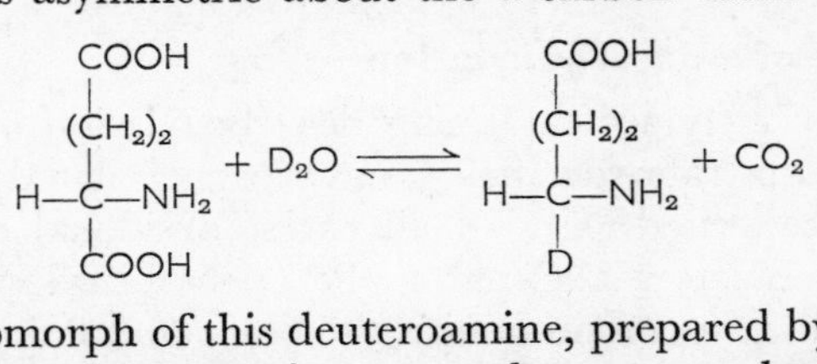

$$
\begin{array}{c}
\text{COOH} \\
| \\
(\text{CH}_2)_2 \\
| \\
\text{H—C—NH}_2 \\
| \\
\text{COOH}
\end{array}
+ D_2O \rightleftharpoons
\begin{array}{c}
\text{COOH} \\
| \\
(\text{CH}_2)_2 \\
| \\
\text{H—C—NH}_2 \\
| \\
\text{D}
\end{array}
+ CO_2
\qquad (32)
$$

The enantiomorph of this deuteroamine, prepared by decarboxylation of α-deuteroglutamate in water, does not exchange deuterium with the solvent, under the same conditions. (α-Deuteroglutamate is prepared by enzyme-catalysed racemisation of D- or L-glutamate in deuterium oxide). Additional support for the stereospecificity of enzyme-catalysed decarboxylation of amino acids comes from experiments in which tyrosine was decarboxylated in D_2O to give R-α-deuterotyramine and α-deuterotyrosine was decarboxylated in H_2O to give S-α-deuterotyramine. The configurations of the products were assigned by comparison of the rates of enzyme-catalysed oxidation (also stereospecific, see section VI.A) of the two isomers with those for oxidation of isomers prepared by unambiguous synthesis[73].

3. Distribution and specificity of α-decarboxylases

Decarboxylases isolated from bacteria act on the amino acids listed in Table 4. Each enzyme appears to be substrate specific and the

TABLE 4.   Common amino acids decarboxylated
          by microorganisms.

Amino acid	Product
Arginine	Agmatine
Aspartic acid	L-Alanine or β-alanine
Cysteic acid	Taurine
Glutamic acid	γ-Aminobutyric acid
Histidine	Histamine
i-Leucine	(2-Methyl)butylamine
Leucine	(3-Methyl)butylamine
Lysine	Cadaverine
Ornithine	Putrescine
Phenylalanine	(2-Phenyl)ethylamine
Tyrosine	Tyramine
Valine	(2-Methyl)propylamine

substrates are in general those found in protein hydrolysates. There is so far no conclusive evidence for decarboxylation of such common amino acids as glycine, serine, alanine, threonine, proline or hydroxy-proline, the sulphur-containing acids or the common acid amides.

In animals, decarboxylation of glutamic acid, histidine, tyrosine, phenylalanine and tryptophane and hydroxy derivatives of the aromatic amino acids is reasonably well established. The enzymes catalysing decarboxylation of glutamic acid and of histidine are sub-strate specific but there is no certain evidence for the existence of specific aromatic amino acid decarboxylases.

a. *Glutamic decarboxylase.* The activity of this enzymye has been demonstrated in the brain[74]. The decarboxylation product (γ-amino-butyric acid) is known to reduce the excitability of nerve cells and has been tentatively identified as the central nervous inhibitor in lobster brain[75], and as the peripheral inhibitor in crustacean muscle.

b. *Histidine decarboxylase.* This catalyses the formation of histamine which is liberated when cells are damaged and which has powerful physiological effects. Although histamine causes contraction of the smooth muscle present in the walls of large blood vessels and one would predict a consequent rise in blood pressure, it also causes dilation of the minute peripheral blood capillaries. The overall affect of administration of histamine is a fall in blood pressure with attendant side effects. The enzyme is present in special components of the blood called mast cells.

c. *The aromatic amino acid decarboxylase.* This catalyses a number of reactions leading to physiologically important amines. Some of these reactions are given in Figures 19a and b. The enzyme has been partly purified from pig kidney[76] and identified in liver[77]. It shows a constant ratio of activities, with respect to decarboxylation of tyrosine, tryptophane and phenylalanine, 3,4-dihydroxyphenylalanine and 5-hydroxytryptophane, up to one hundred-fold purification. Histidine is also decarboxylated by the same enzyme but not as efficiently as by the specific histidine decarboxylase of mast cells.

The specific action of serotonin, the amine produced by decarboxylation of 5-hydroxytryptophane, in stimulating cerebral activity, suggests that a specific decarboxylase for this substrate may be present in brain tissue. The amine itself cannot pass the so-called blood–brain barrier, so must be formed *in situ*. A deficiency of serotonin results in mental depression. The serotonin produced by the kidney decarboxylase and carried in the blood may be concerned with control of blood pressure since it is a powerful vasoconstrictor. The two amines tyramine and tryptamine are both vasopressors.

Norepinephrine, (Figure 19a) and its *N*-methyl derivative epinephrine (adrenalin) are examples of so-called 'chemical transmitters' which are substances released at the end of nerve fibres in response to electrical impulses passing down the fibres. (Electrical impulses, at any rate in the peripheral nervous system, pass down nerves by transverse migration of ions across the nerve membrane.) Their function is to excite neighbouring cells. Where the neighbouring cell is another nerve cell, the junction is called a *synapse*; where the neighbouring cell is a muscle cell, a *neuro-muscular junction* is said to be formed. Norepinephrine and epinephrine are the transmitters at most of the endings of nerve fibres belonging to the sympathetic system affecting smooth and cardiac muscle and glands and, as such, have diverse physiological effects, e.g. increase in rate and force of heart beat, dilation of the bronchi, constriction of peripheral vessels etc. In addition, the same

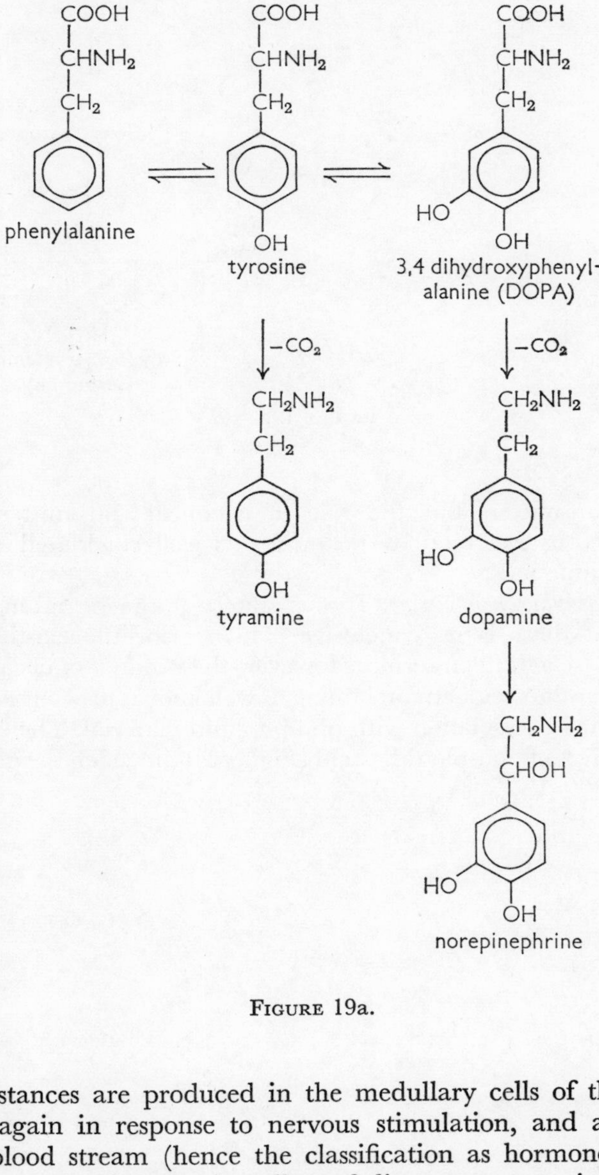

FIGURE 19a.

two substances are produced in the medullary cells of the adrenal glands, again in response to nervous stimulation, and are carried in the blood stream (hence the classification as hormones) to sites at which they reinforce the effect of direct nervous stimulation of effector organs. Epinephrine also affects carbohydrate metabolism in the liver. It is possible that both these substances are also central

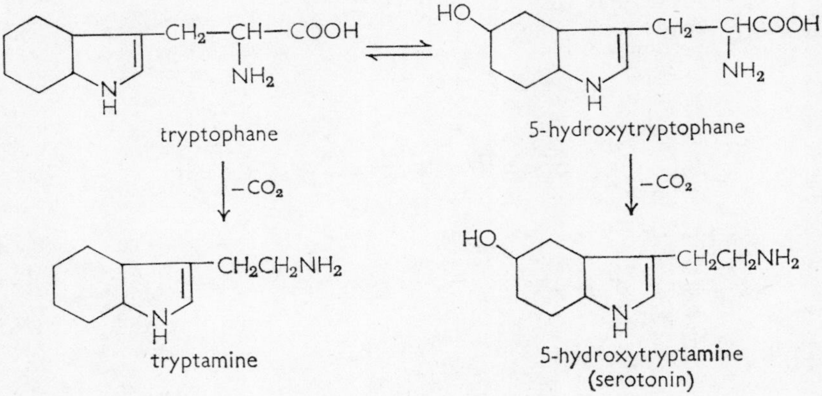

FIGURE 19b.

nervous transmitters but the role of chemical transmitters in the central nervous system is by no means as well elucidated as it is at peripheral junctions.

d. *Decarboxylation of serine.* It was stated at the beginning of this section that there is no conclusive evidence for the existence of a decarboxylase acting on serine. However, the product of decarboxylation of this amino acid, ethanolamine, is well known to occur naturally, in particular in association with phospholipid material. The cephalins are derivatives of phosphatidic acid (Figure 20) in which the phosphate

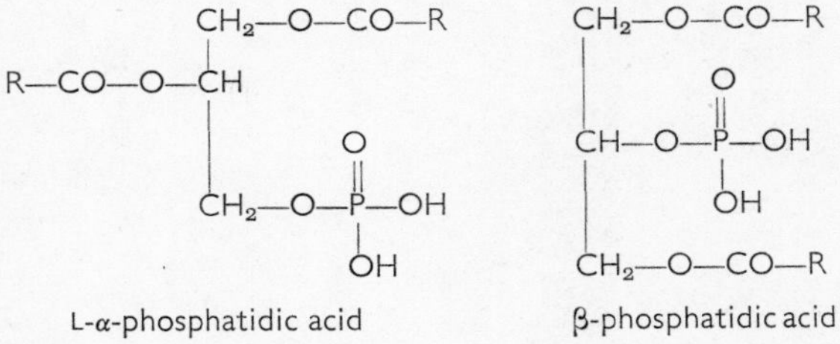

FIGURE 20.

group is esterified with ethanolamine. The lecithins have the same general structure as the cephalins but the phosphate group is esterified with choline in place of ethanolamine. Choline, a quaternary nitrogen base, is itself derived from ethanolamine, and the acetyl derivative of choline is independently important because it is the chemical transmitter involved in synaptic transmission of nerve impulses in both the sympathetic and parasympathetic systems, and is the post ganglionic transmitter for parasympathetic nerves and for some sympathetic nerves. It is also the transmitter at the neuro-muscular junction of the somatic (voluntary) system. Acetylcholine acts at those junctions outside the central nervous system which do not respond to epinephrine or norepinephrine.

Recent work in the general field of lipid biochemistry has revealed that although the carbon skeleton of serine is extensively used in the biosynthesis of ethanolamine (and hence choline and acetylcholine), decarboxylation of free serine does not occur[78, 79], nor is the ethanolamine formed from serine found free. It appears that the hydroxyl group of serine is first esterified with phosphatidic acid by reaction with cytidine diphosphate diglyceride (Figure 21). Phosphatidyl

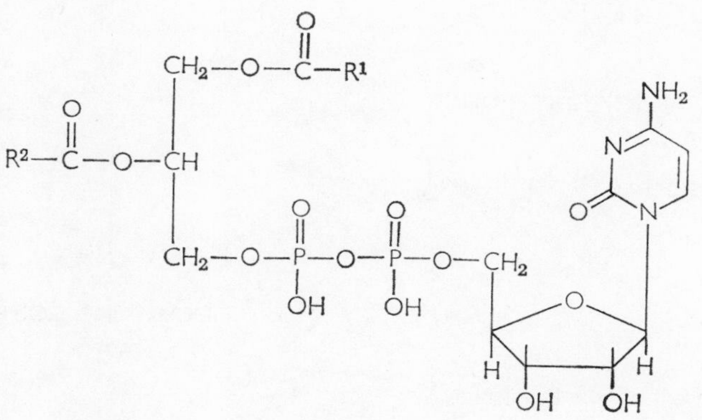

FIGURE 21. Cytidine diphosphate diglyceride.

serine is then decarboxylated[80, 81] to phosphatidyl ethanolamine, methylation of which produces phosphatidyl choline (a lecithin) and choline is eventually released by enzyme-catalysed hydrolysis. The reactions are summarised in Figure 22, with the steps involved in reconversion of choline to serine through a betaine which is successively demethylated to glycine and the glycine hydroxymethylated to serine.

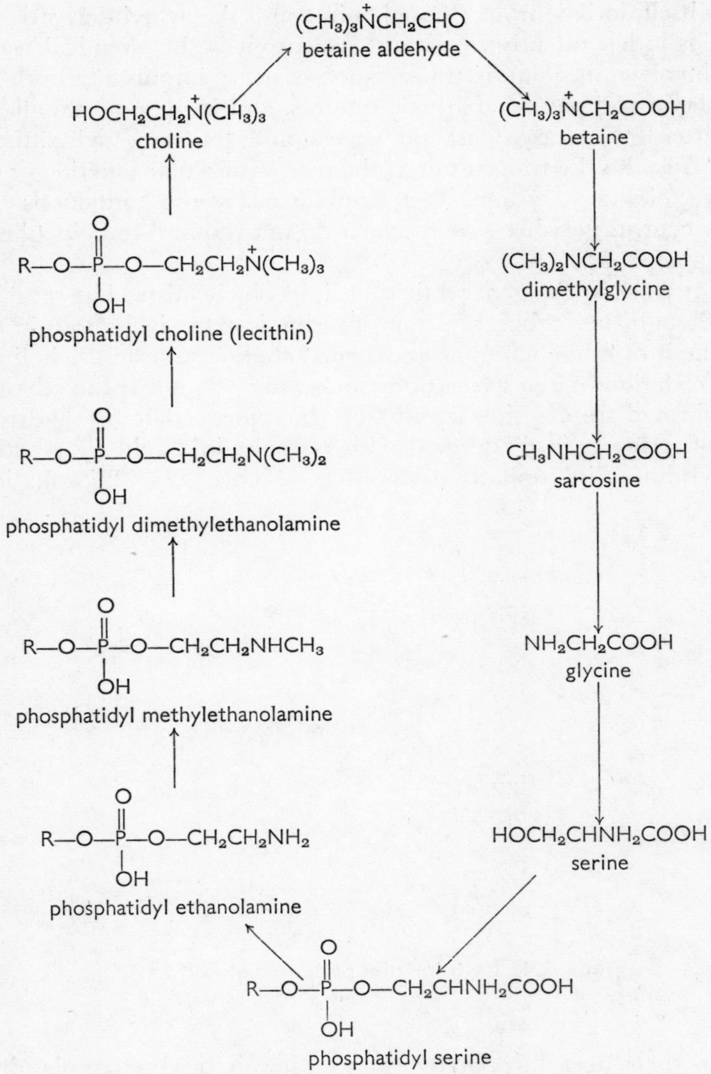
$(CH_3)_3\overset{+}{N}CH_2CHO$
betaine aldehyde
$HOCH_2CH_2\overset{+}{N}(CH_3)_3$
choline
$(CH_3)_3\overset{+}{N}CH_2COOH$
betaine
$R—O—P(=O)(OH)—O—CH_2CH_2\overset{+}{N}(CH_3)_3$
phosphatidyl choline (lecithin)
$(CH_3)_2NCH_2COOH$
dimethylglycine
$R—O—P(=O)(OH)—O—CH_2CH_2N(CH_3)_2$
phosphatidyl dimethylethanolamine
CH_3NHCH_2COOH
sarcosine
$R—O—P(=O)(OH)—O—CH_2CH_2NHCH_3$
phosphatidyl methylethanolamine
NH_2CH_2COOH
glycine
$R—O—P(=O)(OH)—O—CH_2CH_2NH_2$
phosphatidyl ethanolamine
$HOCH_2CHNH_2COOH$
serine
$R—O—P(=O)(OH)—O—CH_2CHNH_2COOH$
phosphatidyl serine

FIGURE 22. Interconversion of serine and choline.

D. Purines and Pyrimidines

I. General

The purine and pyrimidine nucleotides, which are the basic units of the macromolecular nucleic acids, are as important to the functioning of living organisms as are the amino acids. The relevance of these compounds to the present discussion is two-fold. Firstly, three of the five most common nucleotides have amino substituents in the purine or pyrimidine rings (Figure 23), the amino groups coming from the amide group of glutamine or possibly from ammonia, and secondly, the nitrogen atoms of the heterocyclic purine and pyrimidine bases originate variously from aspartate, ammonia (in the form of carbamyl phosphate, see section VI.B) and again the amide group of glutamine. Much information was obtained, in the years 1950–59, concerning the biosynthetic routes to the nucleotides, largely from the results of isotope studies. The fact that purines and pyrimidines are not essential constituents of the diet and, moreover, that dietary purines and pyrimidines are not, in general, incorporated into tissue nucleic acids, suggests immediately that biosynthesis of these compounds may not occur through the free bases. It is now well established that the two types of hetero rings are made at the nucleotide level. Both purine nucleotides originate in inosinic acid, the nucleotide of 6-hydroxypurine, and the pyrimidines originate in uridine-5′-phosphate, a nucleotide of 2,6-dihydroxypyrimidine.

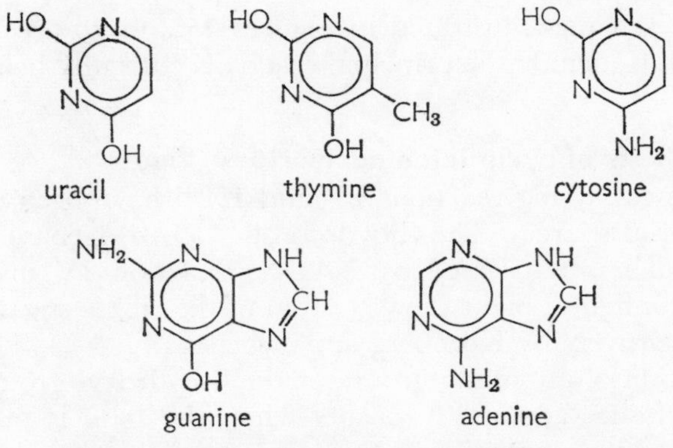

FIGURE 23. Common purine and pyrimidine bases.

Although the origin of each atom in the two ring systems is now known (Figure 24) the postulated routes are not unequivocally established. This is hardly surprising in view of the number of steps apparently involved and the inherent instability of some of the postulated intermediates. The problem of isolating and characterising the enzymes involved is immense and the frequency with which the

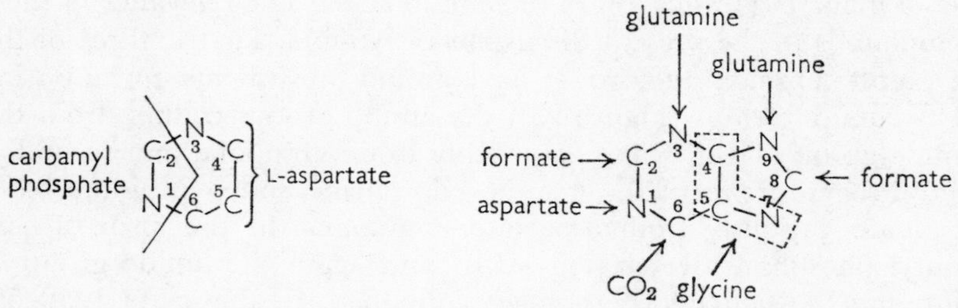

FIGURE 24.   Origins of purine and pyrimidine ring systems.

ubiquitous nucleotide ATP (see section V.B) appears to be associated with intermediate reactions suggests that it will be some years before the detailed mechanism of purine or pyrimidine biosynthesis is understood.

In view of the tentative nature of current theories in this field, mostly arising from work carried out before 1960, no attempt will be made here to give a detailed account of the biosynthesis of purines and pyrimidines. The subject has been included in the recent edition of Meister's book—*Biochemistry of the Amino Acids*[46], and more extensively reviewed at an earlier date[82].

The two schemes illustrated in Figures 25 and 26 are very much simplified but include the principal reactions currently believed to be involved.

2. Biosynthesis of pyrimidine nucleotides (Figure 25)

Enzymes catalysing reactions I[83] and III[84] have been crystallised from bacterial sources. The dihydroorotic dehydrogenase is a flavoprotein which is reoxidised by NAD^+. Reaction IV involving inversion of configuration at the $C_{(1)}$ atom of the ribose sugar, has been studied extensively by Kornberg and associates[85].

Substitution of an amino group for the 6-hydroxyl group of uracil to form cytosine apparently occurs when the base is present as a nucleoside triphosphate. There is some argument as to whether the amino group originates in ammonia[86] or glutamine[87]. The introduction of a methyl group in the $C_{(5)}$ position of the base to form thymine occurs at the nucleotide level, that is after reduction of the sugar ring to form 2-deoxyribose, the sugar found in thymidylic acid.

3. Biosynthesis of purine nucleotides (Figure 26a)

The purine ring system is built, stepwise, onto the $C_{(1)}$ atom of ribose-5-phosphate. Reaction I presumably involves inversion of

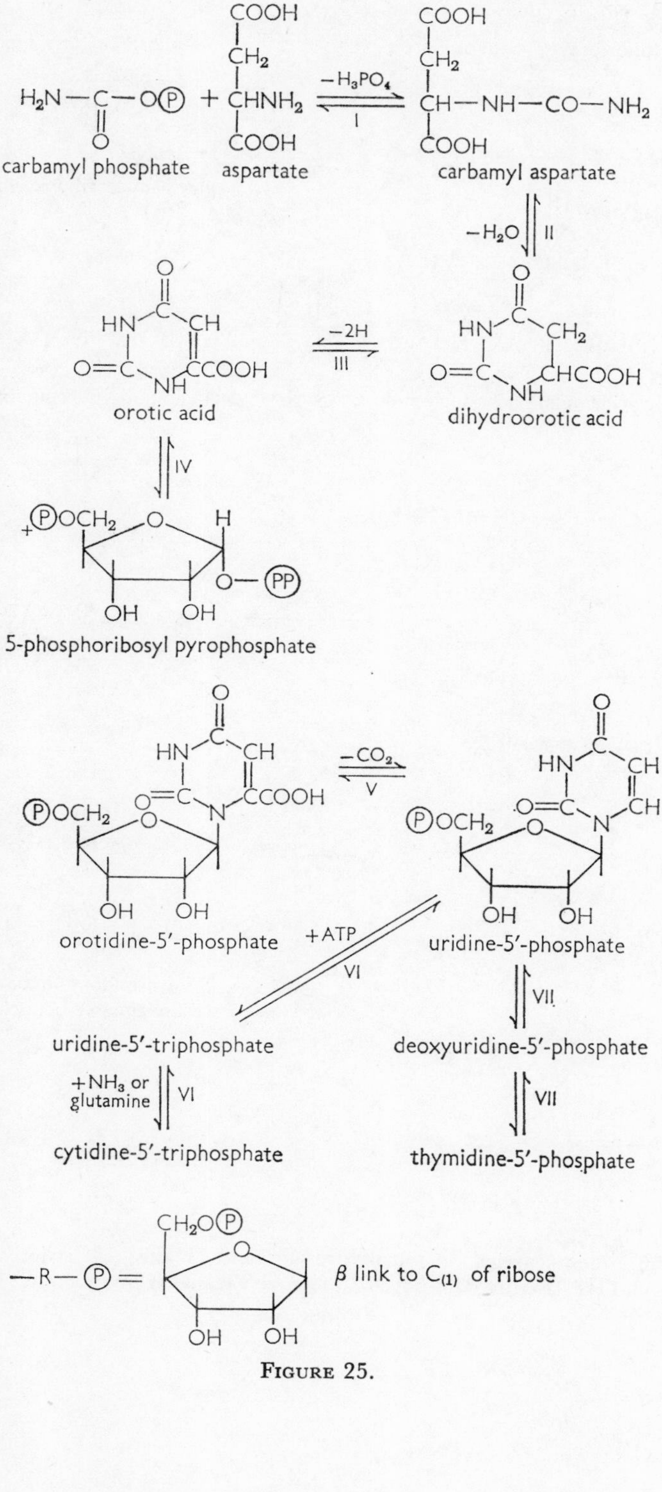

FIGURE 25.

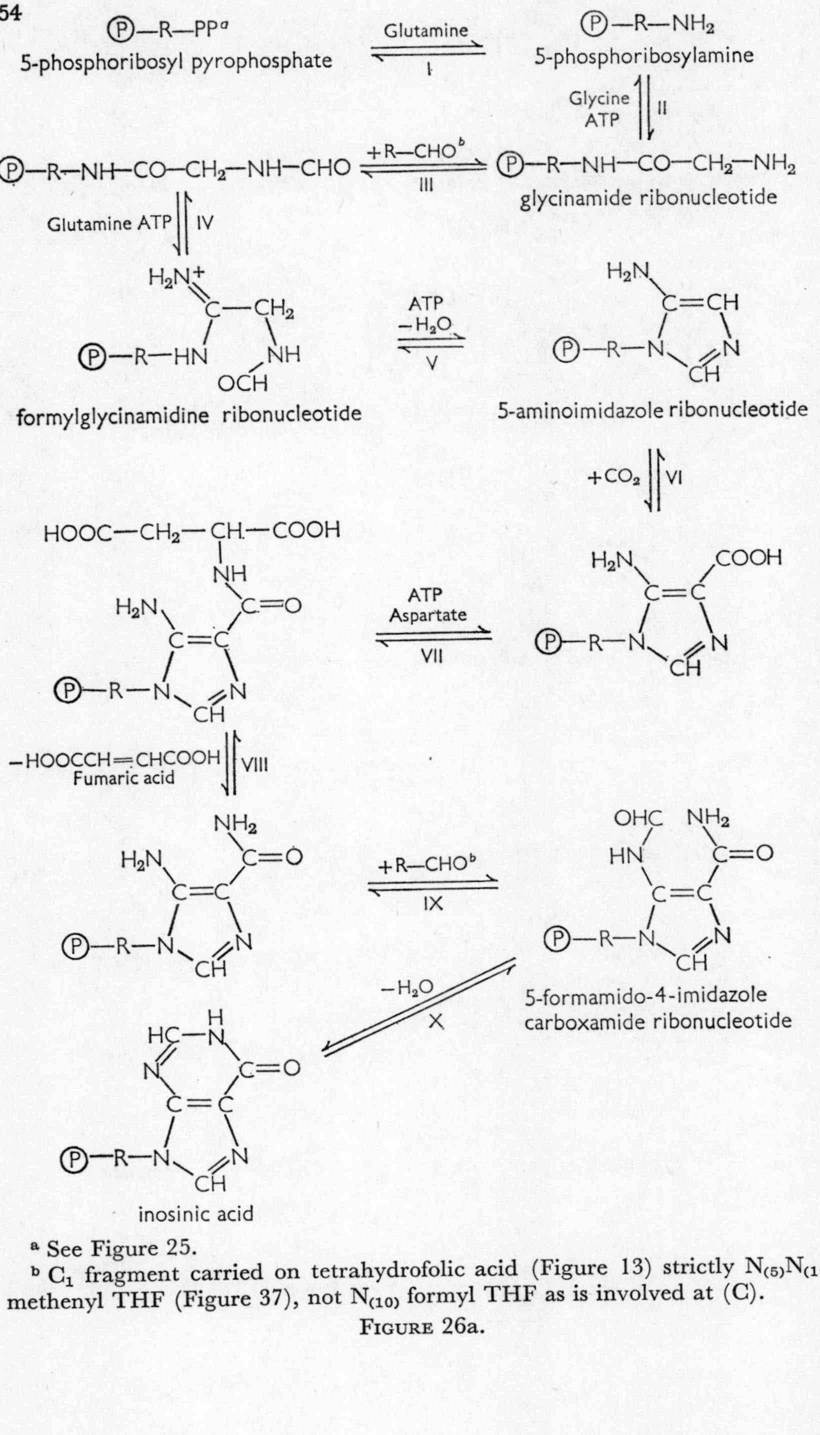

554

$\textcircled{P}$—R—PP^a
5-phosphoribosyl pyrophosphate

$\xrightarrow{\text{Glutamine}}$ I

$\textcircled{P}$—R—NH$_2$
5-phosphoribosylamine

Glycine ATP ‖ II

$\textcircled{P}$—R—NH—CO—CH$_2$—NH—CHO $\underset{\text{III}}{\xrightarrow{+R-CHO^b}}$ $\textcircled{P}$—R—NH—CO—CH$_2$—NH$_2$
glycinamide ribonucleotide

Glutamine ATP ‖ IV

formylglycinamidine ribonucleotide

$\xrightarrow[\text{V}]{\substack{\text{ATP} \\ -H_2O}}$

5-aminoimidazole ribonucleotide

$+CO_2$ ‖ VI

HOOC—CH$_2$—CH—COOH

$\xrightarrow[\text{VII}]{\substack{\text{ATP} \\ \text{Aspartate}}}$

—HOOCCH=CHCOOH ‖ VIII
Fumaric acid

$\xrightarrow[\text{IX}]{+R-CHO^b}$

5-formamido-4-imidazole carboxamide ribonucleotide

$\xrightarrow[\text{X}]{-H_2O}$

inosinic acid

^{a} See Figure 25.

^{b} C$_1$ fragment carried on tetrahydrofolic acid (Figure 13) strictly N$_{(5)}$N$_{(10)}$ methenyl THF (Figure 37), not N$_{(10)}$ formyl THF as is involved at (C).

FIGURE 26a.

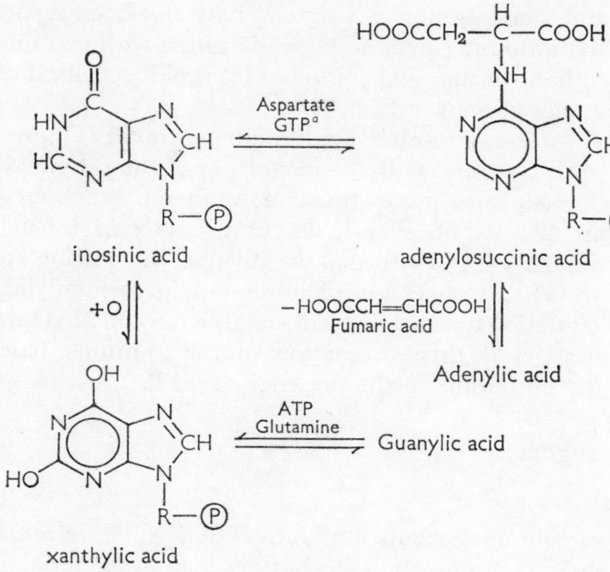

^{a} Guanosine triphosphate.

FIGURE 26b. Conversion of inosinic acid to adenylic and guanylic acids.

configuration at the $C_{(1)}$ atom, but the ribosylamine intermediate has not been isolated, probably because of its chemical reactivity. Synthetic 5-phosphoribosylamine is converted to the second intermediate (reaction II) in the presence of glycine, ATP and an enzyme isolated from pigeon liver[88]. Enzymes catalysing reactions III, IV and V have been partly purified from liver[89, 90]. Carboxylation of 5-amino-imidazole ribonucleotide (reaction VI) occurs in the presence of high concentrations of bicarbonate buffer. The reaction is strongly inhibited by the product[91]. Reaction VII is blocked in sulphonamide-inhibited cultures of *Eschiarischia coli* and the 5-amino-4-imidazole carboxamide ribonucleotide accumulates under these conditions[92]. The purine-N_3 is introduced by a condensation reaction with aspartate (reaction VII) presumably to form an imine which undergoes protropic rearrangement to the *N*-succinyl compound shown. The same type of reaction is involved in the conversion of inosinic acid to adenylic acid (see below). Cleavage of the *N*-succinyl compound releasing fumaric acid (reaction VIII) is apparently catalysed by an enzyme (adenylosuccinase) which has been partly purified from yeast[93]. The

enzymes catalysing reactions IX and X have not been separated nor
has the formamido intermediate been identified with certainty. However, the synthetic formamido compound is rapidly cyclised to inosinic
acid by enzymes present in liver[94].

In the conversion of inosinic acid to adenylic acid (Figure 26b) the
amino group is introduced by condensation of the 6-hydroxyl group
with aspartate to form adenylosuccinic acid (cf. reactions VII and
VIII, Figure 26a). The overall reaction, catalysed by an enzyme
isolated from *E. coli*[95] is alleged to involve the purine nucleoside
triphosphate GTP. Conversion of inosinic acid to guanylic acid involves first oxidation to xanthilic acid, catalysed by an NAD-dependent
dehydrogenase, and then amination of the xanthine base, at the
2-position, by glutamine in the presence of ATP.

E. Amino Sugars

I. General

A number of amino sugars and *N*-acyl derivatives of amino sugars
occur widely[96, 97] in mucopolysaccharides, e.g. in bacterial cell walls,
the skeletal structure of arthropods (chitin), animal cartilage and
bone, in bacterial antibiotics[98] and on the exterior of red blood cells
where they are concerned with the antigen–antibody reactions, on the
basis of which individuals are classified into blood groups. The
mucopolysaccharides are crosslinked polymers containing more than
one sugar derivative and frequently occurring in association with
protein and polypeptide material. Hyaluronic acid, for example,
found in skin and bone, is a crosslinked polymer of glucuronic acid
and *N*-acetylglucosamine; cartilage contains chondroitin, a polymer
of glucuronic acid and *N*-acetylgalactosamine sulphate. Some bacterial cell walls contain crosslinked polymers of *N*-acetylglucosamine,
alternate carbohydrate units having a lactyl side-chain substituted on
the $C_{(3)}$ of the ring (*N*-acetylmuramic acid), and such cell walls are
attacked by the enzyme lysozyme, mentioned in section IV. The
lactyl side-chain of *N*-acetylmuramic acid is frequently linked to a
peptide chain commonly containing both D- and L-alanine, D-glutamic
acid and L-lysine[99] (Figure 27).

A group of compounds, called the sialic acids, are *N*- and *O*-acyl
derivatives of neuraminic acid, itself a derivative of D-mannosamine
(Figure 28). The sialic[99] acids are widely distributed (e.g. in egg yolk,
human semen and in blood serum) both free and bound to the
polysaccharide part of most, if not all, mucoproteins. They are closely

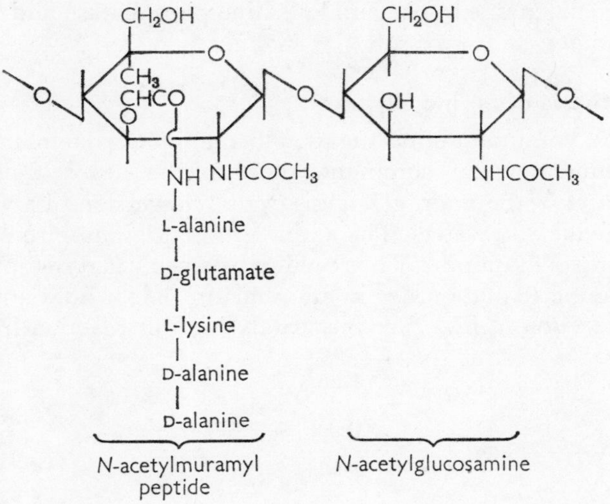
CH₂OH
CH₂OH
O
O
CH₃
CHCO
O
O
OH
O
NH NHCOCH₃
NHCOCH₃
L-alanine
D-glutamate
L-lysine
D-alanine
D-alanine
N-acetylmuramyl peptide
N-acetylglucosamine

FIGURE 27. Type of polysaccharide structure found in some bacterial cell walls.

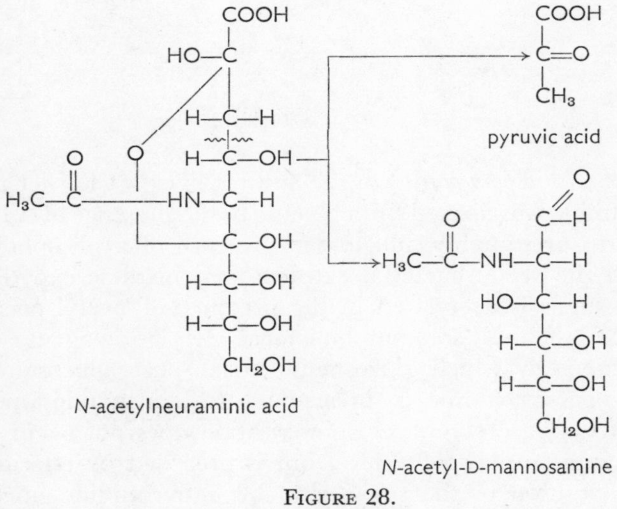
COOH
HO—C
H—C—H
H—C—OH
H₃C—C
HN—C—H
C—OH
H—C—OH
H—C—OH
CH₂OH
N-acetylneuraminic acid
COOH
C=O
CH₃
pyruvic acid
C—H
H₃C—C—NH—C—H
HO—C—H
H—C—OH
H—C—OH
CH₂OH
N-acetyl-D-mannosamine

FIGURE 28.

concerned with preventing agglutination of blood cells, (the response to bacterial or virus infection or the introduction of any foreign material which produces an antigen–antibody reaction). There is now evidence that neuraminic acid derivatives also occur in association with some enzymes, e.g. placental alkaline phosphatase and prostatic acid phosphatase.

2. Biosynthesis of amino sugars

The most common amino sugars, glucosamine, galactosamine and mannosamine, have a common origin in the hexose-6-phosphate intermediates of the main glycolysis cycle (conversion of glycogen to carbon dioxide and water). The amino group originates from the acid amide group of glutamine. The reaction between fructose-6-phosphate and glutamine (equation 33) is 'unusual' in that it does not involve ATP (see section V.B). Enzymes catalysing this reaction have been

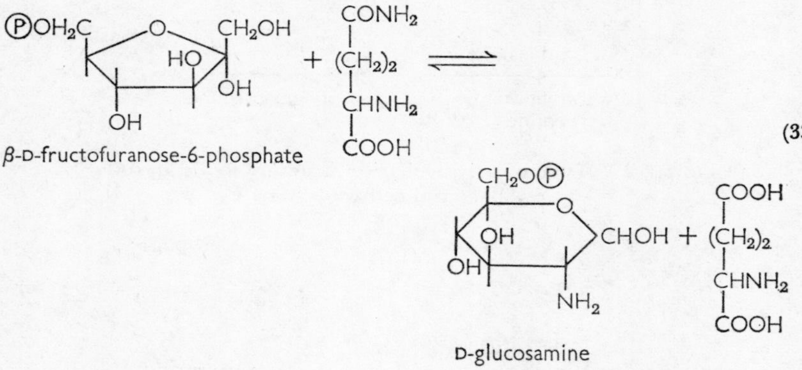

β-D-fructofuranose-6-phosphate

D-glucosamine

$$(33)$$

isolated from *Neurospora crassa*, *E. coli* and rat liver[100]. Earlier reports that glucosamine was formed from glucose in the presence of ATP and glutamine are presumably due to the presence of a phosphohexose isomerase in the crude bacterial extracts used in these experiments. The amino sugar is acetylated in the presence of acetyl coenzyme A (Figure 29), the most common biological acetylating agent.

The reactions by which glucosamine (or *N*-acetylglucosamine) is converted to galactosamine, by inversion at $C_{(4)}$, or mannosamine, by inversion at $C_{(2)}$, are complex. Epimerisation does not occur at the simple hexosamine phosphate level but is preceded by esterification at the $C_{(1)}$ position of the sugar ring by a pyrimidine nucleoside triphosphate (uridine or thymidine triphosphate) to form a nucleoside diphosphoamino (or *N*-acylamino) sugar (Figure 30). Enzymes catalysing epimerisation of uridine (or thymidine) diphosphate esters of glucosamine, to form the nucleoside diphosphate ester of galactosamine, or of *N*-acetylglucosamine, to form the nucleoside diphosphate esters of *N*-acetylgalactosamine or *N*-acetylmannosamine, have been reported [101-103].

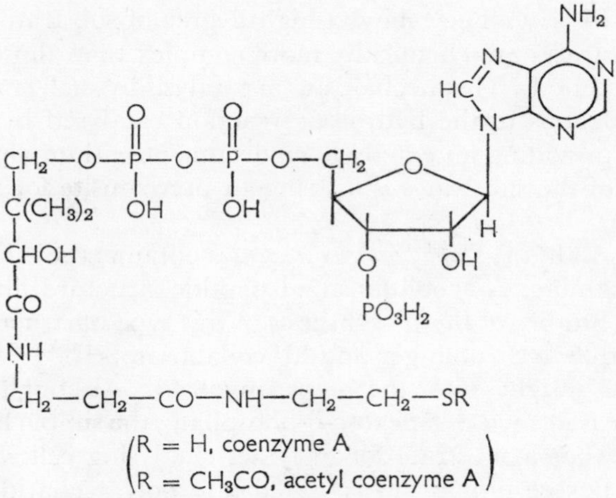

FIGURE 29.

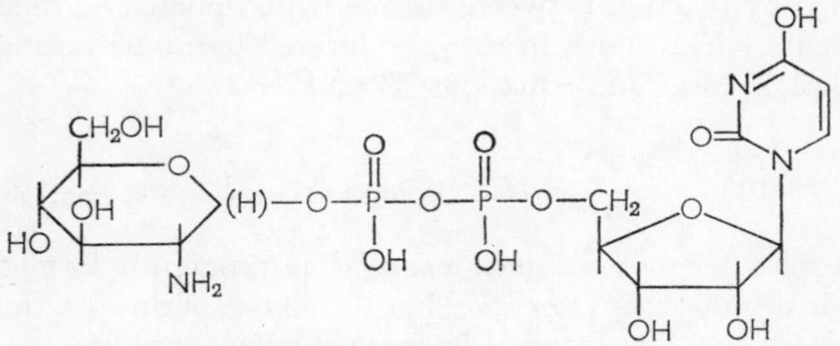

FIGURE 30. Uridine diphosphoglucosamine.

Recent work on the structure and biosynthesis of bacterial cell walls has given some indication of the way in which the crosslinked polysaccharide components are formed. There are several justifications for including here a brief account of work in this field. Firstly, certain antibiotics (e.g. the penicillins, bacitracin A, novabiocin) are known specifically to inhibit bacterial cell-wall synthesis. If, as seems likely, the inhibition is of an enzyme-catalysed reaction, a knowledge of the particular substrates involved might lead to the design of more powerful bactericidal agents. Secondly, the particular type of cell wall to be described is the substrate for the enzyme lysozyme already discussed in section 2. Thirdly, the enzymes catalysing certain of the steps to be outlined below are of considerable interest from a mechanistic point of view. They show a high degree of substrate specificity and the substrates are chemically more complex than those discussed in previous sections. The mechanisms of catalysis by such enzymes may well be analogous to the hydrolytic reaction catalysed by lysozyme, the protein providing an extensive environment within which correct orientation of the substrates is a stringent prerequisite for reaction to occur.

The cell wall of *Staphylococcus aureus* contains the alternate *N*-acetylglucosamine, *N*-acetylmuramyl peptide structure illustrated in Figure 27.* Studies of the biosynthesis of this type of structure, which are largely due to Strominger and his collaborators[104-106] have given considerable insight into the steps involved. Although *N*-acetylglucosamine is formed as a hexose-6-phosphate, the simple hexosamine derivative is not a substrate for enzymes catalysing cell-wall biosynthesis. The hexose unit is incorporated into polysaccharide from the uridine diphosphate ester of *N*-acetylglucosamine. The nucleoside diphosphate esters of glucosamine or *N*-acetylglucosamine (Figure 30) are formed by reaction between uridine triphosphate and the hexose-1-phosphates which must, in turn, be formed by isomerisation of the hexose-6-phosphates (cf. equations 33 and 34).

Hexosamine-1-phosphate + UTP $\rightleftharpoons$

$$\text{Uridine diphosphohexosamine} + \text{Pyrophosphate} \quad (34)$$

The second hexose unit, muramic acid, is apparently formed at the nucleoside diphosphate ester level but the course of this reaction is not well understood. The lactyl side-chain may well be introduced by condensation of uridine diphospho-*N*-acetylglucosamine and phosphoenol pyruvate, liberating inorganic phosphate. The pentapeptide side-chain of the muramyl peptide unit is formed by stepwise addition of the first three amino acids, to uridine diphospho-*N*-acetylmuramic acid, in the presence of ATP and either magnesium or manganese ions, one mole each of ADP and inorganic phosphate being formed for each amino acid added, and final addition of the dipeptide D-alanyl-D-alanine, again in the presence of ATP. The separate enzymes

---

* The figure illustrates a β 1:4 glycosidic link between successive hexose units but it is possible that β 1:6 links are also present in cell-wall material from different microorganisms.

catalysing each addition have been partly purified[104, 105]. Only the completed nucleoside diphospho-N-acetylmuramyl pentapeptide acts as a substrate for cell-wall synthesis. Particulate, cell-free preparations from *Staphylococcus aureus* which catalyse a reaction between the nucleoside diphosphate esters of N-acetylglucosamine and N-acetyl-muramyl pentapeptide to form 'cell-wall-like' material have been obtained[106].

The metabolism of N-acetylmannosamine and its relation to the formation of the sialic acids (derivatives of neuraminic acid) are not well understood. The subject has been reviewed by Roseman[107].

F. Miscellaneous Amino Compounds and Derivatives

There are a number of amino compounds, of immense biological importance, which cannot be classified with the compounds so far discussed. It is intended, in this section, to discuss the possible origins of some of these compounds and also to give a brief account of the biosynthesis of certain important heterocyclic systems which are derived from the simple amino compounds already discussed.

I. Vitamins[108]

The familiar word vitamin derives from 'vital amine', the term being introduced in 1912 following the recognition that certain diseases could be prevented by addition to the diet of some compounds originally believed to be amines. Later it was revealed that not only are these compounds not all amines but some contain no nitrogen at all, such as the 'fat-soluble' A, D, K and E groups and vitamin C. The water-soluble vitamins (B group), all contain nitrogen, and appear to function, in more or less well-understood ways, as cofactors for enzyme-catalysed metabolic reactions, hence the limitation of the present discussion to members of this group and their derivatives.

a. *Vitamin B_1 (thiamine)*. (Figure 31a). The pyrimidine and thiaz-ole rings of thiamin are formed separately but the origins of the two rings are not yet known. It is presumed that the pyrimidine ring is formed from orotic acid but no information is yet available concerning the introduction of the 4-amino group. Two possible routes to the thiazole ring are given in Figure 31b. The two ring systems are linked by a reaction between the pyrimidine pyrophosphate and the thiazole monophosphate. The product is hydrolysed to free thiamine before conversion to a pyrophosphate, in which form vitamin B_1 is a cofactor for an enzyme catalysing the reversible decarboxylation of pyruvate.

Imine formation between pyruvate and the amino substituent in the pyrimidine ring was at one time suggested as a possible step in decarboxylation but this idea has since been abandoned. The reaction is now believed to involve nucleophilic attack on the carbonyl carbon of pyruvate by an anion formed by loss of a proton from the $C_{(2)}$ position in the thiazole ring. Details of the mechanism need not concern us since no amino group is involved in either the substrate or the cofactor.

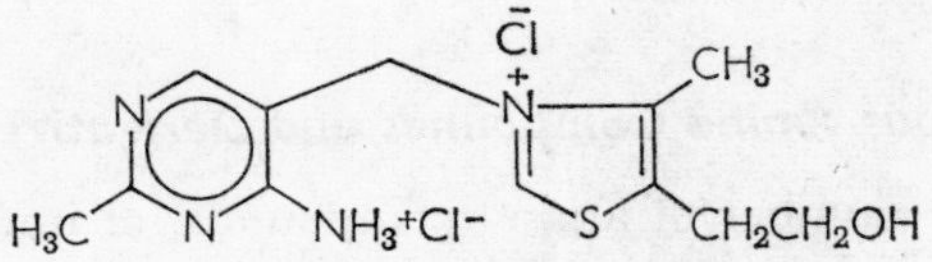

FIGURE 31a. Thiamin.

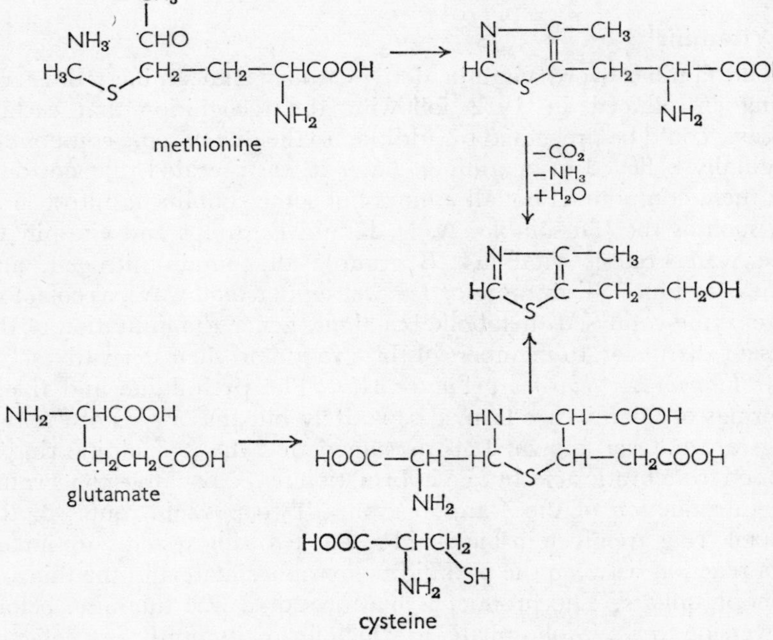

FIGURE 31b.

b. *Vitamin B$_2$ (riboflavine)*. The two nitrogen-containing rings of the isoalloxazine system of riboflavine (Figure 32) are derived from adenine. The routes to the dimethylbenzo ring and the ribityl side-chain are unknown. The two derivatives of riboflavine (Figure 2c) which occur as the cofactors in flavoprotein enzymes catalysing oxidation–reduction reactions (e.g. amino acid oxidases) act as hydrogen acceptors. The addition of two hydrogen atoms across the conjugated double bonds between $N_{(1)}$ and $N_{(10)}$ gives the reduced flavin ring shown in Figure 2d.

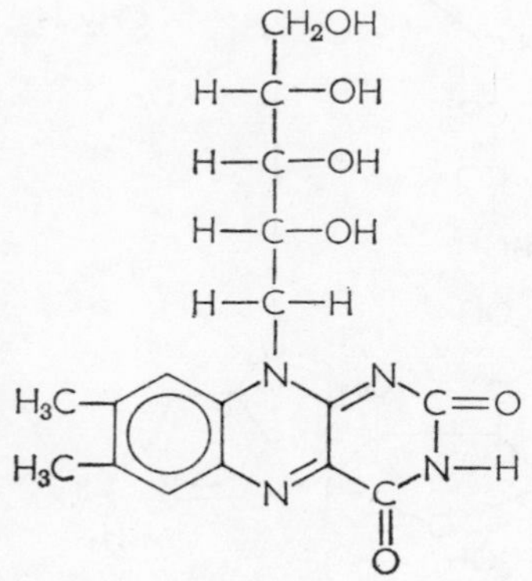

FIGURE 32. Riboflavine.

c. *Vitamin B$_6$ (pyridoxine)*. (Figure 5a). No information is yet available on the biosynthesis of pyridoxine. The phosphorylated derivatives of pyridoxine (see Figures 5b and c) which act as cofactors for enzymes catalysing certain reactions of α-amino acids, have already been discussed (sections IV.C and V.C).

d. *Vitamin B$_{12}$ (cyanocobalamin)*. The somewhat formidable structure of Vitamin B$_{12}$ is given in Figure 33. The 5:6 dimethylbenzimidazole nucleotide unit is probably joined, through the 1-amino-2-propanol bridge formed by decarboxylation of threonine, to the preformed corrin nucleus. The corrin nucleus is itself a modification of the porphyrin ring system which occurs commonly, for example in haemoglobin, chlorophyll and in the cytochromes. Compounds containing the tetrapyrrole ring system probably have a common origin (see section V.F.2). The acid amide groups in the corrin side-chains may be formed by transamidation between glutamine and the

19+c.a.g.

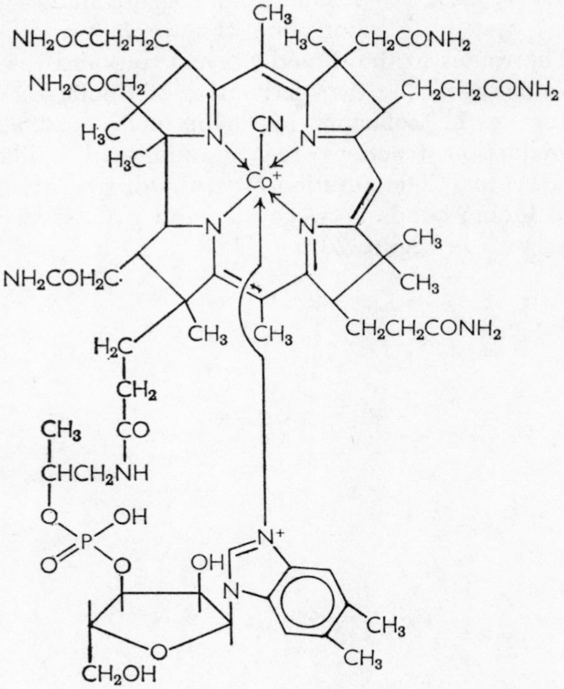

FIGURE 33. Vitamin B_{12}.

acetic acid and propionic acid side-chains which are correctly arranged about the porphyrin rings of uroporphyrin III (Figure 39).

Vitamin B_{12} was first identified as the antipernicious anaemia factor, present in whole liver. Only recently has it been shown that, in common with other members of the B group of vitamins, a derivative of cyanocobalamin (having the cyanide group replaced by adenosine) is a cofactor for certain enzymes, in particular, those catalysing the isomerisation of glutamate to β-methyl aspartate[109] (bacteria) and of the methylmalonyl derivative of coenzyme A (Figure 29) to succinyl CoA[110,111]. It has been suggested that both isomerisations involve cyclic intermediates and that the enzyme–cofactor complexes may stabilise carbanions formed on cleavage of the rings in these intermediates.

e. *Vitamin H (biotin)*. The biosynthetic route to biotin (Figure 34) is unknown, but there is indirect evidence that desthiobiotin is an intermediate and that the side-chain originates in pimelic acid. Biotin is involved in certain decarboxylation reactions and in the deamination of some amino acids, e.g. aspartate, serine and threonine.

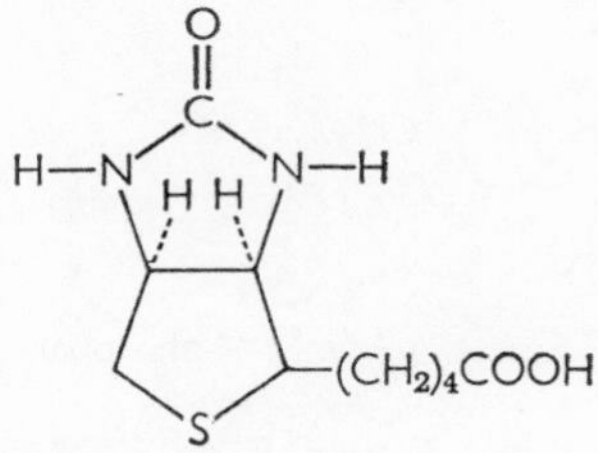

FIGURE 34. Biotin.

f. *Vitamin B_3 (pantothenic acid)*. The precursors of pantothenic acid (Figure 35) are pantoic acid (from α-ketovaleric acid) and β-alanine (a decarboxylation product of aspartate). A bacterial enzyme[112], pantothenic acid synthetase, catalyses the condensation of these two compounds to form pantothenic acid.

Pantothenic acid is a constituent of the CoA molecule (Figure 29), the biosynthesis of which is summarised in Figure 35. Only the sulphydryl group of CoA is known to be directly involved in trans-acetylation reactions. The function of the remainder of the molecule is obscure.

g. *Folic acid*. The structure of folic acid is given in Figure 13. The compound is formed by condensation between *p*-aminobenzoic acid and glutamic acid followed by a second condensation reaction with a substituted pteridine. The pteridine nucleus appears to originate from the purine ring system, probably by a route related to that involved in the formation of the isoalloxazine ring system of riboflavin.

A reduced form of folic acid (tetrahydrofolic acid, FH_4) is active in so-called one-carbon metabolism. A formyl group may be substituted on $N_{(5)}$ or $N_{(10)}$ and the $N_{(5)}$–$N_{(10)}$ methenyl derivative (Figure 36) is also active. Reactions involving the introduction of one-carbon fragments include the conversion of glycine to serine (Section V.C.3), ethanolamine to choline (Section V.C.3) and the introduction of $C_{(8)}$ of the purine nucleus (Section V.D.3).

h. *Nicotinamide*. The route to the precursor of nicotinamide, nicotinic acid, is different in animals and bacteria. In animals (see Figure 37) tryptophane is converted to nicotinic acid by a series of

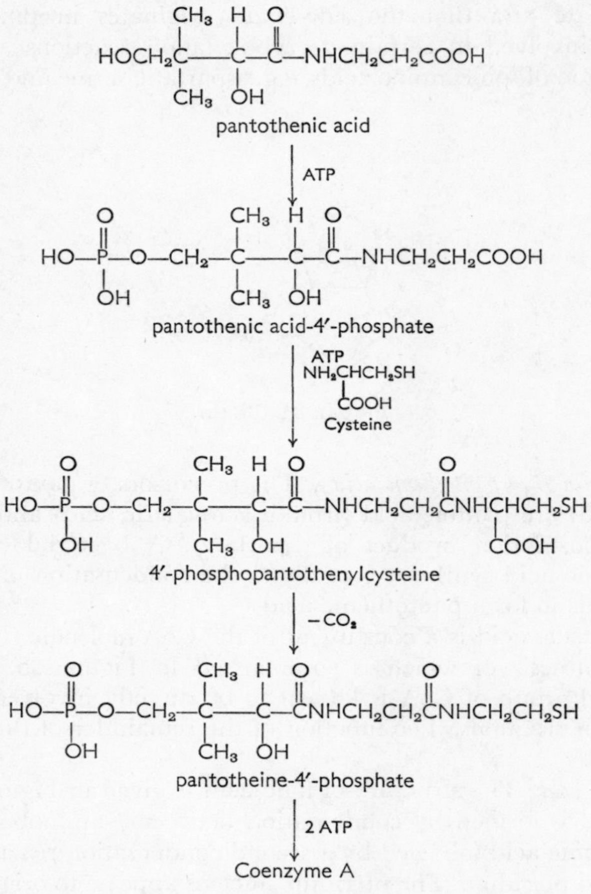

CH_3 H O
HOCH$_2$C—C—C—NHCH$_2$CH$_2$COOH
CH_3 OH

pantothenic acid

ATP

O CH_3 H O
HO—P—O—CH$_2$—C——C—C—NHCH$_2$CH$_2$COOH
OH CH_3 OH

pantothenic acid-4′-phosphate

ATP
NH$_2$CHCH$_2$SH
COOH
Cysteine

O CH_3 H O   O
HO—P—O—CH$_2$—C——C—C—NHCH$_2$CH$_2$CNHCHCH$_2$SH
OH CH_3 OH   COOH

4′-phosphopantothenylcysteine

$-CO_2$

O CH_3 H O   O
HO—P—O—CH$_2$—C——C—CNHCH$_2$CH$_2$CNHCH$_2$CH$_2$SH
OH CH_3 OH

pantotheine-4′-phosphate

2 ATP

Coenzyme A

FIGURE 35.

reactions in which first the five-membered ring is opened, the resulting carbon side-chain in the six-membered ring is oxidised and fragmented; the six-membered ring is then opened and reformed to give a quinolinic acid. In bacteria, the carbon skeleton of nicotinic acid is formed from a three-carbon unit (e.g. glycerol) and a four-carbon dicarboxylic acid[113].

The formation of nicotinamide from nicotinic acid occurs at the nucleotide level[114]. Nicotinic acid first reacts with phosphoribosyl pyrophosphate to form nicotinic acid mononucleotide and inorganic pyrophosphate. The mononucleotide then reacts with ATP to form a

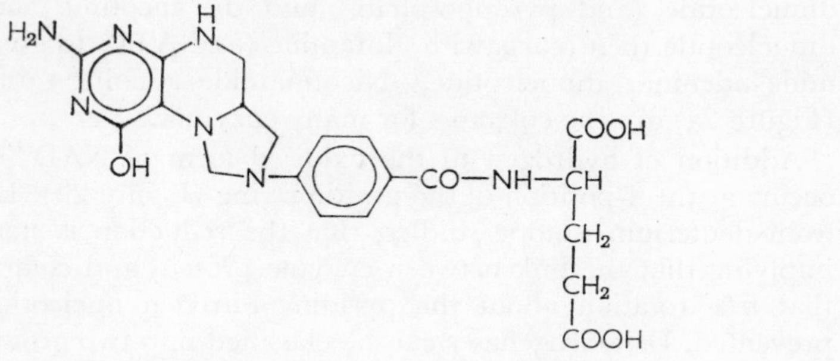

FIGURE 36. $N_{(5)}$, $N_{(10)}$-methyl tetrahydrofolic acid.

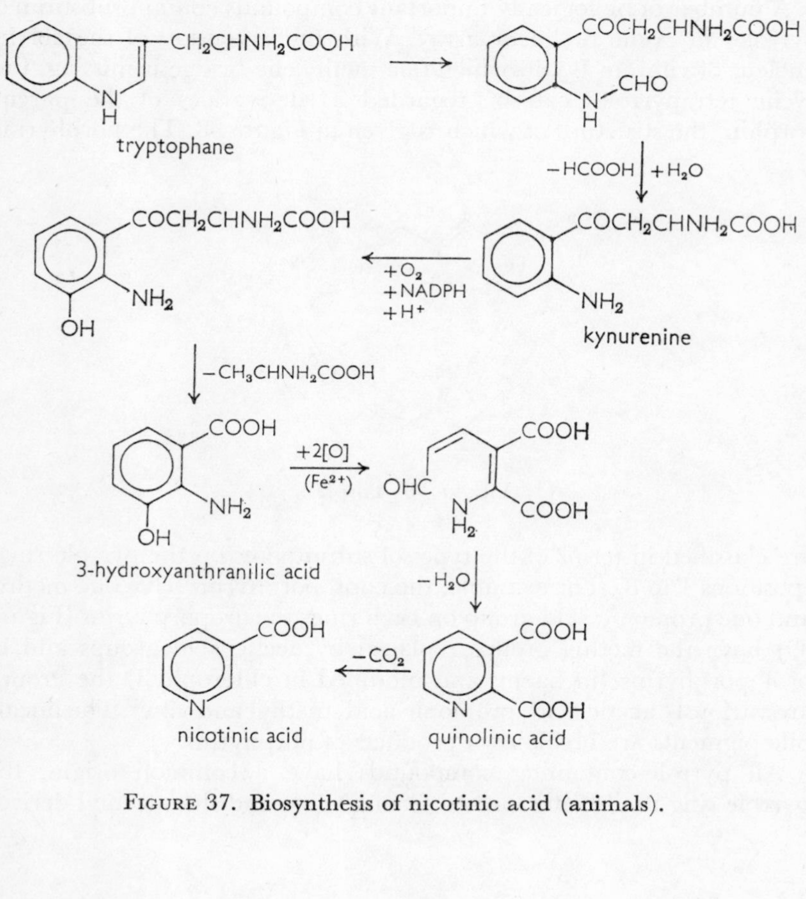

FIGURE 37. Biosynthesis of nicotinic acid (animals).

dinucleotide (and pyrophosphate) and the nicotinic acid–adenine dinucleotide then reacts with glutamine (and ATP) to give nicotinamide–adenine dinucleotide. Nicotinamide–adenine dinucleotides (Figure 2a) are the cofactors for many enzymes.

Addition of hydrogen to the oxidised form of NAD (or NADP) occurs at the 4-position of the pyridine ring (Figure 2b). It is known, from deuterium isotope studies, that the reduction is stereospecific, implying that the link between enzyme protein and cofactor is such that free rotation about the pyridine nitrogen nucleotide bond is prevented. Dehydrogenases can be classified into two groups showing different stereospecificity with respect to reduction of the nicotinamide ring.

2. Pyrrole-containing compounds

A number of biologically important compounds contain substituted pyrroles in cyclic or linear array. With the exception of the corrin nucleus of vitamin B_{12} in which one methylene bridge is missing, the cyclic tetrapyrroles can be regarded as derivatives of the parent porphin, the structure of which is given in Figure 38. The porphyrins

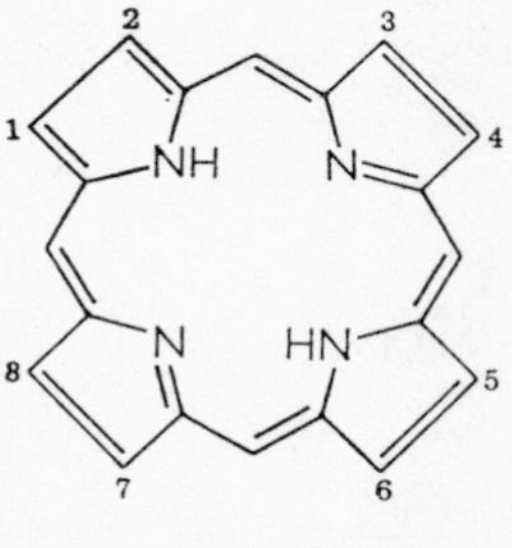

FIGURE 38. Porphin.

are classified in terms of the types of substituents on the pyrrole rings (positions 1 to 8). For example, the coproporphyrins have one methyl and one propionic acid group on each ring; the uroporphyrins (Figure 39) have the methyl groups replaced by acetic acid groups and in protoporphyrins (in haem, and modified in chlorophyll) the groups are variously acetic acid, propionic acid, methyl and vinyl. The linear, bile pigments are breakdown products of porphyrins.

All pyrrole-containing compounds have a common origin, the pyrrole ring itself being formed from glycine and the succinyl derivative of coenzyme A which condense to form α-amino-β-ketoadipic acid. This undergoes α-decarboxylation to form δ-aminolaevulinic acid, two molecules of which condense to give a pyrrole ring which already carries the acetic and propionic acid side-chains found in the uroporphyrins (Figure 39). Subsequent polymerisation of four-substituted pyrroles with the loss of ammonia from the methylamine side-chains gives the reduced precursor of the uroporphyrins. The methyl and vinyl side-chains of copro- and protoporphyrins are formed by decarboxylation of acetic acid side-chains and oxidation and decarboxylation of propionic acid side-chains respectively. The metal ion present in haem compounds and in cytochromes (iron), in chlorophylls (magnesium), and in cyanocobalamine (cobalt), is apparently introduced after the porphyrin (or corrin) nucleus has been formed.

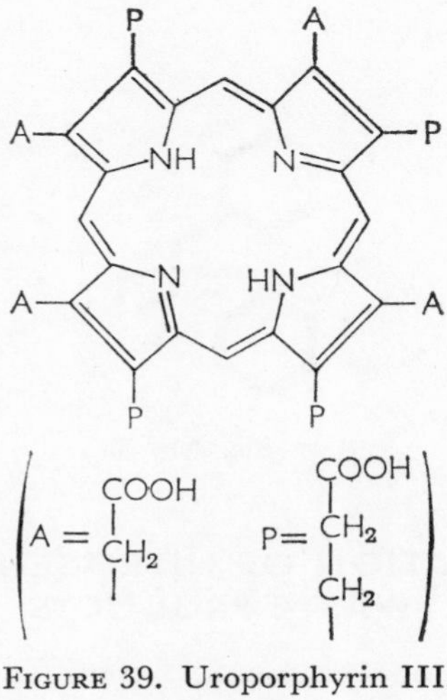

FIGURE 39. Uroporphyrin III.

3. Plant alkaloids

The biosynthesis of plant alkaloids has recently been well reviewed by Bu'Lock[115] and will therefore be considered very briefly. It is not possible to make general statements about the mechanism of biosynthesis of this large and varied class of compounds, partly because of the incomplete state of knowledge of any individual routes. Tracer studies have contributed much information in this field but perhaps more has resulted from hypotheses, based on inductive reasoning, resulting from the observation that certain common units occur within complex organic structures. The alkaloids appear to be derived largely from only a small number of amino acids. For example, ornithine contributes the C_4N unit to the pyrrolidine alkaloids such as hyoscyamine (Figure 40a) while lysine contributes the C_5N units to the piperidine alkaloids such as lupinine (Figure 40b). Alkaloids such as morphine (Figure 40c) and belladine, contain $C_6 C_2 N$ (phenyl-ethylamine units) derived from the aromatic amino acids, phenyl-alanine and tyrosine. The fifth amino acid to contribute significantly to alkaloid biosynthesis is tryptophane, from which are derived the so-called indole alkaloids such as strychnine and yohimbine.

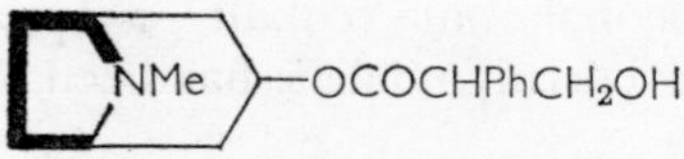

FIGURE 40a. Hyoscyamine.

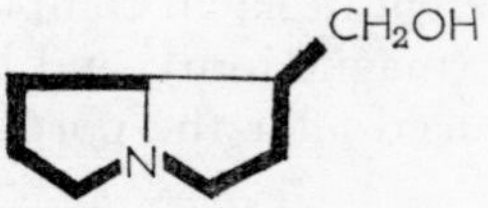

FIGURE 40b. Lupinine.

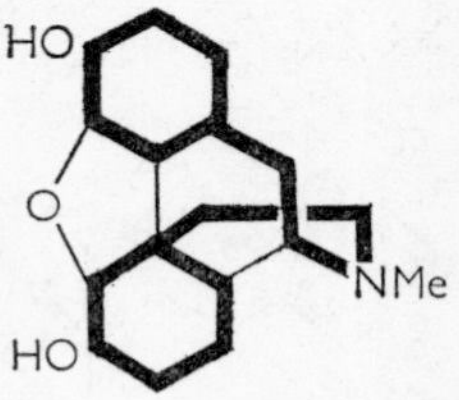

FIGURE 40c. Morphine.

VI. THE FORMATION OF NITROGEN-CONTAINING WASTE PRODUCTS

The reactions of amino compounds, considered so far, have been largely concerned with the production of essential cell constituents from ingested material. The remaining section is concerned with those reactions by which excess amino compounds are converted first to ammonia and then to common excretion products. The division between biosynthetic and degradative reactions is not clear cut as can be seen from the fact that the reaction catalysed by the enzyme glutamic dehydrogenase produces ammonia and has already been discussed in sections II and IV.B. Oxidative deamination of the D-amino acids has also been discussed (section V.A) because the reverse of this reaction appears to be the main route by which D-amino acids are formed in microorganisms but it is also believed to be the main route by which D-amino acids are broken down in animals. Apart from the routes which have already been discussed, two main types of reaction lead to the production of ammonia.

A. Production of Ammonia

I. L-Amino acid oxidases

The original observation by Krebs[116], that homogenates of mammalian liver and kidney catalysed the oxidation of both D- and L-amino acids, was followed by a number of attempts to purify individual amino acid oxidases. Although a crystalline, non-specific D-amino acid oxidase has been isolated from kidney, the only evidence for the existence, in mammals, of a single enzyme catalysing the oxidative deamination of L-amino acids, comes from an early report by Blanchard and coworkers[117]. This particular flavoprotein, isolated from rat kidney, is odd for two reasons: (a) it shows a higher oxidative activity towards some α-hydroxy acids (e.g. L-lactic acid) than to L-amino acids and (b) the catalytic activity is very low compared with most other enzymes. The amounts of this enzyme in kidney homogenates are certainly too small to account for Krebs' original observations of the oxidation of L-amino acids. It is now generally believed that L-amino acid oxidation by mammalian tissue homogenates is due to the coupling of two enzyme-catalysed reactions, the first a transamination between the L-amino acid substrates and α-ketoglutaric acid and the second, deamination of the resulting glutamic acid, catalysed by glutamic dehydrogenase (Section IV.B). This is probably the principal route, in mammals, to the production of ammonia from excess amino acids.

The highest L-amino acid oxidase activity is found in the venoms of a variety of snakes, the enzyme being present to the extent of up to 3% of dried, whole venom in some species. Crystalline enzymes (both flavoproteins) have been isolated from the venoms of water moccasin[118] (*Ancistrodon piscivorus*) and rattle snake[119] (*Crotalus adamanteus*) and shown to be similar with respect to molecular weight (130,000) and flavin content (2 moles FAD/mole protein). The enzymes are not substrate-specific but are more active with the L-isomers of methionine, leucine, tryptophan, phenylalanine and tyrosine than with other L-amino acids. The reactions catalysed by L-amino acid oxidase are formally the same as those discussed in section V.A (reactions 23–26) but, as is the case for the D-amino acid oxidase, the mechanism of

19*

oxidative deamination remains obscure. The function of large quantities of L-amino acid oxidase in snake venom is quite unknown.

L-Amino acid oxidase activity has been detected in a variety of moulds, yeasts and bacteria, but the enzymes from these sources have not been well characterised. A partly purified enzyme from *Proteus vulgaris*[120] is curious in that it does not apparently produce hydrogen peroxide in the presence of molecular oxygen (equations 22–26) and absence of catalase.

2. Amine oxidases[24]

The enzymes catalysing the oxidation of naturally occurring amines, according to equation (35), found in both plants and animals, were

$$RCH_2NH_2 + O_2 + H_2O \rightleftharpoons RCHO + NH_3 + H_2O \qquad (35)$$

originally classed as monoamine oxidases and diamine oxidases. Recent work has shown that, although there are two separate types of enzymes, the difference between them does not lie in the number of amino groups in the substrates. The substrate specificities are, in both cases, more complex. There are two main types of monoamine oxidases, the intracellular, mitochondrial enzyme (present to a significant extent in liver tissue) and the extracellular, soluble enzymes present in the blood serum, particularly that of ruminants. The former are poorly characterised as proteins because of the common problem of solubilising normally particulate enzymes. (A fifty-eight fold purification of the beef mitochondrial enzyme was recently reported[121].) The enzyme acts on simple aliphatic amines (with the exception of methylamine and ethylamine) according to equation (35) and catalyses the oxidation of such physiologically important amines as serotonin, tyramine and norepinephrine (section V.C). It also oxidises secondary and tertiary N-methyl derivatives of amines, e.g. epinephrine and N,N-dimethyltryptamine, producing methylamine and dimethylamine in place of ammonia, and slowly attacks such diamines as cadaverine (formed by decarboxylation of lysine) and some long-chain diamines (C_{14}, C_{16} and C_{18}). Histamine is also oxidised slowly in the presence of the mitochondrial monoamine oxidase. The enzyme is believed to contain functional copper and also, possibly, a flavin cofactor[122]. Final decision on these points must await further purification of the enzyme. The activity of 'monoamine oxidase' is unaffected by reagents known to inhibit pyridoxal phosphate-dependent, enzyme-catalysed reactions.

Soluble amine oxidases have been crystallised from the blood serum of beef[123] and pigs[124]. In both cases the properties of the purified enzyme were notably different from those of the mitochondrial enzymes, both in regard to substrate specificity and cofactor requirement. The beef plasma enzyme, although classified as a monoamine oxidase, does act on certain diamines, in particular spermine and spermidine, (amines found in seminal fluid and apparently derived from putrescine, the decarboxylation product of ornithine). The oxidation products from spermine and spermidine are a dialdehyde and a monoaldehyde respectively[125] (equations 36 and 37). The en-

$$NH_2(CH_2)_3NH(CH_2)_4NH(CH_2)_3NH_2 + 2\,O_2 + 2\,H_2O \longrightarrow$$
$$\text{spermine}$$

$$CHO(CH_2)_2NH(CH_2)_4NH(CH_2)_2CHO + 2\,NH_3 + 2\,H_2O_2 \quad (36)$$

$$NH_2(CH_2)_3NH(CH_2)_4NH_2 + O_2 + H_2O \longrightarrow$$
$$\text{spermidine}$$

$$CHO(CH_2)_2NH(CH_2)_4NH_2 + NH_3 + H_2O \quad (37)$$

zyme crystallised from pig plasma has been called 'benzylamine oxidase'[124] in spite of the fact that it acts equally well with mescaline (2,3,4-trimethoxyphenyl ethylamine) and also oxidises histamine (at approximately half the rate). The enzyme from beef plasma acts on the same monoamine substrates. Unlike the mitochondrial enzymes, neither plasma enzyme will oxidise N-methyl derivatives of amines and this is consistent with the recent finding that the plasma amine oxidases contain tightly bound pyridoxal-5'-phosphate. (The suggestion that pyridoxal phosphate might be involved in amine oxidation was made as long ago as 1949[126].) It is an unusual finding, in that other pyridoxal phosphate-dependent enzymes catalyse reactions of amino acids, rather than amines.* It is now suggested that the reactions catalysed by plasma amine oxidases occur by Schiff's base formation between primary amino groups in the substrates and the aldehyde group of pyridoxal phosphate. The involvement of a pyridoxal cofactor could explain the known sensitivity of plasma amine oxidases to carbonyl reagents such as cyanide. The plasma oxidase is also believed to contain copper (3 g atoms per mole of protein, mol. wt. $\sim 200,000$.) An amine oxidase present in pea seedlings also contains copper[127] and is sensitive to carbonyl reagents, but there is no direct evidence for the presence of bound pyridoxal phosphate.

The enzymic activity originally attributed to a diamine oxidase was first observed in pig kidney and later shown to be fairly widespread in animals, plants and microorganisms. The enzyme catalyses the

* The exception being muscle phosphorylase which contains pyridoxal phosphate, the function of which remains obscure.

oxidation of the diamine decarboxylation products from arginine, (agmatine), lysine, (cadaverine) and ornithine, (putrescine) and the more complex diamines spermine and spermidine. The oxidation products are amine aldehydes which might be expected to cyclise. The amine aldehyde from cadaverine would cyclise to an oxidised precursor of piperidine (equation 38). The enzyme appears to be

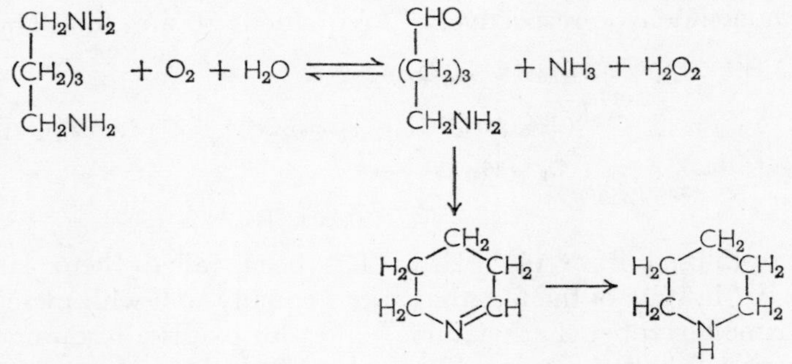

$$
\begin{array}{c} CH_2NH_2 \\ | \\ (CH_2)_3 \\ | \\ CH_2NH_2 \end{array} + O_2 + H_2O \rightleftharpoons \begin{array}{c} CHO \\ | \\ (CH_2)_3 \\ | \\ CH_2NH_2 \end{array} + NH_3 + H_2O_2 \tag{38}
$$

mitochondrial but is more readily solubilised than the mitochondrial monoamine oxidase. The recent report[128] of a substantial purification (1000-fold) of the enzyme from pig kidney suggests also that the enzyme contains both copper and firmly bound pyridoxal phosphate. There has been some controversy regarding the possible identity of the diamine oxidase with histaminase, the enzyme catalysing the oxidation of histamine to imidazole acetaldehyde[129]. (It should be noted that histamine is not a diamine, a fact which underlines the unsatisfactory nature of the original classification of amine oxidases.) The most recent data[128] shows that the ratio of the activities with respect to diamine substrates and histamine, remains constant throughout the purification of the pig kidney 'diamine oxidase'.

3. Miscellaneous reactions

The production of ammonia by non-oxidative processes is of less general importance. Enzymes catalysing non-oxidative deamination are relatively substrate-specific.

a. *Amino acid dehydrases*. The two hydroxyamino acids, serine and threonine, are both dehydrated and deaminated in the presence of specific dehydrases. It has been suggested[130] that the reaction may involve dehydration, to give α, β- or β,γ-unsaturated amino acids, followed by rearrangement, to the corresponding imino acids, and hydrolysis to an α-keto acid and amonia (equation 39). The dehydrases

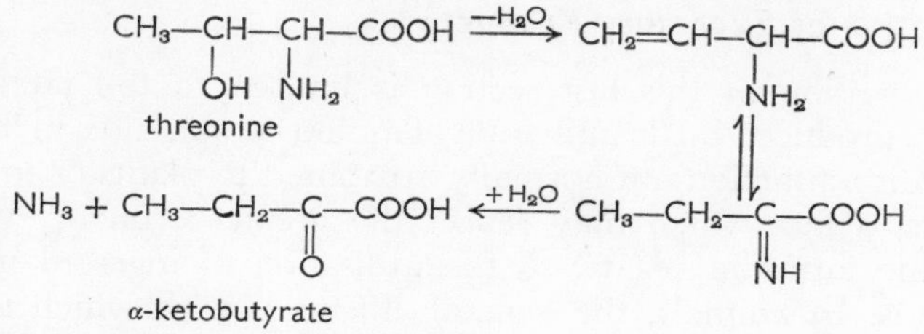

$$\text{CH}_3\text{—CH—CH—COOH} \xrightarrow{-\text{H}_2\text{O}} \text{CH}_2\text{=CH—CH—COOH}$$

threonine

(39)

$$\text{NH}_3 + \text{CH}_3\text{—CH}_2\text{—C—COOH} \xleftarrow{+\text{H}_2\text{O}} \text{CH}_3\text{—CH}_2\text{—C—COOH}$$

α-ketobutyrate

present in *Neurospora crassa*[131], *E. coli*[132] and sheep liver[133] are said to be pyridoxal phosphate dependent. A recent report[134] suggests that the threonine dehydrase of sheep liver may normally be present to variable extents as an inactive complex with serine. Threonine dehydrase activity could be demonstrated after heat treatment to remove serine.

b. *Tryptophanase*. The breakdown of tryptophane to indole, pyruvate and ammonia, is catalysed by a bacterial enzyme which requires pyridoxal phosphate as a cofactor (equation 40).

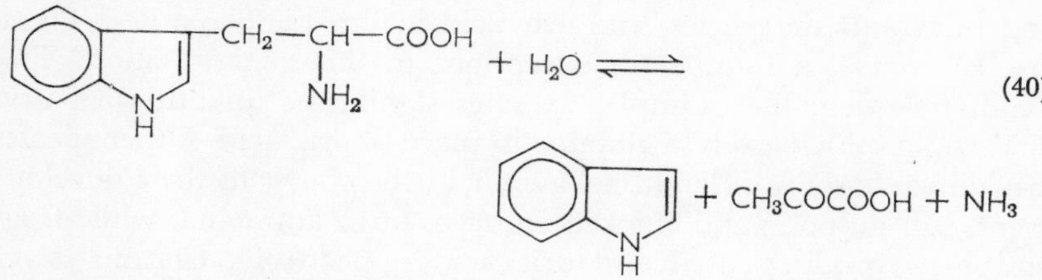

$$+ \text{H}_2\text{O} \rightleftharpoons$$

(40)

$$+ \text{CH}_3\text{COCOOH} + \text{NH}_3$$

c. *Cysteine desulphydrase*[135]. Removal of hydrogen sulphide from cysteine is accompanied by non-oxidative deamination (equation 41). The reaction is analogous to the dehydration and deamination of the hydroxy amino acids and is catalysed by pyridoxal phosphate-dependent enzymes present in some mammalian tissues and some micro-organisms.

$$\text{HSCH}_2\text{—CH—COOH} + \text{H}_2\text{O} \rightleftharpoons \text{H}_2\text{S} + \text{NH}_3 + \text{CH}_3\text{COCOOH} \quad (41)$$

d. *Deamination of purines and pyrimidines*. The amino substituents in the purines, guanine and adenine, and the pyrimidine, cytosine, are not apparently lost by deamination of the free bases. Enzymes catalysing non-oxidative deamination of the corresponding nucleosides have been demonstrated in liver and other tissues, while the enzyme adenylic deaminase, present in muscle, converts the nucleotide adenylic acid to the deaminated analogue, inosinic acid.

B. *Formation of Excretory Products*

The discussion in this last section is limited to the fate of the ammonia, produced by deamination of amino compounds, in animals. Excretion is a function not normally attributed to plants or to micro-organisms, a fact which may reflect the greater economy of these simpler life forms in regard to the utilisation of ingested nitrogen compounds. In animals, the general efficiency with which ingested nitrogenous materials are used to maintain a healthy state of existence is relatively low. In addition, the turnover of such essential nitrogenous compounds as proteins and nucleic acids is relatively rapid. The main problem is the efficient conversion of excess amino compounds to non-toxic products excreted in the urine. The toxicity of ammonia, particularly in the central nervous system of higher animals, has already been noted and it has long been recognised that the level of free ammonia in the tissues, in blood and in urine, is only significant in certain cases of metabolic disorder.

The principal nitrogenous excretion product is urea in mammals and most aquatic species, and uric acid in birds and reptiles. There are, of course, a number of exceptions to this generalisation. The Dalmation dog, for example, excretes significant quantities of uric acid while spiders excrete guanine in place of uric acid. Other species are known to show different behaviour in the course of their development, e.g. tadpoles and chicken embryos form ammonia, while frogs and hens produce urea and uric acid respectively. Certain lower aquatic species lose ammonia directly, a simple diffusion process replacing the complex excretory systems of higher animals.

The main routes by which surplus amino compounds are converted to urea and uric acid, are considered below.

I. Urea formation

It was suggested, by Krebs and Henseleit[136] more than thirty years ago, that urea is formed (primarily in liver tissue), by a cycle of reactions between the basic amino acids, ornithine, citrulline and arginine. The original scheme, given in Figure 41, is now known to be oversimplified. Other intermediates are involved, and enzymes catalysing the individual steps in the reaction have been purified and partly characterised. The reactions involved are as follows:

(a) Formation of carbamyl phosphate from ammonia, carbon dioxide and adenosine triphosphate (ATP). The reaction is catalysed by carbamyl phosphate synthetase which requires two cofactors, i.e.

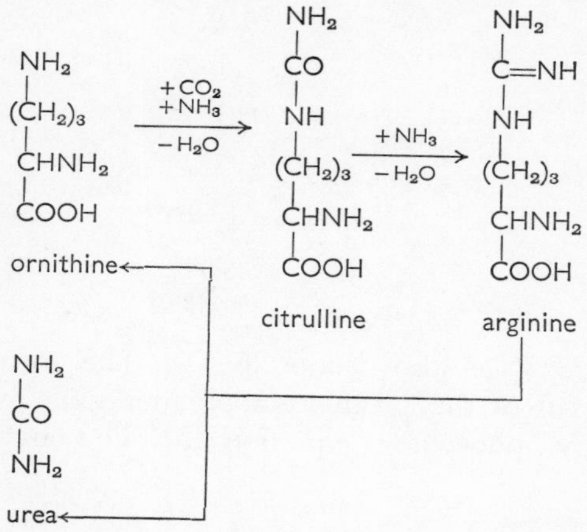

FIGURE 41.   Formation of urea.

magnesium ions and an *N*-acyl derivative of glutamate. Two moles of ATP are broken down in the course of the reaction (equation 42). Equilibrium lies towards carbamyl phosphate formation. The mechanism of the reaction is obscure.

$$CO_2 + NH_3 + 2\,ATP \rightleftharpoons NH_2C(=O)OPO_3H_2 + 2\,ADP + H_3PO_4 \qquad (42)$$

(b) Elimination of inorganic phosphate between carbamyl phosphate and ornithine to give citrulline. The reaction (equation 43) is catalysed by ornithine transcarbamylase which has been purified from liver tissue of rat and ox.

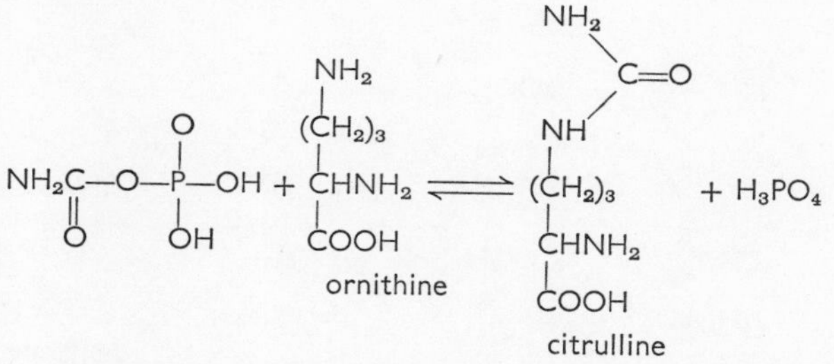

(c) The formation of arginine from citrulline can be represented formally by equation (44). However, the reaction neither occurs in one step nor involves free ammonia. Ratner[137] and her colleagues have shown that arginine formation requires the presence of both aspartate and ATP and that reaction occurs through the intermediate

$$\begin{array}{ccc}
\text{NH}_2 & & \text{NH}_2 \\
\quad\diagdown & & \quad\diagdown \\
\qquad\text{C}{=}\text{O} & & \qquad\text{C}{=}\text{NH} \\
\text{NH}\diagup & & \text{NH}\diagup \\
| & & | \\
(\text{CH}_2)_3 & + \text{NH}_3 \rightleftharpoons & (\text{CH}_2)_3 \qquad + \text{H}_2\text{O} \qquad\qquad (44) \\
| & & | \\
\text{CHNH}_2 & & \text{CHNH}_2 \\
| & & | \\
\text{COOH} & & \text{COOH} \\
\text{citrulline} & & \text{arginine}
\end{array}$$

argininosuccinate, (see also section V.D). The formation of the intermediate involves the production of adenosine monophosphate and inorganic pyrophosphate (equation 45). The enzyme catalysing

$$\begin{array}{cccc}
\text{NH}_2 & & \text{NH} \qquad \text{COOH} \\
\quad\diagdown & \text{COOH} & \qquad\diagdown \\
\qquad\text{C}{=}\text{O} & | & \qquad\text{C}{-}\text{NH}{-}\text{CH} \\
\text{NH}\diagup & \text{NH}_2{-}\text{CH} & \text{NH}\diagup \qquad | \\
| & + \quad | \quad + \text{ATP} \rightleftharpoons & | \qquad\qquad \text{CH}_2 \quad + \text{H}_4\text{P}_2\text{O}_7 + \text{AMP} \\
(\text{CH}_2)_3 & \text{CH}_2 & (\text{CH}_2)_3 \qquad | \\
| & | & | \qquad\qquad \text{COOH} \\
\text{CHNH}_2 & \text{COOH} & \text{CHNH}_2 \\
| & & | \\
\text{COOH} & & \text{COOH} \qquad\qquad\qquad (45) \\
\text{citrulline} & \text{aspartate} & \text{argininosuccinate}
\end{array}$$

this reaction (argininosuccinate synthetase) is inhibited by inorganic pyrophosphate. The presence, in liver, of an inorganic pyrophosphatase is presumed to prevent inhibition, under physiological conditions. Cleavage of argininosuccinate to produce arginine and fumaric acid (equation 46) is catalysed by a second enzyme, argininosuccinase.

$$\begin{array}{ccc}
\text{NH} \qquad \text{COOH} & & \text{NH} \\
\quad\diagdown & & \quad\diagdown \\
\qquad\text{C}{-}\text{NH}{-}\text{CH} & & \qquad\text{C}{-}\text{NH}_2 \qquad \text{COOH} \\
\text{NH}\diagup \qquad | & & \text{NH}\diagup \qquad\qquad | \\
| \qquad\qquad \text{CH}_2 & \rightleftharpoons & | \qquad\qquad + \quad\text{CH} \\
(\text{CH}_2)_3 \qquad | & & (\text{CH}_2)_3 \qquad\qquad \| \quad\quad (46) \\
| \qquad\qquad \text{COOH} & & | \qquad\qquad\qquad \text{CH} \\
\text{CHNH}_2 & & \text{CHNH}_2 \qquad\qquad | \\
| & & | \qquad\qquad\qquad \text{COOH} \\
\text{COOH} & & \text{COOH} \qquad\qquad \text{fumarate} \\
& & \text{arginine}
\end{array}$$

(d) Arginine is hydrolysed to produce urea and regenerate ornithine (equation 47). The enzyme catalysing this reaction (arginase) is activated by some divalent metal ions, e.g. Ni^{2+}, Co^{2+}, Mn^{2+}.

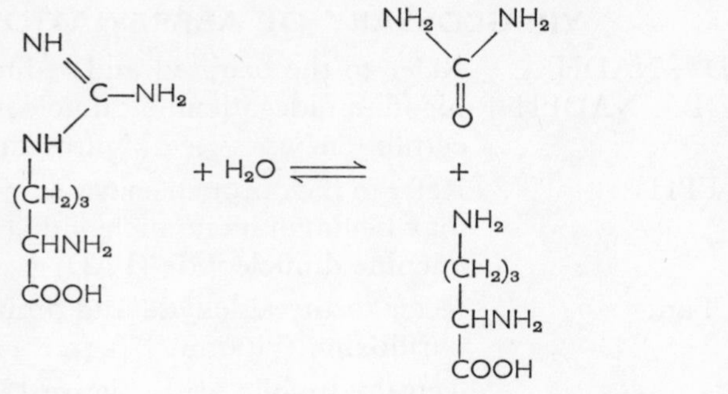

$$
\begin{array}{c}
\text{NH} \\
\diagdown \\
\quad \text{C—NH}_2 \\
\diagup \\
\text{NH} \\
| \\
(\text{CH}_2)_3 \\
| \\
\text{CHNH}_2 \\
| \\
\text{COOH}
\end{array}
\quad + \text{ H}_2\text{O} \ \rightleftharpoons \quad
\begin{array}{c}
\text{NH}_2 \quad \text{NH}_2 \\
\diagdown \diagup \\
\text{C} \\
\| \\
\text{O}
\end{array}
\ + \
\begin{array}{c}
\text{NH}_2 \\
| \\
(\text{CH}_2)_3 \\
| \\
\text{CHNH}_2 \\
| \\
\text{COOH}
\end{array}
\qquad (47)
$$

Although the present discussion is limited to reactions catalysed by liver enzymes, it may be noted that analogous enzymic activities have been demonstrated in both plants and microorganisms.

2. Uric acid formation

Although uric acid is known to be the main nitrogenous excretion product in birds and in reptiles, the route by which it is formed from excess amino compounds is not well understood. Uric acid is also the end product of purine breakdown in man and is excreted as such, while in subprimate mammals, the purine ring is further oxidised to produce first, allantoin and, in some species, eventually urea. The interrelationship between these compounds is illustrated in Figure 42.

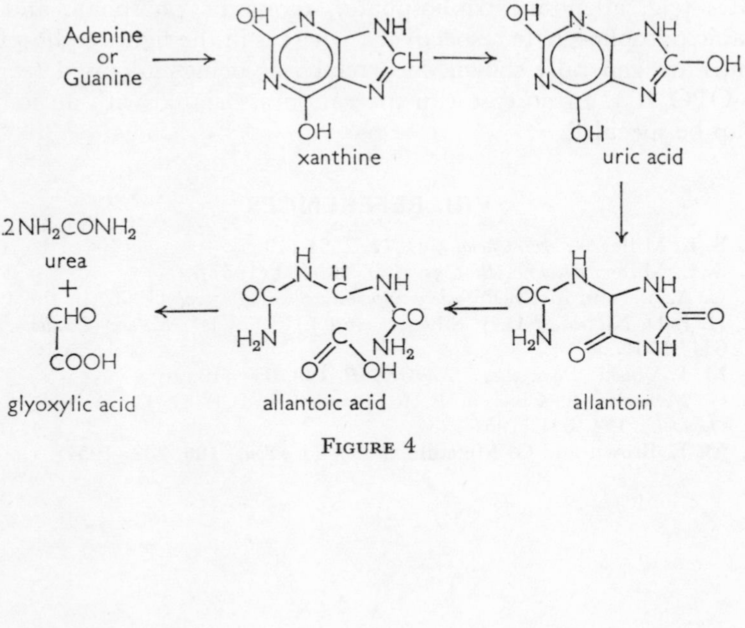

FIGURE 4

VII. GLOSSARY OF ABBREVIATIONS

NAD$^+$, NADH ⎫ Refer to the oxidised and reduced forms of the
NADP$^+$, NADPH ⎬ nicotinamide–adenine dinucleotide cofactors of
certain enzymes (see Figures 2a, b).

FP, FPH$_2$ Refer to flavoprotein enzymes in which the flavin may be flavin mononucleotide (FMN) or flavin-adenine dinucleotide (FAD) (see Figures 2c, d).

Pal, Pam Refer to the aldehyde and amine derivatives of pyridoxine (vitamin B$_6$) (see Figures 5a, b, c).

FH$_4$ Tetrahydrofolic acid (Figure 13).

ATP Adenosine triphosphate (Figure 17).

GTP Guanosine triphosphate.

CoA Coenzyme A (Figure 29).

In Figures 1 and 16, names of amino acids are indicated by the first three letters only. Thus glu represents glutamic acid and phe represents phenylalanine.

The nomenclature used in this chapter for metabolites which contain carboxylic acid groups and for phosphate esters is imprecise since the relevant ionisation states *in vivo* are, in general, ill-defined. α-Keto acids and acidic amino acids are, therefore, referred to as α-ketoglutarate, aspartate etc., but it is not intended to imply that all carboxylic acid groups are necessarily fully ionised under all conditions. In the text, esters of orthophosphoric acid, orthophosphoric acid itself and pyrophosphoric acid are referred to simply as phosphates (e.g. adenosine triphosphate), inorganic phosphate and inorganic pyrophosphate respectively, whereas in the figures, phosphate groups are generally shown, incorrectly, as being unionised (e.g. as X—OPO$_3$H$_2$). In no case can the cation associated with an ionised group be specified.

VIII. REFERENCES

1. S. L. Miller, *J. Am. Chem. Soc.*, **77**, 2351 (1955).
2. S. L. Miller, *Biochim. Biophys. Acta*, **23**, 480 (1957).
3. C. A. Vernon in *Royal Society Symposium on Lysozyme* (1966) in the Press.
4. D. J. D. Nicholas, D. J. Silvester and J. F. Fowler, *Nature (London)*, **189**, 634 (1961).
5. M. I. Volskĭ, *Dokl. Akad. Nauk SSSR*, **128**, 857 (1959).
6. G. Metcalfe, S. Chayen, E. R. Roberts and T. G. G. Wilson, *Nature (London)*, **174**, 841 (1954).
7. M. E. Brown and G. Metcalfe, *Nature (London)*, **180**, 282 (1957).

8. R. H. Burris and P. W. Wilson in *Methods in Enzymology*, Vol. 4 (Eds. S. P. Colowick and N. O. Kaplan), Academic Press, New York, 1957, p. 355.

9. J. E. Carnahan and J. E. Castle, *Ann. Rev. Plant Physiol.* **14**, 125 (1963).

10. D. J. D. Nicholas, *Symp. Soc. Gen. Microbiol.*, **13**, 92 (1963).

11. W. D. P. Stewart, *Nitrogen Fixation in Plants*, University of London, Athlone Press, 1966.

12. L. E. Mortenson, R. C. Valentine and J. E. Carnahan, *Biochem. Biophys. Res. Commun.*, **7**, 448 (1962).

13. L. E. Mortenson, H. F. Mower and J. E. Carnahan, *Bacteriol. Rev.*, **26**, 42 (1962).

14. H. J. Jensen in *Nutrition of the Legumes* (Ed. E. G. Hallsworth), Butterworths, London, 1950, p. 75.

15. D. J. D. Nicholas, P. W. Wilson, W. Heinen, G. Palmer and H. Beinert, *Nature (London)*, **196**, 433 (1962).

16. H. J. Evans and A. Nason, *Plant Physiol.*, **28**, 233 (1953).

17. D. J. D. Nicholas and A. Nason, *J. Bacteriol.*, **69**, 580 (1955).

18. D. J. D. Nicholas and H. M. Stevens, *Nature (London)*, **176**, 1066 (1955).

19. A. Nason, R. G. Abraham and B. C. Averbach, *Biochim. Biophys. Acta*, **15**, 159 (1954).

20. D. Spencer, H. Takahashi and A. Nason, *J. Bacteriol.*, **73**, 553 (1957).

21. A. Medina and D. J. D. Nicholas, *Nature (London)*, **179**, 533 (1957).

22. V. L. Kretovich, A. A. Bundel, M. R. Frasheri and N. V. Borovikova, *Fiziol. Rast.*, **7**, 261 (1960).

23. D. J. D. Nicholas and O. T. G. Jones, *Nature (London)*, **189**, 512 (1960).

24. A. Meister, *Biochemistry of the Amino Acids*, Vol. 1, Academic Press, New York, 1965.

25. E. Marler and C. Tanford, *J. Biol. Chem.*, **239**, 4217 (1964).

26. S. J. Adelstein and B. L. Vallee, *J. Biol. Chem.*, **234**, 824 (1959).

27. J. L. Wood and S. L. Cooley, *Proc. Soc. Exp. Biol. Med.*, **85**, 409 (1954).

28. E. A. Adelberg and H. E. Umbarger, *J. Biol. Chem.*, **205**, 475 (1953).

29. D. F. Silbert, S. E. Jorgensen and E. C. C. Lin, *Biochim. Biophys. Acta*, **73**, 232 (1963).

30. S. P. Bessman, J. Rossen and E. C. Layne, *J. Biol. Chem.*, **201**, 385 (1953).

31. E. M. Scott and W. B. Jakoby, *J. Biol. Chem.*, **234**, 932 (1959).

32. C. Peraino and H. C. Pitot, *Biochim. Biophys. Acta*, **73**, 222 (1963).

33. D. E. Metzler, M. Ikawa and E. E. Snell, *J. Am. Chem. Soc.*, **76**, 637, 648, 653 (1954).

34. D. E. Metzler, *J. Am. Chem. Soc.*, **79**, 485 (1957).

35. B. E. C. Banks, A. Diamantis and C. A. Vernon, *J. Chem. Soc. (London)*, 4235 (1961).

36. R. C. Hughes, W. T. Jenkins and E. H. Fischer, *Proc. Natl. Acad. Sci., U.S.*, **48**, 1615 (1962).

37. V. Scardi, P. Scotto, M. Iaccarino and E. Scarano in *Proceedings of the Symposium on Pyridoxal Catalysis (Rome)*, (Eds. E. E. Snell, P. M. Fasella, A. Braunstein and A. Rossi Fanelli), Pergamon, 1963, p. 167.

38. W. T. Jenkins and I. W. Sizer, *J. Biol. Chem.*, **235**, 620 (1960).

39. H. Lis, P. Fasella, C. Turano and P. Vecchini, *Biochim. Biophys. Acta*, **45**, 529 (1960).

40. W. T. Jenkins, D. A. Yphantis and I. W. Sizer, *J. Biol. Chem.*, **234**, 51 (1959).

41. P. Andrews, *Biochem. J.*, **96**, 595 (1965).

42. S. Doonan, *Ph.D. Thesis*, London, 1966.

43. B. E. C. Banks, A. J. Lawrence, E. M. Thain and C. A. Vernon, *J. Chem. Soc. (London)*, 5799 (1963).

44. B. E. C. Banks, A. J. Lawrence, C. A. Vernon and J. F. Wooton, in *Proceedings of the Symposium on Pyridoxal Catalysis (Rome)*, (Eds. E. E. Snell, P. M. Fasella, A. Braunstein and A. Rossi Fanelli), 1963, p. 197.

45. G. G. Hammes, and P. Fasella, *J. Am. Chem. Soc.*, **84**, 4644 (1962).

46. A. Meister, *Biochemistry of the Amino Acids*, Vol. 2, Academic Press, New York, 1965.

47. H. S. McKee, *Nitrogen Metabolism in Plants*, Clarendon Press, Oxford, 1962.

48. L. Fowden, *Ann. Rev. Biochem.*, **33**, 173 (1964).

49. B. N. Ames and B. L. Horecker, *J. Biol. Chem.*, **220**, 113 (1956).

50. J. A. Miller, *Cancer Res.*, **10**, 65 (1950).

51. G. Hillmann, A. Hillmann-Elies and F. Methfessel, *Nature (London)*, **174**, 403 (1954).

52. E. E. Snell, N. S. Radin and M. Ikawa, *J. Biol. Chem.*, **217**, 803 (1955).

53. W. C. Rose, *Federation Proc.*, **8**, 546 (1949).

54. R. E. Kallio and A. D. Larson in *Amino Acid Metabolism* (Ed. W. D. McElroy and B. Glass), Johns Hopkins Press, Baltimore, Maryland, 1955, p. 616.

55. J. T. Holden, C. Furman and E. E. Snell, *J. Biol. Chem.*, **178**, 789, 799 (1949).

56. L. Glaser, *J. Biol. Chem.*, **235**, 2095 (1960).

57. W. F. Diven, R. B. Johnston and J. J. Schloz, *Biochim. Biophys. Acta*, **67**, 161 (1963).

58. V. Massey, G. Palmer and R. Bennett, *Biochim. Biophys. Acta*, **48**, 1 (1961).

59. K. Yagi, T. Ozawa and M. Harada, *Nature (London)*, **184**, 1938 (1959); *Nature (London)*, **188**, 745 (1960).

60. A. E. Bender and H. A. Krebs, *Biochem. J.*, **46**, 210 (1950).

61. M. Dixon and K. Kleppe, *Biochim. Biophys. Acta*, **96**, 357, 368, 383 (1965).

62. K. Yagi and T. Ozawa, *Biochim. Biophys. Acta*, **81**, 599 (1964).

63. K. Yagi, T. Ozawa and T. Ooi, *Biochim. Biophys. Acta*, **54**, 199 (1961).

64. P. A. Charlwood, G. Palmer and R. Bennett, *Biochim. Biophys. Acta*, **50**, 17 (1961).

65. V. Pamiljans, P. R. Krishnaswamy, G. Durnville and A. Meister, *Biochemistry*, **1**, 153 (1962).

66. C. A. Vernon in *Size and Shape Changes of Contractile Polymers* (Ed. A. Wasserman), Pergamon Press, London, 1960, p. 109.

67. J. M. Ravel, S. J. Norton, J. S. Humphreys and W. Shive, *J. Biol. Chem.*, **237**, 2845 (1962).

68. A. Meister, L. Levintow, R. E. Greenfield and P. A. Abendschein, *J. Biol. Chem.*, **215**, 441 (1955).

69. E. F. Gale, *Advan. Enzymol.*, **6**, 1 (1946).

70. R. Shukuya and G. W. Schwert, *J. Biol. Chem.*, **235**, 1649, 1653 (1960).

71. F. H. Westheimer in *The Enzymes*, Vol. 1 (Eds. P. D. Boyer, H. Lardy and K. Myrbäck), Academic Press, New York, 1959, p. 259.

72. S. Mandeles, R. Koppelman and M. E. Hanke, *J. Biol. Chem.*, **209**, 327 (1954).

73. B. Belleau and J. Burba, *J. Am. Chem. Soc.*, **82**, 5751 (1960).

74. E. Roberts and S. Frankel, *J. Biol. Chem.*, **190**, 505 (1951).

75. E. A. Kravitz, S. W. Kuffler and D. D. Potter, *J. Neurophysiol.*, **26**, 739 (1963).

76. W. Lovenberg, H. Weissbach and S. Udenfriend, *J. Biol. Chem.*, **237**, 89 (1962).

77. J. D. Reid and D. M. Shepherd, *Biochim. Biophys. Acta*, **81**, 560 (1964).

78. J. D. Wilson, K. D. Gibson and S. Udenfriend, *J. Biol. Chem.*, **235**, 3539 (1960).

79. J. Bremer, P. H. Figard and D. M. Greenberg, *Biochim. Biophys. Acta*, **43**, 477 (1960).

80. L. F. Borkenhagen, E. P. Kennedy and L. Fielding, *J. Biol. Chem.*, **236**, P.C. 28 (1961).

81. K. D. Gibson, J. D. Wilson and S. Udenfriend, *J. Biol. Chem.*, **236**, 673 (1961).

82. B. Magasanik in *The Bacteria*, Vol. 3 (Eds. I. C. Gunsalus and R. Y. Stanier), Academic Press, New York, 1962, p. 295.

83. M. Shepherdson and A. B. Pardee, *J. Biol. Chem.*, **235**, 3233 (1960).

84. H. C. Friedmann and B. Vennesland, *J. Biol. Chem.*, **235**, 1526 (1960).

85. H. G. Khorana, J. F. Ferandes and A. Kornberg, *J. Biol. Chem.*, **230**, 941 (1958).

86. I. Lieberman, *J. Biol. Chem.*, **222**, 765 (1956).

87. K. P. Chakraborty and R. B. Hurlbert, *Biochim. Biophys. Acta*, **47**, 607 (1961).

88. S. C. Hartman and J. M. Buchanan, *J. Biol. Chem.*, **233**, 451, 456 (1958).

89. L. Warren and J. M. Buchanan, *J. Biol. Chem.*, **229**, 613 (1957).

90. B. Levenberg and J. M. Buchanan, *J. Biol. Chem.*, **224**, 1005, 1019 (1957).

91. L. N. Lukens and J. M. Buchanan, *J. Am. Chem. Soc.*, **79**, 1511 (1957).

92. G. R. Greenberg and E. L. Spilman, *J. Biol. Chem.*, **219**, 411 (1956).

93. C. E. Carter and L. H. Cohen, *J. Am. Chem. Soc.*, **77**, 499 (1955).

94. L. Warren, J. G. Flaks and J. M. Buchanan, *J. Biol. Chem.*, **229**, 627 (1957).

95. I. Lieberman, *J. Biol. Chem.*, **223**, 327 (1956).

96. M. R. J. Solton, *Ann. Rev. Biochem.*, **34**, 143 (1965).

97. H. R. Perkins, *Bacteriol. Rev.*, **27**, 18 (1963).

98. J. D. Dutcher, *Advan. Carbohydrate Chem.*, **18**, 259 (1963).

99. W. Gottschalk, *Scialic Acids.*, Cambridge University Press, 1960.

100. S. Ghosh, H. J. Blumenthal, E. Davidson and S. Roseman, *J. Biol. Chem.*, **235**, 1265 (1960).

101. L. F. Leloir, *Biochem. J.*, **91**, 1 (1964).

102. S. Kornfeld and L. Glaser, *J. Biol. Chem.*, **237**, 3052 (1962).

103. S. Roseman, *Federation Proc.*, **21**, 1075 (1962).

104. E. Ito and J. L. Strominger, *J. Biol. Chem.*, **237**, 2689, 2696 (1962).

105. E. Ito and J. L. Strominger, *J. Biol. Chem.*, **239**, 210 (1964).

584   Barbara E. C. Banks

106. J. S. Anderson, P. M. Meadow, M. A. Hoskins and J. L. Strominger, *Arch. Biochem. Biophys.*, **116**, 487 (1966).
107. S. Roseman, *Proc. Intern. Congr. Biochem.*, N.Y. VI S5, 467 (1964).
108. S. F. Dyke, *The Chemistry of the Vitamins*, Interscience, London, 1965.
109. H. A. Barker, H. Weissbach and R. D. Smyth, *Proc. Natl. Acad. Sci. U.S.*, **44**, 1093 (1958).
110. H. Eggerer, P. Overath, F. Lynen and E. R. Stadtman, *J. Am. Chem. Soc.*, **82**, 2643 (1960).
111. C. S. Hegre, S. J. Miller and M. D. Lane, *Biochim. Biophys. Acta*, **56**, 538 (1962).
112. T. W. Goodwin, *The Biosynthesis of Vitamins and Related Compounds*, Academic Press, London, 1963, Chap. 5.
113. E. L. R. Stokstad, *Ann. Rev. Biochem.*, **31**, 451 (1962).
114. T. W. Goodwin, *Recent Advances in Biochemistry*, Churchill (London), 1960, p. 161.
115. J. D. Bu'lock, *Biosynthesis of Natural Products*, McGraw Hill, London, 1965.
116. H. A. Krebs, *Biochem. J.*, **29**, 1620 (1935).
117. M. Blanchard, D. E. Green, V. Nocito and S. Ratner, *J. Biol. Chem.*, **155**, 421 (1944); *J. Biol. Chem.*, **161**, 583 (1945).
118. T. P. Singer and E. B. Kearney, *Arch. Biochem. Biophys.*, **29**, 190 (1950).
119. D. Wellner and A. Meister, *J. Biol. Chem.*, **235**, 2013 (1960).
120. P. K. Stumpf and D. E. Green, *J. Biol. Chem.*, **153**, 387 (1944).
121. S. Nara and K. T. Yasunobu, *Pharm. Rev.*, **18**, 144 (1966).
122. H. Blaschko, *Pharm. Rev.*, **18**, 39 (1966).
123. H. Yamada and K. T. Yasunobu, *J. Biol. Chem.*, **237**, 1511 (1962).
124. F. Buffoni and H. Blaschko, *Proc. Roy. Soc. B (London)*, **161**, 153 (1965).
125. C. W. Tabor, H. Tabor and U. Bachrach, *J. Biol. Chem.*, **239**, 2194 (1964).
126. E. Werle and E. Pechmann, *Ann. Chem.*, **562**, 44 (1949).
127. P. J. G. Mann, *Biochem. J.*, **79**, 623 (1961).
128. E. V. Goryachenkova and E. A. Enshova, *Biokhimiya*, **30**, 165 (1965).
129. H. Tabor, *J. Biol. Chem.*, **188**, 125 (1951).
130. A. E. Braunstein in *The Enzymes*, Vol. 2 (Eds. P. D. Boyer, H. Lardy and K. Myrbäck), Academic Press, New York, 1960, p. 113.
131. C. Yanofsky and J. L. Reissig, *J. Biol. Chem.*, **202**, 567 (1953).
132. D. E. Metzler and E. E. Snell, *J. Biol. Chem.*, **198**, 363 (1952).
133. F. W. Sayre and D. M. Greenberg, *J. Biol. Chem.*, **220**, 787 (1956).
134. G. Moss, D. H. Russell, E. M. Thain and C. A. Vernon, *Biochem. J.*, **95**, 3P (1965).
135. C. Fromageot in *The Enzymes*, Vol. 2 (Eds. J. B. Sumner and K. Myrbäck) Academic Press, New York, 1951, p. 248.
136. H. A. Krebs and K. Henseleit, *Z. Physiol. Chem.*, **210**, 33 (1932).
137. S. Ratner in *The Enzymes*, Vol. 6 (Eds., P. D. Boyer, H. Lardy and K. Myrbäck), Academic Press, New York, 1962, p. 495.

CHAPTER **10**

Rearrangements involving amino groups

D. V. BANTHORPE

University College, London

The title reactions can be grouped into several classes depending on the existence, and nature, of intermediates. The following sections summarise the most important examples, but the extent of the literature precludes an exhaustive treatment. Often, especially for minor topics, questions of historicity and priority will be ignored and citation will be made to articles containing discussion of earlier work.

I. REARRANGEMENTS ACCOMPANYING DEAMINATION

A. Ionic Intermediates

In the generally accepted mechanism for deamination of primary alkylamines with nitrosating agents, the transiently formed diazonium ion loses nitrogen, with neither solvent nor neighbouring group assistance to give an unsolvated carbonium ion which is negligibly encumbered by the departing molecule and has no counter ion in the vicinity[1]. This high-energy or 'hot' ion differs from the formally identical but solvated species typically generated in an $S_N 1$ reaction by undergoing a wider spectrum of reactions, some of which, however, are attendant on partial or almost complete solvation[2-7]. Other interpretations require the intervention of diazoalkanes[8] or diazohydroxides[2,9,10], or the direct decomposition at the diazonium ion stage to products[11,12]; but the last two distinctions may be semantic, for the energy profiles for the different mechanisms (exhibiting shallow minima for the various intermediates), must be almost indistinguishable; especially when smoothed out by the thermal energy of the system. Nevertheless, the carbonium ion mechanism is usually considered to accommodate the data most satisfactorily[1].

Much of the evidence for hot ions has been adduced from the study of rearrangements accompanying deamination, and in this review we will assume the existence of such intermediates. Classical ions, which can be represented by one or a series of equilibrating structures in which the charge is largely localised on one carbon and the normal covalencies occur, and also non-classical ions with charge delocalisation and partial bonding are both believed to exist in deamination. The existence of the latter in reactions generally, has been disputed[13], but there is considerable stereochemical evidence, particularly for bridged substrates, that cannot be convincingly interpreted on the classical model[14]; although in many cases a decision as to whether the non-classical species are intermediates or merely transition states is difficult. Unfortunately, kinetic evidence for mesomeric cations, perhaps the best criterion[15], is unavailable because deamination occurs after the rate-determining step of the reaction sequence, but we will follow the literature and often assign non-classical intermediates, on admittedly, minimal evidence, and will represent them using the partial-bond symbolism, despite recent objections to this[16].

Deaminations have usually been conducted in acetous or aqueous media, at 0 to 30°, yielding olefins and either acetates or alcohols respectively. For uniformity, all reactions forming acetates will be referred to alcohol products in the following sections, and the reaction conditions will only be mentioned when mechanistically significant.

B. Acyclic Amines

Deamination of short-chain primary alkylamines with nitrous acid has long been known to give not only an enhanced (25–30%) yield of olefin compared with the analogous S_N1 reactions of halides, but also rearranged products absent from the latter. The proportion of rearranged alcohols falls with increasing homology[17-19] (Table 1) and

TABLE 1. Rearranged products from nitrous deamination of alkylamines[17-19].

R in RNH_2	n-Pr	n-Bu	i-Bu	n-Oct	n-Non	n-Dec
% Rearranged ROH	58	34	75, 25	5	0	0
Isomer formed	iso	s	t, s	s	—	—

(conditions: aqueous or aqueous–acetic acids)

competition of the 1,2-methyl and hydride shifts responsible for the rearrangement with direct S_N1 and $E1$ is very dependent on conditions: e.g. the proportion of rearranged product from n-propylamine falls from 58% in totally aqueous solvent to 31% in aqueous dimethyl-formamide[6, 7]. Later it was appreciated that the driving force of these Wagner–Meerwein-like isomerisations (which can also involve aryl shifts) was the tendency, usually only capable of fruition in hot ions, of the initially formed ion to rearrange to a more stable secondary or tertiary ion. Thus normal and iso substrates yield secondary and tertiary products respectively, whereas secondary substrates never form normal products[20]. Sometimes the hot-ion model seems inadequate[20]—iso:s:t product from isobutylamine changes from $10:19:71$ in water to $23:21:56$ in 7 N aqueous sodium thiocyanate, and the ratio of alkyl thiocyanate and isothiocyanate products also suggests that the anion participates in C–N cleavage to some extent.

Analogous rearrangements accompany related deaminations (reaction 1), in which the similarity of products under comparable conditions suggests a common intermediate[6, 7, 21, 22]; the slight differences being attributable to the (necessarily) different reaction

$$
\begin{array}{l}
\text{n-PrNH}_2 \xrightarrow{\text{HNO}_2, 0°} \\[4pt]
\text{CH}_3\text{CH}_2\text{CHN}_2 \xrightarrow{\overset{+}{\text{H}}, 0°} \\[4pt]
\text{n-PrN(NO)COCH}_3 \xrightarrow{\Delta, 70°}
\end{array}
\longrightarrow \text{n-PrN}_2{}^+ \longrightarrow \text{n-Pr}^+ \longrightarrow \underset{(28\text{–}33\%)}{\text{i-PrOH}} \qquad (1)
$$

temperatures. Only about 1% isoalcohol results from decomposition of diazopropane in ethereal acid[6, 7], probably owing to the formation of specifically orientated ion pairs in these non-solvating conditions (e.g. $\text{R}\overset{+}{\text{N}}_2\overset{-}{\text{O}}\text{Ac}$, when acetic acid was the catalyst) which inhibit rearrangement. In more polar media, ion dissociation simulates the situation in nitrous deamination and permits extensive rearrangement. It is interesting that at the other end of the solvation scale, the nitrous deamination, under completely aqueous conditions, of substrates that initially form stabilised carbonium ions (e.g. t-amines) gives little rearrangement[117]. Here the relatively stable ion initially formed may solvate to form a 'cold' ion before rearrangement can occur; or it may orientate an incipient hydroxide ion from the solvent shell to act as a counter ion at the diazonium-forming stage.

The pyrolyses of N-nitroso-N-acetylamines also probably generate ion pairs in the usual non-polar reaction media, e.g. p-cymene, for little rearrangement is reported[23, 64]. Diazoalkanes, formed by elimination from diazohydroxides, are unlikely intermediates in nitrous deamination, for neither ethylamine nor isobutylamine gives labelled products when the reaction is conducted in deuterated media: these observations also exclude π-complex intermediates[19, 24].

Tracer studies have shown that a proportion of the apparently unrearranged product can result from 1,2-shifts (Table 2). About 90%

TABLE 2.   Rearrangement in deamination of β-substituted ethylamines.

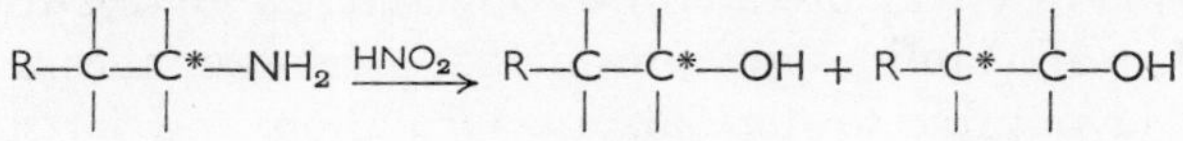

$$R-\overset{|}{\underset{|}{C}}-\overset{|}{\underset{|}{C^*}}-NH_2 \xrightarrow{HNO_2} R-\overset{|}{\underset{|}{C}}-\overset{|}{\underset{|}{C^*}}-OH + R-\overset{|}{\underset{|}{C^*}}-\overset{|}{\underset{|}{C}}-OH$$

$$(A) \qquad\qquad (B)$$

(C* represents ^{14}C)

R	H	Ph	p-MeOC$_6$H$_4$	p-HOCH$_2$CH$_2$OC$_6$H$_4$
% B	1·5	27	45	40
References	24	25, 26	25, 26	27

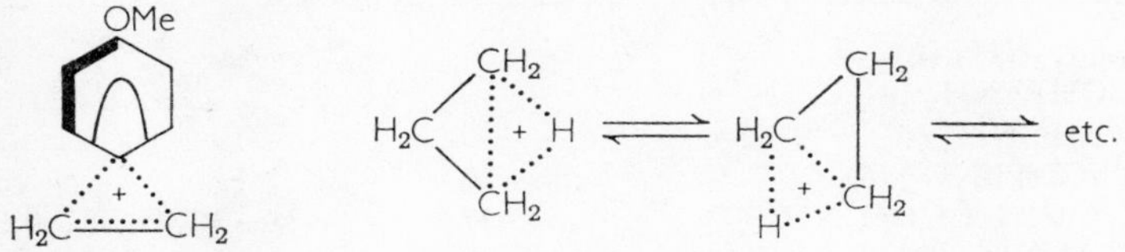

of the p-anisyl compound must have reacted through the bridged ion or transition state **1**, whereas such a route was much less favoured for hydrogen migration. Early investigations of n-propylamine[28, 29, 30]

(1)   (2)

reported 9% rearrangement of ^{14}C tracer from C$_{(1)}$ exclusively to C$_{(3)}$ and so excluded methyl shifts in favour of 1,3- or successive 1,2-hydride migrations. An n.m.r. study[31] of products from deuterated substrate indicated 1,3-migration (reaction 2) and was considered to

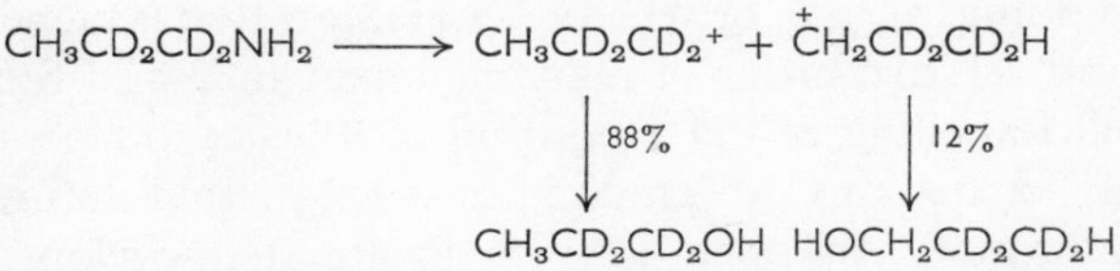

$$CH_3CD_2CD_2NH_2 \longrightarrow CH_3CD_2CD_2{}^+ + \overset{+}{C}H_2CD_2CD_2H$$

$$\Big\downarrow 88\% \qquad\qquad \Big\downarrow 12\% \qquad\qquad (2)$$

$$CH_3CD_2CD_2OH \quad HOCH_2CD_2CD_2H$$

rule out the intermediacy of equilibrating cyclopropanes (**2**)[32]. The existence of the latter are also inconsistent with the ^{14}C studies, but reinvestigation[33] has led to a reappraisal in that not only a lower proportion (about 4%) of tracer rearrangement, but also nearly equal migration to C$_{(2)}$ and C$_{(3)}$, was demonstrated. These, and concurrent studies on deuterated and tritiated substrates[34], are inconsistent with the 1,3-shift occurring entirely through open classical ions, and favour **2** as specific intermediates (rather than transition states) which can give up to 10% cyclopropane in addition to the usual products[35]. Isolation of cyclopropane-d_2 and -d_3 in comparable amounts from decomposition of γ-trideuterated n-propylamine supports this view[36]. Certain other alkylamines give up to 15% cyclopropanes under conditions not favourable to diazoalkane formation and its subsequent carbenoid decomposition, and presumably in all cases proton loss from bridged ions similar to **2** leads to products, but it is difficult to rationalise the relative ease of cyclisation and 1,2- and 1,3-shifts in these compounds in any convincing manner[37].

Reactions of *neo*-pentyl derivatives, based on a skeleton well known for rearrangement tendencies in S_N1, are summarised in Table 3. As

TABLE 3. Deamination of *neo*-pentyl-like structures.

Compound	% Rearrangement	Migrating group	Reference
$Me_3CCH_2NH_2$ (**3**)	100	Me	38
$Me_3CCHPhNH_2$ (**4**)	0	—	38
$Et_3CCHPhNH_2$ (**5**)	~ 30	Et	38
$R_2CPhCHPhNH_2$ (**6**)	~ 50	Ph	38
(p-$MeOC_6H_4$)$_3CCH_2NH_2$ (**7**)	17	p-$MeOC_6H_4$	39
$Ph_3CCH_2NH_2$ (**8**)	28	Ph	39

expected, **3** gives products derived from a tertiary ion, but the first-formed ion from **4** is stabilised by the phenyl group. Such a substituent cannot prevent rearrangement when steric congestion at $C_{(2)}$ is relieved in so doing, as in **5**, or when phenyl migration is possible as in **6**. The size and relative order of rearrangement in **7** and **8** is also surprising. ^{14}C Labelling of the 1-position of **3** leads to *t*-pentanol with tracer solely located at $C_{(3)}$[40], indicating a 1,2-methyl shift rather than hydride migration via open ions or protonated cyclopropanes, in contrast with the situation for n-propylamine (reaction 3). Other differences between the two substrates are that **3** yields no cyclopropane, but gives larger quantities of rearranged olefins[41]. Other tracer studies of alkylamines have demonstrated similar skeletal migrations[42, 43].

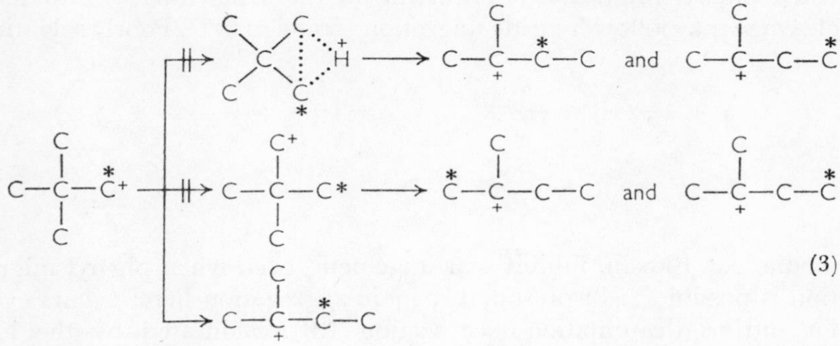

$$(3)$$

C. Semipinacols

1,2-Aminoalcohols (semipinacols) undergo aryl, alkyl, and hydride shifts on treatment with nitrosating agents, and 1,2-diamines behave analogously[44]. The former class gives similar products to those of the corresponding pinacol rearrangement (reaction 4), but with the

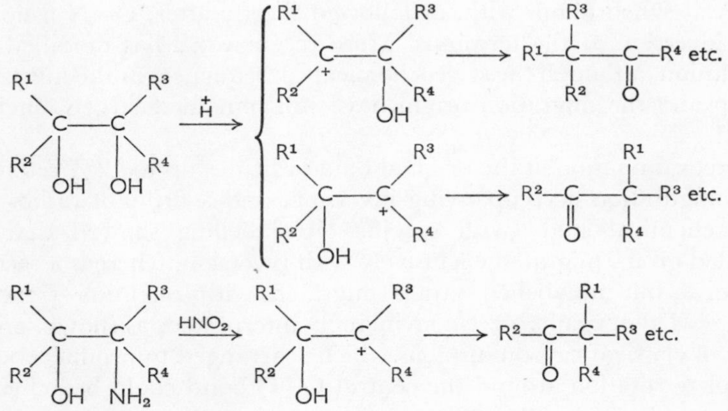

$$(4)$$

important modification of hot-ion formation; although the product distribution from particular substrates in the two reactions may be identical if the hot ion can be solvated before it reacts further[45], and here glycols and epoxides may predominate[44]. 1,3-, 1,4-, and some 1,2-aminoalcohols give up to 20% fragmentation (reaction 5), even

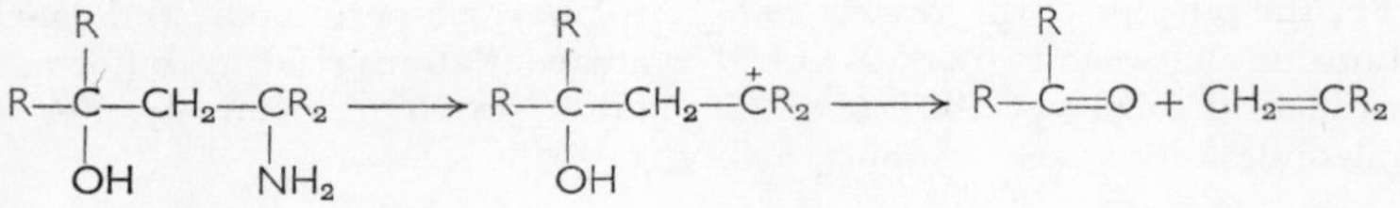

$$(5)$$

when phenyl migration is occurring at the same time[46]; and such cleavage may follow hydride migration (reaction 6)[47]. Poorly solvating

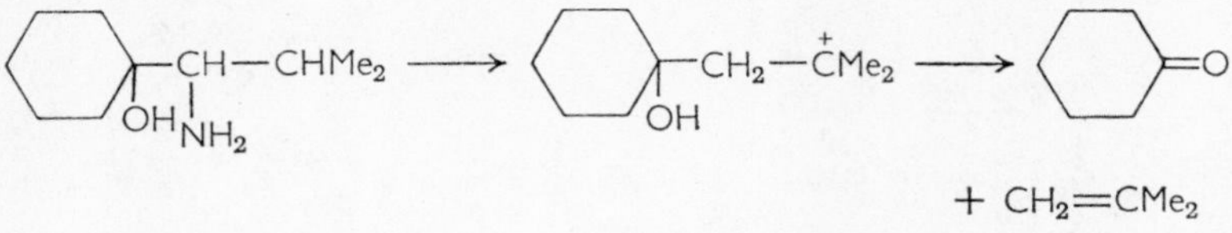

$$+ \ CH_2{=}CMe_2 \quad (6)$$

media, e.g. dioxan, inhibit rearrangement, even when phenyl migration is possible[48, 49]; presumably ionic aggregation here, occurs even for nitrous deamination (see section I.B), facilitated by the low dielectric constant and the orientating effect on the solvation sheath caused by the neighbouring hydroxyl group.

D. Stereochemistry

The fragmentary data[50] available for normal and semipinacolic deamination was once held to indicate strictly stereospecific migration that was synchronous with, or followed shortly after, C—N fission to give inversion at the terminus. More recent work has modified this conclusion, although the stereochemical consequences at the migrating group and the migration origin have still remained largely unelucidated.

A reexamination of the original data indicates up to 12% retention of configuration accompanying inversion, and a brilliant radio- and stereochemical study with specific ^{14}C labelling showed that the labelled group migrated exclusively with inversion, whereas a second, identical but unlabelled group migrated with retention (reaction 7)[51, 52]. These results were convincingly interpreted, as shown, on the basis of classical carbonium ions which rearranged to products before complete rotation around the central C—C bond could be achieved, and non-classical intermediates or rearrangement synchronous with nitrogen loss were excluded. The *anti*-periplanar* relationship between the migrating and leaving groups implied by this scheme is certainly consistent with the last mechanism, but the consensus of opinion, here and in other examples, is that the *anti* relationship enables maximum overlap between the bond electrons of the migrating group and the

---

* For the purpose of this review, *anti*-periplanar, *syn*-periplanar, and *syn*-clinal relationships between two bonds linked to adjacent atoms refer to conformations with dihedral angles of 180°, 0 to 15°, and 15 to 90° respectively. For more complete definitions see reference 398.

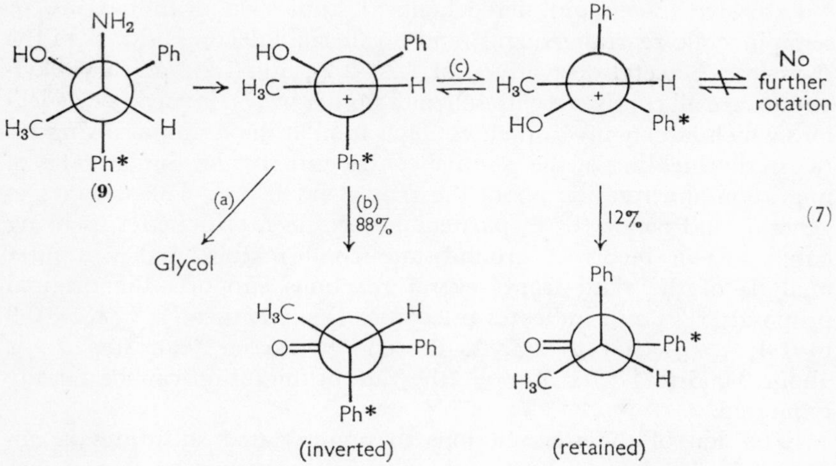

freshly formed empty p orbital on $C_{(1)}$. Rotation of the ion into other conformations than those shown would have given retention with the labelled phenyl having migrated.

Earlier ^{14}C studies (reaction 8) had demonstrated that the ap-

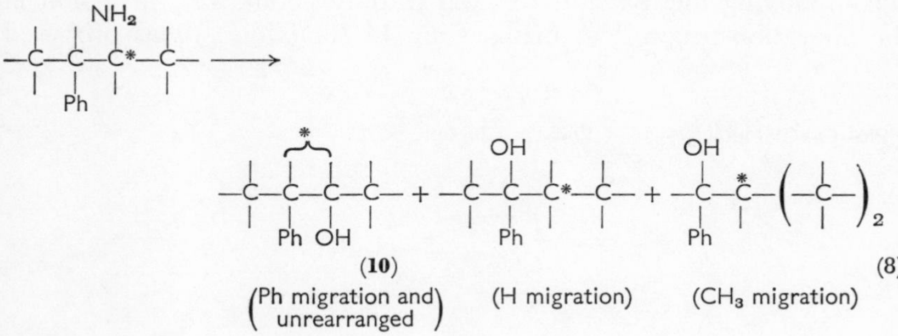

parently unrearranged product **10** had tracer distributed equally between the two carbons indicated[53], and although this was consistent with phenyl migration through a non-classical intermediate similar to **1**, the initial formation of a classical ion was preferred for such an exothermic process as deamination, with (presumed) lack of neighbouring group participation. Much evidence for non-classical phenonium ions is available in S_N1[54], but stereospecific 'equilibrating' rearrangement of the above type is necessary, although not sufficient, evidence for proving such intermediates in deamination.

Extensive tracer and stereochemical studies on deaminations, or semipinacolic rearrangements, and on thermal decompositions of the N-nitroso-N-acetyl derivatives, of 1,2-, 2,2-, and 1,2,2-polyarylalkylamines are all consistent with schemes similar to (7), involving classical ions which largely invert their configuration at the terminus owing to: (a) steric shielding of the positive centre and (b) the similar rates of migration and rotation about the central bond[55-58]. The schemes in general, and factor (b) in particular have been criticised[59] as being based on an incorrect ground-state conformation, but computer analysis of the data from several reactions supports the original proposals[60, 61], and indicates rate ratios (see reaction 7): $k_c/k_b = 0.9$ to 1.8; $k_a/k_b = 1.1$ to 2.3. As the energy barrier[62] for step (c) is about 0.006 to 1.15 kcal/mole, the rate of migration can be readily estimated.

Extension of these conclusions to non-arylated substrates is uncertain as the classical ion may be especially stabilised in these rather unusual structures, and the route utilised may well depend on the reaction conditions. Deamination of 2-amino-3-phenylbutane (**11**), results in stereospecific migration of phenyl with inversion at the origin and terminus (owing to the symmetry of the molecule this rearrangement could only be detected by optical analysis), but the accompanying methyl shift resulted in only about 16% inversion at the migration origin[63]. A bridged ion **12** (reaction 9) was proposed

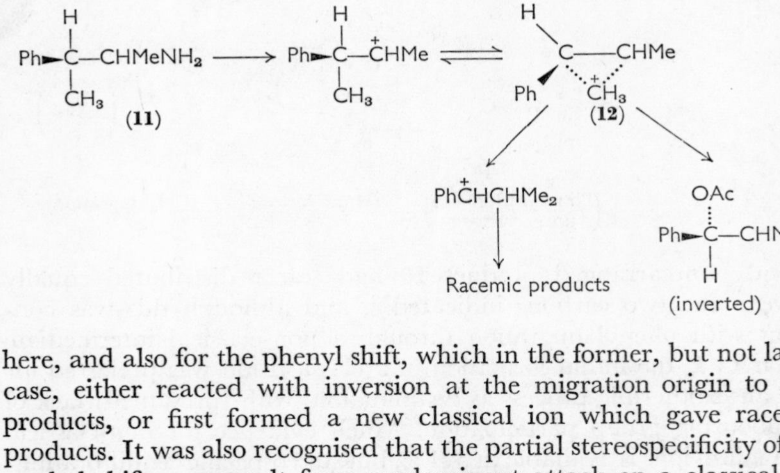

here, and also for the phenyl shift, which in the former, but not latter case, either reacted with inversion at the migration origin to give products, or first formed a new classical ion which gave racemic products. It was also recognised that the partial stereospecificity of the methyl shift may result from predominant attack on a classical ion from the less sterically hindered side. At present a decision is not possible between the two mechanisms, although **12**, in contradistinction to phenonium ions, seems more likely as a transition state than an intermediate. Less ambiguous was the observation that *threo*-**11** gave 32% methyl, 24% phenyl, and 24% hydrogen migration, whereas the *erythro* isomer gave 6, 68, and 20% respectively. These values are in excellent accord with the calculated conformational populations of the substrates, (reaction 10) and with the proposal that an *anti-*

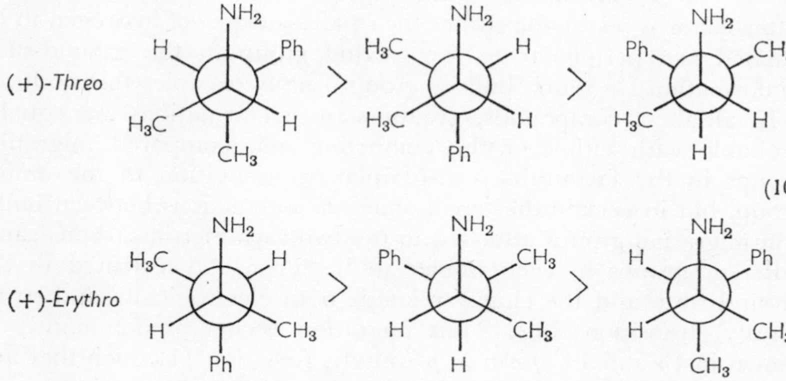

(+)-*Threo*

$$
\begin{array}{ccc}
\text{NH}_2 & \text{NH}_2 & \text{NH}_2 \\
\text{H, Ph, H}_3\text{C, CH}_3 & \text{H}_3\text{C, H, H}_3\text{C, Ph} & \text{Ph, CH}_3, \text{H}_3\text{C, H}
\end{array}
$$

(10)

(+)-*Erythro*

$$
\begin{array}{ccc}
\text{NH}_2 & \text{NH}_2 & \text{NH}_2 \\
\text{H}_3\text{C, H, H, Ph} & \text{Ph, CH}_3, \text{H, H} & \text{H, Ph, H, CH}_3
\end{array}
$$

periplanar relationship between migrating and leaving groups is required. Although this is again consistent with a mechanism where migration is synchronous with nitrogen loss from the diazonium ion, the consensus of opinion again is that it refers to a carbonium ion mechanism in which the migration rate is considerably greater than the rate of rotation about the central bond, *i.e.* analogous to reaction (7), with no rotation of the first-formed ion.

E. Migratory Aptitudes

The ease of migration of groups in the pinacol rearrangement varies in a consistent and interpretable manner over a several hundred-fold range, but little selectivity is shown in migration to a hot ionic centre. Accordingly, the relative migratory aptitude (determined by product analyses of substrates carrying groups to be compared in equivalent positions) for *p*-anisyl:phenyl falls from about 500:1 in the former reaction, to 1·6:1 in semipinacolic deamination, and such values are typical for *p*-substituted aryl groups, irrespective of the polarity of the *para* substituent[4,25]. Discrimination between possible migrating groups is also negligible in deamination of 2,2-diarylethylamines[65]. and *N*-nitroso-*N*-acetylamines[7]; for example ratios for *o*-, *m*-, or *p*-tolyl to phenyl are all 1·2:1. Phenyl has a very much greater migratory tendency in S_N1 than have methyl or hydride ions, in fact the latter have been rarely observed to shift, but in certain deaminations the phenyl:methyl migratory aptitude is only about 6, and phenyl:hydride about 18[25, 26]. Alkyl groups also probably have similar aptitudes in deamination; e.g. $PhCH_2$=CH_3 = 2:1[66]. $\Delta H^{\ddagger}$ for hydride shift in isobutylamine is 2·5 kcal/mole less than that for methyl shift[19], but the ΔS^* factor favours the latter process by some 5 kcal/mole. The latter factor probably illustrates the small tendency of hydrogen to be situated *anti* periplanar to the leaving group in the ground-state conformations: a much bulkier group usually occupies this position.

In all these compounds, ground-state conformations are equally probable with either of the competing and compared migrating groups in the favourable *anti*-periplanar orientation to the amino group, but in certain diastereoisomers *cis* interactions between bulky non-migrating groups attached to the two asymmetric carbons cause different groups in each isomer to be favourably oriented in the ground state and the closely related (both electronically and structurally) transition state. Thus migration occurs predominantly as shown in **13** and **14** (*p*-An = *p*-Anisyl), (reaction 11), such that any

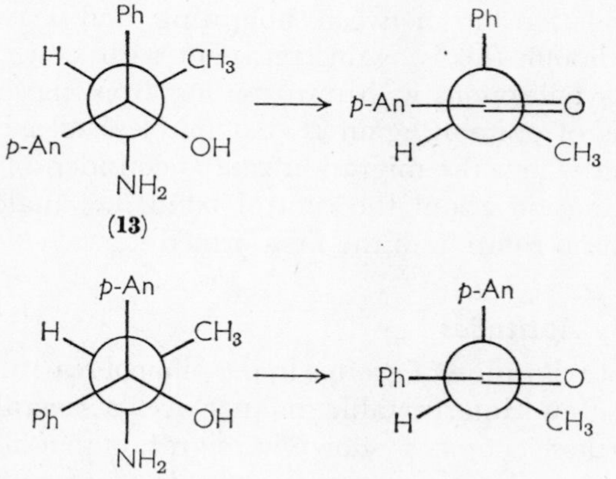

(13)

(11)

(14)

'*cis* effect' of neighbouring aryl and methyl is eliminated[67]. Other unusual relative aptitudes, e.g. Ph > *o*-Tolyl[68], Ph > Naphthyl[69], and *t*-Bu ∼ Ph[70], also reflect the favoured ground-state conformation rather than inherent migrational ability, and again indicate (on the basis of the carbonium ion model) the preference for migration within, rather than rotation of, the initially formed ion.

F. Cycloalkylamines: Ring Expansion

Alicyclic analogues of the above reactions were discovered in 1901 (Demjanov rearrangement of amines) and 1937 (Tiffeneau rearrangement of aminoalcohols), but although scores of examples have been recorded[71], few mechanistic studies are available. These reactions, shown in (12) and (13), usually give good (30–60%) and sometimes excellent (>80%) yields of rearranged alcohols or ketones and

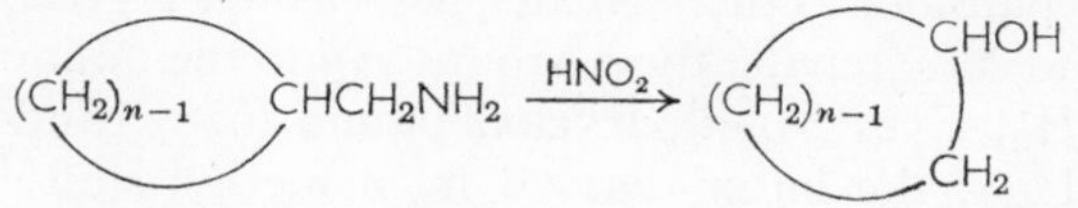

$$(CH_2)_{n-1} \quad CHCH_2NH_2 \xrightarrow{HNO_2} (CH_2)_{n-1} \begin{array}{c} CHOH \\ \\ CH_2 \end{array} \tag{12}$$

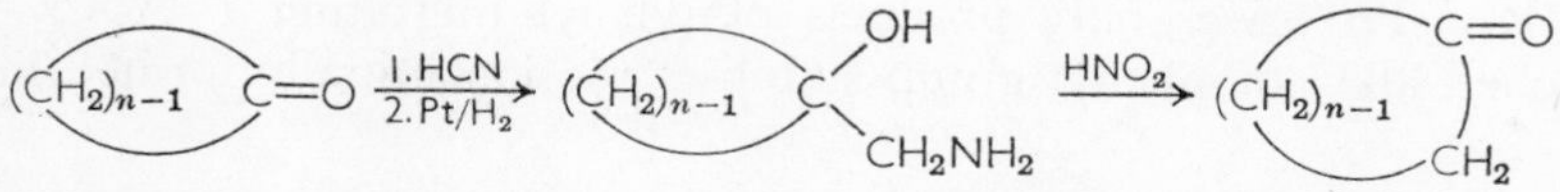

$$(CH_2)_{n-1} \quad C=O \xrightarrow{1.HCN}_{2.Pt/H_2} (CH_2)_{n-1} \begin{array}{c} OH \\ C \\ CH_2NH_2 \end{array} \xrightarrow{HNO_2} (CH_2)_{n-1} \begin{array}{c} C=O \\ \\ CH_2 \end{array} \tag{13}$$

olefins, accompanied by unrearranged products, and three- to twelve-membered carbocyclic rings have been thus expanded: maximum rearrangement occurring with the five to seven membered[71-73]. As a carbon–carbon (ring) bond will predominantly be *anti* periplanar to the leaving group in the substrate, the hindered rearrangement of the larger rings probably reflects an unfavourable entropy factor in the actual migration step, caused by the need to change the conformation of the original ring, when its span is flexed to a new terminus.

The Tiffeneau reaction possesses two properties favouring ring expansion:

(i) No hydrogen is available at $C_{(2)}$ to compete with migration of ring-bond electrons.

(ii) A protonated carbonyl group, $\overset{+}{>}C\!-\!OH$, is formed on ring expansion which is a comparatively stable carbonium ion. Consequently expansion is usually more complete, with less olefinic by-products, than in the Demjanov process. In synthesis, the Tiffeneau method is especially valuable for ring enlargement of ketones (see reaction 13).

These rearrangements are again manifestations of the tendency to form a more stable carbonium ion. Both ring-bond and hydride shifts (in Demjanov conditions) can occur (reaction 14)[74], but the latter are

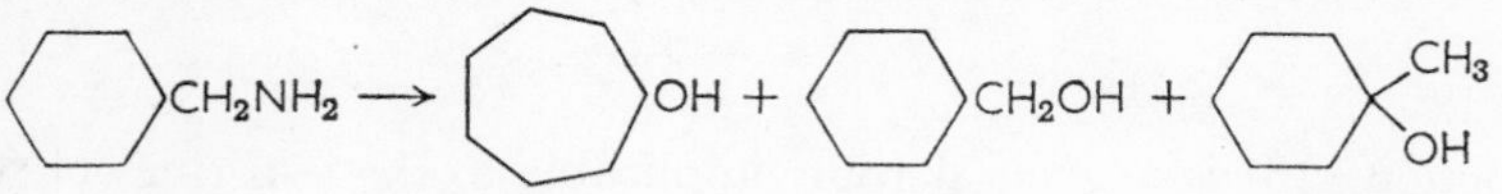

$$\langle \rangle CH_2NH_2 \longrightarrow \langle \rangle OH + \langle \rangle CH_2OH + \langle \rangle \begin{array}{c} CH_3 \\ OH \end{array} \tag{14}$$

(66%)   (33%)   (1%)

usually unimportant as they require an unfavourable ground-state conformation, even though they lead to a tertiary ion. Nevertheless up to 20% tertiary alcohol results from 1-cyclopentylethylamine[75], and hydride shift can be readily demonstrated in non-rearranged products by tracer studies; for example 1-^{14}C-cyclohexylamine gives a mixture of 1-^{14}C- and 2-^{14}C-cyclohexanol[76]. Methyl and phenyl substituents at $C_{(2)}$ can migrate more efficiently than hydrogen in competition with ring expansion, as their greater size ensures a greater proportion of the ground state having these groups *anti* to the amino group. Thus **15** (R = CH$_3$) gives 2-methylcycloheptanol (67%) and 1-ethylcyclohexanol (11%); the latter derived from methyl shift[71,77]; whereas **15** (R = Ph) gives only products of phenyl migration[78]. A $C_{(1)}$-attached alkyl or phenyl group stabilises the ion initially formed in

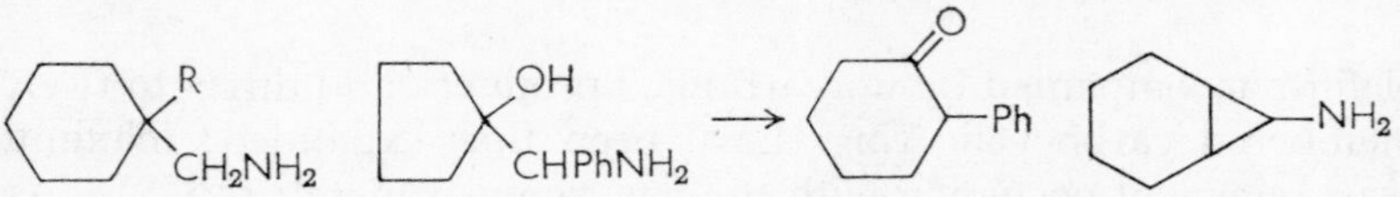

(15)   (16)   (17)

Demjanov deamination and so inhibits ring expansion, even in small rings where strain is released by the latter process[71,75,82]. An intervening methylene group between the ring and the ionic centre (in 3-cycloalkylpropylamines) has the same effect[71]. Tiffeneau reactions are much less affected by such factors, having stronger driving forces, and **16** gives over 50% ring expansion, and the corresponding 1-methylated amine of the cyclohexane series even more[71,83]. Bulky alkyl groups at $C_{(1)}$[84] or the axial/equatorial nature of the aminoethyl group (in conformationally fixed 4-*t*-butyl-1-hydroxy-1-methylamino-cyclohexanes) do not prevent up to 80% ring expansion[47].

Asymmetrically substituted rings can expand in two ways to give position isomers, the proportions of which can usually be rationalised by conformational considerations as to which ring bond is suitably placed for migration (however, see reaction 15)[71]. Demjanov–Tiffeneau reactions can be applied to bicyclics (reaction 15)[71].

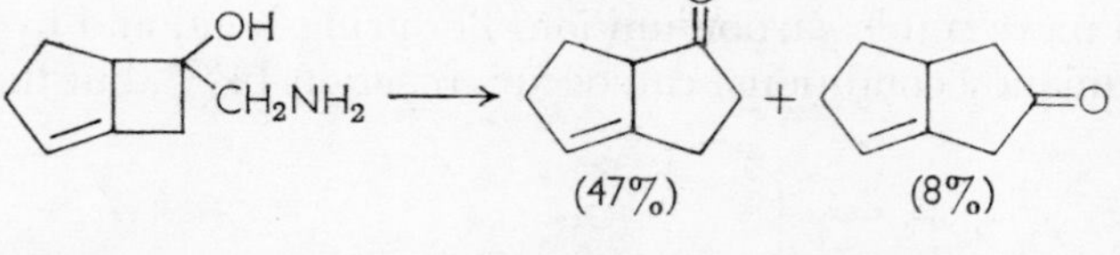

(47%)   (8%)   (15)

Reaction (16) is a general route applicable to the c–d rings of steroids[86,87] and to *N*-, *O*-, and *S*-heterocyclics[71,85]. However, certain

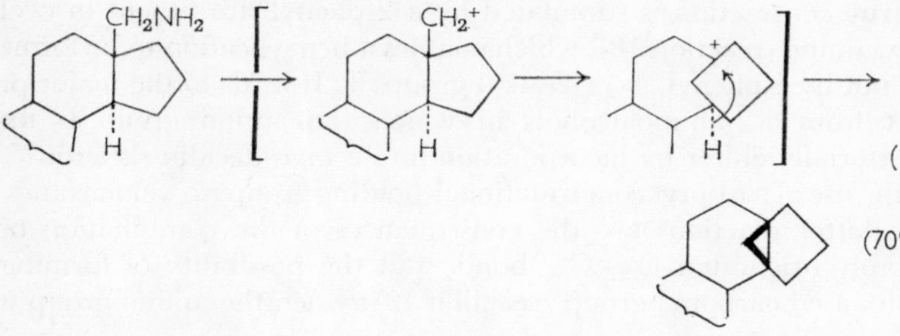

$$(16)$$

$$(70\%)$$

electronegative substituents (2- and 4-Cl, F, and COOEt) inhibit ring expansion in six- and even four-membered rings[71, 79-81], perhaps by favouring direct decomposition in the diazonium ion stage.

G. Cycloalkylamines: Ring Contraction

In a few structures, the most stable ion results from ring contraction, rather than expansion. Cleavage of the strained ion from cyclopropylamine[88] is a particular case, and ^{14}C labelling has elucidated the mechanism (reaction 17)[89]. The bicyclic (17) reacts analogously to

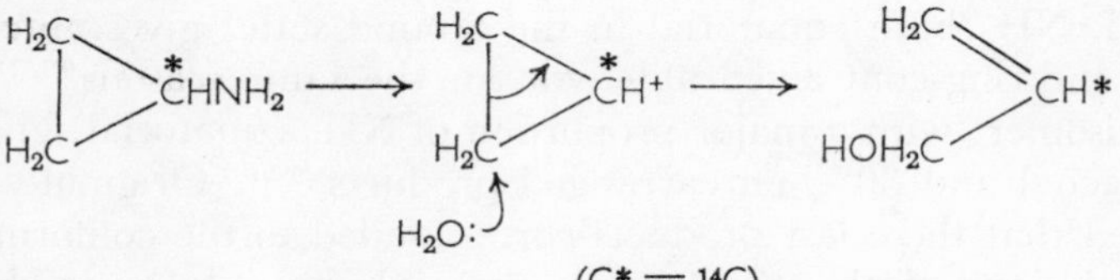

$$(17)$$

$$(C* = {}^{14}C)$$

give 2-hydroxycycloheptene[90], but the ion derived from cyclopropylmethylamine is much more stable (see section I), and the strained 18 gives little rearranged product[91].

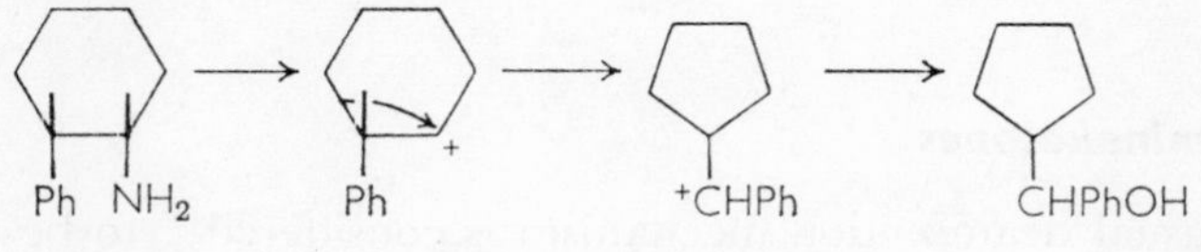

$$(18)$$

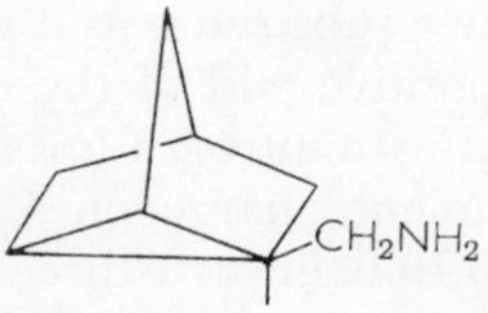

(18)

True contraction is stimulated by a 2-phenyl substituent in cyclo-hexylamine (reaction 18), which enables a benzyl cation to be formed, but not by 4-phenyl, 4- or 2-alkyl groups[79]. It leads to the major product from 1,2-aminoalcohols in which the amino group is fixed equatorially either by incorporation into a *trans*-decalin skeleton[92] or by the use of a *t*-butyl conformational-holding group in cyclohexanes[10]. The latter reactions are the consequences of the combination of a suitably orientated $C_{(2)}$–$C_{(3)}$ bond, and the possibility of forming a protonated carbonyl group (reaction 19); when the amino group was

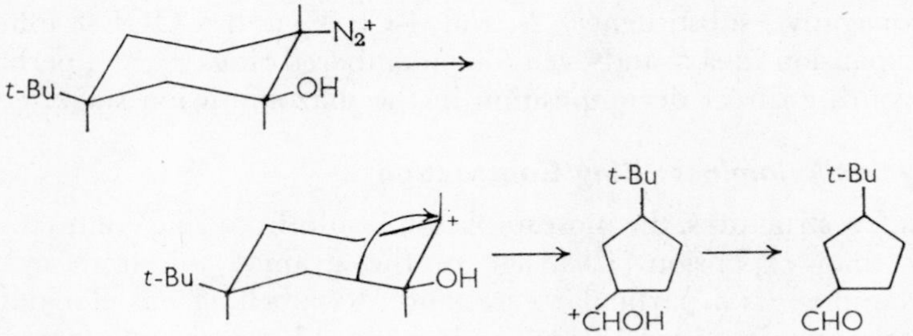

$$\tag{19}$$

fixed axially no contraction took place. *trans*-2-Hydroxycyclohexyl-amine (OH, NH_2 both equatorial in the ground state) gives almost exclusively the ring-contracted aldehyde for the same reasons[93,134], but the *cis* isomer (with a major proportion of NH_2 equatorial) gives 70% contracted and 30% unrearranged products[94-96]. Originally it was believed that these last products corresponded to the conformational populations of the *cis* starting material, but recent work[10] refutes this and interprets products in terms of the equilibrium conformations of diazohydroxides which are supposed to decompose directly to a carbonium ion and to by-pass the diazonium ion. It is not yet clear how general this mechanism is in aliphatic deamination.

H. α-Aminoketones

The usual deamination mechanism is considered[97] to be energetically unfavourable for these substrates, as it would generate a positive centre adjacent to the positive end of the carbonyl dipole, and rearrangement synchronous with nitrogen loss from the diazonium ion is suggested, to allow incipient charge on $C_{(1)}$ to be delocalised. A detailed product study of **19** delineates three types of bond rearrangement (reaction 20)[97]. The cyclopentane carboxylic acid may also be

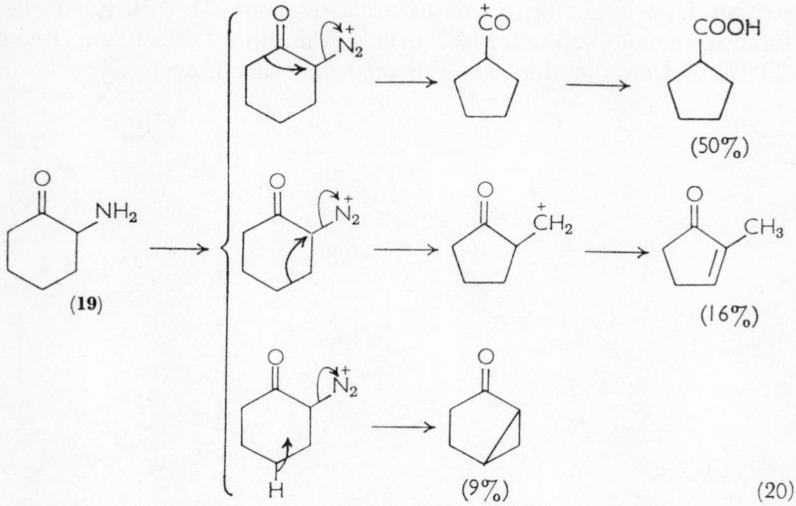

$$(20)$$

formed by direct rearrangement of the ketone hydrate (reaction 21) but a Favorskii-type mechanism (reaction 22) was excluded by the

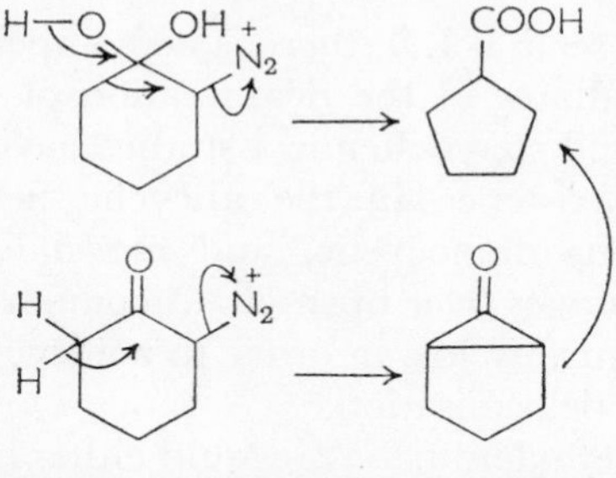

$$(21)$$

$$(22)$$

observation that 2-amino-6,6-dimethylcyclohexanone and **19** both give similar products. 1,3-Transannular elimination of hydrogen to form the bicyclic ketone as in reaction (20), was also demonstrated in a reinvestigation of products from cyclohexylamine which revealed 2% (3.1.0)-bicyclohexane[97, 133]. This may indicate that part, at least, of this substrate decomposes via the diazonium ion with no formation of the hot carbonium ion, as the 2-methyl derivative of **19**, which might be expected to be able to react by the carbonium ion mechanism, gives only unrearranged products[98].

Straight-chain α-aminoketones rearrange analogously to alkylamines by means of methyl shifts[98], but **20** does not undergo 2,6-hydride shift similar to that occurring in the bornylamines (see section I.J). The ring opens to a classical ion **21** with good charge separation, and some 1,3-hydrogen elimination takes place (reaction 23) [97]. Similar reactions are reported for other bicyclics [99].

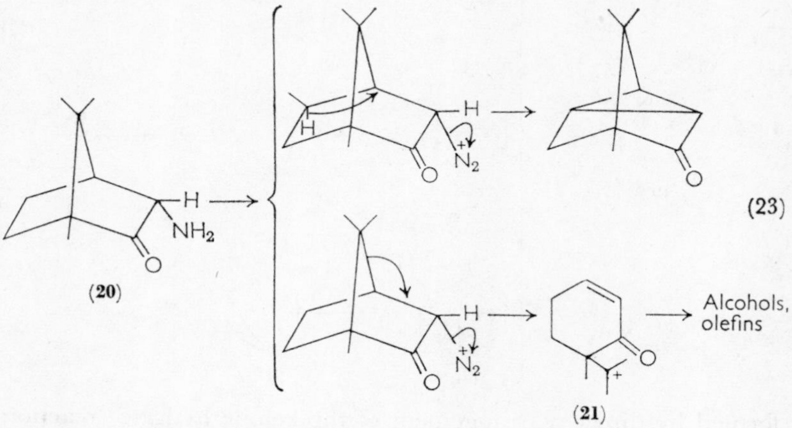

$$(23)$$

I. Non-classical Ions from Monocyclic Amines

As outlined in section I.D, there is no compelling evidence for non-classical intermediates in the deamination of acyclic amines, but a series of tracer and stereochemical studies have led many workers to postulate their existence in the alicyclic field [100, 129]. In forming mesomeric cations monocyclic, and especially bicyclic, ions have important advantages over open-chain compounds as the latter must sustain a large entropy loss in order to achieve the restricted orientation for electron delocalisation.

Cyclopropylmethylamine (**22**) would either be expected to undergo ring expansion to cyclobutyl derivatives or cleavage to allylic products in order to gain advantage from secondary or mesomeric ion formation, but the situation is much more complex. Reaction (24) shows products and tracer distribution from 1-^{14}C-labelled substrate [101]. It is interesting that cyclobutylamine gives the same proportions of products although the equilibrium position of **23** and **24** is nearly 100% in favour of the latter [101], and that 2-phenylated **22** gives 97% ring expansion via a benzylic ion [102]. A simple view of the expansion is that the p orbital of $C_{(1)}$ in the carbonium ion overlaps with one of the 'banana' orbitals of the three-membered ring (reaction 25) such that $C_{(1)}$ of **22** becomes $C_{(2)}$ of **24**, but the tracer distribution shows that an intermediate must intervene in which $C_{(2)}$, $C_{(3)}$, and $C_{(4)}$ of **24** become

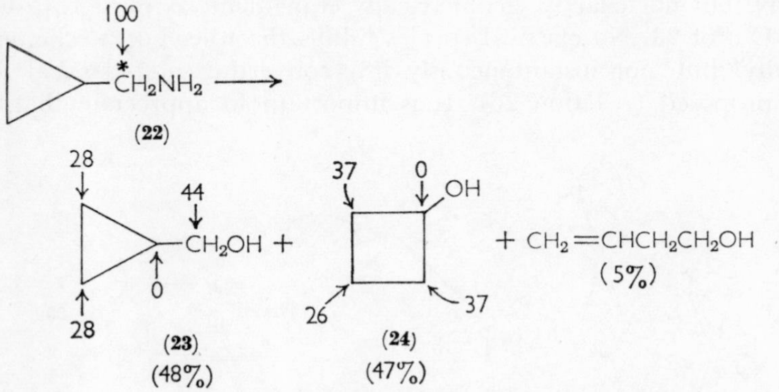

$$(24)$$

(figures give % distribution of tracer)

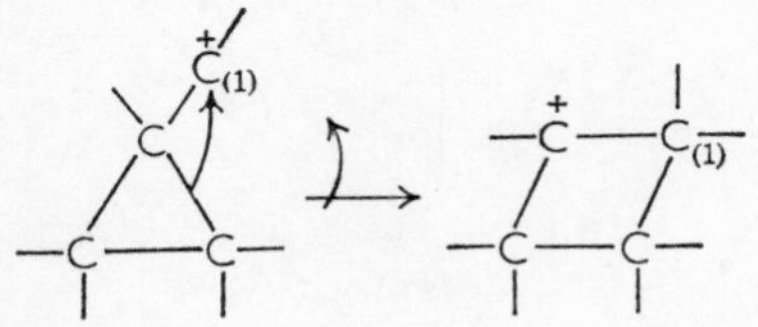

$$(25)$$

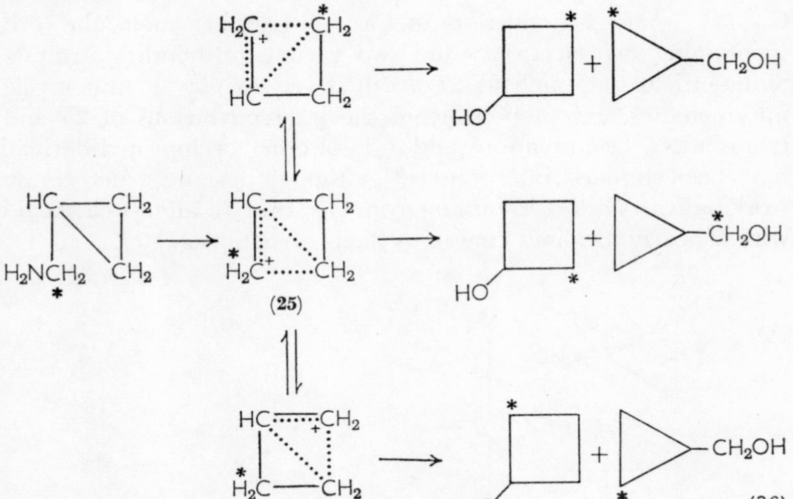

$$(26)$$

$$[C* = {}^{14}C]$$

20*

nearly, but not exactly, geometrically equivalent, as must $C_{(1)}$, $C_{(3)}$, and $C_{(4)}$ of **23**. No classical species fulfills the role, but a scheme of rapidly, but not instantaneously interconverting non-classical ions was proposed (reaction 26). It is important to appreciate that **25**,

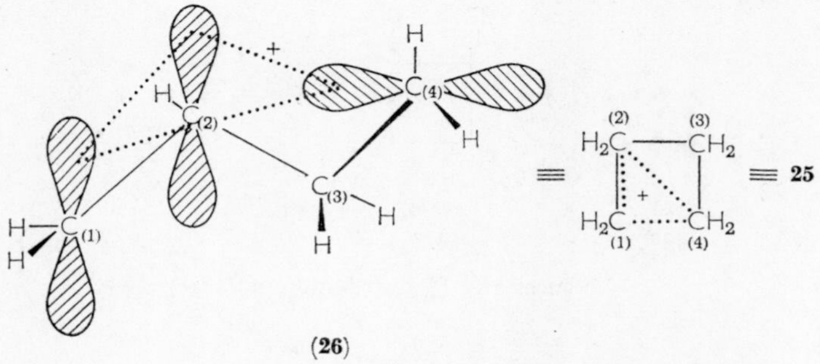

$$(26)$$

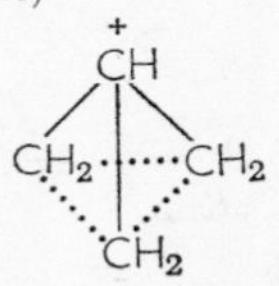

$$(27)$$

for example, is shorthand notation for a complicated orbital diagram: the geometry of this ion is probably **26**, wherein the $2p$ orbitals of $C_{(1)}$, $C_{(2)}$, and $C_{(4)}$ overlap to form a bonding molecular orbital (containing two electrons) and two vacant antibonding orbitals[101]. Symmetrical ions such as **27** are believed to play a minor role as intermediates, except possibly in the interconversions of **25** and its two isomers. Deamination (and S_N1) of other cyclopropyl derivatives have been similarly interpreted[103], although in some structures tracer work indicates little scrambling from $C_{(1)}$, and an almost classical ion, with a nearly localised charge, is likely (reaction 27)[104].

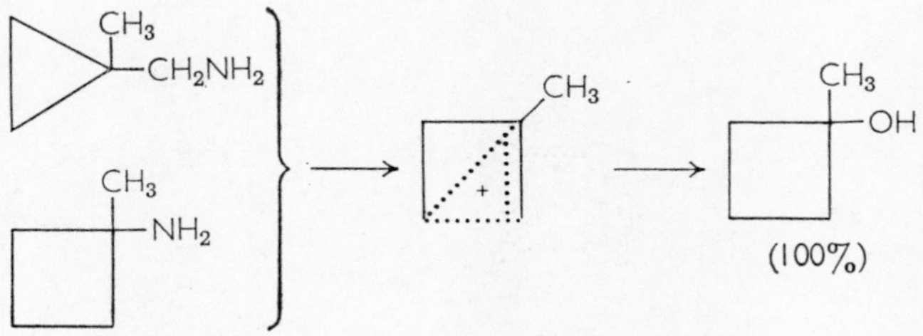

$$(27)$$

An interesting product from 2-cyclopropylethylamine is cyclo-pentanol, which is believed to result from irreversible migration of the ring to $C_{(1)}$ (reaction 28), although the reactions are complicated by

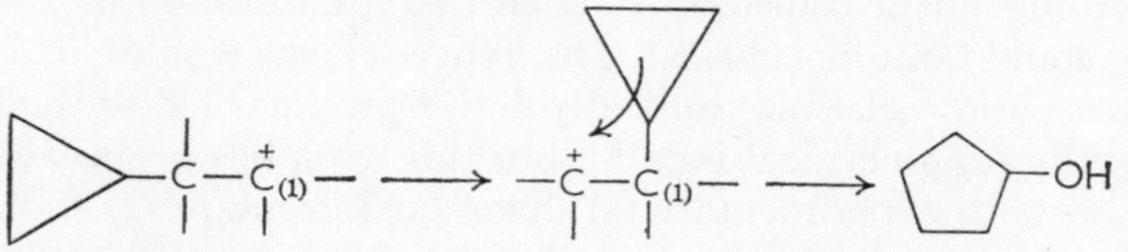

$$\triangleright\!-\!\overset{|}{\underset{|}{C}}\!-\!\overset{+}{\underset{|}{C}}_{(1)}\!-\!\longrightarrow\!-\!\overset{|}{\underset{|}{C}}\!-\!\overset{+}{\underset{|}{C}}_{(1)}\!-\!\longrightarrow\!\bigcirc\!-\!OH \qquad (28)$$

much isotope scrambling from $C_{(1)}$ and probably proceed through non-classical intermediates[105].

J. Non-classical Ions from Bicyclic Amines

Solvolyses and deaminations of *exo*- and *endo*-norbornyl (**28**) and norbornenyl derivatives labelled with ^{14}C at $C_{(2)}$ and $C_{(3)}$ have been held to indicate non-classical intermediates (reaction 29)[106,107]. If **29**

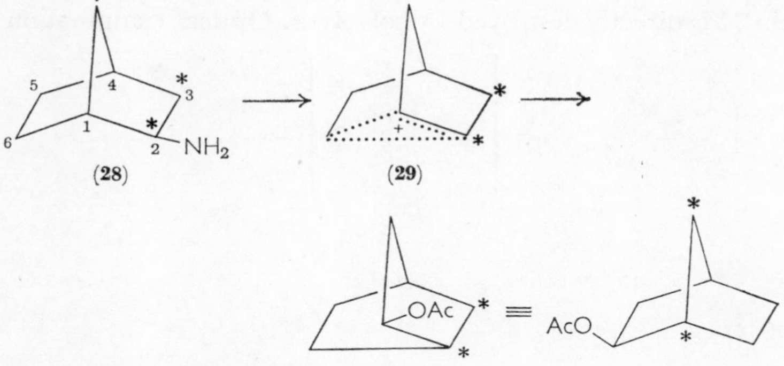

(**28**)   (**29**)

(29)

was the sole intermediate, tracer would appear at $C_{(1)}$ and $C_{(7)}$ in the product, but substantial, although not equivalent amounts were found at all positions except $C_{(4)}$. This distribution was accommodated by postulating a 2,6-hydride shift linking non-classical ions **29** and **30**, via the special non-classical symmetrical species **31** (reaction 30), and

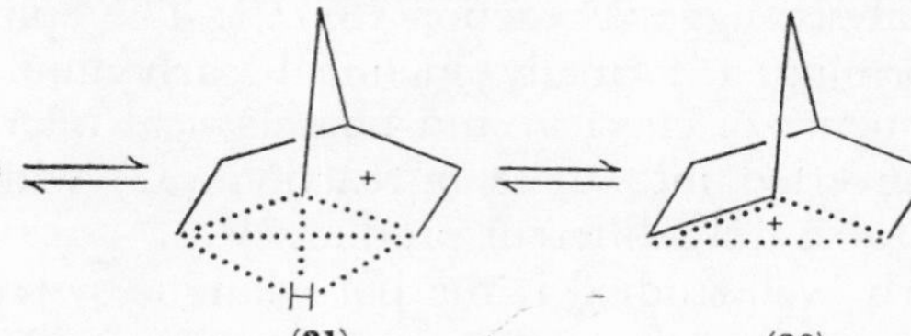

(**31**)   (**30**)

so achieving a system whereby partial or total equilibration of tracer in all positions except $C_{(4)}$ could be realised, before destruction of the intermediates by solvolysis. However, there is no evidence that **31** is anything but a transition state in the equilibration of **29** and **30**, which could both be classical. The above scheme would yield racemic products, and a claim to optically active products led to the proposal of equilibrating classical ions[108], but this work has been validly criticised on both experimental and theoretical grounds[109], and the consensus of opinion favours non-classical intermediates, although not necessarily the existence of **31**[100].

More recently, analyses of tracer distribution or optical activity, and geometric structure (i.e. *endo* or *exo*) of products from various bicyclic amines have presented evidence for non-classical intermediates[100,113]. A typical example is shown in reaction (31)[110]. In both non-solvating (to carbonium ions) acetic acid and solvating aqueous conditions, only *exo* product could be detected and such stereospecificity is difficult to interpret on a classical mechanism[100], wherein **32** is directly destroyed by solvolysis. Optical examination of

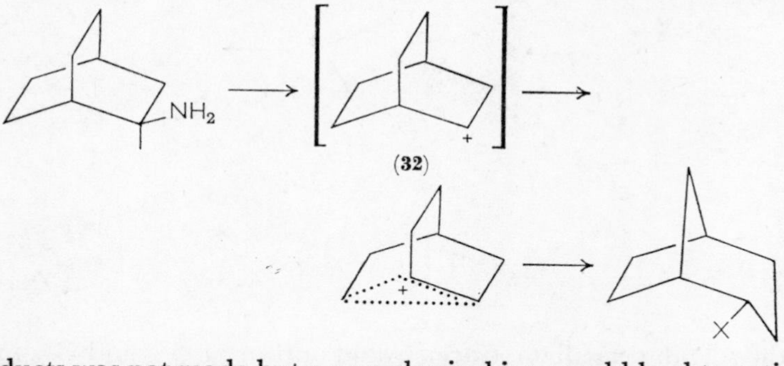

products was not made but a non-classical ion would lead to active *exo* product whereas the *endo–exo* mixture from **32** would be racemic. For other systems it has been adduced that both classical and non-classical ions act as precursors of products: for example **33** gives optically active products and this rules out the symmetrical **34** as sole intermediate. Detailed analysis suggests reaction (32)[111]. The epimeric *exo*-amine, although forming a formally identical carbonium ion, utilises a different sequence of classical and non-classical intermediates which are only converted into those of reaction (32) with difficulty, and which lead to entirely different products[112].

Particularly well studied is the bornylamine system. Both bornylamine (**35**), and its epimer (**36**), give camphene (**37**) and tricyclene

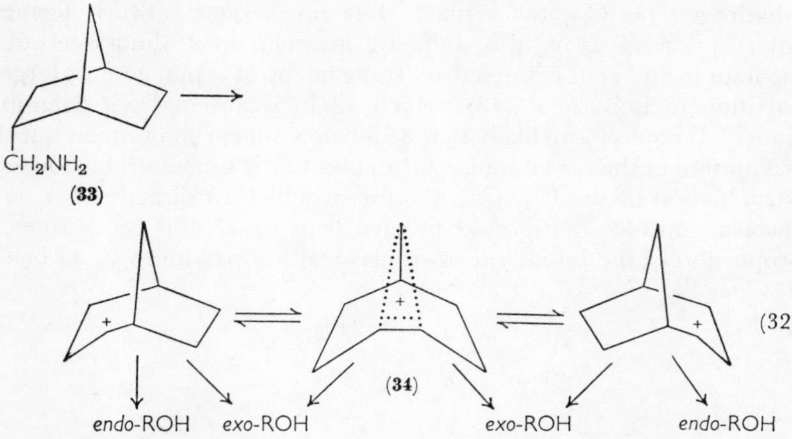
CH2NH2
(33)
+
+
+
(34)
endo-ROH
exo-ROH
exo-ROH
endo-ROH

(32)

(**38**) on deamination in aqueous or acetous solvents[114,115], and the former is optically pure, i.e. there is no evidence for Nametkin or 2,6-hydride shifts. However, **35** gives up to 10% of monocyclic products **39**, whereas its epimer gives none (reaction 33). The corresponding

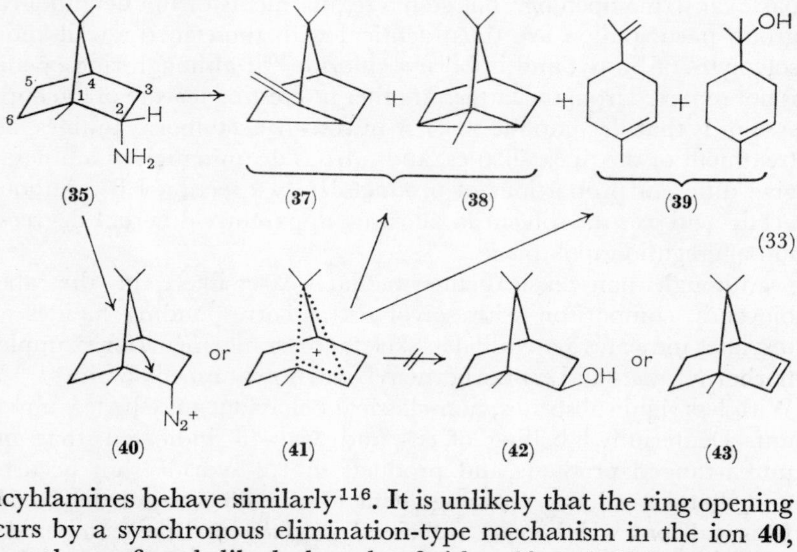
5
4
3
1
2
6
H
NH2
(35)
+
+
+
OH
(37)
(38)
(39)
or
N2+
+
OH
or
(40)
(41)
(42)
(43)

(33)

fencyhlamines behave similarly[116]. It is unlikely that the ring opening occurs by a synchronous elimination-type mechanism in the ion **40**, for as the preferred dihedral angle of either 0° or 180° between the participating bonds cannot be achieved (owing to the rigid geometry) in either the *endo* or *exo* isomer, each isomer should behave similarly. Also, perfect eclipsing (dihedral angle ~0) of the leaving group and a

β-hydrogen on $C_{(3)}$ takes place, but no bornene (**43**) is formed. Intervention of **41** is also unlikely, for such ions almost certainly mediate in the acid-catalysed rearrangement of α-pinene and largely partition to isoborneol (**42**), which again is absent from deamination[100]. It seems more likely that **35** forms a hot carbonium ion but the orientation of the developing p orbital on $C_{(2)}$ is unfavourable for such participation of the $C_{(1)}$–$C_{(6)}$ bond as would lead directly to a non-classical ion, **44**, that would in turn lead to **37** and **38**. Rather, a proportion of the initially formed classical ion partitions to **39** before

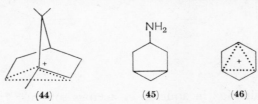

(44)   (45)   (46)

being converted into **44**, and thence to **37** and **38**. In contrast, the incipient carbonium ion from **36** is perfectly orientated for $C_{(1)}$–$C_{(6)}$ bond participation and a smooth transition to products via **44** by-passes any ring opening. The stereo requirements of the neighbouring group participation are thus identical with those in the well-known solvolyses of bornyl and isobornyl chloride[100]; although ring opening is not reported from the latter. Another interesting feature of the bornyl system is that decomposition of *N*-nitroso-*N*-acetylbornylamines, acid treatment of the diazoalkanes, and nitrous deamination of amines, all give different proportions of products[115] (see section I.B). Although acetic acid was the solvent in all cases, apparently different degrees of ion aggregation took place.

Although non-classical intermediates are likely in the above bicyclics, comparison with solvolyses of corresponding halides and tosylates indicates generally less skeletal rearrangement, for example in norbornyl, *endo*- and *exo*-norbornenyl, and norcamphenylamines[100,118]. With less rigid substrates, non-classical behaviour may be less import-ant. Deuterium labelling of *cis*- and *trans*-**45** indicated that only unrearranged products and products of 1,2-hydride shift occurred, and that 1,3-valence rearrangement to give **46** was not detected[119]. Classical ions which react sufficiently rapidly to be encumbered by departing nitrogen account for the non-stereospecificity of products. Here, as opposed to the previously considered bicyclics, the low free energy of decomposition of the classical ion severely limits the time available for acquisition of vibrational energy in the mode producing motion towards formation of **46**. Valence rearrangement is also not detected in systems with the cyclopropyl and ionic centre held in rigid, non-orthogonal positions (reaction 34) [120].

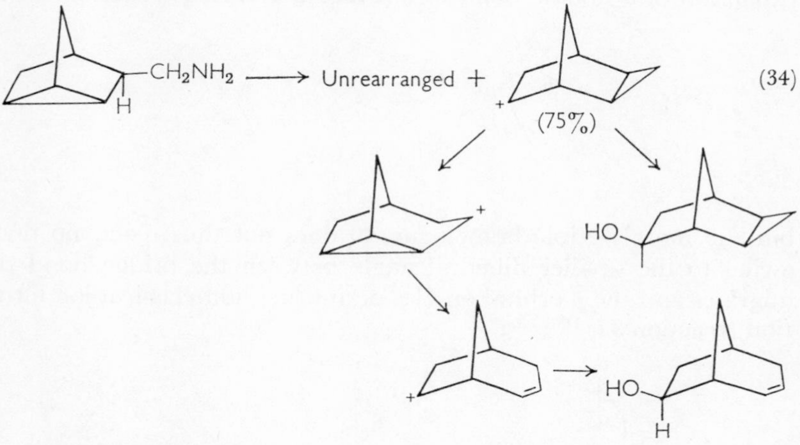

$$\text{(34)}$$

K. Allylamines

Products of nitrous deamination of allylamines in acetic acid [121] indicate that the mesomeric ion found in S_N1 for analogous halides is not generated. Thus primary alcohols are formed rather than the secondary or tertiary isomers expected from facile double-bond rearrangement. Apparently, the hot ion has its p orbital at $C_{(1)}$ orientated at a dihedral angle $>0°$, and probably $\sim 90°$ with the π orbital of the double bond, and solvolysis is more rapid than alignment to ensure overlap. In more aqueous media such alignment, permitted by partial or entire solvation of the carbonium centre, is allowed and products typical of S_N1 are obtained [122].

Ring size controls the dihedral angle in cyclic allylamines. **47** is held with the orbitals overlapping and extensive ring enlargement caused by an allyl-type shift results [123] (reaction 35); but the

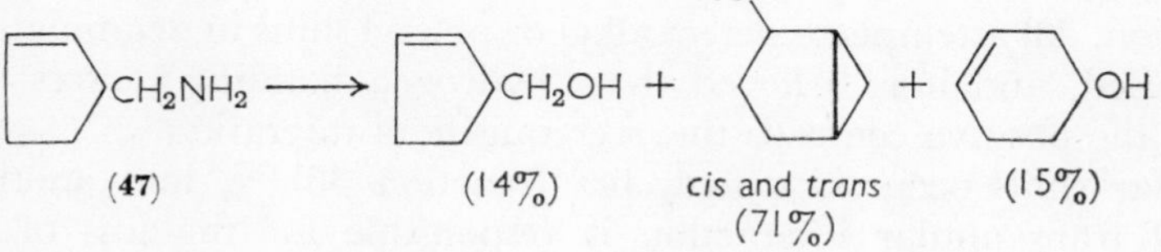

$$\text{(35)}$$

cyclohexyl analogue is more flexible, little overlap is possible in the incipient ion, and only 5% ring expansion is found[124]. Similar rigidity and overlap occurs in bicyclics and **48** undergoes formal expansion of a double bond to a 3-membered ring (reaction 36)[125],

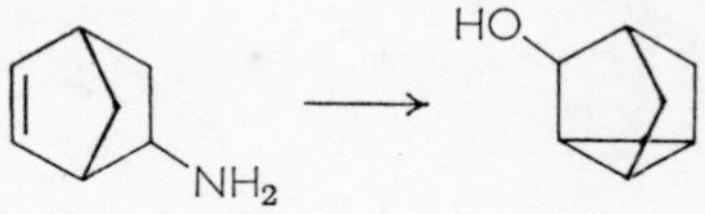

$$\tag{36}$$

but the more flexible homologue **49** does not thus react, no doubt owing to the smaller dihedral angle between the bridge bond that migrates and the p orbital on $C_{(2)}$ permitting non-classical ion formation (reaction 37)[100, 125].

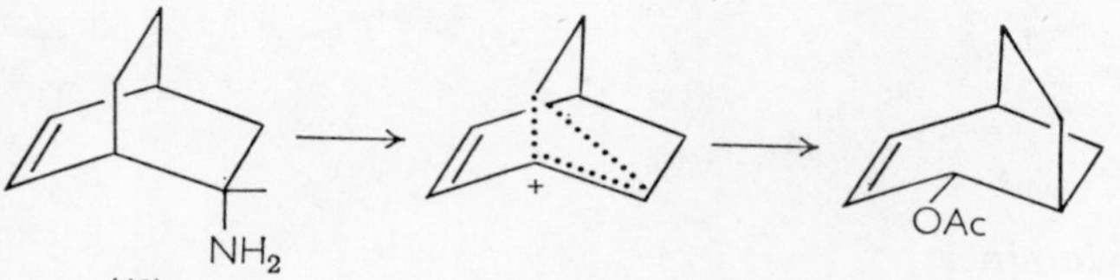

$$\tag{37}$$

L. Transannular Migration

1,3- and 2,6-hydride shifts have already been mentioned as occurring when bicyclic, or occasionally monocyclic, systems allow close approach of non-bonded atoms to an ionic centre. Puckering in rings larger than those previously considered allows stable conformations to have the same proximity relations and 1,5- and 1,6-hydride shifts have been demonstrated by tracer studies in the deamination of cyclononyl- and cyclodecylamines[126]. Similar shifts have been detected in S_N1 of tosylates, under suitable conditions, and here more migration occurs than in deamination. Also there is less rearrangement in olefin products than in the alcohol products of deamination. All attempts to detect alkyl or phenyl shifts in deamination have failed, and it is believed that 'hydrogen bonding' across the ring to the positive centre is the prerequisite of migration.

Similar shifts occur in polycyclics (reaction 38)[127], and another type of transannular interaction is responsible for the loss of all

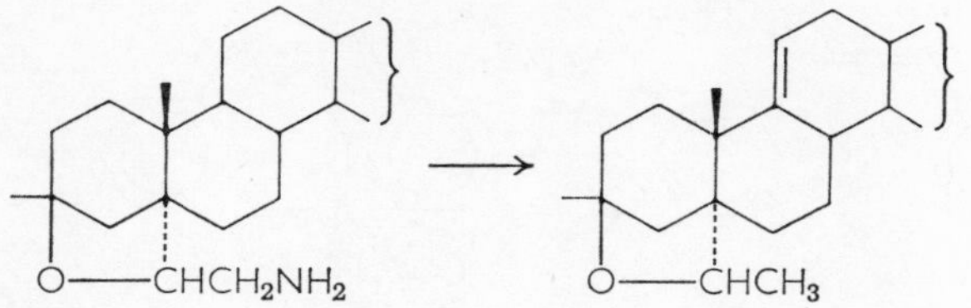

(38)

nitrogen from glutamine on treatment with nitrous acid (reaction 39)[128]. In the latter example cyclisation is faster than solvolysis of

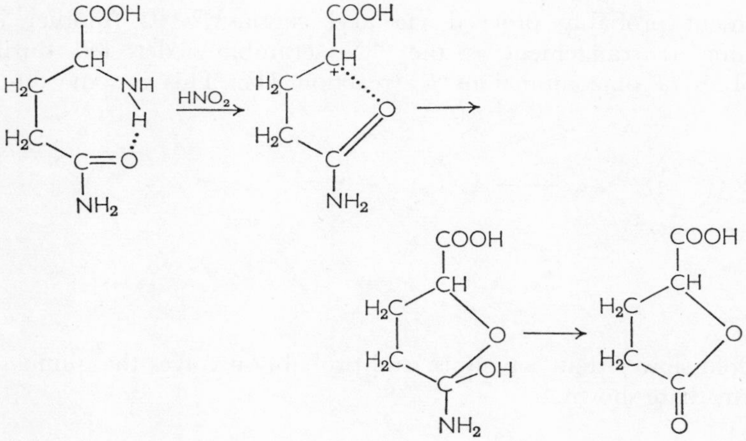

(39)

the first-formed ion because a suitable conformation is held in the substrate by hydrogen bonding between carbonyl oxygen and amino hydrogen, and deamination and cyclisation is more rapid than adoption of another conformation from the original.

M. Aromatic Diazonium Compounds

Aromatic diazonium ions are stabilised by resonance and do not show the exotic behaviour of their aliphatic counterparts, but they are intermediates in a group of cyclisations, known as Pschorr reactions, that result in skeletal rearrangement.

The prototype reaction—(40), in which decomposition of the diazonium salt is catalysed by copper powder in strong acid, is almost certainly a radical process. Other cyclisations, induced by heat

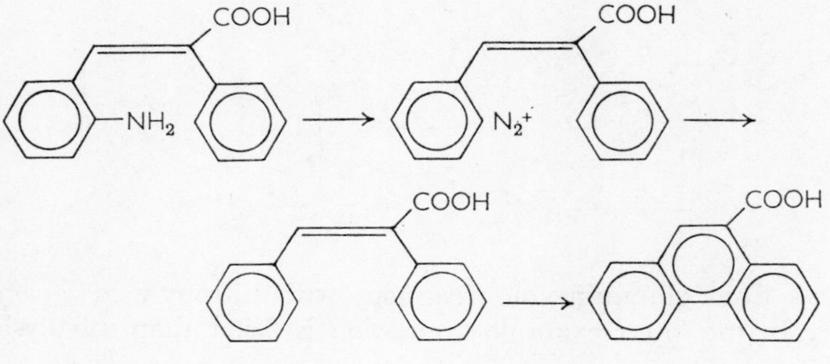

COOH ⟶ COOH ⟶ (40)

COOH ⟶ COOH

treatment probably proceed via aryl cations[130, 131]. Another interesting rearrangement is the ^{15}N scrambling detected during solvolysis of diazonium ions[132] (reaction 41). This is some 30- to

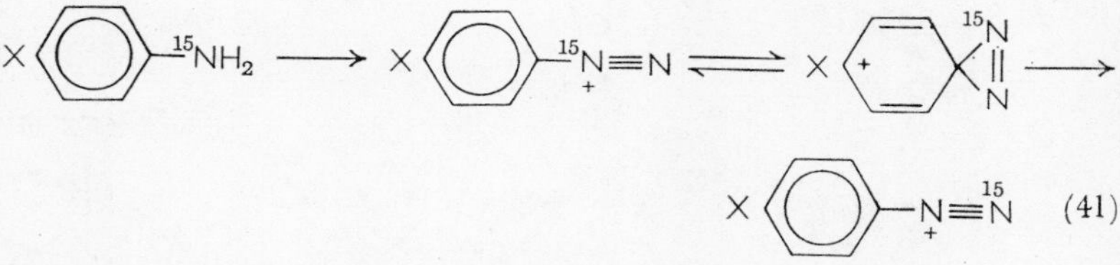

$$X\text{—}C_6H_4\text{—}^{15}NH_2 \longrightarrow X\text{—}C_6H_4\text{—}^{15}N\equiv N^+ \rightleftharpoons X \cdots \overset{^{15}N}{\underset{N}{\|}} \longrightarrow$$

$$X\text{—}C_6H_4\text{—}N\equiv\overset{15}{N}^+ \quad (41)$$

100-fold slower than solvolysis, and probably involves the quinonoid intermediate shown.

II. REARRANGEMENTS OF QUATERNARY COMPOUNDS TO AMINES

A. Stevens Rearrangement

The base-promoted migration of a side-chain of an 'onium compound to what is now recognised as an adjacent carbanionic site was discovered by Stevens in 1928 on attempted Hofmann degradation of **50** (Ar = Ar[1] = Ph) (reaction 42)[136], although a closely related process (reaction 43) had been recorded earlier[135]. Subsequent

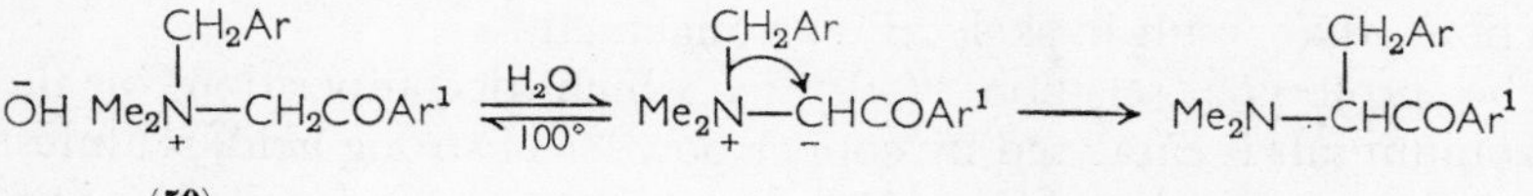

$$\overset{-}{O}H \quad \underset{+}{Me_2N}\text{—}CH_2COAr^1 \overset{H_2O}{\underset{100°}{\rightleftharpoons}} \underset{+}{Me_2N}\text{—}\underset{-}{C}HCOAr^1 \longrightarrow Me_2N\text{—}CHCOAr^1 \quad (42)$$

(with CH_2Ar substituents)

(**50**)

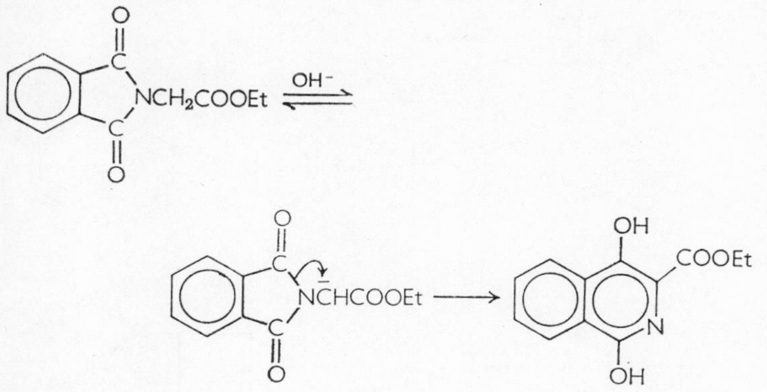

(43)

studies[137-139] (mainly in aqueous media) demonstrated the necessity for proton loss to form a mesomeric carbanion incorporating a carbonyl or allyl group, (see (44)); and the reaction was extended to

$$\underset{+}{Y_2}\overset{R}{\underset{|}{N}}\text{—}\overset{-}{\underset{}{CHX}}\left\{\begin{array}{l} X = \text{acetonyl, phenacyl, allyl;} \\ R = \text{alkyl, 1-phenylethyl, benzyl, phenacyl;} \\ Y = \text{alkyl, } Y_2 = (CH_2)_5 \end{array}\right.$$ (44)

sulphonium compounds. The aptitudes of different migrating groups and termini[140], the influence of unsaturated side-chains[141], and the nature of by-products of elimination and fragmentation[142], have recently been delineated. Strongly basic or aprotic media, e.g. fused sodamide or phenyllithium in ether[139,143], can readily induce even the usually sluggish methyl migration in the absence of an activating carbonyl group (reaction 45) and a unique method of generating a carbanion is shown in reaction (46)[144]. Polycyclics readily rearrange

$$\bar{I}\ \underset{+}{Me_2}\overset{CH_3}{\underset{|}{N}}\text{—}CH_2Ph \xrightarrow[30°]{PhLi} \underset{+}{Me_2}\overset{CH_3}{\underset{|}{N}}\text{—}\overset{-}{\underset{}{CHPh}} \longrightarrow \underset{}{Me_2}\overset{CH_3}{\underset{|}{N}}\text{—}CHPh$$ (45)

$$R_2CN_2 \xrightarrow[\Delta]{Ether} R_2C: \xrightarrow{R^1NMe_2} \underset{+}{Me_2}\overset{R^1}{\underset{|}{N}}\text{—}\overset{-}{\underset{}{CR_2}} \longrightarrow \text{Products}$$ (46)

under such forcing conditions (reaction 47)[145,146], and ring contraction can be achieved (reaction 48)[147] although in certain strained

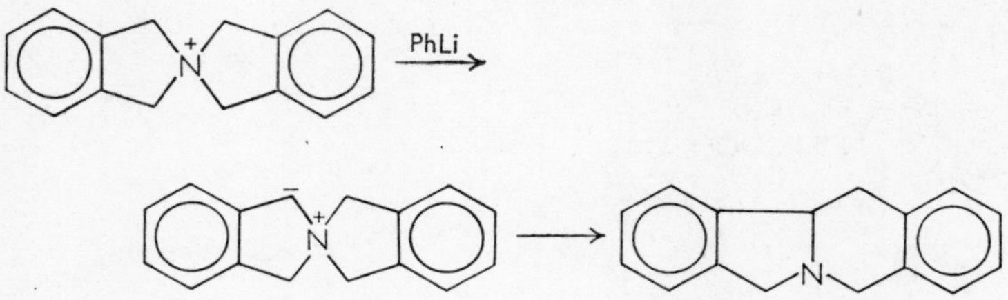

$$(47)$$

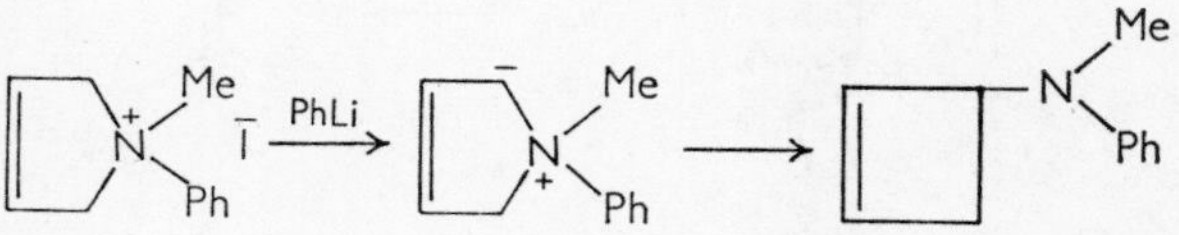

$$(48)$$

substrates rearrangement may be evaded (reaction 49) [145, 146]. The most stable carbanionic centre is involved in all these reactions.

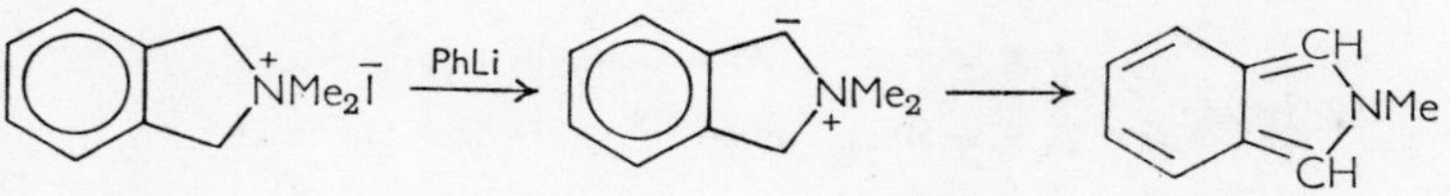

$$(49)$$

The carbanion or ylide mechanism [143] can be regarded as electrophilic migration to a negative centre analogous to the Wagner–Meerwein shift in carbonium ions [137], but an equivalent and possibly more enlightening view is to consider the reaction as an internal nucleophilic substitution, $S_N i$ (cf. **51**) [148]. Such mechanisms were once considered operative in the decompositions of alkyl chlorosulphites, but although an ion-pair mechanism is now preferred for these processes [150],

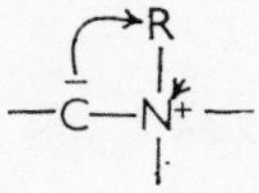

(51) 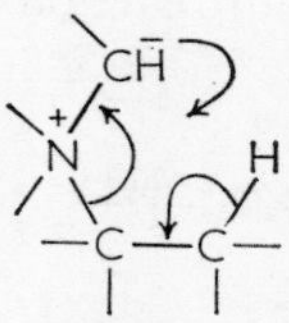   **(52)**

the incidence of *cis* eliminations (cf. **52**), especially in aprotic media [151], shows that the Pauli principle, as applied to the orbitals of the transacting electrons, need not necessarily forbid an analogous rearrangement mechanism.

The mechanism is supported by much data:

(i) Rate is proportional to base concentration in aqueous media[149], although the rough kinetics require repetition as the observation of a limiting rate at high basicities is difficult to reconcile with what should be (in aqueous media) a preequilibrium proton transfer.

(ii) Colours characteristic of carbanions develop during reaction[148].

(iii) Corearrangement of two approximately equally reactive substrates with different migrating groups produces no cross-over products[136, 152] nor intermolecular transfer of ^{14}C tracer when one group was labelled[153].

(iv) 1-Phenylethyl and certain substituted allyl groups migrate with >97% preservation of optical activity[154] and retention of configuration[155, 156].

(v) Substituents in the migrating group of **50** increase the rate in the order: m,p-OMe < H, Me < Cl, Br < NO$_2$, but a weaker and almost reverse order applies to substituents on the phenacyl group. This is consistent with the main effect of substitution being on the migration, rather than on the proton abstraction[137].

An early mechanism (reaction 50), which is a limiting case of S_Ni

$$
\underset{+}{PhCOCH_2\overset{\overset{\displaystyle CH_2Ph}{|}}{N}Me_2} \rightleftharpoons \underset{+}{PhCO\overset{\overset{\displaystyle CH_2Ph}{|}}{\bar{C}}H\overset{}{N}Me_2} \longrightarrow \left[\underset{+}{PhCOCH{=}NMe_2} \overset{\overset{\displaystyle \bar{C}H_2Ph}{}}{} \right] \quad (50)
$$

$$
\text{ion pair}
$$

$$
\downarrow
$$

$$
PhCO\overset{\overset{\displaystyle CH_2Ph}{|}}{C}HNMe_2
$$

in that bond fission is completed before the onset of bond making, accommodates most of these observations, and has been reproposed to account for the 1,2- and 1,4-migrations with the optical consequences shown in reaction (51)[157], and for the dependence of the two routes

$$
\underset{\underset{*}{CH_3\overset{\displaystyle |}{C}HPh}}{CH_2{=}CHCH_2\overset{+}{N}Me_2} \xrightarrow[\text{NaNH}_2]{\text{Liquid NH}_3} \begin{cases} \underset{CH_3\overset{\displaystyle |}{C}HPh}{CH_2{=}CHCHNMe_2} & \left(\begin{array}{l}>90\% \text{ retention of} \\ \text{optical activity}\end{array}\right) \\ \\ \underset{CH_3\overset{\displaystyle |}{C}HPh}{CH_2CH{=}CHNMe_2} & (70\text{–}80\% \text{ retention}) \end{cases} \quad (51)
$$

(* = optically active carbon centre)

on reaction medium. An $S_N i$ mechanism with allylic rearrangement (reaction 52) was rejected for the 1,4-migration as not leading to the

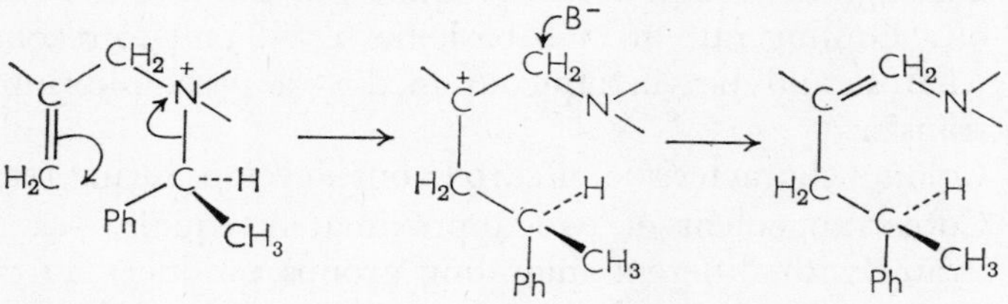

$$(52)$$

expected stereospecific inversion, and was also rejected, by analogy, for the usual 1,2-route.

The validity of these conclusions seems questionable. The original proposal[139] that an ion-pair mechanism could account for production of dibenzyl from **50** by dissociation of the ion pair and attack on unreacted substrate, is unsatisfactory, as toluene formation would have intervened in the protic media used. The stability of a carbanion-containing ion pair in aqueous or protic media in general is doubtful, and the widely different temperature effects[158] on the rates of 1,2- and 1,4-migrations suggests a difference in activation energies that is too large for two competing recombinations of an ion pair. A concerted or step-wise 1,4-shift, mainly with retention (reaction 52) could be favoured in poorly cation-solvating media by lack of stabilisation of the $C_{(1)}$–N bond, and cannot be convincingly ruled out on the presently available data; nor has the possibility of complete or partial racemisation of the substrate before rearrangement been checked. Such racemisation could account for the isolation of inactive products from migrations involving initially optically active allylic groups[156]. The $S_N i$ mechanism seems adequate to account for all the available data. It is also consistent with the highly stereospecific formation of optically active products in reaction (53) in which a compound with

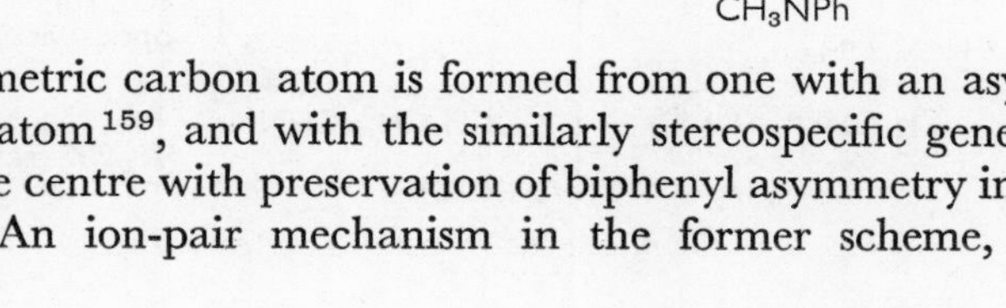

$$(53)$$

an asymmetric carbon atom is formed from one with an asymmetric nitrogen atom[159], and with the similarly stereospecific generation of the amine centre with preservation of biphenyl asymmetry in reaction (54)[160]. An ion-pair mechanism in the former scheme, and the

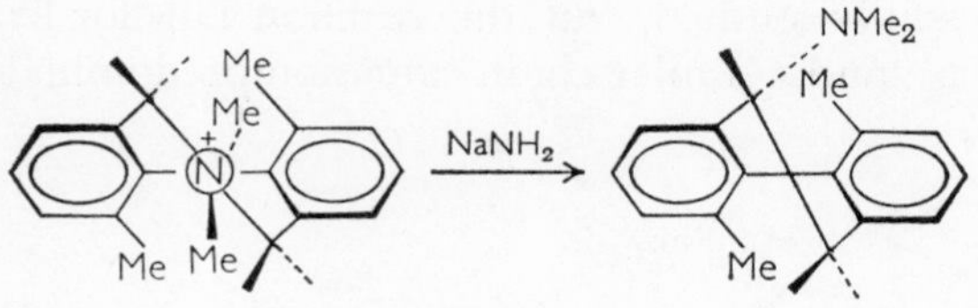

$$(54)$$

analogous formation of a zwitterion in the latter would be unlikely to result in such stereospecificity.

Transannular Stevens rearrangement is possible[186], and even specially activated tertiary amines will sometimes undergo a similar base-promoted reaction[187, 188].

B. Sommelet Rearrangement

In 1937, Sommelet found that a mixture of **53** and phosphorus pentoxide effervesced in sunlight to give a good yield of **54**[161], and ultraviolet irradiation of aqueous substrate later gave the same result[162]. However, the absorption properties of the substrate and glassware suggest that the reaction was thermal, rather than photochemical, and a 'dark' reaction can be achieved at 100–150° under the conditions of Hofmann degradation[161, 163]. Wittig[143], and Hauser[165] and their coworkers subsequently carried out similar dark reactions for many benzylammonium compounds using phenyllithium in ether, or sodamide in liquid ammonia, as catalyst, and suggested an ylide mechanism (reaction 55). The sodamide method

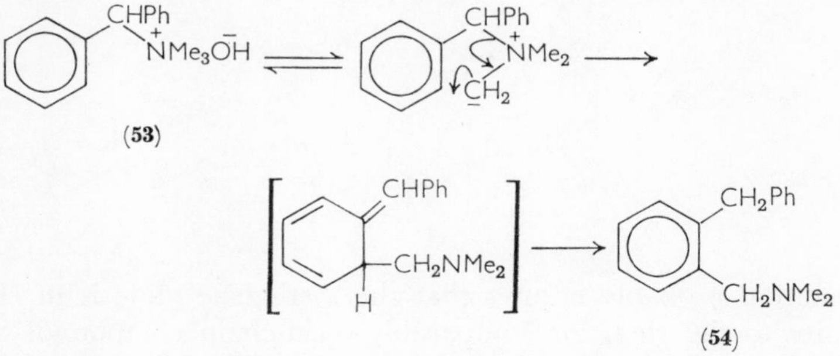

$$(55)$$

is especially convenient: typically some 95% rearrangement occurs within one minute at −33°.

Remethylation of product, and recycling, permits several methyl groups to be introduced at adjacent ring positions[172], and polycyclics[164, 166], and ring-[167] or α-benzyl-substituted substrates[168] have been successfully studied, but the reaction fails for ferrocenes[169, 170] (reaction 56), and a similar chain extension predominates in naphthyl

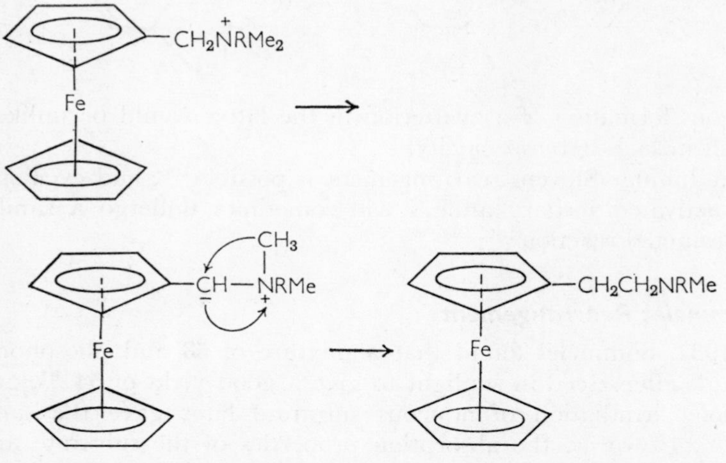

$$(56)$$

derivatives with the 2-position blocked[171]. An interesting ring expansion is shown in reaction (57)[173]. A favourable ground-state

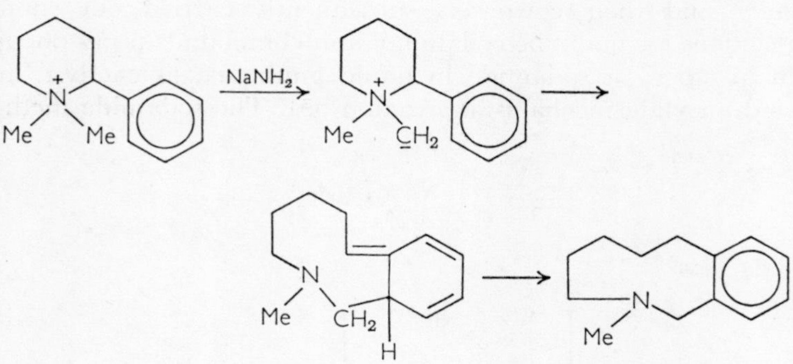

$$(57)$$

conformation probably ensures that the methylene ylide is in close proximity to the ring, for comparable open-chain compounds and the pentacyclic analogue give predominant β-elimination. The latter side-reactions, either $E2$ or α'-β processes on substrates or products of alkylation (reaction 58) often reduce the yield of rearranged product[174].

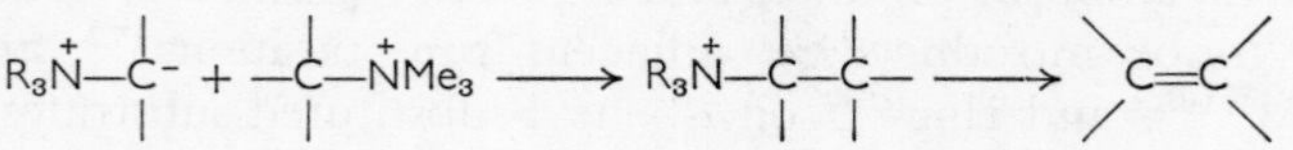

$$(58)$$

The ylide mechanism (the transition state of which has been argued to possess especial stability in view of the delocalisation of electrons possible in the cyclic transfer)[137], has been generally accepted. An ion-pair mechanism[176], similar to that proposed for the Stevens reaction, seems unlikely as the rotation within the pair necessary to lead to *ortho* migration would be expected to produce some *para* migration, which has never been observed. When no *o*-hydrogen is available as in **55**, the expected intermediate **56** can be isolated[63,164,177], and this breaks down on heating or on acid treatment as shown in reaction (59), but attempts to capture the methylene ylide (**57**) as a

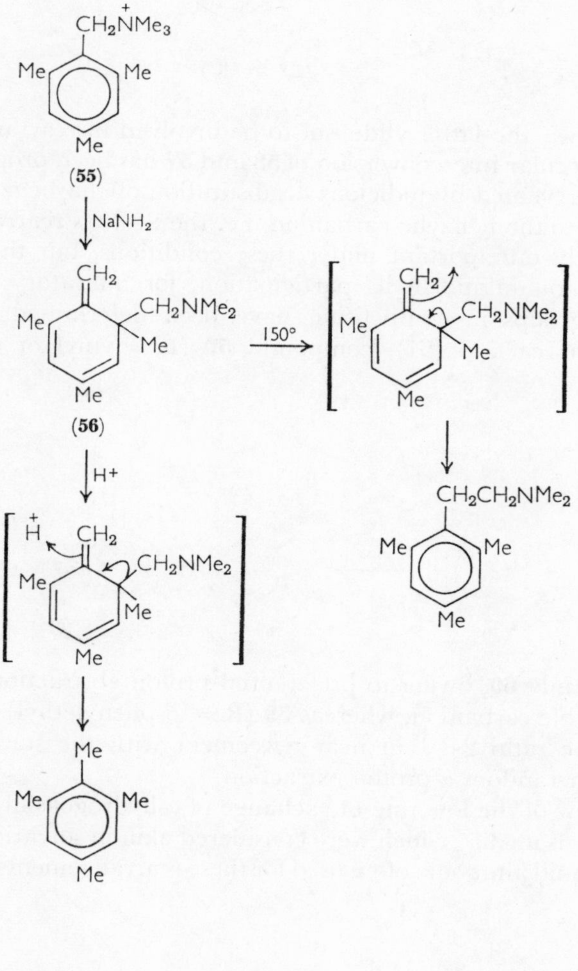

(59)

benzophenone adduct have only trapped the more stable benzyl ylide
(**58**) [178]. As tracer studies (reaction 60) and detailed product analysis [179]

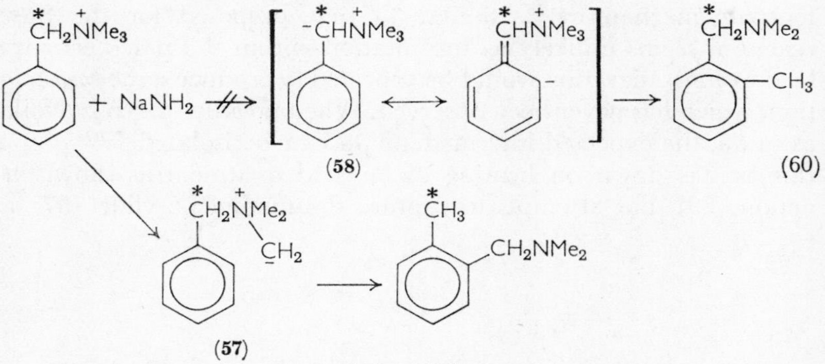

(**58**)

(**57**)

(60)

$(C^* = {}^{14}C)$

have shown the latter ylide not to be involved in rearrangement, an
intramolecular interconversion of **58** and **57** has been proposed, which
can be prevented by judicious α-substitution of the benzyl group [177].
1,2-Shift to the benzylic carbanion, i.e. the Stevens rearrangement, is
apparently unimportant under these conditions, but the methylene
ylide is not unique in its participation, for migratory aptitudes of
certain groups in competition have been determined [175]. Thus as
shown in reaction (61), compound **59** (R = vinyl or cyclopropyl)

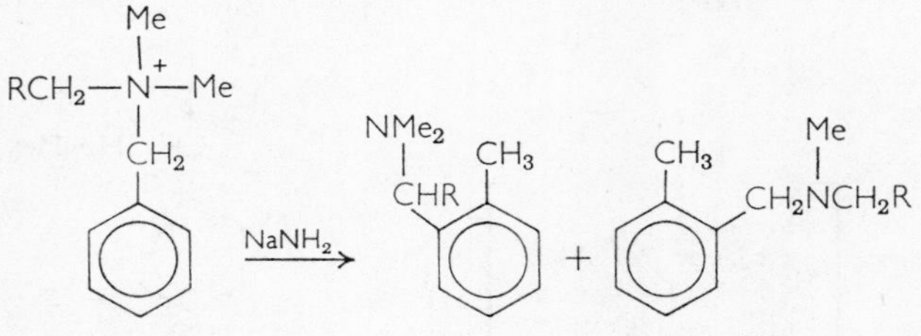

(**59**)

(**60**)

(**61**)

(61)

gives mainly **60**, owing to preferential proton abstraction to give the
more stable carbanion, whereas **59** (R = β-phenylethyl) gives **61** and
60 in the ratio 2·8:1 in near agreement with the statistical value,
(3:1), for random α-proton extraction.

In view of the low rate of exchange of α-hydrogens in 'onium salts
in aqueous media (which were considered akin in solvation properties
to the liquid ammonia often used for these rearrangements), it has been

reasonably suggested[180] that methylene ylides have no independent existence but that proton loss and bonding to the *ortho* position of the aromatic ring are concerted. Product analyses from *p*-monosubstituted benzhydryldimethylammonium compounds also indicate that proton removal may sometimes determine the direction and rate of rearrangement[181].

The Stevens and Sommelet reactions both involve similar substrates and often occur concurrently (reaction 62)[145]. The general trend is

$$(PhCH_2)_2\overset{+}{N}Me_2 \xrightarrow[30°]{PhLi} \begin{cases} \begin{matrix} PhCHNMe_2 \\ | \\ PhCH_2 \end{matrix} \quad \text{(Stevens, 56\%)} \\ \\ PhCHNMe_2 \\ \text{[ring]}\!-\!Me \quad \text{(Sommelet, 36\%)} \end{cases} \quad (62)$$

(62)

that the former is favoured at higher temperatures, although it also predominates under mild conditions when the *ortho* positions are blocked, when bulky substituents exert steric influences[182], and when butyllithium, rather than phenyllithium is the base[180]. The change from Sommelet to Stevens product as the conditions vary (sodamide in liquid ammonia, phenyllithium in ether, fused sodamide in refluxing toluene) is probably a temperature effect[183], for a similar trend occurs on changing from phenyllithium in ethyl ether at 25°, to the same base in butyl ether at 120°[184]. Although the entropy factor is more unfavourable for the Sommelet reaction—a five-membered transition state freezes rotation about three σ bonds whereas only one rotational degree of freedom is lost about the ylide $C_{(1)}$–N in the S_Ni mechanism of the Stevens reaction—this is counteracted at low temperatures by a lower heat of activation. The latter is undoubtedly connected with the 10 delocalised π electrons in the transition state as opposed to 4 for the Stevens reaction, which lower the heat content of the transition state relative to that for the Stevens process; and also with the higher energy level of the initial state of the Sommelet process (relative to the Stevens reaction) which is a result of it being a fully formed or incipient methylene ylide of high instability. The net result is that the Stevens reaction predominates at higher temperatures. These qualitative arguments are supported by rough molecular-orbital calculations[185].

C. Rearrangements of t-Amine Oxides

The Meisenheimer reaction[189] (reactions 63 and 64), which also

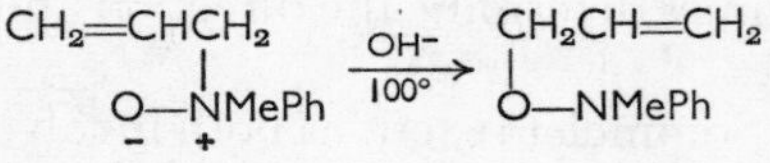

$$CH_2{=}CHCH_2 \quad \xrightarrow[\text{100°}]{OH^-} \quad CH_2CH{=}CH_2 \tag{63}$$
$$\underset{-}{O}{-}\underset{+}{N}MePh \qquad\qquad O{-}NMePh$$

(63)

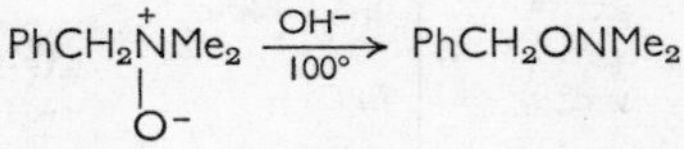

$$PhCH_2\overset{+}{N}Me_2 \xrightarrow[\text{100°}]{OH^-} PhCH_2ONMe_2 \tag{64}$$
$$\underset{O^-}{|}$$

(64)

applies to nitrones[202], is probably intramolecular in view of the first-order (in substrate) kinetics, the lack of cross-over products in co-occurring reactions, and the rearrangement of crotylmethylaniline oxide with allylic shift[190, 191]. The base prevents hydration masking the nucleophilic oxygen, although alkaline solutions of tenaciously hydrated alkyldimethylamine oxides have to be concentrated to a syrup to effect reaction[189, 192].

Compound **63** probably undergoes $S_N i$ with allylic shift, but re-arrangement of $(+)$-α-deutero-**64** leads to extensive racemisation (proved to occur during reaction), and the kinetic effects of ring substituents, e.s.r. measurements, and the photochemical induction of reaction, support a radical-pair mechanism[193,397]. Kinetics of the gas-phase reaction have led to similar conclusions[194], but it is disturbing that the activation energy is much lower than the bond energy of $C_{(1)}$–N^+. These should be approximately equal for a radical non-chain process.

Treatment of amine oxides containing at least one nitrogen-attached methyl group with acetic anhydride, results in the Polonovski reaction (reaction 65)[195]; product studies on which indicate the involvement of radicals[196]. The reaction is also catalysed by Fe(III) and by many transition-metal complexes (reaction 66)[197]. These reactions have been studied as models for the bacterial degradation of nicotine which involves N-oxidation as the first step[198]. N-Arylamine oxides give considerable amounts of o-acetylarylamines on treatment with acetic anhydride, presumably via cyclisation from the N-acetate[199], and similar migrations are common for heterocyclic

N-oxides[200]. A related, but little studied rearrangement, useful in alkaloid synthesis, is reaction (67)[201,396].

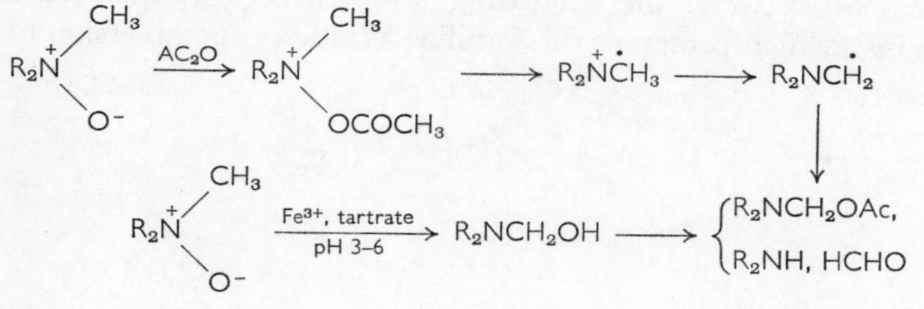

$$R_2\overset{+}{N}\underset{O^-}{\overset{CH_3}{|}} \xrightarrow{AC_2O} R_2\overset{+}{N}\underset{OCOCH_3}{\overset{CH_3}{|}} \longrightarrow R_2\overset{+}{N}\overset{\cdot}{C}H_3 \longrightarrow R_2N\overset{\cdot}{C}H_2 \qquad (65)$$

$$R_2\overset{+}{N}\underset{O^-}{\overset{CH_3}{|}} \xrightarrow[pH\ 3-6]{Fe^{3+},\ tartrate} R_2NCH_2OH \longrightarrow \begin{cases} R_2NCH_2OAc, \\ R_2NH,\ HCHO \end{cases} \qquad (66)$$

$$\xrightarrow{HCl,\ Ac_2O} \qquad (67)$$

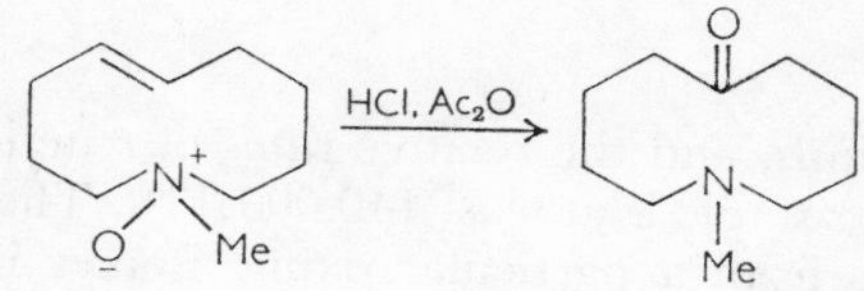

III. REARRANGEMENTS WITH MIGRATION TO AN ELECTRON-DEFICIENT NITROGEN

A. Beckmann Rearrangement

The vast literature concerning this reaction has been recently reviewed[203,204], and references to recent work only will be cited. The rearrangement (reaction 68), is promoted by many Brönsted and Lewis

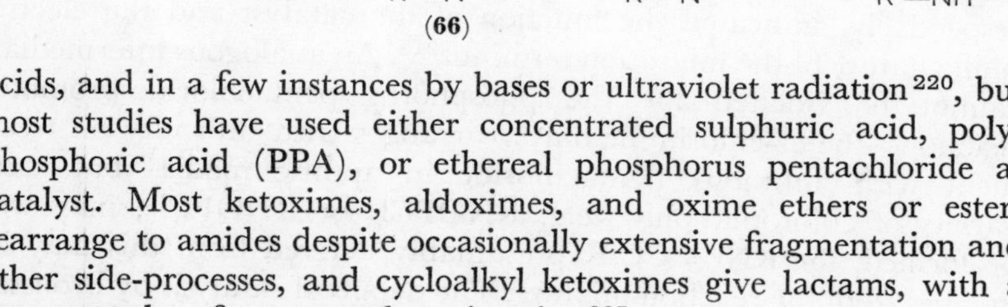

$$R^1{-}\underset{N{-}OH}{\overset{C{-}R^2}{\|}} \xrightarrow{Acid} R^1{-}\underset{N{-}X}{\overset{C{-}R^2}{\|}} \longrightarrow \left[\underset{\mathbf{(65)}}{R^1\overset{C{-}R^2}{\underset{N\cdots X}{\|}}} \right] \longrightarrow \qquad (68)$$

$$\underset{\mathbf{(66)}}{\overset{+C{-}R^2}{\underset{R^1{-}N}{\|}}} \longrightarrow R^2{-}\underset{R^1{-}N}{\overset{C{-}OH}{\|}} \longrightarrow R^2{-}\underset{R^1{-}NH}{\overset{C{=}O}{|}}$$

acids, and in a few instances by bases or ultraviolet radiation[220], but most studies have used either concentrated sulphuric acid, polyphosphoric acid (PPA), or ethereal phosphorus pentachloride as catalyst. Most ketoximes, aldoximes, and oxime ethers or esters rearrange to amides despite occasionally extensive fragmentation and other side-processes, and cycloalkyl ketoximes give lactams, with a few examples of transannular migration[205].

Product analyses from ketoximes assigned as *anti* or *syn* isomers on the basis of synthesis or cyclisation have shown an *anti*-periplanar requirement between the migrating and leaving groups (reactions 68 and 69), which preserves the familiar Walden-type inversion at the

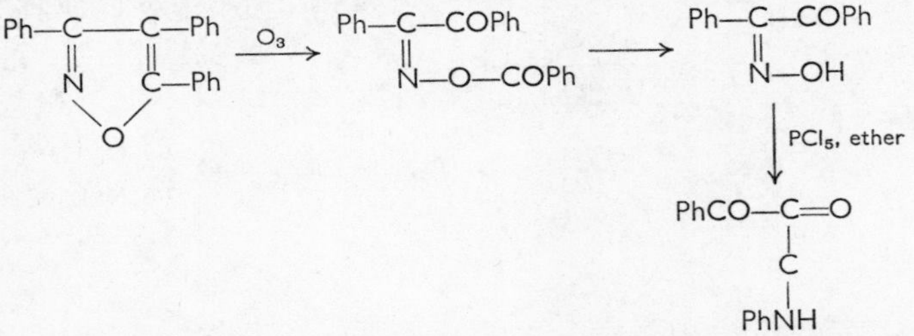

migration terminus, and the relative rates of reaction of *anti* and *syn* isomers in one example is at least 140,000:1[209]. The formation of two isomeric amides from a particular oxime isomer in rearrangements catalysed by protic acids is attributable to an equilibration of the *anti* and *syn* forms of substrate before rearrangement. Optically active migrating groups retain their configuration and ^{18}O-tracer studies demonstrate that oxygen is introduced intermolecularly, as in reaction (68). Species **66** can readily be trapped by addition of a nucleophile, e.g. azide ion or amine.

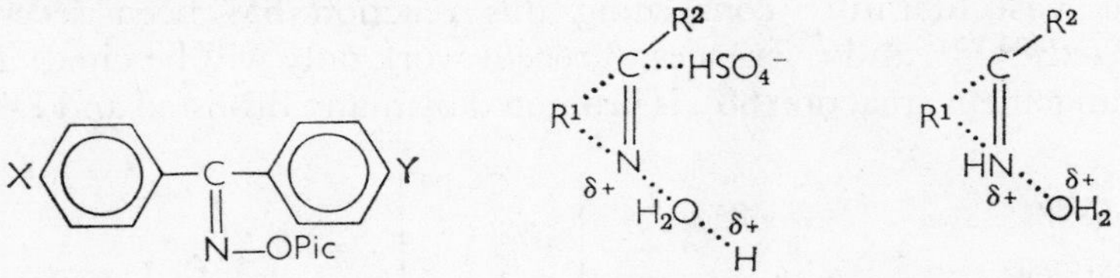

The early discovery that *O*-picryl ethers (**67**) can be rearranged thermally in the absence of catalyst, and the large and small increases in rate caused by electron-donating X and Y substituents respectively, indicated the function of the catalyst and the electrophilic nature of the migration terminus[212]. An analogous intermediate cannot be isolated for the phosphorus pentachloride-promoted reaction—the *N*-chlorimine only rearranges with difficulty on treatment with antimony pentachloride in aprotic media—but some variety of chlorophosphite, e.g. $RNOPCl_4$ or $RNOPCl_2$, may well be formed, for $RN = CClR$ presumably derived from this may be isolated from the reaction mixture. The details of catalysis by Brönsted acids are obscure. Acid salts of oximes can be readily prepared, but only rearrange at elevated temperatures, but N-protonation, rather than O-protonation is almost certain in these, and proton shift may be necessary to initiate rearrangement; analogies are the BF_3- and $SbCl_5$-catalysed reactions which undergo rate-determining transfer of catalyst from nitrogen to oxygen. The rate in concentrated sulphuric acid depends approximately on the h_0, or better the j_0, acidity functions, but levels off at high acidities and detailed mechanistic interpretations vary. An oxime sulphonic acid may form in the rate-determining step or a second proton may enter the transition state (cf. **68** and **69**). Reactions promoted by hydrochloric, and possibly hydrofluoric, acid show autocatalysis by the product, and mechanism (70) is likely. A similar scheme has been proposed for other catalysts[210], but cannot accommodate the generally observed acid dependence and stereochemistry[204].

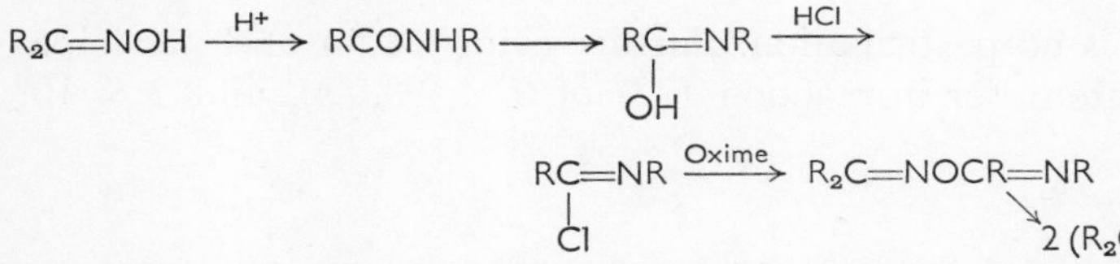

$$R_2C{=}NOH \xrightarrow{H^+} RCONHR \longrightarrow RC{=}NR \xrightarrow{HCl}$$
$$\underset{OH}{|}$$

$$RC{=}NR \xrightarrow{Oxime} R_2C{=}NOCR{=}NR$$
$$\underset{Cl}{|} \qquad\qquad \searrow 2\ (R_2CONH_2) \quad (70)$$

An ionic intermediate, **70**, was originally proposed, but concerted fission and rearrangement is necessary to explain the stereochemistry, and species **65** is generally assumed to be a transition state or intermediate, although products derived from an ion, **70**, are obtained if the normal rearrangement is hindered by structural rigidity (reaction 71)[206,241]. A bridged intermediate or transition state, **71**, is indicated

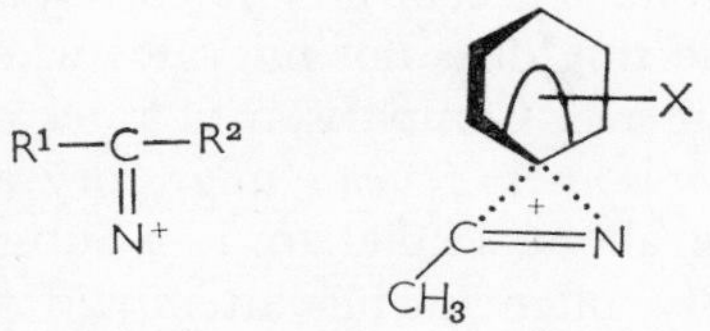

$$R^1{-}\underset{\underset{N^+}{\|}}{C}{-}R^2$$

(**70**)   (**71**)

by the Hammett ρ value $(-4\cdot1)$ obtained for a series of m- and p-substituted *anti*-acetophenone oxime picrates[207,208]. This is similar to the values obtained in aromatic substitution proceeding through a bridged Wheland intermediate, and is nearer the value obtained for solvolysis of 2-aryl-2-methyl-1-propylbrosylates $(\rho = -3\cdot0)$, in which aryl participation is believed to occur, than that for the corresponding reactions of 1-aryl-2-methyl-2-propyl chlorides $(\rho = -1\cdot1)$ in

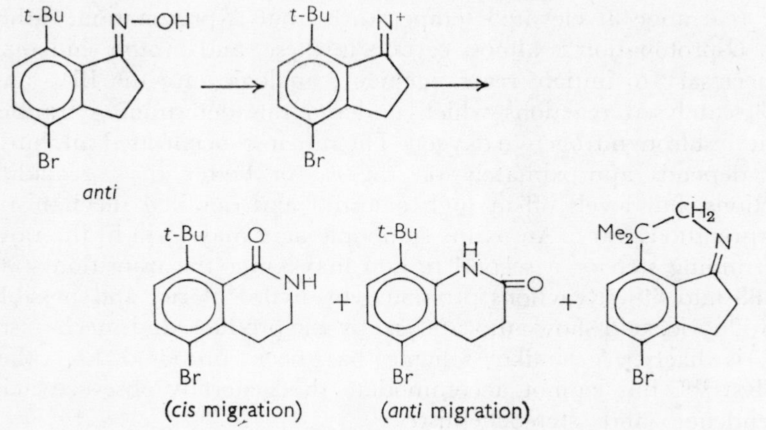

$$(71)$$

which there is no participation. Further evidence for aryl participation is the rate order in reaction (72) of 0, 1, 93,000, and 2×10^6

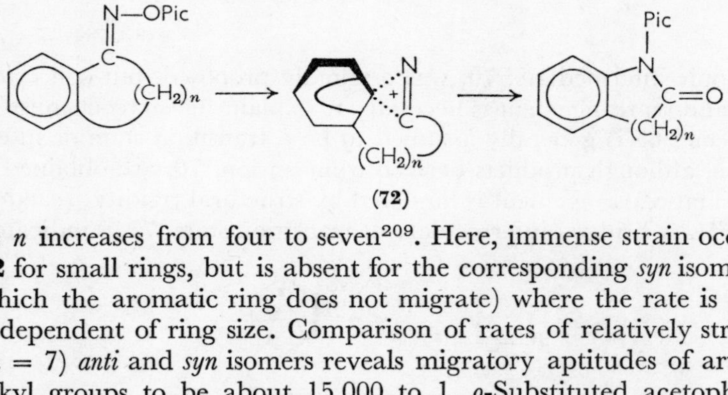

$$(72)$$

as n increases from four to seven[209]. Here, immense strain occurs in **72** for small rings, but is absent for the corresponding *syn* isomers (in which the aromatic ring does not migrate) where the rate is almost independent of ring size. Comparison of rates of relatively strainless ($n = 7$) *anti* and *syn* isomers reveals migratory aptitudes of aryl and alkyl groups to be about 15,000 to 1. *o*-Substituted acetophenone oximes react rapidly, often during attempted preparation, as steric inhibition of conjugation of the oxime double bond and the aromatic ring destabilises the substrate. Studies of salt effects have been held to indicate the rate-determining formation of an ion pair comprising a cation similar to **71**[211], but the detailed interpretation of these kinetics has been disputed[204].

An interesting adaption of the usual is reaction (73) possessing a common intermediate, species **73**, for rearrangement and the accompanying fragmentation[213], which has been isolated as a component of an ion pair. This may well be a general mechanism. Similar frag-

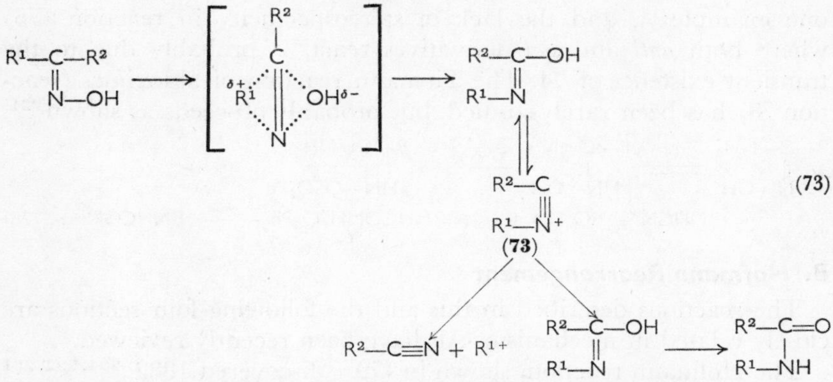

$$R^1-\underset{\underset{N-OH}{|}}{C}-R^2 \longrightarrow \left[\begin{array}{c} R^2 \\ \overset{\delta^+}{R^1} \overset{C}{\underset{N}{\|}} OH^{\delta-} \end{array}\right] \longrightarrow \begin{array}{c} R^2-C-OH \\ R^1-N \end{array}$$

(73)

$$\begin{array}{c} R^2-C \\ R^1-N^+ \\ \textbf{(73)} \end{array}$$

$$R^2-C\equiv N + R^{1+} \qquad \begin{array}{c} R^2-C-OH \\ R^1-N \end{array} \longrightarrow \begin{array}{c} R^2-C=O \\ R^1-NH \end{array}$$

mentation and recombination of suitably substituted oximes (reaction 74) has been detected (in strongly acidic media which permit

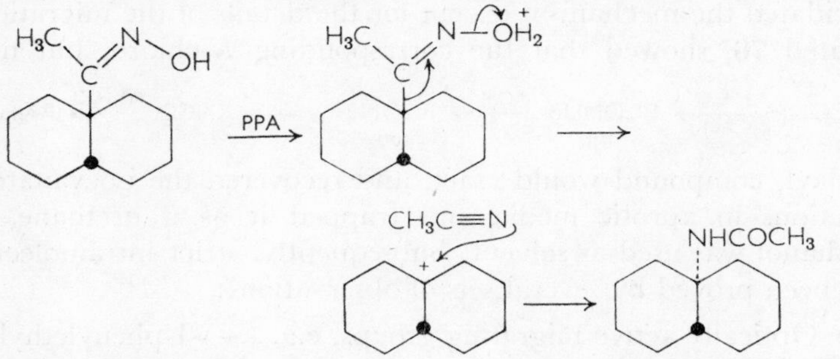

(74)

the existence of carbonium ions) by 'cross-over' and stereochemical studies, and by isolation of free nitrile. Under similar conditions Ritter condensation of the nitrile with a tertiary carbonium centre (derived from an alcohol) gives identical products[214, 215].

Related rearrangements, restricted to substrates with suitably placed hydrogen atoms, are reactions (75) (the Neber reaction)[216], (76)[218], and (77)[219]. The intermediate **75** has been isolated in at least

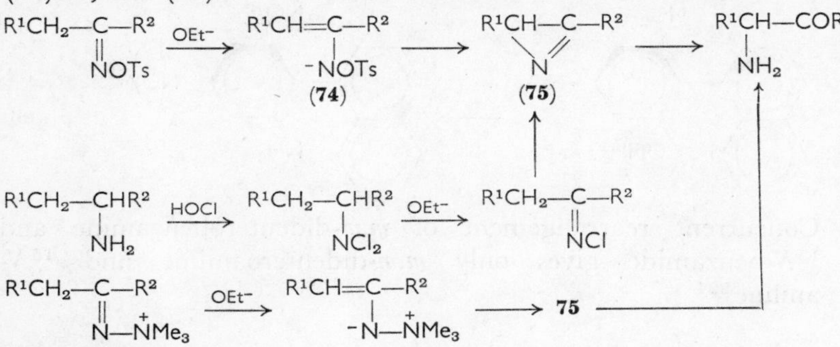

$$R^1CH_2-\underset{\underset{NOTs}{\|}}{C}-R^2 \xrightarrow{OEt^-} \underset{\underset{\textbf{(74)}}{-NOTs}}{R^1CH=C-R^2} \longrightarrow \underset{\underset{\textbf{(75)}}{N}}{R^1CH-C-R^2} \longrightarrow \underset{\underset{NH_2}{|}}{R^1CH-COR^2}$$

(75)

$$R^1CH_2-\underset{\underset{NH_2}{|}}{CHR^2} \xrightarrow{HOCl} R^1CH_2-\underset{\underset{NCl_2}{|}}{CHR^2} \xrightarrow{OEt^-} R^1CH_2-\underset{\underset{NCl}{\|}}{C}-R^2$$

(76)

$$R^1CH_2-\underset{\underset{N-\overset{+}{N}Me_3}{\|}}{C}-R^2 \xrightarrow{OEt^-} R^1CH=\underset{\underset{-N-\overset{+}{N}Me_3}{|}}{C}-R^2 \longrightarrow \textbf{75}$$

(77)

21+C.A.G

one example[217], and the lack of stereospecificity in reaction (75) where both *anti* and *syn* derivatives react, is probably due to the transient existence of **74**. The Tiemann reaction of aldoximes (reaction 78) has been rarely studied, but probably proceeds as shown[221].

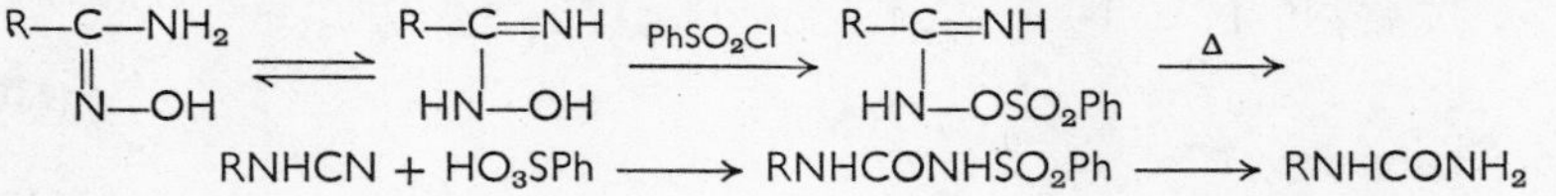

$$R—C—NH_2 \rightleftharpoons R—C{=}NH \xrightarrow{PhSO_2Cl} R—C{=}NH \xrightarrow{\Delta}$$
$$\underset{N—OH}{\|} \qquad \underset{HN—OH}{|} \qquad \underset{HN—OSO_2Ph}{|}$$

$$RNHCN + HO_3SPh \longrightarrow RNHCONHSO_2Ph \longrightarrow RNHCONH_2 \quad (78)$$

B. Hofmann Rearrangement

The reactions described in this and the following four sections are closely related in mechanism. All have been recently reviewed.

The Hofmann reaction, shown in (79) (discovered 1882)[204, 222, 251] gives excellent (80–90%) yields with lower alkyl and aryl amides, but side-reactions intrude with higher homologues. The discoverer elucidated the mechanism except for the details of the migration. He isolated **76**, showed that the corresponding *N*-chloro, but not the

$$RCONH_2 \xrightarrow[60°]{Br_2} RCONHBr \xrightarrow{OH^-} RCO\bar{N}Br \longrightarrow RNCO \xrightarrow{H^+,H_2O} RNH_2 \quad (79)$$
$$\qquad\qquad (76) \qquad\qquad\quad (77)$$

N-alkyl, compound would react, and recovered the isocyanate from reactions in aprotic media and trapped it as a urethane, when methanol was used as solvent. Subsequently, strict intramolecularity has been proved by several sets of observations:

(i) Optically active migrating groups, e.g. (+)-1-phenylethyl, show ~96% retention of configuration in products, the loss of activity being traced to slight racemisation of substrate under the strongly basic conditions[223].

(ii) Bridgehead amides of (2.2.1)-bicycloheptanes readily react when inversion would be sterically prohibited[204].

(iii) Biphenyl assymetry is preserved during rearrangement (reaction 80)[204].

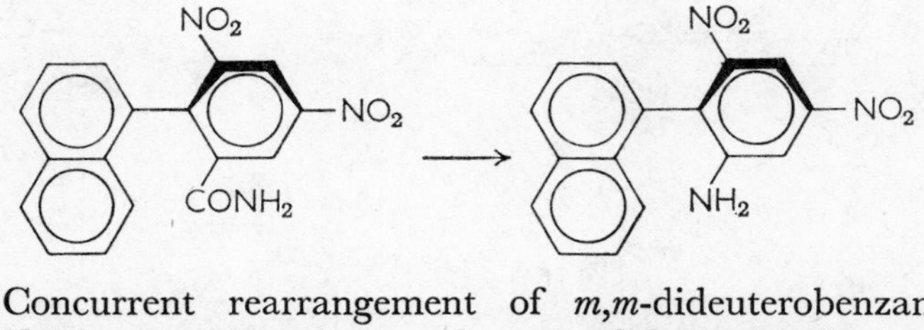

$$(80)$$

(iv) Concurrent rearrangement of *m,m*-dideuterobenzamide and ^{15}N-benzamide gives only *m,m*-dideuteroaniline and ^{15}N-aniline[224].

Consequently, the reaction may be used to assign configurations to acids, by conversion into amines of known configuration[225].

Details of the migration to nitrogen are obscure. Early studies on the related Curtius rearrangement failed to detect radical intermediates by trapping techniques, and formation of free carbonium ions is inconsistent with the lack of Wagner–Meerwein rearrangement in products from $Ph_3CCH_2CONH_2$, and similar substrates[204]. The rate acceleration ($\rho = -2\cdot5$) found in substituted benzamides with *m*- and *p*-electron-releasing substituents, and the converse with electron-attracting groups, favours migration concerted with loss of halide ion from **77**. A two-step migration with the initial formation of the

$$\begin{array}{c} O \\ \parallel \\ R-C-\ddot{N}: \end{array}$$
(78)

azene (**78**) would be expected to be accompanied by formation of hydroxamic acid resulting from radical attack on water, but this could not be detected in the Curtius process, even by sensitive colorimetric techniques[226], although analogous radicals from the decompositions of hydrazoic acid, phenylazide, and benzenesulphonyl-azide could be readily demonstrated. Despite these observations, it has been argued that an azene intermediate cannot be ruled out[204, 227, 228] in this and the related rearrangements, and obviously much work on different substrates is necessary before any general, or even restricted, conclusions can be made.

C. Lossen Rearrangement

Most of the above observations and evidence for mechanism also apply to this reaction of acylhydroxamic acids (reaction 81)[229],

$$RCONHOCOR^1 \xrightarrow[\text{or OH}^-]{\Delta} RCO\bar{N}OCOR^1 \longrightarrow RNCO \longrightarrow RNH_2 \quad (81)$$

which was discovered in 1872. It is first order, intramolecular, and shows similar substituent effects in aromatic R ($\rho = -2\cdot6$), as the Hofmann and Beckmann reactions; for the same reasons *ortho* substituents facilitate reaction, irrespective of their polarity. An additional datum here, which implicates N—OAc cleavage in the rate-determining step, is the retardation ($\rho = +0\cdot87$) exhibited by *meta* or *para* electron-releasing groups in R^1, and the acceleration promoted by corresponding groups of opposite polarity.

D. Curtius Rearrangement

This reaction of acyl azides (reaction 82) (discovered 1890) may be

$$ROCl + \bar{N}_3 \atop RCONHNH_2 + HNO_2 \Bigg\} \longrightarrow RCON_3 \xrightarrow{\Delta} RNCO \longrightarrow \text{Urethanes,}$$
$$\text{amines, etc.} \quad (82)$$

conducted in aprotic, alcoholic, or aqueous media to give isocyanates, urethanes, or amines respectively in excellent (often 80%) yields[204, 230, 251]. Benzenesulphonyl azides react similarly on heating or under ultraviolet irradiation[234].

Intramolecularity has been demonstrated as in the previous two reactions, but whilst electron-releasing *meta* substituents in aromatic substrates increase the rate of rearrangement, and *ortho* groups of any polarity behave similarly, a difference is that all *para* substituents decrease the rate compared with the unsubstituted compound. Uniquely here, the substrate is stabilised by considerable conjugation between the aromatic and triazo groups which more than counteracts the migratory ability even of *para* electron-releasing groups[231]. The details of the migration are again obscure but tracer studies have confirmed the absence of far-spanning migrations (reaction 83)[232].

$$ArCONHNH_2 + H^{15}NO_2 \longrightarrow ArCO\overset{-}{N}\!\!-\!\!\overset{+}{N}\!\!\equiv\!\!^{15}N \longrightarrow ArNH_2 + N\!\equiv\!^{15}N \quad (83)$$

The ease of rearrangement of acyl azides compared with their alkyl counterparts may indicate that the former are concerted, for the latter reactions are insensitive to polar influences, and may well involve an azene intermediate (see section III.F).

Both Brönsted and Lewis acid catalysis have been observed[204, 233] and probably coordination to oxygen rather than nitrogen puts a positive charge on the middle nitrogen of the triad to give a labile species similar to that in the Schmidt reaction.

E. Schmidt Rearrangement

This is a group of reactions (discovered 1923) resulting from treatment of a carbonyl compound with sodium azide and concentrated acid, usually in an inert solvent (reaction 84)[204, 235, 251]. With

$$RCOR^1 \xrightarrow[40°]{H^+} R\!-\!\overset{OH}{\underset{+}{\underset{|}{C}}}\!-\!R^1 \xrightarrow{N_3H} R\!-\!\overset{OH}{\underset{HN\!-\!N_2^+}{\underset{|}{C}}}\!-\!R^1 \longrightarrow R\!-\!\overset{OH}{\underset{N\!-\!N_2^+}{\underset{\|}{C}}}\!-\!R^1 \longrightarrow$$

$$(79) \qquad\qquad (80) \qquad\qquad (84)$$

$$\overset{+}{\underset{R\!-\!N}{\underset{\|}{C}}}\!-\!R^1 \longrightarrow \overset{+}{\underset{R\!-\!NH}{\underset{|}{HO}}}\!=\!C\!-\!R^1 \longrightarrow O\!=\!\overset{}{\underset{R\!-\!NH}{\underset{|}{C}}}\!-\!R^1$$

$$(81) \qquad\qquad (82)$$

$$R = \text{alkyl, aryl}; \quad R^1 = \begin{cases} H \\ \text{alkyl, aryl} \\ OH \end{cases} : \text{Product} \begin{cases} RNHCHO \\ RNHCOR^1 \\ [RNHCOOH] \end{cases} \longrightarrow RNH_2 + CO_2 \qquad (85)$$

carboxylic acids as substrates, amines are formed directly (90% yields), although an isocyanate has been isolated in one example[236]. No other intermediates have been characterised and the mechanism has been deduced from secondary evidence and analogy. Although extensively used in synthesis[239], the reaction is less flexible than its counterparts owing to the forcing conditions necessary, and the several steps sometimes give good scope for side-reactions. Bonded nitrogen can migrate to the nitrogen terminus in guanyl azides[237] and phenyl bonded to phosphorus, sulphur, and other metalloids also migrates[238]. 2,4,6-Trimethylbenzoic acid reacts under much milder conditions than does benzoic acid, and an acyl ion, RCO^+, may be formed[243]. It is not possible to decide whether benzoic acid reacts through a very small concentration of a similar ion or via the more plentiful but less reactive conjugate acid $RC(OH)_2{}^+$.

The reaction is catalysed by protons (h_0 dependence) and Lewis acids, and is strictly intramolecular by the criteria of section III.B. Substituent effects are similar to those in the Hofmann and Lossen processes and a similar concerted mechanism is likely. Despite the forcing conditions, no Wagner–Meerwein shifts in suitably chosen migrating groups have been detected by ^{14}C-tracer studies[240]. Although **79** could directly form **82**, the route shown is necessary to accommodate (i) the ineffectiveness of alkyl azides as substitutes for hydrazoic acid, and (ii) the formation of tetrazoles by 1,3-addition of **81** to azide ion, when the latter is in excess[204,242].

Use of asymmetric ketones enables the detection of a *trans*-migratory requirement as in the Beckmann reaction. Thus the tendency for the larger group to migrate in such substrates, i.e. Ph > Me, Et > Me, is a consequence of favoured formation of **80** with the larger group and the leaving group situated *trans*, and the same irreversible pre-rearrangement step accounts for the identical migratory aptitudes in *para*-monosubstituted benzophenones, notwithstanding the polarity of the substituents.

Reaction (86) is related to the Schmidt reaction[244,245].

$$Ph_2C(N_3)_2 \xrightarrow{H^+} Ph_2C{=}N{-}N_2{}^+ \longrightarrow PhCONHPh$$

$$\Big\uparrow HNO_2 \qquad\qquad (86)$$

$$PH_2C{=}NNH_2$$

F. Azide Rearrangements

Adaptions of the previous method are shown in reactions (87) and (88) [204,245]. The rates of these rearrangements follow h_0 and a concentrated acid is required both to produce carbonium ions necessary to form organic azides and (as in the Curtius process) to protonate the latter to initiate reaction. A good Hammett dependence, $\rho = -2\cdot0$, is found for suitable m- and p-substituted aromatic substrates, and the migration is almost certainly concerted in these examples. Qualitatively similar results are obtained from separately prepared azides and concentrated mineral acid.

$$R_3COH \xrightarrow[HN_3]{H_2SO_4} R_3CN_3 \xrightarrow{H^+} R_3CNHN_2{}^+ \longrightarrow R_2C{=}NR \qquad (87)$$

$$R_2C{=}CR_2 \xrightarrow[HN_3]{H_2SO_4} R_2CHCR_2N_3 \xrightarrow{H^+} R_2CHCR_2NHN_2{}^+ \longrightarrow R_2CHCR{=}NR \quad (88)$$

Ethyl azide, formed by any of these techniques, gives products (CH_3NH_2 and $HCHO$) derived from both methyl and hydrogen migration, but n-hexyl azide and higher homologues only give the latter type of rearrangement. Product studies from different s-heptyl azides show the migration tendencies: n-Am > Me, n-Bu > Et, and 1-methylcycloalkyl azides expand the ring to give heterocyclics rather than undergo methyl shift [246]. In general, the smallest group has least tendency to migrate, probably owing to the populations of the different ground-state conformations controlling the availability of migrating groups in the concerted (or very nearly so) rearrangement, as in deamination.

Ultraviolet irradiation of alkyl azides induces rearrangements best accounted for on a radical mechanism (reaction 89) [247]. Route (a)

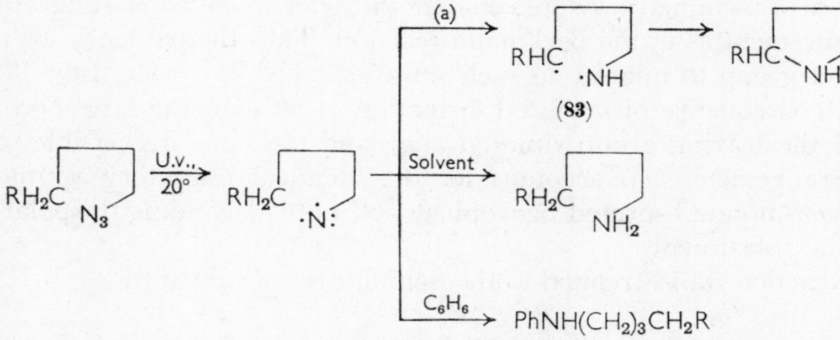

predominated when the reactive nitrogen and a hydrogen, five atoms removed, could adopt the conformation shown, and was developed into an elegant synthetic route, e.g. of proline and connessine (the Barton reaction). The diradical **83** was originally proposed to account for racemic products from optically active substrate—an insertion of a singlet azene into the neighbouring bond was expected to retain activity in products—but recent studies on other systems have demonstrated active products[227], and no universal mechanism seems likely. Azenes probably also intervene in the silver-catalysed bicyclisation of N-chloroazacyclononane[248]. Irradiation of substituted trityl azides (Ar_3CN_3) readily gave rearranged products but the rates of rearrangements and migratory aptitudes of various p-substituted aryl groups were identical. Again a triplet nitrene was postulated as an intermediate which showed negligible selectivity as a migration terminus. Presumably sufficient activation was achieved by absorption of radiation to eliminate any necessity for aryl participation in a concerted process of migration and loss of nitrogen[249]. Considerable evidence for the intermediacy of such triplet azenes has been presented[227] and they can be trapped by dimethylsulphoxide in the ultraviolet-induced decomposition of benzoyl azide[250].

In contrast, thermal decomposition of these trityl azides in inert solvents at 170–190° is sensitive to *para* substitution, although the rates of the anisyl and nitro compounds only differ ten-fold[249]. Analysis of the range of $\Delta H^{\ddagger}$ and ΔS^{*} values indicates that inductive effects cannot be wholly responsible, and a small amount of aryl participation seems likely, although a singlet nitrene has been proposed[204, 227] that does show some discrimination towards available β-linked groups, despite its great electrophilicity.

G. Stieglitz Rearrangements

A group of reactions of trityl derivatives discovered by Stieglitz ($\sim$ 1916) are shown in reactions (90a) to (90c)[204]. Such reactions have not been reported for other classes probably owing to synthetic difficulties; for only in trityl and a few similar types does steric hindrance

$$Ar_3CNHOH \xrightarrow{PCl_5,\ ether} Ar_3CNHO\overset{+}{P}Cl_3 \longrightarrow Ar_2C{=}NAr \qquad (90a)$$

$$Ar_3CNHOCOPh \xrightarrow{OH^-} Ar_3C\overset{-}{N}OCOPh \longrightarrow Ar_2C{=}NAr \qquad (90b)$$

$$\left.\begin{array}{l} Ar_3CNHCl \\ Ar_3CNCl_2 \end{array}\right\} \xrightarrow{OH,\ \Delta} Ar_3C\overset{-}{N}Cl \longrightarrow Ar_2C{=}NAr \qquad (90c)$$

to the nitrogen atom generate conditions where reaction at the oxygen, either during rearrangement (90a), or in preparation of substrates (90b), can predominate. Intermediates have not been isolated, but the mechanisms are probably as shown. p-Substituted compounds obey a Hammett relationship ($\rho = -1\cdot2$) similar to that of the thermal reactions of trityl azides previously discussed, and the reactions are presumably intramolecular.

IV. INTERMOLECULAR AROMATIC REARRANGEMENTS

An important group of acid-catalysed 'rearrangements' involves fission and intermolecular recombination (reaction 91), rather than

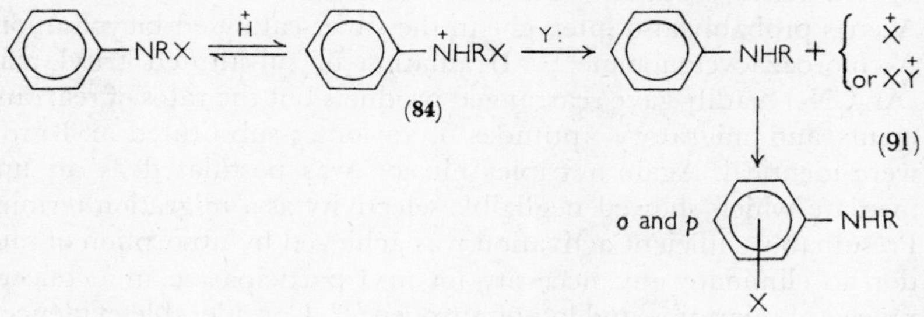

(84)

(91)

o and p

intramolecular migration. A widely applied criterion for intermolecularity has been trapping of the electrophile X with an additive, e.g. anisole or aniline, but this procedure is valueless as the scavenger may react with the substrate, with an intermediate of an intramolecular process, or with side-products accompanying, but quite unrelated to, rearrangement. Better tests for intermolecularity are the scrambling into product of added X^+ (or a species convertible into it under the reaction conditions) containing isotopic tracer, the demonstration of coupling of the separately prepared fission products to give rearrangement products in the same proportions as obtained from 'rearrangement' under similar conditions, and the predominant formation of *para* rather than *ortho* products. Application of these criteria and detailed kinetic studies are sadly lacking for the following reactions.

A. Diazoaminobenzene Rearrangement

(reaction 91; R = H; X = N=NAr)

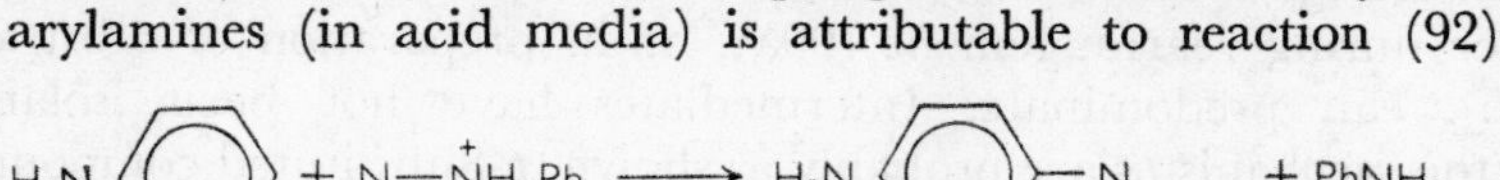

This reaction, promoted by ethanolic hydrochloric acid at 0–20°, was discovered by Griess and Martius in 1886, and is important in dyestuffs manufacture[252]. Little controversy has arisen over mechanism, which has since 1885 been accepted as fission to a diazo compound or diazonium ion (characterised by coupling with scavenger) and amine (isolable)[253], and recoupling. The marked catalytic effect of arylamines (in acid media) is attributable to reaction (92)[254], where the base usurps the role of water or ethanol. In aniline as solvent the rate was always first order in substrate, but with strong acids it was predominantly first order in acid and with weak acids second order; the former rate expression follows from reaction (92) in which aniline (present in large excess) is implicated, and the latter from a mechanism where the more nucleophilic anion of the acid can intervene to give a diazo compound, $RN{=}NX$, in a step with a rate proportional to $[H^+][X^-]$, i.e. to $[HX]^2$. Reaction (92) has also been supported by ^{15}N and product studies[253, 255], but the role of covalent diazo compounds in aqueous media awaits elucidation.

B. Hofmann-Martius Rearrangement

(reaction 91; R = alkyl, H; X = alkyl)

Pyrolyses of N-alkylarylamine hydrochlorides at 200–300° lead mainly to p-alkylated derivatives (discovered 1871)[271]. Reactions of N-methyl heterocyclics (Ladenburg reaction)[257], and Hofmann degradations of certain 'onium hydroxides[256] give similar products. Formation of alkyl halide (always with unrearranged carbon skeleton) and olefins (sometimes rearranged) reduces the yields unless sealed vessels are used, and these observations coupled with the frequent formation of products with rearranged alkyl substituents (e.g. N-isobutylaniline gives isobutyl bromide, isobutylene, and p-t-butylaniline) suggests nucleophilic attack by halide ion to give alkyl halide followed by carbonium ion formation (in the polar molten salt) and para substitution[258].

Consistent with this, halides and olefins condense with anilines under simulated conditions to form para products[259], and ^{14}C-tracer studies show no hydride shift on migration of straight-chain and isopropyl groups[260]. (+)-N-1-Phenylethylaniline possesses a migrating group that is appreciably stable as a cation, and here rearrangement gives quite different isomer ratios from alkylation of aniline with styrene or 1-phenylethylchloride[264]. Extensive racemisation of products eliminates an intramolecular route, and presumably the normal pathway is employed involving relatively free, non-paired cations. These product studies and the general dependence of products on the particular acids used[262,263], rule out a direct S_N1 heterolysis of the protonated substrate and subsequent para coupling[261].

Rearrangement is also catalysed by colbalt(II), zinc, and other metal chlorides[265], with the important differences that neither olefin,

21*

alkyl halide nor product with rearranged alkyl group, have been detected[266], and the *ortho : para* product ratio differs from that of the amine salt pyrolyses[262, 264]. A Friedel-Crafts-type process (reaction 93) is likely[261].

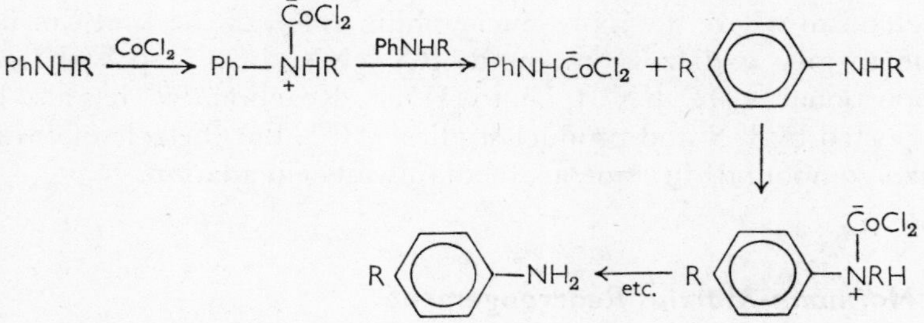

$$PhNHR \xrightarrow{CoCl_2} Ph-\overset{+}{\underset{}{N}}HR \xrightarrow{PhNHR} PhNH\bar{C}oCl_2 + R\text{-}C_6H_4\text{-}NHR \tag{93}$$

An intramolecular π-complex mechanism (see section V.a) has been proposed for the proton-catalysed reactions[267] to accommodate the alleged lack of polyalkylation such as was expected in the inter-molecular route due to the activating effect of alkyl groups on ben-zenoid reactivity. In fact, polyalkylation *does* occur to a small extent[266], but the previous argument ignores the fact that the most reactive *para* position is first blocked, and that the *ortho* positions, probably for steric reasons, are much less susceptible to attack. All the available evidence refutes such a mechanism.

C. Fischer-Hepp Rearrangement (reaction 91; R = alkyl; X = NO)

The rearrangements of N-nitrosoamines catalysed by hydrochloric or hydrobromic acids usually in aqueous ethanol or acetic acid at 20°, were discovered in 1886. N-nitrosodiphenylamines also react, especially with Lewis acids (AlCl$_3$ etc.) in aprotic media[268, 269], and N-nitrosotetrahydroquinoline gives migration into the aromatic ring[251]. The great catalytic efficiency of halogen acids (compared with other mineral acids), the facile cross-nitrosation of additives, the improved yields of products on adding sodium nitrite, and the isola-tion of halogenated side-products (with chlorine derived from reaction of nitrous and hydrochloric acids) all support a nucleophilic attack by halide ion to form nitryl halide (NOHal) which attacks the ring, but a satisfactory kinetic study is lacking. Nitrosoamines with t-alkyl or other bulky N-substituents denitrosate to amines on attempted reaction[270, 271], probably owing to steric inhibition of the initial S_N2 step.

D. Orton Rearrangement (reaction 91; R = COCH$_3$; X = Br, Cl)

This rearrangement to mainly *para*, but often appreciable quantities (~40%) of *ortho* products, was discovered by Bender in 1886, but was chiefly elucidated by Orton[272]. Reactions of N-chloroacetanilides in aqueous or ethanolic media are uniquely catalysed by hydrochloric acid with rates proportional to [HCl]^2, i.e. to [H$^+$][Cl$^-$], although other halogen acids promote non-isomeric transformations. For example, hydrobromic acid gives *o*- and *p*-bromo derivatives; other mineral acids are largely ineffective. These observations, the isolation of acetanilides and chlorine from rearranging solutions, the cross-chlorination of additives, and the identity of products of rearrangement and of chlorination of acetanilides[274], suggests an initial nucleophilic attack by chloride ion on the N-bonded chlorine forming fission products which combine intermolecularly. The formation of chlorine via hydrolysis to hypochlorous acid is less likely[258]. This mechanism is consistent with detailed kinetic analysis[258] and with the discovery that use of ^{36}Cl-labelled substrate or catalysing acid, leads to products with the chlorine containing approximately the radioactivity calculated from pooling the organic chloride with inorganic chloride[273].

Rearrangements catalysed by carboxylic acids in aprotic media, e.g. chlorobenzene, give mainly *para* product, but the mechanism is obscure. An intramolecular process[275], of the π-complex type[267, 280], was proposed, as free halogen could not be detected in quantities sufficient to account for reaction; but kinetic evidence for intermolecularity with XOCOR as the halogen carrier has since been presented[276, 277]; and the predictions of π-complex theory are inconsistent with the ready isomerisations of 2,6-disubstituted N-bromoacetanilides in which extension of the π orbitals to nitrogen would be sterically inhibited[278, 279]. ^{14}C- and ^{82}Br-tracer studies of concurrent rearrangements of two N-bromoacetanilides, and of the kinetics of cross-bromination of added anisole, suggest that intra complex reactions occur in aggregates consisting of brominating agent, bromine acceptor, and acid catalyst[281, 282]. Such aggregates are very reasonable for polar reactions in non-polar media, and account for the observed complex kinetics at least qualitatively.

Rearrangements of N-chloroacetanilides induced by ultraviolet radiation or benzoyl peroxide in carbon tetrachloride involve free-radical intermolecular chlorination[283], as probably do similar reactions of N-chloro-N-alkylarylamines[284] and N-chloroarylsulphanilides[285].

E. Phenylhydroxylamine Rearrangement

Although formally similar to the previous processes, this reaction (Bamberger, 1894) which is often carried through in one step by electrolytic reduction of nitrobenzenes in acid conditions[251], involves nucleophilic attack at the *para* (or less favourably the *ortho*) positions by the solvent (reaction 94). Redox reactions catalysed by oxygen

$$ \text{NHOH} \underset{}{\overset{\overset{+}{H}}{\rightleftharpoons}} \text{NHOH}_2^+ \xrightarrow{\text{Slow}} $$

$$ =NH \longrightarrow HO\text{---}NH_2 \tag{94} $$

lead to azoxy products[286], and non-catalytic thermal reactions[287] and transformations of *N*- or *O*-sulphonic esters[290] and *N,O*-diacetyl derivatives[291] have been recorded. An interesting unstudied analogue is the interconversion of phenylhydrazine and *p*-phenylenediamine catalysed by hydrochloric acid[296].

There is no of evidence for a proposed[267] intramolecular π-complex mechanism, and reaction (94) accommodates the dependence of rate on acidity[288], the formation of *p*-chloro and *p*-ethoxy compounds in the presence of hydrochloric acid and ethanol[286, 289], and the total incorporation of ^{18}O from labelled aqueous solvent[288]. A cationic intermediate, rather than concerted fission and *para* substitution, is suggested by the non-dependence of rate of formation of chloro compounds on chloride ion concentration[288].

Interest has recently been stimulated by the discovery that carcinogenic arylamines are degraded *in vivo* by *N*-hydroxylation and rearrangement, and a microsome fraction from rat liver converts 3-aminobiphenyl into 3-amino-4-hydroxybiphenyl via this route[292].

F. Chattaway Rearrangement (reaction 91; X = R = acyl)

The interconversion of *N,N*- and *N,p*-diacylarylamines was found (1915) to be promoted by zinc chloride and other Lewis acids, but has been seldom studied. A Friedel-Crafts type of mechanism is probable[293] for this and for the related rearrangements of *N*-aryl-sulphonylamines[294]. *N*-Acylarylamines form the *para* isomer under ultraviolet irradiation[295], but the proposed intramolecular shift involving radicals caged in a solvent shell (Frank–Rabinovitch effect) seems less likely than an intermolecular process.

V. INTRAMOLECULAR AROMATIC REARRANGEMENTS

A. Nitramine Rearrangement

The acid-catalysed rearrangement of N-nitroarylamines (Bamberger, 1893) is formally similar to reaction (91) (R = H, alkyl; X = NO$_2$), but complete intramolecularity has been demonstrated for several substrates in the range 0·1 M to 16 M acid in several solvents. For example:

(i) *ortho* Products predominate (typically 70–90%) over the whole range[297–299] whereas intermolecular nitration would give predominantly *para* products.

(ii) Nitration of added scavenger is rare[297,298], but is attributable (with 2,4-dinitro and chloro-substituted nitroanilines) to electrophilic attack by protonated substrate[301].

(iii) No tracer enters either *ortho* or *para* products when N-nitroaniline[297,302] or N-nitronaphthylamines[300] are rearranged in the presence of ^{15}N-labelled nitrate or nitrite ion over a large acidity range; and no cross-transfer of tracer occurs when the former substrate labelled in the nitro group, and p-methyl-N-nitroaniline are simultaneously rearranged[303]. Cross-nitration during a similar rearrangement of N-methyl-N-nitroaniline and its labelled p-fluoro derivative[304] is inconclusive in the absence of appropriate controls[300].

(iv) Possible products of fission, for example NO$_2{}^+$ and amine at high acidities, either do not react under appropriate conditions[300,301] or react but give a different isomer ratio from products of rearrangement[301]. The latter observation (see (95)) eliminates

Products (%)

	o	m	p
Nitration of PhN$\overset{+}{H}_3\overset{-}{N}O_3$	6	34	59
Rearrangement of PhNHNO$_2$	93	0	7

85% H$_2$SO$_4$: H$_2$O at 10°   (95)

both Bamberger's original theory and a recent variant[305] that aromatic C-nitration proceeds via N-nitration and rearrangement, and the absence of *meta* rearrangement product confirms that no intermolecular component intrudes.

The dependence of rate on Hammett's h_0 function and the large solvent isotope effects (k_{D_2O}/k_{H_2O}) for several N-nitroamines[297,298,306] suggest preequilibrium protonation of substrate, but details of the subsequent migration are disputed. What is clear is that some structural feature denied to other N-substituted amines here permits an intramolecular route. Three mechanisms are extant:

(i) The π-bond mechanism: A scheme, applicable to intramolecular rearrangements in general, has been vigorously but uncritically advocated[261, 267, 281, 307] (despite its unsatisfactory molecular-orbital foundation[297, 307]). In this, protonated substrate hetero-lyses to NO_2^+ which bonds to the π orbital of its counter-fragment and migrates without becoming kinetically free. However, fission in this direction is ruled out at low acidities by the excellent Hammett σ^+-ρ relationship ($\rho = -3\cdot7$) shown by a series of p-substituted N-nitroanilines[306, 308], and heterolysis to an aromatic dication and nitrite ion appears universal as indicated by the general formation of nitrous acid, diazonium ions, quinones, and imines as side-products, and the non-occurrence of nitrate ion, derived from NO_2^+ [297, 298]. Also the π complex must collapse to σ complexes *en route* to products, identical with those occurring in aromatic nitration. Consequently the theory cannot accommodate the appreciable o- and p-deuterium isotope effects on products that are observed, for such isotope effects do not occur in nitration processes[297, 298], and furthermore it predicts migration into either ring of naphthyl derivatives (and perhaps *meta* and deaminated products) whereas only the 2- and 4-nitro isomers are formed[298].

(ii) The caged-radical mechanism: reaction (96) has been pro-posed[299, 306] in which caged radicals either recombine intra-molecularly to products, or dissociate to allow an intermolecular component. The sole evidence was the formation of amine and nitrous acid, the former presumably by reduction of the dis-

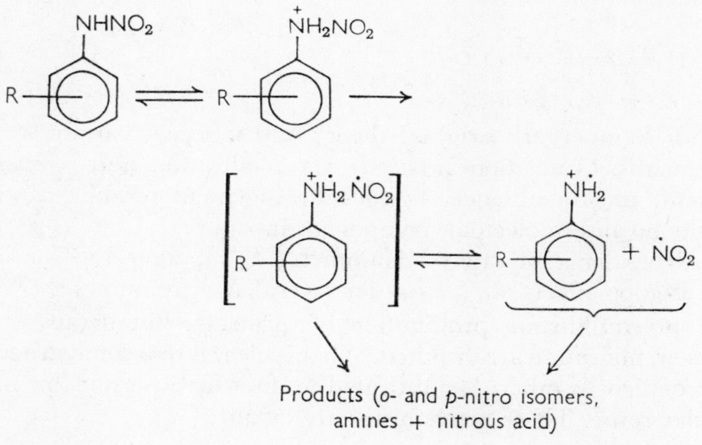

(96)

Products (o- and p-nitro isomers, amines + nitrous acid)

sociated species, and the action of radical inhibitors (α-naphthol, hydroquinone) in diminishing the yield of products by destroying the intermolecular component. Tracer evidence for the latter component is illusory (see above) and the inhibitors may also react with nitrous acid or nitryl chloride (which are known to be present in the reaction products) and so may well suppress rearrangements proceeding through N-nitrosation, Fischer-Hepp migration, and oxidation, that involve no radicals. In this mechanism the completely intramolecular reactions, of N-nitroaniline for example, would have to show 100% primary recombination of radicals, an efficiency never approached even for neutral species[307, 309]. Also, product isotope effects cannot be accommodated for similar reasons as in the π-bond mechanism, and *meta* products, and migration of naphthyl derivatives into either ring, neither of which are observed, would be predicted. Furthermore, electron spin resonance techniques or tests for initiation of polymerisation provide no evidence for radicals[300]. Reaction through an ion pair **86**, was considered unlikely by the proponents of this theory.

(iii) The cyclic transition-state mechanism: neither of the above theories accounts for the different behaviour of N-nitro as compared with other N-substituted compounds, for, if radicals or π complexes could readily be formed from the former, they should equally well occur with the latter. This is contrary to experience. A mechanism, shown in its latest form[298] as reaction (97), envisaging *para* migration through a reactive σ-bonded intermediate[302], accommodates all the presently available data[297, 298], and has analogies with the Claisen rearrangement of allyl aryl ethers, and the thermal rearrangements of N-allylamines[315]. **85** and **87** break down to o- and p-nitro isomers directly, in highly exothermic processes that by-pass σ complexes and so show product isotope effects; and uncoupling of bond making at $C_{(2)}$ from the initial bond fission leads to **86** which dissociates to side-products[297]. For some substrates, extensive solvent attack on **85** and **87** results in nitrous acid and tars derived from aminophenols.

Nitramine rearrangements occur in quinoline[310], pyridine[311], and thiazole[312] derivatives, and although migration into both rings is reported for the first type, ^{15}N studies in the pyridine series indicate an intramolecular pathway[313]. Thermal and photochemical reactions of N-nitroanilines and naphthylamines give 30 to 50% yields of

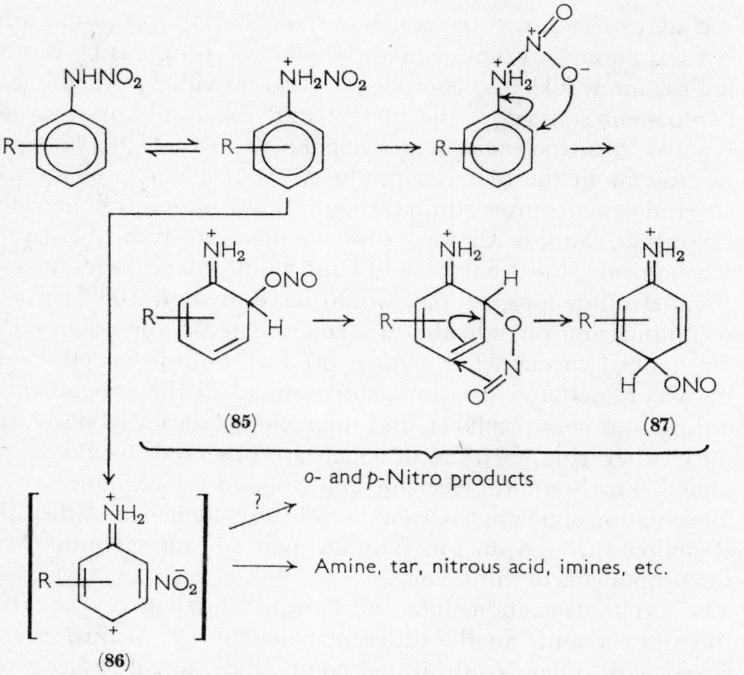

$$(97)$$

rearrangement products together with tars, and the lack of evidence for radicals[314] makes cyclic processes, similar to reaction (97), likely.

B. Benzidine Rearrangement

Acid treatment of hydrazobenzene (reaction 98) (Hofmann 1863) gives almost exclusively the $C_{(4)}-C'_{(4)}$-linked benzidine (**88**) and $C_{(4)}-C'_{(2)}$-linked diphenyline (**89**); although in conditions where proton acceptors show low activity, e.g. 40–90% sulphuric acid–water[316] or ethereal hydrogen chloride[317], both the parent hydrazo compound (**90**) and ring-substituted derivatives may form $C_{(2)}-N'$, $C_{(4)}-N'$, and $C_{(2)}-C'_{(2)}$-linked products (o- and p-semidines and o-

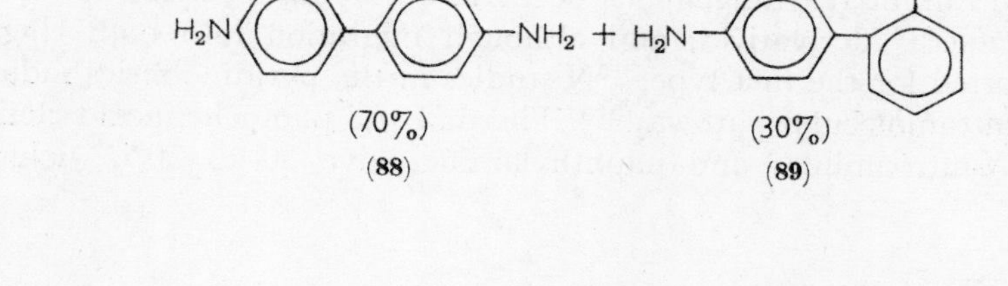

$$(98)$$

benzidines) or disproportionate to yield monocyclic amines and azo compound[318, 319]. The last reaction predominates with p,p'-disubstituted substrates and is generally favoured by moderate acidities and elevated temperatures[320].

Most of the early product studies used the convenient reduction of azo compounds with stannous chloride in acid (although many other reductants and even halogens can effect reductive rearrangement) but intrusion of tin complexes cannot be ruled out—although the one relevant kinetic analysis of reduction by titanous salts indicates hydrazo intermediates[321]. Reaction of separately prepared substrate under kinetically controlled conditions is essential for mechanistic investigations.

Early studies showed that semidines were not intermediates in the formation of aminobiphenyls for they were stable under the conditions typical for the rearrangement, and no intermediates, except mono- and possibly diprotonated substrate, have been detected. All rearrangements that have been studied are strictly intramolecular under all investigated conditions. This is shown by the inability to detect, by product isolation or by sensitive radiochemical and chromatographic techniques, any products B.B or A.A from the rearrangement of unsymmetrical substrates A.B, or products A.B from concurrently performed reactions of A.A and B.B[322-324]. The substrates must react at comparable rates for the last test to be valid. Specific examples are the failure to detect tracer-containing **91** in reaction (99).

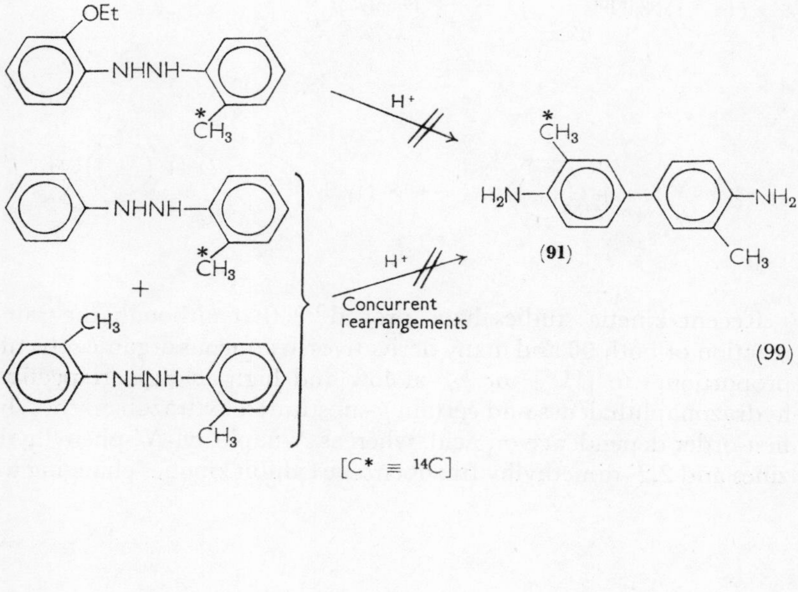

$$[C^* \equiv {}^{14}C]$$

(99)

The influence of ring substituents on products is complex and the 150 or so recorded examples[318, 325] have only recently been systematised and rationalised[325]. Ejection of certain groups, e.g. p-COOH, p-SO$_3$H, and p-Hal, may occur to varying extents but, this aside, the following statement of orientational principles is possible:

(i) A $C_{(4)}$ substituent leads to $C_{(2)}$-N' linking if it is strongly electron donating, and to $C_{(2)}$–$C'_{(4)}$ linkage if it is not.

(ii) A $C_{(4)}$ substituent showing both inductive electron attraction and mesomeric electron donation (e.g. a halogen) leads to both of the above modes of linking.

(iii) A $C_{(2)}$ electron-donating substituent orients to $C_{(4)}$–$C'_{(4)}$ linking with a strength paralleling that of its electron donation, or to $C_{(4)}$–$C'_{(2)}$ linking if the $C'_{(4)}$ position is blocked.

(iv) Substitution at $C_{(3)}$ and/or $C_{(5)}$, $C'_{(3)}$, and $C'_{(5)}$ favours $C_{(2)}$–$C'_{(2)}$ linkage.

Reaction (100) gives typical examples.

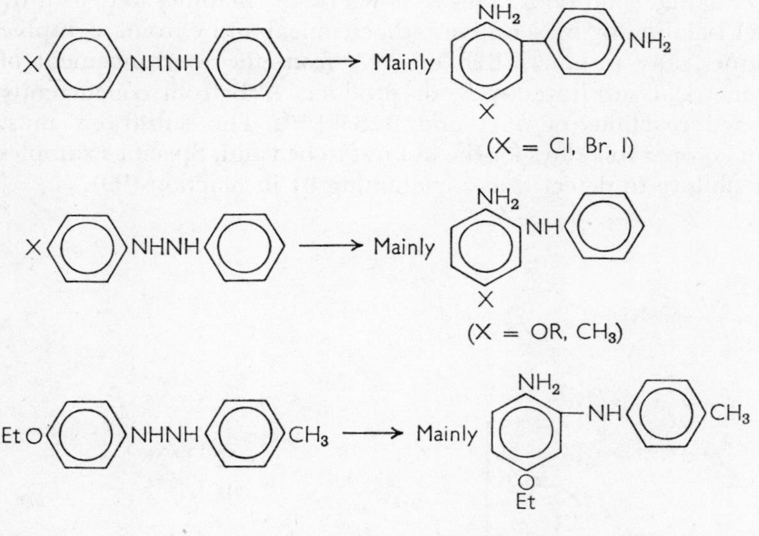

$$(100)$$

Recent kinetic studies have proved[325] that although the rate of reaction of both **90** and many derivatives in aqueous organic solvents is proportional to [H$^+$]^2 or h_0^2 at low and high acidities respectively, hydrazonaphthalenes and certain p-substituted hydrazobenzenes show first-order dependence on acid, whereas N-naphthyl-N'-phenylhydrazines and 2,2'-dimethylhydrazobenzene exhibit kinetics changing from first to second order with increasing acidity and quantitatively obeying the rate law (equation 101). General acid catalysis is not detected, and

$$k_{obs} = k_2[S][H^+] + k_3[S][H^+]^2 \tag{101}$$
$$(S = substrate)$$

is conceptually excluded[325], and solvent isotope effects (k_{D_2O}/k_{H_2O}) of 2·3 to 4·8, depending on the order in acid, clearly indicate that either one or two protons are transferred in preequilibrium.

Several mechanisms have been proposed to account for these, and other, data. Some[326] completely ignore the stereochemical problem of bonding the 4,4'-positions (which even in the *cis* form of substrate are 4·3 Å apart) without complete N–N fission and geometrical reorganisation leading to intermolecular and disproportionation processes. Others[327] cannot account for products and product isotope effects[325], and either are completely unphysical[328, 329] or are based on fallacious reasoning concerning rates and equilibria[330, 331]. But three can be seriously considered as meeting the stereochemical challenge:

(i) The caged-radical mechanism: although often mentioned, the formation and intramolecular recombination of caged ion-radicals ($ArNH_2^+ \cdot$) from diprotonated substrates has only recently been seriously advocated as an inconsequential conclusion from certain experimental observations[332]. Long ago[333] this theory had been made unlikely by the observation that although tetraphenylhydrazines would rearrange they did not do so under conditions where dissociation into radicals occurred; and the theory, based as it is on the idea of homolysis promoted by adjacent-charge repulsion, cannot account for the coexistence of, and steady transition between, one and two proton routes[325]. Nor can it account for the widely different products from different substrates. The reactivity of the fragments should be largely independent of ring substituents, and semidines would be always expected to predominate as the odd-electron density is greatest on the nitrogen atom[261, 281, 307].

Physical and chemical tests have not detected radicals[325], and the low probability of primary recombination of radical-ions of like charge would not lead to quantitative and intramolecular rearrangement. Radical-ions may be formed in certain compounds (e.g. **92**, $n = 4$) when rearrangement (and dissociation) are structurally inhibited, but with longer internitrogen bridges a normal type of migration occurs (reaction 102)[334].

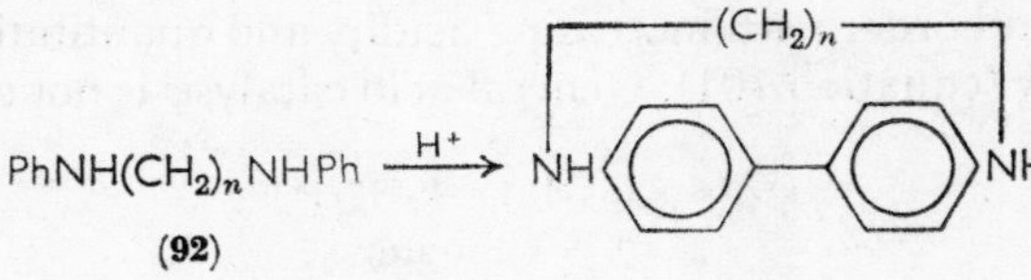

$$\text{PhNH(CH}_2)_n\text{NHPh} \xrightarrow{\text{H}^+} \text{NH} \underset{}{\bigcirc} \text{---} \underset{}{\bigcirc} \text{NH} \qquad (102)$$

(92)

(ii) The π-bond mechanism: this theory[261], analogous to that proposed for the nitramine rearrangement, has little theoretical justification[307,335], but is superficially attractive[336]. Unfortunately, except in attenuated form[307], it does not fit the observed facts and has no predicative value[325].

π-Complex formation was initially envisaged to result from a monoprotonated substrate. The discovery of second-order and transitional-order kinetics led to an amended mechanism in which the π complex was broken up with formation of rearrangement products by the attack (which could be rate determining) of a second proton. However, this required general acid catalysis for the second step which is not observed, and also a rate law (equation 103) that was found subsequently to be completely at variance with experiment (cf. equation 101).

$$k_{obs} = \frac{A[S][H^+]^2}{1 + B[H^+]} \qquad (103)$$

(S = substrate; A, B constants)

In order to accommodate the later kinetic data, π-complex formation from the diprotonated substrate had to be introduced[307], despite a previous theoretical demonstration[261] that this was improbable. As the earlier versions of the theory also could not account for the kinetic order of the reaction of hydrazonaphthalene and of semidine- and diphenyline-forming rearrangements[325], or for the products and kinetics for compounds with bulky *para* substituents[337], and furthermore, led to incorrect general rules for products, there seems little justification for its continued advocacy. The latest version of the theory has been stripped of its quantitative content, and is little more than a statement that the reaction is intramolecular and has some kind of (unspecified) interaction between the different potential fragments, as they readjust their relative positions.

(iii) The polar transition-state mechanism: this, the most satisfactory theory at the present time[325], recognises that the transition states of these rearrangements are unique, on account of the far-reaching structural changes involved, in bearing little electronic or geometrical resemblance to either reactants or products. The di- or monoprotonated substrate heterolyses (reactions 104 and 105), such that an incipient dication with one positive change

$$\text{Products} \qquad (104)$$

$$\text{Products} \qquad (105)$$

anchored on nitrogen and the other delocalised to the 2-, and (chiefly) 4-position of the ring, or a monocation with charge delocalised similarly, is formed in proximity with an incipient neutral species capable of undergoing electrophilic attack. The N–N fission results in lowered force constants for the $N—C_{(1)}$ and $N'—C_{(1)}$ bonds[338], permitting bond angles quite different from normal, and a concerted movement of the electronically distinct quasi fragments into coplanarity is accompanied by weakly directed electrostatic interaction (facilitated by the low apparent dielectric constant of the medium at these molecular distances) at the potential bonding sites. This develops into well-directed covalency formation and internal electrophilic substitution. A detailed analysis[325], taking into account the polarity and steric effects of ring substituents and the possibility of lateral displacement and rotation of the rings leads to a plausible rationalisation of all data concerning substituent effects on rate and products, salt and solvent effects, and kinetic and product isotope effects. Complete dissociation to an aromatic dication and amine is feasible when, for example, di-*p*-substitution inhibits rearrangement, and the former species may have been detected as a powerful unidentified oxidising agent formed during re-arrangement of *p,p'*-dimethylhydrazobenzene[339].

Heating hydrazobenzene with methyl iodide in the absence of acid gives tetramethylbenzidine[340], which may be formed either via quaternisation of the nitrogen of the substrate or by a purely thermal rearrangement followed by *N*-methylation. The thermal route is well known in a variety of solvents at 60–100°, and kinetic and product studies[341] indicate that hydrogen bonding leads to a quasi acid-catalysed intramolecular process in protic media. In aprotic solvents, semidines often predominate and a caged-radical process, which is still strictly intramolecular[342], may occur. Thermal rearrangements are undoubtedly responsible for reactions allegedly promoted by certain unusual 'catalysts' at elevated temperatures: e.g. refluxing **90** with phosgene in xylene is reported to induce rearrangement[343].

Benzidine-like rearrangements are reported on acid treatment of suitable imidazoles[344], pyridine[345] and thiazole hydrazocompounds[346], but not ferrocene[347] derivatives. Related, probably intramolecular, rearrangements of *p*-quinamines[348] and *O,N*-diarylhydroxylamines[349], (reactions 106 and 107) have been arbitrarily assigned π-bond mechanisms, whereas reactions (108) and (109) are probably inter-molecular, although the observed formation of cross-over products counts for little[251].

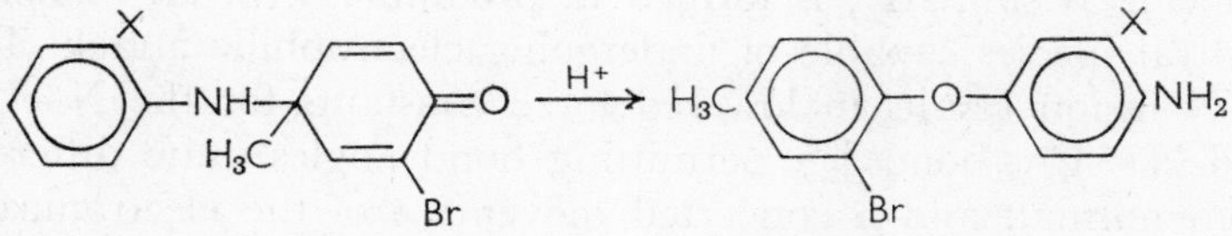

$$\text{(106)}$$

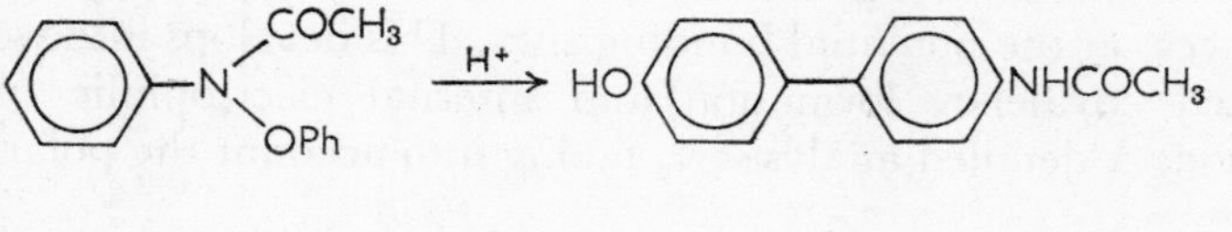

$$\text{(107)}$$

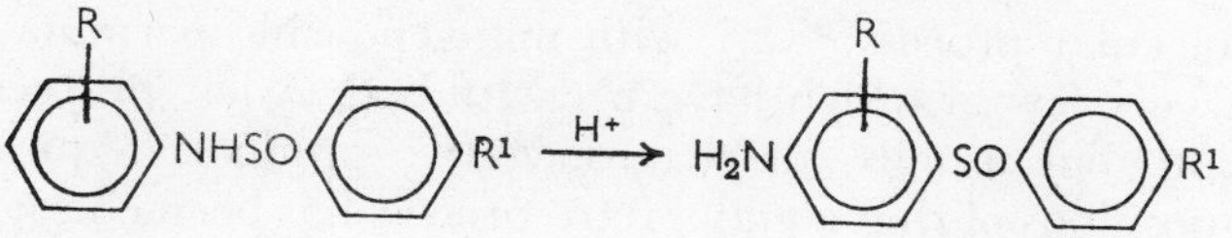

$$\text{(structure with NHSO, } R, R^1\text{)} \xrightarrow{H^+} \text{(structure H}_2N, SO, R^1\text{)} \quad (108)$$

$$\text{(structure with NHS, NO}_2, R\text{)} \xrightarrow{\Delta}$$

$$\text{(structure S, NH}_2, NO_2\text{)} + \text{(structure NH}_2, S, NO_2\text{)} \quad (109)$$

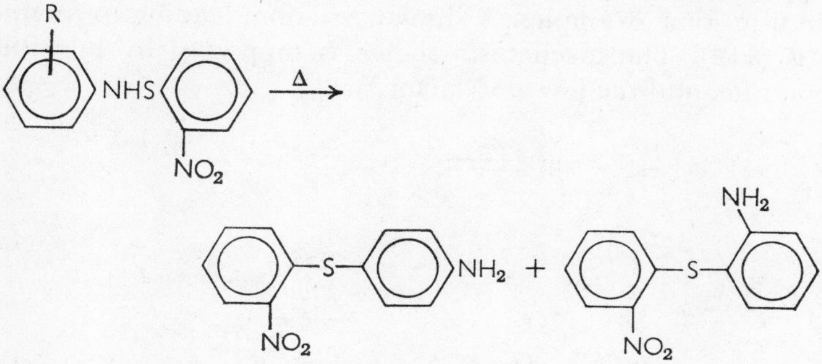

C. Chapman and Smiles Rearrangements

The thermal transformation of aryl N-arylbenzimidates into N-aroyldiphenylamines (reaction 110), systematically studied by

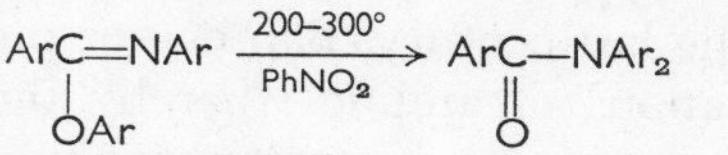

$$\underset{\underset{\text{OAr}}{|}}{\text{ArC}}\!\!=\!\!\text{NAr} \xrightarrow[\text{PhNO}_2]{200–300°} \underset{\underset{\text{O}}{\|}}{\text{ArC}}\!\!-\!\!\text{NAr}_2 \quad (110)$$

Chapman ($\sim$1925)[350, 351], provides an intramolecular (as proved by lack of cross over in concurrent reactions) route to diphenylamines. Electron-attracting substituents (NO_2, Cl) accelerate the internal nucleophilic substitution at the aryl ring by the nitrogen. Related thermal and acid-promoted displacements at an aliphatic centre have been studied[352].

The Smiles reaction (discovered 1931) (reaction 111) is a similar

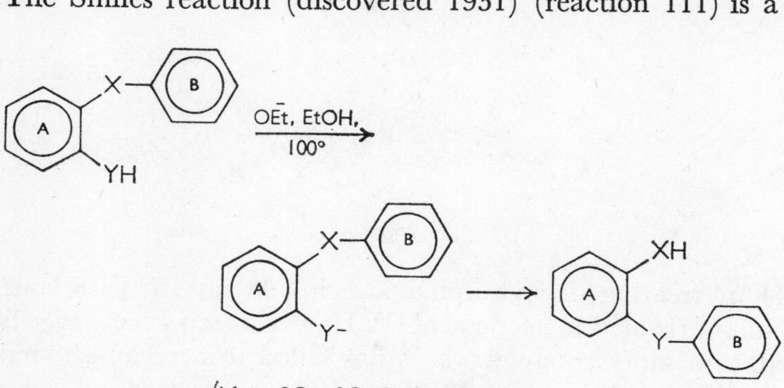

$$\text{(structure A, X, B, YH)} \xrightarrow[100°]{\bar{O}Et,\ EtOH,}$$

$$\text{(structure A, X, B, Y}^-\text{)} \longrightarrow \text{(structure A, XH, Y, B)} \quad (111)$$

$$\left(\begin{array}{l} X = SO_2,\ SO,\ S,\ O,\ SO_2O; \\ Y = NR,\ NOAc,\ NH,\ CONH,\ SO_2NH \end{array}\right)$$

intramolecular process[251, 353] with many possible permutations of Y and X, very few of which have been studied. A few proceed without activation, but usually an electronegative group is required at the 2- or 4-positions of ring B and a retro process has been characterised at pH 2 to 6 in some examples. A similar reaction, leading to fragmentation, is (112). The mechanism shown is supported by substituent effects on rate, and the low ΔS^* factor[354].

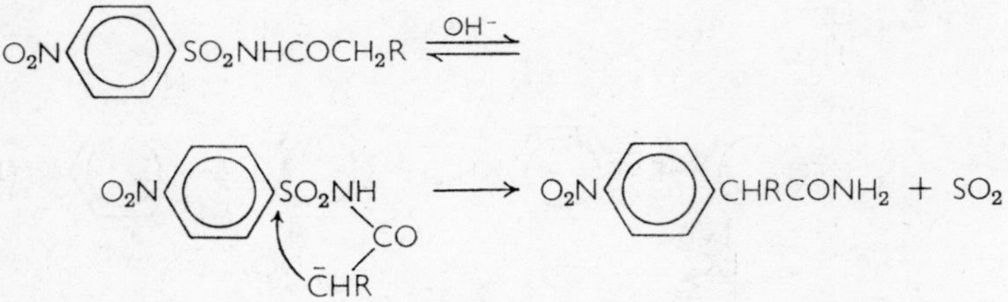

$$O_2N\langle\bigcirc\rangle SO_2NHCOCH_2R \xrightleftharpoons{OH^-}$$

$$O_2N\langle\bigcirc\rangle SO_2NH \longrightarrow O_2N\langle\bigcirc\rangle CHRCONH_2 + SO_2 \quad (112)$$

VI. OTHER REARRANGEMENTS

A. Rearrangement of Phenylsulphamic Acid and Related Compounds

At the end of the last century, Bamberger put forward a theory of 'indirect sulphonation' of aniline whereby the 'baking' of aniline sulphate gave rise to a series of rearrangements that can be dissected as reaction (113). In support he claimed to demonstrate steps II

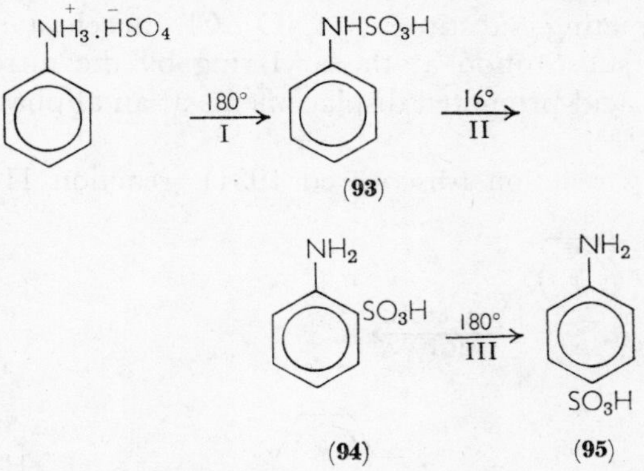

and III by treating phenylsulphamic acid (93) at 16° with concentrated sulphuric acid in acetic acid[355]. However step I has never been claimed, and more recent work[356] has failed to accomplish step II under any conditions: disproportionation occurred in all solvents except in dioxan at 100° when the conversion of 93 into 95 predominated. However, treatment, of phenylhydroxylamine with sulphur dioxide did yield **94**, in addition to **93**, and intramolecular *ortho* rearrangement similar to that proposed for the nitramine rearrangement of an initially formed sulphitoamine (**96**) is probable[355].

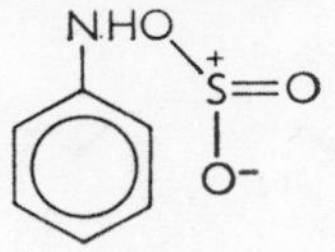

(**96**)

Tracer studies, using ^{34}S, indicate that the thermal isomerisation of 1-naphthylsulphamic acid to (mainly) 1-aminonaphthalene-2-sulphonic acid is intermolecular in the presence of sulphuric acid[357]— presumably involving fission to amine and sulphuric acid, and recombination—but largely intramolecular in its absence[358]. The situation is confused, for example a sequence analogous to **94→93→95** has been demonstrated in the naphthalene series[359], and requires further study. As before, observation of cross-sulphonation of additives during these reactions[360, 362] has no mechanistic significance.

A sulphamic acid rearrangement has been reported in the thiazole senes[361], and reactions (114) and (115), of unknown mechanism, are similar[362,363].

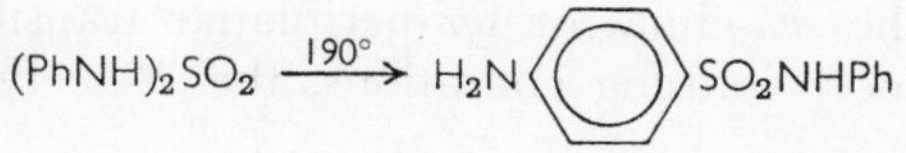

$$(PhNH)_2SO_2 \xrightarrow{190°} H_2N-\!\!\langle\bigcirc\rangle\!\!-SO_2NHPh \qquad (114)$$

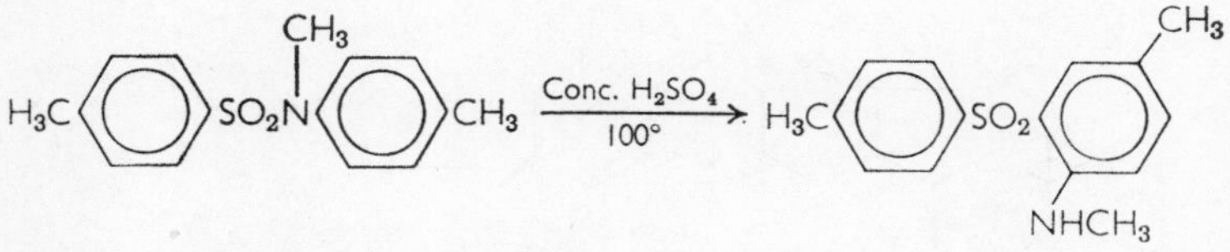

B. Acyl Migrations to and from Nitrogen

An $O \rightarrow N$ acyl shift (reaction 116) was discovered in 1883 and found not to occur with the acid salt, *meta* or *para* isomers, or the

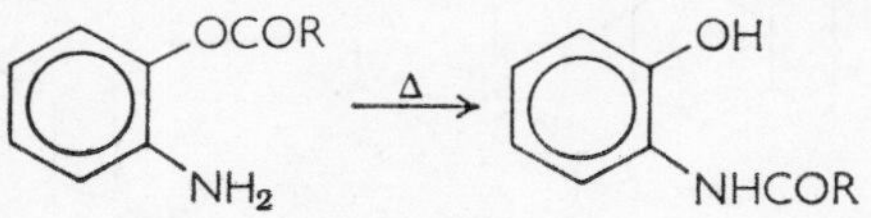

corresponding α-amino ether. Such reactions have since been found to be base promoted in contrast to the acid-catalysed reverse $N \rightarrow O$

shift. Detailed kinetics for these types of catalysis have not been reported, but two mechanisms, (117) and (118), (for clarity the role of catalyst is omitted) leading to retention or inversion at the oxygenated carbon are reasonable[364]. **97**, rather than the oxazoline

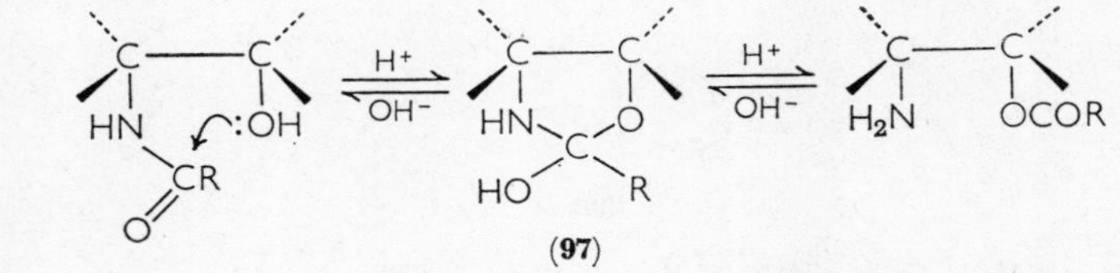

(117)

(97)

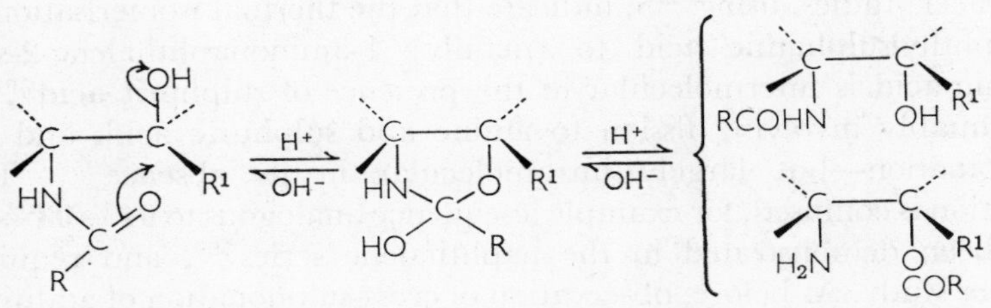

(118)

derived from it by dehydration, is probably an intermediate[365].

Reaction (117) accounts for the complete retention in configuration in rearrangements of *N*-benzoylephedrines, (**98**) and (**99**), in dilute ethanolic acid. **98** reacts some 70-fold faster than its isomer owing to *cis* effects in the *syn*-clinal or *syn*-periplanar transition states of the latter[364]. In more ionising conditions, the *N*-acetyl analogue of **99**

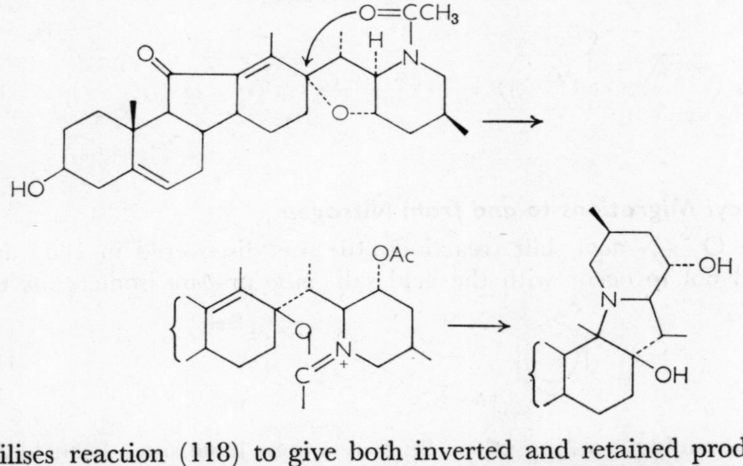

(119)

utilises reaction (118) to give both inverted and retained products. The same result is found for *o*-substituted benzamides when the ratio of inversion to retention increases as the *ortho* group is made larger. The retro $O \rightarrow N$ shift gives retained configurations in products from both isomers, under all investigated conditions. *cis-N*-Acetyl-2-aminocyclopentanol, in which the two reacting groups are held rigidly eclipsed, reacts as in (117)[365, 366], but the *trans* isomer only undergoes irreversible reaction with inversion under forcing conditions. The benzoates of both *cis*- and *trans*-2-aminocyclohexanols give $N \rightarrow O$ and $O \rightarrow N$ acyl shifts with retention, as the flexible geometry allows either to assume a conformation with the reacting groups *syn*-clinal[367].

Similar acyl shifts occur for derivatives of the β-hydroxyamino acids, serine and threonine. As incorporation of either of these into a polypeptide chain is equivalent to forming an *N*-derivative, it is not surprising that the amino groups of these acid residues are among the first to be liberated during mild acid hydrolysis of proteins[368]. Analogous shifts occur in glucosamines[369], steroids[370], and particularly alkaloids. Of interest in the last class is the ready transannular shift in **100**, but not in **101**, under specified conditions[371] (apparently reaction 118 is here inapplicable), and the far-reaching reorganisations in jervine acetate (reaction 119)[372]. $S \rightarrow N$ (reaction 120) and $N \rightarrow N$ acyl shifts have been reported at near neutral pH's in certain structures[373,374].

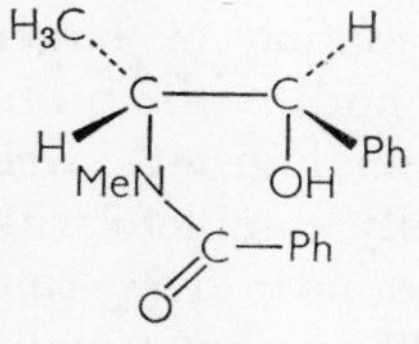

(98)

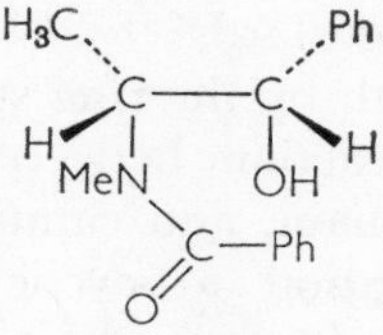

(99)

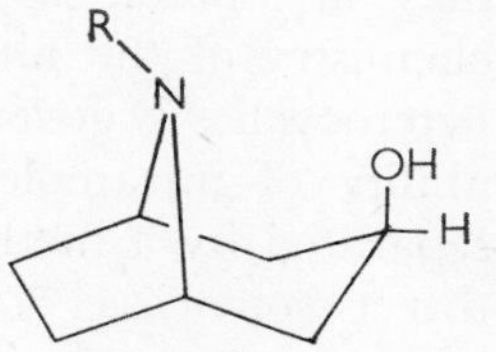

(100)

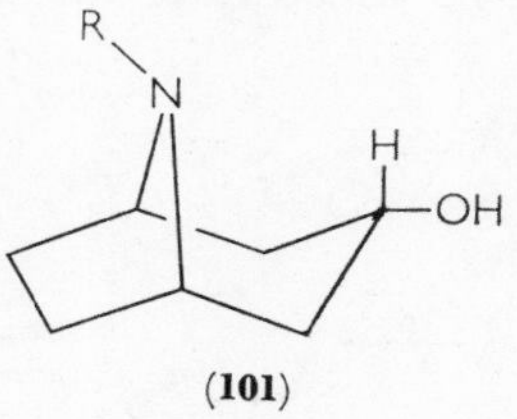

(101)

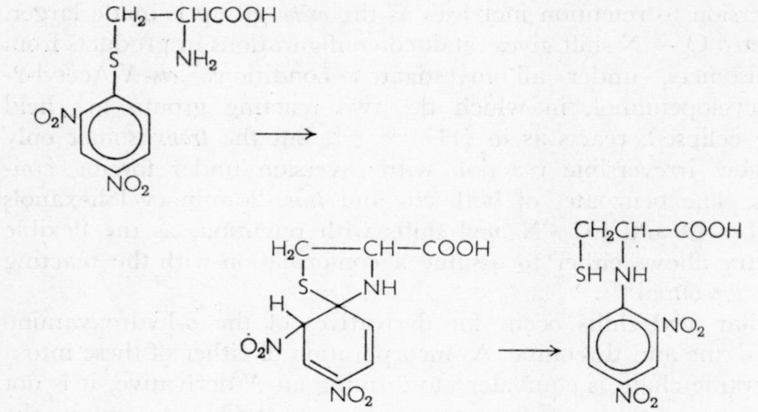

$$CH_2-CHCOOH \quad (with\ S,\ NH_2\ substituents\ on\ O_2N\text{-}NO_2\ benzene\ ring) \longrightarrow$$

$$\underset{NO_2}{\underset{|}{\overset{H_2C-----CH-COOH}{\underset{S}{|}\ \ \underset{NH}{|}}}} \quad \longrightarrow \quad \underset{NO_2}{\overset{CH_2CH-COOH}{\underset{SH\ \ NH}{|\ \ \ |}}} \quad (120)$$

C. Rearrangements Involving S_Ni Displacements by Amino Groups

Solvolysis of $X(CH_2)_nNH_2$ (X = Halogen, $n = 2$ to 6) under a variety of conditions is much faster (except for $n = 3$) than that of the corresponding alkyl halides, and is unaffected by added base or silver salts. Amino group participation in the rate-determining step occurs (reaction 121) to form a cyclic species which, for $n = 2$ or 3,

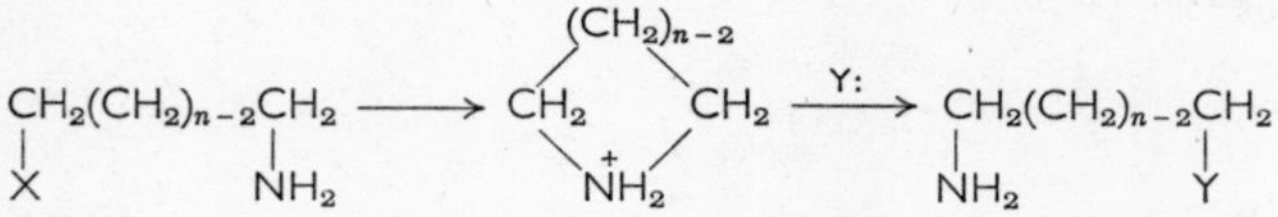

$$\underset{X}{\overset{CH_2(CH_2)_{n-2}CH_2}{|}} \longrightarrow \underset{\overset{+}{NH_2}}{\overset{(CH_2)_{n-2}}{CH_2}} \overset{Y:}{\longrightarrow} \underset{NH_2}{\overset{CH_2(CH_2)_{n-2}CH_2}{|}} \quad Y \quad (121)$$

rapidly ring opens; although 5- and 6-membered ring compounds may be isolable as quaternary salts. Many examples exist in the chemistry of the nitrogen mustards[251, 375-377]. Formation of larger heterocyclics is governed both by the ring stability and by the probability of intramolecular reaction between terminal groups well separated by a methylene chain, and cannot usually complete with direct solvolysis[376]. Application to cyclic compounds gives ring expansion (reaction 122)[251], but bicyclics, e.g. **102**, cannot usually sustain the strain involved in forming the intermediate and do not react[376].

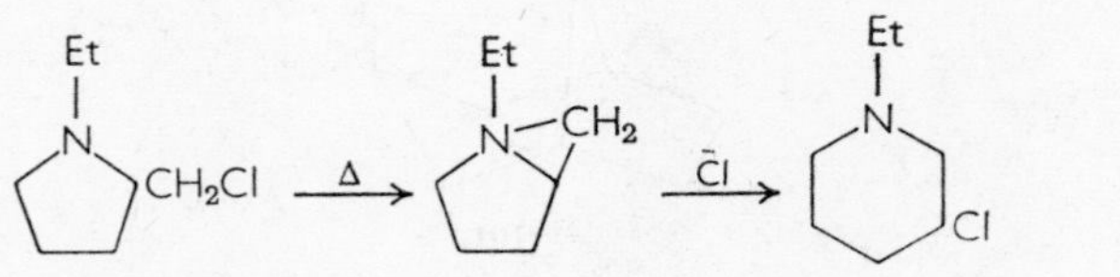

$$\underset{}{\overset{Et}{\underset{N}{|}}CH_2Cl} \overset{\Delta}{\longrightarrow} \underset{}{\overset{Et}{\underset{N}{|}}-CH_2} \overset{\bar{C}l}{\longrightarrow} \underset{Cl}{\overset{Et}{\underset{N}{|}}} \quad (122)$$

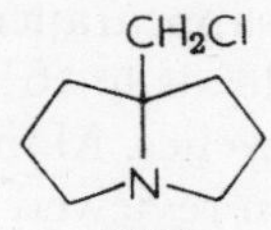

$$CH_2Cl$$

(102)

In many cases a primary substrate goes over to a secondary product, but this is not evidence for opening of the cyclic immonium ion to give a carbonium ion before product formation, it is merely an indication of the relative stabilities of product and substrate, and kinetic analysis and the application of the principle of microscopic reversibility indicates that the immonium ion is attacked by anion or solvent in a S_N2 step[378].

Amides possess feebly nucleophilic nitrogen atoms, but can readily be activated for cyclisation and rearrangement by base[379] (reaction 123).

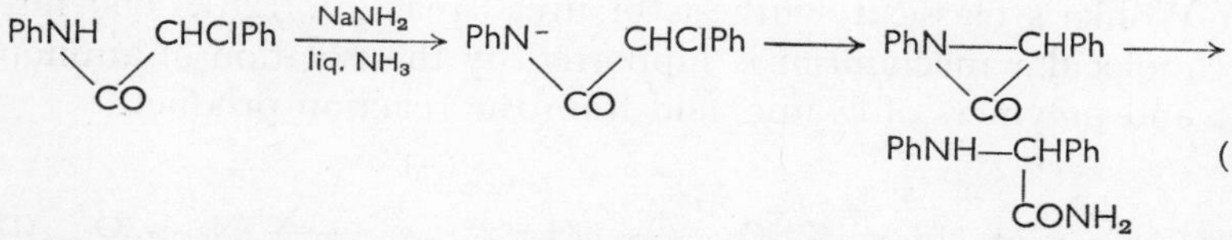

$$PhNH \quad CHClPh \xrightarrow[\text{liq. NH}_3]{\text{NaNH}_2} PhN^- \quad CHClPh \longrightarrow PhN\!-\!CHPh \longrightarrow$$
$$\underset{CO}{} \qquad\qquad \underset{CO}{}$$

$$PhNH\!-\!CHPh \qquad (123)$$
$$\underset{CONH_2}{|}$$

Transannular reactions (sometimes followed by ring opening) (reactions 124, 125, and 126) are well known[380-382]. The first type can

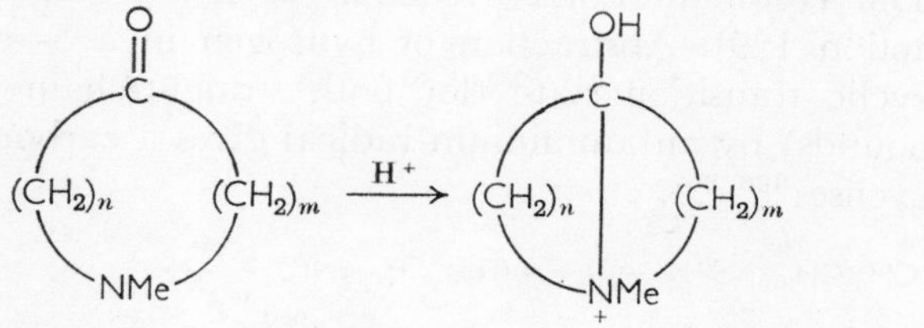

$$\text{(CH}_2)_n \quad \text{(CH}_2)_m \xrightarrow{H^+} \text{(CH}_2)_n \quad \text{(CH}_2)_m \qquad (124)$$

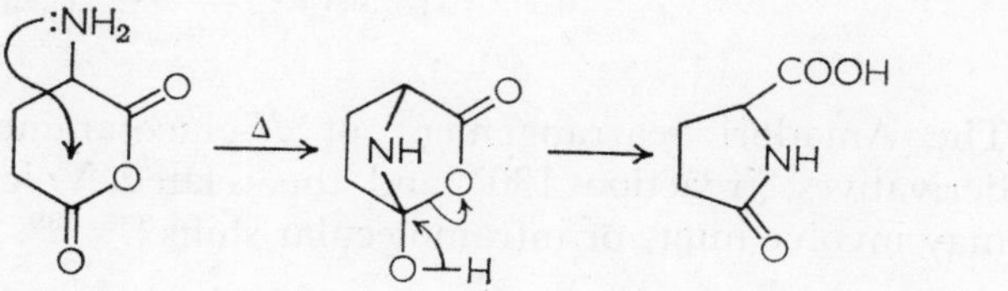

$$(125)$$

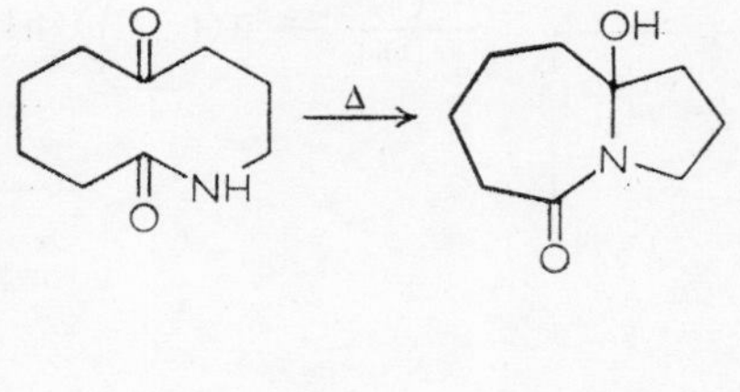

$$(126)$$

readily be followed by the disappearance of the carbonyl vibration in the infrared spectrum and leads to (6.6.0), (6.5.0), and (5.5.0), but not (7.6.0) or (7.5.0) bicyclics. Many similar cyclisations of biochemical importance have been reviewed[383]. An interesting analogous reaction is the 'amino nitrile' rearrangement (reaction 127)[384, 385].

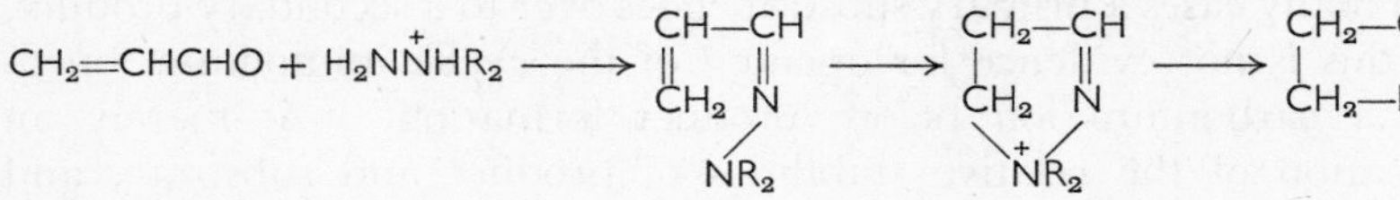

$$CH_2{=}CHCHO + H_2N\overset{+}{N}HR_2 \longrightarrow \begin{matrix} CH{-}CH \\ \| \quad \| \\ CH_2 \quad N \\ \diagup \\ NR_2 \end{matrix} \longrightarrow \begin{matrix} CH_2{-}CH \\ | \quad \| \\ CH_2 \quad N \\ \diagdown\overset{+}{N}R_2\diagup \end{matrix} \longrightarrow \begin{matrix} CH_2{-}C{\equiv}N \\ | \\ CH_2{-}NR_2 \end{matrix} \quad (127)$$

D. Miscellaneous Rearrangements

Some of the more important of the remaining reactions that come within our terms of reference are grouped here. Few have been adequately studied from the mechanistic viewpoint, and none fall into the previously considered classes.

(i) Wöhler's classical synthesis of urea (reaction 128). The intermolecular mechanism is supported by the isolation of ammonia and polymers of cyanic acid from the reaction product.

$$NH_4CNO \overset{\Delta}{\longrightarrow} NH_3 + HCNO \longrightarrow NH{=}C\overset{\displaystyle OH}{\underset{\displaystyle NH_2}{\diagup}} \longrightarrow (NH_2)_2CO \quad (128)$$

(ii) The Hofmann–Löffler reaction of N-halogenated amines (reaction 129). Abstraction of hydrogen in a 5- or 6-membered cyclic transition state (for both straight-chain or cyclic compounds) by an amminium radical gives a carbon radical which cyclises[386, 387].

$$RCH_2(CH_2)_nNR^1X \longrightarrow RCH_2(CH_2)_n\overset{\cdot}{N}R^1 \overset{H^+}{\underset{\Delta}{\longrightarrow}}$$

$$RCHX(CH_2)_n\overset{+}{N}H_2R^1 \overset{OH^-}{\longrightarrow} RCH(CH_2)_nNR^1 \quad (129)$$

$$(n = 3, 4)$$

(iii) The Amadori rearrangement of N-glucosamines to fructose derivatives, (reaction 130) and the related Voigt reaction[251], may involve inter- or intramolecular shifts[388, 389].

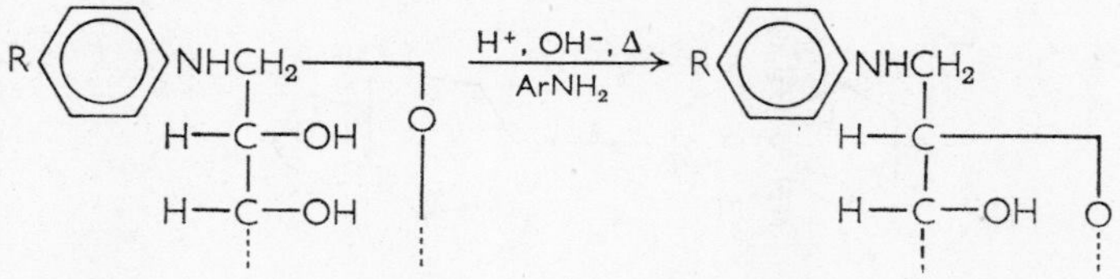

$$(130)$$

(iv) The 'amino ketone' rearrangement (reaction 131) involves intramolecular O-migration but intermolecular introduction of the amine moiety, as shown by 'crossing-over' of the latter during simultaneous rearrangement of two substrates[390].

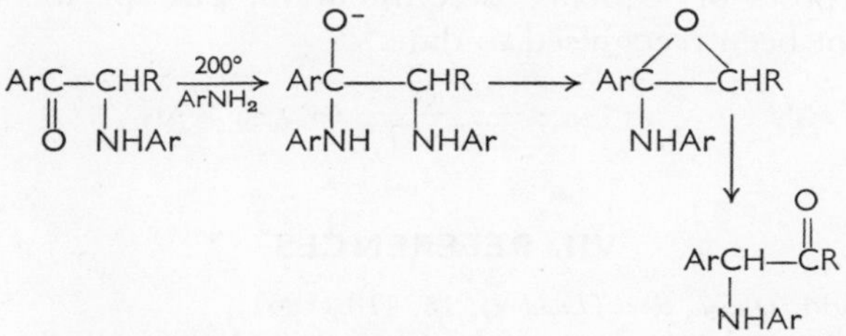

$$\text{ArC—CHR} \xrightarrow[\text{ArNH}_2]{200°} \text{ArC——CHR} \longrightarrow \text{ArC——CHR} \qquad (131)$$

with the scheme showing O^-, $NHAr$, $ArNH$, $NHAr$ and the final product

$$\text{ArCH—CR}$$
$$\text{NHAr}$$

(v) The Dimroth reaction (132) is facilitated by electron-withdrawing groups in the nucleus[391], and occurs for purines, pyrimidines[392], triazoles, and pteridines[391]. Similar ring fission and

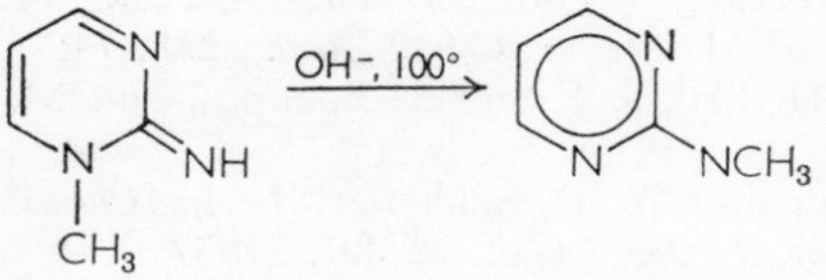

$$\xrightarrow{\text{OH}^-,\ 100°} \qquad (132)$$

(with pyrimidine ring bearing CH_3 and NH rearranging to NCH_3)

recyclisation accounts for the well-known 'hydantoin exchange' in amino acid derivatives[393] and the formation of nicotinic acid in the enzymic degradation of tryptophan[394]. A reverse type of rearrangement via cyclisation and ring opening is reaction (133) leading to sequence inversion in peptides[383, 395]. Such a

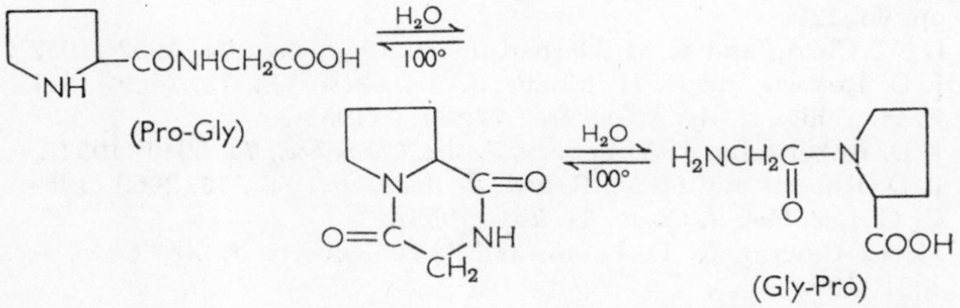

$$\text{—CONHCH}_2\text{COOH} \xrightleftharpoons[100°]{\text{H}_2\text{O}}$$

(Pro-Gly)

$$\xrightleftharpoons[100°]{\text{H}_2\text{O}} \text{H}_2\text{NCH}_2\text{C—N} \qquad (133)$$

with C=O, O=C, NH, CH$_2$ intermediate and COOH on product

(Gly-Pro)

phenomenon has not been demonstrated in the course of the determination of primary structure of typical polypeptides, but is well known for simpler peptides.

(vi) Transpeptidation: a variety of peptides have been shown to undergo interchange of units on treatment with certain proteolytic enzymes (reaction 134)[383]. Again, the same phenomenon is theoretically possible during the degradation of polypeptides for purposes of sequence determination, but specific examples have not been recognised to date.

$$\text{Gly-Val} + \text{Ala-Leu} \underset{25^\circ, \text{ pH } 7.2}{\overset{\text{Enzyme}}{\rightleftharpoons}} \text{Gly-Leu} + \text{Ala-Val} \qquad (134)$$

VII. REFERENCES

1. J. H. Ridd, *Quart. Rev. (London)*, **15**, 418 (1961).
2. P. S. Bailey and J. G. Burr, *J. Am. Chem. Soc.*, **75**, 2951 (1953).
3. D. J. Cram and J. E. McCarty, *J. Am. Chem. Soc.*, **79**, 2866 (1957).
4. D. Y. Curtin and M. C. Crew, *J. Am. Chem. Soc.*, **76**, 3719 (1954).
5. A. W. Fort and J. D. Roberts, *J. Am. Chem. Soc.*, **78**, 584 (1956).
6. R. Huisgen and C. Rüchardt, *Ann. Chem.*, **601**, 1 (1956).
7. R. Huisgen and C. Rüchardt, *Ann. Chem.*, **601**, 21 (1956).
8. L. S. Cieresko and J. G. Burr, *J. Am. Chem. Soc.*, **74**, 5431 (1952).
9. G. Darzens and M. Meyer, *Compt. Rend.*, **233**, 749 (1951).
10. M. Cherest, H. Felkin, J. Sicher, F. Sipos, and M. Tichy, *J. Chem. Soc.*, 2513 (1965).
11. A. Streitwieser and W. D. Schaeffer, *J. Am. Chem. Soc.*, **79**, 2888 (1957).
12. A. Streitwieser, *J. Org. Chem.*, **22**, 861 (1957).
13. H. C. Brown, *Chem. Brit.*, **2**, 199 (1966).
14. J. A. Berson in *Molecular Rearrangements*, (Ed. P. de Mayo) Interscience, New York, 1963, p. 111.
15. P. v. R. Schleyer, *J. Am. Chem. Soc.*, **86**, 1856 (1964).
16. M. J. S. Dewar and A. P. Marchand, *Ann. Rev. Phys. Chem.*, **16**, 321 (1965).
17. F. C. Whitmore and R. S. Thorpe, *J. Am. Chem. Soc.*, **63**, 1118 (1941).
18. D. W. Adamson and J. Kenner, *J. Chem. Soc.*, 838 (1934).
19. L. G. Cannell and R. W. Taft, *J. Am. Chem. Soc.*, **78**, 5812 (1956).
20. H. Zollinger, *Azo and Diazo Chemistry*, Interscience, New York, 1961, pp. 96, 129.
21. D. Y. Curtin and S. M. Gerber, *J. Am. Chem. Soc.*, **74**, 4052 (1952).
22. J. D. Roberts and R. H. Mazur, *J. Am. Chem. Soc.*, **73**, 2509 (1951).
23. E. H. White, *J. Am. Chem. Soc.*, **77**, 6011 (1955).
24. J. D. Roberts and J. A. Yancey, *J. Am. Chem. Soc.*, **74**, 5943 (1952).
25. J. D. Roberts and C. M. Regan, *J. Am. Chem. Soc.*, **75**, 2069 (1953).
26. C. C. Lee, *Can. J. Chem.*, **31**, 761 (1953).
27. M. C. Caserio, R. D. Levin, and J. D. Roberts, *J. Am. Chem. Soc.*, **87**, 5651 (1965).
28. J. D. Roberts and M. Halmann, *J. Am. Chem. Soc.*, **75**, 5759 (1953).
29. O. A. Reutov and T. N. Shatkina, *Tetrahedron*, **18**, 237 (1962).
30. O. A. Reutov and T. N. Shatkina, *Izv. Akad. Nauk SSSR, Otd. Khim. Nauk*, 2038 (1961).
31. G. J. Karabatsos and C. E. Orzech, *J. Am. Chem. Soc.*, **84**, 2838 (1962).

32. P. S. Skell and I. Starer, *J. Am. Chem. Soc.*, **84**, 3962 (1962).
33. C. C. Lee and J. E. Kruger, *J. Am. Chem. Soc.*, **87**, 3986 (1965).
34. C. C. Lee, J. E. Kruger, and E. W. C. Wong, *J. Am. Chem. Soc.*, **87**, 3985 (1965).
35. M. S. Silver, *J. Org. Chem.*, **28**, 1686 (1963).
36. A. A. Aboderin and R. L. Baird, *J. Am. Chem. Soc.*, **86**, 2300 (1964).
37. M. S. Silver, *J. Am. Chem. Soc.*, **82**, 2971 (1960).
38. A. Brodhag and C. R. Hauser, *J. Am. Chem. Soc.*, **77**, 3024 (1955).
39. W. A. Bonner and T. A. Putkey, *J. Org. Chem.*, **27**, 2348 (1962).
40. G. J. Karabatsos, C. E. Orzech, and S. Meyerson, *J. Am. Chem. Soc.*, **86**, 1994 (1964).
41. M. S. Silver, *J. Am. Chem. Soc.*, **83**, 3482 (1961).
42. W. H. Saunders, *J. Am. Chem. Soc.*, **78**, 6127 (1956).
43. A. W. Fort and R. E. Leary, *J. Am. Chem. Soc.*, **82**, 2497 (1960).
44. H. Söll in *Methoden der Organischen Chemie*, Vol. II(2), 4th edn. (Ed. Houben-Weyl), Thieme, Stuttgart, 1958, p. 133.
45. Y. Pocker, *Chem. Ind. (London)*, 322 (1959).
46. J. English and A. D. Bliss, *J. Am. Chem. Soc.*, **78**, 4057 (1956).
47. H. Favre and D. Gravel, *Can. J. Chem.*, **41**, 1452 (1963).
48. H. Felkin, *Compt. Rend.*, **234**, 2203 (1952).
49. H. Felkin, *Bull. Soc. Chim. France*, 1582 (1960).
50. H. I. Bernstein and F. C. Whitmore, *J. Am. Chem. Soc.*, **61**, 1324 (1939).
51. B. M. Benjamin, H. J. Schaeffer, and C. J. Collins, *J. Am. Chem. Soc.*, **79**, 6160 (1957).
52. B. M. Benjamin, P. Wilder, and C. J. Collins, *J. Am. Chem. Soc.*, **83**, 3654 (1961).
53. W. A. Bonner and D. D. Tanner, *J. Am. Chem. Soc.*, **80**, 1447 (1958).
54. A. Streitweiser, *Solvolytic Displacement Reactions*, McGraw Hill, New York, 1963, p. 170.
55. C. J. Collins, *Quart. Rev. (London)*, **14**, 357 (1960).
56. C. J. Collins, *Advan. Phys. Org. Chem.*, **2**, 46 (1964).
57. J. W. Huffmann and R. P. Eliot, *J. Org. Chem.*, **30**, 365 (1965).
58. C. J. Collins, J. B. Christie, and V. F. Raaen, *J. Am. Chem. Soc.*, **83**, 4267 (1961).
59. J. C. Martin and W. G. Bentude, *J. Org. Chem.*, **24**, 1902 (1959).
60. C. J. Collins and B. M. Benjamin, *J. Am. Chem. Soc.*, **85**, 2519 (1963).
61. C. J. Collins, B. M. Benjamin, and M. H. Lietze, *Ann. Chem.*, **687**, 150 (1965).
62. E. B. Wilson, *Proc. Natl. Acad. Sci.*, **43**, 816 (1957).
63. C. R. Hauser and D. N. van Eenam, *J. Am. Chem. Soc.*, **79**, 6274 (1957).
64. E. H. White, *J. Am. Chem. Soc.*, **77**, 6014 (1955).
65. B. M. Benjamin and C. J. Collins, *J. Am. Chem. Soc.*, **78**, 4952 (1956).
66. P. Warrick and W. H. Saunders, *J. Am. Chem. Soc.*, **84**, 4095 (1962).
67. D. Y. Curtin and M. C. Crew, *J. Am. Chem. Soc.*, **77**, 354 (1955).
68. V. F. Raaen and C. J. Collins, *J. Am. Chem. Soc.*, **80**, 1409 (1958).
69. A. McKenzie and W. S. Dennler, *J. Chem. Soc.*, **125**, 2105 (1924).
70. H. O. House and G. J. Grubb, *J. Am. Chem. Soc.*, **81**, 4733 (1959).
71. P. A. S. Smith and D. R. Baer, *Org. Reactions*, **11**, 157 (1960).
72. B. Tchoubar, *Bull. Soc. Chim. France*, 164 (1949).

73. L. I. Zakharin and V. V. Norneva, *Zh. Org. Khim.*, **1**, 1602 (1965).
74. P. A. S. Smith and D. R. Baer, *J. Am. Chem. Soc.*, **24**, 6135 (1952).
75. P. A. S. Smith, D. R. Baer, and S. N. Ege, *J. Am. Chem. Soc.*, **76**, 4564 (1954).
76. O. A. Reutov and T. N. Shatkina, *Dokl. Akad. Nauk SSSR*, **142**, 835 (1962).
77. R. Kotari, *J. Org. Chem.*, **30**, 350 (1965).
78. J. Diamond, W. F. Bruce, and F. T. Tyson, *J. Org. Chem.*, **30**, 1840 (1965).
79. D. V. Nightingale, J. D. Kerr, J. A. Gallacher, and M. Maienthal, *J. Org. Chem.*, **17**, 1017 (1952).
80. M. Maxim, M. Avram, and C. D. Nenitescu, *Acad. Rep. Populare Romine, Studii Cercetari Chim.*, **8**, 187 (1960); *Chem. Abstr.*, **55**, 9327h (1961).
81. J. Baer, *J. Org. Chem.*, **23**, 1560 (1958).
82. I. Elphimoff-Felkin and B. Tchoubar, *Compt. Rend.*, **231**, 1314 (1950).
83. I. Elphimoff-Felkin and B. Tchoubar, *Compt. Rend.*, **233**, 964 (1951).
84. I. Elphimoff-Felkin and Y. Gault, *Compt. Rend.*, **246**, 1871 (1958).
85. L. N. I. Putokin and V. S. Egorova, *Dokl. Akad. Nauk SSSR*, **96**, 293 (1954).
86. J. Hora, V. Cerny, and F. Sorm, *Tetrahedron Letters*, 501 (1962).
87. W. G. Dauben and P. Lang, *Tetrahedron Letters*, 453 (1962).
88. C. Dupin and R. Fraisse-Jullien, *Bull. Soc. Chim. France*, 1993 (1964).
89. E. J. Corey and R. F. Atkinson, *J. Org. Chem.*, **29**, 3703 (1964).
90. J. E. Hodgkins and R. J. Flores, *J. Org. Chem.*, **28**, 3356 (1963).
91. H. Hart and R. A. Martin, *J. Am. Chem. Soc.*, **82**, 6362 (1960).
92. R. Anliker, O. Rohr, and H. Heusser, *Helv. Chim. Acta*, **38**, 1171 (1955).
93. G. E. McCasland, *J. Am. Chem. Soc.*, **73**, 2293 (1951).
94. M. Mousseron and R. Granger, *Bull. Soc. Chim. France*, 850 (1947).
95. P. B. Talukdar and P. E. Fanta, *J. Org. Chem.*, **24**, 555 (1959).
96. J. W. Huffmann and J. E. Engle, *J. Org. Chem.*, **24**, 1844 (1959).
97. O. E. Edwards and M. Lesage, *Can. J. Chem.*, **41**, 1542 (1963).
98. R. Granger, H. Tecker, and A. Massiau, *Compt. Rend.*, **250**, 4378 (1960).
99. D. E. Applequist and J. P. Kleiman, *J. Org. Chem.*, **26**, 2178 (1961).
100. J. A. Berson, reference 14, p. 168.
101. R. H. Mazur, W. N. White, D. A. Semenov, C. C. Lee, M. S. Silver, and J. D. Roberts, *J. Am. Chem. Soc.*, **81**, 4390 (1959).
102. C. Dupin and R. Fraisse-Jullien, *Bull. Soc. Chim. France*, 1997 (1964).
103. M. S. Silver, M. C. Caserio, H. E. Rice, and J. D. Roberts, *J. Am. Chem. Soc.*, **83**, 3671 (1961).
104. E. F. Cox, M. C. Caserio, M. S. Silver, and J. D. Roberts, *J. Am. Chem. Soc.*, **83**, 2719 (1961).
105. G. E. Cartier and S. C. Bunce, *J. Am. Chem. Soc.*, **85**, 932 (1963).
106. J. D. Roberts, C. C. Lee, and W. H. Saunders, *J. Am. Chem. Soc.*, **77**, 3034 (1955).
107. J. D. Roberts, C. C. Lee, and W. H. Saunders, *J. Am. Chem. Soc.*, **76**, 4501 (1954).
108. E. J. Corey, J. Casanova, P. A. Vatakencherry, and R. Winter, *J. Am. Chem. Soc.*, **85**, 169 (1963).
109. J. A. Berson and A. Remanick, *J. Am. Chem. Soc.*, **86**, 1749 (1964).
110. H. L. Goering and M. F. Sloan, *J. Am. Chem. Soc.*, **83**, 1992 (1961).
111. J. A. Berson and P. Reynolds-Warnhoff, *J. Am. Chem. Soc.*, **86**, 595 (1964).
112. J. A. Berson and D. Willner, *J. Am. Chem. Soc.*, **86**, 609 (1964).

113. J. Meinwald, P. G. Gassmann, and J. J. Hurst, *J. Am. Chem. Soc.*, **84**, 3722 (1962).
114. W. Hückel and P. Rieckmann, *Ann. Chem.*, **625**, 1 (1959).
115. D. G. Morris, *Ph.D. Thesis*, University of London, 1964.
116. W. Hückel and U. Ströle, *Ann. Chem.*, **585**, 182 (1954).
117. J. Burgess, *Ph.D. Thesis*, University of London, 1953.
118. D. S. Noyce, *J. Am. Chem. Soc.*, **72**, 924 (1950).
119. E. J. Corey and R. L. Dawson, *J. Am. Chem. Soc.*, **85**, 1782 (1963).
120. R. R. Sauers and J. A. Beisler, *Tetrahedron Letters*, 2181 (1964).
121. D. Semenov, C. H. Shih, and W. G. Young, *J. Am. Chem. Soc.*, **80**, 5472 (1958).
122. R. H. De Wolfe, and W. G. Young, *Chem. Rev.*, **56**, 753 (1956).
123. M. Hanack and H. J. Schneider, *Tetrahedron*, **20**, 1863 (1964).
124. J. Cologne and H. Daunis, *Compt. Rend.*, **251**, 1080 (1960).
125. W. C. Wildman and D. R. Saunders, *J. Am. Chem. Soc.*, **76**, 946 (1954).
126. V. Prelog and J. G. Traynham in *Molecular Rearrangements*, (Ed. P. de Mayo), Interscience, New York, 1963, p. 593.
127. W. Nagata, T. Sugasawa, Y. Hayase, and K. Sasakura, *Proc. Chem. Soc.*, 241 (1964).
128. A. T. Austin and J. Howard, *Chem. Ind. (London)*, 1413 (1959).
129. W. Hückel, *J. Prakt. Chem.*, **28**, 27 (1965).
130. P. H. Leake, *Chem. Rev.*, **56**, 27 (1956).
131. D. F. De Tar. *Org. Reactions*, **9**, 410 (1957).
132. E. S. Lewis and J. M. Insole, *J. Am. Chem. Soc.*, **86**, 32 (1964).
133. M. Hanack and G. H. Allmendinger, *Chem. Ber.*, **97**, 1669 (1964).
134. M. Godchot and M. Mousseron, *Compt. Rend.*, **198**, 2000 (1934).
135. K. Gabriel and M. Colman, *Chem. Ber.*, **33**, 980, 2630 (1900).
136. T. S. Stevens, E. M. Creighton, A. B. Gordon, and M. MacNicol, *J. Chem. Soc.*, 3193 (1928).
137. C. K. Ingold, *Structure and Mechanism in Organic Chemistry*, Bell, London 1953, p. 524.
138. F. Möller in reference 44, p. 905.
139. T. Thomson and T. S. Stevens, *J. Chem. Soc.*, 1932 (1932).
140. A. T. Babayan, M. G. Indzhikyan, and R. A. Aivazova, *Zh. Obshch. Khim.*, **33**, 1773 (1963).
141. A. T. Babayan, M. G. Indzhikyan, and G. B. Bagdasaryan, *Zh. Obshch. Khim.*, **34**, 411 (1964).
142. G. C. Jones, W. Q. Beard, and C. R. Hauser, *J. Org. Chem.*, **28**, 199 (1963).
143. G. Wittig, R. Mangold, and G. Felletschin, *Ann. Chem.*, **560**, 116 (1948).
144. W. R. Bamford and T. S. Stevens, *J. Chem. Soc.*, 4675 (1952).
145. G. Wittig, H. TenHaeff, W. Schoch, and G. Koenig, *Ann. Chem.*, **572**, 1 (1951).
146. G. Wittig, G. Koenig, and K. Clauss, *Ann. Chem.*, **593**, 127 (1955).
147. G. Wittig and H. Sommer, *Ann. Chem.*, **594**, 1 (1955).
148. C. R. Hauser and S. W. Kantor, *J. Am. Chem. Soc.*, **73**, 1437 (1951).
149. T. Thomson and T. S. Stevens, *J. Chem. Soc.*, 55 (1932).
150. C. E. Boozer and E. S. Lewis, *J. Am. Chem. Soc.*, **75**, 3182 (1953).
151. D. V. Banthorpe, *Elimination Reactions*, Elsevier, Amsterdam, 1963, p. 101.

662   D. V. Banthorpe

152. M. G. Indzhikyan and A. T. Babayan, *Izv. Akad. Nauk Arm. SSR Ser. Khim. Nauk*, **10**, 411 (1957); *Chem. Abstr.*, **52**, 16256f (1958).
153. R. A. W. Johnstone and T. S. Stevens, *J. Chem. Soc.*, 4487 (1955).
154. A. Campbell, A. H. J. Houston, and J. Kenyon, *J. Chem. Soc.*, 93 (1947).
155. J. H. Brewster and M. W. Kline, *J. Am. Chem. Soc.*, **74**, 5179 (1952).
156. B. J. Millard and T. S. Stevens, *J. Chem. Soc.*, 3397 (1963).
157. F. Jenny and J. Druey, *Angew. Chem. Intern. Ed. Engl.*, **1**, 155 (1962).
158. H. Hellmann and G. M. Scheytt, *Ann. Chem.*, **654**, 39 (1962).
159. R. M. Hill and T. H. Chan, *J. Am. Chem. Soc.*, **88**, 866 (1966).
160. K. Mislow and J. Joshua, *J. Am. Chem. Soc.*, **87**, 666 (1965).
161. M. Sommelet, *Compt. Rend.*, **205**, 56 (1937).
162. A. J. Beckwith, *Ph.D. Thesis*, University of London, 1955.
163. E. D. Hughes and C. K. Ingold, *J. Chem. Soc.*, 69 (1933).
164. G. E. Hilbert and L. A. Pinck, *J. Am. Chem. Soc.*, **60**, 494 (1938).
165. S. W. Kantor and C. R. Hauser, *J. Am. Chem. Soc.*, **73**, 4122 (1951).
166. C. R. Hauser, D. N. van Eenam, and P. L. Bayless, *J. Org. Chem.*, **23**, 358 (1958).
167. W. Q. Beard and C. R. Hauser, *J. Org. Chem.*, **25**, 334 (1960).
168. W. H. Puterbaugh and C. R. Hauser, *J. Am. Chem. Soc.*, **86**, 1394 (1964).
169. C. R. Hauser, J. K. Lindsay, and D. Lednicer, *J. Org. Chem.*, **23**, 358 (1958).
170. Y. A. Ustynok and E. G. Perevalova, *Izv. Akad. Nauk SSSR*, 62 (1964).
171. C. R. Hauser, D. N. van Eenam, and P. L. Bayless, *J. Org. Chem.*, **23**, 354 (1958).
172. C. R. Hauser and D. N. van Eenam, *J. Am. Chem. Soc.*, **79**, 6280 (1957).
173. C. R. Hauser and G. C. Jones, *J. Org. Chem.*, **27**, 3572 (1962).
174. F. N. Jones and C. R. Hauser, *J. Org. Chem.*, **27**, 4020 (1962).
175. C. L. Bumgardener, *J. Am. Chem. Soc.*, **85**, 73 (1963).
176. D. J. Cram, *Fundamentals of Carbanion Chemistry*, Academic Press, New York, 1965, p. 272.
177. C. R. Hauser and D. N. van Eenam, *J. Am. Chem. Soc.*, **79**, 5512 (1957).
178. W. H. Puterbaugh and C. R. Hauser, *J. Am. Chem. Soc.*, **86**, 1105 (1964).
179. N. J. Jones and C. R. Hauser, *J. Org. Chem.*, **26**, 2979 (1961).
180. A. R. Lepley and R. H. Becker, *J. Org. Chem.*, **30**, 3888 (1965).
181. W. Q. Beard and C. R. Hauser, *J. Org. Chem.*, **26**, 371 (1961).
182. C. R. Hauser, W. Q. Beard, and D. N. van Eenam, *J. Org. Chem.*, **26**, 2062 (1961).
183. C. R. Hauser, S. W. Kantor, and W. R. Braser, *J. Am. Chem. Soc.*, **75**, 2660 (1953).
184. G. Wittig and H. Streib, *Ann. Chem.*, **584**, 1 (1953).
185. H. Zimmerman in *Molecular Rearrangements* (Ed. P. De Mayo), Interscience, New York, 1963, p. 407.
186. J. Schmitz, *Helv. Chim. Acta*, **38**, 1712 (1955).
187. H. Dahn and U. Solms, *Helv. Chim. Acta*, **34**, 907 (1951).
188. W. F. Cockburn, R. A. W. Johnstone, and T. S. Stevens, *J. Chem. Soc.*, 3340 (1960).
189. J. Meisenheimer, H. Greeske, and A. Willmersdorf, *Chem. Ber.*, **55**, 513 (1922).
190. A. H. Wragg, T. S. Stevens, and D. M. Astlee, *J. Chem. Soc.*, 4057 (1958).

191. R. F. Kleinschmidt and A. C. Cope, *J. Am. Chem. Soc.*, **66**, 1929 (1944).
192. A. C. Cope and P. H. Toule, *J. Am. Chem. Soc.*, **71**, 3423 (1949).
193. U. Schoellkopf, M. Patsch, and H. Shaefer, *Tetrahedron Letters*, 2512 (1964).
194. G. P. Shulman, P. Ellgen, and M. Conner, *Can. J. Chem.*, **43**, 3459 (1965).
195. M. Polonovski and M. Polonovski, *Bull. Soc. Chim. France*, **41**, 1190 (1927).
196. J. C. Craig, F. D. Dwyer, A. N. Glazer and E. C. Horning, *J. Am. Chem. Soc.*, **83**, 1871 (1961).
197. C. C. Sweeley and E. C. Horning, *J. Am. Chem. Soc.*, **79**, 2620 (1957).
198. J. C. Craig, N. Y. Mary, N. L. Goodman, and L. Wolff, *J. Am. Chem. Soc.*, **86**, 3866 (1964).
199. R. Huisgen, F. Bayerlin, and W. Heydkamp, *Chem. Ber.*, **92**, 3223 (1959).
200. J. H. Markgraf and C. G. Carson, *J. Am. Chem. Soc.*, **86**, 3699 (1964).
201. R. D. Haworth and W. H. Perkins, *J. Chem. Soc.*, 1769 (1926).
202. A. C. Cope and A. C. Haven, *J. Am. Chem. Soc.*, **72**, 4896 (1950).
203. W. Z. Heldt and L. G. Donaruma, *Org. Reactions*, **11**, 1 (1960).
204. P. A. S. Smith in *Molecular Rearrangements* (Ed. P. de Mayo), Interscience, New York, 1963, p. 457.
205. N. Tokura, R. Tada, and K. Suzuki, *Bull. Chem. Soc. Japan*, **32**, 654 (1959)
206. P. T. Lansbury and N. R. Mancuso, *J. Am. Chem. Soc.*, **88**, 1205 (1966).
207. R. Huisgen, J. Witte, H. Walz, and W. Jira, *Ann. Chem.*, **604**, 191 (1957).
208. P. J. McNulty and D. E. Pearson, *J. Am. Chem. Soc.*, **81**, 612 (1959).
209. R. Huisgen, J. Witte, and I. Ugi, *Chem. Ber.*, **90**, 1844 (1957).
210. A. Stephen and B. Staskin, *J. Chem. Soc.*, 980 (1956).
211. W. Z. Heldt, *J. Am. Chem. Soc.*, **80**, 5927 (1958).
212. B. Jones, *Nature*, **157**, 519 (1946).
213. C. A. Grob, H. P. Fischer, W. Raudenbusch, and J. Zergenyi, *Helv. Chim. Acta*, **47**, 1003 (1964).
214. R. K. Hill, R. T. Conley, and O. T. Chortyk, *J. Am. Chem. Soc.*, **87**, 5646 (1965).
215. R. T. Conley and R. J. Lange, *J. Org. Chem.*, **28**, 210 (1963).
216. C. O'Brien, *Chem. Rev.*, **64**, 81 (1964).
217. M. J. Hatch and D. J. Cram, *J. Am. Chem. Soc.*, **75**, 38 (1953).
218. G. H. Alt and W. S. Knowles, *J. Org. Chem.*, **25**, 2047 (1960).
219. P. A. S. Smith and G. E. Most, *J. Org. Chem.*, **22**, 358 (1957).
220. J. H. Amin and P. de Mayo, *Tetrahedron Letters*, 1585 (1963).
221. M. W. Partridge and H. A. Turner, *J. Pharm. Pharmacol.*, **5**, 103 (1953).
222. E. S. Wallis and J. F. Lane, *Org. Reactions*, **3**, 267 (1946).
223. A. Campbell and J. Kenyon, *J. Chem. Soc.*, 25 (1946).
224. T. J. Prosser and E. L. Eliel, *J. Am. Chem. Soc.*, **79**, 2544 (1957).
225. J. A. Berson and D. A. Ben-Efraim, *J. Am. Chem. Soc.*, **81**, 4094 (1959).
226. C. R. Hauser and S. W. Kantor, *J. Am. Chem. Soc.*, **72**, 4284 (1950).
227. R. A. Abramovitch and B. A. Davis, *Chem. Rev.*, **64**, 149 (1964).
228. K. K. Brower, *J. Am. Chem. Soc.*, **83**, 4370 (1961).
229. F. Mathis, *Bull. Soc. Chim. France*, D9 (1953).
230. P. A. S. Smith, *Org. Reactions*, **3**, 337 (1946).
231. Y. Yukawa and T. Tsuno, *J. Am. Chem. Soc.*, **79**, 5530 (1957).
232. A. A. Bothner-By and L. Friedman, *J. Am. Chem. Soc.*, **73**, 5391 (1951).
233. E. Fahr and Z. Neumann, *Angew. Chem.*, **77**, 591 (1965).
234. W. Lwowski and E. Scheiffele, *J. Am. Chem. Soc.*, **87**, 4357 (1965).

235. H. Wolff, *Org. Reactions*, **3**, 307 (1946).
236. K. G. Rutherford and M. S. Newman, *J. Am. Chem. Soc.*, **79**, 213 (1957).
237. E. Leiber, R. A. Henry, and W. G. Finnegan, *J. Am. Chem. Soc.*, **75**, 2023 (1953).
238. W. T. Reichle, *Inorg. Chem.*, **3**, 402 (1964).
239. S. Uyeo, *Pure Appl. Chem.*, **7**, 269 (1963).
240. C. C. Lee, G. P. Slater, and J. W. T. Spinks, *Can. J. Chem.*, **35**, 276 (1957).
241. P. T. Lansbury and N. R. Marcuso, *Tetrahedron Letters*, 2445 (1965).
242. P. A. S. Smith and E. P. Antoniades, *Tetrahedron Letters*, 210 (1960).
243. A. T. Blomquist and R. D. Spencer, *J. Am. Chem. Soc.*, **70**, 30 (1948).
244. S. Götzky, *Chem. Ber.*, **64**, 1555 (1931).
245. D. E. Pearson, K. N. Carter, and C. M. Greer, *J. Am. Chem. Soc.*, **75**, 5905 (1953).
246. K. Dietzsch, *J. Prakt. Chem.*, **27**, 34 (1965).
247. D. H. R. Barton and L. R. Morgan, *J. Chem. Soc.*, 622 (1962).
248. O. E. Edwards, D. Vocell, J. W. ApSimon, and F. Hague, *J. Am. Chem. Soc.*, **87**, 678 (1965).
249. W. H. Saunders and E. A. Caress, *J. Am. Chem. Soc.*, **86**, 861 (1964).
250. L. Horner and A. Christmann, *Chem. Ber.*, **96**, 388 (1963).
251. F. Möller in *Methoden der Organischen Chemie*, Vol. II(2), 4th ed. (Ed. Houben-Weyl), Thieme, Stuttgart, 1958, p. 876.
252. Reference 20, p. 182.
253. H. V. Kidd, *J. Org. Chem.*, **2**, 198 (1937).
254. H. Goldschmidt, S. Johnsen, and E. Overwein, *Z. Physik. Chem.* (*Leipzig*), **110**, 251 (1924).
255. K. Clausius and H. R. Weisser, *Helv. Chim. Acta.*, **35**, 1524 (1952).
256. D. A. Archer and H. Booth, *Chem. Ind.* (*London*), 1570 (1962).
257. J. H. Brewster, *Org. Reactions*, **7**, 135 (1953).
258. Reference 137, p. 615.
259. W. J. Hickinbottom, *J. Chem. Soc.*, 404 (1937).
260. M. J. Molera, J. M. Gamboa, M. Del Val Cob, and N. Ortin, *Anales Real Soc. Españ. Fis. Quim.* (*Madrid*), **59B**, 319 (1963).
261. M. J. S. Dewar in *Molecular Rearrangements* (Ed. P. de Mayo), Interscience, New York, 1963, p. 295.
262. Y. Ogata, T. Tabuchi, and K. Yoshida, *Tetrahedron*, **20**, 747 (1964).
263. G. J. Russell, R. D. Topson, and J. Vaughan, *Chem. Commun.*, 529 (1965).
264. H. Hart and J. R. Kosak, *J. Org. Chem.*, **27**, 116 (1962).
265. J. Reilly and W. J. Hickinbottom, *J. Chem. Soc.*, **117**, 103 (1920).
266. W. J. Hickinbottom, *J. Chem. Soc.*, 1119 (1937).
267. M. J. S. Dewar, *Electronic Theory of Organic Chemistry*, Oxford, London, 1949, p. 300.
268. O. K. Nikiforova, *Tr. Khim-Met. Inst. Akad. Nauk SSSR*, **1**, 53 (1953); *Chem. Abstr.*, **49**, 8158i (1955).
269. O. Czeija, *Austrian Pat.*, 163,203 (1949); *Chem. Abstr.*, **46**, 9125a (1952).
270. W. J. Hickinbottom, *J. Chem. Soc.*, 1070 (1933).
271. J. Willenz, *J. Chem. Soc.*, 1677 (1955).
272. K. J. P. Orton, F. G. Soper and G. Williams, *J. Chem. Soc.*, 998 (1928).
273. A. R. Olson and J. C. Hornel, *J. Org. Chem.*, **3**, 76 (1938).
274. C. Beard and W. J. Hickinbottom, *Chem. Ind.* (*London*), 1421 (1957).

275. R. P. Bell and P. V. Danckwerts, *J. Chem. Soc.*, 1774 (1939).
276. G. C. Israel, A. W. N. Tuck, and F. G. Soper, *J. Chem. Soc.*, 547 (1945).
277. J. C. Chan and G. C. Israel, *J. Chem. Soc.*, 196 (1960).
278. M. J. S. Dewar and J. M. W. Scott, *J. Chem. Soc.*, 1445 (1957).
279. M. J. S. Dewar and J. M. W. Scott, *J. Chem. Soc.*, 2676 (1957).
280. J. M. W. Scott, *Can. J. Chem.*, **38**, 2441 (1960).
281. M. J. S. Dewar in *Theoretical Organic Chemistry*, Butterworths, London, 1959, p. 217.
282. J. M. W. Scott and J. C. Martin, *Can. J. Chem.*, **43**, 732 (1965).
283. K. N. Ayad, C. Beard, R. F. Garwood, and W. J. Hickinbottom, *J. Chem. Soc.*, 2981 (1957).
284. P. Haberfield and D. Paul, *J. Am. Chem. Soc.*, **87**, 5502 (1965).
285. C. M. Veerma and R. C. Srivastava, *Ind. J. Chem.*, **3**, 266 (1965).
286. E. Bamberger, *Ann. Chem.*, **441**, 207 (1925).
287. R. Kijana and S. Minomura, *Rev. Phys. Chem. Japan*, **22**, 9 (1952).
288. H. E. Heller, E. D. Hughes, and C. K. Ingold, *Nature*, **168**, 909 (1951).
289. Y. Yukawa, *J. Chem. Soc. Japan*, **71**, 603 (1950).
290. E. Boyland and R. Nery, *J. Chem. Soc.*, 5217 (1962).
291. L. Horner and H. Steppan, *Ann. Chem.*, **606**, 24 (1957).
292. E. Boyland and R. Nery, *Biochem. J.*, **91**, 362 (1964).
293. J. F. J. Dippy and V. Moss, *J. Chem. Soc.*, 2205 (1952).
294. A. Mustafa and M. I. Ali, *J. Am. Chem. Soc.*, **77**, 4593 (1955).
295. D. Elad, D. V. Rao, and V. I. Stenberg, *J. Org. Chem.*, **30**, 3252 (1956).
296. E. Koenigs and J. Freund, *Chem. Ber.*, **80**, 143 (1947).
297. D. V. Banthorpe, E. D. Hughes, and D. L. H. Williams, *J. Chem. Soc.*, 5349 (1964).
298. D. V. Banthorpe and J. A. Thomas, *J. Chem. Soc.*, 7149 (1965).
299. W. N. White, D. Lazdins, and H. S. White, *J. Am. Chem. Soc.*, **86**, 1517 (1964).
300. D. V. Banthorpe, J. A. Thomas, and D. L. H. Williams, *J. Chem. Soc.*, 6135 (1965).
301. E. D. Hughes and G. Jones, *J. Chem. Soc.*, 2678 (1950).
302. S. Brownstein, C. A. Bunton, and E. D. Hughes, *J. Chem. Soc.*, 4356 (1958).
303. B. A. Geller and L. N. Dubrova, *Zh. Obshch. Khim.*, **30**, 2646 (1960).
304. W. N. White and J. T. Golden, *Chem. Ind. (London)*, 138 (1962).
305. P. H. Gore, *J. Chem. Soc.*, 1436 (1957).
306. W. N. White, J. R. Klink, D. Lazdins, C. Hathaway, J. T. Golden, and H. S. White, *J. Am. Chem. Soc.*, **83**, 2024 (1961).
307. M. J. S. Dewar and A. P. Marchand, *Ann. Rev. Phys. Chem.*, **16**, 381 (1965).
308. J. R. Klink, *Dissertation Abstr.*, **25**, 6962 (1965).
309. G. S. Hammond and R. C. Newmann, *J. Am. Chem. Soc.*, **85**, 1501 (1963).
310. J. B. Dickey, E. B. Towne, and G. F. Wright, *J. Org. Chem.*, **20**, 499 (1955).
311. H. Jensch, *Ann. Chem.*, **568**, 73 (1950).
312. S. Kasman and A. Towers, *Can. J. Chem.*, **34**, 1261 (1956).
313. B. A. Geller and L. S. Samsvat, *Zh. Obshch. Khim.*, **34**, 613 (1964).
314. D. V. Banthorpe and J. A. Thomas, *J. Chem. Soc.*, 7158 (1965).
315. S. Markinkiewicz, J. Green, and P. Mamalis, *Tetrahedron*, **14**, 208 (1961).
316. J. Z. Allan and V. Chmatal, *Collection Czech Chem. Commun.*, **29**, 531 (1964).

317. V. O. Lukashevich and L. G. Krolik, *Dokl. Akad. Nauk SSSR*, **129**, 117 (1959).
318. P. Jacobsen, *Ann. Chem.*, **428**, 76 (1922).
319. M. Vecera, *Chem. Listy*, **52**, 1373 (1958).
320. S. Hashimoto and J. Sunamoto, *Kogyo Kagaku Zasshi*, **67**, 2090 (1964).
321. Reference 20, p. 301.
322. G. W. Wheland and J. R. Schwartz, *J. Chem. Phys.*, **17**, 425 (1949).
323. D. H. Smith, J. R. Schwartz, and G. W. Wheland, *J. Am. Chem. Soc.*, **74**, 2282 (1952).
324. D. V. Banthorpe, *J. Chem. Soc.*, 2413 (1962).
325. D. V. Banthorpe, E. D. Hughes, and C. K. Ingold, *J. Chem. Soc.*, 2864 (1964).
326. R. Robinson, *J. Chem. Soc.*, 220 (1941).
327. S. Brownstein, C. A. Bunton, and E. D. Hughes, *Chem. Ind.* (*London*), 981 (1956).
328. L. Festandig, *Tetrahedron Letters*, 1235 (1963).
329. S. Shamin-Ahmad and H. Hasan, *J. Ind. Chem. Soc.*, **29**, 955 (1952).
330. V. O. Lukashevich, *Dokl. Akad. Nauk. SSSR*, **159**, 1095 (1964).
331. V. O. Lukashevich, *Dokl. Akad. Nauk SSSR*, **133**, 115 (1960).
332. M. Vecera, M. Synek, and F. Sterba, *Collection Czech. Chem. Commun.*, **25**, 1992 (1960).
333. H. Wieland, *Ann. Chem.*, **392**, 127 (1912).
334. G. Wittig, P. Börzel, F. Neumann, and G. Klau, *Ann. Chem.*, **691**, 109 (1966).
335. J. Snyder, *J. Am. Chem. Soc.*, **84**, 340 (1962).
336. J. S. Clovis and G. S. Hammond, *J. Org. Chem.*, **28**, 3290 (1963).
337. H. J. Shine and J. P. Stanley, *Chem. Commun.*, 394 (1965).
338. D. L. Hammick and S. F. Mason, *J. Chem. Soc.*, 220 (1946).
339. G. S. Hammond and J. S. Clovis, *J. Org. Chem.*, **28**, 3283 (1963).
340. A. Pongratz and H. Wurstner, *Chem. Ber.*, **73**, 423 (1940).
341. D. V. Banthorpe and E. D. Hughes, *J. Chem. Soc.*, 2849 (1964).
342. D. V. Banthorpe, *J. Chem. Soc.*, 2860 (1964).
343. E. Klauke and O. Bayer, *Ger. Pat.*, 1,154,091 (1963); *Chem. Abstr.*, **60**, 457e (1964).
344. T. Pyl, H. Lahmer, and H. Beyer, *Chem. Ber.*, **94**, 3217 (1961).
345. H. Beyer, H. J. Haase, and W. Wildgrube, *Chem. Ber.*, **91**, 247 (1958).
346. H. Beyer and H. J. Haase, *Chem. Ber.*, **90**, 66 (1957).
347. A. N. Nesmeyanov, E. G. Perevalova, and T. V. Nikitina, *Dokl. Akad. Nauk SSSR*, **138**, 1118 (1961).
348. B. Miller, *J. Am. Chem. Soc.*, **86**, 1127 (1964).
349. J. R. Cox and M. F. Dunn, *Tetrahedron Letters*, 985 (1963).
350. J. W. Schulenberg and S. Archer, *Org. Reactions*, **14**, 1 (1965).
351. K. B. Wiberg and B. I. Rowland, *J. Am. Chem. Soc.*, **77**, 2205 (1955).
352. R. M. Roberts and F. A. Hussein, *J. Am. Chem. Soc.*, **82**, 1950 (1960).
353. J. F. Bunnett and R. E. Zahler, *Chem. Rev.*, **49**, 273 (1951).
354. R. Dohmori, *Chem. Pharm. Bull.* (*Tokyo*), **12**, 601 (1964).
355. Reference 137, p. 628.
356. G. Illuminati, *J. Am. Chem. Soc.*, **78**, 2603 (1956).
357. P. Megson, *Dissertation Abstr.*, **18**, 793 (1958).

358. E. A. Shilov, M. N. Bogdanov, and A. E. Shilov, *Dokl. Akad. Nauk SSSR*, **92**, 93 (1953).
359. V. V. Kozlov, *Zh. Obshch. Khim.*, **27**, 1146 (1957).
360. R. L. Lantz and P. M. J. Obellianne, *Compt. Rend.*, **238**, 2243 (1954).
361. C. R. Hurd and N. Kharasch, *J. Am. Chem. Soc.*, **68**, 657 (1946).
362. F. L. Scott and O. J. J. Broderick, *Chem. Ind. (London)*, 1058 (1962).
363. S. Searles and S. Nukina, *Chem. Rev.*, **59**, 1077 (1959).
364. L. H. Welsh, *J. Am. Chem. Soc.*, **71**, 3500 (1949).
365. E. E. van Tamelen, *J. Am. Chem. Soc.*, **73**, 5773 (1951).
366. G. Fodor and J. Kiss, *J. Chem. Soc.*, 1589 (1952).
367. G. Fodor and J. Kiss, *J. Am. Chem. Soc.*, **72**, 3495 (1950).
368. P. Desnuelle and A. Casal, *Biochem. Biophys. Acta*, **2**, 64 (1948).
369. B. R. Baker and P. E. Schaub, *J. Am. Chem. Soc.*, **75**, 3864 (1953).
370. L. Labler and F. Sorm, *Collection Czech. Chem. Commun.*, **25**, 265 (1960).
371. G. Fodor and K. Nador, *J. Chem. Soc.*, 721 (1953).
372. O. Wintersteiner and M. Moore, *J. Am. Chem. Soc.*, **75**, 4938 (1953).
373. H. Burchfield, *Nature*, **181**, 49 (1958).
374. W. B. Wright, H. J. Brabander, and R. A. Hardy, *J. Org. Chem.*, **26**, 2120 (1961).
375. B. Capon, *Quart. Rev. (London)*, **18**, 45 (1964).
376. A. Streitwieser, *Chem. Rev.*, **56**, 571 (1956).
377. P. L. Southwick, A. K. Colter, R. J. Owellan, and Y. C. Lee, *J. Am. Chem. Soc.*, **84**, 4299 (1962).
378. J. Hine, *Physical Organic Chemistry*, 2nd ed., McGraw Hill, New York, 1962, p. 146.
379. S. Sarel and A. Greenberger, *J. Org. Chem.*, **23**, 330 (1958).
380. N. J. Leonard, T. W. Milligan, and T. L. Brown, *J. Am. Chem. Soc.*, **82**, 4075 (1960).
381. W. E. Hanby, S. G. Waley, and J. Watson, *J. Chem. Soc.*, 3239 (1950).
382. L. A. Cohen and B. Witkop, *J. Am. Chem. Soc.*, **77**, 6595 (1955).
383. L. A. Cohen and B. Witkop in *Molecular Rearrangements* (Ed. P. de Mayo), Interscience, New York, 1963, p. 965.
384. W. E. Rosen, *Tetrahedron Letters*, 481 (1962).
385. B. V. Joffe and K. N. Zelenin, *Zh. Obshch. Khim.*, **33**, 3231 (1963).
386. M. E. Wolff, *Chem. Rev.* **63**, 55 (1963).
387. R. S. Neale, M. R. Walsh, and N. L. Marcus, *J. Org. Chem.*, **30**, 3683 (1965).
388. H. Simon, *Z. Naturforsch.*, **18B**, 419 (1963).
389. F. Micheel and I. Dijong, *Ann. Chem.*, **658**, 120 (1962).
390. K. L. Nelson, C. J. C. Robertson, and J. J. Duval, *J. Am. Chem. Soc.*, **86**, 684 (1964).
391. D. J. Brown and J. S. Harper, *J. Chem. Soc.*, 5542 (1955).
392. I. Wemper, G. B. Brown, T. Ueda, and J. J. Fox, *Biochem.*, **4**, 57 (1965).
393. E. Ware, *Chem. Rev.*, **46**, 403 (1950).
394. A. H. Mehler, *J. Biol. Chem.*, **218**, 241 (1956).
395. H. Schmidt, *Angew. Chem.*, **73**, 253 (1961).
396. R. D. Haworth and W. H. Perkins, *J. Chem. Soc.*, 445 (1926).
397. U. Schoellkopf, M. Patsch, and H. Shaefer, *Ann. Chem.*, **683**, 42 (1965).
398. W. Klyne and V. Prelog, *Experientia*, **16**, 521 (1960).

CHAPTER **11**

Protection of the amino group

Y. WOLMAN

Department of Organic Chemistry, The Hebrew University of Jerusalem, Israel

I. INTRODUCTION

The amino group undergoes readily oxidation, alkylation, acylation and a variety of other reactions, so there is a need to protect it while other parts of the molecule are undergoing chemical changes. This can be done by converting the amino group to a derivative which is stable under the reaction conditions to be employed and from which the amino group can be regenerated without affecting the rest of the molecule.

The great interest in amino-protecting groups is due mainly to the significant developments in peptide syntheses during the last fifteen years. This chapter is not going to deal with specific aspects and problems of peptide synthesis (racemization, preferential protection and selective removal of the protecting groups) nor with any other specific synthetic problems (like the syntheses of nucleotides, amino sugars and their derivatives etc.). It will deal only with the formation, use and removal of various amino-protecting groups.

In recent years a review dealing with amino-protecting groups[1] and few articles devoted to the specific problems of the amino-protecting groups in peptide syntheses[2-5] have been published. We shall try mainly to concentrate on recent developments and on some aspects which had not yet received sufficient coverage.

II. ACYL-TYPE PROTECTING GROUPS

The acyl derivatives are formed very easily (usually by reaction of the amine with the acyl chloride or anhydride, employing the Schotten–Baumann procedure[6]) and a good number of them have been used to protect various amines. While all of them can be removed with varying degrees of ease, using acidic or basic hydrolysis, some protecting groups can be removed by other means such as oxidation (formyl, phlorethyl), reduction (formyl, dibenzylphosphonyl) or electrolysis (phloretyl, benzoyl).

A. Formyl Derivatives

The formyl derivatives are obtained from the amines by the action of a mixture of 98% formic acid and acetic anhydride[7]. It seems that either formic anhydride or a mixed anhydride of formic and acetic acid is formed as an intermediate and acts as the acylating agent. In some cases formyl derivatives were prepared via the aminolysis of ethyl formate[8] or p-nitrophenyl formate[9].

The formyl group is removed by solvolysis in methanol or in a 1:1 mixture of dioxane–water containing a small excess of hydrochloric acid[7]. It can also be removed by oxidation using a two- to three-fold excess of hydrogen peroxide[10]. The reaction proceeds through the formation of the corresponding carmabic acid derivatives (reaction 1). It is interesting to note that the formyl group is susceptible

$$HCONHR \xrightarrow{H_2O_2} HOCONHR \longrightarrow CO_2 + RNH_2 \qquad (1)$$

to reductive cleavage and could be removed by catalytic hydrogenation using tetrahydrofuran (containing hydrochloric acid) as a solvent[11]. Nothing is known, as yet, about the mechanism of the reaction, but while formylamino acid esters yield the corresponding amino acid ester hydrochloride quantitatively, no amino acids could be recovered from the hydrogenolysis of formylamino acids. It has been shown lately that the formyl group could be removed in 30–70% yield from various disubstituted formamides by the action of sodium hydride[12]. The reaction probably proceeds by a concerted mechanism in which the hydride ion abstracts the formyl proton while carbon monoxide is eliminated simultaneously (equation 2).

$$HCONRR^1 + NaH \longrightarrow RR^1NNa + CO + H_2 \qquad (2)$$

B. Acetyl and Benzoyl Derivatives

Acetyl and benzoyl derivatives of various amines have been prepared using the acyl chloride or the anhydride as the acylating agent. Those groups are removed usually by acidic or basic hydrolysis. Compounds of biological importance (peptides, oligosacharides, nucleotides etc.), are usually destroyed under these conditions. Recently it has been shown that the benzoyl group can be removed by electrolytic cleavage[13]. This is a reductive cleavage, in contrast to the oxidative electrolytic cleavage of the phlorethyl group (see section II.J). In several cases the acetyl group could be removed from acylamino sugars by hydrazinolysis without destruction of the glycosidic linkage (e.g. with derivatives of D-glucosamines[14]).

The acetyl and benzoyl derivatives have been used as well as the formyl derivatives in the synthesis of various amino compounds employing acetamido[15], benzamido[16] and formamido malonate[17] respectively. In the synthesis of the 64 possible ribotrinucleotides which are derived from the four major ribonucleotides, the amino group of two of the heterocyclic bases, adenine (1) and cytosine (2), was protected by an acetyl or a benzoyl group[18]. The acyl derivatives in this

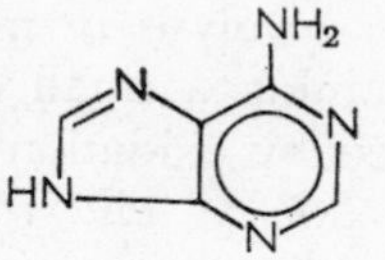

(1) 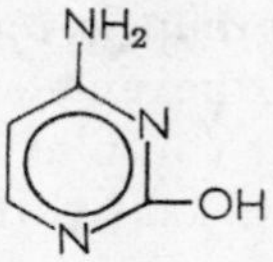　　　　　　　(2)

case are very sensitive toward acidic and basic hydrolysis and can be removed under mild conditions[19, 20]. The hydrolysis of various acylamino derivatives of deoxycytidine-5-phosphate (3) was studied[19],

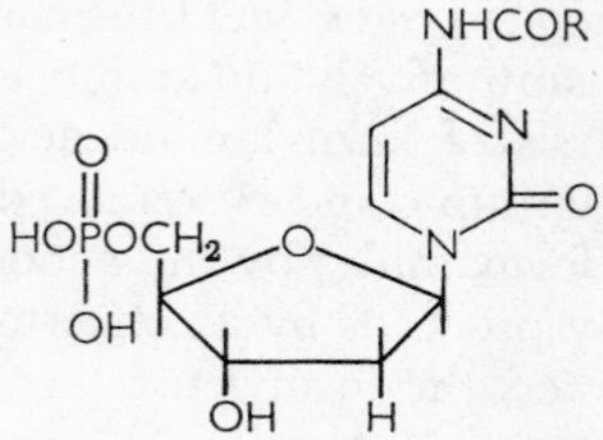

(3)

(R = p-CH$_3$OC$_6$H$_4$, C$_6$H$_5$, CH$_3$)

and it was found that the time which is required for 50% removal of the protecting group varies with each acyl group, the order being: p-methoxybenzoyl > benzoyl > acetyl. The acetyl group is removed very easily by treatment with aqueous ammonia for a few minutes.

C. Trifluoroacetyl Derivatives

The trifluoroacetyl group is obtained by acylation of the amine with trifluoroacetic anhydride[21] or by the aminolysis of ethyl trifluorothiolacetate (CF$_3$COSEt)[22], phenyl trifluoroacetate[23] or methyl trifluoroacetate[24]. The trifluoroacetyl group can be cleaved easily by very mild alkaline agents, such as dilute sodium hydroxide[21], dilute ammonia solution[21] or 0·2 N barium hydroxide[25]. The trifluoroacetyl group is so sensitive toward basic hydrolysis that it can be removed by using a weakly basic polyamine ion exchange resin (e.g. IR–4B)[22].

In the last two years the trifluoroacetyl group was used as an amino-protecting group for the synthesis of steroid glycosides of amino sugars[26] and amino sugar nucleosides[27].

D. Acetoacetyl Derivatives

The acetoacetyl derivatives are obtained by the reaction of the amine with diketene[28]. In the case of amino acid ester hydrochlorides the presence of one equivalent of sodium methoxide or of a tertiary base is necessary. The group is removed by applying the conditions of the Knorr synthesis of pyrazol derivatives from various β-keto esters and their derivatives[29]. The protecting group is removed in acetic acid by using an equimolar amount of phenylhydrazine (reaction 3).

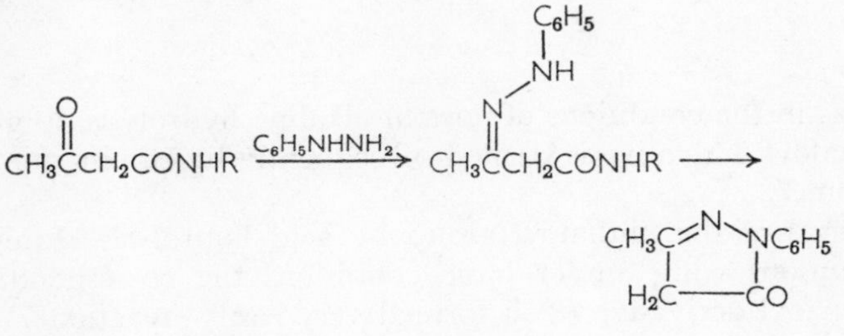

$$CH_3\overset{O}{\overset{\|}{C}}CH_2CONHR \xrightarrow{C_6H_5NHNH_2} CH_3\overset{\overset{\overset{C_6H_5}{\underset{|}{N}}}{\underset{\|}{N}}}{C}CH_2CONHR \longrightarrow$$

$$\overset{CH_3C\underset{\underset{H_2C\underline{\hspace{0.5cm}}CO}{|}}{\overset{N}{\diagdown}}NC_6H_5}{} + RNH_2 \quad (3)$$

E. Diacetyl Derivatives

Acyclic N,N-diacyl derivatives are hydrolyzed so easily that they are generally of no practical use, although they have been used very occasionally[30]. In contrast, the cyclic imides, e.g. N-substituted phthalimides, are very stable and have been widely used.

The phthaloyl group can be introduced by heating an alkyl halide with potassium phthalimide (4) alone or in a non-polar solvent (equation 4)[31], or by heating the amine with phthalic anhydride[32, 33]

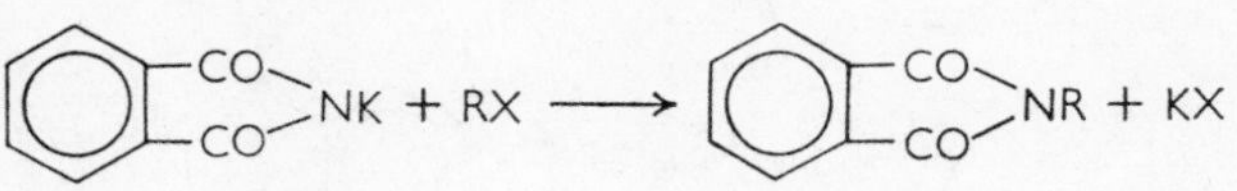

$$\text{(4)} \quad NK + RX \longrightarrow NR + KX \quad (4)$$

(4)

or with carbethoxythiobenzoic acid (5) (equation 5)[34]. The phthaloyl derivatives can also be obtained, without heating and under very mild conditions, by reaction of the amine with N-carbethoxyphthalimide

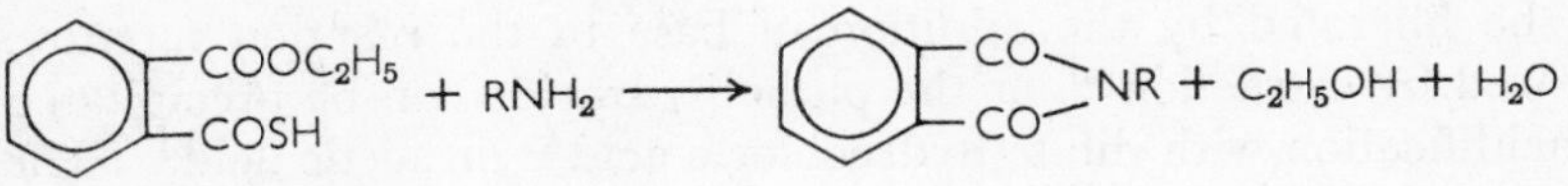

$$\text{(5)} \quad \begin{array}{c} COOC_2H_5 \\ COSH \end{array} + RNH_2 \longrightarrow NR + C_2H_5OH + H_2O \quad (5)$$

(5)

(6) at room temperature in aqueous solution (reaction 6)[35]. The reaction proceeds by the opening of the five-membered ring of 6 by

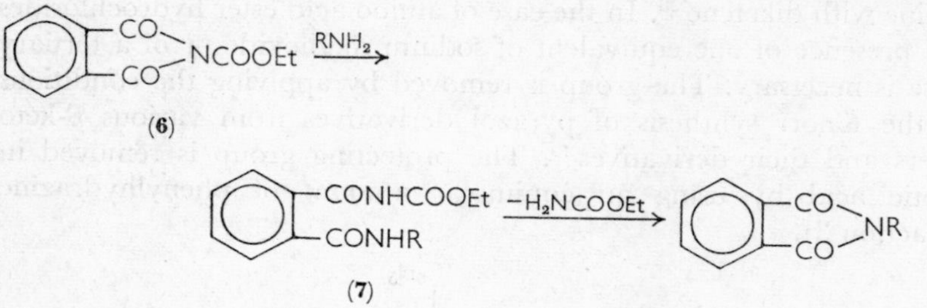

(6)

(7)

the amine, as in the conditions of normal alkaline hydrolysis, to give 7. The phthaloyl derivative is formed subsequently by elimination of urethane from 7.

The removal of the phthaloyl group by acid hydrolysis requires drastic conditions while under basic conditions the corresponding o-carboxybenzoyl derivative (8) is formed very easily (reaction 7)[36].

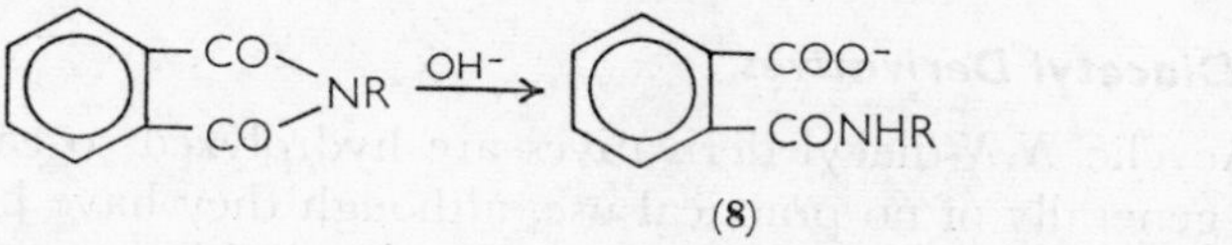

(7)

(8)

Ring closure of 8 to recover the starting material can be achieved by treatment with thionyl chloride[37] or methanolic HCl[38]. The phthaloyl group is removed by hydrazinolysis, using hydrazine hydrate, to yield the amine and phthalhydrazide (9) (equation 8)[39].

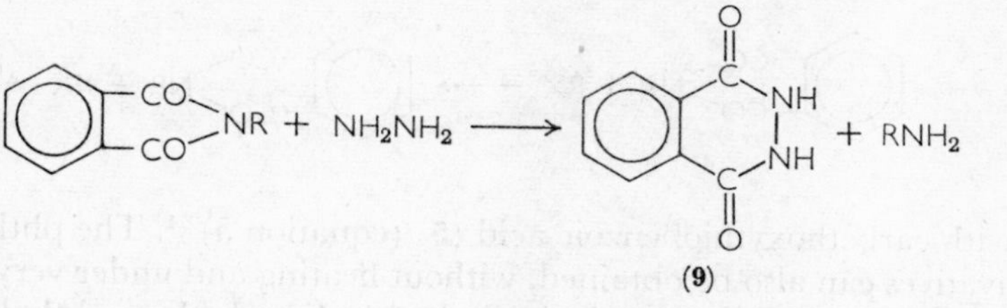

(9)

Due to the acidic character of 9 the amine is obtained as its salt, and can be liberated by the addition of base to the reaction mixture, followed by extraction[40], or the phthalhydrazide can be precipitated by acidification with dilute hydrochloric acid[41] or acetic acid[42]. The hydrazinolysis can also be carried out with excess of phenylhydrazine in the presence of a tertiary base[43]. The phthaloyl group can also be removed by the action of copper acetate–basic copper carbonate in aqueous solution[42].

The phthaloyl group has been used for a long time as a protecting group in the course of the synthesis of various amines[31]. The most important example of its use is in the synthesis of penicillin V (10)[44].

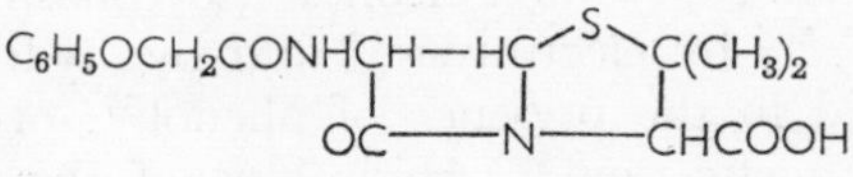

$$C_6H_5OCH_2CONHCH—HC\overset{S}{\diagup}C(CH_3)_2$$
$$OC———N———CHCOOH$$

(10)

F. Amidines

Imidoesters (11) react with various amines in nonaqueous solutions[45] or in aqueous solution at low temperatures and at pH near neutrality[46] to yield the corresponding amidines (12) (reaction 9).

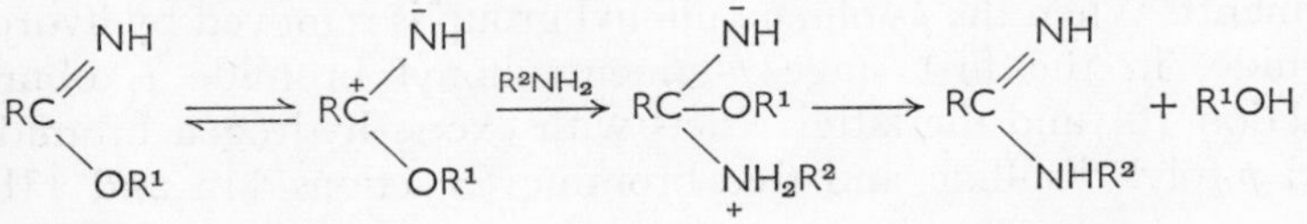

(11)   (12)

$$(9)$$

The reaction proceeds through a mechanism similar to the aminolysis of esters. It has been shown that the reaction between glycine and methyl benzimidate at pH 9·5 in water–dimethylformamide (7:3) solution yields benzamidinoacetic acid quantitatively[46].

Amidines react with nucleophiles like ammonia, amines or hydrazine, in a displacement reaction[45]. Application of this reaction for the removal of the acetamidine group was described in the literature[47]. Side-reactions are avoided by running the reaction at room temperature for 8 hours, employing as the nucleophile an ammonia–ammonium acetate buffer (pH 11·3).

G. Sulfonyl Derivatives

Aryl- and alkylsulfonyl protecting groups have been used in the synthesis of secondary amines from primary ones[48-50]. The sulfonyl group increases the acidity of the nitrogen–hydrogen bond so that the sodium salt of the sulfonyl amide is obtained easily. This salt can be alkylated in aqueous media to give the secondary amide derivative which in turn yields the secondary amine on removal of the protecting group[51].

The p-toluenesulfonyl (tosyl) group is the best known protecting group of this class. These derivatives (like all other sulfonyl derivatives) are obtained easily by reaction of p-toluenesulfonyl chloride with the amine in alkaline media[52], but their removal is not a straightforward reaction. The most widely used methods for the removal of the p-toluenesulfonyl group are cleavages by sodium in liquid ammonia[53], by sodium in butanol[50] or amyl alcohol[54], by phosphonium iodide in hydriodic acid[55,56], by electrolysis[13] and by hydrolysis with 48% hydrobromic acid in the presence of phenol[57], or by 30% hydrobromic acid in acetic acid[58]. Hydrolysis of the p-toluenesulfonyl group with hydrochloric acid requires high concentration of the acid, elevated temperatures and long periods of boiling (e.g. with 25% hydrochloric acid for 12–36 hours)[59]. These reactions are neither simple reductive cleavages nor simple hydrolyses, but involve the cleavage of the carbon–sulfur and the nitrogen–sulfur bond as well as various oxidation–reduction reactions of the sulfur-containing fragments. When the p-toluenesulfonyl group is removed by hydrogen bromide, in the first stage p-toluenesulfonyl bromide is obtained (reaction 10) and the latter reacts with excess hydrogen bromide to yield p-tolyl disulfide and free bromine (reactions 11a and 11b)[59].

$$p\text{-}CH_3C_6H_4SO_2NHR + HBr \longrightarrow p\text{-}CH_3C_6H_4SO_2Br + RNH_2 \qquad (10)$$

$$p\text{-}CH_3C_6H_4SO_2Br + 4\,HBr \longrightarrow p\text{-}CH_3C_6H_4SBr + 2\,H_2O + 2\,Br_2 \qquad (11a)$$

$$p\text{-}CH_3C_6H_4SBr \longrightarrow p\text{-}CH_3C_6H_4SSC_6H_4CH_3\text{-}p + Br_2 \qquad (11b)$$

The bromine which usually reacts with the liberated amine could be removed from the reaction mixture by adding phenol to it[57]. In the case of the reductive cleavage the picture is even more complicated. It has been shown in the cleavage by means of sodium and alcohol that reductive cleavage as well as alcoholysis proceeds at the same time[60]. Removal of the p-toluenesulfonyl group from p-toluene-sulfonyl-L-glutamic acid by means of sodium in liquid ammonia yields 14–15% of p-thiocresol, 70–75% of sulfur dioxide and 86% of toluene (reaction 12). It has been suggested that the reaction proceeds via two different paths[61]: path (a) which is the major course of the reaction involves a carbon–sulfur bond cleavage to yield toluene, sulfur dioxide and the free amine; path (b) involving the cleavage of a nitrogen–sulfur bond, gives p-toluenesulfinic acid and the free amine. The p-toluenesulfinic acid itself reacts with the reductive agent giving mainly p-thiocresol as well as a small amount of toluene and sulfur dioxide through carbon–sulfur bond cleavage.

The benzylsulfonyl group has also been used as a protecting group in the synthesis of secondary amines from primary ones. The preparation of the benzylsulfonyl derivatives is very similar to that of the

$$RNH_2 + SO_2 + C_6H_5CH_3$$

$$p\text{-}CH_3C_6H_4SO_2NHR \xrightarrow{\text{(a)}}$$

$$\xrightarrow{\text{(b)}} RNH_2 + p\text{-}CH_3C_6H_4SO_2H \longrightarrow p\text{-}CH_3C_6H_4SH$$
$$\downarrow$$
$$C_6H_5CH_3 + SO_2$$

(12)

p-toluenesulfonyl derivatives. The benzylsulfonylamides are hydrolyzed by hydrochloric acid more easily than the corresponding p-toluenesulfonyl derivatives to give the free amine, benzyl chloride and sulfur dioxide[48]. The benzylsulfonyl group, like benzyl derivatives generally, could be removed relatively easily by reduction. Although the benzylsulfonyl group could not be removed by catalytic hydrogenation in the presence of platinium oxide[62], it can be removed by reduction using Raney nickel under mild conditions at room temperature and at atmospheric pressure[62-64].

H. Sulfenyl Derivatives

Various aromatic sulfenyl chlorides, e.g. phenylsulfenyl chloride[65], triphenylmethylsulfenyl chloride[66] and o-nitrophenylsulfenyl chloride[67], react with amines to yield the corresponding sulfenyl derivatives (reaction 13). The acylation is carried out under neutral

$$RSCl + R^1NH_2 \longrightarrow RSNHR^1 + HCl$$
$$(R = C_6H_5, (C_6H_5)_3C, o\text{-}NO_2C_6H_4)$$

(13)

conditions or in the presence of triethylamine. Under strongly alkaline conditions no acylation takes place and the corresponding disulfide and the thio ester of the sulfonic acid are formed, (e.g. in the case of o-nitrophenylsulfenyl chloride, the disulfide (**13**) and the thio ester (**14**) are obtained[68]). The nitrogen–sulfur bond of the sulfenyl amide is very sensitive toward acidic cleavage and the sulfenyl group can be

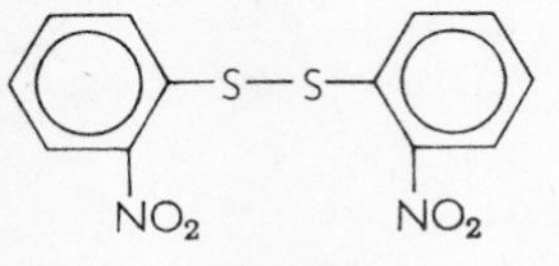

(**13**)

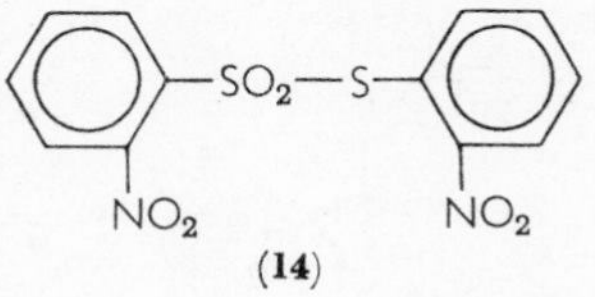

(**14**)

removed by using two equivalents of hydrogen chloride in nonaqueous media[66]. Other means for the removal of the sulfenyl group are desulfurization employing Raney nickel[69], or displacement reaction with highly polarizable nucleophiles: hydrogen cyanide[70], sulfur dioxide[70], thioacetamide[70], thiophenol[71] and thioglycolic acid[71].

The use of triphenylmethylsulfenyl, *o*-nitrophenylsulfenyl and 2,4-dinitrophenylsulfenyl as protecting groups in peptide synthesis, has been reported recently[72-74].

I. Phosphonyl Derivatives

Dibenzyl- and *p*-substituted dibenzylphosphonyl groups have been used to protect the amino group[75, 76]. Those derivatives are obtained by reaction of the corresponding dibenzylphosphonyl chloride with the amine in pyridine. The benzyl or *p*-substituted benzyl groups are removed by catalytic hydrogenation to yield the corresponding phosphamic acid which is very acid sensitive. Treating the phosphamic acid with aqueous solution (pH 4) at 25° results in the formation of free phosphoric acid and the amine, within five to ten minutes, in quantitative yield (reaction 14)[76]. It has been shown that the di-

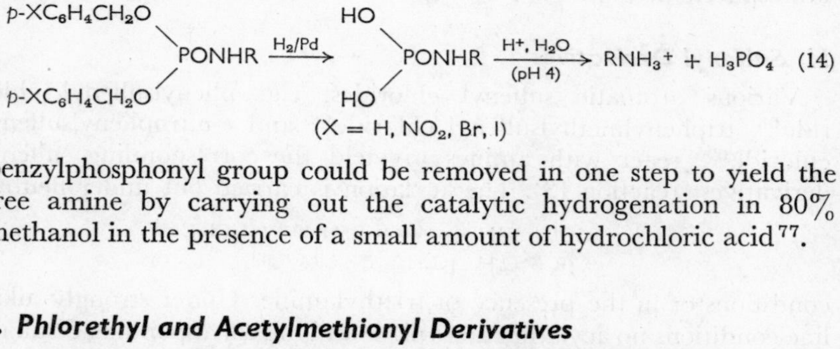

$$\begin{array}{c} p\text{-}XC_6H_4CH_2O \\ \diagdown \\ PONHR \xrightarrow{H_2/Pd} \\ \diagup \\ p\text{-}XC_6H_4CH_2O \end{array} \quad \begin{array}{c} HO \\ \diagdown \\ PONHR \xrightarrow[(pH\,4)]{H^+,\,H_2O} RNH_3{}^+ + H_3PO_4 \quad (14) \\ \diagup \\ HO \end{array}$$

$$(X = H,\ NO_2,\ Br,\ I)$$

benzylphosphonyl group could be removed in one step to yield the free amine by carrying out the catalytic hydrogenation in 80% methanol in the presence of a small amount of hydrochloric acid[77].

J. Phlorethyl and Acetylmethionyl Derivatives

The phlorethyl (β-(p-hydroxyphenyl)propionyl **15**) and the acetyl-methionyl (**16**) protecting groups are both removed in the same way,

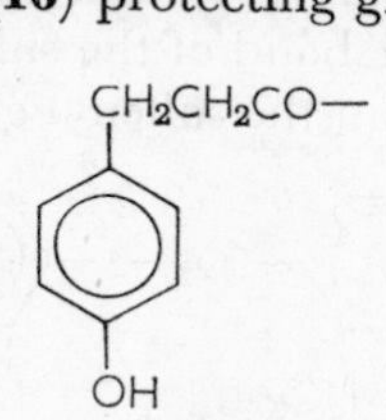

(15) 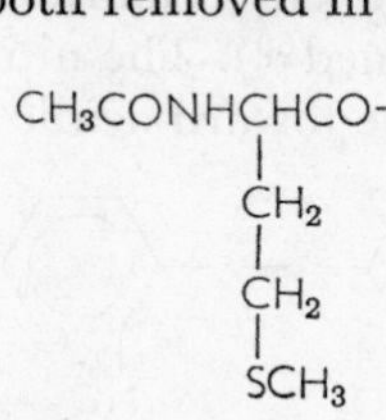   (16)

via participation of the carbonyl oxygen by 1,5-interaction to yield the corresponding iminolactones, which in aqueous solution give the free amine and the lactone[78-80]. This 1,5-interaction is known to occur with other amino derivatives too (e.g. acetylated allylglycine derivatives[81], indole-3-propionyl derivatives[82]).

It has been shown that phlorethylglycine or 3,5-dibromophorethylglycine liberate glycine upon the addition of *N*-bromosuccinimide (NBS) with the formation of a dibromospirodienone lactone (**17**) (reaction 15)[79]. The amine could be obtained in over 90% yield by

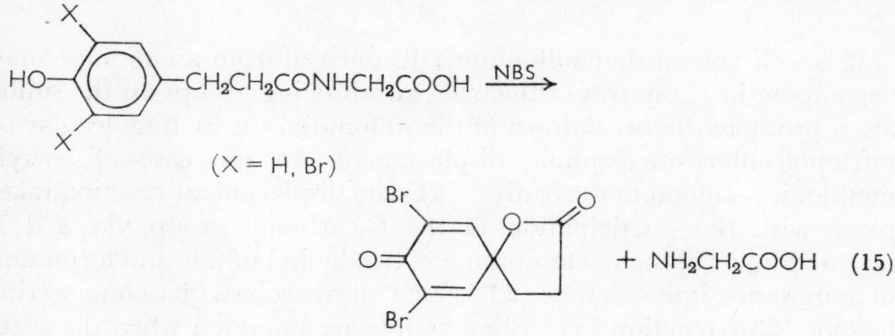
HO
X
X
CH₂CH₂CONHCH₂COOH
NBS
(X = H, Br)
Br
Br
O
O
O
+ NH₂CH₂COOH
(15)

(**17**)

using dilute acetic acid as the reaction media[83]. The reaction proceeds through the formation of the tribromodienone (**18**) followed by a displacement of the $C_{(1)}$-bromo atom by the carbonyl group to give the iminolactone (**19**) which upon hydrolysis yields the amine and the dibromospirolactone (**17**) (reaction 16). The phlorethyl group can be

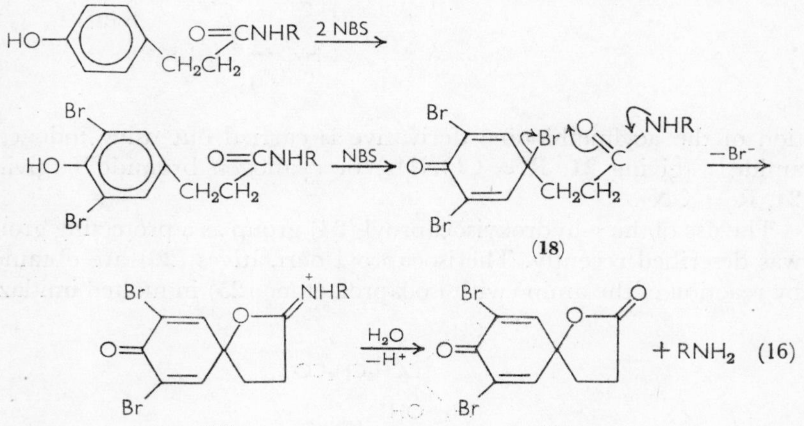
HO
O=CNHR
CH₂CH₂
2 NBS
Br
Br
NBS
Br
NHR
C
−Br⁻
(18)
NHR
H₂O
−H⁺
+ RNH₂
(16)
(19)
(17)

removed also by electrolytic cleavage[84]. To avoid destruction of the amine formed at the anode, the electrolysis is conducted at the lowest practical voltage and in moderately acidic medium. Carrying out the electrolysis of phlorethylglycine at pH 1 to 2 using a voltage of 7 volts yields 70% glycine and spirodienone lactone (20).

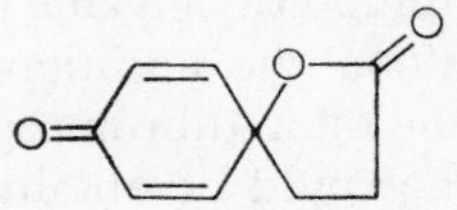

(20)

It is well known that sulfonium salts derived from methionine may decompose in many ways. Electron-withdrawing groups on the sulfur atom promote the breakdown of the sulfonium salt by bimolecular or intramolecular nucleophilic displacement. In the case of acetyl-methionine sulfonium derivatives (21) the displacement reaction takes place with the participation of the C-carbonyl group via a 1,5-interaction, resulting in elimination of the dialkyl sulfide and formation of homoserine iminolactone (22) which on hydrolysis gives homoserine lactone (23) (reaction 17)[80]. Best results are obtained when the reac-

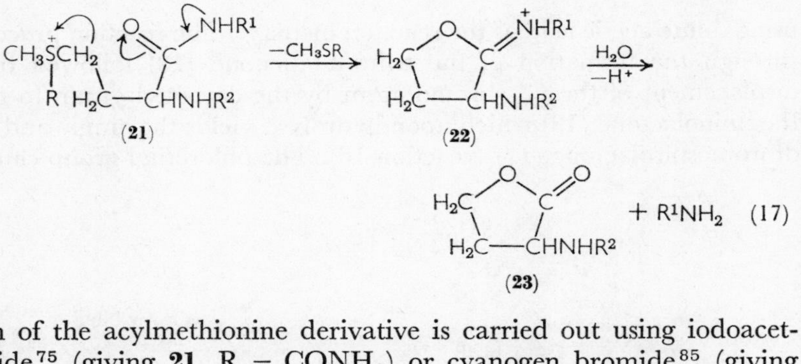

tion of the acylmethionine derivative is carried out using iodoacet-amide[75] (giving 21, R = $CONH_2$) or cyanogen bromide[85] (giving 21, R = CN).

The use of the γ-hydroxyisocaproyl (24) group as a protecting group was described recently. The isocaproyl derivatives (26) are obtained by reaction of the amine with isocaprolactone (25) in molten imidazol

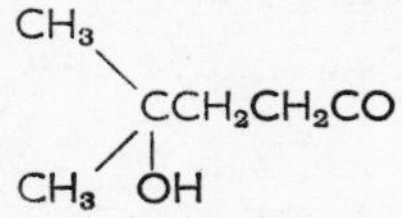

(24)

and could be removed by means of 50% trifluoroacetic acid at room temperature[86]. It seems that the reaction proceeds through the formation of the carbonium ion (26a) from which the iminolactone (27) is obtained by a 1,5-interaction of the carbonyl group with the positive carbon, followed by hydrolysis of the iminolactone (27) to yield the free amine and the lactone (25) (reaction 18).

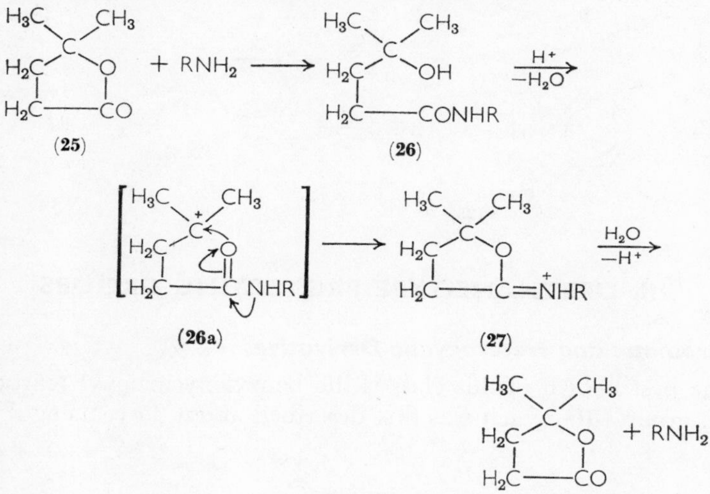

K. o-Nitrophenoxyacetyl and Chloroacetyl Derivatives

The observation that o-nitrophenoxyacetamide gives spontaneously upon reduction with ferrous sulfate the lactam of o-aminophenoxyacetic acid (28) and ammonia[87] was employed for the use of the o-nitrophenoxyacetyl group as a protecting group. Removal of the group is effected by reduction followed by short boiling of the aqueous solution containing the o-aminophenoxyacetyl derivative to yield the free amine and the lactam (28) (reaction 19)[84].

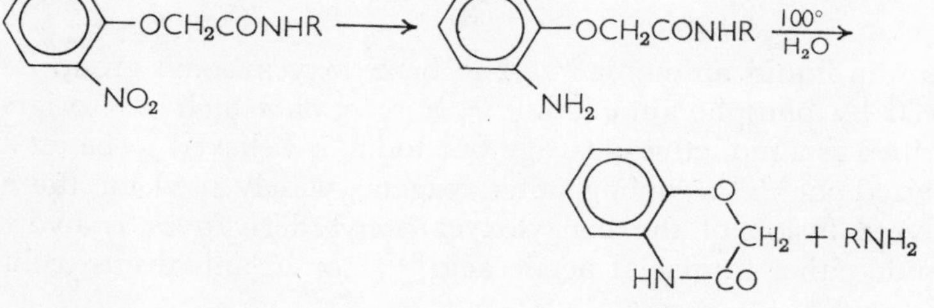

Similarly, the chloroacetyl group is cleaved by reaction with *o*-phenylenediamine[88], giving **29** which upon heating yields the lactam (**30**) and the free amine (reaction 20).

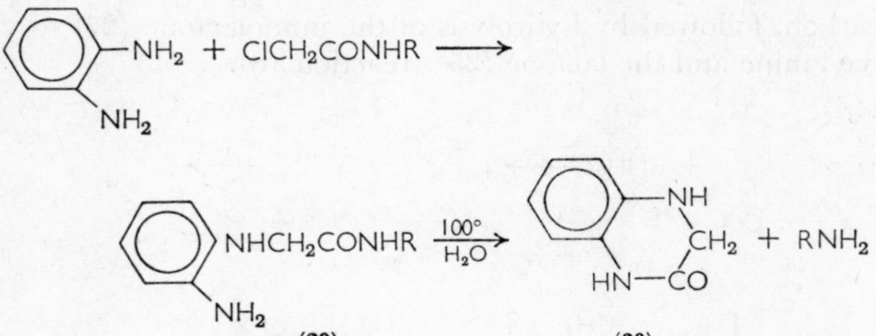

III. URETHANE-TYPE PROTECTING GROUPS

A. *Aromatic and Heterocyclic Derivatives*

The best known of this class is the benzyloxycarbonyl (carbobenzoxy) group (**31**) which was first described about 35 years ago[89] and

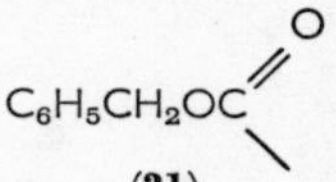

$$C_6H_5CH_2OC\!\!\!\diagup\!\!\!\!\!\!\!\overset{O}{\diagdown}$$

(**31**)

is still widely in use. Its introduction had opened new horizons in the field of peptide synthesis. The benzyloxycarbonyl derivatives result from the interaction of benzyloxycarbonyl chloride with the amine in basic media. The protecting group is sensitive toward reductive and acid-catalyzed fission. Reductive cleavage may occur either by catalytic hydrogenation[89], where the reaction proceeds via the formation of the carbamic acid (**32**) (reaction 21), or by chemical reduction (e.g. using

$$C_6H_5CH_2OCONHR \longrightarrow C_6H_5CH_3 + HOCONHR \longrightarrow CO_2 + RNH_2 \quad (21)$$

(**32**)

sodium in liquid ammonia[90]). The benzyloxycarbonyl group is also cleaved by phosphonium iodide[91], a reaction which was originally described as a reductive cleavage but today is believed to be an acid-catalyzed one[92, 93]. Among other reagents widely used for the acid-catalyzed fission of the benzyloxycarbonyl derivatives is hydrogen bromide either in glacial acetic acid[93-95] or in anhydrous trifluoro-acetic acid[96].

Various modified benzyloxycarbonyl groups have been proposed as amine-protecting groups (p-nitrobenzyloxycarbonyl[97, 98], p-chlorobenzyloxycarbonyl[95, 99], p-methylbenzyloxycarbonyl[100], p-bromobenzyloxycarbonyl[100], p-methoxybenzyloxycarbonyl[101, 102], 3,5-dimethoxybenzyloxycarbonyl[103], 1-(naphthylmethyl)oxycarbonyl (**33**)[100], p-(p-methoxyphenylazo)benzyloxycarbonyl (**34**)[104] and benzhydryloxycarbonyl (**35**)[105]) but none of them was widely used. The synthesis

$$\text{—CH}_2\text{OCO—} \qquad \text{CH}_3\text{OC}_6\text{H}_4\text{—N}{=}\text{N—C}_6\text{H}_4\text{CH}_2\text{OCO—} \qquad \text{Ph}_2\text{CHOCO—}$$

(**33**)   (**34**)   (**35**)

of the various modified derivatives is carried out from the amine with the corresponding chloroformate ($ROCOCl$) or azidoformate ($ROCON_3$) employing the Schotten–Baumann procedure. In certain cases the derivatives are obtained by reaction of the aryl-substituted alcohol with the isocyanate derived from the corresponding amine (reaction 22). Unblocking is effected by the same general methods

$$ArOH + OCNR \longrightarrow ArOCONHR \qquad (22)$$

which are employed for the removal of the benzyloxycarbonyl group. The cleavage of some of the modified benzyloxycarbonyl groups could be promoted by ultraviolet irradiation[103, 106]. It was known for some time that the p-nitro- and p-chlorobenzyloxycarbonyl derivatives are more stable, and p-methoxybenzyloxycarbonyl derivatives are much less stable, toward acidic cleavage than the corresponding unsubstituted derivatives. This phenomenon which was known qualitatively from preparative studies was recently confirmed quantitatively by kinetic studies[107, 108].

Until recently very little was known about the kinetics and mechanism of the acid-catalyzed fission of the benzyloxycarbonyl group. It was accepted[92, 93] that the cleavage involved prior protonation of the carbamate (although the precise site of the protonation was not clear), followed by fission of the benzyl–oxygen bond (reaction 23). Lately

$$C_6H_5CH_2OCONHR \overset{H^+}{\rightleftharpoons} C_6H_5CH_2\overset{+}{O}CONHR \longrightarrow C_6H_5CH_2Br + CO_2 + RNH_2$$
$$(23)$$

two studies appeared dealing with the detailed mechanism and the kinetics of the reaction[107, 108]. The first stage of the cleavage is protonation of the carbonyl oxygen of the carbamate. Depending upon the acid used and the substituents in the benzyl group the reaction proceeds either via a S_N2 nucleophilic displacement at the benzyl group (equation 25) or through the formation of a benzylcarbonium ion (equation 26).

$$H^+ + R\!-\!O\!-\!\overset{O}{\overset{\|}{C}}\!-\!NHR^1 \longrightarrow R\!-\!O\!-\!\overset{\overset{+}{O}H}{\overset{\|}{C}}\!-\!NHR^1 \qquad (24)$$

$$X^-R\!-\!O\!-\!\overset{\overset{+}{O}H}{\overset{\|}{C}}\!-\!NHR^1 \longrightarrow RX + HOOCNHR^1 \longrightarrow CO_2 + R^1NH_2 \quad (25)$$

$$R\!-\!O\!-\!\overset{\overset{+}{O}H}{\overset{\|}{C}}\!-\!NHR^1 \longrightarrow R^+ + HOOCNHR^1 \xrightarrow{\;X^-\;} CO_2 + R^1NH_2 + RX \quad (26)$$

The furfuryloxycarbonyl (**36**) protecting group[107,109] is very

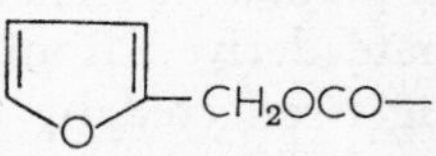

(**36**)

sensitive toward acidic cleavage, its rate of fission ($k_1 = 10^{-2}$/sec) being nearly the same as that of the corresponding p-methoxybenzyl-oxycarbonyl derivative ($k_1 = 1.5 \times 10^{-2}$/sec) which is the most acid-sensitive compound[107].

The phenyloxycarbonyl and p-tolyloxycarbonyl groups have been described as amine-protecting groups[95]. Contrary to all other ure-thane-type derivatives which have been described in this section the phenyloxy- and p-tolyloxycarbonyl derivatives are stable towards acidic cleavage. Their acid-catalyzed fission proceeds only using forced reaction conditions (high temperatures and long times). These groups can be removed by reductive cleavage using catalytic hydro-genation (which in this case requires a very long reaction time) or by chemical means, such as treatment with sodium in liquid ammonia.

The 8-quinolyloxycarbonyl (**37**) protecting group is susceptible toward metal ion-promoted hydrolysis[110]. The group is removed by Cu^{2+} or Ni^{2+} in neutral solution. The reaction proceeds via the forma-tion of a stable coordination complex **38** which upon hydrolysis yields the derivative of the oxine anion (**39**) (reaction 27).

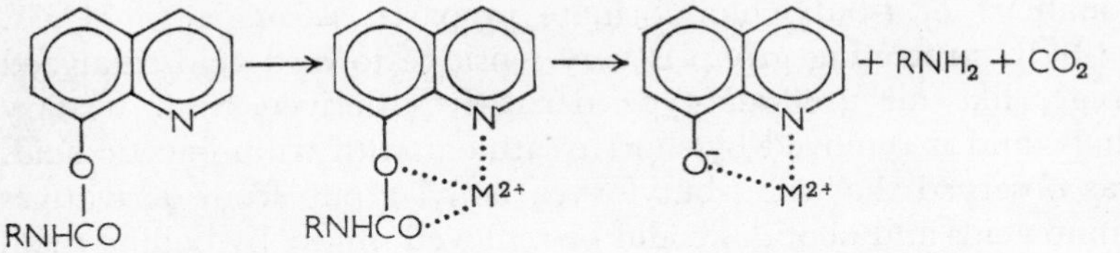

$$+ RNH_2 + CO_2 \quad (27)$$

(37)   (38)   (39)

B. Aliphatic and Alicyclic Derivatives

Various aliphatic and alicyclic urethane-type derivatives have been used to protect the amino group (allyloxycarbonyl (40)[111], cyclopentyloxycarbonyl (41)[101], 1-methylcyclopentyloxycarbonyl (42)[112], cyclohexyloxycarbonyl (43)[101], (diisopropyl)methyloxycarbonyl (44)[101], t-butyloxycarbonyl (45)[101,113], t-amyloxycarbonyl (46)[112,114], 9-methyl-9-fluoronyloxycarbonyl (47)[112] and adamantyloxycarbonyl (48)[115].

$CH_2=CHCH_2OCO—$

(40)

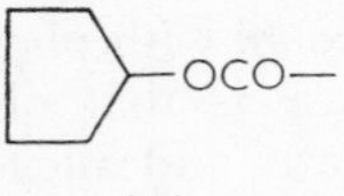

(41)

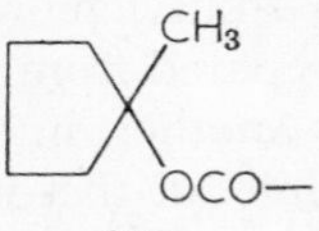

(42)

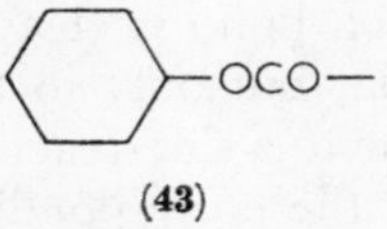

(43)

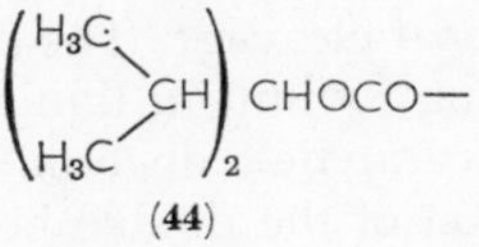

(44)

$(CH_3)_3COCO—$

(45)

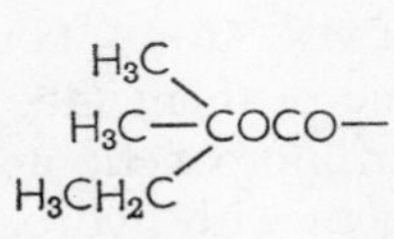

(46)

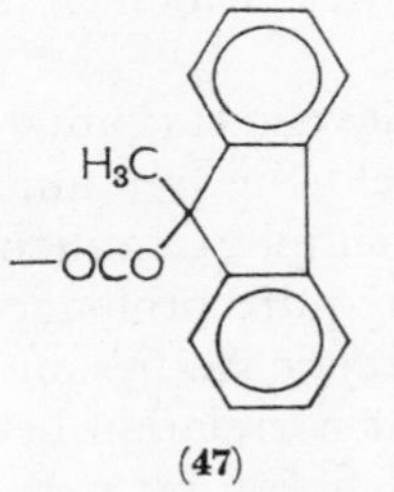

(47)

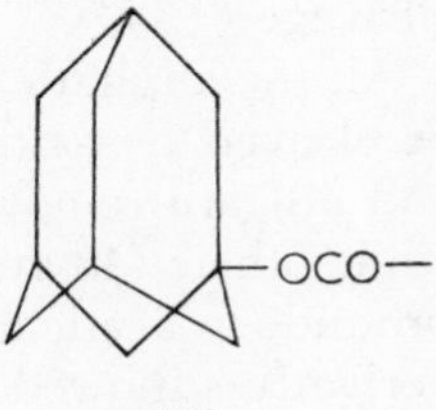

(48)

The best known group of this class is the t-butyloxycarbonyl(t-BOC) (45). Since its introduction as an amino-protecting group in peptide synthesis about seven years ago[116] it became second in use only to the benzyloxycarbonyl group. Due to the instability of t-butylchloroformate the t-butyloxycarbonyl derivatives are obtained by reaction of the corresponding isocyanate with t-butanol[101] or by the action of the amine with t-butyl-p-nitrophenyl carbonate[113], t-butylazidoformate[117], t-butylcyanoformate[118], t-butyl N-hydroxysuccinimide carbonate[119] or t-butylchloroformate prepared *in situ* at $-74°$[120]. The t-BOC protecting group is very sensitive toward acid-catalyzed cleavage, like all urethane-type derivatives derived from tertiary alcohols, and is removed by short treatment with trifluoracetic acid. It was observed that the t-butyloxycarbonyl group from derivatives of amino acids and peptides could be removed simply by boiling them in water for an hour or two[121]. Although there is a need for a free carboxylic group in the molecule, the reaction is not acid catalyzed (e.g. addition of acetic acid does not have any appreciable effect on the reaction rate), but may be, at least partially, autocatalytic. The t-butyloxycarbonyl group could even be removed by heating the solid derivative for a short time at $150-160°$[121].

The t-amyloxycarbonyl derivatives which are obtained by the reaction of t-amylchloroformate with amines[114], can be cleaved as easily as the corresponding t-butyloxycarbonyl derivatives. In contrast to t-butylchloroformate, t-amylchloroformate is stable and can be prepared from t-amyl alcohol with phosgene.

Another interesting case is that of the allyloxycarbonyl group; while all the other aliphatic and alicyclic carbamate derivatives are resistant to reductive cleavage this group can be removed both by catalytic or by chemical cleavage. The allyl–oxygen bond is cleaved quantitatively by using sodium in liquid ammonia, but only to the extent of $65-70\%$ by catalytic reduction. This is due to a side-reaction in which the reduction of the double bond yields the corresponding propyloxycarbonyl derivative which is resistant toward reductive cleavage[111].

All the aliphatic and alicyclic carbamate derivatives (40–48) could be cleaved by acids to yield the free amines. Some of them (45, 46 and 48) are removed by simple treatment with trifluoroacetic acid, while others (43 and 44) require prolonged treatment with hydrogen bromide in acetic acid. Kinetic studies on various alicyclic carbamate derivatives showed a linear correlation between the logarithms of the rate constants of the acid fission by means of hydrogen bromide in acetic acid, and between the solvolysis of the corresponding alicyclic tosylates[122]. It was concluded that the cleavage of the alicyclic carbamates proceeds by solvolysis of the protonated compounds.

The β-tolylsulphonylethyloxycarbonyl (49)[123] group is sensitive to alkaline cleavage. The protecting group is introduced by reaction of

$$p\text{-}CH_3C_6H_4SO_2CH_2CH_2OCO-$$
(49)

β-tolylsulphonyloxycarbonyl chloride with the amine and can be removed by treating it with ethanolic potassium hydroxide. The mechanism of the removal (reaction 28) is identical with that of the elimination of arylsulphonyl group from 1,2-diarylsulphonylalkanes[124].

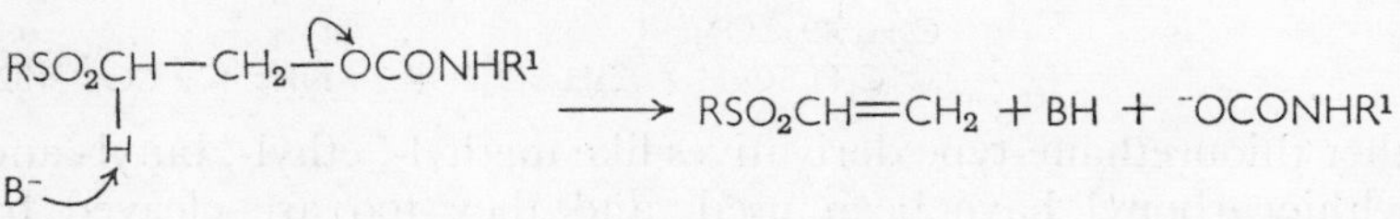

$$RSO_2CH-CH_2-OCONHR^1 \longrightarrow RSO_2CH=CH_2 + BH + {}^-OCONHR^1 \tag{28}$$

C. Thiourethane-type Derivatives

The thiocarbamate derivatives can be obtained by reaction of the aryl- or alkylthiocarbonyl chloride with two equivalents of an amine in an organic solvent (equation 29)[125]. When the alkylation is carried

$$RSCOCl + 2 H_2NR^1 \longrightarrow RSCONHR^1 + RNH_2 \cdot HCl \tag{29}$$

out in aqueous solution, aminolysis of the thioester bond occurs and the substituted urea derivatives are obtained (reaction 30)[126].

$$RSCOCl \xrightarrow{H_2NCHR^1COO^-} RSCONHCHR^1COO^- \xrightarrow{H_2NCHR^1COO^-}$$
$$\qquad\qquad {}^-OOCR^1HCHNCONHCHR^1COO^- + RSH \tag{30}$$

Phenylthiocarbonyl derivatives are stable toward acid hydrolysis but could be removed easily by basic hydrolysis[125]. Their cleavage occurs by the action of lead acetate in methanol or by treating the thiocarbamate with dilute alkali in the presence of basic lead carbonate or lead hydroxide[127]. In the case of various phenylthiocarbonyl derivatives of dipeptide esters the removal of the protecting group by means of lead acetate does not yield the desired dipeptide esters, but hydantoins, in very good yields[128]. The reaction proceeds with the elimination of phenylmercaptan and the formation of an isocyanate which reacts with a neighboring nitrogen to yield the hydantoin (reaction 31). The thiocarbamate group could be removed by

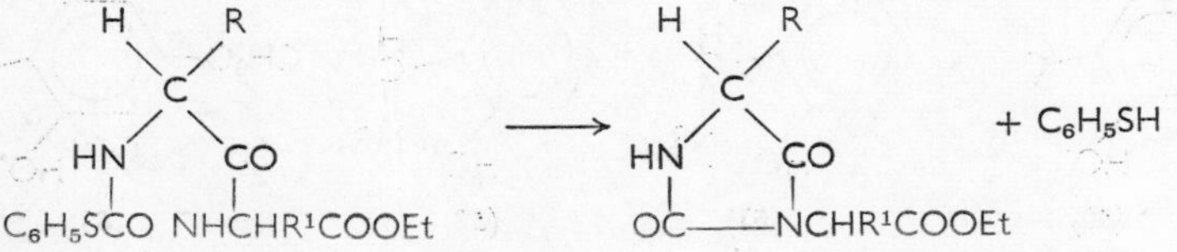

$$\text{(structures)} \longrightarrow \text{(structures)} + C_6H_5SH \tag{31}$$

oxidation with perbenzoic acid when besides the desired amine, carbon dioxide, sulfonic and sulfinic acids are obtained (reaction 32)[129].

$$2 C_6H_5SCONHR \xrightarrow{C_6H_5COOOH} \begin{matrix} C_6H_5SO_2CONHR \\ | \\ O \\ | \\ C_6H_5SO_2CONHR \end{matrix} \xrightarrow{H_2O}$$

$$C_6H_5SO_3H + C_6H_5SO_2H + 2 RNH_2 + 2 CO_2 \quad (32)$$

Other thiourethane-type derivatives like methyl-, ethyl-, butyl- and benzylthiocarbonyl have been used, and they too are cleaved by oxidation with perbenzoic acid[126].

IV. ARYLIDENE-TYPE PROTECTING GROUPS

A. Schiff Bases of Aromatic Aldehydes and Ketones

Schiff bases of benzaldehyde have found a very limited use due to their instability. The benzylidene derivatives of various primary amines have been used for the synthesis of secondary amines (e.g. N-methyl-β-phenylethylamine from β-phenylethylamine (reaction 33)[130]). Benzylidene derivatives of amino acids[131] are known, but

$$C_6H_5CH_2CH_2NH_2 \longrightarrow C_6H_5CH_2CH_2N{=}CHC_6H_5 \longrightarrow$$

$$C_6H_5CH_2CH_2\overset{+}{N}(CH_3){=}CHC_6H_5\ I^- \longrightarrow C_6H_5CH_2CH_2NHCH_3 \quad (33)$$

attempts to use them for the protection of amino groups in peptide synthesis fail because of their instability[132, 133]. However, the benzylidene group has been used to protect the ε-amino group of lysine during the acylation of the α-amino group[134].

Several other aromatic aldehydes form more stable Schiff bases which have found some use as amino-protecting groups. These Schiff bases are derived from salicylaldehyde (50)[135], 5-chlorosalicylaldehyde (51)[136], 2-hydroxy-1-naphthaldehyde (52)[136, 137] and p-methoxysalicylaldehyde (53)[138]. These Schiff bases could be cleaved by treatment with acid but they are resistant toward reductive cleavage by catalytic hydrogenation (contrary to the benzylidene

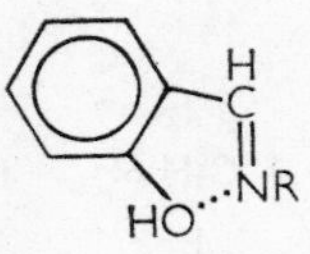

(50) 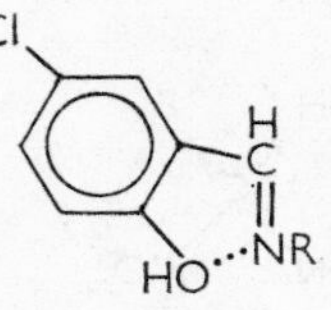   (51) 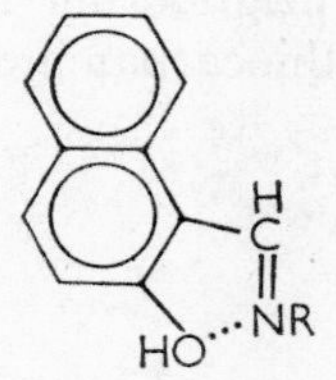   (52) 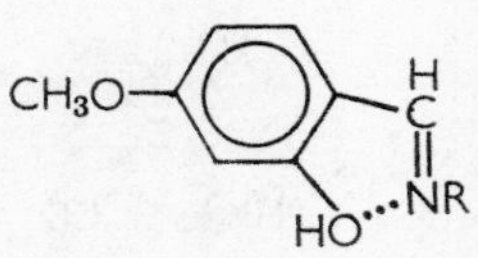   (53)

derivatives)[136]. The stability of o-hydroxybenzylidene derivatives compared with the corresponding benzylidene derivatives is caused by the hydrogen bond between the nitrogen and the hydroxylic group.

Schiff bases derived from ketones have been rarely used; one described example is the synthesis of cycloserine starting with benzophenone oxime[139].

B. Derivatives of 1,3-Dicarbonyl Compounds

Amines react with β-dicarbonyl compounds such as acetylacetone, benzoylacetone or cyclopentan-2-one-1-carboxylic acid ethyl ester to form enamines. These enamines (54) like the o-hydroxybenzylidene derivatives are stabilized by hydrogen bonding. The protecting group is removed by treating the corresponding derivative with dilute acid[140].

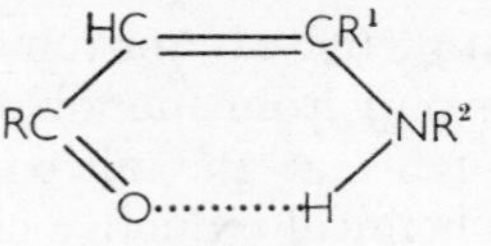

(54)

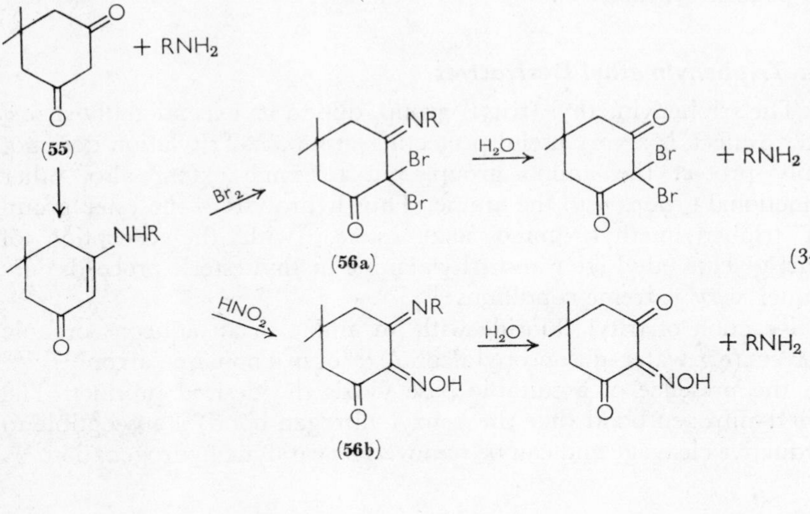

5,5-Dimethylcyclohexane-1,3-dione (dimedone) (55) reacts with amines to form enamine derivatives[141] which are stable toward acids and bases. Those derivatives are set apart from the enamines which are obtained by the reaction between amines and other β-dicarbonyl compounds since cleavage takes place only after bromination[142] or nitrosation[141, 143] at $C_{(2)}$ (reaction 34); the azomethine intermediates (**56a, 56b**) are hydrolyzed immediately by the acid present to yield the free amine.

V. ALKYL- AND ARYL-TYPE PROTECTING GROUPS

A. Benzyl and Dibenzyl Derivatives

Mono- and dibenzyl derivatives are obtained easily by reaction of the amine with benzyl chloride in the presence of base[144, 145]. The monobenzyl derivatives can also be obtained by selective hydrogenolysis of the corresponding benzylidene[144] or dibenzyl[146] derivatives.

The benzyl group is removed by reductive cleavage of the benzyl–nitrogen bond by means of catalytic hydrogenation. While the first benzyl group is removed from dibenzyl derivatives very easily (using palladium black or palladium on barium sulfate as a catalyst at 25° and atmospheric pressure), the ease of removal of the second benzyl group (or the benzyl group from monobenzyl derivatives) depends upon the other substituents at the nitrogen atom[147]. The benzyl group is removed by a chemical reductive cleavage too, using sodium in liquid ammonia[53].

The benzyl and dibenzyl groups are used mainly as an amino-protecting group during the synthesis of primary and secondary amines. It has been shown that the use of N-benzylamino acids or N-benzylamino acid esters instead of the corresponding amino acids or amino acid esters, prevents racemization during the coupling stage in peptide synthesis[148].

B. Triphenylmethyl Derivatives

The triphenylmethyl (trityl) group, due to its exceptionally strong steric effect, is a very useful protecting group. N-Tritylation does not only protect the amino group, but to some extent also, other functional groups α to the amine. Thus hydrolysis of the ester group of triphenylmethyl amino acid esters, (with the exception of tritylglycine ethyl ester and tritylalanine methyl ester), proceeds only under very extreme conditions[149, 150].

Reaction of trityl chloride with an amine in an aqueous–organic solvent (e.g. water–diisopropyl alcohol)[150] or in a nonaqueous one[151,152] in the presence of a suitable base yields the desired product. The trityl–nitrogen bond (like the benzyl–nitrogen bond) is susceptible to reductive cleavage and can be removed by catalytic hydrogenation[150].

It is also very sensitive to acids and can be removed under very mild conditions (15% acetic acid at 30° for 80 minutes)[153]. The sensitivity toward acids is increased by using the mono-*p*-methoxytriphenyl-methyl group, which is removed by treatment with 6% acetic acid at 22° for 60 minutes[154].

C. Dinitrophenyl Derivatives

Amines react readily with 2,4-dinitrofluorobenzene in the presence of sodium bicarbonate to form the corresponding dinitrophenyl derivative (equation 35)[155]. These compounds like most dinitro-

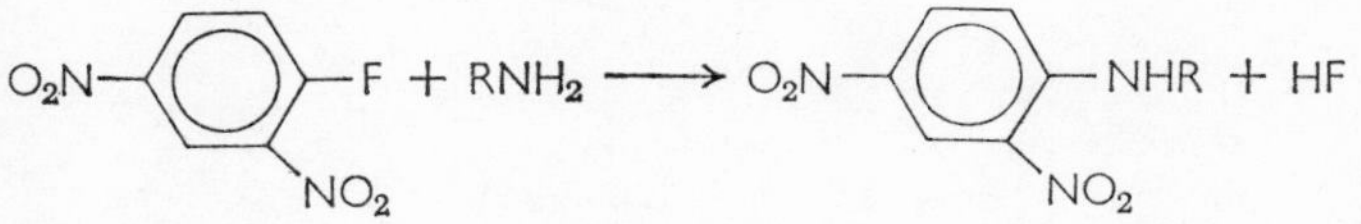

$$O_2N-\left\langle\bigcirc\right\rangle-F + RNH_2 \longrightarrow O_2N-\left\langle\bigcirc\right\rangle-NHR + HF \qquad (35)$$
$$NO_2 \qquad\qquad NO_2$$

phenyl derivatives are susceptible to nucleophilic displacement re-actions which may liberate the free amine. The displacement (in the case of *N*-2,4-dinitrophenyl 2'-amino-2'-deoxy sugar derivatives) could be carried out under very mild conditions using a Dowex 1 ion-exchange resin in the hydroxyl form as the nucleophile[156,157].

D. Trialkylsilyl Derivatives

Trialkylsilazanes are obtained by reaction of trialkylchlorosilane with an amine (reaction 36) (e.g. trimethylchlorosilane with an amino

$$R_3SiCl + R^1NH_2 \longrightarrow R_3SiNHR^1 + 3 HCl \qquad (36)$$

acid ester[158]). Because of their high sensitivity towards hydrolysis, the trialkylsilazanes have not yet found practical use as amine-protecting groups.

VI. PROTECTION BY PROTONATION

Protection of the amino group by protonation is of importance only in very special cases. Some 'activated' amino acids and peptides can be obtained only as their salts, since the free compounds form diketo-piperazines or polymers[159]. One amino group in diamino mono-carboxylic acids can be protected in certain cases by protonation (salt formation). In neutral solution the amino acids are present as a Zwitterionic inner salt, therefore reaction of benzyl phenyl carbon-ate[160] or ethyl trifluorothiolacetate[22] with lysine in neutral solution yields the $N_{(\epsilon)}$-acylated derivatives. Reaction of lysine with trifluoro-acetic anhydride in presence of trifluoroacetic acid yields the $N_{(\alpha)}$ derivative[161] due to the protonation of the $N_{(\epsilon)}$ under the reaction conditions.

α-Amino alcohols can be oxidized to the corresponding α-amino acids if the amino group is protonated first (e.g. the oxidation of 2-aminopropanol to alanine by means of potassium permanganate[162]).

N-Carboxyanhydrides (**57**) usually react with amines to give a

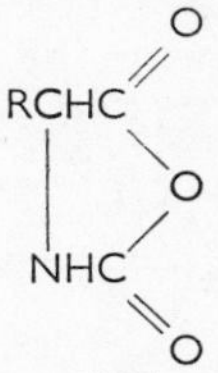

(**57**)

complex mixture of products (including simple amines, polypeptides, the amino acid from which the anhydride is derived, as well as carbamic acids, hydantoins, ureido derivatives etc.)[163]. A clean-cut reaction is observed and the corresponding amides are obtained in 60–90% yield when the reaction is carried out between the *N*-carb-oxyanhydride and a weakly basic amine under suitable conditions. 6-Aminopenicillic acid (**58**) reacts with *N*-carboxy-D-phenylglycine anhydride (**59**) in aqueous solution at pH 4·9 to give the dipeptide (**60**) in nearly quantitative yield[164]. At this pH half of the 6-amino-penicillic acid ($pK_2 = 4·92$) has its amino group in the form of the free base and capable of reacting with the anhydride (**59**) while only

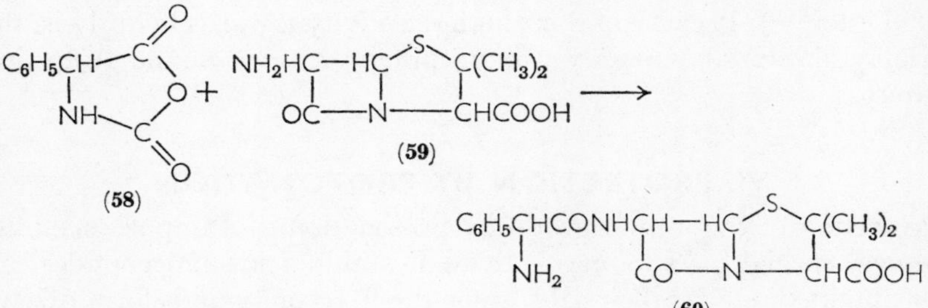

one molecule out of ten thousand of the hydrolyzed *N*-carboxy-anhydride ($pK_2 = 9·03$) is in the form of the free base, and the rest of the molecules are protonated and unavailable for further reaction. Similarly, the hydrochlorides of *p*-anisidine and *p*-chloroaniline react with N-carboxyphenylalanine anhydride in dimethylformamide to give the corresponding amides in 50–70% yield[165]. Recently it was observed that a clean-cut reaction takes place between the anhydride (**57**) and any amino acid or peptide when using suitable conditions[166]. It seems that the amino group is protected as the carbamate anion ($^-$OOCHNR) under the reaction conditions (in a buffer of pH 10 as the solvent at 0°).

VII. PROTECTION BY CHELATE FORMATION

This kind of protection is limited to amino groups which are in the α- or β-position to a carboxylic group.

It has been shown that copper ions give complexes with α- or β-amino acids. In neutral or basic solution these complexes have a chelate structure (**61**) which due to its stability protects the amino

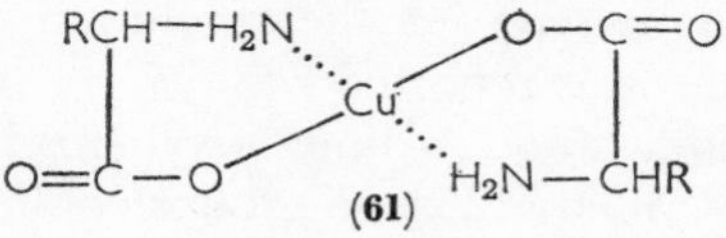

(61)

group[167]. Other ions may form a chelate structure too, but the superiority of the cupric ions over others, lies both in the ease of the complex formation and their removal. Carbamylation, acylation and guanidization at the $N_{(\omega)}$ of α, γ-diaminobutyric acid, ornithine and lysine[168], as well as the acylation of the phenolic hydroxyl group of tyrosine[169], using the copper complexes, were reported.

It is interesting to note that all attempts to work with the chelate in nonaqueous media or to protect the α-amino group during alkylation were unsuccessful.

VIII. ADDENDUM

Since the completion of this chapter some more work of interest have appeared.

The specific problems of the amino-protecting groups in peptide syntheses were discussed in a recent book[170]. Various haloacetyl groups have been used in order to protect the amino group of cytosine (**2**)[171]. The ease of removal of the haloacetyl group is: trifluoroacetyl > trichloroacetyl > dichloroacetyl > chloroacetyl > acetyl. The trifluoroacetyl group could be removed even by treatment with aqueous bicarbonate solution for a few seconds. A new method for the introduction of the phthaloyl group by reaction of phthalic acid with ethoxyacetylene in the presence of the desired amine has been described[172]. A modified sulfenyl protecting group, 2,3,4-trichlorophenylsulfenyl, that could be removed by treatment with boiling aqueous acetic acid was reported[173]. The p-nitrophenylsulfenyl group could be removed by displacement with various aliphatic thiols: ethanethiol, propane-2-thiol, butane-2-thiol and hydrogen sulfide[174,175].

The synthesis of benzyloxycarbonyl (31) derivatives by the aminolysis of various active esters of benzyl carbonate has been reported[176]. The use of N-hydroxypiperidyl p-methoxybenzyl carbonate and N-hydroxypiperidyl t-butyl carbonate for the synthesis of p-methoxybenzyloxycarbonyl and t-butyloxycarbonyl (45) derivatives, respectively, has been described[177]. The benzhydryloxycarbonyl group (35) could be removed by the action of boron trifluoride in glacial acetic acid[178]. The use of easily cleaved, optically active, protecting groups of the urethane type for the resolution of racemic primary and secondary amines was reported[179].

The use of p-methoxy- and 2,4-dimethoxybenzyl as amino-protecting groups has been described[180]; these groups could be removed by the action of trifluoroacetic acid. The trimethylsilyl group was used to protect the amino group during various organometallic transformations[181].

IX. REFERENCES

1. J. F. W. McOmie in *Advances in Organic Chemistry*, Vol. 3 (Ed. R. A. Raphael), Interscience Publishers, New York, 1963, pp. 203–215.
2. J. Rudinger, *Pure Appl. Chem.*, **7**, 335 (1963).
3. R. A. Boissonnas in *Advances in Organic Chemistry*, Vol. 3 (Ed. R. A. Raphael), Interscience Publishers, New York, 1963, pp. 159–190.
4. K. Hofmann and P. G. Katsoyannis in *The Proteins*, Vol. 1 (Ed. H. Neurath), Academic Press, New York, 1963, pp. 59–69.
5. E. Schröder and K. Lubke, *The Peptides*, Vol. 1, Academic Press, New York, 1965, pp. 3–51.
6. C. Schotten, *Chem. Ber.*, **21**, 2235 (1888); E. Baumann, *Chem. Ber.*, **19**, 3218 (1886).
7. J. C. Sheehan and D. D. H. Yang, *J. Am. Chem. Soc.*, **80**, 1154 (1958).
8. K. Hofmann, E. Stutz, G. Spühler, H. Yajima and E. T. Schwartz, *J. Am. Chem. Soc.*, **82**, 3727 (1960).
9. K. Okawa and S. Hase, *Bull. Chem. Soc. Japan*, **36**, 754 (1963).
10. G. Losse and W. Zönnchen, *Ann. Chem.*, **636**, 140 (1960).
11. G. Losse and D. Nadolski, *J. Prakt. Chem.*, **24**, 118 (1964).
12. J. C. Power, R. Seidner and T. G. Parson, *Tetrahedron Letters*, 1713 (1965).
13. L. Horner and H. Neumann, *Chem. Ber.*, **98**, 3462 (1965).

14. M. Fujinaga and Y. Matsushima, *Bull. Chem. Soc. Japan*, **39**, 185 (1966).
15. R. Locquin and V. Cerchez, *Bull. Soc. Chim. France*, **49**(4), 42 (1931).
16. F. Knop and H. Oesterlin, *Z. Physiol. Chem.*, **170**, 186 (1927).
17. M. Conrad and A. Schulze, *Chem. Ber.*, **42**, 729 (1909).
18. R. Lohrmann, D. Soll, H. Hayatsu, E. Ohtsuka and H. G. Khorana, *J. Am. Chem. Soc.*, **88**, 819 (1966).
19. H. G. Khorana, A. F. Turner and J. P. Vuzsolyi, *J. Am. Chem. Soc.*, **83**, 686 (1961).
20. R. Lohrmann and H. G. Khorana, *J. Am. Chem. Soc.*, **86**, 4188 (1964).
21. F. Weygand and E. Csendes, *Angew Chem.*, **64**, 136 (1952).
22. E. E. Schallenberg and M. Calvin, *J. Am. Chem. Soc.*, **77**, 2779 (1955).
23. F. Weygand and A. Röpsch, *Chem. Ber.*, **92**, 2095 (1959).
24. F. Weygand, B. Kolb, A. Prox, M. A. Tilak and I. Tomida, *Z. Physiol. Chem.*, **322**, 38 (1960).
25. F. Weygand and W. Swodenk, *Chem. Ber.*, **90**, 639 (1957).
26. H. Newman, *J. Org. Chem.*, **30**, 1287 (1965).
27. M. L. Wolfrom and H. B. Bhat, *Chem. Commun.*, 146 (1966).
28. F. D'Angeli, F. Filira and E. Scoffone, *Tetrahedron Letters*, 605 (1965).
29. L. Jacobs in *Heterocyclic Compounds*, Vol. 5 (Ed. R. C. Elderfield), John Wiley and Sons, New York, 1957, p. 116.
30. J. C. Sheehan and E. J. Corey, *J. Am. Chem. Soc.*, **74**, 4555 (1952).
31. S. Gabriel, *Chem. Ber.*, **20**, 2224 (1887).
32. J. H. Billman and W. F. Harting, *J. Am. Chem. Soc.*, **70**, 1473 (1948).
33. A. K. Bose, F. Greer and C. C. Price, *J. Org. Chem.*, **23**, 1335 (1958).
34. K. Balenović and B. Gaspert, *Chem. Ind.* (*London*), 624 (1960).
35. G. H. L. Nefkens, G. I. Tesser and R. J. F. Nivard, *Rec. Trav. Chim.*, **79**, 688 (1960).
36. J. C. Sheehan, D. W. Chapman and R. W. Roth, *J. Am. Chem. Soc.*, **74**, 3822 (1952).
37. B. Helferich and H. Böshagen, *Chem. Ber.*, **92**, 2813 (1959).
38. F. E. King and D. A. A. Kidd, *J. Chem. Soc.*, 3315 (1949).
39. H. R. Ing and R. H. F. Manske, *J. Chem. Soc.*, 2348 (1926).
40. H. J. Barber and W. R. Wragg, *Nature*, **158**, 514 (1946).
41. J. C. Sheehan and V. S. Frank, *J. Am. Chem. Soc.*, **71**, 1856 (1949).
42. W. Grassmann and E. Schulte-Uebbing, *Chem. Ber.*, **83**, 244 (1950).
43. I. Schumann and R. A. Boissonnas, *Helv. Chim. Acta*, **35**, 2237 (1952).
44. J. C. Sheehan and K. R. Henery-Logan, *J. Am. Chem. Soc.*, **81**, 3089 (1959).
45. R. L. Shriner and F. W. Neumann, *Chem. Rev.*, **35**, 351 (1944).
46. M. J. Hunter and M. L. Ludwig, *J. Am. Chem. Soc.*, **84**, 3491 (1962).
47. M. L. Ludwig and R. Byrne, *J. Am. Chem. Soc.*, **84**, 4160 (1962).
48. T. B. Johnson and J. A. Ambler, *J. Am. Chem. Soc.*, **36**, 372 (1914).
49. W. H. Carothers, C. F. Bickford and G. J. Hurwitz, *J. Am. Chem. Soc.*, **49**, 2908 (1927).
50. G. Wittig, W. Joos and P. Rathfelder, *Ann. Chem.*, **610**, 180 (1957).
51. O. Hinsberg, *Chem. Ber.*, **23**, 2962 (1890).
52. E. Fischer and W. Lipschitz, *Chem. Ber.*, **48**, 360 (1915).
53. V. du Vigneaud and O. K. Behrens, *J. Biol. Chem.*, **117**, 27 (1937).
54. C. C. Howard and W. Marckwald, *Chem. Ber.*, **32**, 2031 (1899).

55. E. Fischer, *Chem. Ber.*, **48**, 93 (1915).
56. R. Schönheimer, *Z. Physiol. Chem.*, **154**, 203 (1926).
57. H. R. Snyder and R. E. Heckert, *J. Am. Chem. Soc.*, **74**, 2006 (1952).
58. H. Ohle and G. Haeseler, *Chem. Ber.*, **69**, 2324 (1936).
59. R. S. Schreiber and R. L. Shriner, *J. Am. Chem. Soc.*, **56**, 1618 (1934).
60. D. Klamann and G. Hofbauer, *Chem. Ber.*, **86**, 1246 (1953).
61. J. Kovacs and U. R. Ghatak, *J. Org. Chem.*, **31**, 119 (1966).
62. H. B. Milne and C. H. Peng, *J. Am. Chem. Soc.*, **79**, 639 (1957).
63. K. Onodera, S. Kitaoka and H. Ochiai, *J. Org. Chem.*, **27**, 156 (1962).
64. M. L. Wolfrom and R. Wurmb, *J. Org. Chem.*, **30**, 3058 (1965).
65. T. Zincke and F. Farr, *Ann. Chem.*, **391**, 57 (1912).
66. H. Lecher and F. Holschneider, *Chem. Ber.*, **57**, 755 (1924).
67. J. H. Billman and E. O'Mahony, *J. Am. Chem. Soc.*, **61**, 2340 (1939).
68. J. Goerdeler and A. Holst, *Angew. Chem.*, **71**, 775 (1959).
69. J. Meienhofer, *Nature*, **205**, 73 (1965).
70. W. Kessler and B. Iselin, *Helv. Chim. Acta*, **49**, 1330 (1966).
71. A. Fontana, F. Marchiori, L. Moroder and E. Scoffone, *Tetrahedron Letters*, 2985 (1966).
72. L. Zervas, D. Borovas and E. Gazis, *J. Am. Chem. Soc.*, **85**, 3660 (1963).
73. L. Zervas and C. Hamalidis, *J. Am. Chem. Soc.*, **87**, 99 (1965).
74. A. Fontana, F. Marchiori and L. Moroder, *Ric. Sci. Suppl.*, **36**, 261 (1966).
75. S. O. Li and R. E. Eakin, *J. Am. Chem. Soc.*, **77**, 1866 (1955).
76. L. Zervas and P. G. Katsoyannis, *J. Am. Chem. Soc.*, **77**, 5351 (1955).
77. A. Cosmatos, I. Photaki and L. Zervas, *Chem. Ber.*, **94**, 2644 (1961).
78. E. J. Corey and L. F. Haefele, *J. Am. Chem. Soc.*, **81**, 2225 (1959).
79. G. L. Schmir, L. A. Cohen and B. Witkop, *J. Am. Chem. Soc.*, **81**, 2228 (1959).
80. W. B. Lawson, E. Gross, C. M. Foltz and B. Witkop, *J. Am. Chem. Soc.*, **84**, 1715 (1962).
81. N. Izumiya, J. F. Francis, A. V. Robertson and B. Witkop, *J. Am. Chem. Soc.*, **84**, 1702 (1962).
82. A. Patchornik, W. B. Lawson, E. Gross and B. Witkop, *J. Am. Chem. Soc.*, **82**, 5923 (1960).
83. G. L. Schmir and L. A. Cohen, *J. Am. Chem. Soc.*, **83**, 723 (1961).
84. L. Farber and L. A. Cohen, *Biochemistry*, **5**, 1027 (1966).
85. E. Gross and B. Witkop, *J. Biol. Chem.*, **237**, 1856 (1962).
86. T. Wieland, Ch. Lamperstorfer and Ch. Birr, *Makromol. Chem.*, **92**, 277 (1966).
87. W. A. Jacobs and M. Heidelberger, *J. Am. Chem. Soc.*, **39**, 2418 (1917).
88. R. W. Holley and D. Holley, *J. Am. Chem. Soc.*, **74**, 3069 (1952).
89. M. Bergmann and L. Zervas, *Chem. Ber.*, **65**, 1192 (1932).
90. R. H. Sifferd and V. du Vigneaud, *J. Biol. Chem.*, **108**, 753 (1935).
91. C. R. Harington and T. H. Mead, *Biochem. J.*, **29**, 1602 (1935).
92. N. F. Albertson and F. C. McKay, *J. Am. Chem. Soc.*, **75**, 5323 (1953).
93. D. Ben Ishai and A. Berger, *J. Org. Chem.*, **17**, 1564 (1952).
94. G. W. Anderson, J. Blodinger and A. D. Welcher, *J. Am. Chem. Soc.*, **74**, 5309 (1952).
95. R. A. Boissonnas and G. Preitner, *Helv. Chim. Acta*, **36**, 875 (1953).
96. S. Guttmann and R. A. Boissonnas, *Helv. Chim. Acta*, **42**, 1257 (1959).

97. F. H. Carpenter and D. T. Gish, *J. Am. Chem. Soc.*, **74**, 3818 (1952).
98. D. T. Gish and F. H. Carpenter, *J. Am. Chem. Soc.*, **75**, 950 (1953).
99. L. Kisfaludy and S. Dualszky, *Acta. Chim. Acad. Sci. Hung.*, **24**, 301 (1960).
100. D. M. Channing, P. B. Turner and G. T. Young, *Nature*, **167**, 487 (1951).
101. F. C. McKay and N. F. Albertson, *J. Am. Chem. Soc.*, **79**, 4686 (1957).
102. F. Weygand and K. Hunger, *Chem. Ber.*, **95**, 1, (1962).
103. J. W. Chamberlin, *J. Org. Chem.*, **31**, 1658 (1966).
104. R. Schwyzer, P. Sieber and K. Zatskó, *Helv. Chim. Acta*, **41**, 491 (1958).
105. R. G. Hiskey and J. B. Adams, Jr., *J. Am. Chem. Soc.*, **87**, 3969 (1965).
106. J. A. Barltrop, and P. Schofield, *J. Chem. Soc.*, 4758 (1965).
107. K. Blàha and J. Rudinger, *Collection Czech. Chem. Commun.*, **30**, 585 (1965).
108. R. B. Homer, R. B. Moodie and H. N. Rydon, *J. Chem. Soc.*, 4403 (1965).
109. H. Jeschkeit, G. Losse and K. Neubert, *Chem. Ber.*, **99**, 2803 (1966).
110. E. J. Corey and R. L. Dawson, *J. Am. Chem. Soc.*, **84**, 4899 (1962).
111. C. M. Stevens and R. Watanabe, *J. Am. Chem. Soc.*, **72**, 725 (1950).
112. W. J. Bailey and J. R. Griffith, *Polymer Preprints*, **5**, 266 (1964).
113. G. W. Anderson and A. C. McGregor, *J. Am. Chem. Soc.*, **79**, 6180 (1957).
114. S. Sakakibara, M. Shin, M. Fujino, Y. Shimonishi, S. Inove and N. Inukai, *Bull. Chem. Soc. Japan*, **38**, 1522 (1965).
115. W. L. Haas, E. V. Krumkalns and K. Gerzon, *J. Am. Chem. Soc.*, **88**, 1988 (1966).
116. R. Schwyzer, P. Sieber and H. Kappeler, *Helv. Chim. Acta*, **42**, 2622 (1959).
117. L. A. Carpino, *J. Am. Chem. Soc.*, **79**, 98 (1957).
118. L. A. Carpino, *J. Org. Chem.*, **29**, 2820 (1964).
119. M. Frankel, L. Ladkany, C. Gilon and Y. Wolman, *Tetrahedron Letters*, 4765 (1966).
120. R. B. Woodward, K. Heusler, J. Gosteli, P. Naegeli, W. Oppolzer, R. Ramage, S. Ranganathan and H. Vorbruggen, *J. Am. Chem. Soc.*, **88**, 852 (1966).
121. W. J. Bailey and J. R. Griffith, *Polymer Preprints*, **5**, 279 (1964).
122. L. Blàha and J. Rudinger, *Collection Czech. Chem. Commun.*, **30**, 599 (1965).
123. A. T. Kader and C. J. M. Stirling, *Proc. Chem. Soc.*, 363 (1962).
124. A. T. Kader and C. J. M. Stirling, *J. Chem. Soc.*, 3686 (1962).
125. M. H. Rivier, *Bull. Soc. Chim. France*, **1**(4), 733 (1907).
126. J. Kollonitsch, J. Gábor and A. Hajos, *Chem. Ber.*, **89**, 2293 (1956).
127. G. C. H. Ehrensvard, *Nature*, **159**, 500 (1947).
128. A. Lindenmann, N. H. Khan and K. Hofmann, *J. Am. Chem., Soc.* **74**, 476 (1952).
129. J. Kollonitsch, A. Hajos and J. Gábor, *Chem. Ber.*, **89**, 2288 (1956).
130. H. Decker and P. Becker, *Ann. Chem.*, **395**, 362 (1913).
131. M. Bergmann and L. Zervas, *Z. Physiol. Chem.*, **152**, 282 (1926).
132. T. Wieland and W. Schäfer, *Ann. Chem.*, **576**, 104 (1952).
133. E. Gazis, B. Bezas, G. C. Stelakatos and L. Zervas in *Peptide Symposium*, (Ed. G. Young), Macmillan, New York, 1963, p. 17.
134. B. Bezas and L. Zervas, *J. Am. Chem. Soc.*, **83**, 719 (1961).
135. J. N. Williams, Jr. and R. M. Jacobs, *Biochem. Biophys. Res. Commun.*, **22**, 695 (1966).
136. F. C. McIntire, *J. Am. Chem. Soc.*, **69**, 1377 (1947).
137. Z. E. Jolles and W. T. J. Morgan, *Biochem. J.*, **34**, 1183 (1940).

138. L. Zervas and S. Konstas, *Chem. Ber.*, **93**, 435 (1960).
139. H. Bretschneider and W. Vetter, *Monatsh. Chem.*, **90**, 799 (1959).
140. E. Dane, F. Drees, P. Konrad and T. Dockner, *Angew. Chem.*, **74**, 873 (1962).
141. P. Haas, *J. Chem. Soc.*, **95**, 421 (1909).
142. B. Halpern and L. B. James, *Nature*, **202**, 592 (1964).
143. B. Halpern and A. D. Cross, *Chem. Ind. (London)*, 1183 (1965).
144. L. Velluz, G. Amiard and R. Heymes, *Bull. Soc. Chim. France*, 1012 (1954).
145. L. Velluz, G. Amiard and R. Heymes, *Bull. Soc. Chim. France*, 201 (1955).
146. H. J. Haas, *Chem. Ber.*, **94**, 2442 (1961).
147. W. H. Hartung and R. Simonoff in *Organic Reactions*, Vol. 7 (Ed. R. Adams), John Wiley and Sons, New York, 1953, p. 275.
148. G. C. Stelakatos and M. Argyropoulos, *Chem. Commun.*, 270 (1966).
149. G. Amiard, R. Heymes and L. Velluz, *Bull. Soc. Chim. France*, 97 (1956).
150. L. Zervas and D. M. Theodoropoulos, *J. Am. Chem. Soc.*, **78**, 1359 (1956).
151. A. Hillmann-Elies, G. Hillmann, and H. Jatzkewitz, *Z. Naturwiss.*, **8b**, 445 (1953).
152. G. Amiard, R. Heymes and L. Velluz, *Bull. Soc. Chim. France*, 191 (1955).
153. R. Schwyzer and W. Rittel, *Helv. Chim. Acta.*, **44**, 159 (1961).
154. Y. Lapidot, N. deGroot, M. Weiss, R. Peled and Y. Wolman, *Biochim. Biophys. Acta*, **138**, 241 (1967).
155. F. Sanger, *Biochem. J.*, **39**, 507 (1945).
156. P. E. Lloyd and M. Stacey, *Tetrahedron*, **9**, 116 (1960).
157. M. L. Wolfrom, H. G. Garg and D. Horton, *J. Org. Chem.*, **30**, 1556 (1965).
158. L. Birkhofer and A. Ritter, *Ann. Chem.*, **612**, 22 (1958).
159. E. Katchalski, M. Sela, H. I. Silman and A. Berger in *The Proteins*, Vol. 2, 2nd ed., (Ed. H. Neurath), Academic Press, New York, 1964, pp. 425–426.
160. H. Zahn and H. R. Falkenburg, *Ann. Chem.*, **636**, 117 (1960).
161. F. Weygand and R. Geiger, *Chem. Ber.*, **89**, 647 (1956).
162. J. H. Billman, E. E. Parker and W. T. Smith, *J. Biol. Chem.*, **180**, 29 (1949).
163. E. Katchalski, M. Sela, H. I. Silman and A. Berger in *The Proteins*, Vol. 2, 2nd ed., (Ed. H. Neurath), Academic Press, New York, 1964, pp. 413–416.
164. N. H. Grant and H. F. Alburn, *J. Am. Chem. Soc.*, **86**, 3870 (1964).
165. Y. Knobler, S. Bittner and M. Frankel, *J. Chem. Soc.*, 3941 (1964).
166. R. G. Denkewalter, H. Schwam, R. G. Stracham, T. E. Beesley, D. F. Veber, E. F. Schoenewaldt, H. Barkemeyer, W. J. Paleveda Jr., T. A. Jacob and R. Hirschmann, *J. Am. Chem. Soc.*, **88**, 3163 (1966).
167. A. C. Kurtz, *J. Biol. Chem.*, **122**, 477 (1937).
168. A. C. Kurtz, *J. Biol. Chem.*, **180**, 1253 (1949).
169. B. G. Overell and V. Petrow, *J. Chem. Soc.*, 232 (1955).
170. M. Bodanszky and M. A. Ondetti, *Peptide Synthesis*, Interscience Publishers, New York, 1966, pp. 21–41.
171. R. S. Goody and R. T. Walker, *Tetrahedron Letters*, 289 (1967).
172. G. R. Banks, D. Cohen, G. E. Pattenden and J. A. G. Thomas, *J. Chem. Soc.* (C), 126 (1967).

173. St. Guttmann in *Peptides*, Proceedings of the 6th European Symposium Athens, September 1963 (Ed. L. Zervas), Pergamon Press, Oxford, 1966, p. 116.
174. A. Fontana, F. Marchiori, L. Moroder and E. Scoffone, *Gazz. Chem. Ital.*, **96**, 1313 (1966).
175. D. Brandenburg, *Tetrahedron Letters*, 6201 (1966).
176. Y. Wolman, D. Ladkany and M. Frankel, *J. Chem. Soc.* (C), 689 (1967).
177. J. H. Jones and G. T. Young, *Chem. Ind.* (*London*), 1722 (1966).
178. R. G. Hiskey and E. L. Smithwick Jr., *J. Am. Chem. Soc.*, **89**, 437 (1967).
179. L. A. Carpino, *Chem. Commun.*, 858 (1966).
180. F. Veygand, W. Steglich, J. Bjarnson, R. Akktar and N. M. Khan, *Tetrahedron Letters*, 3483 (1966).
181. D. R. M. Walton, *J. Chem. Soc.* (C), 1706 (1966).

CHAPTER 12

Tetraaminoethylenes

DAVID M. LEMAL

Dartmouth College, Hanover, New Hampshire, U.S.A.

I. INTRODUCTORY NOTE

Tetraaminoethylenes are a particularly fascinating class of amines. Their double bond bristling with the best (uncharged) electron-donating groups available, these substances in many ways represent the antithesis of the more familiar tetracyanoethylene. Thus they are powerful nucleophiles, π bases, and reducing agents. Were these their only outstanding properties, they might be viewed merely as ultra-powerful enamines, but another factor non-existent in the chemistry of simple enamines profoundly influences synthetic approaches to, and reactions of, peraminoethylenes. This factor is the unusual

701

stability of diaminocarbenes, more accurately described as ylids, which constitute in a formal sense the two halves of the ethylene. Hence much of peraminoethylene chemistry is genuinely novel in character; it is novel in point of time as well, for the field is still in an early stage of development.

II. METHODS FOR SYNTHESIS

A. From a Perhaloethylene

The first tetraaminoethylene (1) was synthesized in 1950 by Pruett[1] by the reaction of dimethylamine with chlorotrifluoroethylene in a bomb. Vigorous in its early stages, the reaction requires heating

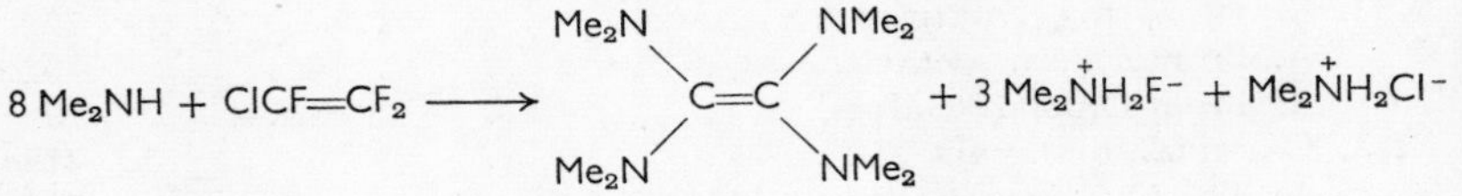

$$8 \; Me_2NH + ClCF{=}CF_2 \longrightarrow \underset{Me_2N}{\overset{Me_2N}{}}\!\!\!\!C{=}C\!\!\!\!\underset{NMe_2}{\overset{NMe_2}{}} + 3 \; Me_2\overset{+}{N}H_2F^- + Me_2\overset{+}{N}H_2Cl^-$$

(1)

to 70–80° for completion[2]. This approach fails for the next higher homolog, tetrakis(diethylamino)ethylene, presumably for steric reasons. The addition–elimination sequence can be accomplished only twice, giving 2-chloro-1,1-bis(diethylamino)-2-fluoroethylene. It has been reported, in fact, that despite attempts with both cyclic and acyclic secondary amines, no tetrakis(dialkylamino)ethylene other than 1 could be synthesized via Pruett's method[3,4]. Chlorotrifluoroethylene was apparently the only perhaloolefin used in these experiments.

An exception is the fused bicyclic olefin (2) prepared in low yield from N,N'-dimethylethylenediamine[5]. It is not entirely clear why 2 is preferred over its isomer 3 in this reaction; the result is a fortunate

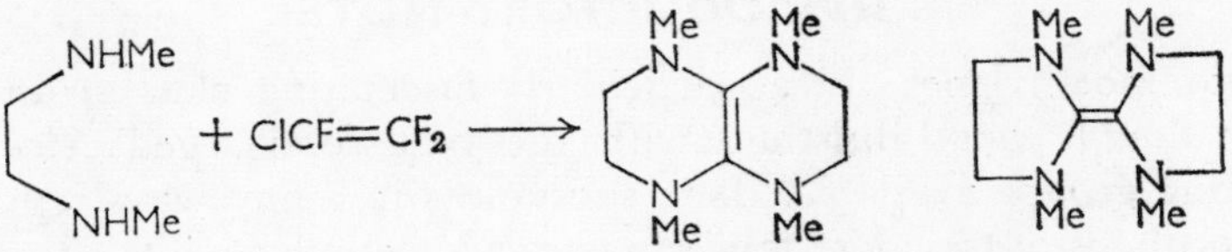

(2)   (3)

one, however, because presently available alternative routes to tetra-aminoethylenes are well adapted to generating ring systems akin to 3 and incapable of yielding fused systems like 2. Though the Pruett method is clearly very limited in scope, one can hope that it will succeed with other short-chain secondary diamines.

B. From Dipolar 'Halves'

After Pruett's brief account, the next report concerning a tetra-aminoethylene appeared a decade later from the laboratory of H.-W. Wanzlick[6a], who became a major contributor to the field. Heated in collidine, the condensation product 4 of chloral and N,N'-diphenylethylenediamine had yielded a high-melting substance whose Raman spectrum displayed a strong band at 1640 cm^{-1} (absent in the infrared spectrum) and whose microanalytical data and chemical

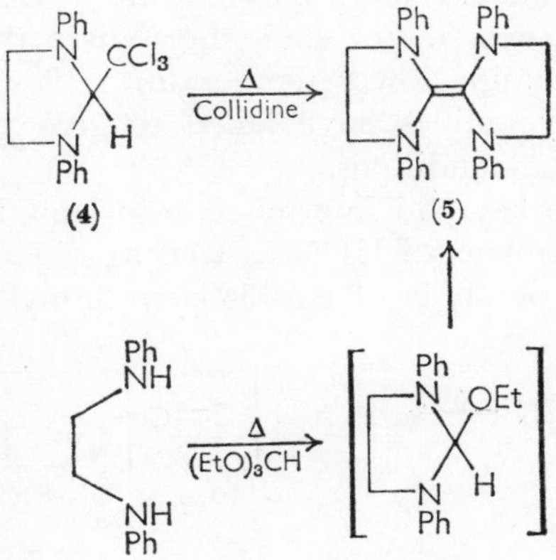

behavior established its structure as 5. A more facile synthesis based on the same principle soon followed, viz. reaction of the diamine with ethyl orthoformate[6b]. In this instance the elements of ethanol, instead of chloroform, are lost from the intermediate 6 and the re-sulting ylid (7) is isolated as its dimer. Generated in a different manner,

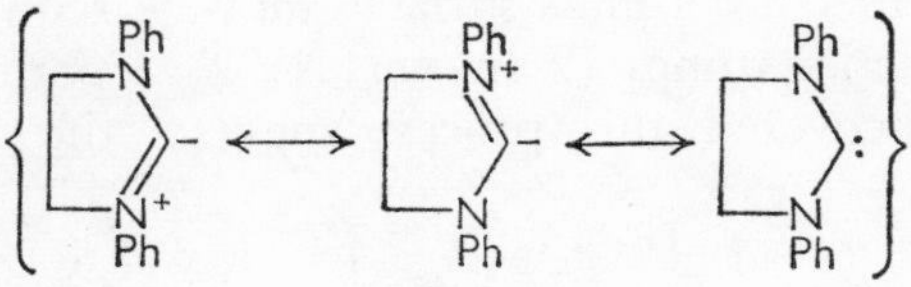

to be discussed, the methoxy analog of 6 was actually isolated and then thermally decomposed to 5[7]. The Berlin workers used the ethyl orthoformate method to prepare a whole array of tetrakis(aralkyl-amino)ethylenes[8], and a group at duPont synthesized a long series of tetrakis(dialkylamino)ethylenes using an ingenious variation on this theme[9,10]. In this approach the dimethyl acetal of N,N-dimethyl-formamide is simply heated with a secondary amine and the methanol and dimethylamine generated in the resulting series of equilibria (equations 1-3) are removed by distillation. Secondary amines with

$$R_2NH + Me_2NCH(OMe)_2 \rightleftharpoons R_2NCH(OMe)_2 + Me_2NH \qquad (1)$$

$$R_2NCH(OMe)_2 + R_2NH \rightleftharpoons (R_2N)_2CHOMe + MeOH \qquad (2)$$

$$2\,(R_2N)_2CHOMe \rightleftharpoons (R_2N)_2C{=}C(NR_2)_2 + 2\,MeOH \qquad (3)$$

high basicity and small steric requirements give excellent conversions, particularly at relatively high temperatures (125–250°) with the amine in generous excess. For such short-chain diamines as N,N'-dimethylethylenediamine lower temperatures (80–125°) suffice, but hindered secondary amines have failed to give peraminoethylenes regardless of reaction conditions.

Formation of the key ylid intermediate in equation (3) can result from loss of the elements of HCN as well as of ROH or $CHCl_3$, as illustrated by the pyrolysis of α,α-bis(pyrrolidinyl)acetonitrile (8)[9].

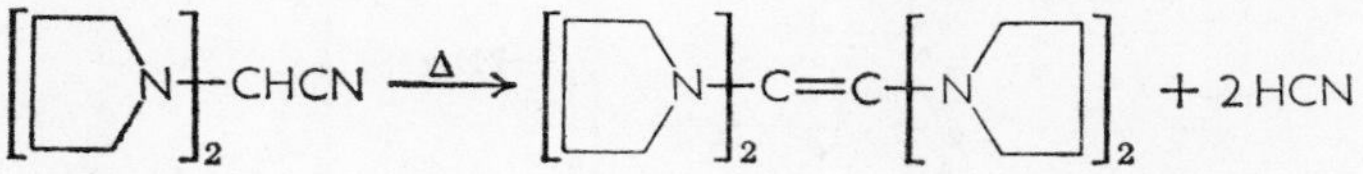

(8)

Each of the three kinds of 'leaving molecule' can be represented as HY, the conjugate acid of a moderately good leaving group; one may suggest that in each of the syntheses discussed HY is lost via ionization to a formamidinium salt (9) followed by proton transfer to the anion. Recently tetrakis(dimethylamino)ethylene (1) has been prepared in high yield by pyrolysis of tris(dimethylamino)methane[10b]. Since the dimethylamide ion is a very poor leaving group, trace proton sources, as provided by a glass surface, may be required. The great disparity in reaction times (2 hr versus 72) reported by the two groups who discovered this process suggests the involvement of

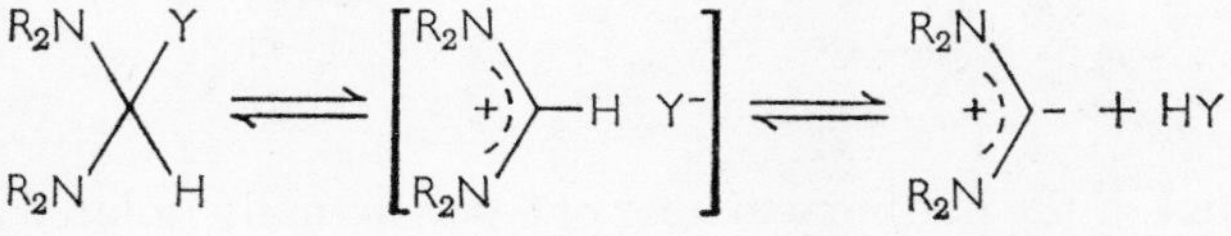

(9)

catalysis. In any event, the notion that a formamidinium ion intervenes in all of the reactions under discussion points to yet another synthetic variation, viz. the reaction of a formamidinium salt with a hindered base. Here the anion Y^- is relatively inert, but choice of an external base which is sufficiently hindered not to add (or at least not to form a stable adduct) permits proton removal at room temperature or below $(\mathbf{10} \rightarrow \mathbf{5})$ * [5, 11].

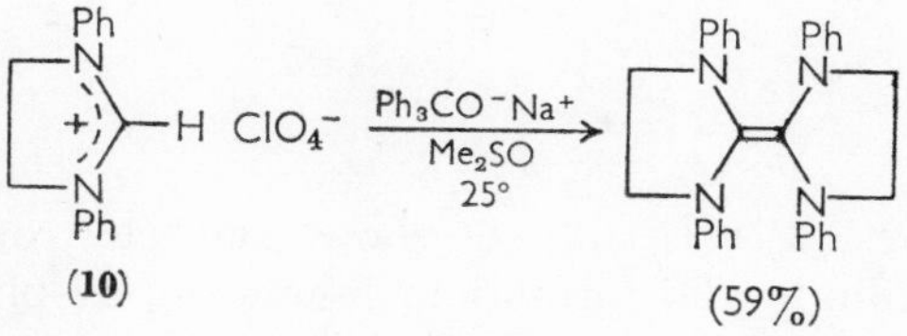

$$(\mathbf{10}) \qquad \qquad (59\%)$$
$$(\mathbf{5})$$

Ultimately, then, all of the peraminoethylene syntheses from dipolar 'halves' have depended upon the acidity of the central proton of a formamidinium ion. Thus all were foreshadowed by the pioneering work of Breslow on the mechanism of thiamine action [14, 15] and, more remotely, by Hammick's study of the decarboxylation of certain heterocyclic acids [16a]. The Hammick reaction, illustrated in equation (4), led its discoverer to conclude that decarboxylation yields dipolar

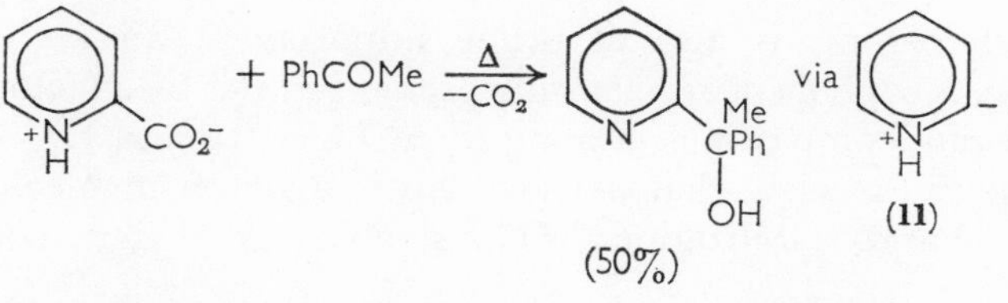

$$(50\%)$$

intermediates such as **11** (though he believed that subsequent attack on a carbonyl compound was effected by the conjugate base of the ylid) [16b]. In the late fifties, Breslow demonstrated that imidazolium and particularly thiazolium salts (e.g. **12**) suffer deuterium exchange at $C_{(2)}$ under extraordinarily gentle conditions, again via dipolar intermediates such as **13**.

* Tetramethylformamidinium salts are reported not to react even with a variety of very powerful bases [12, 13], but solubility problems may be primarily responsible for this result. With methyllithium fairly high yields of methane, but very little **1**, are obtained [12].

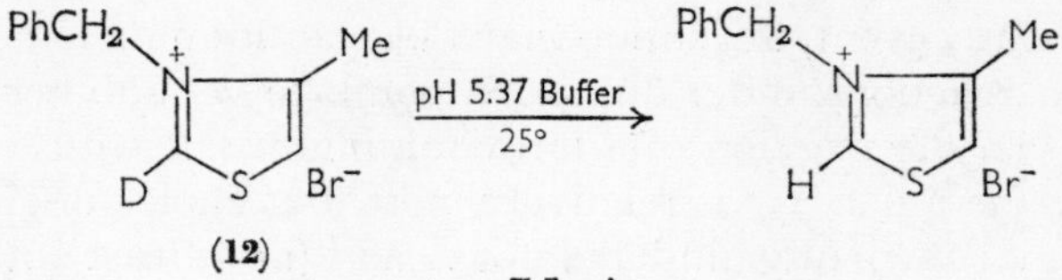

(12)

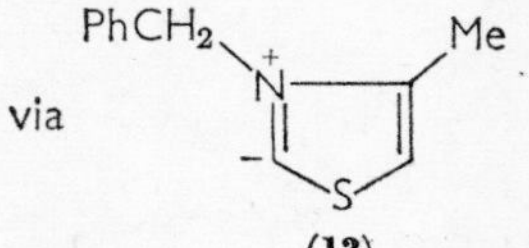

$t_{\frac{1}{2}} = 7.5$ min

via

(13)

Although Breslow was able to show that electron-withdrawing substituents facilitate ylid formation, it was impossible to assess the relative stabilizing influence of (1) coulombic attraction, (2) inductive electron withdrawal by the ring heteroatoms and substituents, (3) d–σ interaction involving the filled sp^2 orbital at $C_{(2)}$ and a vacant d orbital on sulfur, and (4) resonance involving a carbenoid form. Olofson and his coworkers' recent thoroughgoing investigation of deuterium exchange in a wide variety of 5-ring heterocycles has provided some clues to this complex problem[17-20]. Not surprisingly, the presence of a positive charge in the system undergoing deprotonation has an enormous influence on the rate of base-catalyzed exchange: typically ten powers of ten compared to uncharged systems. By virtue of their electronegativity, both the total number and proximity of ring heteroatoms is also of prime importance; an α-nitrogen for example, is capable of accelerating exchange rates by a factor of $\sim 10^5$. Several rate comparisons, among them **14** vs. **15** and $H_{(5)}$ vs. $H_{(3)}$ of **16** support the view that d–σ overlap is a powerful mechanism for anionic charge stabilization[18]. The significance of these comparisons

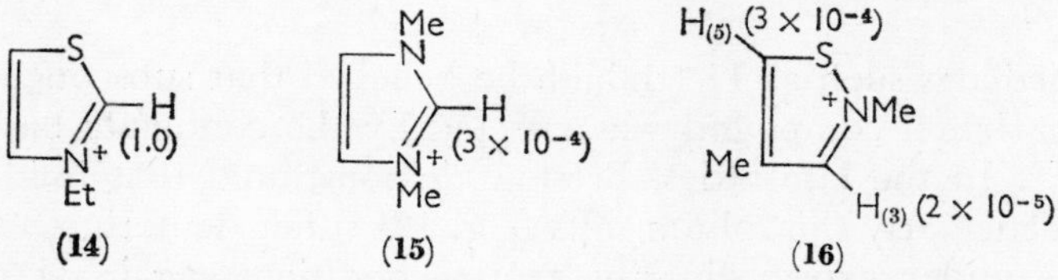

(14)   **(15)**   **(16)**

(numbers in parentheses are relative rates of exchange at 31°)

is heightened by the fact that electronegativity, and in the latter instance, Coulombic considerations, would have predicted slower exchange for the protons with neighboring sulfur than those with neighboring nitrogen. Incidentally, similar exchange experiments with neutral heterocycles have revealed more dramatic rate differences, again favoring CH's flanked by sulfur over those flanked by nitrogen[19]. Hence in anions, where the demand for stabilization is greater, d–σ overlap apparently plays an even more important role than in the corresponding ylids.

There is evidence that the contribution of carbenoid resonance forms (cf. **7**) in certain heterocyclic ylids is of minor importance at best[20]. In particular, Hammett rho values for deuterium exchange in a series of 1-aryl-4-ethyltetrazolium salts (**17**) were found to be rather small: $+0.76$ in 3.8 N DCl, $+1.38$ in the less polar medium 9.0 N

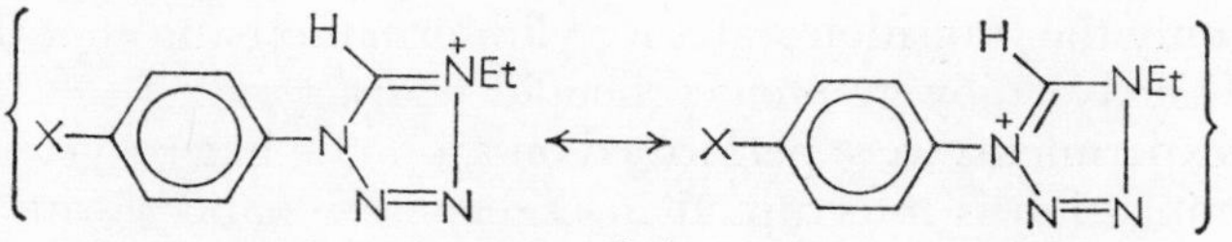

(**17**)

F_3CCO_2D. Were **18** an important contributor to the electronic structure of the ylid intermediate the lone-pair electrons of the aryl nitrogen should be much more available than in the starting cation for interaction with the substituent X; i.e. resonance form **19** should be more significant than the rho values suggest. Moreover, an exchange-rate comparison of triazolium salts (**20**) and (**21**) points to

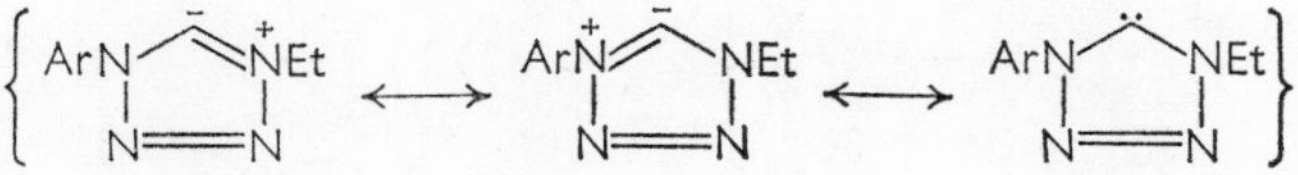

(**18**)

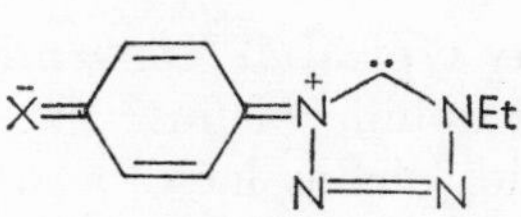

(**19**)

the same conclusion[20]. It will be noted that the ability of these structures to stabilize the ylid is closely matched, with the exception that a carbene contributor can be written for the ylid from **20** but not from **21**. The latter salt exchanges only one order of magnitude faster than the former, an unimpressive enhancement by comparison with those attributable to Coulombic, inductive and d–σ overlap effects. Olofson is careful to note that the experiments involving **17**, **20**, and **21** can

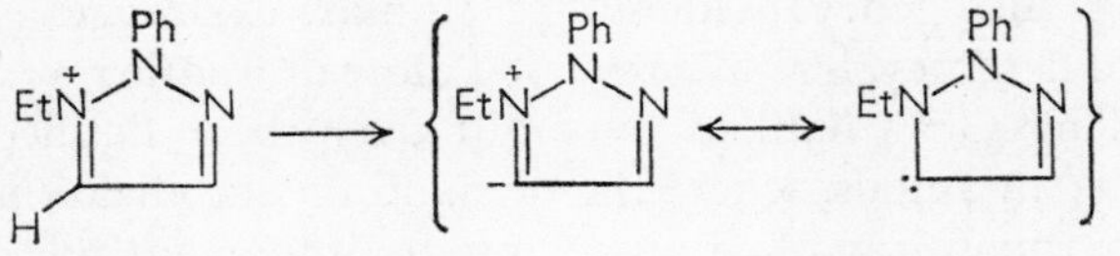

(20)

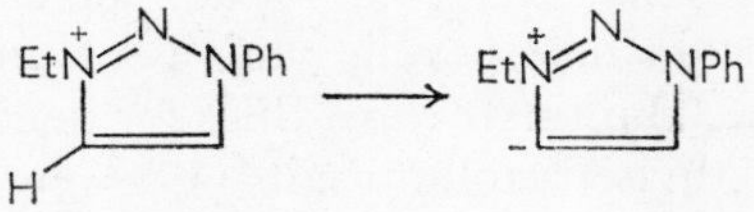

(21)

examine only the transition states for ylid formation and that the ylids themselves may enjoy greater carbenoid character.

These experiments were performed on aromatic heterocycles, whose cyclic conjugation is interrupted in carbene resonance forms for the ylids. Carbenoid character may be perceptibly more significant in non-aromatic analogs of these ylids. Moreover, an ylid's carbenoid character may be expected to depend heavily on the nature of the heteroatoms present. In the sequence $N > O > S$ the heteroatom should assume a smaller net positive charge, and the ylid carbon a correspondingly smaller negative charge. It is significant that dithio-carbenes (22), in dramatic contrast to their nucleophilic nitrogen

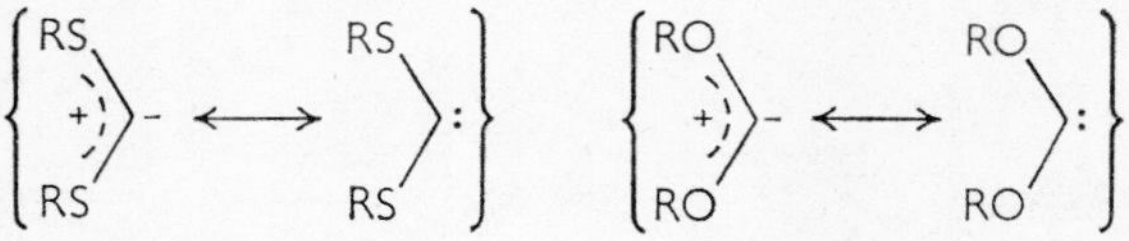

(22)

(23)

analogs, possess the ability typical of 'conventional' carbenes to form cyclopropanes with nucleophilic olefins[21, 22]. Attempts to intercept dialkoxycarbenes (23) with olefins have been unsuccessful to date[23].

All performed in protic media, the extensive exchange experiments discussed above carry intriguing implications for aprotic solvents. Here ylid concentrations are not suppressed nor are their dimers cleaved as a consequence of solvent acidity, so it should be possible to prepare a host of yet-unknown heteroatom-substituted ethylenes by the ylid route. In addition to the peraminoethylenes synthesized via ylids, in fact, several 1,2-diamino-1,2-dithioethylenes[24-28] and tetrathioethylenes[29] have already been prepared by deprotonation of thiazolium and 1,3-dithiolium salts, respectively.

The gentleness of the conditions which suffice for deprotonation of these salts in aprotic media (rapid reaction with a tertiary amine at temperatures from 0° to ambient) is made striking by the fact that triethylamine in aprotic solvents failed to transform 1,3-diphenyl-imidazolinium perchlorate (**10**) into 'Wanzlick's dimer' (**5**) even at elevated temperatures[5b, 30b]. Attempts to intercept the air-sensitive dimer with oxygen to give 1,3-diphenyl-2-imidazolidone were fruitless, so the possibility that dimer was being formed, but reversibly, is very doubtful. Comparison of **14** and **15**, and similar contrasts, provide a partial explanation, but deductions about deprotonation rates in aprotic solvents from exchange rates in protic solvents may be hazardous. As an illustration, the recalcitrant **10** in D_2O solution suffers instantaneous exchange when a drop of triethylamine is added[5b]. The implications are that hydrogen bonding is already present in the transition state for deprotonation in D_2O (**24** is a possible representation) and that it greatly stabilizes nitrogen ylids such as **7**, species in

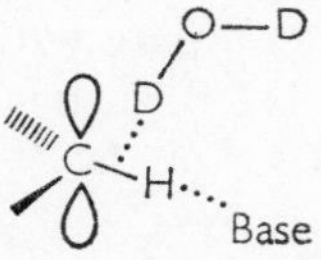

(**24**)

which much negative charge appears at carbon. Ylids containing other heteroatoms may prove to be less sensitive to the hydrogen-bonding capability of the solvent. These ideas definitely require more experimental support, but at the moment it appears that relative exchange rates in protic solvents provide only a rough (albeit useful) guide for selecting deprotonating conditions for aprotic media.

Several years ago Balli developed an ingenious method for generating ylids which did not depend at any stage on deprotonation of a cation[31, 32]. Treatment of certain azidinium salts with azide ion yielded 3 moles of nitrogen and an ylid (equation 5) which could be

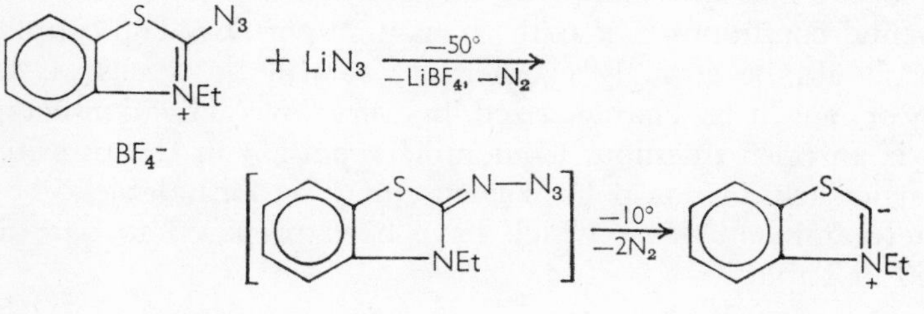

(5)

trapped with a variety of electrophiles. No dimers of the ylids have been obtained in experiments of this type, however. Oxidation of hydrazones such as **25** yields products which have lost the exocyclic nitrogens, and again the ylid, formed via an unstable diazo compound, is believed to intervene[25b].

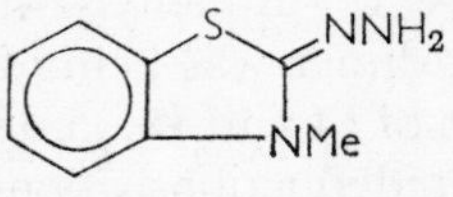

(**25**)

Finally, mechanisms for dimer formation from ylids deserve comment. Since Breslow's investigations it has been clear that the kinds of ylids under discussion are good nucleophiles, so attack of an ylid on its cationic precursor, followed by proton loss, is an attractive scheme (path a). It is not yet known how effectively simple coupling of pairs of ylids (path b) can compete with path (a) (though work of Quast and Hünig (Section III.B.3) has shed much light on this

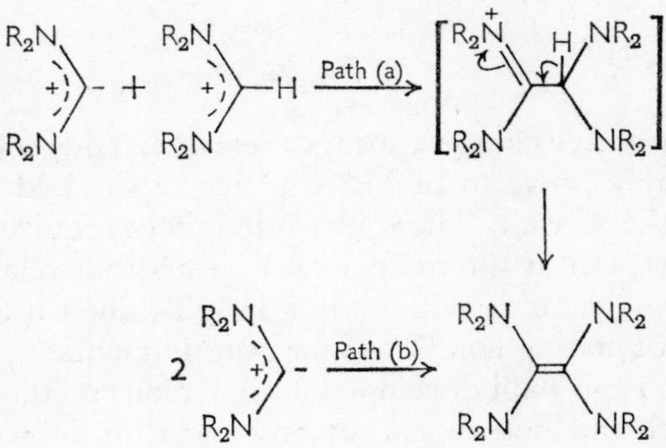

question for ylids from thiazolium salts). The relatively high reactivity of these species may maintain their concentrations so low under most experimental conditions that path (b) is unimportant. The coupling reaction should be strongly exothermic, as will be discussed, but whether or not it is characterized by very low activation-energy barriers is an open question. Coulombic repulsion in the transition state, particularly in non-polar solvents, may be formidable.

The tetraaminoethylenes which have been prepared to date are presented in Table 1.

TABLE 1. Tetraaminoethylenes.

$$\begin{array}{ccc} R_2N & & NR_2 \\ & C{=}C & \\ R_2N & & NR_2 \end{array}$$

R_2N	M.p. (b.p.)	Method (yield)	References[a]
Me_2N	0 (59/0·9)	F (54–65) G (80–82)	1, 2 10[b]
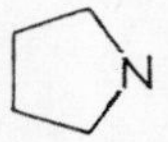 (pyrrolidin-1-yl)	91–93	A (98) D	3, 4, 9, 33 9
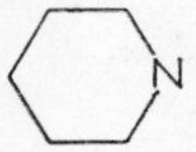 (piperidin-1-yl)	59–61	A (89)	3, 4, 9
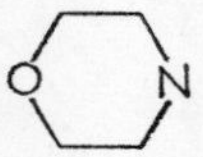 (morpholin-4-yl)	170–171	A (41)	3, 4, 9
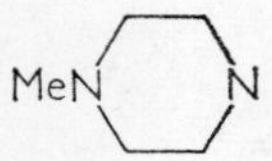 (4-methylpiperazin-1-yl)	79–80·5	A (53)	3, 4, 9
(aziridin-1-yl)		A	3
(azetidin-1-yl)		A	3
(azepan-1-yl)		A	3
(2,3-dihydro-1H-pyrrol-1-yl)		A	3
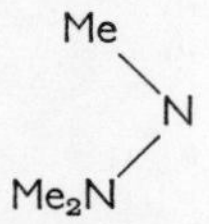 $Me_2N{-}N(Me){-}$		A	3
$(Me_2N)_2C{=}N$		A (17)	4
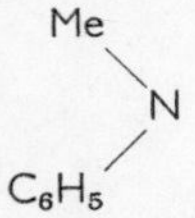 $C_6H_5{-}N(Me){-}$	255–257 (dec.)	B (36)(40)[b]	34

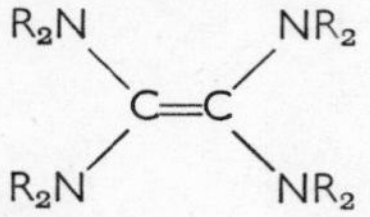

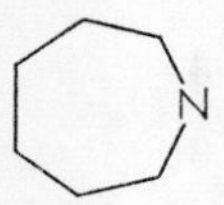

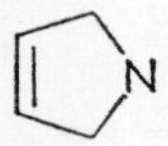

R	M.p. (b.p.)	Method (yield)	References[a]
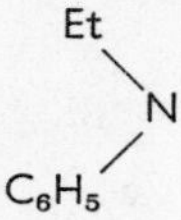 Et–N–C6H5 (N with Et and C6H5)	180–182	B (3)[b]	34

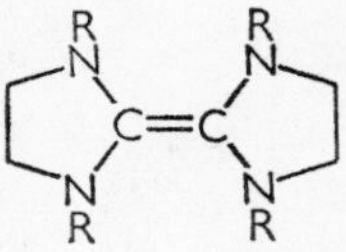

R	M.p. (b.p.)	Method (yield)	References[a]
Me	57	A (83)	3, 4, 9
Et	48	A (82)	3, 4, 9
Me and Et	(80–82/0·25)	A (77)	3, 4, 9
n-Pr	(108–110/0·20)	A (75)	9
n-Bu	(125–129/0·20)	A (73)	9
$n\text{-}C_6H_{13}$	(185/0·20)	A (77)	9
$C_6H_5CH_2$	162–164	A (100, crude)	4, 9
C_6H_5	285(dec.)[c] 294(dec.)[d]	B (90), c (90) E (59)	8, 6, 35 5, 11
$p\text{-}CH_3C_6H_4$	~230(dec.) 281–282(dec.)[d]	B (80), c (50) E (69)	8 5, 11
$p\text{-}CH_3OC_6H_4$	~190(dec.)	B (75), c (70)	8
$p\text{-}ClC_6H_4$	~220(dec.)	B (90), c (45)	8
$p\text{-}O_2NC_6H_4$	~290(dec.)	c (45)	8
$\alpha\text{-}C_{10}H_7$	244–248(dec.)	B (90)	8

Others[e]

	M.p. (b.p.)	Method (yield)	References[a]
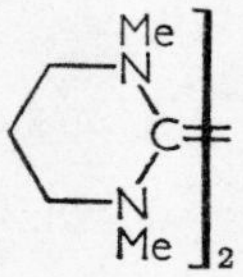	(104–105/8·5)	A (34)	3, 4, 9

	M.p. (b.p.)	Method (yield)	References[a]
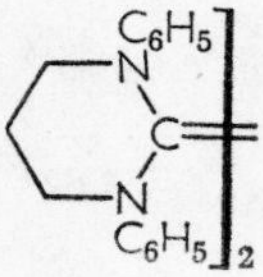	~240(dec.)	B	8

	M.p. (b.p.)	Method (yield)	References[a]

| 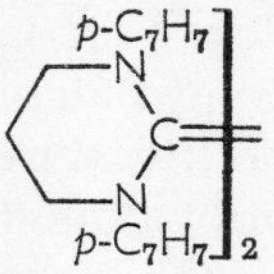 | ~230(dec.) | B (10) | 8 |

| 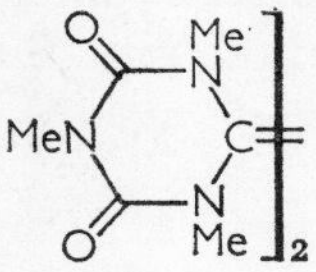 | >350° | E (53)[f] | 37 |

| 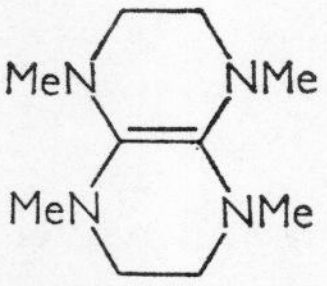 | 56–57·4(~65/0·3) | F | 5 |

A. Amine was heated with N,N-dimethylformamide dimethyl acetal.
B. Amine was treated with hot ethyl orthoformate.
C. Aminal of chloral was heated in collidine.
D. Aminal of glyoxylonitrile was pyrolyzed.
E. Formamidinium salt was deprotonated.
F. Amine was allowed to react with chlorotrifluoroethylene.
G. Orthoamide was pyrolyzed.
[a] For a number of compounds, only selected references dealing with their preparation are cited.
[b] Sulfuric acid was used as a catalyst here.
[c] M.p. was determined under nitrogen.
[d] M.p. was determined *in vacuo*.
[e] 1,1′-Dimethoxy-3,3′dimethyl-$\Delta^{2,2'}$-bi(benzimidazolidine) has been postulated as an intermediate by S. Takahashi and H. Kano, *Tetrahedron Letters*, 3789 (1965). Urry and Sheeto[36] have proposed that methylaminotris(dimethylamino)ethylene is an intermediate in certain oxidations of the fully methylated compound.
[f] Pyrolysis at 300° of 1,3,5-trimethyl-6-methoxyhexahydro-*s*-triazine-2,4-dione gave the ethylene in 38% yield, and pyrolysis at 160° of 2,4-dimethoxy-*s*-triazine yielded the same compound (23%).

III. CHEMICAL BEHAVIOR

The reactions of peraminoethylenes can be divided into two main classes, one in which the polyaminoolefin functions as a nucleophile and another in which it functions as an electron donor, or reducing agent. The distinction is by no means clear in all cases, but those reactions which result in covalent bonding between reactants will be considered under the first category and those which do not under the second. In order to discuss the first class mechanistically, it is necessary to face up to the question whether or not tetraaminoethylenes are capable of fragmenting, as originally suggested by Wanzlick[6a], into the ylid 'halves' from which most of them are synthesized.

A. The Question of Dissociation

For a variety of reasons, among them the relatively high thermo-dynamic stability of such ylids and the marked tendency for tetra-aminoethylenes to give reaction products derived from only half of the molecule, the dissociation theme pervades most of the literature on these compounds in the form either of an assumption or of a question. Since nearly all of the many transformations of the 'Wanzlick dimer' (5) have been interpreted in terms of initial dissociation into the

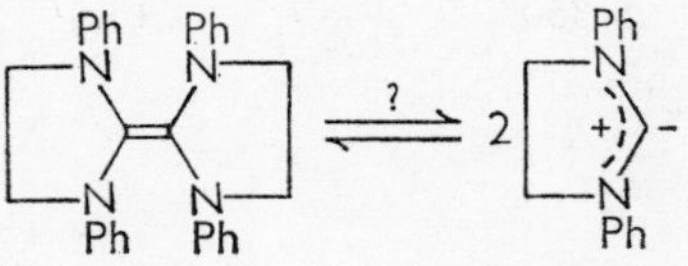

reactive fragment 7, a crossover experiment involving 5 and its very close relative 26 was performed. It was expected that heating an equimolar mixture of 5 and 26 (refluxing xylene, two hours) would give the crossed ethylene (27) in yields approaching 50% if dissociation were occurring. As it was not possible to analyze the insoluble, air-sensitive mixture of ethylenes directly, this product was oxidized by silver nitrate in excellent yield to a mixture of the corresponding stable, colored dinitrate salts. The n.m.r. singlet corresponding to the

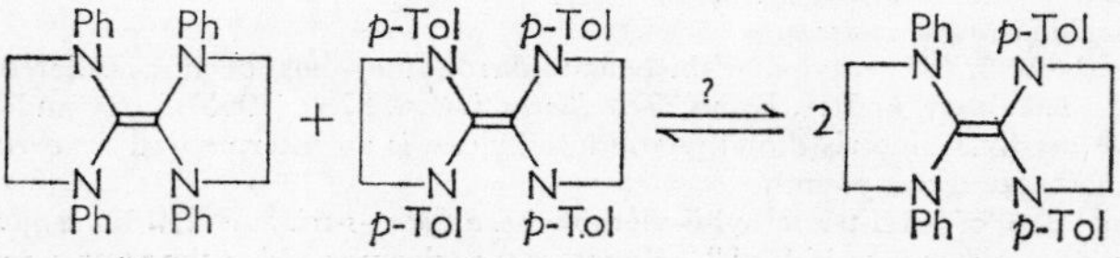

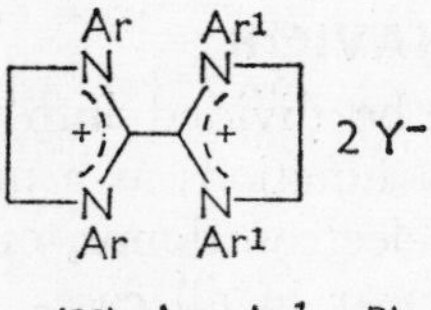

(28) Ar, Ar¹ = Ph
(29) Ar, Ar¹ = p-Tol
(30) Ar = Ph, Ar¹ = p-Tol

methylene protons of 29 (Y = NO₃) was found 5 Hz (at 60 MHz) upfield from that of 28 (Y = NO₃), and a mixture of 28–30 (Y = NO₃, prepared from a mixture of the two imidazolinium perchlorates with sodium triphenylmethoxide, followed by silver nitrate oxidation) revealed an additional pair of methylene resonances belonging to **30**, matched in area and centered between the signals for **28** and **29**. Their existence testifies to the remarkable fact that the methylene protons in one half of the molecule 'feel' the presence or absence of *p*-methyl groups in the other half. In contrast to this 'authentic' mixture of **28–30**, the mixture of dinitrate salts from the crossover experiment showed only the signals attributable to **28** and **29**.* Thus a reinterpretation of the chemistry of **5** became necessary.

Seeking to answer the question of dissociation for the more reactive tetrakis(dimethylamino)ethylene (**1**), Wiberg and Buchler studied its reactions with a variety of Lewis acids[39, 40]. In no case was an adduct of the ylid 'halves' isolated. Diborane and boron trifluoride yielded salts with the structure **31**. Zinc chloride gave a 1:1 complex, and trimethylboron failed to yield any complex at temperatures up to 150°. Though the results with the first three Lewis acids are inconclusive with regard to a dissociation equilibrium, the last experiment certainly inspires serious doubt about its existence even at elevated temperatures. Powerful evidence to support Wiberg and Buchler's

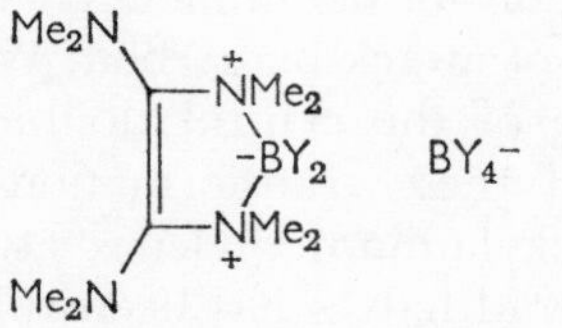

(**31**) Y = H,F

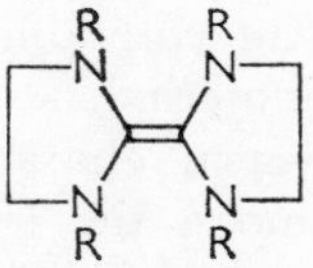

(**32**) R = Et
(**33**) R = n-Bu

suspicions was provided by crossover experiments of Winberg and his coworkers[9]. Mixtures of **32** and **1** and of **32** and **33** were heated at 175°, and no crossed products were detectable by vapor chromatography. Consistent with this result are simple HMO calculations which suggest that dissociation of **1** would be endothermic by about 80 kcal/mole[41].

In recent years dissociation of olefins other than tetraaminoethylenes into carbenoid 'halves' has repeatedly been suggested or alleged to occur[42]. Steric hindrance in the olefins and/or unusual ability of the substituents to stabilize divalent carbon have been offered as explanations for the proposed fragmentations. Though definitive contrary

---

* Vacuum sublimation at 230° of a mixture of **5** and **26** is claimed to yield a mixture containing about 5% of the crossed product[38].

evidence, such as that which crossover experiments can provide, is unavailable, these examples should be viewed with scepticism in the opinion of this reviewer. It is true that ketenes yield carbenes upon pyrolysis, but the conditions are violent and one of the divalent carbon fragments is noted for its stability!

Finally, there is a caveat to be heeded in using crossover experiments as a touchstone for thermal dissociation. If properly conceived and executed, those which 'fail' are trustworthy, but those which yield crossed products can be ambiguous. In the **5–26** crossover experiments, for example, highly pure starting materials are required if crossing over via a catalyzed, not a purely thermal process, is to be eliminated[5].

B. Nucleophilic Reactivity

Like simple enamines, tetraaminoethylenes might be expected to display nucleophilic reactivity both at nitrogen and at carbon. Attack by nitrogen is discernible in a few reactions of the tetrakis(dimethylamino) compound (**1**), notably those with BF_3 and BH_3, previously cited. All other nucleophilic reactions of peraminoethylenes reported to date are interpretable in terms of attack by carbon. With tetrakis-(aralkylamino)ethylenes, cleavage of the central double bond is a nearly universal characteristic of these transformations, and it is common among the tetrakis(dialkylamino)ethylenes as well. Since dissociation of the ethylenes into ylid halves has been ruled out, it is proposed that such reactions conform, with exceptions to be noted, to the mechanism presented in equations (6–9) (where E^+ represents an electrophile, charged or neutral)[11]. Equation (7) finds analogy in Breslow's classic investigation of thiamine, which revealed that ylids

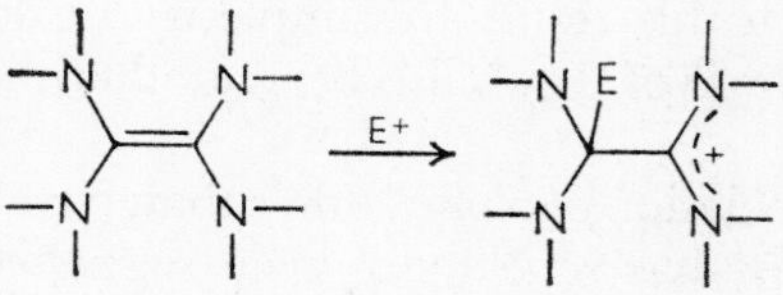

$$(6)$$

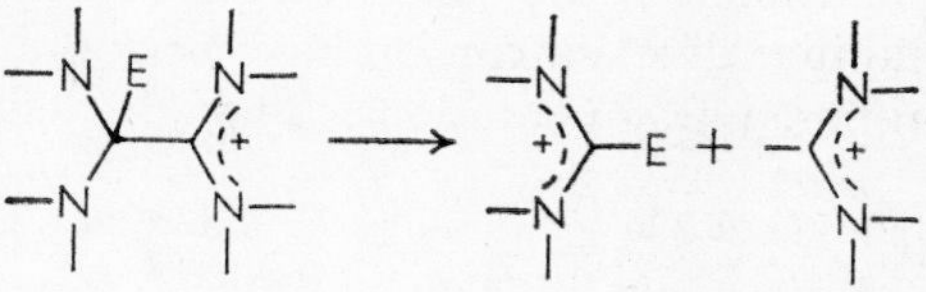

$$(7)$$

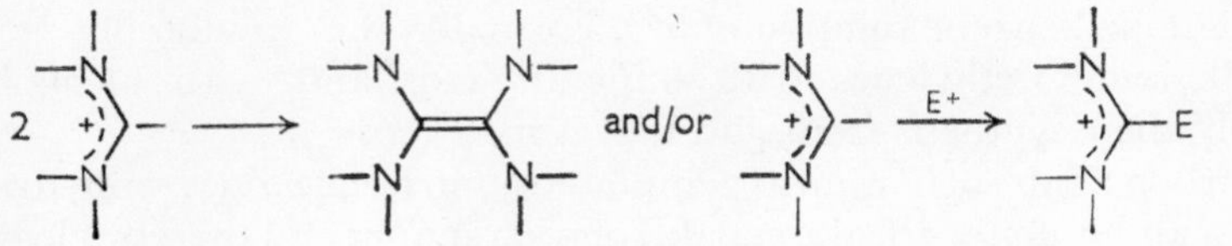

$$2 \overset{\underset{\displaystyle |}{-N}}{\underset{\underset{\displaystyle |}{-N}}{+\!\cdot\!\rangle\!-}} \longrightarrow \overset{\underset{\displaystyle |}{-N} \quad \underset{\displaystyle |}{N-}}{\underset{\underset{\displaystyle |}{-N} \quad \underset{\displaystyle |}{N-}}{\rangle\!=\!\langle}} \quad \text{and/or} \quad \overset{\underset{\displaystyle |}{-N}}{\underset{\underset{\displaystyle |}{-N}}{+\!\cdot\!\rangle\!-}} \overset{E^+}{\longrightarrow} \overset{\underset{\displaystyle |}{-N}}{\underset{\underset{\displaystyle |}{-N}}{+\!\cdot\!\rangle\!-E}} \tag{8}$$

$$\overset{\underset{\displaystyle |}{-N}}{\underset{\underset{\displaystyle |}{-N}}{+\!\cdot\!\rangle\!-E}} \longrightarrow \text{Stable products.} \tag{9}$$

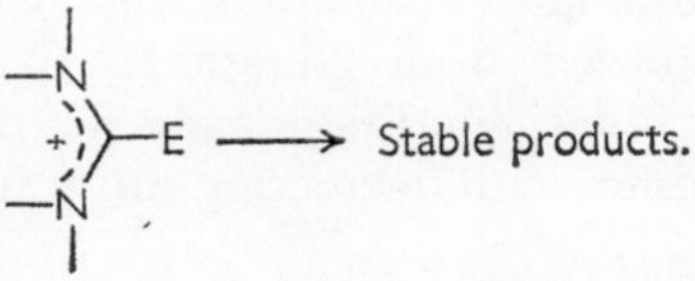

of the sort under discussion are effective not only as nucleophiles but also as leaving groups[14, 43]. The relative importance of the two parts of equation (8) will depend greatly, of course, on the reactivity of the electrophile E^+.

I. Attack at hydrogen

The prototype for equations (6–9) is the cleavage of tetrakis-(aralkylamino)ethylenes into formamidinium salts by mineral acids ($E^+ = H^+$), e.g. $5 \rightarrow 34$[44]. Addition of hot water to a pyridine

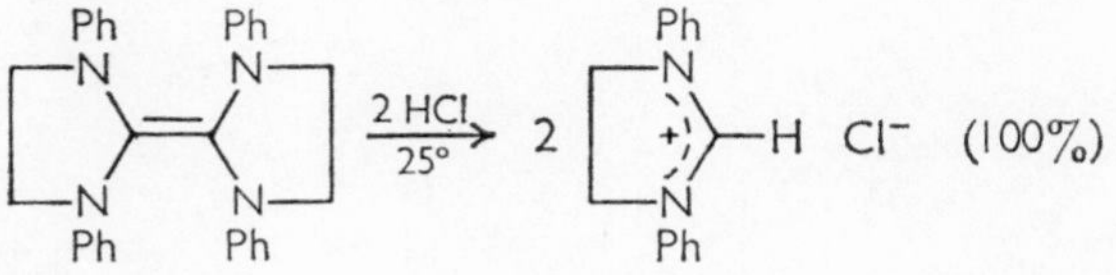

$$(5) \xrightarrow[25°]{2\,HCl} 2\,(34) \quad (100\%)$$

solution of **5** transforms it essentially quantitatively into 2 moles of N-formyl-N,N'-diphenylethylenediamine (**37**)[35]. In this instance the counter ion accompanying the intermediate imidazolinium cation (**35a**) is hydroxide, with the result that covalent bonding and ring opening ensue. With an alcohol the final step in this sequence is blocked, and indeed Wanzlick and Schikora obtained **36b** by warming a

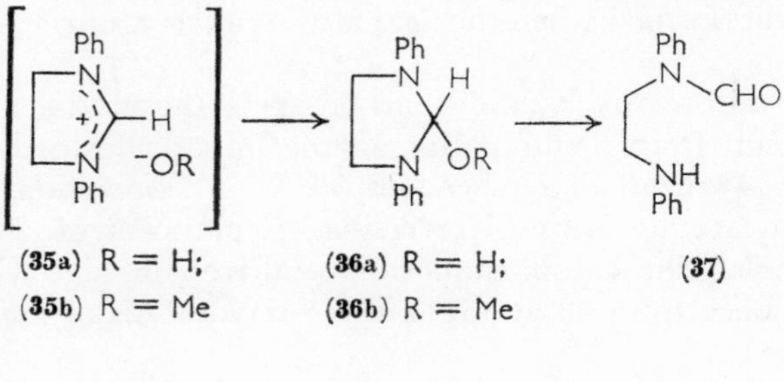

(35a) R = H;   (36a) R = H;   (37)
(35b) R = Me   (36b) R = Me

David M. Lemal

methanolic benzene solution of **5**[44]. Formally, the reaction of tetrakis-(aralkylamino)ethylenes with mineral acids and with alcohols is simply their synthesis from ylid halves in reverse.

Carbon acids such as nitro compounds, nitriles, ketones, anhydrides, esters, and sulfones are also capable of protonating and thereby cleaving these bases with ultimate formation of new carbon–carbon bonds[35, 45, 46]. As expected from equation (6), and as suggested by the variation in reaction conditions used for preparing **38** to **42**, ease of reaction diminishes with decreasing substrate acidity. Whereas

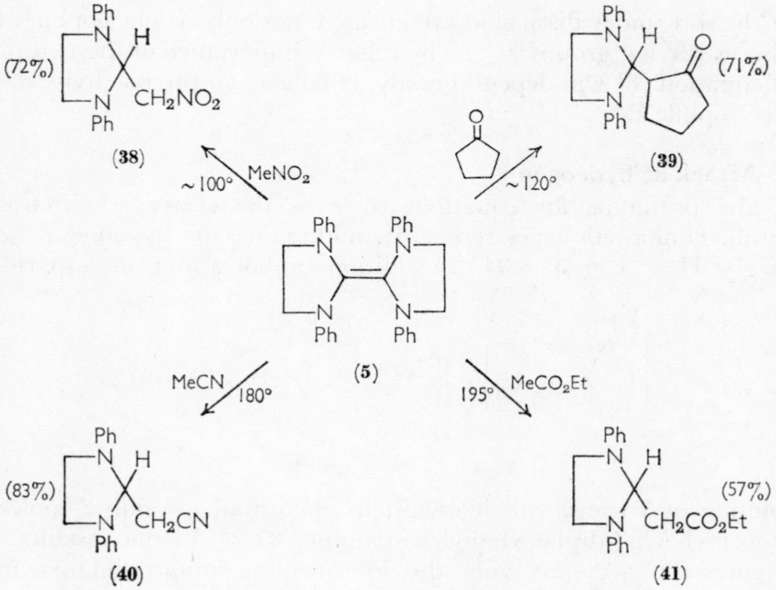

the first part of equation (8), ylid dimerization, must be completely suppressed in the presence of mineral acid, it may consume much or all of the ylid formed when the acid is as weak as a nitrile, sulfone, or ester.

With certain relatively acidic carbon acids, the products which are isolated result from β-elimination in the initially formed imidazolidines (e.g. **42** from desoxybenzoin)[45b, 46]. The reaction of **5** with ethyl phenylacetate is freely reversible, reminiscent of the situation with alcohols. The equilibrium can be driven in high yield either toward or back from **43** at 130–140°[45b]. Hydrolysis of the imidazol-

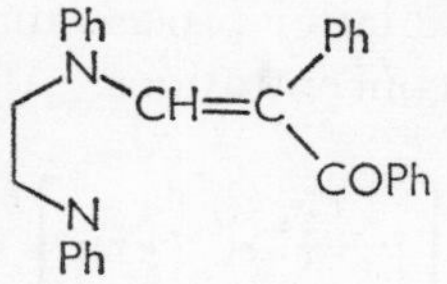

(42) 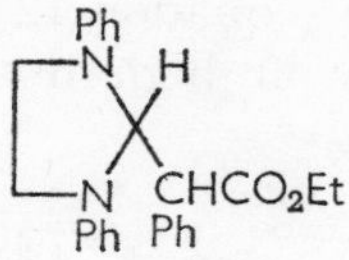 (43)

idines prepared from **5** and carbon acids can yield *C*-formyl compounds (e.g. equation 10)[35]. In some instances, however, hydrolytic cleavage has occurred in the opposing sense, generating the *N*-formyl compound (**37**) and the original active hydrogen compound (e.g. equation 11)[45b].

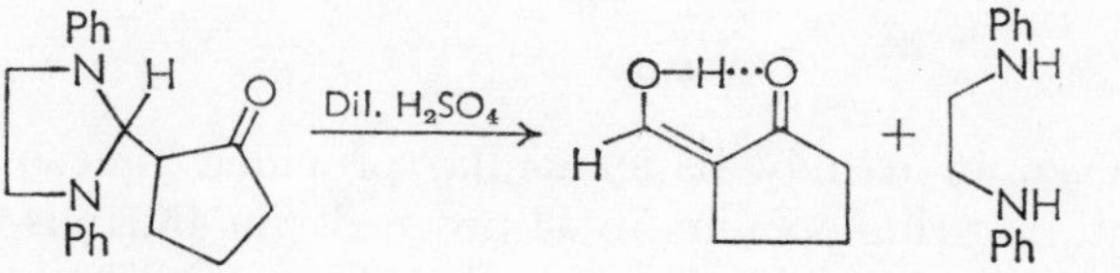

$$\xrightarrow{\text{Dil. H}_2\text{SO}_4} \qquad + \qquad \qquad (10)$$

$$\xrightarrow[\text{aqueous solvents}]{\text{Moist alumina or}} \qquad + \text{ MeCO}_2\text{Et} \qquad (11)$$

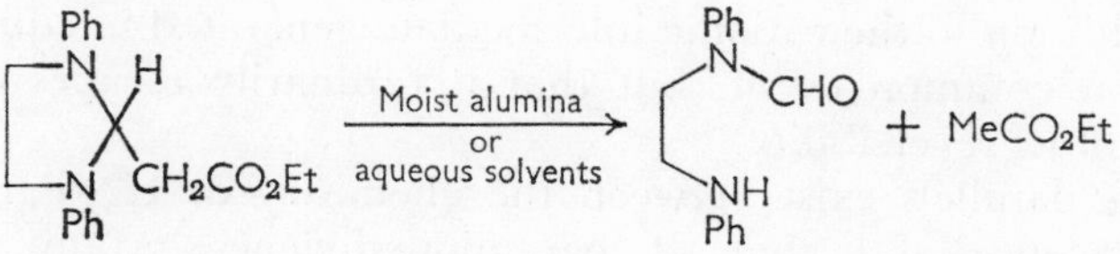

(37)

A discussion of the reactions of **5** with proton donors would be incomplete without mention of its hydrolysis by strong alkali[7]. In sharp contrast to cleavage by aqueous pyridine, which yields 2 moles of the *N*-formyl compound (**37**) as noted earlier, the same solvent containing potassium hydroxide transforms **5** into the amide (**44**) in 90% yield. The weak base–strong base contrast is understandable in

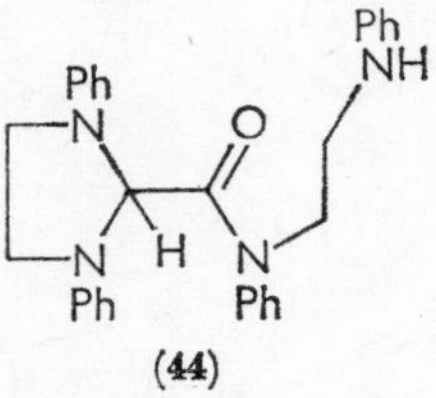

(44)

terms of initial protonation by water (equation 6) with subsequent competition between fragmentation (equation 7) and addition of hydroxide ion (equation 12)[5, 47]. The latter process may assume importance only at high hydroxide concentrations. Alternatively, it

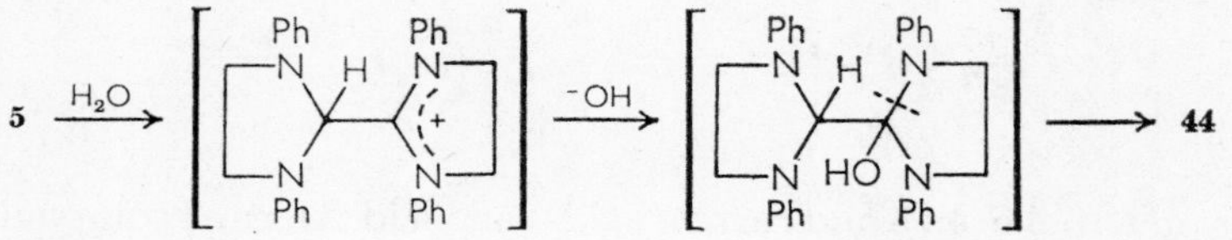

$$5 \xrightarrow{\text{H}_2\text{O}} \left[\begin{array}{c} \text{Ph} \quad \text{Ph} \\ \boxed{\begin{array}{c}\text{—N} \quad \text{H} \quad \text{N—} \\ \text{—N} \qquad \text{N—} \\ \text{Ph} \qquad \text{Ph}\end{array}} \end{array} \right] \xrightarrow{-\text{OH}} \left[\begin{array}{c} \text{Ph} \quad \text{Ph} \\ \boxed{\begin{array}{c}\text{—N} \quad \text{H} \quad \text{N—} \\ \text{—N} \quad \text{HO} \quad \text{N—} \\ \text{Ph} \qquad \text{Ph}\end{array}} \end{array} \right] \longrightarrow 44 \qquad (12)$$

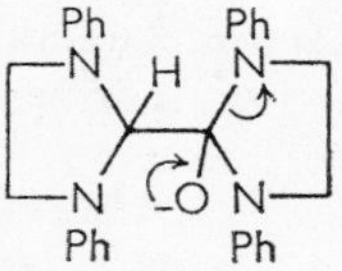

(45)

may occur even in weakly basic media, provided that it happens reversibly and that ring opening to **44** proceeds via **45** (thus requiring a second hydroxide ion). This thought raises the possibility that nucleophilic addition to the cationic intermediate generated in equation (6) may be a common event, but that it ordinarily escapes detection because of its reversibility.

Strong parallels exist between the chemistry of 1,2-diamino-1,2-dithioethylenes and that of peraminoethylenes. Many cleavage reactions analogous to equations (6–9) are recognized[24, 25, 27b], but it is also common for one of the central C—C bonds to remain intact as in the reaction of **5** with strong alkali. With 2,2-bi(3-methylbenzthiazolidine), for example, ammonium iodide yields the salt **46**, dimethyl sulfate yields **47** and methanol (presumably via initial protonation) gives the adduct **48**[24]. In comparison with this ethylene, the

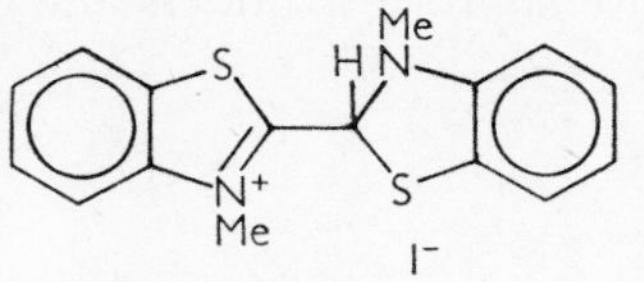

(46)   I⁻   (47)   MeSO₄⁻

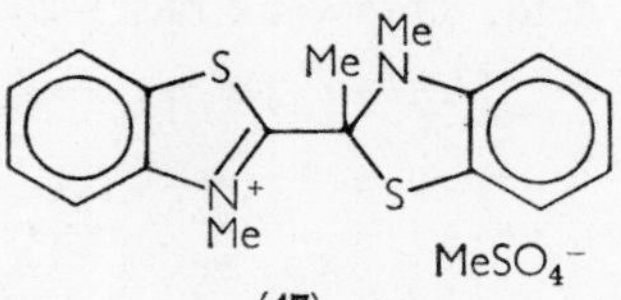

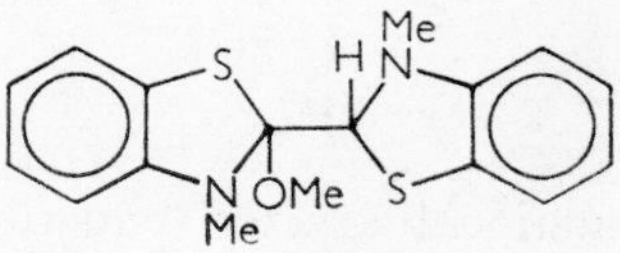

(48)

greater tendency of the crowded peraminoethylene (5) to fragment may be attributable in part to steric relief and a more favorable entropy of activation for cleavage.

With the behavior of 5 in mind, one encounters a number of surprises among the reactions of aliphatic peraminoethylene (1) with proton donors. Hydrolysis in water alone is reported to be slow even at 160°[48],*, due in part presumably to its low solubility. In this single instance the cleavage products expected by analogy to 5, N,N-dimethylformamide and dimethylamine, are obtained, and the mechanism may parallel that for 5. Though methanolysis has been claimed to be even more sluggish[48], this has been challenged[49]; in any event, the products have not been characterized. Reactions of 1 with strong acids do not fit the pattern of equations (6–9), and they are quite complex. Excess dilute hydrochloric acid attacks 1 readily in the cold, yielding not the formamidinium chloride, but glyoxylic acid dimethyl amide (49, 47%), octamethyloxamidinium chloride (50, Y = Cl, isolated as the hexafluorophosphate, 22%), dimethylamine, and other unidentified substances[48]. The tremendous difference,

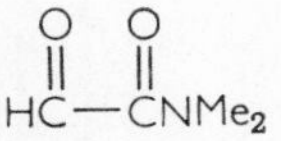

(49)

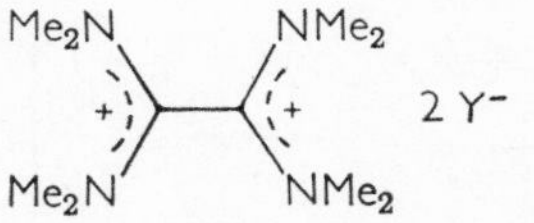

(50)

both in rate and products, between 'neutral' and acid hydrolysis may imply that the latter involves a *di*protonated intermediate. Based on this idea, a mechanism for formation of 49 is suggested in equation (13). Besides suffering hydrolysis, dication 51 should be

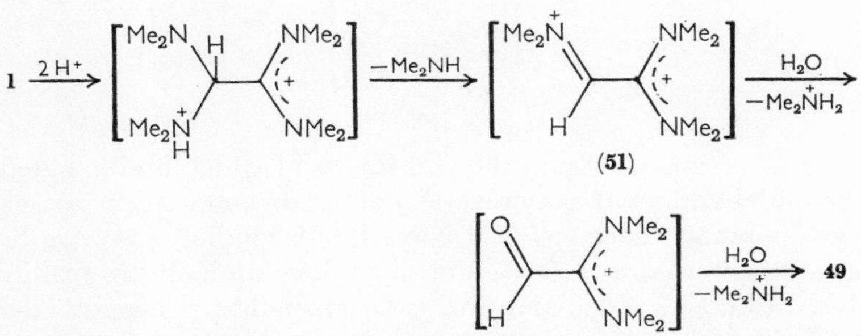

(51)

(13)

* The hydrolysis of 1,1-dipiperidinoethylene, giving piperidine and N-acetylpiperidine, in water at room temperature, makes an interesting comparison[13].

capable of accepting two electrons from **1**, thereby accounting for the oxamidinium salt which is found. An alternative scheme for arriving at **49** has been postulated by Wiberg and Buchler[48].

Oxamidinium salts are very common by-products when tetra-aminoethylenes (even of the aralkyl type) are treated with acids, Brønsted or Lewis. With Brønsted acids, reduction to molecular hydrogen is feasible thermodynamically, but it has been ruled out in at least one instance[41]. The oxidizing agent is air in some reactions of this type (section III.B.5), but care has been taken to exclude it in others, for example the reaction of bi(imidazolidine) (**5**) with boron trifluoride in methylene chloride[5b]. Much remains to be learned about the mechanisms of these processes.

2. Attack at carbon

Unlike the ketones and esters whose reactions with **5** have already been considered, aromatic aldehydes lack α-hydrogens; hence they suffer attack at the carbonyl carbon yielding 2-aroylimidazolidines (**54**)[7, 35, 46, 50]. Here E^+ of equation (6) becomes the entire aldehyde molecule, and equation (14) represents the 'general mechanism' as

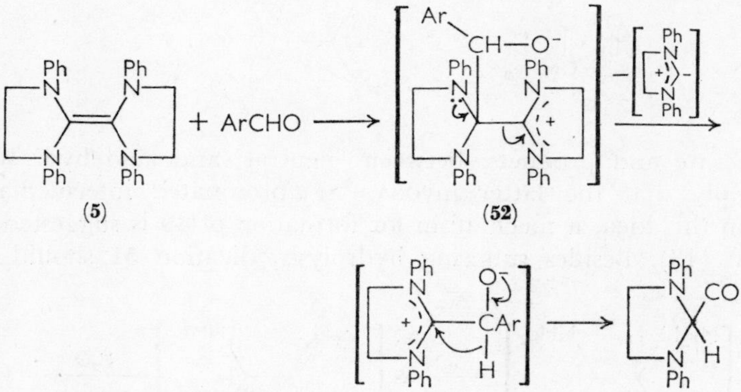

$$(5) \qquad\qquad (52) \qquad\qquad (14)$$

$$(53) \qquad\qquad (54)$$

applied to this reaction. Again the ylid fragment which is eliminated from the initial adduct **52** may revert to dimer, or more likely, attack aldehyde giving **53**. Since yields of 2-aroylimidazolidines (**54**) can be excellent (90% for Ar = C_6H_5), and since these aminals are readily hydrolyzed to α-keto aldehydes, this two-step pathway has potential value in synthesis.

Interestingly, small quantities of benzoin and major amounts of furoin (**56**, Ar = Ph, C_4H_3O) accompany the ketones (**54**) in the

product from the benzaldehyde and furfural reactions, respectively.
These results are reminiscent of Breslow's thiamine studies, wherein

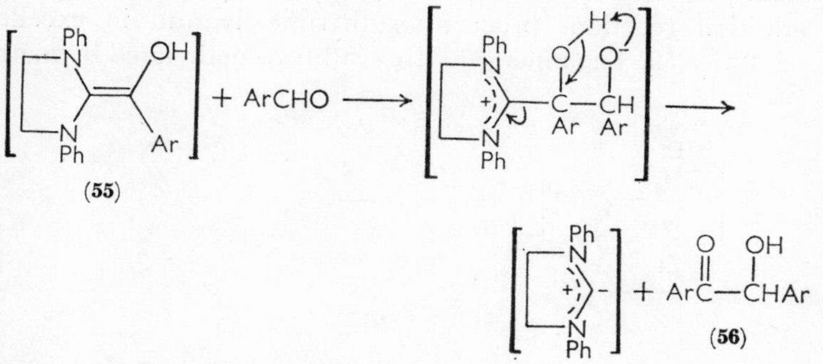

(55)   + ArCHO ⟶

$$\begin{bmatrix} \ \end{bmatrix} + ArC\!\!\overset{O}{\underset{\|}{}}\!\!-\!\!\overset{OH}{\underset{|}{}}\!CHAr \quad (15)$$

(56)

a variety of thiazolium and imidazolium ions (acting via their ylids)
were found to catalyze the benzoin condensation[14]. The formation of
benzoin and furoin, with **5** as catalyst, indicates that **53** and/or **54**
tautomerize to **55**, which proceeds to the benzoin à la Breslow
(equation 15).

Acid chlorides react with **5** to yield 2-acylimidazolinium chlorides[45a]. The ethylene is capable of conjugate as well as direct addition
to a carbonyl group, as exemplified by transformation of the α,β-
unsaturated ketone (**57**) into the furan (**58**)[46]. Viewed in terms of the
'general mechanism', this reaction occurs as shown in equation (16).

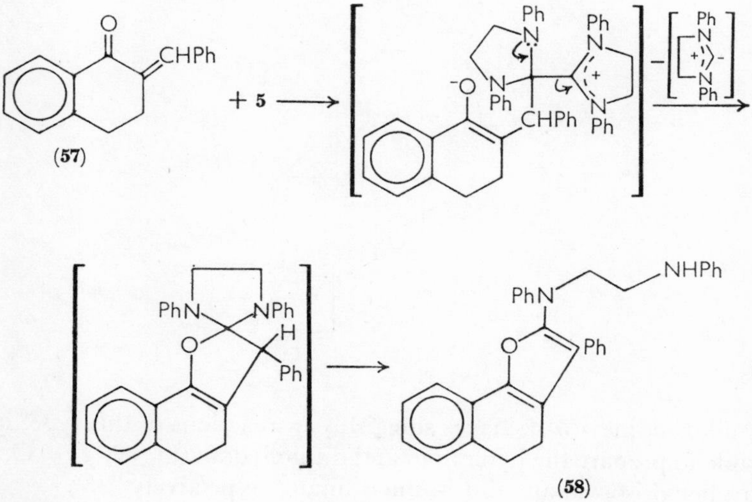

(57)   + 5 ⟶

(58)   (16)

Winberg and Coffman have recently discovered that aliphatic peraminoethylenes react with carbon disulfide and with aryl isothiocyanates to yield a new class of 1,3-dipolar species[51]. Typically, these addition reactions proceed exothermically and in excellent yield, e.g. **32** → **59**. The inner salts are readily S-methylated by methyl

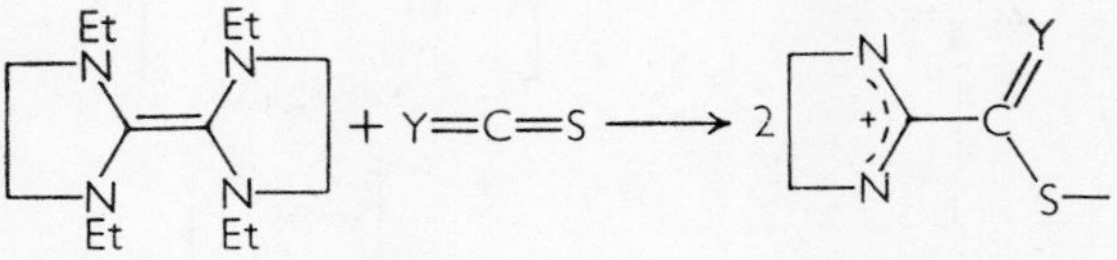

(**32**)   (**59**)

(Y = S, PhN)

iodide, and those derived from aryl isothiocyanates are quite reactive toward a variety of 1,3-dipolarophiles, as illustrated for **59** (Y = C_6H_5N) in equation (17). If alkyl rather than aryl isothiocyanates are combined with an aliphatic $\Delta^{2,2'}$-bi(imidazolidine), the reaction fails to stop at the inner-salt stage and proceeds directly by 1,3-dipolar addition to the spiroheterocyclic. Though Wanzlick's

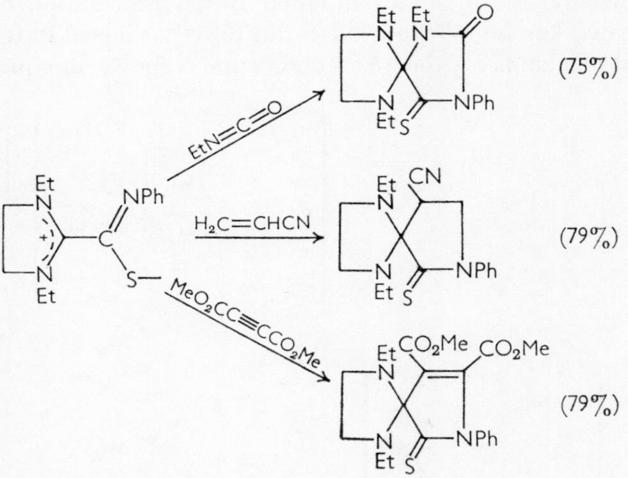

(75%)

(79%)   (17)

(79%)

bi(imidazolidine) (**5**) behaves sluggishly in reactions of this type, it is possible to prepare the hydantoin and dithiohydantoin (**60**, Y = O, S) with phenyl isocyanate and isothiocyanate, respectively[38, 51].

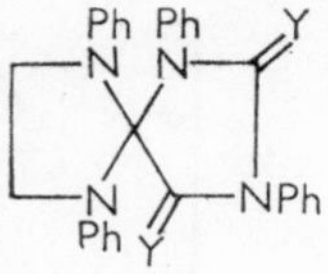

$$(60)\ Y = O, S$$

Treatment of **1** with methyl iodide is reported to yield octamethyl-oxamidinium iodide (**50**, Y = I), tetramethylammonium iodide, and the iodide of betaine dimethyl amide[49]. The course of this remarkable transformation has not been elucidated.

3. Attack at nitrogen

Since the terminal nitrogen atom of electronegatively substituted azides and diazo compounds displays distinctly electrophilic properties, attack at this site by peraminoethylenes comes as no surprise. Both Wanzlick[52] and Reimlinger[53] have discovered that **5** reacts with diazofluorene to give the azine (**61**).

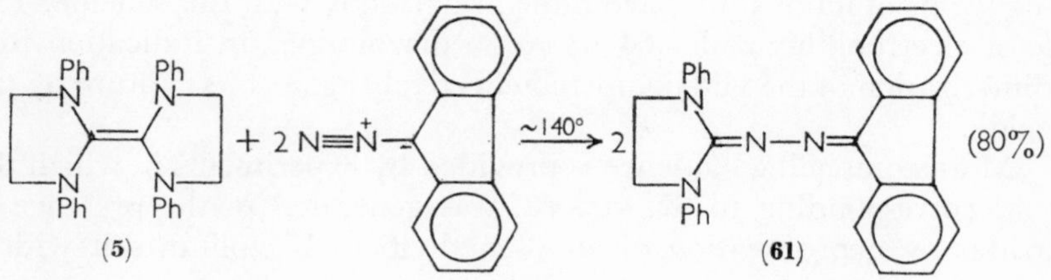

$$(5) \quad + 2\ N\!\equiv\!\overset{+}{N}\!-\!\!\!\!\quad \xrightarrow{\sim 140°} 2 \quad (61) \quad (80\%)$$

In order to understand peraminoethylene–azide reactions, it will be useful to consider first the ingenious work of Quast and Hünig on diaminodithioethylenes[25]. Since known cleavage reactions of both types of ethylene gave products traceable to either the ethylene or its ylid halves, it had not been possible to evaluate the relative contribution of the two parts of equation (8). The German researchers then discovered that electrophilic azides yield products with diaminodithioethylenes which differ from those with the corresponding ylids. As in enamine–azide reactions studied by Fusco[54], a cyclo adduct forms from azide and ethylene and then decomposes to an imine and a diazo compound (equation 18). In the present instance the labile diazo compound loses nitrogen (if indeed it has an independent

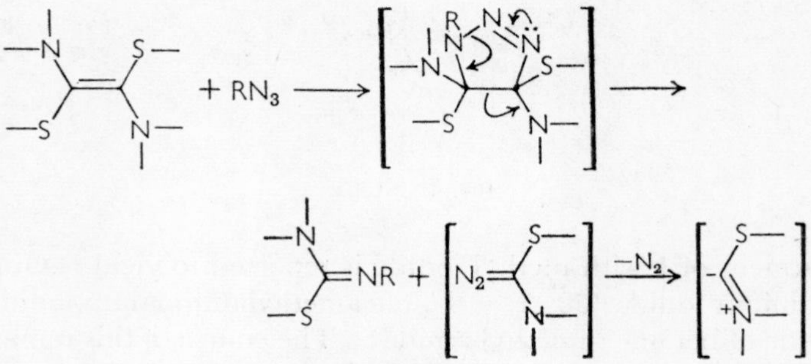

$$(18)$$

existence) to give the familiar ylid; this is intercepted by another molecule of azide to yield a stable diazoimine (equation 19). A series of azides, all electrophilic but to a widely varying degree, reacted

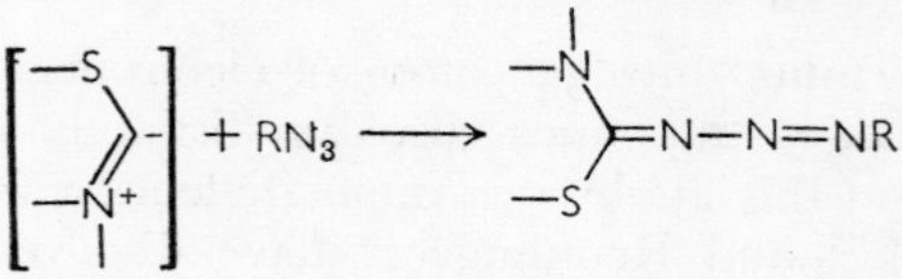

$$(19)$$

with **62** to give in each instance at least one mole of nitrogen and varying yields of imine and diazoimine in harmony with this scheme. The least electrophilic azide (**64**) gave no diazoimine, an indication that dimerization of the ylid competed effectively against its capture by the azide.

More compelling evidence is provided by experiments in which the ylid corresponding to **62**, viz. **63**, was generated in the presence of azides by deprotonation of an *N*-methylbenzthiazolium salt with a

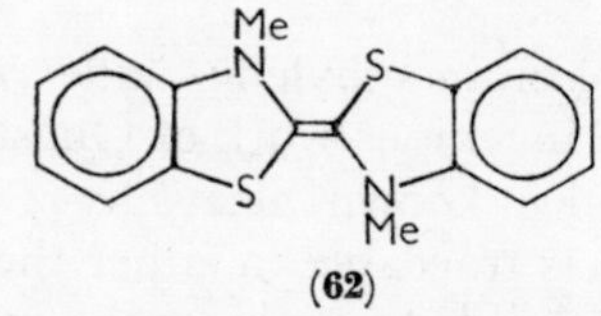

(**62**)   (**63**)

(double bond configuration unknown)

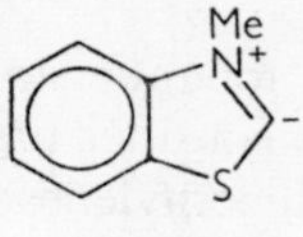

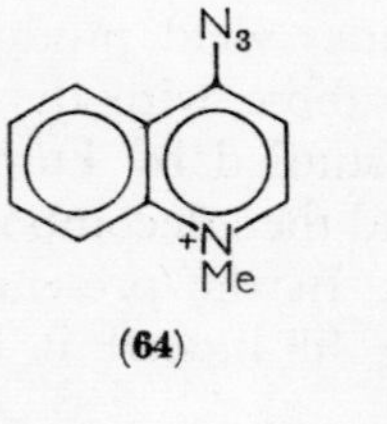

(**64**) 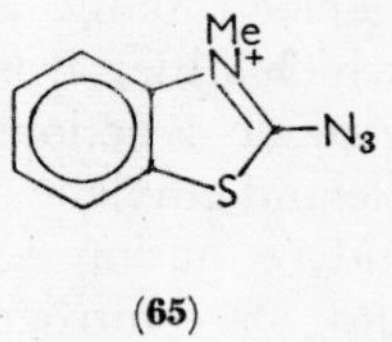   (**65**)

tertiary amine at room temperature. With the relatively weak electrophile tosyl azide in 7-fold excess, the yields of imine were about 60% and those of nitrogen slightly higher, but *no* diazoimine was detectable. Apparently tosyl azide, even in large excess, cannot compete with the electrophilic benzthiazolium ion for ylid (**63**), which is transformed accordingly into **62** as shown in equation (20). This rationale accounts for the good yields of imine and nitrogen as well as the absence of diazoimine. In sharp contrast, the powerfully

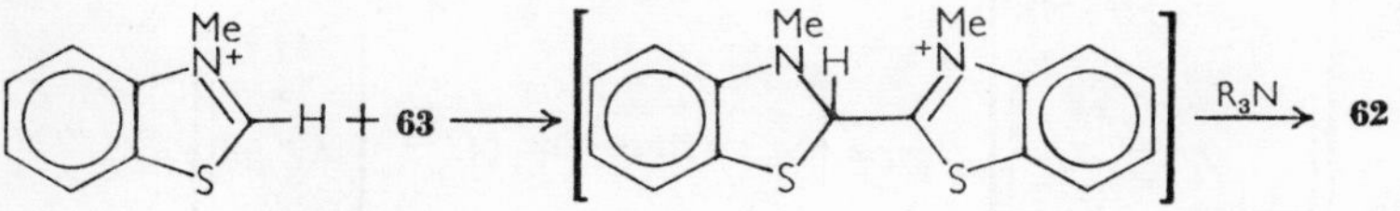

$$\xrightarrow{R_3N} \mathbf{62} \qquad (20)$$

electrophilic azide (**65**) in only 100% excess gave 72–78% diazoimine, little nitrogen, and no imine. Clearly this azide intercepts the ylid with high efficiency, thus precluding the dimer formation (equation 20) which is a necessary prelude to imine formation (cf. the earlier-discussed experiments of Balli[31, 32]).

Side-reactions involving the ylids in Quast and Hünig's experiments have not been identified, but a major side-reaction of the ethylenes is oxidation to their radical-cations and/or -dications.

With these experiments as background, the results of Winberg and Coffman with peraminoethylenes appear clear-cut. Even in the cold **32** reacts with tosyl, diphenylphosphonyl, and *p*-nitrophenyl azides to yield the yellow-to-orange diazoiminoimidazolidines (**66**)[51]. The reaction with tosyl azide also produces the white imino compound

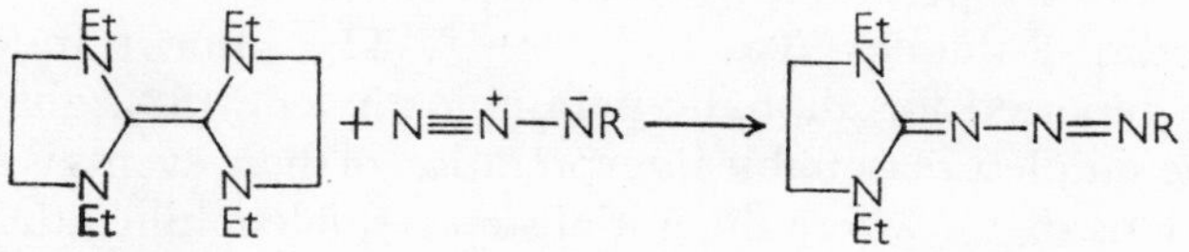

(R = *p*-TolSO₂, Ph₂P(O), *p*-O₂NC₆H₄)

(**67**), which presumably arises via cycloaddition to give **69** followed by cleavage (equation 21). Exclusive operation of this scheme would require that the yield of imine equal or exceed that of diazoimine, as it invariably does in Quast and Hünig's reactions of **62** with azides. If fragmentation of **68**, formed en route to/or by ring opening of **69**, also occurs (equation 22), however, this requirement disappears.

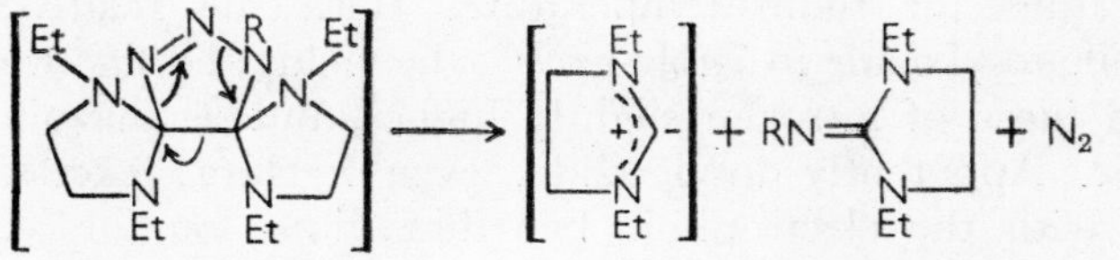

$$\text{(69)} \longrightarrow \text{(67)} \quad + N_2 \tag{21}$$

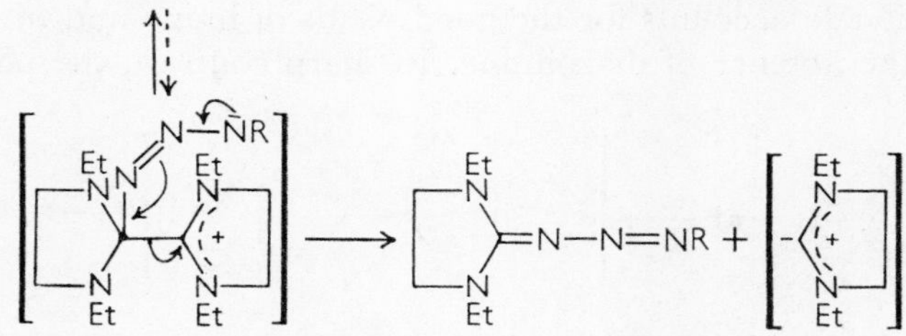

$$\text{(68)} \longrightarrow \text{(66)} \tag{22}$$

Since no imines were reported from the reaction of diphenylphos-phonyl and *p*-nitrophenyl azides with **32**, decomposition of **68** apparently predominates over that of **69**. In other words, the reaction of a particular peraminoethylene and azide may conform strictly to the general mechanism outlined in equations (6–9), but with another pair of reactants the alternative fragmentation process observed by Quast and Hünig may compete with or supersede equation (7).

4. Attack at sulfur

The much-studied tetraaminoethylene (**5**) is oxidized slowly, but even at room temperature, by elemental sulfur to yield 1,3-diphenyl-imidazolidine-2-thione (**70a**, 85%)[45a, 52]. The same transformation has been reported for aliphatic peraminoethylene (**32**), giving **70b**[9]. Again the simplest reasonable interpretation of these events is expressed by equations (6–9), where the initial step is nucleophilic attack on the S_8 ring, with ring opening. Initiation by electron transfer from the ethylene to sulfur, however, is also an attractive possibility.

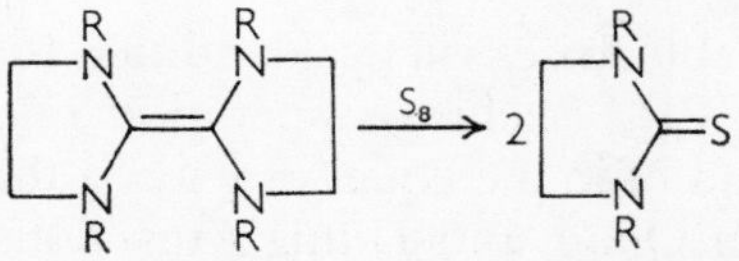

$$\xrightarrow{S_8} 2$$

(**5**) R=Ph   (**70a**) R=Ph
(**32**) R=Et   (**70b**) R=Et

5. Attack at oxygen

The reaction of peraminoethylenes with oxygen has a number of unusual features, one of which is spectacular as well. In the first publication on tetraaminoethylenes, Pruett[1] noted the brilliant, prolonged blue–green chemiluminescence which is generated when air comes in contact with tetrakis(dimethylamino)ethylene (**1**)[55]. Winberg's group has found that the phenomenon is rather general among aliphatic peraminoethylenes[9], though there is no report of oxyluminescence among their aryl-substituted analogs. Before considering further the light-emission process, it is appropriate to turn to the reaction products. The sole reported products from air oxidation of peraminoethylenes (**5**) (in non-hydroxylic solvents) and **32** are the imidazolidones (**71a**) and (**b**), respectively. Though the analogous

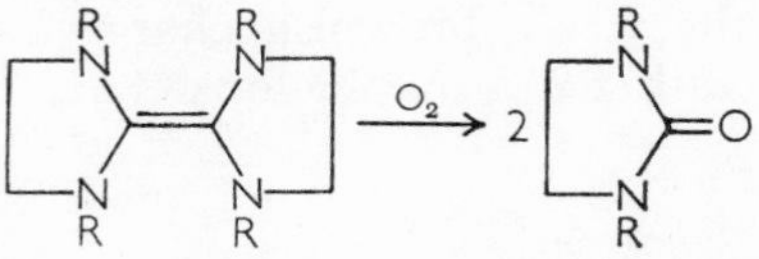

(**5**) R=Ph
(**32**) R=Et

(**71a**) R=Ph
(**71b**) R=Et

product, tetramethylurea, is obtained in air oxidations of **1** under certain conditions, one finds in addition a bewildering mixture whose composition is affected dramatically by the reaction conditions[2, 36, 41]. By far the most complete study of the oxidation products has been carried out by Urry and Sheeto[36], whose major results will be summarized.

When performed neat or in a variety of aprotic solvents, autoxidation of **1** yields **72–75**, and the reaction is strongly catalyzed by hydroxylic compounds. Layered on water or aqueous solutions of inorganic

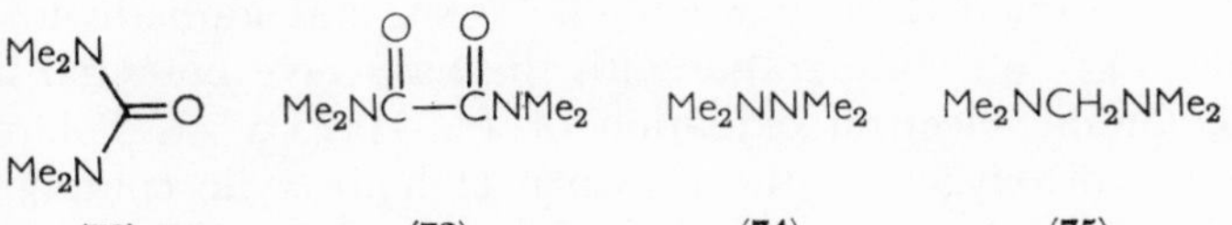

(**72**)   (**73**)   (**74**)   (**75**)

salts, **1** gives the oxamidinium hydroxide hydroperoxide (**50**, Y = OH, OOH), which is slowly transformed into **73, 75, 76**, formate, and dimethylamine. In methanol and certain aprotic solvents containing lithium chloride or water the two reaction pathways exist in competition. With proper choice of medium and counter ion, the dication

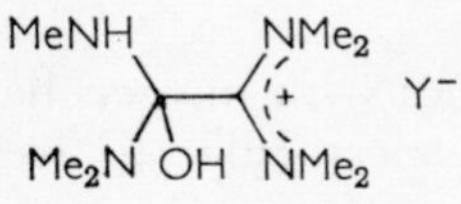

(76) 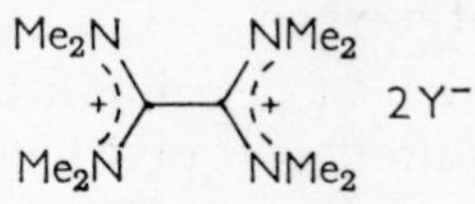   (50)

crystallizes as the oxidation progresses. Even aralkyl peramino-ethylene (5) yields the corresponding oxamidinium acetate when air oxidized in acetic acid containing sodium acetate[36]. In harmony with the proposed scheme, octamethyloxamidinium dibromide reacts with sodium peroxide in aqueous solution to give the same array of products formed by air oxidation of 1 in polar media.

The counterpart of 50 (Y = OH, OOH) in an aprotic medium is the zwitterion 77 or its cyclic tautomer 78. Precisely how such species are formed from the weak, brownish charge-transfer complex[56] of molecular oxygen with 1 is a matter for debate. Spin-forbidden as it

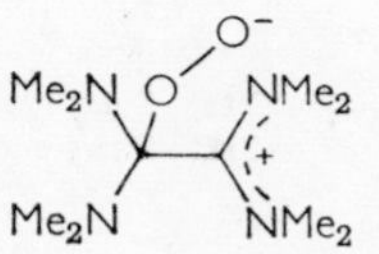

(77) 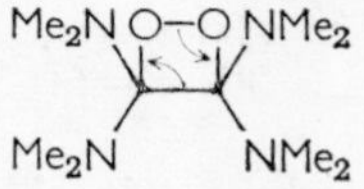   (78)

is, concerted cycloaddition to yield 78 is not an appealing possibility, and the rate acceleration of 200 times which accompanies a change in solvent from decane to dioxane points to a polar transition state[36]. This need not imply direct formation of 77 (also spin-forbidden), however, since initial electron transfer from ethylene to oxygen might occur, giving a radical-ion pair which could couple subsequently. Half-wave potentials for reduction of oxygen to superoxide ion $(O_2^{\overline{\cdot}})$ in dimethyl sulfoxide $(-0.73$ v) and dimethylformamide $(-0.80$ v)[57] are comparable with the half-wave potential in aceto-nitrile for one-electron oxidation of 1 $(-0.75$ v)[58]. Polarographic reduction of oxygen in the presence of hydroxylic compounds is a two-electron process yielding $^-$OOH and $^-$OR and occurring at less negative potentials, a fact which may underlie the large catalytic effect of these substances on the oxidation of 1[59].

With regard to the ultimate oxidation products, two molecules of 72 could be formed in aprotic media by fragmentation of 78 as indicated. There is evidence that 73–75 arise via free-radical pathways[49], but the details are not yet understood. Urry and Sheeto have explained formation of oxamide (**73**) and dimethylamine in aqueous media by hydrolysis of the oxamidinium dication. The properties of **76**, whose structure is supported by considerable evidence, are no less surprising than its formation. It is reported to be recovered unchanged after two days in refluxing 6 N hydrochloric acid. According to Urry and Sheeto the first step in the sequence leading to **76** is proton loss from the oxamidinium ion (**50**) to give immonium ion (**79**); deuterium exchange experiments support this proposal. Hydrolysis of the

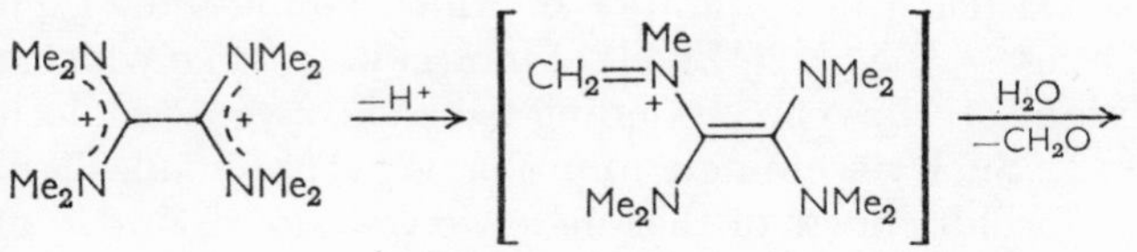

immonium compound would yield formaldehyde (which is oxidized under the reaction conditions to the observed formate) and a new tetraaminoethylene (**80**). The latter could transfer two electrons to dication **50** to yield **81**, a process which could well be exothermic since coplanarity is more easily achieved in the new dication. Addition of hydroxide ion to **81** would complete the sequence.

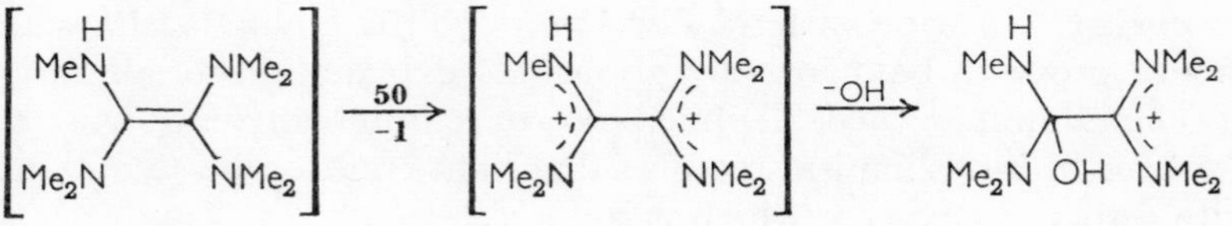

Returning now to the matter of chemiluminescence, the radiating species appears to be the ethylene itself in the lowest excited singlet state. This conclusion follows from the near identity of the fluorescence and oxyluminescence spectra at room temperature and from the blue shift which characterizes each when the temperature is lowered[60]. The oxidation kinetics are first order in oxygen and ethylene and variable order in alcohol catalyst due to hydrogen-bonding association of the latter. The kinetics of luminescence, obtained from emission intensity measurements as a function of time, are similar, except that ethylene concentration appears as the second power (and that quenching terms involving alcohol and oxygen are involved)[59]. The calculated chemical efficiency (efficiency of generating excited **1**)

24*

approximately equals the ethylene molarity except at high concentrations of **1** (>0.1 M), where it drops off. Since fluorescence quantum yields are rather high (0.35 in n-decane), it is clear that emission quantum yields can also be substantial.

The nature of the step which leads to excited **1** (**1***) remains shrouded in mystery, but at least there exists evidence which militates against attractive possibilities. A recent proposal of Khan and Kasha[61] has shed much light on the remarkable fact that most chemiluminescent reactions involve molecular oxygen. Their work and others' have established the existence of a number of 'double-molecule' excited states ($2[^1\Delta g]$, $[^1\Delta g + {}^1\Sigma g^+]$, $2[^1\Sigma g^+]$) of molecular oxygen which can arise by collision of the corresponding singlet oxygen molecules ($[^1\Delta g]$ or $[^1\Sigma g^+]$). Such a seemingly improbable event is made feasible by the extreme forbiddenness of the singlet oxygen-to-ground transition. One or more of the double-molecule excited states are sufficiently far above the ground state (75 kcal/mole for $2[^1\Sigma g^+]$) to transfer energy to a wide variety of acceptors, so the Khan–Kasha scheme should prove to be a broadly applicable explanation for the origins of chemiluminescence. Applied to tetraaminoethylene oxidation, however, this mechanism requires that light emission be second order with respect to oxygen, which it is not[59].

A second tempting possibility is a mechanism which has been proposed for the chemiluminescent oxidation of certain imidazole[62], acridane[63], and indole[64], derivatives. Its essence is expressed in equation (23), where a 4-ring peroxide collapses, leaving one of the fragments in an electronically excited state. This scheme is attractive

$$\tag{23}$$

because ring strain, the weakness of the O—O bond, and the high resonance energy of carbonyl groups combine to make the fragmentation tremendously exothermic. The selection rules of Woodward and Hoffmann[65] notwithstanding, it is difficult to predict whether this retro '2 + 2 cycloaddition' should proceed concertedly or via rupture of the strained O—O bond to a discrete diradical, which subsequently fragments. It is also unclear whether or not excited-state product is to be expected on theoretical grounds[66]. Whatever the details may be, this chemiluminescence mechanism could be adapted to the autoxidation of **1** simply by postulating electronic energy transfer from excited tetramethylurea, formed by fragmentation of peroxide (78) ($-\Delta H$ estimated to be well in excess of 100 kcal/mole), to a molecule of 1. Unfortunately, Fletcher and Heller found no square quenching terms for oxygen over a 1500-fold pressure range[67]. Known to quench 1* efficiently, oxygen should do the same for tetramethylurea in its lowest singlet excited state, for that state should lie above the lowest excited triplet state of oxygen ($E_T \cong 100$ kcal/mole)[68]. Thus the Fletcher–Heller result casts serious doubt on a scheme for chemiluminescence which requires both 1* and excited urea. It is reasonable to suppose, nonetheless, that urea formation is intimately related to light production, not only because of the energy it liberates but also because oxyluminescence fails to occur under reaction conditions which do not yield urea, and because imidazolidone (71b) is the only product reported from the chemiluminescent autoxidation of tetraaminoethylene (32)[9].

Chandross and Sonntag[69] and also Hercules[70] discovered that electron transfer between two species in solution can yield an electronically excited product, provided of course that the transfer reaction be sufficiently exothermic. Suppose that a short-lived pair of radical-ions, 82 and 83, is generated by electron transfer from 1 to peroxide

(82)   (83)

(78). Cleavage of the central bond of 82 would yield a molecule of the urea and one of its 'ketyl', an extremely powerful electron donor. Electron transfer to the neighboring cation could generate 1*; for that matter, the transfer might be concerted with collapse of 82. A variant on this proposal involves covalent-bond formation between radicals 82 and 83, yielding 84 or even 85, followed by concerted fragmentation into two urea molecules plus 1*. The reaction of the oxamidinium ion (50) with concentrated aqueous base (see Section III.B.6), which may well proceed via 84 and/or 85, is not reported to be chemiluminescent; hence the electron-transfer interpretation is a more inviting, albeit still highly speculative, possibility. It is noteworthy that the 1,4-dioxane, 85, is a very crowded molecule, and the isomeric 1,2-dioxane would experience fearful non-bonded repulsions. Formation of the latter substance, whose fragmentation might well yield 1*, is thus most unlikely.

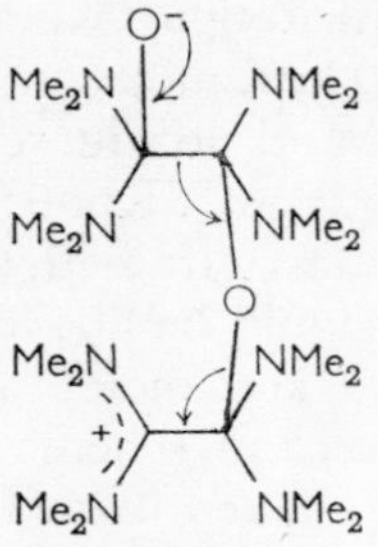

Me_2N NMe_2
Me_2N NMe_2
Me_2N NMe_2
Me_2N NMe_2

(84)

NMe_2 NMe_2
Me_2N NMe_2
Me_2N NMe_2
NMe_2 NMe_2

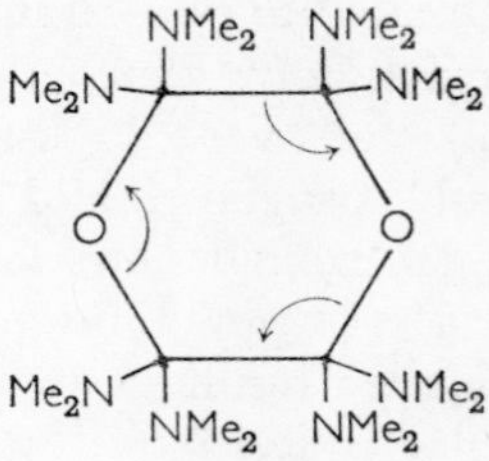

(85)

Finally, the great reactivity of peraminoethylenes toward oxygen has suggested a variety of uses for them: scavenging oxygen, as in gasolines[4,33], detecting traces of oxygen by chemiluminescence[41], determining molecular oxygen quantitatively by measuring the amount of carbonyl compound formed[4,33], and generating electrical current in fuel cells, e.g. for space travel[71]. A more deep-seated reaction of aliphatic peraminoethylenes with oxygen and other powerful oxidizing agents, namely combustion, also merits attention. These compounds are useful as high-energy rocket fuels because they combine high heats of combustion (~ 9000 cal/g) for simple $\Delta^{2,2'}$-bi-(imidazolidines) with specific impulses comparable to those of hydrazine and methylhydrazine with dinitrogen tetroxide, nitric acid, and other rocket motor oxidizing agents[4,33].

6. Related reactions of oxamidinium ions

In principle, the same kind of intermediate obtained by electrophilic attack on a peraminoethylene is accessible by nucleophilic attack on the corresponding oxamidinium cation[5,47] (equations 6 and 24). A single example illustrating this idea has already appeared in the work of Urry and Sheeto on autoxidation of **1**[36]. Other reactions of this type are presented below; a number of them will be seen to conform to the pattern of equations (7–9).

The salts **28** and **29** ($Y = NO_3$) were rapidly cleaved by aqueous base to the corresponding imidazolidones and N-formyl compounds in high yield (equation 25)[5,30]. Had addition of a second hydroxide ion and double ring opening occurred in analogy to the strong-base hydrolysis of **5** (equation 12), an oxamide would have been formed. The lack of competition from this process may be attributable to base catalysis of the fragmentation of **86**, a possibility which does not exist

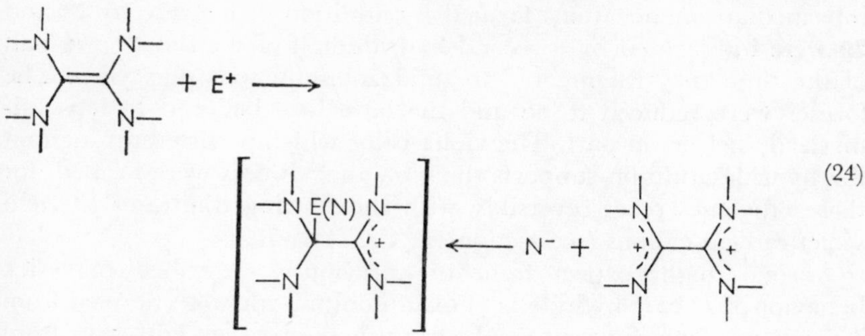

$$(24)$$

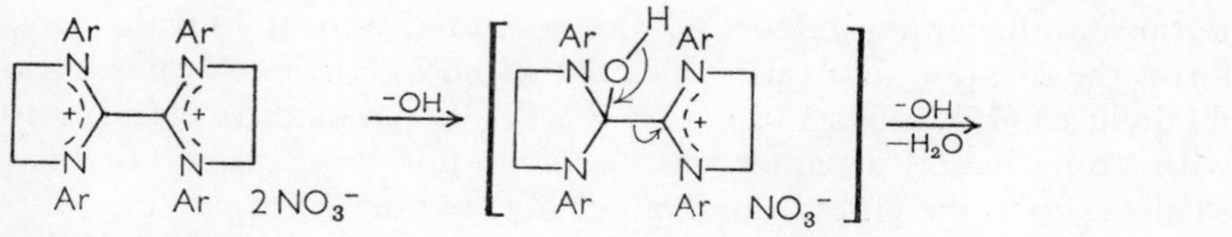

(28) Ar = Ph

(29) Ar = p-Tol

$$(25)$$

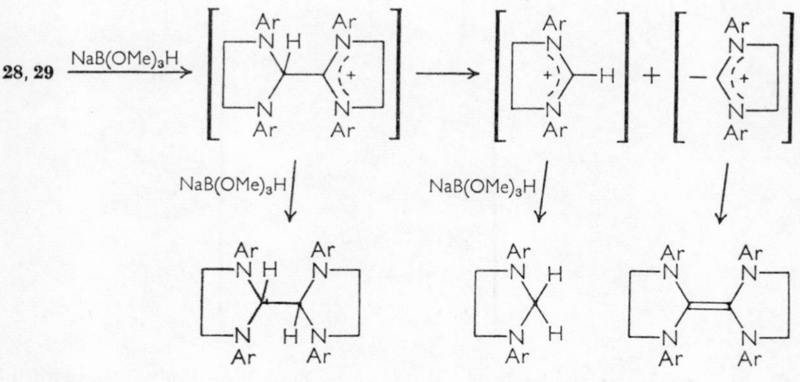

when H takes the place of OH in that cation. Reduction with sodium trimethoxyborohydride of the same dinitrates gave the ethanes **87** (~50%) and the imidazolidines **88** (~25%)[5]. Apparently the intermediate monocations formed by addition of hydride to **28** and **29** were intercepted by a second hydride half of the time*, and half of the time they fragmented to imidazolinium ions plus ylids. The former were reduced to **88** and the latter are believed to have dimerized, at least in part. The violet color which persisted throughout the hydride addition supports the view that dimers were formed, for these ethylenes react reversibly with the starting dinitrates to yield violet radical-cations (see section III.C).

As one might expect from the previously described maverick behavior of **1**, base hydrolysis of oxamidinium salts (**50**) derived from **1** produces quite different results from those observed with salts from tetrakis(aralkylamino)ethylenes. Concentrated aqueous alkali transforms the dication in a rather clean reaction to ethylene **1** plus tetramethylurea (**72**), molar ratio $1:2$[36, 72]. Treating a mixture of dications with strong base, Carpenter was unable to find any crossed ethylene, so cleavage to the ylid in the manner of equation (7) apparently does not occur[73]. Resistance to such cleavage has been evident in other reactions of **1** (see sections III.B.1, 2, and 5). The contrast with tetrakis(aralkylamino)ethylenes may be traceable in part to steric assistance for scission in the aralkyl compounds and in part to the lesser 'leaving ability' of the aliphatic ylid (see section II.B). The opposing argument could be advanced, however, that aliphatic are better than aralkyl amino groups at stabilizing the cationic fragment generated in the cleavage process. In any event, a rationalization of the base hydrolysis products from **50** is given in equation (26).

$$\mathbf{50} \xrightarrow{\ ^-OH\ } \begin{bmatrix} Me_2N & OH & NMe_2 \\ & \diagdown | \diagup \\ & | & +\!\! \\ Me_2N & & NMe_2 \end{bmatrix} \xrightarrow{\ ^-OH\ } \begin{bmatrix} Me_2N & O^- & NMe_2 \\ & \diagdown | \diagup \\ & | & +\!\! \\ Me_2N & & NMe_2 \end{bmatrix} \xrightarrow{\ Dimerization\ }$$

$$(89)$$

$$\begin{bmatrix} Me_2N & O^- & NMe_2 \\ & \diagdown | \diagup \\ Me_2N & | & NMe_2 \\ & O \\ Me_2N & | & NMe_2 \\ & \diagup | \diagdown \\ Me_2N & +| & NMe_2 \end{bmatrix} \quad or\ \mathbf{85} \longrightarrow 2\,\mathbf{72} + \mathbf{1} \quad (26)$$

$$(84)$$

* The diiodide derived from $\Delta^{2,\,2'}$-bi(3-methylbenzthiazoline) gives an adduct with two moles of methoxide ion (central bond retained)[24].

It will be noted that formation of the dipolar species **89** requires two hydroxide ions, so it is not too surprising that the reaction takes a totally different course in dilute base. Here the products are tetramethyl-oxamide (**73**), bis(dimethylamino)methane (**75**), cation **76**, dimethyl-amine, formaldehyde, and **1**[36]. With the exception of the last two, these products were also obtained in the reaction of dication **50** with aqueous sodium peroxide (see section III.B.5), and their formation may be rationalized in the same manner. The identification of formaldehyde and **1** in this experiment supports Urry and Sheeto's interpretation of the peroxide reaction. Both products were proposed in this scheme, but both were shown to be oxidized rapidly by aqueous sodium peroxide.

C. Electron-donor Reactivity

As noted earlier, a division of the reactions of tetraaminoethylenes into those which reflect nucleophilic character and those which reflect electron-donor ability must be somewhat arbitrary. The criterion by which the assignments have been made is the presence or absence of new covalent bonds in the product. The reader must bear in mind, however, that a reaction which ultimately results in covalent-bond formation may well proceed via one or more electron-transfer steps (air oxidation is a possible example), and that a reaction whose net result is electron transfer may involve covalently bonded intermediates (halogenation might be such a process). It is convenient to subdivide the known 'electron-donor' reactions of peraminoethylenes into those which form *bona fide* salts and those which form complexes, despite the fact that the borderline, like most borderlines, is fuzzy.

I. Oxamidinium salt formation

Pruett[1] was the first to observe that the action of a halogen on a peraminoethylene (**1**) yielded a salt-like dihalide*. Carrying out simi-lar experiments with tetrakis(aralkylamino)ethylene (**5**), Kawano[30] characterized the dication and identified the oxidation-state inter-mediate between ethylene and dication. When such oxidizing agents as iodine or silver salts were added to acetonitrile solutions of **5** the intense violet color characteristic of radical-cation **90** ($\lambda_{max}^{CH_3CN}$ 517 mμ,

---

* For the preparation of corresponding compounds from a 1,2-diamino-1,2-dithioethylene, see reference 24.

$\epsilon \sim 6800$) developed and then faded as soon as the second equivalent had been introduced (equation 27). Solutions of **90** displayed an intense e.s.r. signal, but one surprising for its narrowness and lack of

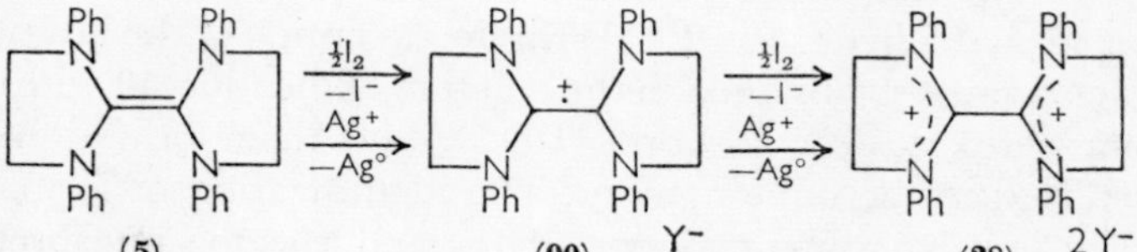

$$(27)$$

well-resolved hyperfine structure (features whose significance is still not clearly understood)[5b, 30, 58]. The same e.s.r. spectrum and violet color were observed when solutions of **5** and **28** were mixed. Using a method based on solubilities, Lokensgard[5b] estimated that K_{eq} in acetonitrile for the reaction **5** + **28** ($Y = NO_3$) $\rightleftharpoons$ 2 **90** exceeds 25. Though never isolated, the radical-cation was stable for long periods of time in solutions that were protected from the air. Acetonitrile solutions of the radical-cation were found to obey Beer's Law over a 100-fold concentration range, so dimerization of the radical is negligible in this medium. In addition to the electronic and Coulombic factors which would oppose dimer formation, the steric barrier is formidable.

Some interesting features characterize the behavior of **91**[74], a relative of **28** in which the aryl groups have been interposed between

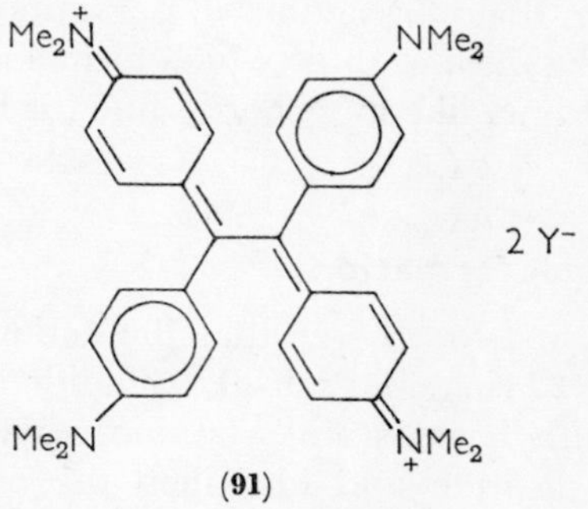

the nitrogens and the central carbons. In water **91** dinitrate and diiodide display nearly identical optical properties, and neither shows e.s.r. absorption. Switching to ethylene chloride does not alter the e.s.r. situation or change the optical spectrum of the dinitrate significantly, but the diiodide in this solvent shows an intense e.s.r. signal and a drastically modified optical spectrum (green solution, whereas the others are purple). Treatment of tetrakis(p-dimethylaminophenyl)-ethylene with iodine in ethylene chloride gives a solution of the same composition ($> 20\%$ free radical). These data are interpretable in terms of the electron-transfer equilibria shown in equation (28). In water the diiodide salt is stable, but the poorly solvating ethylene chloride opposes charge separation sufficiently to drive the equilibria

$$(Ar = Me_2NC_6H_4) \tag{28}$$

significantly to the left. Stabilization of the cationic charge by nitrogen requires quinoidal resonance forms in **91** but not in **28**; thus the lack of evidence for electron transfer in the ground state of **28** diiodide is not surprising. Its orange–red color suggests charge transfer by photoexcitation, however, since the corresponding dinitrate and diperchlorate are canary yellow. Finally, **28** diiodide in acetonitrile is smoothly reduced to the ethylene **5** by metallic silver[30] ($E°$ for $AgI + e \rightarrow Ag + I^-$ is -0.40 v).

Aliphatic peraminoethylenes are, understandably, even more powerful as electron donors than their aralkyl counterparts[75], and again both radical-cation and stable dication oxidation states are known[77]. In Figure 1 are presented the HMO energy levels calculated by Wiberg[41] for the tetraaminoethylene system as a function of the Coulomb integral chosen for nitrogen (α_N). It will be noted that the highest filled orbital, b_{1u}, is antibonding except at very high values of α_N (probably appropriate only to the dication[76]), thus accounting for the remarkable donor properties of these compounds. Using the value ($\alpha + 1.5\beta$) for α_N, Wiberg has calculated an array of physical data for the system **1**, **83**, and, **50** making comparisons with experimental data where possible. The appearance potential of **83** in the mass spectrum of **1** is 6.5 ev, comparable to the ionization potential calculated from charge-transfer spectra for the powerful electron donor N,N,N',N'-tetramethyl-p-phenylenediamine (6.35–6.6 ev)[41,78]. Polarographic half-wave potentials for oxidation of **1** in acetonitrile ($\sim 4 \times 10^{-4}$ M) are -0.75 and -0.61 v. vs. s.c.e. (equation 29); these values yield an estimate of 2.3×10^2 for K_{eq} in the

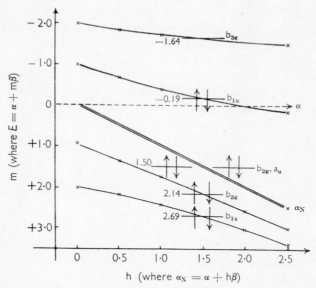

FIGURE 1. HMO Energy levels of tetrakis(dimethylamino)ethylene (**1**)[41].

$$\text{reaction } \mathbf{1} + \mathbf{50} \rightleftharpoons 2\ \mathbf{83}\ [58].$$ (29)

reaction **1** + **50** ⇌ 2 **83**[58]. The dication can be reduced to **1** by zinc dust distillation ($E°$ for $Zn^{2+} + 2e \rightarrow Zn$ is $-1·01$ v. vs. s.c.e.)[41, 77b].

In sharp contrast with the poorly resolved e.s.r. spectrum of the violet radical-cation **90**, that of the orange **83** ($\lambda_{max}^{CH_3CN}$ 385 mμ) reveals about 300 lines from which these coupling constants were obtained: a_N 4·85, $a_{H_{(1)}}$ 2·84, and $a_{H_{(2)}}$ 3·28 G[58]. The nitrogen coupling constant is considered reasonable for an approximately coplanar structure, and the existence of two proton coupling constants corresponding to equal numbers of protons indicates that rotation about the central C—N bonds is slower than $\sim 10^6$/sec (no change in line breadth between 0 and 120°). Hindered rotation is to be expected about both the central C—N and C—C bonds since the calculated π bond orders for each are about 0.5 [41].

The presence of two equally intense signals in the p.m.r. spectrum of the colorless oxamidinium ion (**50**) ($\delta = 2.90$, 3.18 p.p.m., aceto-nitrile, external TMS) signifies that rotation about the central C—N bonds in this molecule is slow even on the n.m.r. time scale. Both simple Hückel calculations and estimates from vibrational spectra point to high C—N and substantial C—C π bond orders; together with the p.m.r spectrum these data support a planar or near-planar structure for the dication [41].

Ethylene **1** displays a single sharp line in its p.m.r. spectrum ($\delta = 2.32$ p.p.m., external TMS), indicating that rotation about the central C—N bonds is rapid on the n.m.r. time scale. The require-ment of a center of symmetry imposed by the lack of C=C stretching absorption in the infrared spectrum of **1** (strong Raman band at 1630 cm^{-1}) [60] is, of course, consistent with a propeller-like structure as well as a planar one, and the steric barrier to coplanarity is con-siderable. Bond-order calculations, from Hückel theory and vibra-tional spectra, indicate that nearly all of the twist in a non-planar structure would involve the C—N bonds rather than the C—C bond [41].

Peraminoethylene (**1**) is oxidized to the oxamidinium ion by poly-haloalkanes [77b, 79]. Ease of reaction increases in the sequence F < Cl < Br < I, and increases with the number of halogens at-tached to a given carbon. The major products from the halo com-pounds are those in which a halogen has been replaced by hydrogen or in which a pair of vicinal halogens has been lost. Ionic processes have been proposed to account for these reactions, as illustrated for carbon tetrachloride. Electron transfer from or nucleophilic attack by **1** results in heterolysis of a Cl—C bond, producing the trichloromethyl anion; this species accepts a proton from the weakly acidic oxami-dinium ion to give chloroform, or loses chloride ion to yield dichloro-carbene (trapped as dichloronorcarane in a related experiment with bromotrichloromethane). Products of radical coupling or radical attack on solvent were not detected in the reactions of **1** with a variety of polyhaloalkanes.

2. Complex formation

Dozens of charge-transfer complexes of aliphatic peraminoethylenes with organic acceptor molecules of widely differing electron affinity have been prepared, primarily by Winberg and his coworkers[3, 80]. They range in properties from the unisolated, diamagnetic yellow complexes formed between 1 and the weak acceptors tetrachloroethylene and styrene to high-melting, deeply colored, strongly paramagnetic solids such as the blue–black 1:2 complex of tetrakis-(N-pyrrolidino)ethylene with 7,7,8,8-tetracyanoquinodimethane, for which structure 92 is probably a close approximation. Typical stoi-

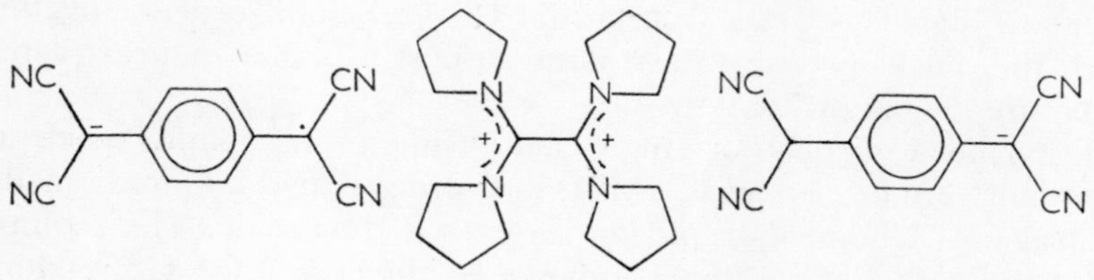

(92)

chiometry for the peraminoethylene complexes ranges from 1:1 to 1:3, donor-to-acceptor. A number of the paramagnetic complexes show very low electrical resistivities.

Spectral data have been reported for only a few of these substances, notably the 1:2 complexes of 1 with sym-trinitrobenzene[81] and tetracyanoethylene[41, 77b,*]. The former is an air-sensitive, black–red solid (charge-transfer maxima at 545, ε 210 and 442 mμ, ε 300 in acetonitrile), which is diamagnetic in solution and reported to be slightly paramagnetic (weak e.s.r. signal at g = 2) in the crystalline state. A sandwich structure with 1 inside has been proposed for this complex. In acetonitrile solution its p.m.r. spectrum reveals only two sharp signals, δ = 2·08 and 8·35 p.p.m. (external TMS), corresponding to the protons of donor and acceptor, respectively. This implies either a rapid equilibrium between the sandwich and its components or, less likely, a loose enough sandwich structure to permit rapid rotation about the central C—N bonds of the donor molecule. Unlike the sym-trinitrobenzene complex, the black–violet 1:2 complex of 1 with tetracyanoethylene can be represented fairly accurately as a salt analogous to 92. Its infrared spectrum contains bands characteristic of the dication 50 and TCNE radical-anion. The optical spectrum reveals the presence of approximately two TCNE radical-anions per mole. In acetonitrile the p.m.r. spectrum shows the two signals charac-

* Despite earlier reports[7, 35], the reaction of 5 with TCNE also yields a 1:2 charge-transfer compound[38].

teristic of **50**, but greatly broadened and shifted to higher field under the influence of the paramagnetic counter ion. The e.s.r. spectrum (acetonitrile) displays the sharp nonet characteristic of TCNE radical-anion in dilute solution ($\sim 10^{-4}$ M), but as the concentration is raised the line breadth increases and the hyperfine splitting disappears. At $\sim 0\cdot 1$ M a single line 5 gauss in width is observed, signifying very rapid scrambling of the free-electron's environment. These results have been interpreted[41] to mean that even in a salt-like peraminoethylene complex with a very powerful π acid, electron transfer from acceptor back to donor is facile*, as is true for the TCNE-N,N',N',N'-tetramethyl-p-phenylenediamine complex[82]. Thus ionic structures such as **92** may represent the complexes correctly in dilute solution in polar solvents, but may exaggerate the degree of electron transfer in the solid state. Regarding the optical spectra of such complexes, charge-transfer absorption very likely leads to excited states less polar than the ground states[41] (as in the pyridinium iodides, for example)[83]; charge-transfer transitions in peraminoethylene complexes with weak acceptors presumably give excited states more polar than the ground states.

In drawing conclusions about structure from the magnetic properties of charge-transfer complexes, one must bear in mind that both dia- and paramagnetism are subject to more than one interpretation. The former need not signify a lesser degree of electron transfer than the latter, for strong spin coupling resulting in a singlet ground state can occur even when transfer is essentially complete[84]. A case in point is the complex of N,N,N',N'-tetramethyl-p-phenylenediamine with 7,7,8,8-tetracyano-p-quinodimethane (**93**), which is clearly ionic in the crystalline state as judged from absorption spectra. Nonetheless, the

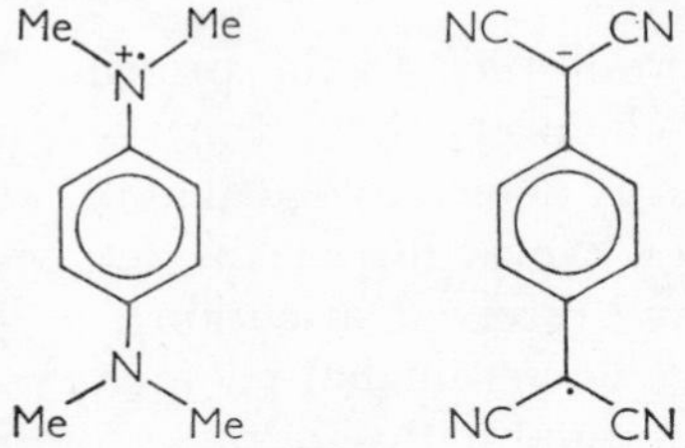

(**93**)

---

* It would be helpful to know, however, how large a part of the concentration effect on the e.s.r. spectrum is traceable to spin–spin and exchange interaction involving only the radical-anions.

intensity and temperature dependence of its e.s.r. spectrum reveal that its ground state is singlet, lying 1·7 kcal/mole below a triplet state[85]. A complex containing unpaired spins may display either doublet (if the spins are essentially independent) or triplet paramagnetism: indeed, complexes are known which can individually exist in singlet, doublet, or triplet states depending upon the temperature[86]. Though many tetraaminoethylene complexes have been identified in the literature as diamagnetic or paramagnetic, very little has been reported regarding e.s.r. spectra and their variation with such parameters as state, solvent, concentration, and temperature. The magnetic properties of the charge-transfer compounds with powerful acceptors appear to be well worth more detailed investigation.

The paramagnetic complexes derived from aliphatic peramino-ethylenes possess the property, unusual for organic compounds, of strong, broad, light absorption in the near-infrared region, typically from 0·5 to 2·0 μ. Combined with their high stability, this characteristic makes them valuable as pigments in writing inks which are suitable for thermographic reproduction[3]. Most fountain-pen inks, and particularly ball-point inks, are not copied satisfactorily by thermographic methods because they are nearly transparent in the infrared. Peraminoethylene complexes also have promise as pigments for impact printing. When a sheet of paper impregnated with donor is laid upon one impregnated with acceptor and pressure is applied, as with a stylus, an image created by complex formation appears on the bottom sheet. The many other uses of peraminoethylene-strong π-acid complexes suggested by their intense colors, their stability, and their paramagnetism will not be enumerated here.[3]

IV. PAST AND FUTURE

During the period from 1960 to the present, the broad outlines of tetraaminoethylene chemistry have emerged, accompanied by an already impressive array of technological applications for the ethylenes and their derivatives. Nevertheless, the areas of peraminoethylene chemistry which have received attention are liberally strewn with puzzling questions, and yet-uncharted areas constitute a challenge made lively by the unusual nature of these compounds.

Some unexplored aspects of their synthesis from ylid halves will serve as illustration. Among divalent carbon derivatives, where these ylids formally belong, their stability is relatively great. Can they then be generated along with benzene by thermal decomposition of tropone aminals[87, 88]? Will they be formed when phenylcyclopropanone aminals are irradiated, just as methylene[89] and dichlorocarbene[90] are produced by photolysis of the corresponding cyclopropanes? If so, flash photolytic techniques may make possible a kinetic study of the theoretically interesting dimerization of ylids. Can the ylids be trapped as ligands in coordination complexes[91], and will such compounds change into peraminoethylene complexes?

May this record of progress in peraminoethylene chemistry rapidly become outdated.

V. ACKNOWLEDGMENTS

The author wishes to express his appreciation to Drs. H. Quast and S. Hünig, R. A. Olofson, H.-W. Wanzlick, N. Wiberg, and J. W. Buchler, H. E. Winberg, A. N. Fletcher, and C. A. Heller, W. H. Urry, and J.-J. Vorsanger for providing him with unpublished information, and, in a number of instances, helpful counsel. He is indebted as well to Professor R. B. Woodward, Doz. Dr. R. W. Hoffman, Dr. J. P. Lokensgard, and Dr. S. D. McGregor for valuable discussions, and to the National Institutes of Health for financial support during the preparation of the manuscript.

VI. REFERENCES

1. R. L. Pruett, J. T. Barr, K. E. Rapp, C. T. Bahner, J. D. Gibson, and R. H. Lafferty, Jr., *J. Am. Chem. Soc.*, **72**, 3646 (1950).
2. N. Wiberg and J. W. Buchler, *Z. Naturforsch.* **19b**, 5 (1964).
3. H. E. Winberg, *U.S. Pat.*, 2,239,518 (1966).
4. H. E. Winberg, *U.S. Pat.*, 3,239,519 (1966).
5. (a) R. A. Lovald, *Ph.D. Dissertation*, University of Wisconsin, 1965.
   (b) D. M. Lemal, R. A. Lovald, K. I. Kawano, and J. P. Lokensgard, *J. Am. Chem. Soc.*, in press.
6. (a) H.-W. Wanzlick and E. Schikora, *Angew. Chem.*, **72**, 494 (1960).
   (b) H.-W. Wanzlick and H.-J. Kleiner, *Angew. Chem.*, **73**, 493 (1961).
7. H.-W. Wanzlick, *Angew. Chem. Intern. Ed. Engl.* **1**, 75 (1962). A review of 'nucleophilic carbene' chemistry.
8. H.-W. Wanzlick, F. Esser, and H.-J. Kleiner, *Chem. Ber.*, **96**, 1208 (1963).
9. H. E. Winberg, J. E. Carnahan, D. D. Coffman, and M. Brown, *J. Am. Chem. Soc.*, **87**, 2055 (1965).
10. W. J. Chambers, *U.S. Pat.*, 3,141,906 (1964). An acid catalyst is employed in this particular procedure.
11. D. M. Lemal, R. A. Lovald, and K. I. Kawano, *J. Am. Chem. Soc.*, **86**, 2518 (1964).
12. N. Wiberg and J. W. Buchler, *Z. Naturforsch.*, **19b**, 953 (1964).
13. H. Böhme and F. Soldan, *Chem. Ber.*, **94**, 3109 (1961).
14. R. Breslow, *J. Am. Chem. Soc.*, **80**, 3719 (1958).

15. (a) R. Breslow, *Chem. Ind.* (*London*), 893 (1957).
    (b) *J. Am. Chem. Soc.* **79**, 1762 (1957).
    (c) R. Breslow and E. McNelis, *J. Am. Chem. Soc.*, **81**, 3080 (1959).
    (d) *J. Am. Chem. Soc.* **82**, 2394 (1960).
16. (a) P. Dyson and D. L. Hammick, *J. Chem. Soc.*, 1724 (1937); 809 (1939).
    (b) A recent study of the decarboxylation of heterocyclic betaines has produced a series of interesting Hammick-like reactions (H. Quast and E. Frankenfeld, *Angew. Chem. Intern. Ed. Engl.*, **4**, 691 (1965)).
17. R. A. Olofson, W. R. Thompson, and J. S. Michelman, *J. Am. Chem. Soc.*, **86**, 1865 (1964); R. A. Olofson and J. S. Michelman, *J. Org. Chem.*, **30**, 1854 (1965); R. A. Olofson, R. A. Coburn, J. M. Landesberg, and D. S. Kemp, unpublished results.
18. R. A. Olofson and J. M. Landesberg, *J. Am. Chem. Soc.*, **88**, 4263 (1966). The earlier work of Haake and Miller and of Häfferl, Lundin, and Ingraham is discussed in this paper.
19. R. A. Olofson, J. M. Landesberg, K. N. Houk, and J. S. Michelman, *J. Am. Chem. Soc.*, **88**, 4265 (1966).
20. R. A. Olofson, A. C. Rochat, and W. R. Thompson, unpublished results.
21. U. Schöllkopf and E. Wiskott, *Angew. Chem.*, **75**, 725 (1963). Trapping of dithiocarbenes by the nucleophilic triphenylphosphine has been reported by D. M. Lemal and E. H. Banitt, *Tetrahedron Letters*, 245 (1964).
22. W. Kirmse, *Carbene Chemistry*, Academic Press, New York, 1964, Chap. 9.
23. D. M. Lemal, E. P. Gosselink, and S. D. McGregor, *J. Am. Chem. Soc.*, **88**, 582 (1966).
24. J. Metzger, H. Larivé, R. Dennilauler, R. Baralle, and C. Gaurat, *Bull. Soc. Chim. France*, 2857 (1964).
25. (a) H. Quast and S. Hünig, *Angew. Chem. Intern. Ed. Engl.*, **3**, 800 (1964).
    (b) *Chem. Ber.*, **99**, 2017 (1966).
26. J.-J. Vorsanger, *Bull. Soc. Chim. France*, 119 (1964); W. Friedrich, H. Kehr, K. Kröhnke, and P. Schiller, *Chem. Ber.*, **98**, 3808 (1965).
27. (a) H.-W. Wanzlick and H.-J. Kleiner, *Angew. Chem. Intern. Ed. Engl.*, **3**, 65 (1964).
    (b) H.-W. Wanzlick, H.-J. Kleiner, I. Lasch, and H. U. Füldner, *Angew. Chem. Intern. Ed. Engl.*, **4**, 126 (1965). A correction of the earlier paper is noted in the latter.
28. For an alternative approach to 1,2-diamino-1,2-dithioethylenes, see W. R. Brasen, *U.S. Pat.*, 3,122,539 (1964).
29. H. Prinzbach, H. Berger, and A. Lüttringhaus, *Angew. Chem. Intern. Ed. Engl.*, **4**, 435 (1965).
30. (a) D. M. Lemal and K. I. Kawano, *J. Am. Chem. Soc.*, **84**, 1761 (1962).
    (b) K. I. Kawano, *Ph.D. Dissertation*, University of Wisconsin, 1965.
31. H. Balli, *Angew. Chem.*, **70**, 442 (1958).
32. H. Balli, *Angew. Chem.*, **76**, 995 (1964).
33. H. E. Winberg, *U.S. Pat.*, 3,239,534 (1966).
34. I. Hagedorn, K. E. Lichtel, and H. D. Winkelmann, *Angew. Chem. Intern. Ed. Engl.*, **4**, 702 (1965); I. Hagedorn and K. E. Lichtel, *Chem. Ber.*, **99**, 524 (1966).
35. H.-W. Wanzlick and E. Schikora, *Chem. Ber.*, **94**, 2389 (1961).
36. W. H. Urry and J. Sheeto, *Photochem. Photobiol.*, **4**, 1067 (1965).

37. A. Piskala, *Tetrahedron Letters*, 2587 (1964).
38. H.-W. Wanzlick, private communication.
39. N. Wiberg and J. W. Buchler, *J. Am. Chem. Soc.*, **85**, 243 (1963).
40. N. Wiberg and J. W. Buchler, *Chem. Ber.*, **96**, 3000 (1963).
41. N. Wiberg, *Habilitationsschrift*, University of Munich, 1966.
42. V. Franzen and H. Joschek, *Ann. Chem.*, **633**, 7 (1960). E. Bayer and B. Krämer, *Chem. Ber.*, **97**, 1057 (1964). R. W. Hoffmann and H. Häuser, *Tetrahedron Letters*, 1365 (1964), the latter authors now favor an alternative interpretation (private communication). Reference. 52 J.-J. Vorsanger, *Bull. Soc. Chim. France*, 1772 (1966).
43. See also F. G. White and L. L. Ingraham, *J. Am. Chem. Soc.*, **84**, 3109 (1962).
44. E. Schikora, *Ph.D. Thesis*, F. U. Berlin, 1961. Quoted by H.-W. Wanzlick in reference 7.
45. (a) H.-W. Wanzlick, H. Ahrens, B. König, and M. Riccius, *Angew. Chem. Intern. Ed. Engl.*, **2**, 560 (1963).
    (b) H.-W. Wanzlick and H. Ahrens, *Chem. Ber.*, **97**, 2447 (1964).
    (c) *Ann. Chem.* **693**, 176 (1966); *Chem. Ber.* **99**, 1580 (1966); H.-W. Wanzlick, *Ger. Pat.*, 1,215,127 (1966).
46. H.-W. Wanzlick and H.-J. Kleiner, *Chem. Ber.*, **96**, 3024 (1963).
47. H.-W. Wanzlick, B. Lachmann, and E. Schikora, *Chem. Ber.*, **98**, 3170 (1965).
48. N. Wiberg and J. W. Buchler, *Z. Naturforsch.*, **19b**, 9 (1964).
49. W. H. Urry, private communication.
50. H.-W. Wanzlick, E. Schikora, H.-J. Kleiner, and M. Riccius, *Angew. Chem.*, **73**, 765 (1961).
51. H. E. Winberg and D. D. Coffman, *J. Am. Chem. Soc.*, **87**, 2776 (1965).
52. H.-W. Wanzlick and B. König, *Chem. Ber.*, **97**, 3513 (1964).
53. H. Reimlinger, *Chem. Ber.*, **97**, 3503 (1964).
54. R. Fusco, G. Bianchetti, D. Pocar, and R. Ugo, *Chem. Ber.*, **96**, 802 (1963).
55. K. D. Gundermann, *Angew. Chem. Intern. Ed. Engl.*, **4**, 566 (1965); F. McCapra, *Quart. Rev.*, **20**, 485 (1966). Reviews of chemiluminescence.
56. C. A. Heller and A. N. Fletcher, *J. Phys. Chem.*, **69**, 3313 (1965).
57. D. L. Maricle and W. G. Hodgson, *Anal. Chem.*, **37**, 1562 (1965).
58. K. Kuwata and D. H. Geske, *J. Am. Chem. Soc.*, **86**, 2107 (1964).
59. A. N. Fletcher and C. A. Heller, *J. Catalysis*, **6**, 263 (1966); private communication.
60. H. E. Winberg, J. R. Downing, and D. D. Coffman, *J. Am. Chem. Soc.*, **87**, 2054 (1965).
61. A. U. Khan and M. Kasha, *J. Am. Chem. Soc.*, **88**, 1574 (1966).
62. E. H. White and M. J. C. Harding, *J. Am. Chem. Soc.*, **86**, 5686 (1964).
63. F. McCapra and D. G. Richardson, *Tetrahedron Letters*, 3167 (1964).
64. F. McCapra, D. G. Richardson, and Y. C. Chang, *Photochem. Photobiol.*, **4** 1111 (1965); F. McCapra and Y. C. Chang, *Chem. Commun.*, 522 (1966).
65. R. Hoffmann and R. B. Woodward, *J. Am. Chem. Soc.*, **87**, 2046 (1965).
66. A helpful discussion of some of the considerations involved is provided by H. C. Longuet-Higgins and E. W. Abrahamson, *J. Am. Chem. Soc.*, **87**, 2045 (1965).
67. A. N. Fletcher and C. A. Heller, *Photochem. Photobiol.*, **4**, 1051 (1965).
68. Electronic energy transfer is discussed in N. Turro, *Molecular Photochemistry*, W. A. Benjamin, Inc., New York, New York, 1965, Chap. 5.

69. E. A. Chandross and F. I. Sonntag, *J. Am. Chem. Soc.*, **86**, 3179 (1964); *J. Am. Chem. Soc.*, **88**, 1089 (1966), and references contained therein.
70. D. M. Hercules, *Science*, **145**, 808 (1964).
71. D. H. Geske, quoted in reference 41.
72. J. P. Paris, *Photochem. Photobiol.*, **4**, 1059 (1965).
73. W. Carpenter, quoted in reference 59.
74. D. H. Anderson, R. M. Elofson, H. S. Gutowsky, S. Levine, and R. B. Sandin, *J. Am. Chem. Soc.*, **83**, 3157 (1961); R. M. Elofson, D. H. Anderson, H. S. Gutowsky, R. B. Sandin, and K. F. Schulz, *J. Am. Chem. Soc.*, **85**, 2622 (1963).
75. Voltaic cells based on this ability have been constructed by W. R. Wolfe, Jr., *U.S. Pat.*, 3,156,587 (1964).
76. A. Streitwieser, Jr., *Molecular Orbital Theory for Organic Chemists*, John Wiley & Sons, Inc., New York, 1961, Chap. 5.
77. (a) N. Wiberg and J. W. Buchler, *Angew. Chem.*, **74**, 490 (1962); (b) *Chem. Ber.*, **96**, 3223 (1963).
78. G. Briegleb and J. Czekalla, *Z. Electrochem.*, **63**, 6 (1959).
79. W. Carpenter, *J. Org. Chem.*, **30**, 3082 (1965).
80. For coordination complexes with transition metals, see E. L. Muetterties, *J. Am. Chem. Soc.*, **82**, 1082 (1960); reference 40; N. R. Fetter and D. W. Moore, *Can. J. Chem.*, **42**, 885 (1964). With certain metal carbonyl derivatives, oxamidinium salts are obtained (R. B. King, *Inorg. Chem.*, **4**, 1518 (1965)).
81. N. Wiberg and J. W. Buchler, *Chem. Ber.*, **97**, 618 (1964).
82. W. Liptay, G. Briegleb, and K. Schindler, *Z. Electrochem.*, **66**, 331 (1962).
83. E. M. Kosower, *J. Am. Chem. Soc.*, **80**, 3253 (1958); E. M. Kosower, *Molecular Biochemistry*, McGraw-Hill, Inc., New York, 1962, pp. 180–189.
84. D. B. Chestnut and W. D. Phillips, *J. Chem. Phys.*, **35**, 1002 (1961).
85. M. Kinoshita and H. Akamatu, *Nature*, **207**, 291 (1965), and references contained therein.
86. F. Schneider, K. Möbius, and M. Plato, *Angew. Chem. Intern. Ed. Engl.*, **4**, 856 (1965). A review on e.s.r. applications.
87. At ~100° 7-(dimethylamino)tropilidene yields, in addition to double-bond isomers, several percent of benzene and, apparently, dimethylaminocarbene. A. P. ter Borg, E. Razenberg, and H. Kloosterziel, *Rec. Trav. Chim.*, **84**, 1230 (1965); **85**, 774 (1966).
88. The proposed fragmentation is reminiscent of the thermal decomposition of certain norbornadienone ketals, which yield dimethoxycarbene and benzene derivatives (R. W. Hoffmann and H. Häuser, *Tetrahedron Letters*, 197 (1964); *Tetrahedron*, **21**, 891 (1965). D. M. Lemal, E. P. Gosselink, and A. Ault, *Tetrahedron Letters*, 579 (1964). Reference 23). Indeed, pyrolysis of tropone dimethyl ketal gives benzene in 40% yield (R. W. Hoffmann, private communication).
89. D. B. Richardson, L. R. Durrett, J. M. Martin, Jr., W. E. Putnam, S. C. Slaymaker, and I. Dvoretzky, *J. Am. Chem. Soc.*, **87**, 2763 (1965).
90. M. Jones, Jr., W. H. Sachs, A. Kulczycki, Jr., and F. J. Waller, *J. Am. Chem. Soc.*, **88**, 3167 (1966).
91. For transition metal complexes of carbenes, see F. D. Mango and I. Dvoretzky, *J. Am. Chem. Soc.*, **88**, 1654 (1966), and references contained therein.

Author Index

This author index is designed to enable the reader to locate an author's name and work with the aid of the reference numbers appearing in the text. The page numbers are printed in normal type in ascending numerical order, followed by the reference numbers in brackets. The numbers in *italics* refer to the pages on which the references are actually listed.

If reference is made to the work of the same author in different chapters, the above arrangement is repeated separately for each chapter.

26*

Subject Index